METHODS IN MOLECULAR BIOLOGY™

Series Editor
John M. Walker
School of Life Sciences
University of Hertfordshire
Hatfield, Hertfordshire, AL10 9AB, UK

For further volumes:
http://www.springer.com/series/7651

Electron Crystallography of Soluble and Membrane Proteins

Methods and Protocols

Edited by

Ingeborg Schmidt-Krey

School of Biology, School of Chemistry and Biochemistry, Georgia Institute of Technology, Atlanta, GA, USA

Yifan Cheng

Department of Biochemistry and Biophysics, University of California San Francisco, San Francisco, CA, USA

Humana Press

Editors
Ingeborg Schmidt-Krey
School of Biology
School of Chemistry and Biochemistry
Georgia Institute of Technology
Atlanta, GA, USA

Yifan Cheng
Department of Biochemistry and Biophysics
University of California San Francisco
San Francisco, CA, USA

ISSN 1064-3745 ISSN 1940-6029 (eBook)
ISBN 978-1-4939-5876-4 ISBN 978-1-62703-176-9 (eBook)
DOI 10.1007/978-1-62703-176-9
Springer New York Heidelberg Dordrecht London

Printed on acid-free paper

Humana Press is a brand of Springer
Springer is part of Springer Science+Business Media (www.springer.com)

Preface

This volume is aimed at biological scientists intending to gain structural information on membrane proteins or soluble proteins by electron crystallography. Membrane proteins may be reconstituted and induced to form two-dimensional (2D) crystals in a near-native lipid environment. Soluble proteins may also be induced to form ordered 2D arrays. From such 2D crystals, both images and electron diffraction patterns can be acquired with an electron microscope. The basic principle of electron crystallography is to calculate a 3D density map by combining the amplitudes obtained from electron diffraction patterns with the experimental phases calculated from images. The technology for studying membrane proteins is very well developed and has produced a number of atomic models of membrane proteins in a lipid environment. The book is focused on comprehensive experimental protocols, and its chapters cover the entire range of techniques used in electron crystallography, from protein sample preparation, 2D crystallization, and screening in negative stain over electron cryo-cryomicroscopy (cryo-EM) and data processing to modeling of conformational changes. The reader will be guided through each step, including sample preparation and initial low-resolution studies, which require only a minimum of equipment and in-house expertise, to an overview of planning and using a state-of-the-art cryo-EM facility for high-resolution data collection and processing. Additional chapters provide perspective on past, present, and future challenges as well as complementary methods.

It is our hope that this book will guide both laboratories new to the methods as well as state-of-the-art facilities in the nontrivial techniques used in electron crystallography. Protocols will be applicable to laboratories wishing to gain initial structural information on, for example, the number of subunits on a protein, over secondary structural features to atomic resolution information with the aim of structure–function studies.

In addition to biological scientists, a considerable number of physicists have moved to apply their skills to biological problems. This book will provide valuable guidelines for understanding the differences encountered when working with beam-sensitive organic samples from specimen preparation, handling, and analysis to low-dose techniques.

The introduction (Chapter 1) covers the importance of the method to structural biology and provides a historical overview of the development of cryo-EM and electron crystallography in terms of both advances in methods development as well as biological applications and findings. At the foundation of all structural studies lies a sample that is properly purified in large-enough quantities. Furthermore, the structural and functional integrity needs to be maintained, which still poses a challenge for membrane proteins. Chapter 2 describes protocols for membrane protein overexpression and purification based on three challenging examples. Following the biochemical sample preparation, the next step is to induce 2D crystallization, which is most commonly achieved through dialysis and described in Chapter 3. Here, considerations important to strategy as well as basic protocols are discussed. Chapter 4 describes the method of 2D crystallization by the monolayer technique, which provides a valuable extension of electron crystallography to soluble proteins. Chapter 5 then focuses on the screening by transmission electron microscopy of negatively stained samples, essential to identifying and optimizing 2D crystallization conditions. Once large and well-ordered 2D crystals are produced, data collection by Cryo-EM is the next step (Chapter 6). For this, the specimen needs to be frozen in a thin layer of vitreous buffer under the most favorable circumstances (Chapter 7) to collect the best quality image and electron diffraction data. Careful image data collection (Chapter 8) is absolutely critical to obtaining the highest resolution information possible and plays an equal role in the number of useful images collected, controlling the timeframe of the project significantly. Chapter 9 outlines detailed protocols for collecting electron diffraction patterns. Next is the extraction of the structural information from the cryo-EM data by processing. Image processing is described in Chapters 10 and 11, providing many critical details to the understanding of the individual steps. As for any method in structural biology, data obtained needs to be comprehensively and meticulously assessed, showing the validity of the model of the structure (Chapter 12). Once a model is available, structures may be modeled in terms of higher resolution and conformational changes (Chapter 13). Comparison of structures of related proteins and docking, allow for further analysis and interpretation on the structural and functional level. One of the newest methodological advances in terms of phasing electron diffraction data by molecular replacement is described in Chapter 14.

In addition to protocols for these essential steps of electron crystallography, the book also covers a number of recent technological development efforts, particularly the technology of automation. Such technological advances are critical for the future of electron crystallography. Chapter 15 describes applying automated methods to 2D crystallization and a promising cyclodextrin protocol as well as several additional, detailed procedures that give insights adding to and complementing Chapter 3. Chapters 16 and 17 outline automation of grid handling and imaging at the electron microscope, and Chapter 18 provides guidelines on the automation of image processing.

The strength of electron crystallography is always related to the instrumentation. A variety of set-ups of minimal to state-of-the-art cryo-EM facilities are available and can be arranged, depending on needs, funding, and space. Chapter 19 gives guidance on selection and maintenance of all critical instrumentation. Chapter 20 provides a detailed overview of past and possible future developments of instrumentation benefiting electron crystallography.

Further chapters describe methods that may be used on their own or in conjunction with electron crystallography. Tubular crystals and helical arrays may be obtained by methods similar to 2D crystallization and provide the advantage of not having a missing cone as well as that images of tilted samples are not necessary. An overview of the handling of such samples is given in Chapter 21. Single particle electron microscopy (Chapter 22) is a sister cryo-EM technique to electron crystallography and has contributed a wealth of structural information on both soluble and membrane proteins. Electron tomography can reveal essential details on ordered structures that might not be accessible by electron crystallography, due to, e.g., paracrystalline samples such as described in Chapter 23. Atomic force microscopy provides structural information on membrane proteins within 2D crystals that is highly complementary to electron crystallography and can reveal a wealth of additional information (Chapter 24). X-ray crystallography and nuclear magnetic resonance spectroscopy (NMR) are traditional structural methods that are continuing to break new ground and often go hand in hand with the cryo-EM methods. X-ray crystallography and NMR are covered in Chapters 25 and 26, respectively. Chapter 27 outlines just a few examples of the wealth of structural and functional information revealed by electron crystallography of soluble and membrane proteins. Chapter 28 provides a comprehensive description of protocols to induce 2D crystallization of the two soluble protein complexes, carboxysome shell proteins, and HIV CA. Chapter 29 describes the processing of electron diffraction data.

While an impressive amount of methods development has been accomplished and automation is in the process of breaking further ground for electron crystallography, Chapter 30 introduces and highlights the great potential of electron crystallography in terms of biological information that cannot be revealed by other methods.

We are immensely grateful to our authors for sharing their expertise and insights in preparing this book. Particular thanks are due to our series editor, John M. Walker, Professor Emeritus in the School of Life Sciences at the University of Hertfordshire, for sharing his invaluable experience and his kindness in doing so. It was an absolute pleasure to work with our editor David C. Casey at Springer. We would also like to express our sincere thanks to our Springer production editor Lesley Poliner and Raja Dharmaraj at SPi.

Atlanta, GA, USA ***Ingeborg Schmidt-Krey***
San Francisco, CA, USA ***Yifan Cheng***

Contents

Preface . . . . *v*
Contributors . . . . *xi*

1 Introduction to Electron Crystallography 1
Werner Kühlbrandt

2 Practical Aspects in Expression and Purification of Membrane Proteins for Structural Analysis 17
Kutti R. Vinothkumar, Patricia C. Edwards, and Joerg Standfuss

3 Two-Dimensional Crystallization of Membrane Proteins by Reconstitution Through Dialysis 31
Matthew C. Johnson, Tina M. Dreaden, Laura Y. Kim, Frederik Rudolph, Bridgette A. Barry, and Ingeborg Schmidt-Krey

4 Monolayer Two-Dimensional Crystallization of Membrane Proteins 59
Luc Lebeau and Catherine Vénien-Bryan

5 Screening for Two-Dimensional Crystals by Transmission Electron Microscopy of Negatively Stained Samples 73
Tina M. Dreaden, Maureen Metcalfe, Laura Y. Kim, Matthew C. Johnson, Bridgette A. Barry, and Ingeborg Schmidt-Krey

6 Low Dose Techniques and Cryo-Electron Microscopy 103
Yoshinori Fujiyoshi

7 Grid Preparation for Cryo-Electron Microscopy 119
Nobuhiko Gyobu

8 Recording High-Resolution Images of Two-Dimensional Crystals of Membrane Proteins 129
Agustin Avila-Sakar, Xueming Li, Shawn Q. Zheng, and Yifan Cheng

9 The Collection of High-Resolution Electron Diffraction Data 153
Tamir Gonen

10 Image Processing of 2D Crystal Images 171
Marcel Arheit, Daniel Castaño-Díez, Raphaël Thierry, Bryant R. Gipson, Xiangyan Zeng, and Henning Stahlberg

11 Merging of Image Data in Electron Crystallography 195
Marcel Arheit, Daniel Castaño-Díez, Raphaël Thierry, Priyanka Abeyrathne, Bryant R. Gipson, and Henning Stahlberg

12 Evaluation of Electron Crystallographic Data from Images of Two-Dimensional Crystals 211
Vinzenz M. Unger

13 Modeling, Docking, and Fitting of Atomic Structures to 3D Maps from Cryo-Electron Microscopy 229
Gregory S. Allen and David L. Stokes

14 Phasing Electron Diffraction Data by Molecular Replacement: Strategy for Structure Determination and Refinement 243
Goragot Wisedchaisri and Tamir Gonen

15 High-Throughput Methods for Electron Crystallography. 273
David L. Stokes, Iban Ubarretxena-Belandia, Tamir Gonen, and Andreas Engel

16 Automated Grid Handling and Image Acquisition for Two-Dimensional Crystal Screening 297
Anchi Cheng

17 Automation of Data Acquisition in Electron Crystallography 307
Anchi Cheng

18 Automation of Image Processing in Electron Crystallography. 313
Marcel Arheit, Daniel Castaño-Díez, Raphaël Thierry, Bryant R. Gipson, Xiangyan Zeng, and Henning Stahlberg

19 Choice and Maintenance of Equipment for Electron Crystallography 331
Deryck J. Mills and Janet Vonck

20 Future Developments in Instrumentation for Electron Crystallography. 353
Kenneth H. Downing

21 Tubular Crystals and Helical Arrays: Structural Determination of HIV-1 Capsid Assemblies Using Iterative Helical Real-Space Reconstruction 381
Peijun Zhang, Xin Meng, and Gongpu Zhao

22 Single Particle Electron Microscopy 401
Wilson C.Y. Lau and John L. Rubinstein

23 Electron Tomography of Paracrystalline 2D Arrays. 427
Hanspeter Winkler, Shenping Wu, and Kenneth A. Taylor

24 High-Resolution Imaging of 2D Outer Membrane Protein F Crystals by Atomic Force Microscopy 461
Dimitrios Fotiadis and Daniel J. Müller

25 Determination of Soluble and Membrane Protein Structures by X-Ray Crystallography. 475
Raquel L. Lieberman, Mary E. Peek, and J. Derrick Watkins

26 Solution Nuclear Magnetic Resonance Spectroscopy. 495
James J. Chou and Remy Sounier

27 Structure–Function Insights of Membrane and Soluble Proteins Revealed by Electron Crystallography 519
Tina M. Dreaden, Bharanidharan Devarajan, Bridgette A. Barry, and Ingeborg Schmidt-Krey

28 Lipid Monolayer and Sparse Matrix Screening for Growing Two-Dimensional Crystals for Electron Crystallography: Methods and Examples 527
Mark Yeager, Kelly A. Dryden, and Barbie K. Ganser-Pornillos

29 Processing of Electron Diffraction Patterns with the XDP Program 539
Kaoru Mitsuoka

30 Future Directions of Electron Crystallography 551
Yoshinori Fujiyoshi

Index *569*

Contributors

PRIYANKA ABEYRATHNE • *C-CINA, Biozentrum, University Basel, Basel, Switzerland*

GREGORY S. ALLEN • *Department of Cell Biology, Skirball Institute of Biomolecular Medicine, New York University School of Medicine, New York, NY, USA*

MARCEL ARHEIT • *C-CINA, Biozentrum, University Basel, Basel, Switzerland*

AGUSTIN AVILA-SAKAR • *The W.M. Keck Advanced Microscopy Laboratory, Department of Biochemistry and Biophysics, University of California San Francisco, San Francisco, CA, USA*

BRIDGETTE A. BARRY • *School of Chemistry and Biochemistry, Georgia Institute of Technology, Atlanta, GA, USA*

DANIEL CASTAÑO-DÍEZ • *C-CINA, Biozentrum, University Basel, Basel, Switzerland*

ANCHI CHENG • *Department of Cell Biology, The National Resource for Automated Molecular Microscopy, The Scripps Research Institute, La Jolla, CA, USA*

YIFAN CHENG • *Department of Biochemistry and Biophysics, University of California San Francisco, San Francisco, CA, USA*

JAMES J. CHOU • *Jack and Eileen Connors Structural Biology Laboratory, Department of Biological Chemistry and Molecular Pharmacology, Harvard Medical School, Boston, MA, USA*

BHARANIDHARAN DEVARAJAN • *School of Biology, Georgia Institute of Technology, Atlanta, GA, USA*

KENNETH H. DOWNING • *Lawrence Berkeley National Laboratory, Life Science Division, Berkeley, CA, USA*

TINA M. DREADEN • *School of Chemistry and Biochemistry, Georgia Institute of Technology, Atlanta, GA, USA*

KELLY A. DRYDEN • *Department of Molecular Physiology and Biological Physics, University of Virginia School of Medicine, Charlottesville, VA, USA*

PATRICIA C. EDWARDS • *MRC Laboratory of Molecular Biology, Cambridge, UK*

ANDREAS ENGEL • *Department of Pharmacology, Case Western Reserve University, Cleveland, OH, USA*

DIMITRIOS FOTIADIS • *Institute of Biochemistry and Molecular Medicine, University of Bern, Bern, Switzerland*

YOSHINORI FUJIYOSHI • *Department of Basic Biology, Cellular and Structural Physiology Institute (CeSPI), Nagoya University, Furo-cho, Chikusa, Nagoya, Japan*

BARBIE K. GANSER-PORNILLOS • *Department of Molecular Physiology and Biological Physics, University of Virginia School of Medicine, Charlottesville, VA, USA; Department of Cell Biology, The Scripps Research Institute, La Jolla, CA, USA*

BRYANT R. GIPSON • *Department of Computer Sciences, Rice University, Houston, TX, USA*

TAMIR GONEN • *Janelia Farm Research Campus, Howard Hughes Medical Institute, Ashburn, VA, USA*

NOBUHIKO GYOBU • *Biomedicinal Information Research Center (BIRC), Tokyo, Japan*

MATTHEW C. JOHNSON • *School of Biology, Georgia Institute of Technology, Atlanta, GA, USA*

LAURA Y. KIM • *School of Biology, Georgia Institute of Technology, Atlanta, GA, USA*

WERNER KÜHLBRANDT • *Max-Planck-Institute of Biophysics, Frankfurt, Germany*
WILSON C.Y. LAU • *Molecular Structure and Function Program, Departments of Biochemistry and Medical Biophysics, The Hospital for Sick Children, The University of Toronto, Toronto, ON, Canada*
LUC LEBEAU • *Laboratoire de Conception et Application de Molécules Bioactives, CNRS, Université de Strasbourg, Faculté de Pharmacie, BP, Illkirch, France*
XUEMING LI • *The W.M. Keck Advanced Microscopy Laboratory, Department of Biochemistry and Biophysics, University of California San Francisco, Route du Rhin, BP, Illkirch, France*
RAQUEL L. LIEBERMAN • *School of Chemistry and Biochemistry, Georgia Institute of Technology, Atlanta, GA, USA*
XIN MENG • *Department of Structural Biology, University of Pittsburgh School of Medicine, Pittsburgh, PA, USA*
MAUREEN METCALFE • *School of Biology, Georgia Institute of Technology, Atlanta, GA, USA*
DERYCK J. MILLS • *Max-Planck-Institute of Biophysics, Frankfurt, Germany*
KAORU MITSUOKA • *Biomedicinal Information Research Center (BIRC), National Institute of Advanced Industrial Science and Technology (AIST), Tokyo, Japan*
DANIEL J. MÜLLER • *ETH Zürich, Biosystems Science and Engineering (BSSE), Basel, Switzerland*
MARY E. PEEK • *School of Chemistry and Biochemistry, Georgia Institute of Technology, Atlanta, GA, USA*
JOHN L. RUBINSTEIN • *Molecular Structure and Function Program, Departments of Biochemistry and Medical Biophysics, The Hospital for Sick Children, The University of Toronto, Toronto, ON, Canada*
FREDERIK RUDOLPH • *School of Biology, Georgia Institute of Technology, Atlanta, GA, USA*
INGEBORG SCHMIDT-KREY • *School of Biology, School of Chemistry and Biochemistry, Georgia Institute of Technology, Atlanta, GA, USA*
REMY SOUNIER • *Jack and Eileen Connors Structural Biology Laboratory, Department of Biological Chemistry and Molecular Pharmacology, Harvard Medical School, Boston, MA, USA*
HENNING STAHLBERG • *C-CINA, Biozentrum, University Basel, Basel, Switzerland*
JOERG STANDFUSS • *Laboratory of Biomolecular Research, Paul Scherrer Institute, Villigen PSI, Switzerland*
DAVID L. STOKES • *Department of Cell Biology, Skirball Institute of Biomolecular Medicine, New York University School of Medicine, New York, NY, USA; Division of Cryo-Electron Microscopy, New York Structural Biology Center, New York, NY, USA*
KENNETH A. TAYLOR • *Institute of Molecular Biophysics, Florida State University, Tallahassee, FL, USA*
RAPHAËL THIERRY • *C-CINA, Biozentrum, University Basel, Basel, Switzerland*
IBAN UBARRETXENA-BELANDIA • *Department of Structural and Chemical Biology, Mt. Sinai School of Medicine, New York, NY, USA*
VINZENZ M. UNGER • *Department of Molecular Biosciences, Chemistry of Life Processes Institute, Northwestern University, Evanston, IL, USA*
CATHERINE VÉNIEN-BRYAN • *IMPMC Institut de Minéralogie et de Physique des Milieux Condensés, Univ Paris 06, CNRS, UMR 7590, Paris Cedex, France*
KUTTI R. VINOTHKUMAR • *MRC Laboratory of Molecular Biology, Cambridge, UK*
JANET VONCK • *Max-Planck-Institute of Biophysics, Frankfurt, Germany*
J. DERRICK WATKINS • *School of Chemistry and Biochemistry, Georgia Institute of Technology, Atlanta, GA, USA*

HANSPETER WINKLER • *Institute of Molecular Biophysics, Florida State University, Tallahassee, FL, USA*

GORAGOT WISEDCHAISRI • *Janelia Farm Research Campus, Howard Hughes Medical Institute, Ashburn, VA, USA*

SHENPING WU • *Department of Biochemistry & Biophysics, University of California San Francisco, San Francisco, CA, USA*

MARK YEAGER • *Department of Molecular Physiology and Biological Physics, University of Virginia School of Medicine, Charlottesville, VA, USA; Department of Cell Biology, The Scripps Research Institute, La Jolla, CA, USA; Division of Cardiovascular Medicine, Department of Medicine, University of Virginia Health System, Charlottesville, VA, USA*

XIANGYAN ZENG • *Department of Mathematics and Computer Science, Fort Valley State University, Fort Valley, GA, USA*

PEIJUN ZHANG • *Department of Structural Biology, University of Pittsburgh School of Medicine, Pittsburgh, PA, USA*

GONGPU ZHAO • *Department of Structural Biology, University of Pittsburgh School of Medicine, Pittsburgh, PA, USA*

SHAWN Q. ZHENG • *The W.M. Keck Advanced Microscopy Laboratory, Department of Biochemistry and Biophysics, University of California San Francisco, San Francisco, CA, USA*

Chapter 1

Introduction to Electron Crystallography

Werner Kühlbrandt

Abstract

From the earliest work on regular arrays in negative stain, electron crystallography has contributed greatly to our understanding of the structure and function of biological macromolecules. The development of electron cryo-microscopy (cryo-EM) then lead to the first groundbreaking atomic models of the membrane proteins bacteriorhodopsin and light harvesting complex II within lipid bilayers. Key contributions towards cryo-EM and electron crystallography methods included specimen preparation and vitrification, liquid-helium cooling, data collection, and image processing. These methods are now applied almost routinely to both membrane and soluble proteins. Here we outline the advances and the breakthroughs that paved the way towards high-resolution structures by electron crystallography, both in terms of methods development and biological milestones.

Key words: History of electron crystallography, Development of electron cryo-microscopy, Membrane protein structure, cryo-EM, 2D crystallization

1. Introduction

Since its origins almost 50 years ago, electron crystallography of two-dimensional (2D) crystals has become an invaluable and versatile tool in structural biology, along with X-ray crystallography, NMR spectroscopy and the two more recently developed electron microscopy techniques in biology, single-particle electron cryo-microscopy and electron tomography. The first structure of a protein to be determined at high resolution by electron crystallography was bacteriorhodopsin, but others soon followed. With one exception, all were of proteins that form highly ordered crystalline sheets, whereas most membrane proteins prefer to form vesicular 2D arrays that tend to be less well ordered. Electron

Ingeborg Schmidt-Krey and Yifan Cheng (eds.), *Electron Crystallography of Soluble and Membrane Proteins: Methods and Protocols*, Methods in Molecular Biology, vol. 955, DOI 10.1007/978-1-62703-176-9_1,

crystallography is particularly suitable for studying the structure and conformational changes of membrane proteins in their native lipid environment. Developments in electron microscopy (EM) instrumentation, 2D crystallization, and 2D crystal screening are bound to increase the range of proteins, especially eukaryotic membrane proteins, to be investigated by this technique in future.

2. Early Beginnings

The principles of electron crystallography were established by Aaron Klug and David DeRosier, who in the 1960s and 1970s examined the structure of regular arrays of viral proteins in the electron microscope. Protein crystallography with X-rays had just had its first triumphs in resolving the structures of myoglobin (1) and hemoglobin (2, 3) at atomic resolution—pioneering achievements that ushered in a new era in molecular biology, and were recognized by the Nobel prize in chemistry to Max Perutz and John Kendrew in 1962. Working door to door to the protein crystallographers at the Cambridge MRC LMB, Klug and DeRosier realized that the principles of crystallography could also be applied to processing electron microscopic images of crystalline arrays (Fig. 1). All early work was carried out with negatively stained samples. Many of the first specimens to be studied were helical arrays, such as T4 polyheads, bacteriophage tails (4), or actin filaments decorated with myosin heads (5). X-ray fiber diffraction of helical arrays had been used by Rosalind Franklin, Maurice Wilkins, Francis Crick, and James Watson (the latter two also at the MRC LMB) to work out the structure of the DNA double helix, and the principles were the same for Fourier transforms of helical arrays on electron micrographs. The polyheads were wide enough to be flattened on the carbon support film and could be treated as 2D crystals (6, 7). The calculation of Fourier transforms required the most powerful computers available, and progress in image processing paralleled the development of better and faster computers. Still, even in the late 1970s, calculating the Fourier transform of a scanned image area of $1{,}024 \times 1{,}024$ pixels was no small undertaking that required the services of a "main frame" computer, and was best carried out over night. The program was started in the evening, and in the morning a printout of the transform on many meters of paper was arranged on the laboratory floor, and inspected from an elevated vantage point, such as a chair, to spot the reflections on the reciprocal lattice.

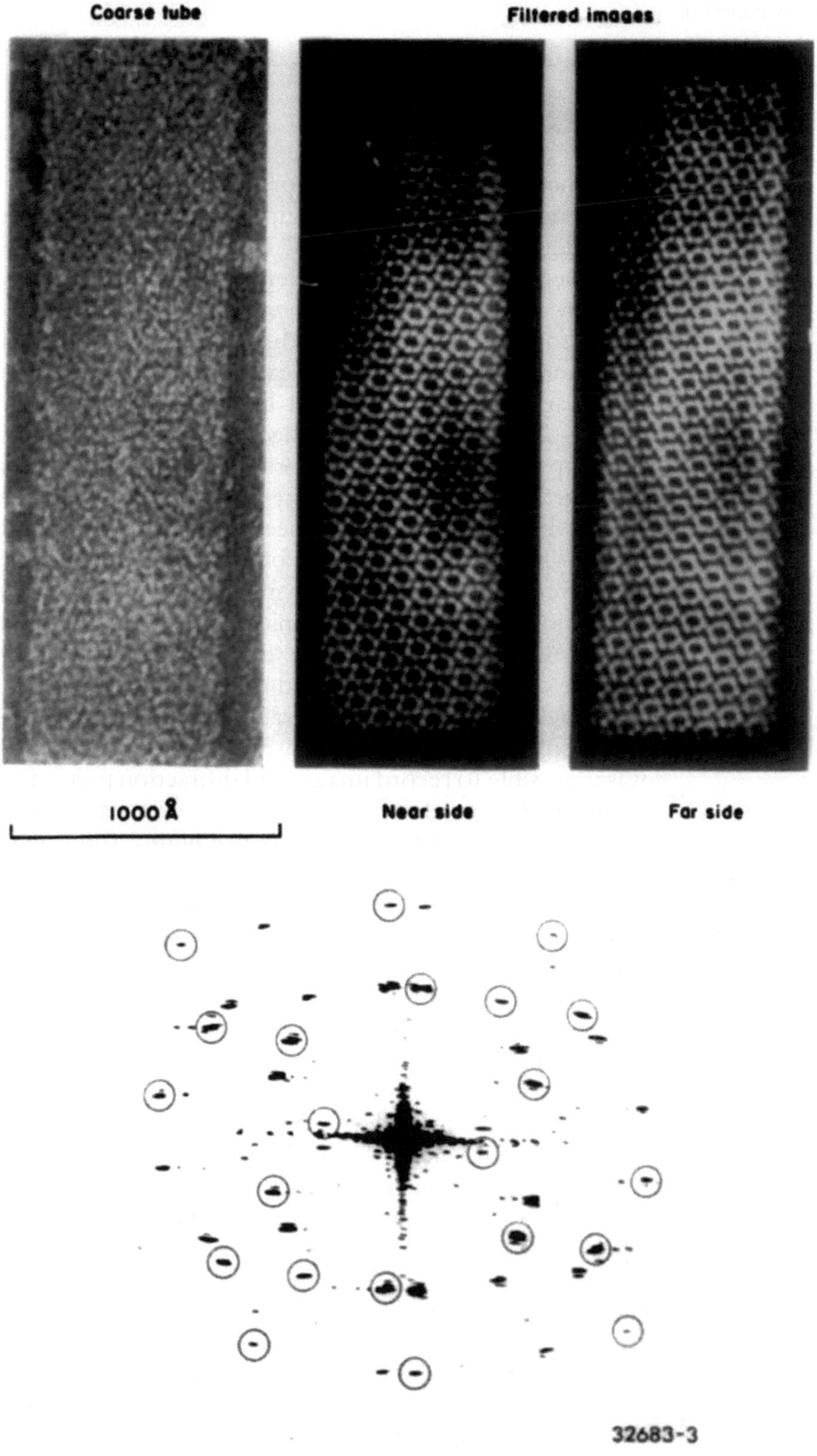

Fig. 1. Electron micrograph of negatively stained T4 polyhead tube, and optically filtered images of the two component layers of the flattened tube. Polyheads are polymers of T4 bacteriophage capsid hexamers, produced by a mutant that lacks the connector protein between T4 head and tail. The tube is 750 Å wide. The optical transform of the electron micrograph is shown below. Reflections from the near-sided lattice are circled (from ref. (7)).

3. Bacteriorhodopsin

It was not long before the first proper 2D crystals were examined by the new technique, by a couple of young scientists who also worked at the Cambridge MRC LMB, next to Perutz, Kendrew, DeRosier, and Klug. Richard Henderson, who had done his PhD in X-ray protein crystallography, and Nigel Unwin, an electron microscopist originally from the Cambridge materials science department, got together to examine the structure of purple membrane, the first (and, as it turns out, still one of the most perfect) 2D protein crystal to be analyzed. The fact that bacteriorhodopsin, the protein that formed the 2D lattice, was a membrane protein added to the interest and the excitement—no other membrane protein had ever been obtained in crystalline form before. The crystalline nature of purple membrane was known from freeze fracture electron microscopy and X-ray powder diffraction (8), but electron microscopy of isolated, unstained purple membranes on carbon film was at first disappointing, because the membranes cracked in the high vacuum of the electron microscope due to dehydration. Only when washed with a dilute solution of glucose to preserve them in a hydrated state was it possible to record images and diffraction patterns—at room temperature, as electron cryo-microscopy had not yet been invented. Electron diffraction revealed high-resolution peaks, but only when the electron dose was reduced to extremely low levels of 1 e/Å^2 or less to avoid radiation damage. In 1975, a map at 7 Å resolution of bacteriorhodopsin was obtained by electron diffraction and image processing of unstained purple membranes tilted in the electron microscope up to 57° (9, 10). The resolution in the direction perpendicular to the crystal plane was lower, due to the limited tilt range, which results in a missing cone of data around the *z*-axis. This affects all structures determined by electron crystallography of planar lattices. In the case of bacteriorhodopsin and other mainly α-helical membrane proteins this does not matter much, however, as most of the structural information is contained in the region of Fourier space covered by tilting to 60°. The 7 Å map of bacteriorhodopsin was the first three-dimensional (3D) structure of a membrane protein, and indeed the first detailed structure of any biological macromolecule obtained by electron microscopy (Fig. 2). The electron diffraction patterns had shown early on (10) that the membranes were highly ordered, and that it was therefore possible to determine the structure of bacteriorhodopsin at high resolution using electrons instead of X-rays. But it took another 15 years, and two major technical advances, before this became a reality.

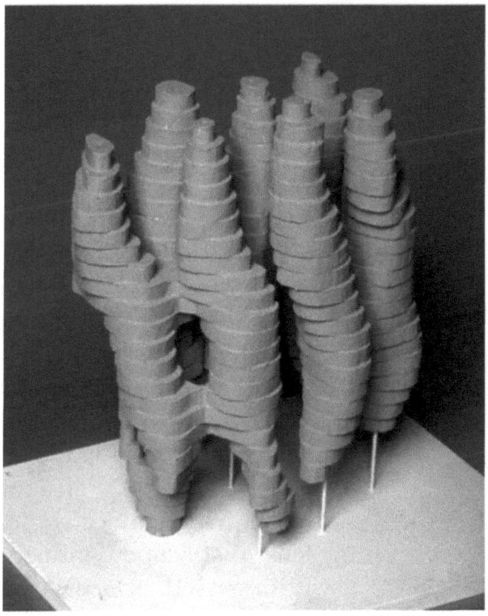

Fig. 2. Balsawood model of bacteriorhodopsin at 7 Å resolution, the first protein structure determined by electron crystallography. The 3D map was generated by processing low-dose images and electron diffraction patterns of glucose-preserved purple membranes recorded at room temperature. The 7 α-helices of the bacteriorhodopsin protomer running roughly perpendicular to the plane of the membrane are clearly resolved (from ref. (9), 1975, with permission from MacMillan Journals).

4. The Advent of Electron Cryo-Microscopy

The first of these technical advances was the development of electron cryo-microscopy. Already in the early 1970s, Robert Glaeser in Berkeley and his PhD student Kenneth Taylor were devising contraptions to freeze biological samples on EM grids, transfer them into the electron microscope at liquid-nitrogen temperature, and to keep them at low temperature in the microscope for imaging or electron diffraction (11). The samples they studied first were thin crystals of catalase, which diffracted electrons to high resolution. For more than 10 years, the JEOL 100B in the Berkeley Donner lab remained the only electron microscope in the world in which high-resolution electron micrographs and diffraction patterns of biological specimens could be recorded at low temperature almost routinely. The lowest temperature that could be used was −100°C because below this, ice contamination built up due to the relatively poor vacuum. A few years later, Jacques Dubochet and his team at the EMBL Heidelberg devised a way to vitrify samples in holey carbon films by rapid freezing in liquid ethane, which avoids the formation of damaging ice crystals. The EM400 microscope produced by the Philips company had a much better vacuum, and a double-blade cold trap also developed at EMBL prevented ice contamination. The Heidelberg team succeeded in preparing vitrified specimens of icosahedral viruses, and to image them at low temperature in the electron microscope (12). For biological samples, vitrification in buffer was a big improvement on air-drying in glucose, which only the most indestructible proteins such as bacteriorhodopsin survive. However, sugar embedding is still the most frequently used preparation technique for 2D crystals on continuous support films, using trehalose rather than glucose or tannin (see below) combined with freezing in liquid ethane or liquid nitrogen—very rapid cooling rates as for vitrification are not needed for sugar-embedded samples.

Dubochet and colleagues experimented with a prototypic superconducting objective lens, a marvel of technology in which the lens coil of superconducting wire, as well as the specimen itself, were immersed in a cryostat cooled with liquid helium (13). The Dietrich lens was built by the company Siemens, who however soon stopped the production of electron microscopes, which they deemed unprofitable. Only three lenses of this type were built. The one in the microscope operated by Fritz Zemlin at the Fritz-Haber-Institute in Berlin later yielded the best images of bacteriorhodopsin that diffracted to 2.8 Å resolution (14), but it took 70 different manual operations to record each single micrograph. A more user-friendly electron microscope in which the sample, but not the whole lens, was cooled with liquid helium in a top-entry stage was developed by Yoshinori Fujiyoshi at Osaka University (15), and commercialized by the company JEOL. This instrument has yielded more structures of 2D crystals at high resolution than any other. Liquid-helium cooling is an advantage for imaging 2D crystals on continuous support film, because it delays the onset of radiation damage roughly by a factor of two in radiation dose, compared to liquid nitrogen. Consequently, electron doses of 20 e/Å^2 or more can be used at liquid helium temperature to record high-resolution data. However, for vitrified samples on holey carbon films used for single-particle processing or electron tomography, liquid-helium cooling can be problematic. The conductivity of water, carbon or protein, which is already low at ambient or liquid nitrogen temperature, becomes even lower in samples cooled with liquid helium. Electrostatic charging and beam-induced movement is then a serious problem, unless special precautions are taken, as for example in case of the nicotinic acetylcholine receptor (16). Charging and beam-induced movement are much reduced by spot-scan imaging (17), which is used routinely in some inveterate electron crystallography laboratories, but for imaging crystal lattices it requires parallel illumination at small spotsize, which is achieved by using a small, 10 or 20 μm, C2 aperture.

Side-entry cryo-transfer stages for routine use at liquid nitrogen temperature did not arrive on the scene until later. Standard side-entry cold stages have not yielded nearly as many high-resolution structures, because they are mechanically less stable and prone to thermal drift.

5. High-Resolution Image Processing

Another technical development had to happen before structures of biological macromolecules could be determined by electron microscopy at close-to-atomic resolution: image processing routines needed to be improved so that the high-resolution phase information could be retrieved from the scanned film negatives. Film is still an excellent medium for recording high-resolution images of 2D crystals (or indeed any biological object), because the number of pixels of CCD cameras is limited and their detective quantum efficiency (DQE) is poor, especially at high resolution and at high voltages. Compared to a sheet of EM film, the sensitive area of even an 8 K CCD camera is small. However, the large, solid-state CMOS pixel detectors that are now coming on line are set to finally replace film, also for 2D crystals (see also Chapter 20).

High-resolution crystallographic image processing involves several rounds of lattice unbending (18), because 2D protein lattices, although more perfect than 3D crystals in terms of mosaicity, are subject to distortions. The lattice distortions cause a broadening of reflections in the Fourier transform. The reflections are then more difficult to detect, especially at high resolution. This is less of a problem in electron diffraction, for which large (1 μm, more for large unit cells), well-ordered 2D crystals are needed. More scattered electrons then contribute to a diffraction peak, and even the broadened high-resolution reflections are intense enough to be measurable. This makes processing of electron diffraction patterns more straightforward than high-resolution image processing. For highly tilted 2D crystals, specimen flatness is a major issue, as reflections become progressively blurred in the direction perpendicular to the tilt axis, due to slight curvature or local surface roughness of the support film, but the problem can be overcome by careful specimen preparation (19, 20). Apart from lattice unbending, major improvements in crystallographic image processing included an objective assessment of reflection amplitudes and phases in terms of figures of merit, and a quantitative treatment of background noise. With these improvements, a 3.5 Å map of bacteriorhodopsin was obtained, and an atomic model built—the first of a biological macromolecule derived by electron microscopy (21, 22). In the 15 years it took to progress from the

7 Å map to the first atomic model of bacteriorhodopsin, a whole new set of techniques and programs had been devised. The new method of high-resolution electron crystallography was now ready to be applied to other 2D crystals.

6. More High-Resolution Structures from Crystalline Sheets

Already while the experimental and computational procedures for high-resolution electron crystallography were being put in place, researchers in various laboratories around the world were growing well-ordered 2D crystals of other proteins. One of these was the light-harvesting complex of plant photosystem-II (LHC-II), like bacteriorhodopsin an interesting membrane protein. LHC-II is isolated from leaves of green plants easily and in large quantities. Untypically for a membrane protein, it crystallizes readily in two and three dimensions (23, 24), even though, unlike bacteriorhodopsin, it does not form crystalline arrays in vivo. In 2D crystals of LHC-II, the crystal plane contains a twofold axis of symmetry so that the orientation of the molecules on the lattice alternates up and down. Many 2D crystals grown in vitro from isolated, purified protein have an in-plane twofold axis (even more common is a twofold screw axis). The 2D crystals of LHC-II, being grown in the presence of detergent, did not take well to washing with glucose, and the success rate with frozen-hydrated specimens was low, but a dilute tannin solution proved to be excellent for preserving the crystalline order for data collection at low temperature (25). Data were recorded and processed by the procedures that had been developed for bacteriorhodopsin, and a combination of amplitudes from electron diffraction patterns collected at home at liquid nitrogen temperature, and phases from images recorded with Fujiyoshi's liquid helium-cooled instrument in Osaka, yielded a 3.4 Å map (26). This was the second near-atomic resolution structure of a protein determined by electron microscopy.

Next came α,β-tubulin, the main component of the microtubule cytoskeleton in eukaryotic cells. Tubulin is not a membrane protein, but the only known soluble or filamentous protein that forms highly ordered 2D sheets, in this case in the presence of zinc ions (27). The 3.7 Å resolution electron crystallographic structure of tubulin (28) remains unique, as there are still no 3D crystals suitable for X-ray crystallography, whereas there are now higher-resolution X-ray structures of bacteriorhodopsin at 1.55 Å (29) and of LHC-II at 2.5 Å (30). It is worth nothing though that these two X-ray structures were solved by molecular replacement with EM data, in the case of the 2.5 Å resolution X-ray structure of LHC-II by molecular replacement with the EM map rather than the atomic model.

The aquaporins, a class of eukaryotic membrane proteins forming water channels in the plasma membrane, appear to be particularly suitable for electron crystallography, perhaps because they are fairly rigid, square-shaped tetramers that produce highly ordered 2D crystals readily. Some aquaporins form regular arrays already in the cell membrane, which is always a good indication that a membrane protein might be induced to form large, highly ordered 2D crystals. Accordingly, well-ordered 2D crystals were grown, and structures solved, of aquaporin-1 at 3.8 Å (31), and of aquaporin-0 at 1.9 Å resolution (32). A recent study of lipid-protein interactions in 2D crystals of aquaporin-0 grown in two different lipids revealed that the lipid adapts to the protein, rather than the other way round (33). Aquaporin-0 holds the resolution record for a protein structure by electron crystallography, but it was determined by phasing electron diffraction data by molecular replacement with the X-ray structure of aquaporin-1 (34), rather than with image phases. For phasing by image processing the resolution limit is currently around 3 Å (35).

Another example of a membrane protein that has given a high-resolution 3D structure by electron crystallography is the microsomal prostaglandin E synthase at 3.5 Å resolution (36), published soon after the 4 Å X-ray structure of the similar 5-lipoxygenase-activating protein (37). It should be said, however, that a 3.5 Å resolution electron crystallographic map is usually significantly better than an X-ray map at the same nominal resolution, because of the higher quality of the phase information obtained by image processing.

Excellent, well-ordered crystalline sheets have also been grown of several bacterial outer membrane porins, such as PhoE (38), OmpF (39), and OmpG (40), which are pore-forming β-barrels. Projection maps or low-resolution 3D maps of these porins have been published, but so far there are no atomic models derived by electron crystallography, because the β-barrel outer membrane porins generally crystallize well in 3D, and are then solved more easily by X-ray crystallography.

7. Tubular 2D Crystals

Many more membrane proteins form crystalline tubular vesicles rather than sheet-like crystals, most likely because the open hydrophobic edge around a crystalline sheet is energetically unfavorable, so it tends to close into a cylinder, unless the protein has a strong predisposition to form planar layers. This appears to be the case for bacteriorhodopsin, LHC-II, and the aquaporins. If the protein does not have this predisposition—which would be difficult to predict, and even more difficult to influence—it usually forms tubular vesicles that are most often 0.5–1 μm wide. Tubular crystals of these dimensions are easily distorted by surface forces, which precludes helical processing. They flatten on the support film and can then be treated as two superposed 2D lattices, as in the case of the T4 polyheads (Fig. 1). However, the fact that they form crystalline vesicles rather than open-edged sheets means that the forces between the proteins on the 2D lattice are too weak to resist vesicle closure. Probably for this reason, most 2D crystals of this type are less well ordered than sheet-like crystals.

An exception is the nicotinic acetylcholine receptor (nAChR), a ligand-gated ion channel from the postsynaptic membrane, so far the only high-resolution structure of a membrane protein to have been determined from tubular crystals. The receptor forms ~1,000 Å-wide tubes, with the 2D lattice on the wall of a cylinder. The tubes are narrow and strong enough not to be flattened by surface forces in the thin layer of buffer on a holey support film, and therefore can be processed as helical arrays, so the grid does not need to be tilted in the electron microscope. On the other hand, individual tubes are too narrow for electron diffraction, so the structure had to be determined without the aid of high-resolution diffraction amplitudes. Nonetheless, helical processing of images of a large number of tubular crystals recorded at liquid helium temperature have yielded the structure of nAChR in several different states, most recently at 4 Å resolution (16), a real tour de force of crystallographic electron image processing.

To date, none of the flattened tubular crystals have produced a map of similar quality. The best collapsed tubular vesicles, such as the bacterial sodium/proton antiporter NhaA (41) or ATP synthase c-rings (42) go to ~4 Å resolution, but the 3D maps of these proteins (42, 43) have not resolved side chains. Yet the 3D structure of ATP synthase c-ring provided sufficient information to build a model (42) with significant agreement with the later high resolution structure by X-ray crystallography, for which the EM structure provided the necessary phasing data. More often, 3D maps from this type of crystal go to ~6–8 Å resolution in-plane. At this resolution, transmembrane α-helices show up nicely, and the maps can serve as an excellent template for molecular modelling, if a high-resolution structure of a related protein is available, as in the recent example of the archaeal sodium/proton antiporter NhaP1 (44). The comparatively poor resolution of 3D maps from these planar-tubular arrays is all the more frustrating since single-particle processing of icosahedral (45) or helical viruses (46) has recently achieved resolutions as good as, or even better than, those obtained by image processing of the best 2D arrays. This means that the viruses must be more rigid and more highly ordered than collapsed tubular crystals, even though the latter contain many more copies of the protein to be averaged on the 2D lattice. The most likely reason is that, due to weak lattice forces, the crystal contacts are not strong enough to prevent small rotations or distortions of individual molecules. In principle these rotations and distortions can be detected by cross correlation with a reference and then corrected, as in lattice unbending, but at present the signal/noise ratio of electron micrographs is too low to do this for individual proteins, as would be necessary to correct rotations in 3D, or to sort out distorted protein molecules. Perhaps electron-optical phase plates that should improve the signal/noise ratio of the image, or new detectors with a quantum efficiency better than film will make this possible in future. At present, lattice distortions can be reliably corrected only in two dimensions, with a reference comprising several unit cells, which accounts for the limited resolution achieved with less-than-perfect 2D crystals. The key factor limiting resolution thus appears to be the molecular mass of the structural unit that can be considered as rigid. The viruses and helical structures that have given high-resolution maps usually consist of rigid structural units greater than 1 MDa. The size of mosaic domains in crystalline sheets that give high-resolution structures are also larger than ~1 MDa.

8. Thin 3D Crystals

While membrane proteins tend to form 2D lattices, this is not the case for soluble or filamentous proteins, which normally crystallize in three dimensions. Still, many of these proteins form thin, plate-like crystals that are several unit cells thick. These crystals can be highly ordered and are, in principle, suitable for structure determination to high resolution by electron crystallography. For example, the earliest electron diffraction patterns of biological samples were recorded of thin crystals of catalase, a soluble enzyme, either with glucose-embedded samples at room temperature (10), or with frozen grids (11). Likewise, many membrane proteins form multilayers of 2D crystals, which are sometimes highly ordered. However, at this writing, there are more than 60,000 protein structures in the protein data bank that have been determined by X-ray crystallography, but not a single one by electron crystallography of thin 3D crystals (as mentioned above, tubulin was solved with 2D sheets). How can this be? It is certainly not due to lack of trying. There have been heroic efforts, most notably with crotoxin (47), a small phospholipase from snake venom, and sarcoplasmic calcium ATPase (48), a membrane calcium pump of substantial importance. Both form thin 3D crystals that diffract electrons to 2.2 Å and 3.5 Å, respectively, and have yielded phases by image processing. But structure determination with such thin 3D crystals is complicated because the phases and amplitudes of diffraction peaks depend on the number of unit cells in the third dimension, i.e., on the exact thickness of each individual crystal, which is difficult to measure. This presents a serious obstacle especially when the crystals are tilted. Although these difficulties can be overcome in principle, the associated practical problems proved to be insurmountable, at least on a timescale to make electron crystallography of small 3D crystals competitive with X-ray crystallography. Small crystals grew bigger, phase problems were solved, and in the end both structures were determined by X-ray crystallography (49, 50) before suitable methods were in place for doing this with electrons. The diffraction of sub-micron sized 3D protein crystals is now being studied with extremely intense X-ray pulses from free electron lasers. It will be interesting to see if this new approach can compete with X-ray crystallography of larger protein crystals, which has become highly streamlined and stupendously successful in the past 20 years.

9. Advantages of Electron Crystallography for Membrane Proteins

Electron crystallography has been particularly useful for studying membrane proteins, which have an inherent tendency to form 2D lattices in the plane of a lipid bilayer. A few of them, such as bacteriorhodopsin, aquaporin-0 and connexins, already form crystalline arrays in vivo. In a 2D crystal there are fewer degrees of freedom for a membrane protein, and therefore, the chances of success are greater for 2D than for 3D crystallization of membrane proteins. At the same time, there are fewer parameters to vary compared to 3D crystallization, which makes it more difficult to enhance the order of 2D crystals by optimizing conditions for crystal growth. Consequently, 2D crystals are often less well ordered, because there are fewer crystal contacts, and those that are present tend to be hydrophobic and less strong than the polar or ionic contacts in a 3D protein crystal. In 3D crystallization of membrane proteins, carefully optimized conditions often improve the crystal quality from intermediate resolution (7–10 Å) to 3.0 Å or better. This has rarely, if ever, been possible with 2D crystals: good crystals tend to be highly ordered at the outset, but those that are of inferior quality do not often improve much, even with considerable effort. On the other hand, 3D crystallization of membrane proteins is still a slow and tedious process, and this is not likely to change in the foreseeable future. Even when 3D crystals are obtained, they are rarely very good to begin with, and improving their quality to the point where they are suitable for structure determination by X-ray crystallography usually takes years. During this time, little or no structural information is forthcoming. By comparison, 2D crystals often form quite readily and yield important, usually unique information about the structure and function of the protein at every stage, even at comparatively low resolution. Not least for this reason, electron microscopy of 2D crystals will remain an important technique for membrane proteins, especially for eukaryotic membrane proteins, which are difficult to produce and purify in quantities required for 3D crystallization.

2D crystals of membrane proteins have another important advantage. In the 2D lattice, the protein is surrounded by lipids, as in its own native environment. Not only does this mean that the protein is more stable than in detergent solution—2D crystals of membrane proteins often can be kept for months or years without deteriorating—but the protein also behaves as if it were in a membrane. In the fluid, but tenacious membrane lipid environment, movements and conformational changes can occur unhindered, whereas lattice contacts between the hydrophilic, surface exposed protein regions would either prevent such movements in 3D crystals, or these contacts would break. In 2D crystals, conformational changes between different functional states, which are implicit in the molecular mechanism of many membrane proteins, especially transporters, receptors, and ion channels, can be studied comparatively easily, whereas in X-ray crystallography, each new conformation presents in principle a new crystallization problem. In 2D crystals, conformational changes of the protein can be triggered directly on the EM grid, for example by laser flashes in bacteriorhodopsin (51), exposure to chemical ligands in nAChR (52), or incubation at different ionic or pH conditions in sodium/proton antiporters (53), followed in each case by rapid freezing in liquid ethane or liquid nitrogen to fix the new conformation. Studies of this kind may be possible only with 2D crystals, where each protein is freely accessible from the surrounding medium, and can move in the membrane environment without disrupting lattice contacts in the third dimension such as in 3D crystals.

10. Outlook

What would it take to make electron crystallography of 2D crystals as successful as X-ray protein crystallography? Computational procedures for electron crystallography have become largely routine. There is room for improvement, especially with respect to automation and user-friendliness, but the new 2dx suite by Stahlberg and colleagues has improved the situation significantly. There have been major improvements in electron microscope design in the past decade, to the point where single-particle processing at near-atomic resolution is now possible, and phase plates and better detectors may improve image contrast to enable single-unit cell unbending in 3D.

The main limitation is the availability of 2D crystals suitable for structure analysis at high resolution. Major bottlenecks are 2D crystal growth itself, and identifying promising 2D crystals in the crystallization volume. Devices for 2D crystallization by dialysis in a multi-well format have been designed and tested, for example in the laboratory of Andreas Engel in Basel, but sample volumes are still quite large compared to 3D crystallization screens. Not nearly as many conditions can be tried, also because for further evaluation, at least one EM grid has to be prepared and examined for each condition. Although in future it may become possible to screen for 2D crystals by light microscopy or spectroscopy, at present the success or failure of a 2D crystallization experiment can be assessed reliably only by electron microscopy. In theory, it is possible to do this automatically by electron diffraction, but few 2D crystals have been sufficiently large and well ordered for this approach. Moreover, electron diffraction requires preparation conditions that need to be carefully optimized for each protein. Smaller, less well-ordered 2D crystals can be spotted in the Fourier transform of images recorded on CCD, but attempts to automate this process have so far met with little success. Other steps that do not lend themselves easily to automation are the preparation of EM grids, and mounting them into a specimen holder for screening in the electron microscope, although progress has been made on both fronts. Still, for the time being, 2D crystal screening is most effectively done by eye in the electron microscope, and requires the skills of an experienced electron microscopist, as does specimen preparation and grid mounting. This is a serious disadvantage compared to X-ray protein crystallography, where 3D crystals can be visualized by light microscopy, and discriminated from other objects by comparatively straightforward edge detection techniques, or by following the growth of individual crystals. However, in principle all steps for 2D crystallization and 2D crystal screening can, and no doubt will, be automated once there is sufficient demand.

Today, high-resolution electron cryo-microscopy of 2D crystals is an essential technique for studying the structure and dynamics of membrane proteins in their native lipid environment. This book provides a unique and uniquely useful guide with contributions from many experts in the field. The book is ideal reading material both for beginners seeking advice on how to get started, and for experts wanting to inform themselves on detailed aspects of the method. It also serves as an excellent survey of complementary techniques. No similarly comprehensive volume covering all aspects of electron crystallography has ever been compiled, and therefore, this book will be and remain a standard reference in the field for many years to come.

References

1. Kendrew JC, Bodo G, Dintzis HM, Parrish RG, Wyckoff H, Phillips DC (1958) A three-dimensional model of the myoglobin molecule obtained by x-ray analysis. Nature 181: 662–666
2. Perutz MF, Miurhead H, Cox JM, Goaman LC, Mathews FS, McGandy EL, Webb LE (1968) Three-dimensional Fourier synthesis of horse oxyhaemoglobin at 2.8 A resolution: (1) x-ray analysis. Nature 219:29–32
3. Perutz MF, Muirhead H, Cox JM, Goaman LC (1968) Three-dimensional Fourier synthesis of horse oxyhaemoglobin at 2.8 A resolution: the atomic model. Nature 219:131–139
4. King J (1968) Assembly of the tail of bacteriophage T4. J Mol Biol 32:231–262
5. Moore PB, Huxley HE, DeRosier DJ (1970) Three-dimensional reconstruction of F-actin, thin filaments and decorated thin filaments. J Mol Biol 50:279–295
6. Finch JT, Klug A, Stretton AO (1964) The structure of the "Polyheads" of T4 bacteriophage. J Mol Biol 10:570–575
7. Yanagida M, DeRosier DJ, Klug A (1972) Structure of the tubular variants of the head of bacteriophage T4 (polyheads). II. Structural transition from a hexamer to a 6+1 morphological unit. J Mol Biol 65:489–499
8. Blaurock AE, Stoeckenius W (1971) Structure of the purple membrane. Nat New Biol 233: 152–155
9. Henderson R, Unwin PN (1975) Three-dimensional model of purple membrane obtained by electron microscopy. Nature 257: 28–32
10. Unwin PN, Henderson R (1975) Molecular structure determination by electron microscopy of unstained crystalline specimens. J Mol Biol 94:425–440
11. Taylor KA, Glaeser RM (1974) Electron diffraction of frozen, hydrated protein crystals. Science 186:1036–1037
12. Adrian M, Dubochet J, Lepault J, McDowall AW (1984) Cryo-electron microscopy of viruses. Nature 308:32–36
13. Dietrich I, Formanek H, Fox F, Knapek E, Weyl R (1979) Reduction of radiation damage in an electron microscope with a superconducting lens system. Nature 277:380–381
14. Baldwin JM, Henderson R, Beckman E, Zemlin F (1988) Images of purple membrane at 2.8 A resolution obtained by cryo-electron microscopy. J Mol Biol 202:585–591
15. Fujiyoshi Y, Mizusaki T, Morikawa H, Yamagashi H, Aoki Y, Kihara H, Harada Y (1991) Development of a superfluid helium stage for high-resolution electron microscopy. Ultramicroscopy 38:241–251
16. Unwin N (2005) Refined structure of the nicotinic acetylcholine receptor at 4 A resolution. J Mol Biol 346:967–989
17. Downing KH (1991) Spot-scan imaging in transmission electron microscopy. Science 251:53–59
18. Henderson R, Baldwin J, Downing KH, Lepault J, Zemlin F (1986) Structure of purple membrane from Halobacterium Halobium: recording, measurement and evaluation of electron micrographs at 3.5 A resolution. Ultramicroscopy 19:147–178
19. Butt HJ, Wang DN, Hansma PK, Kühlbrandt W (1991) Effect of surface roughness of carbon support films on high resolution electron diffraction of two-dimensional protein crystals. Ultramicroscopy 36:307–318
20. Gyobu N, Tani K, Hiroaki Y, Kamegawa A, Mitsuoka K, Fujiyoshi Y (2004) Improved specimen preparation for cryo-electron microscopy using a symmetric carbon sandwich technique. J Struct Biol 146:325–333
21. Henderson R, Baldwin JM, Ceska TA, Zemlin F, Beckmann E, Downing KH (1990) A model for the structure of bacteriorhodopsin based on high resolution electron cryo-microscopy. J Mol Biol 213:899–929
22. Grigorieff N, Ceska TA, Downing KH, Baldwin JM, Henderson R (1996) Electron-crystallographic refinement of the structure of bacteriorhodopsin. J Mol Biol 259:393–421
23. Kühlbrandt W (1984) 3-Dimensional structure of the light-harvesting chlorophyll a/b-protein complex. Nature 307:478–480
24. Kühlbrandt W (1987) Three-dimensional crystals of the light-harvesting chlorophyll a/b protein complex from pea chloroplasts. J Mol Biol 194:757–762
25. Wang DN, Kühlbrandt W (1991) High-resolution electron crystallography of light-harvesting chlorophyll a/b protein complex in three different media. J Mol Biol 217:691–699
26. Kühlbrandt W, Wang DN, Fujiyoshi Y (1994) Atomic model of plant light-harvesting complex by electron crystallography. Nature 367: 614–621
27. Baker TS, Amos LA (1978) Structure of the tubulin dimer in zinc-induced sheets. J Mol Biol 123:89–106
28. Nogales E, Wolf SG, Downing KH (1998) Structure of the alpha beta tubulin dimer by electron crystallography. Nature 391:199–203

29. Luecke H, Schobert B, Richter HT, Cartailler JP, Lanyi JK (1999) Structure of bacteriorhodopsin at 1.55 angstrom resolution. J Mol Biol 291:899–911
30. Standfuss J, Terwisscha van Scheltinga A, Lamborghini M, Kühlbrandt W (2005) Mechanisms of photoprotection and non-photochemical quenching in pea LHC-II at 2.5 Å resolution. EMBO J 24:919–928
31. Murata K, Mitsuoka K, Hirai T, Walz T, Agre P, Heymann JB, Engel A, Fujiyoshi Y (2000) Structural determinants of water permeation through aquaporin-1. Nature 407:599–605
32. Gonen T, Cheng YF, Sliz P, Hiroaki Y, Fujiyoshi Y, Harrison SC, Walz T (2005) Lipid-protein interactions in double-layered two-dimensional AQPO crystals. Nature 438:633–638
33. Hite RK, Li Z, Walz T (2010) Principles of membrane protein interactions with annular lipids deduced from aquaporin-0 2D crystals. EMBO J 29:1652–1658
34. Sui H, Han BG, Lee JK, Walian P, Jap BK (2001) Structural basis of water-specific transport through the AQP1 water channel. Nature 414:872–878
35. Henderson R (1995) The potential and limitations of neutrons, electrons and X-rays for atomic resolution microscopy of unstained biological molecules. Q Rev Biophys 28: 171–193
36. Jegerschöld C, Pawelzik SC, Purhonen P, Bhakat P, Gheorghe KR, Gyobu N, Mitsuoka K, Morgenstern R, Jakobsson PJ, Hebert H (2008) Structural basis for induced formation of the inflammatory mediator prostaglandin E2. Proc Natl Acad Sci USA 105: 11110–11115
37. Ferguson AD, McKeever BM, Xu S, Wisniewski D, Miller DK, Yamin TT, Spencer RH, Chu L, Ujjainwalla F, Cunningham BR, Evans JF, Becker JW (2007) Crystal structure of inhibitor-bound human 5-lipoxygenase-activating protein. Science 317:510–512
38. Jap BK (1988) High-resolution electron diffraction of reconstituted PhoE porin. J Mol Biol 199:229–231
39. Sass HJ, Buldt G, Beckmann E, Zemlin F, van Heel M, Zeitler E, Rosenbusch JP, Dorset DL, Massalski A (1989) Densely packed beta-structure at the protein-lipid interface of porin is revealed by high-resolution cryo-electron microscopy. J Mol Biol 209:171–175
40. Behlau M, Mills DJ, Quader H, Kühlbrandt W, Vonck J (2001) Projection structure of the monomeric porin OmpG at 6 Å resolution. J Mol Biol 305:71–77
41. Williams KA, Geldmacher-Kaufer U, Padan E, Schuldiner S, Kühlbrandt W (1999) Projection structure of NhaA, a secondary transporter from Escherichia coli, at 4.0 A resolution. EMBO J 18:3558–3563
42. Vonck J, Krug von Nidda T, Meier T, Matthey U, Mills DJ, Kühlbrandt W, Dimroth P (2002) Molecular architecture of the undecameric rotor of a bacterial Na+-ATP synthase. J Mol Biol 321:307–316
43. Williams K (2000) Three-dimensional structure of the ion-coupled transport protein NhaA. Nature 403:112–115
44. Goswami P, Paulino C, Hizlan D, Vonck J, Yildiz O, Kühlbrandt W (2011) Structure of the archaeal Na+/H+ antiporter NhaP1 and role of its uncleaved N-terminal signal sequence. EMBO J, 30:439–449
45. Zhang X, Jin L, Fang Q, Hui WH, Zhou ZH (2010) 3.3 A cryo-EM structure of a nonenveloped virus reveals a priming mechanism for cell entry. Cell 141:472–482
46. Sachse C, Chen JZ, Coureux PD, Stroupe ME, Fandrich M, Grigorieff N (2007) High-resolution electron microscopy of helical specimens: a fresh look at tobacco mosaic virus. J Mol Biol 371:812–835
47. Jeng TW, Chiu W (1983) Low dose electron microscopy of the crotoxin complex thin crystal. J Mol Biol 164:329–346
48. Shi D, Hsiung HH, Pace RC, Stokes DL (1995) Preparation and analysis of large, flat crystals of Ca(2+)-ATPase for electron crystallography. Biophys J 68:1152–1162
49. Toyoshima C, Nakasako M, Nomura H, Ogawa H (2000) Crystal structure of the calcium pump of sarcoplasmic reticulum at 2.6 A resolution. Nature 405:647–655
50. Marchi-Salvador DP, Correa LC, Magro AJ, Oliveira CZ, Soares AM, Fontes MR (2008) Insights into the role of oligomeric state on the biological activities of crotoxin: crystal structure of a tetrameric phospholipase A2 formed by two isoforms of crotoxin B from Crotalus durissus terrificus venom. Proteins 72:883–891
51. Subramaniam S, Henderson R (2000) Molecular mechanism of vectorial proton translocation by bacteriorhodopsin. Nature 406: 653–657
52. Unwin N (1995) Acetylcholine receptor channel imaged in the open state. Nature 373:37–43
53. Appel M, Hizlan D, Vinothkumar KR, Ziegler C, Kühlbrandt W (2009) Conformations of NhaA, the Na/H exchanger from Escherichia coli, in the pH-activated and ion-translocating states. J Mol Biol 386:351–365

Chapter 2

Practical Aspects in Expression and Purification of Membrane Proteins for Structural Analysis

Kutti R. Vinothkumar, Patricia C. Edwards, and Joerg Standfuss

Abstract

A surge of membrane protein structures in the last few years can be attributed to advances in technologies starting at the level of genomes, to highly efficient expression systems, stabilizing conformational flexibility, automation of crystallization and data collection for screening large numbers of crystals and the microfocus beam lines at synchrotrons. The substantial medical importance of many membrane proteins provides a strong incentive to understand them at the molecular level. It is becoming obvious that the major bottleneck in many of the membrane projects is obtaining sufficient amount of stable functional proteins in a detergent micelle for structural studies. Naturally, large effort has been spent on optimizing and advancing multiple expression systems and purification strategies that have started to yield sufficient protein and structures. We describe in this chapter protocols to refold membrane proteins from inclusion bodies, purification from inner membranes of *Escherichia coli* and from mammalian cell lines.

Key words: Expression systems, *Escherichia coli*, Inclusion bodies, HEK cells, Rhodopsin

1. Introduction

The primary requirement of any successful structural project is a necessity for abundant source material. Only a handful of membrane proteins are found in such high amounts so that they can be directly isolated from the native membranes (1). Most membrane proteins are expressed in low abundance hence there is a need to identify an optimal expression system that produces sufficient material for structure determination but at the same time preserving their function. Due to their inherent hydrophobic nature, over-expression of membrane proteins is not straightforward and often results in proteins targeted to the wrong cellular compartments, misfolded

Ingeborg Schmidt-Krey and Yifan Cheng (eds.), *Electron Crystallography of Soluble and Membrane Proteins: Methods and Protocols*, Methods in Molecular Biology, vol. 955, DOI 10.1007/978-1-62703-176-9_2,

and therefore not functional and/or degraded (2). Over the years several different heterologous over-expression systems have been identified facilitating structure determination of a large variety of membrane proteins obtained from bacterial, yeast, insect, and human cells (3). A recent addition includes cell-free translation of membrane proteins that has shown promise in obtaining functional proteins (4) and in one case structure (5). We would like to recommend the readers the following reviews (3, 6–8) and book chapters in Methods in Molecular Biology (9) for in depth description of different expression systems available for membrane proteins.

The process to obtain a good quality protein for structural studies can be divided into two steps: (1) finding a suitable host for expression and (2) extraction of protein from cellular membranes and subsequent purification. The three examples of membrane proteins described here include protein expressed as inclusion bodies or in the inner membranes of *Escherichia coli* and in HEK cells that have resulted in crystals and eventually structure (10–14). Expression of proteins as inclusion bodies results in large yield, however a need to refold often limits the final amount of protein for structural analysis. Structures of a number of β-barrel membrane proteins refolded from inclusion bodies have been successfully determined (15). There are also numerous examples of α-helical membrane proteins that have been refolded from inclusion bodies and shown to be functional but their structures yet to be determined (16–19). The expression into *E. coli* inner membranes has so far been the most successful for membrane proteins from prokaryotic source, and the relative ease with which huge quantity of protein can be produced makes it still the most attractive system. Recent success in determining structure of eukaryotic membrane proteins with material obtained from COS or HEK cells are promising (13, 14, 20). These include the well-studied bovine rhodopsin and trimeric Rh protein of the ammonia transporter family. Although rhodopsin can be isolated in abundance from bovine retinas, the advantage of over-expression can be exemplified with the ability to make mutants that are medically relevant or that stabilize a particular conformation (27). The procedures described here is applicable to any membrane protein of interest since most proteins upon extraction from membranes are enriched using specific affinity tags, however depending on the expression system certain modifications need to be made or depending on protein additional steps/factors may be required for purification.

2. Materials

2.1. Protein Expression

1. Expression vectors: OmpG without leader sequence or an affinity tag is expressed from a T7 promotor in a pET26b vector (see Note 1). GlpG is expressed with a C-terminal his

tag in pET25b vector. Rhodopsin is expressed in pACMVtetO vector (21).

2. Strains: *E. coli* C41 or C43 (see Note 2).
   Human embryonic kidney (HEK) cells (HEK-293S-GnTI$^-$, a mutant cell line for restricted and homogenous glycosylation) (22).
3. Antibiotics: The stock solutions of antibiotics are ampicillin (100 mg/mL), kanamycin (100 mg/mL), blasticidin (5 mg/mL), geneticin (50 mg/mL), tetracycline (2 mg/mL). Tetracycline stock is made in ethanol, the rest of the antibiotics can be dissolved in water.
   Final concentrations of antibiotics used: ampicillin (100 μg/mL), kanamycin (50 μg/mL), blasticidin (5 μg/mL), geneticin –G418 (200 μg/mL), tetracycline (2 μg/mL).
4. Media for expression in *E. coli*:
   LB medium:
   - 10 g Bacto tryptone.
   - 5 g Bacto yeast extract.
   - 5 g NaCl.

   2XYT:
   - 16 g Bacto tryptone.
   - 10 g Bacto yeast extract.
   - 5 g NaCl.
   - 2 g Glucose.
5. Media for propagation and expression in HEK-293S cells:
   Dulbecco's modified Eagle's medium/F12 (DMEM) supplemented with 10% fetal calf serum (FCS), PenStrep (Gibco), Geneticin-G418 and blasticidin.
   Freestyle medium (Invitrogen) supplemented with 5% FCS and PenStrep. (Alternatively, Protein expression medium from Gibco with 10% newborn calf serum, NBCS can be used. NBCS is more cost effective).
   IPTG (isopropyl-β-d-1-thiogalactopyranoside):
   A 1 M stock is prepared in water and stored at –20°C.
6. Detergents:
   20% stock solutions of detergents *n*-Octyl-β-d-glucopyranoside, *n*-Nonyl-β-d-glucopyranoside, *n*-Decyl-β-d-maltopyranoside (from Anatrace) are prepared and stored at –20°C. Triton X-100 is obtained from Sigma.
7. Buffers and reagents:
   1M Tris–HCl pH 8, 1 M HEPES pH 7, 1 M glycine pH 8, 1 M sodium bicarbonate pH 8.5, 1 M Mes pH 5.0, 1 M ammonium bicarbonate, 1 M sodium acetate pH 4.0, 5 M sodium chloride, 1 M sodium azide, 1 M Tris-HCl pH 6.8, Phosphate-buffered saline (PBS), chymotrypsin (Sigma), Bradford reagent (Sigma), sodium butyrate (Sigma, made fresh each time

before use), 1D4, known as GC R1, antibody is bought from the University of British Columbia. http://rho1D4.com, 11-cis retinal (a gift from Rosalie Crouch, which is dissolved in ethanol. See also Note 16).

8. Lysis buffer: 25 mM Tris–HCl, pH 8 with Roche complete protease inhibitor cocktail (four tablets for 200 mL).
9. Urea solution:
   Fresh solution of urea is prepared just before the start of purification. Typically 1–2 L of 8 M stock is prepared.
10. Transfection: Lipofectamine 2000 obtained from Invitrogen.
11. Chromatography materials: DEAE sepharose fast flow, Q-sepharose fast flow, CNBR-activated sepharose 4B, Superdex 200 (10/300 GL) obtained from GE healthcare.

2.2. Protein Analysis

1. Protein gel electrophoresis system (Invitrogen) (see Note 3).
2. 4× protein sample buffer (0.24 M Tris–HCl, pH 6.8, 8% Sodium dodecyl sulfate (SDS), 40% glycerol, 0.04% (w/v) bromophenol blue, 5% 2-mercaptoethanol).
3. Running buffer (see Note 4).
4. Molecular weight standards (SIGMA wide range marker or equivalent).
5. Gel staining solution (Coomassie blue R250 staining solution in 50% methanol, 10% acetic acid).
6. Destaining solution (10% (v/v) ethanol, 10% (v/v) acetic acid).

3. Methods

3.1. OmpG

3.1.1. Protein Expression

1. Transform BL21 (DE3) C41 competent cells with OmpG plasmid, plate cells on LB agar plates with kanamycin and incubate overnight at 37°C.
2. Inoculate a single colony in 25 mL LB medium containing kanamycin in a 250 mL flask and grow for 12–16 h at 37°C with shaking (200 rpm).
3. Dilute the culture into 2 L of 2XYT medium with kanamycin in a 5 L flask and incubate at 37°C with shaking (180 rpm).
4. At an OD600 of 0.6, add 0.5 mM IPTG and allow cell growth for 3–4 h.
5. Harvest the cells by centrifugation (4,000×*g* for 10 min at 4°C).

3.1.2. Cell Disruption and Harvesting of Inclusion Bodies

1. Resuspend the cells in 50 mL of lysis buffer.
2. Break the cells by two passages through an emulsiflex (see Note 5) at 15,000 kpi.

3. Collect inclusion bodies (IB) by centrifugation for 20 min at 6,000 × *g* at 4°C.
4. Wash IB in 25 mM Tris–HCl, pH 8, 1 M urea, and 1% Triton-X 100 and centrifuge for 20 min at 6,000 × *g* at 4°C.
5. Resuspend IB in 25 mM Tris–HCl, pH 8 and 8 M urea (see Note 6).

3.1.3. Purification and Refolding of OmpG

1. Equilibrate 10 mL of DEAE sepharose fast flow with 25 mM Tris–HCl, pH 8 and 8 M urea (urea buffer).
2. Apply IB to DEAE column and wash extensively with 150 mL urea buffer.
3. Remove weakly bound protein by washing the column with 50 mL of urea buffer containing 50 mM and 165 mM NaCl in successive steps.
4. Elute OmpG with a buffer containing 300 mM NaCl in 25 mM Tris–HCl, pH 8 and 8 M urea.
5. Dilute OmpG in 75 mM *n*-octyl glucoside (OG) such that the final concentration of urea is ~3 M and final protein concentration is ~0.4 mg/mL (see Note 7).
6. Incubate the diluted mixture at RT for 12–16 h.
7. Load 5–10 μL of refolded OmpG in a SDS-PAGE gel and monitor the efficiency of refolding (see Note 8 and Fig. 1a).
8. Urea is removed by dialysis against a buffer with 25 mM Tris–HCl, pH 8 and 25 mM OG.
9. Alternatively, refolded OmpG can be loaded onto an anion-exchange column and urea removed (see Note 9), this enriches the folded product. Refolded OmpG can then be reconstituted back into lipid bilayers to obtain 2D crystals (Fig. 1b, c) or for 3D crystals (10).

3.2. Expression and Purification of GlpG

3.2.1. Protein Expression

1. Transform BL21 (DE3) C41 competent cells with GlpG plasmid, plate cells on LB agar plates with ampicillin and incubate overnight at 37°C.
2. Inoculate a single colony in 5 mL LB medium containing ampicillin in a 20 mL glass tube and grow for 8–10 h at 37°C with shaking (200 rpm).
3. Dilute the culture into 300 mL of LB medium with ampicillin in a 1 L flask and incubate at 37°C with shaking (200 rpm).
4. Dilute 25 mL of cells in 1 L 2XYT medium with ampicillin in a 2 L flask and incubate at 37°C with shaking (200 rpm).
5. At an OD600 of 0.6, add 0.5 mM IPTG. At this point, reduce the growth temperature to 24°C and allow the cells to grow overnight.
6. Harvest cells by centrifugation (4,000 × *g* for 10 min at 4°C).

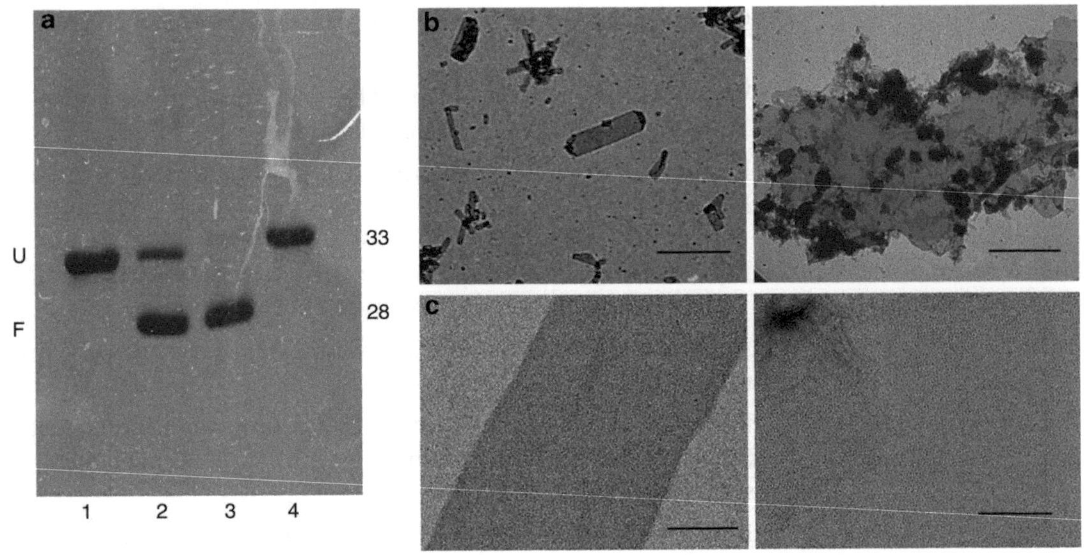

Fig. 1. (**a**) OmpG purified and refolded from inclusion bodies: A SDS PAGE gel showing the differential migratory pattern of OmpG. *Lane 1* shows OmpG in 8 M urea and migrates at ~33 kDa. *Lane 2* shows a mixture obtained after dilution of OmpG in 75 mM Octyl glucoside and removal of urea, with the folded species migrating faster at ~28 kDa. A further enrichment in an ion-exchange column results in removal of unfolded protein as shown in *lane 3*. The same sample when heated at 95°C for 5 min prior to loading in the gel results in complete denaturation of OmpG as shown in *lane 4*. Equal amount of protein was loaded in all lanes. (**b**, **c**) Two-dimensional (2D) crystals of refolded OmpG: Refolded OmpG in detergent is mixed with lipids and gradual removal of detergent by dialysis results in the formation of 2D crystals. Panel (**b**) shows low magnification CCD images of negatively stained 2D crystals of OmpG with two distinct morphologies either as isolated tubes or stacked membranes (scale bar—1 μM). Panel (**c**) shows the crystal lattice of these crystals as viewed in higher magnification (scale bar—100 nm).

3.2.2. Cell Disruption and Preparation of Membrane Fraction

1. Resuspend cells in 250 mL of lysis buffer and homogenize completely.
2. Break the cells by two passages through an emulsiflex (see Note 5) at 15,000–20,000 kpi.
3. Remove unbroken cells by centrifugation for 20 min at 6,000 × *g* at 4°C.
4. Distribute the supernatant into polypropylene tubes suitable for Ti45 rotors (Beckmann) and collect the membrane fraction by centrifugation at 100,000 × *g* at 4°C for 120 min.
5. Resuspend the membrane fraction in 25 mM Tris–HCl, pH 8 and estimate the total protein concentration (see Note 10).
6. Aliquot the membranes at a concentration of ~40–50 mg/mL and store at −80°C.

3.2.3. Solubilization and Purification of GlpG

1. Thaw the membranes in a warm water bath. Dilute the membranes with 25 mM Tris–HCl, pH 8 to a final concentration of 5 mg/mL (see Note 10).
2. Add *n*-β-decyl-maltoside (DM) from a stock of 20% to a final concentration of 1.25% to initiate the solubilization.

3. Allow membrane solubilization at RT for 30–45 min with continuous stirring.
4. Separate the unsoluble fractions by ultracentrifugation at 100,000 × *g* for 30 min at 4°C.
5. Pre-equilibrate 4–5 mL of Ni-NTA column with DM buffer (25 mM Tris pH 8, 0.3 M NaCl, 0.2% DM) containing 10 mM imidazole (see Note 11).
6. Add imidazole and NaCl to final concentrations of 10 mM and 0.3 M, respectively, to the detergent soluble fraction (see Note 12).
7. Pass the supernatant through the Ni-NTA column and collect the unbound fraction by gravity flow.
8. Wash the Column with 20× DM buffer containing 10 and 30 mM imidazole in succession to remove nonspecifically bound proteins (see Note 12).
9. Elute GlpG with 0.2 M imidazole in DM buffer and estimate the protein concentration with Bradford (23) reagent (Sigma).
10. Add chymotrypsin (Sigma) at a ratio of 1:50 (w/w) and incubate at RT for 36 h to remove the soluble N-terminal domain of GlpG (see Note 13).
11. Exchange buffer to remove salts by concentration and dilution on a Vivaspin concentrator with a 10 kDa cutoff.
12. Pass the concentrated protein through Q-sepharose fast flow column pre-equilibrated with 25 mM Tris–HCl, pH 8 and 0.2% DM.
13. Collect the flow through, which contains the majority of N-terminally truncated GlpG.
14. Concentrate the protein to 0.5 mL and load on to Superdex-200 column pre-equilibrated with 0.5% *n*-nonyl-β-glucoside.
15. Collect peak fractions (~13 mL) of GlpG and concentrate with a Vivaspin concentrator with a 10 kDa cutoff. This procedure yields ~0.5–1 mg of truncated GlpG from a liter of culture and both 2D (Fig. 2) and 3D crystals can be obtained (11, 12).

3.3. Rhodopsin

3.3.1. Preparation of Stable Cell Lines

1. Grow HEK293-GnTI-cells (21) in DMEM/F12 medium supplemented with 10% FCS, blasticidin, at 37°C in an environment with 5% CO_2 (see Note 14).
2. Day 1; split the cells into a 10 cm plate with medium free of antibiotic and plate cells such that they will attain 90% confluency for transfection, the next day.
3. Day 2; mix rhodopsin DNA (20 μg per transfection) gently with 1.5 mL serum and antibiotic-free media. Add 60 μL Lipofectamine 2000 separately to 1.5 mL serum-free media and leave at RT for 5 min. Combine the DNA and Lipofectamine

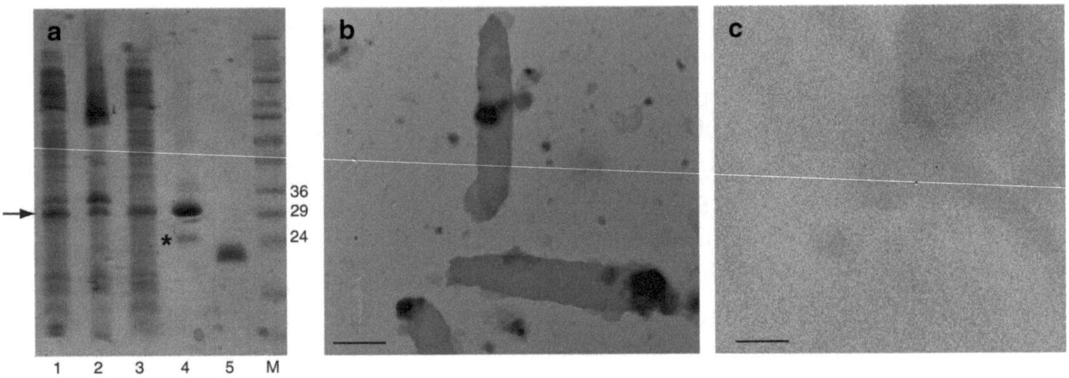

Fig. 2. (**a**) Purification of GlpG from inner membrane of *E. coli*: GlpG, a rhomboid protease is homologously expressed in *E. coli* in the inner membranes. *Lanes 1–3* show the total membrane proteins, unsoluble fraction after detergent extraction and proteins extracted by detergent, respectively. Equal amount of protein (15 μg) was loaded. The band marked with *arrow* shows the GlpG full-length protein. A prominent band in the membrane fraction when visualized by Coomassie stain is a good indicator of efficient expression. A comparison of the unsoluble and soluble fractions shows that >80% of GlpG has been extracted by decyl maltoside. In *lane 4*, GlpG purified using a his-tag in decyl maltoside but left at 4°C for 2 days shows the appearance of a proteolysed product (*marked with asterisk*). *Lane 5* shows the N-terminally truncated GlpG after ion exchange and gel filtration columns. *Lane M* denotes the sigma wide range marker used in this gel. (**b**, **c**) Two-dimensional crystals of GlpG: N-terminally truncated GlpG in detergent was mixed with *E. coli* polar lipids and detergent removal by dialysis results in 2D crystals (12). Panel (**b**) shows a low magnification view of negatively stained GlpG 2D crystals (scale bar—2 μM). Panel (**c**) shows the crystal lattice as viewed in higher magnification (scale bar—100 nm). Panels (**b**) and (**c**) are reproduced with permission from ref. 12.

together, mix gently, and leave for 20 min. Add this mixture in drops to the plate of cells and rock the plate back and forth to spread the mixture.

4. Day 3; split the cells tenfold into media containing blasticidin.
5. Day 4; replace media with DMEM/F12 containing blasticidin and geneticin (G418) (250 μg/mL), and 20% conditioned media (see Note 15).
6. Cells transfected with rhodopsin survive in the presence of G418 and start to grow. After ~2 weeks, small colonies can be seen growing. At this stage, these colonies can be picked clonally or all cells surviving G418 selection can be harvested and expanded.

3.3.2. Expression in HEK Cells

1. Expand cells containing rhodopsin under tetracycline inducible promotor as adherent cells to confluency.
2. Collect the confluent cells, count and seed in Freestyle media (Invitrogen) plus 5% FCS at 0.5×10^6 cells/mL. Typically, cells from five 75-cm^2 flasks are used to expand to a 300 mL suspension culture. Cells stay in suspension rotating on an orbital shaker at 125 rpm.

3. After 2–3 days this initial suspension culture is diluted into ~1 L media resulting in 0.5×10^6 cells/mL in a Wave bioreactor (GE Healthcare). The bag is supplied with 8% CO_2. The rock rate rises between 10 and 25 as the bag is filled, while the angle rises from 5 to 7.
4. Allow cells to grow to a density of ~2×10^6 cells/mL and dilute to 0.5×10^6 cells/mL every 2–3 days with Freestyle medium plus 5% FCS.
5. At 9 L induce rhodopsin expression by adding a final concentration of tetracycline (2 μg/mL) and sodium butyrate (5 mM) diluted in 1 L of medium.
6. Allow cell growth for 72 h after induction, cell density is typically $2–3 \times 10^6$ cells/mL.
7. Harvest the cells by centrifugation at $3,000 \times g$ for 10 min at 4°C.
8. Wash cell pellets with PBS buffer containing protease inhibitor cocktail.
9. Freeze cell pellets in liquid nitrogen and store at −80°C.

3.3.3. Preparation of 1D4 Column

1. 1D4 is dialyzed against three changes of 5 mM NH_4HCO_3 and then lyophilized (optional).
2. To 1 g of CNBr-activated sepharose 4B, add 50 mL of 1 mM HCl. Leave rotating until the column material has swelled and is in homogeneous suspension. 1 g should swell to 3.5 mL in a few minutes.
3. Wash the resin with 50 mL of 1 mM HCl.
4. Wash with 50 mL 0.1 M $NaHCO_3$, 0.5 M NaCl, pH 8.5.
5. Dissolve antibody in 0.1 M $NaHCO_3$, 0.5 M NaCl, pH 8.5.
6. Add 14 mg antibody to 3.5 mL of sepharose. Leave over night at 4°C.
7. Wash with 50 mL 0.2 M glycine, pH 8.0.
8. Incubate in 50 mL 0.2 M glycine, pH 8.0 at RT for 2 h.
9. Wash with 50 mL 0.1 M $NaHCO_3$, 0.5 M NaCl, pH 8.5.
10. Wash with 50 mL 0.1 M NaOAc, 0.5 M NaCl, pH 4.0.
11. Wash with 50 mL 0.1 M $NaHCO_3$, 0.5 M NaCl, pH 8.5.
12. Wash with PBS/10 mM sodium azide and store in the same solution at 4°C.

3.3.4. Purification of Rhodopsin and Reconstitution with Retinal

1. Thaw cell pellets and resuspend in PBS buffer containing protease inhibitor cocktail.
2. Initiate solubilization by the addition of 1.25% DM and incubate for 1 h at 4°C.

3. Centrifuge at 100,000 × *g* at 4°C for 30 min to remove insoluble fractions.
4. Incubate supernatant with 1D4 antibody column pre-equilibrated with PBS containing DM (0.125%) for 2–4 h at 4°C.
5. Wash the matrix with PBS/0.125% DM. All subsequent steps are done in dim red light.
6. At this point, add 50 μM 11-cis retinal to the matrix and incubate overnight at 4°C (see Note 16).
7. Wash the matrix with PBS and 0.125% DM to remove excess retinal.
8. Protein is exchanged to detergent OG (1%) in 10 mM HEPES, pH 7.0 (see Note 17).
9. Elute rhodopsin with peptide TETSQVAPA (80 μM) resembling the C-terminus of rhodopsin, in 10 mM HEPES pH 7 and 1% OG buffer. Repeat elution three times.
10. Concentrate the protein to 0.5 mL with a Vivapsin 30 kDa cut-off concentrator.
11. Load reconstituted recombinant rhodopsin onto a Superdex-200 column pre-equilibrated with 10 mM Mes pH 5.0, 0.1 M NaCl and 1% OG.
12. Collect peak fractions and concentrate to 10–15 mg/mL for crystallization trials. Typically, 0.2–0.4 mg of recombinant rhodopsin with correctly bound retinal and homogenous glycosylation (Fig. 3) can be obtained from a liter of suspension culture.

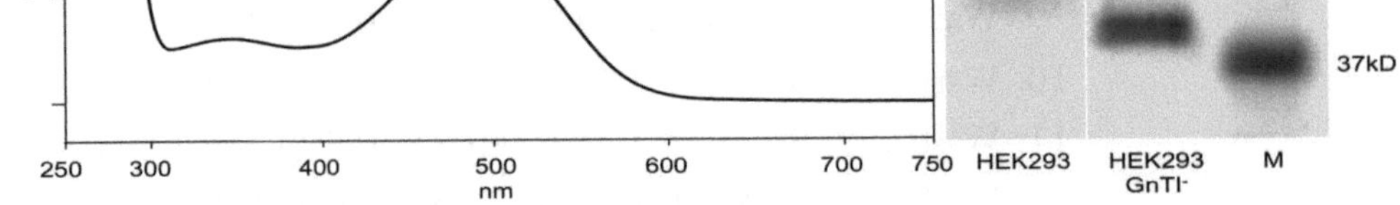

Fig. 3. (**a**) The UV-VIS absorption spectrum of purified recombinant rhodopsin shows the characteristic peak at 280 nm from protein and that at 500 nm from the protonated 11-*cis*-retinal ligand. The ratio of these two peaks is often used to access the quality of a rhodopsin preparation as misfolded protein fails to bind retinal and increases the relative 280 nm contribution. HEK293-GnTI⁻ cell expressed rhodopsin routinely shows a ratio that is very close to the 1.6 ratio observed for native rhodopsin purified from bovine retinas indicating that nearly all protein expressed are in a functional form. (**b**) SDS PAGE comparing rhodopsin expressed in HEK293S and HEK293S-GnTI⁻ cells. The homogenous glycosylation of HEK293S-GnTI⁻ cell expressed rhodopsin leads to a faster and more even migration compared to the complex glycosylation pattern of HEK293S cell expressed protein.

4. Notes

1. For expression in *E. coli*, good expression is typically obtained with any of the pET vectors available from Novagen. The choice of the tags depends on protein and the best option is identified by trials. Use of T7 promoter (as is the case with most pET vectors) sometimes is detrimental to expression with protein ending in inclusion bodies. As an alternative, controlled expression can be achieved with arabinose inducible pBAD system or similar other vectors. For a detailed list of vectors and strains, please refer to Junge et al. 2008 and Zoonens & Miroux 2010 (28).
2. We have had greater success with BL21 derivatives C41 and C43 than the parent strains as explained by Miroux and Walker (24). However, some membrane proteins do not express at high levels, which could be due to codon usage or inefficient insertion into membranes.
3. Invitrogen offers a range of gels and buffer systems. Bis-tris gels in combination with Mes buffer is the most commonly used system. However, we have observed for many membrane proteins (in particular those that are very hydrophobic) the best results are achieved with a gradient gel run with Tris-glycine buffer.
4. Membrane proteins are not typically heated prior to loading in a gel, as they tend to aggregate (beta-barrel membrane proteins can be heated, see Note 8). Since they are not completely denatured by SDS, the ratio of SDS bound to the protein differs from that of a soluble protein. This results in faster migration of membrane proteins typically 4–8 kDa lower than the size estimated from the amino acid sequence.
5. Lysis of *E. coli* cells and preparation of the membrane fraction is a critical step to remove unfolded proteins. When compared to yeast, *E. coli* cells are easily and efficiently disrupted using high-pressure systems such as Emulsiflex C3/C5 from Avestin or Constant cell disruption system.
6. Guanidium hydrochloride (GuHCl) at a concentration of 6 M can be used instead of urea. Remember to dialyze out GuHCl prior to loading in a gel.
7. The critical micelle concentration (CMC) of detergent micelles differs depending on environment such as ionic strength, additives, and temperature. For instance, the CMC of octyl glucoside in water is 25 mM but in 8 M urea the CMC increases to 42.5 mM. Thus it is important to take into account the concentration of urea when attempting to refold a membrane protein.

8. Beta-barrel membrane proteins show different migratory behavior on SDS PAGE gel depending on whether the sample is heated. When boiled at 95°C for 5 min, beta-barrel membrane proteins are completely denatured and the mobility closely matches that of the estimated mass from the amino acid sequence. If the sample is not boiled, the protein migrates faster. This difference in mobility can be used to monitor the efficiency of refolding.
9. Different approaches can be utilized for refolding that include gradual dilution of denaturant in an appropriate refolding buffer, exchange of buffer by dialysis or refolding on the column. Success depends on the protein of choice. We found that gradual dilution worked best for OmpG in terms of the efficiency and yield of refolded protein. A database comprising details of refolding of various proteins can be found at http://refold.med.monash.edu.au.
10. Estimation of total protein concentration in the membranes is a critical step, since it determines the amount of detergent to be used for solubilization. Use of less detergent as a result of underestimating the protein concentration can result in inefficient extraction of protein from membranes. A modified Lowry's protocol that uses acid precipitation and detergent gives a reliable estimate (25).
11. Application of detergent soluble fraction to Ni-NTA resin can be performed by batch method or directly on the column. The source of Ni-NTA can sometimes be crucial on the yield of the protein. We use Ni-NTA from Qiagen and regenerate it to be used multiple times.
12. Addition of imidazole and sodium chloride is optional and depends on efficiency of protein binding to the resin. Use of low concentration of imidazole in the binding step is beneficial in preventing nonspecific binding. The concentration of imidazole required for washing weakly bound proteins and for elution of target protein largely depends on protein of interest. As an initial step, a gradient of imidazole can be used to determine these concentrations. AcrB, a multi-drug transporter is a very common contaminant from *E. coli* that binds to Ni-NTA with very high affinity. Care must be taken to remove AcrB in subsequent steps.
13. In the case of GlpG, removal of N-terminal domain by a protease is a crucial step for obtaining well-diffracting crystals. An expression construct devoid of the N-terminal domain produces crystals that diffract poorly.
14. Heterogenous glycosylation of proteins expressed in eukaryotic cells is often detrimental in obtaining good crystals. With rhodopsin it was essential to use this cell line (HEK-293S-GnTI$^-$)

in combination with stabilizing mutations to obtain well-diffracting crystals (14). In some cases it may however be necessary to remove glycosylation once the protein is correctly folded to facilitate crystal growth. In such cases enzymatic digestion of HEK-293S-GnTI$^-$ expressed proteins with EndO H or PNGase F has been used very successfully (26).

15. Conditioned media is harvested from near confluent cells and filtered that contains conditioning factors. Media is stored at −20°C and used for selection of stably transfected cell lines.
16. Synthetic 9-cis retinal (Sigma) can be used instead of 11-cis retinal. Both these ligands are inverse agonists of rhodopsin with similar potency.
17. Exchange of detergent to octyl glucoside and addition of lipids is a crucial step to obtain diffracting crystals of recombinant rhodopsin (14).

Acknowledgments

VK acknowledges the European Union for a Marie-Curie Intra-European Fellowship and the Medical Research Council for a Career Development fellowship. The work on OmpG was done by VK at the Max-Planck Institute of Biophysics, Frankfurt as part of Ph.D. thesis. This work was further supported by the Swiss National Science Foundation (SNSF) grants 31003A_132815 and 31003A_141235 to JS. We thank Gebhard Schertler at the PSI and Daniel Oprian at the Brandeis University for discussion and advice on the expression of rhodopsin.

References

1. Vinothkumar KR, Henderson R (2010) Structures of membrane proteins. Q Rev Biophys 43:65–158
2. Tate CG, Haase J, Baker C, Boorsma M, Magnani F, Vallis Y, Williams DC (2003) Comparison of seven different heterologous protein expression systems for the production of the serotonin transporter. Biochim Biophys Acta 1610:141–153
3. Junge F, Schneider B, Reckel S, Schwarz D, Dotsch V, Bernhard F (2008) Large-scale production of functional membrane proteins. Cell Mol Life Sci 65:1729–1755
4. Junge F, Haberstock S, Roos C, Stefer S, Proverbio D, Dotsch V, Bernhard F (2010) Advances in cell-free protein synthesis for the functional and structural analysis of membrane proteins. N Biotechnol 28:262–271
5. Chen YJ, Pornillos O, Lieu S, Ma C, Chen AP, Chang G (2007) X-ray structure of EmrE supports dual topology model. Proc Natl Acad Sci USA 104:18999–19004
6. Sorensen HP, Mortensen KK (2005) Advanced genetic strategies for recombinant protein expression in Escherichia coli. J Biotechnol 115:113–128
7. Midgett CR, Madden DR (2007) Breaking the bottleneck: eukaryotic membrane protein expression for high-resolution structural studies. J Struct Biol 160:265–274
8. Lee JK, Stroud RM (2010) Unlocking the eukaryotic membrane protein structural proteome. Curr Opin Struct Biol 20:464–470
9. Mus-Veteau I (2010) Heterologous expression of membrane proteins for structural analysis. Methods Mol Biol 601:1–16

10. Yildiz O, Vinothkumar KR, Goswami P, Kuhlbrandt W (2006) Structure of the monomeric outer-membrane porin OmpG in the open and closed conformation. EMBO J 25: 3702–3713
11. Wang Y, Zhang Y, Ha Y (2006) Crystal structure of a rhomboid family intramembrane protease. Nature 444:179–180
12. Vinothkumar KR (2011) Structure of rhomboid protease in a lipid environment. J Mol Biol 407:232–247
13. Standfuss J, Xie G, Edwards PC, Burghammer M, Oprian DD, Schertler GF (2007) Crystal structure of a thermally stable rhodopsin mutant. J Mol Biol 372:1179–1188
14. Standfuss J, Edwards PC, D'Antona A, Fransen M, Xie G, Oprian DD, Schertler GF (2011) The structural basis of agonist-induced activation in constitutively active rhodopsin. Nature 471:656–660
15. Hiller S, Abramson J, Mannella C, Wagner G, Zeth K (2010) The 3D structures of VDAC represent a native conformation. Trends Biochem Sci 35:514–521
16. Standfuss J, Kuhlbrandt W (2004) The three isoforms of the light-harvesting complex II: spectroscopic features, trimer formation, and functional roles. J Biol Chem 279:36884–36891
17. Baneres JL, Martin A, Hullot P, Girard JP, Rossi JC, Parello J (2003) Structure-based analysis of GPCR function: conformational adaptation of both agonist and receptor upon leukotriene B4 binding to recombinant BLT1. J Mol Biol 329:801–814
18. Baneres JL, Mesnier D, Martin A, Joubert L, Dumuis A, Bockaert J (2005) Molecular characterization of a purified 5-HT4 receptor: a structural basis for drug efficacy. J Biol Chem 280:20253–20260
19. Park SH, Prytulla S, De Angelis AA, Brown JM, Kiefer H, Opella SJ (2006) High-resolution NMR spectroscopy of a GPCR in aligned bicelles. J Am Chem Soc 128:7402–7403
20. Gruswitz F, Chaudhary S, Ho JD, Schlessinger A, Pezeshki B, Ho CM, Sali A, Westhoff CM, Stroud RM (2010) Function of human Rh based on structure of RhCG at 2.1 A. Proc Natl Acad Sci USA 107:9638–9643
21. Reeves PJ, Kim JM, Khorana HG (2002) Structure and function in rhodopsin: a tetracycline-inducible system in stable mammalian cell lines for high-level expression of opsin mutants. Proc Natl Acad Sci USA 99: 13413–13418
22. Reeves PJ, Callewaert N, Contreras R, Khorana HG (2002) Structure and function in rhodopsin: high-level expression of rhodopsin with restricted and homogeneous N-glycosylation by a tetracycline-inducible N-acetylglucosaminyltransferase I-negative HEK293S stable mammalian cell line. Proc Natl Acad Sci USA 99:13419–13424
23. Bradford MM (1976) A rapid and sensitive method for the quantitation of microgram quantities of protein utilizing the principle of protein-dye binding. Anal Biochem 72: 248–254
24. Miroux B, Walker JE (1996) Over-production of proteins in Escherichia coli: mutant hosts that allow synthesis of some membrane proteins and globular proteins at high levels. J Mol Biol 260:289–298
25. Peterson GL (1977) A simplification of the protein assay method of Lowry et al. which is more generally applicable. Anal Biochem 83:346–356
26. Chang VT, Crispin M, Aricescu AR, Harvey DJ, Nettleship JE, Fennelly JA, Yu C, Boles KS, Evans EJ, Stuart DI, Dwek RA, Jones EY, Owens RJ, Davis SJ (2007) Glycoprotein structural genomics: solving the glycosylation problem. Structure 15:267–273
27. Deupi X, Edwards P, Singhal A, Nickle B, Oprian D, Schertler G, Standfuss J (2012) Stabilized G protein binding site in the structure of constitutively active metarhodopsin-II 109:119–124
28. Zoonens M, Miroux B (2010) Expression of membrane proteins at the Escherichia coli membrane for structural studies. Methods Mol Biol 601:49–66

Chapter 3

Two-Dimensional Crystallization of Membrane Proteins by Reconstitution Through Dialysis

Matthew C. Johnson, Tina M. Dreaden, Laura Y. Kim, Frederik Rudolph, Bridgette A. Barry, and Ingeborg Schmidt-Krey

Abstract

Studies of membrane proteins by two-dimensional (2D) crystallization and electron crystallography have provided crucial information on the structure and function of a rapidly growing number of these intricate proteins within a close-to-native lipid bilayer. Here we provide protocols for planning and executing 2D crystallization trials by detergent removal through dialysis, including the preparation of phospholipids and the dialysis setup. General factors to be considered, such as the protein preparation, solubilizing detergent, lipid for reconstitution, and buffer conditions are discussed. Several 2D crystallization conditions are highlighted that have shown great promise to grow 2D crystals within a surprisingly short amount of time. Finally, conditions for optimizing order and size of 2D crystals are outlined.

Key words: Two-dimensional crystallization, 2D crystals, Membrane protein, Dialysis, Lipid, Detergent, Electron crystallography

1. Introduction

Once a membrane protein has been sufficiently purified and is available in adequate amounts, two-dimensional (2D) crystallization arguably constitutes the major bottleneck in electron crystallography. Yet data is emerging, especially as the focus of methods development of a number of laboratories is shifting to 2D crystallization conditions and automation (see Chapters 15 and 16), that it is indeed in many cases possible to grow 2D crystals quite easily and quickly (1–7). It appears to be generally accepted that 1,2-dimyristoyl-*sn*-glycero-3-phosphocholine (DMPC) (5, 6) at low lipid-to-protein ratios (LPR) is frequently critical in obtaining 2D crystals,

Ingeborg Schmidt-Krey and Yifan Cheng (eds.), *Electron Crystallography of Soluble and Membrane Proteins: Methods and Protocols*, Methods in Molecular Biology, vol. 955, DOI 10.1007/978-1-62703-176-9_3,

and that the salt concentration in the dialysis buffer will likely have an impact on crystal size and/or morphology (2, 4). New advances in membrane protein over-expression and purification, resulting in a larger yield to be used in automated 2D trials, as well as a shifting focus onto 2D crystallization of many more membrane proteins to take advantage of the state-of-the-art electron cryo-microscopy (cryo-EM) and image processing methods, will likely function towards eliminating this bottleneck even further.

The goals of 2D crystallization are in general twofold: to obtain (A) crystals of high quality and (B) a crystal size of at least 1 μm. While the main aim is to crystallize the active protein, inhibitors might be necessary to induce crystallization and can even prove advantageous in understanding structure–function aspects such as for some ATPases (8–12).

Dialysis-mediated reconstitution (Fig. 1) of membrane proteins into lipid bilayers is one of the most frequently utilized methods of 2D crystallization (13). Monolayer 2D crystallization is used less frequently, but has shown great success for a number of proteins available in small amounts or more amenable towards this approach as outlined in Chapter 4. Dialysis against detergent-free buffer results in the membrane proteins being embedded in a continuous lipid bilayer, which advantageously resembles the native protein environment. Briefly, the method involves mixing of detergent-solubilized lipids with detergent-solubilized, usually completely delipidated proteins at low LPRs, followed by detergent removal through dialysis. Under optimized conditions, reconstitution into proteoliposomes and ordering into a 2D crystal results. The aim of these experiments is to obtain pure protein in highly ordered arrays at an optimal diameter of 1 μm or larger, which can be used for cryo-EM data collection for a 3D reconstruction. While upon first consideration the number of parameters involved in the formation of well-ordered crystals may appear daunting, surprisingly few factors are critical in obtaining both well-ordered and large crystals. These may include protein purity, lipid type, LPR, detergent concentration, salt type and concentration, dialysis time, and temperature (4–6). The model for this type of 2D crystallization is typically described as occurring in a two-stage or three-stage mechanism in which lipid bilayer formation occurs simultaneously with protein insertion into the membrane, simultaneously with order, or followed by rearrangement into an ordered 2D lattice (14). During crystallization trials, selected parameters are adjusted until the conditions are optimized for data collection. Frequently the number of critical parameters for obtaining both large and well-ordered 2D crystals is in the range of 2–3 conditions, provided that the protein is of sufficient purity and stability (6). This limits the time invested in 2D crystallization trials significantly, despite the importance of careful screening (6). On more rare occasions, 4–6 factors will be critical in inducing 2D crystallization (15).

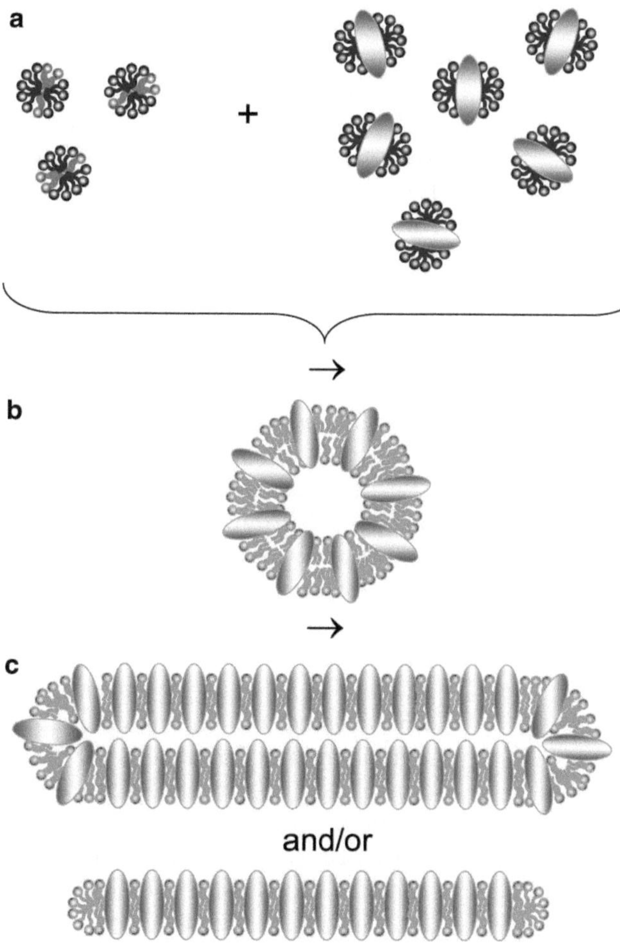

Fig. 1. Schematic representation of cross-sections of (**a**) the detergent–lipid mixture that is added to the purified protein in detergent; (**b**) a proteoliposome after dialysis, ideally containing no or minimal amounts of detergent; (**c**) 2D crystals lying flat on an EM grid. Two ordered layers are found in the case of vesicles or planar-tubular 2D crystals (*top*), while a sheet is composed of only one ordered layer (*bottom*) and a helical crystal will retain its tubular shape on the grid.

The controlling step in 2D crystallization of reconstituted membrane proteins is thought to be the transition from mixed micelles to proteoliposomes, which is to a large extent governed by dialysis time and the speed of detergent removal, which is in turn closely related to temperature (3). The ideal temperature for crystallization depends partly on the phase transition temperature of the lipid(s) in use. Generally, crystallization is more likely to occur in the lipid fluid phase (4), although crystallization below the phase transition temperature (PPT) has been shown to occur (6). Exploitation of lipid phase transition through temperature transitions during dialysis by slowly ramping the temperature above and then below the PTT has been shown to be helpful (4, 16).

The dialysis temperature must also take the stability of the protein into account.

A variety of methods are suitable for 2D crystallization, including dialysis using tubing such as used for simple dialysis, buttons or glass capillaries (14). Significant success has been achieved through the use of automated, continuous-flow devices, which allow high-throughput screening of crystallization conditions (2, 17–19). Chapters 15–18 describe highly promising automation approaches for 2D crystallization that are already showing success for a number of projects. In addition, a number of the excellent protocols in Chapter 15 are applicable to the basic dialysis approach described here. The procedure described in this chapter, which entails the use of dialysis tubing or buttons in glass beakers, does not allow for as high of a throughput, but provides excellent reproducibility and ease of use. Furthermore, the method is cheap, does not require specialized equipment, and can be used for samples from low (~100 μl/week) to unlimited availability.

The goal of this chapter is to convey the steps involved in setting up 2D crystallization experiments by dialysis as well as considerations regarding an initial strategy for a new project. While many of the steps are quite simple to carry out and do not require any sophisticated equipment, careful handling of the setup as well as adjustment of parameters are critical to success. Here, we discuss preparation of phospholipids to be used in dialysis, choice of dialysis method, and selection of parameters to be tested. Furthermore, guidance is given on identifying suitable starting factors and considerations for crystallizing any membrane protein of choice. It should be emphasized that while the choice of 2D crystallization parameters is the key to obtaining structural information, proper screening to identify ordered arrays is just as critical as outlined in Chapter 5. Our hope is that in the near future it will be possible to describe the 2D crystallization process in much greater detail to be able to give even more comprehensive and directed guidelines.

2. Materials

A key factor in successful and reproducible testing of 2D crystallization conditions is the usage of clean technique, especially concerning unwanted lipid contamination. Gloves worn at all times to avoid contamination, properly cleaned and thoroughly rinsed glassware, clean surfaces and avoidance of dust particles is a necessity. Residual salts on glassware from tap water can alter the salt concentration in the dialysis buffer, and any remaining detergent can also affect the crystallization conditions or structural and functional integrity of the protein. Moreover, the copious amounts of

Fig. 2. Examples of dialysis devices: (a) dialysis tubing, (b) "hockey stick" (≥40 μL), (c) slide-a-lyzer (≥100 μL), (d) dialysis cup (≥20 μL), and (e) dialysis button (≥5 μL). The dialysis membrane is indicated by *dark gray* shading and is indicated by an *arrow* for the "hockey stick".

natural lipids present on human skin can easily transfer to any aspect of the dialysis sample or setup and also alter conditions. If one shares a lab space or materials with labmates, likewise encourage the importance of glove use and cleanliness.

2.1. Dialysis Setup

1. Dialysis membrane tubing with a 12–14,000 Da molecular weight cut-off (Spectra/Por from Spectrum Laboratories, 10 mm flat width) cut to a length of ~8–9 cm (see Note 1). While the method of dialysis with tubing and buttons will be described in detail here, these can be substituted with other types of dialysis devices, including Slide-A-Lyzers, cups, hockey sticks, and automated dialysis devices (see Chapter 15). Depending on the specimen and conditions, usually related to sample volume and availability, a device is chosen (Fig. 2; see Note 2). Alternatively, dialysis buttons of 10–100 μl or a dialysis apparatus (see Chapter 15) are used. The dialysis buttons are particularly well suited for volumes of 5–50 μl.

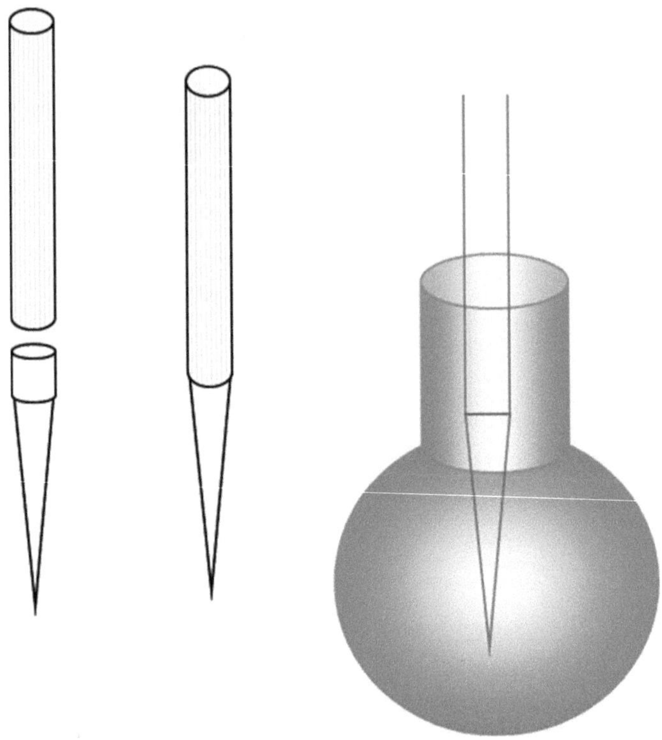

Fig. 3. Chloroform is evaporated from a round-bottom flask for preparation of the lipid stock solutions. A micropipette tip is inserted into Tygon tubing, the Nitrogen or Argon gas pressure carefully adjusted to prevent agitation of the chloroform, and the micropipette tip is held above the surface of the chloroform for evaporation.

2. For use with dialysis buttons: O-rings and tool (usually a golf "T") for securing the dialysis membrane smoothly to the buttons.
3. Two standard 35 mm polypropylene tubing closures (e.g., Spectra/Por type from Spectrum Laboratories) for securing the ends of tubing with the sample inside. Although these clips are designed to float, rotation of the clips by 180° relative to one another prevents surface flotation (Fig. 3).
4. Any heating/cooling incubator for temperature-controlled dialysis experiments (e.g., Memmert brand precision cooled incubator models ICP 200, 400, and 500 provide controlled temperature conditions: although variable by model, temperatures can range from 0 to 60°C). Depending on the temperature requirements for the experiment, a temperature-consistent air-conditioned room can also be used for dialyses conducted in the range of 18–25°C. In addition, a cold room or refrigerator can be used when the dialysis must be conducted at low temperatures in the range of 4–8°C in the case of thermally sensitive proteins.

5. One 400-ml glass beaker per dialysis condition.
6. Squares of aluminum foil (~15 cm × 15 cm), Parafilm, or a large Petri dish top to cover the top of the dialysis beakers during dialysis.
7. Fresh ultrapure water for equilibration and hydration of the dialysis tubing.
8. 1-l beaker for equilibration and hydration of the dialysis tubing.

2.2. Dialysis Buffer

1. Buffer components as used during the final purification step and for protein storage. 250 ml (see Note 3) of detergent-free buffer in a 400 ml beaker is more than sufficient per 10–100 μl of sample.
2. Glycerol (see Note 4).
3. 1 mM NaN_3 (see Note 5).
4. Optional, depending on protein: ligand or inhibitor (see Note 6).

2.3. Protein

1. Detergent-solubilized membrane protein of interest, purified to the highest degree possible (see Note 7). Aliquot solubilized protein samples at appropriate concentration (0.5—1 mg/ml) into 75–100 μl aliquots (dependent on individual trial volume) and rapidly freeze by plunging into liquid N_2. These samples can then be stored at −80°C. This method of flash freezing prevents slow freeze-thaw cycles, which can compromise structural integrity. An individual aliquot of 75–100 μl (or smaller) will be used per single dialysis experiment. Some proteins might require immediate use for 2D crystallization trials after purification and without freezing, if protein stability cannot be maintained under any storage conditions.
2. Any nonstructurally essential co-purified lipids should be removed from the purified protein (see Note 8).
3. The protein remains solubilized in detergent until dialysis. At least in initial trials, the detergent used in the purification should be kept since the protein will be stable in this detergent (see Subheading 2.4 and Note 9).
4. The optimal protein concentration for dialysis is currently estimated to be roughly in the range of 0.5–1 mg/ml (5, 6, 14). Due to the importance of the LPR parameter in crystallization (see Subheading 2.5), the protein concentration should be measured carefully, precisely, and always by the same method (see Note 10).
5. Optional, depending on protein concentration: additional volumes of solubilizing detergent. If the protein is at a higher concentration than desired, this is added to a small volume of buffer to reduce the protein concentration before dialysis (see Note 11).

2.4. Detergent

1. As outlined above, the detergent type and concentration used during the protein purification will be the starting point (see Note 12).
2. The CMC of a detergent determines the rate, and subsequent length, of dialysis (see Note 13).
3. If the detergent concentration is excessively high, such as several-fold the CMC, in the purified sample, then it can be reduced to avoid a long dialysis time, possibly in combination with BioBeads. Although, the concentration should not be altered at the expense of protein stability or activity.
4. Detergent can be removed by the addition of small successively added amounts of hydrophobic BioBeads (Bio-Rad), which enable fast absorption of both high and low CMC detergents by the polystyrene beads (20). The rate of removal can be controlled by adjustment of the quantity of beads added, although reproducibility can become problematic with this method (see Note 14).

2.5. Lipids

2.5.1. Lipid Stock Solution Preparation

1. Round bottom flask with a volume of 10–20 ml, carefully cleaned to especially avoid any residual detergent and lipid from previous uses.
2. Nitrogen or Argon gas cylinder with regulator and Tygon tubing.
3. Clean micropipette tips (approximately 100–200 μl volume or smaller) to attach to tubing for directed evaporation; ~2 mm of the tip can be removed with a razor blade or scalpel and the tip may be held in place with parafilm. The wide end of the pipette tip may also be removed with a razor blade, depending on the inner diameter of the Tygon tubing.
4. Detergent-buffer mixture, which is chosen depending on protein stability and solubilization assays of different detergents. We found 0.5% sodium-deoxycholate to work well for both solubilization of lipids as well as 2D crystallization of a large range of membrane proteins (6).
5. Lipid in chloroform (Avanti Polar Lipids, see Note 15), ideally at a concentration of 1–25 mg/ml depending on LPR range; use of smaller concentrations is helpful for lower LPRs as slightly larger volumes are then pipetted and pipetting errors are avoided/minimized. Alternatively, the lipid is solubilized to produce a lower concentration (see Note 16).
6. Eppendorf tubes for preparation of aliquots.
7. Small sonicator bath with wire basket for holding the round-bottom flask.
8. Fumehood for handling of chloroform.

9. Gloves to be worn at all times of handling lipids and glassware, etc.
10. Freezer (−20 or −80°C).

2.5.2. Lipid

1. Lipid stock solutions in detergent.

3. Methods

Two-dimensional crystallization trials by dialysis require several steps. While the setup is very simple and reproducible when executed carefully, planning of especially the first 2D crystallization trial requires consideration, and/or discussion with collaborators, of four main factors: the protein to be crystallized in 2D, the detergent used for solubilizing the protein, the lipid to be used for reconstitution/crystallization, and the dialysis buffer. These factors will be critical in making decisions for the entire 2D crystallization setup. After planning the initial strategy, aliquots of lipid stock solutions are prepared, which can in some cases, depending on detergent type and concentration compatibility with each protein, be used for multiple projects. Additional preparation steps for the dialysis setup itself involve accurate measurement of protein concentration, and preparation of dialysis membranes as well as buffers. The dialysis setup itself is relatively short and includes mixing of components followed by an appropriate dialysis time for sufficient detergent removal, which commonly spans at least several days. Further parameters for optimizing the conditions are outlined.

3.1. Devising an Initial Strategy for 2D Crystallization Trials

3.1.1. The Protein

1. Availability: The first question to be considered is how much protein is available and whether it can be obtained at regular intervals. The concentration and amount of the protein obtainable on a regular basis will determine the feasibility of the 2D crystallization approach, the number of experiments that can be conducted in each set of trials, and the range of conditions that can be tested. The minimum available amount of a protein to be considered for this approach will depend on the willingness of the investigator to, e.g., calculate waiting time for new protein and thus delays into the overall timeframe. Thus, a recommendation is to obtain minimally 0.3 ml of the protein at a given/chosen concentration (see below) every 1–2 weeks. A supply of less protein will make it worthwhile to first focus on improving the reproducible protein yield.
2. Concentration: The protein concentration achievable on a regular basis needs to be sufficiently high to make 2D crystallization trials feasible. Furthermore, determination of the

concentration is necessary for calculating LPRs. While a protein concentration of 0.5–1.0 mg has been used most frequently in successful crystallization trials (5, 6), concentration is rarely the determining factoring 2D crystallization, and lower concentrations (ideally not less than 0.2 mg/ml) of an interesting sample might have to suffice until protein production can be optimized. While different methods will frequently show a discrepancy in protein concentration determined due to the presence of detergent, consistency in the use of one method should minimize this concern and provide reproducible relative values. The reproducibility of the determination of protein concentration, even if the value should overall be relative, will be critical in calculating the amount of lipid to be added in terms of the lipid-to-protein ratio (LPR).

3. Molecular weight and, if known, number of subunits: The size of the protein is used for calculating the LPR. As the LPR frequently plays the most critical role in 2D crystallization, questions can be answered at this stage to start within a roughly realistic range and/or to at least obtain initial reconstitution. The molecular weight of the protein, and possibly information on the number of subunits, will provide preliminary guidance for calculations on an estimated minimum of lipid required for reconstitution (see Note 17). If this starting estimate should be too low and protein precipitation occurs, a range of higher LPRs are tested. Molar LPRs make it easier to visualize how many lipid molecules are used and are easy to describe in terms of numbers when one or only a few lipid molecules per protein molecule are used (see Note 18). Weight ratios rather than molar ratios are used by the majority of laboratories though, which makes it important to point out the type of ratio used.
4. Flexibility/conformations: Information on structural flexibility requiring stabilization, inhibitors, etc. can be a determining factor in inducing 2D crystallization (8–12) and possibly provide additional structure–function information (see Note 19).
5. Temperature sensitivity: Usually at this point, information on stability of the protein at various temperatures is available from purification data and/or biochemical studies. Temperature requirements, however, will change as soon as lipid is added to the detergent-solubilized membrane protein, which usually stabilizes the protein dramatically (6).

3.1.2. The Detergent

1. Type: As the solubilizing detergent is removed during the dialysis, the type of detergent generally appears to have less of an effect on the 2D crystallization. Thus, it is recommended to work with the solubilizing detergent for at least the first few trials. We have, however, not found it necessary to change the type of detergent for any of our projects as all solubilizing

detergents used led to 2D crystallization (see Note 20). Others, such as Vinothkumar et al. (see Chapter 2), found an improvement in size and quality of their 2D crystals after a detergent exchange, possibly due to the use of detergents with lower critical micellar concentrations (CMC) for both the protein and the lipid solubilization.

2. Concentration and CMC: Knowledge of the detergent type, CMC, and concentration will be essential for deciding on the length of the dialysis. The time needed for successful crystallization depends on a number of factors, first and foremost the CMC of the detergent to be dialyzed, with lower CMC detergents taking significantly more time for removal. In general, dialysis of at least a few days to 1 week or more may be necessary. A detergent with a high CMC, such as CHAPS (3-[(3-cholamidopropyl)dimethylammonio]-1-propanesulfonate), depending on the concentration used, might require 1–4 days of dialysis for removal (21), while detergents with lower CMCs, such as Triton X-100 and DDM (*n*-dodecyl-β-d-maltoside) will require at least 1 week (6, 10, 21–23). The first estimate for length of dialysis might need to be reconsidered, depending on protein stability under the chosen conditions or completeness of detergent removal (see also Chapter 5; see Note 21).
3. For at least the first 2D crystallization trials, it is recommended to keep the solubilizing detergent at the concentration determined to be most optimal to retain protein stability and function. A later increase or decrease can be tested for an influence on 2D crystal size (6, 22).
4. Compatibility of detergent with protein: If available, a list of detergents tested during the purification trials will be helpful should the solubilizing detergent of the lipid be different and possibly not compatible with retaining protein stability.

3.1.3. The Lipid

Lipid has to be considered both in the context of possible co-purified lipid with the protein and the lipid to be added for reconstitution in terms of the LPR. The LPR is the major crystallization parameter to be tested, as it is most frequently the deciding factor in 2D crystallization (5, 6, 22). Optimal molar LPRs are often in the range of 1–30, but can vary widely from protein to protein up to LPRs in the hundreds for some proteins.

The choice of lipid, whether native or endogenous, is often a significant factor in 2D crystallization. Some proteins can only be induced to crystallize *in situ* within native lipid bilayers (24), in which case it is preferable to isolate and improve upon the naturally present crystalline arrays, if possible. Success in crystallization has also been observed using copurified lipids with and without additional endogenous lipid (25). One frequent choice is the use of mixed phospholipids, either separately purified from native membranes or from sources such as *E. coli*, and it is often the case that suitable crystals can be produced in these heterogeneous membranes. Screening of multiple lipids within native batches indicates that only one or two types of lipids are usually necessary, and that while some proteins are relatively insensitive to lipid type, others rely heavily on the presence of specific, usually native, lipids to crystallize (26). In some cases different crystal packing occurs with different LPRs as well as with lipids of different acyl chain lengths (1, 15, 22). The oxidation of multiple unsaturated phospholipids (such as are often found in mixed lipid preparations) (27) is undesirable, and may be avoided through the use of synthetic monounsaturated lipids. Of particular interest is the success of DMPC, either alone or with one additional lipid [often POPC (1-palmitoyl-2-oleoyl-sn-glycero-3-phosphocholine)]. In fact, it was found that the use of DMPC has most frequently resulted in 2D crystals of membrane proteins (5, 6).

1. Co-purified lipid: Co-purified lipids can be identified by thin layer chromatography (TLC), mass spectrometry, or by simply dialyzing the protein without the addition of extrinsic lipids. In the latter case, liposomes or membrane structures of various morphologies will be observed during screening by electron microscopy (EM) after dialysis of samples with excessive co-purified lipid (see Chapter 5), which might require further purification steps, such as washes with detergent, to remove lipid. Varying amounts of co-purified lipids with protein purification batches will likely require adjustments of the LPR for each 2D crystallization experiment (15). The co-purified lipid should be absolutely minimal, as excessive amounts of lipid will "dilute" the protein within the proteoliposome and prevent growth of large crystals.
2. Choice of reconstitution lipid: Most successful 2D crystallizations of membrane proteins were grown with DMPC (5), and thus it is highly recommended to start all 2D crystallization trials with this phospholipid. As certain membrane proteins require mixtures or lipids from the native source, these strategies might be tried at later stages, or in early experiments in cases where reconstitution with DMPC should not be successful.
3. Lipid-to-protein ratio: A range of molar LPRs including 0, 3, 15, and 30 are a good starting point for the first crystallization trial. An LPR of 0 allows for possible co-purified lipid to be identified quickly and with high sensitivity by EM. The LPR = 3 is within the range of which we have mostly frequently been able to induce 2D crystallization (unpublished data). An LPR of 15 might be necessary for a larger membrane protein or simply provides, together with an LPR of 30, the safe end of

the LPR range as a control to ensure that proteoliposomes will form under the buffer and dialysis conditions used. Another factor to consider is that the final measured LPR in 2D crystals is sometimes slightly different than the initial crystallization conditions due to lipid that was not incorporated into the crystals, as well as noncrystallizable protein that remains in solution. This metastability between crystalline and noncrystalline lipids may lead to a loss in long-term stability of crystals, and it may be necessary to remove noncrystalline lipids through the use of phospholipase or detergent cocktails (28). The LPR may be calculated most reliably for lipid stocks that have been prepared in chloroform, as opposed to desiccated lipids that are very hygroscopic, easily leading to weighing errors (see Note 16).

3.1.4. The Dialysis Buffer

Several aspects of the composition and character of the dialysis buffer influence the 2D crystallization of a membrane protein. In initial trials, the conditions of the buffer used in the final purification step are a good starting point, as the protein is expected to be stable and active in this environment.

1. Choice of buffer: The dialysis buffer for the first trials should be identical, or closely match, the buffer used for the last step of the purification protocol (and/or protein storage) as this was very likely chosen based on maximal protein stability.
2. Modification of buffer components: Possibly very expensive buffer components can be omitted or exchanged in the dialysis buffer.
3. Glycerol: The addition of 20% glycerol has been a factor in obtaining particularly large crystals of a number of proteins (6, 15, 21–23) but not all. Thus it is worthwhile to investigate the effect of glycerol on the size of proteoliposomes under the given conditions.
4. Salt: The choice and quantity of salt can have a significant impact on the size and morphology of 2D crystals and is discussed below. Typically, the salt(s) used in the protein purification buffer are tested first. Combinations of various salts, including both mono- and divalent cationic salts, may also be considered.
5. Temperature: Unless the protein is highly sensitive to higher temperatures, 24°C has been successful in the 2D crystallization of a number of membrane proteins and is conveniently above the phase transition temperature of DMPC. Even with a highly sensitive membrane protein a trial at this temperature will be worthwhile though, as the presence of lipids is highly stabilizing to many membrane proteins (6).

3.2. Pre-dialysis

3.2.1. Preparation of Lipid–Detergent Stock

It is critical to wear gloves for the following steps to avoid lipid contamination from the large amounts of different lipids present on hands. Steps 1–6 are performed in a well-ventilated fumehood.

1. Prepare detergent-buffer solution for solubilization of lipids. Depending on the scale of the preparation, 2 ml of solution might be sufficient. The total volume is determined based on the range of LPRs to be used in 2D crystallization trials. A final lipid concentration of 1 mg/ml will allow for easy pipetting without pipetting errors of molar LPRs even below 1. For this protocol we will describe the preparation of 1 ml of 1 mg/ml of stock solution. For LPRs above 15 at protein concentrations of 1 mg/ml or higher, a 25 mg/ml lipid stock solution can be ideal, especially in cases where the solubilizing detergent is very different between the protein and the lipid. This allows for pipetting of smaller volumes and prevents a significant disturbance to the system.
2. Open lipid vial with a razor blade or brake off the top of the vial with a thick layer of paper to protect gloved hands from broken glass.
3. To avoid evaporation of the chloroform and thus concentration of the lipid, quickly pipette 1 ml of the solution into a round-bottom flask (Fig. 3). This is conventionally done with a Hamilton syringe (see Note 22). Ensure that the chloroform–lipid mixture stays pooled in one location at the bottom of the flask to avoid resolubilization of only part of the lipid and thus inaccurate LPRs. For this purpose the flask is held upright, positioned in a cork support, or stabilized in aluminum foil shaped as a stand.
4. Test a stream of nitrogen gas by carefully and slowly directing the tubing with the attached pipette tip on the back of a gloved hand or against a piece of paper; adjust the flow to the point where a gentle stream can be directed against the solution in the flask without changing the location of the detergent–lipid mixture; at this point the nitrogen gas flow can be carefully upregulated to increase the rate of chloroform evaporation. Avoid any movement or excessive spreading of the solution as this will make complete solubilization of the dried lipid with detergent solution of the same or a lower volume easier.
5. Carefully evaporate all chloroform, which may take from a few minutes up to roughly 10 min, depending on the stream of nitrogen gas.
6. Stop the stream of nitrogen (see Note 23).
7. Very carefully pipette 1 ml of detergent-buffer solution into the round-bottom flask to cover the lipid. Take particular care to avoid pipetting errors as any discrepancies in the final lipid concentration will lead to errors in LPR identification,

reproducibility of crystallization conditions, and/or reproducible preparation of future lipid stocks.

8. The detergent-buffer solution in the round-bottom flask will start to solubilize the dried lipid. The flask is either first lightly tapped with a finger to advance the solubilization process, or it is immediately placed in a sonicator bath for up to roughly 5 min. Heating of the solution should be monitored by touch every few minutes (see Note 24).
9. Upon complete solublization, the solution will become completely clear/transparent (see Note 25). Certain lipids and lipid mixtures will have a transparent, colored appearance.
10. Once solubilized, the lipid–detergent mixture is aliquoted into 10 μl aliquots in Eppendorf tubes and stored at either −20 or −80°C. The aliquots may be smaller or larger in volume, depending on the number of crystallization experiments set up, the lipid concentration, and the LPR. Individual tubes of lipid aliquots can now be conveniently removed from storage for immediate use in 2D crystallization trials (see Note 26).

3.2.2. Dialysis Setup Using Dialysis Tubing

The protocol below outlines the steps involved in using dialysis tubing. The protocol for the alternative use of dialysis buttons is outlined in Subheading 3.2.3.

1. Cut dialysis tubing to a length of preferably ≥9 cm (see Note 27) with a clean razor blade, place the dialysis tubing in buffer or ultrapure water for approximately 30 min to hydrate the dialysis membrane. During this and all following steps, wear gloves and be mindful of surface contact at all times to control contamination, especially from any unwanted lipids that may be present on the skin or in the laboratory environment.
2. Prepare 250 ml of buffer per 2D crystallization condition in glassware that has been extensively cleaned and rinsed to remove any residual detergent. For each condition, add 250 ml of dialysis buffer to a 400 ml beaker (see Note 28).
3. On ice, mix the purified and detergent-solubilized protein (see Note 29) suspension with the lipid stock of choice to obtain a final sample volume of 75–100 μl in an Eppendorf tube (see Note 30). Vortex samples for 5 s.
4. Incubate samples on ice for 10 min, followed by another 5 s of vortexing.
5. Fasten one dialysis clip ≥5 mm, preferably 1 cm, from one end of the dialysis tubing, and strongly flick the tubing in an arc (holding the clip end) in order to remove as much water or buffer solution from inside the tubing as possible. Carefully pipette the protein–detergent–lipid mixture into the open, longer end of the tubing and apply a second dialysis clip at a

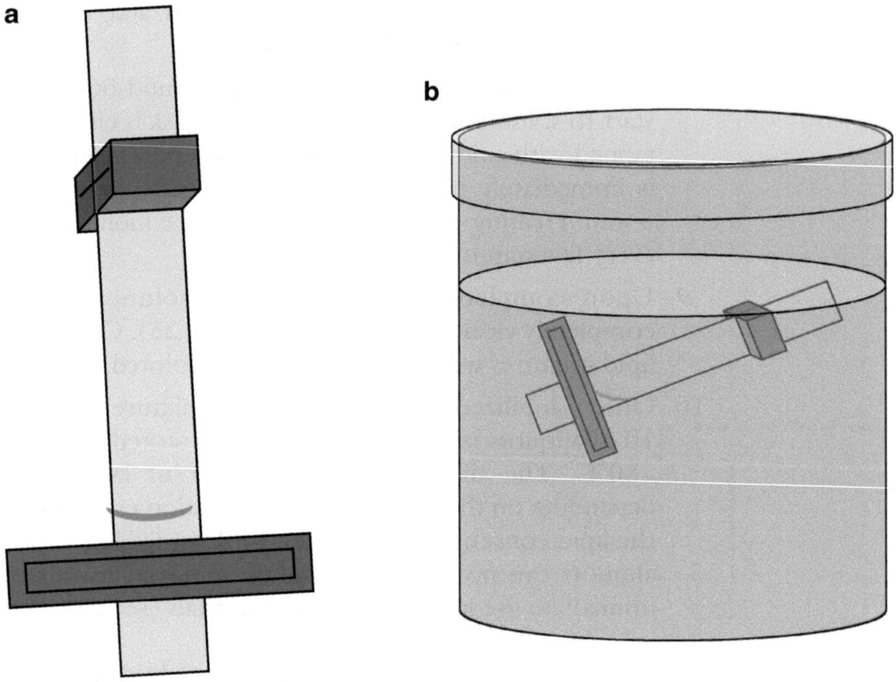

Fig. 4. (**a**) The dialysis clips are rotated by 90° in respect to each other to prevent contact of the dialysis membrane with the air–buffer interface. Most importantly, the dialysis tubing is ~9 cm long to prevent the loss of sample volume that frequently occurs with shorter tubing. (**b**) The rotation of the clips ensures that the sample remains immersed in the dialysis buffer.

90° angle to the first clip, at least 5 mm from the opposite end of the tubing. The angle of the clips will ensure that the tubing stays submerged in the dialysis buffer (see Note 31). Next, pipette the protein–lipid suspension directly into the lower end of the tubing, near the bottom clip (see Note 32).

6. Place the properly sealed dialysis tubing in the preferred volume of dialysis buffer (Fig. 4; see Note 33). Cover the beaker with aluminum foil, Parafilm, or the lid of a large Petri dish, and place it in an incubator or a laboratory with a consistent temperature (see Note 34).

3.2.3. Dialysis Setup Using Dialysis Buttons

1. Dialysis tubing is prepared and hydrated as in Subheading 3.2.2, except for that a shorter piece of tubing of approximately twice the diameter of a dialysis button is used per condition. If available, tubing of a wider flat width than the button diameter is preferable to avoid a fold from running across the dialysis surface.
2. After hydration, follow steps 2–4 in Subheading 3.2.2.
3. Pipette the sample into thoroughly cleaned and rinsed dialysis buttons. The amount pipetted into the button should exceed

the button capacity by approximately 10%, or at least several microliters, to avoid inclusion of air bubbles under the dialysis membrane. If sufficient protein is available, a button volume of at least 10–20 μl is recommended to be able to prepare a minimum of two grids per condition for screening.

4. The dialysis button is placed on a clean surface such as a laboratory paper towel or a Petri dish, and a sample volume with an excess of 10–20% of the button volume is pipette without bubbling into the dialysis button.
5. The membrane is placed on top of the button.
6. The O-ring is placed on the golf "T" tool. The tool is carefully pushed down to avoid bubbles in the dialysis chamber of the button, after which the O-ring is pushed down into the groove of the button.
7. Each button is placed into one beaker with at least 100 ml of dialysis buffer (see Note 33), which should be far in excess of the required amount for dialysis, such as the 250 ml recommended for the dialysis tubing.
8. Place the properly sealed dialysis buttons in the preferred volume of buffer (see Note 33). Cover the beaker with aluminum foil, Parafilm, or the lid of a large Petri dish, and place it in an incubator or a laboratory with a consistent temperature (see Note 34).

3.3. Dialysis

1. Visually examine the dialysis mixture within the dialysis setup for precipitate once a day (see Note 35). This is more important while examining the initial reconstitution conditions rather than later when parameters are merely fine-tuned and only minimally adjusted.

3.4. Post-dialysis

1. Remove the dialysis tubing from the buffer solution. Hold the membrane in a similar orientation to what was described above: one clip held between the thumb and fourth finger and the other between the index and third fingers. Cut the membrane with a clean razor blade above the level of the sample. If air was allowed to remain in the tubing, it will provide a convenient place to cut that is free of sample, thus limiting sample loss and contamination during this part of the procedure (Fig. 5).
2. Insert the tip of a pipette into the very bottom of the dialysis tubing and remove as much of the sample as possible. The volume of the sample may have changed slightly. Pipette the sample into a microcentrifuge tube. Alternatively for dialysis buttons, the buttons are removed from the beakers and placed into a Petri dish. The membrane can simply be pierced with the tip of the plastic micropipette for sample removal.

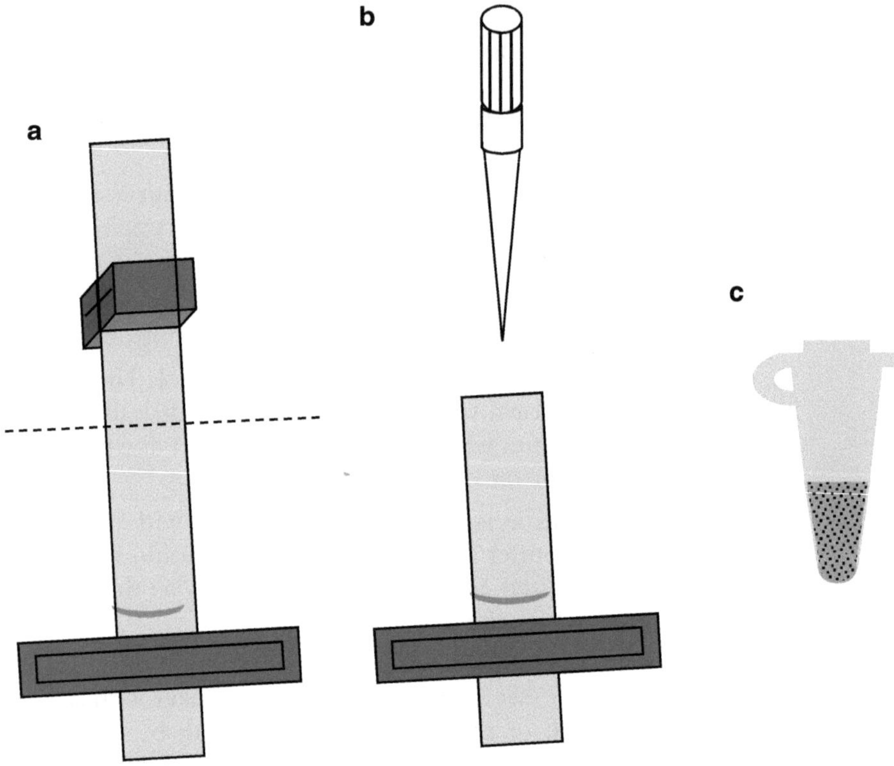

Fig. 5. (**a**) The dialysis tubing is held upright for sampling. Then (**b**) the tubing is cut between the clips, and (**c**) the sample is pipetted into a microcentrifuge tube.

3. Set aside the beakers with the dialysis buffer. Especially at the early stages of crystallization screening, when the time needed to sufficiently dialyze the detergent may not be known, it is important to check for the continued presence of detergent (see Note 36). In cases where samples are evaluated to still contain detergent, they can simply be pipetted into new dialysis tubing and placed in the preserved beakers with buffer.
4. There are three options for storage of samples to be screened for 2D crystals. Long-term storage of 2D crystals depends partly on the specific properties of the proteins under study, but in general membrane proteins that have been reconstituted into a membrane bilayer are much more stable than when in a detergent-solubilized state, and may be stored at room temperature. It is also helpful to control long-term loss of water due to evaporation by sealing the storage tube with parafilm. More fragile 2D crystals of membrane proteins are stored at 4°C. Highly sensitive samples may require that the microcentrifuge tubes are plunged into liquid nitrogen for storage at −80°C or in liquid nitrogen dewars/freezers (see Note 37).

5. Ideally two grids of each condition are immediately prepared for screening by EM as outlined in Chapter 5. This is highly critical as samples might change with time, especially upon incomplete detergent removal.

3.5. Subsequent Dialysis Trials

As 2D crystallization is still very much a step-by-step approach relying on feedback from evaluating negatively stained samples by EM, with the requirement to factor in variables such as sample availability and the time required for screening by EM, strategies will vary significantly.

The first 2D trials are evaluated for different levels of success by EM of negatively stained samples (see Chapter 5).

1. Length of dialysis in terms of sufficient detergent removal: See Note 21 for determination of the length of dialysis.
2. Reconstitution is the first achievement. In cases of insufficient or no reconstitution, a dramatically higher LPR will show whether the protein is reconstituted under the chosen lipid and buffer conditions. Should reconstitution be problematic, a change of lipid or the use of a lipid mixture can often make a crucial difference. Salt and temperature could also have an effect on reconstitution. While we are generally successful in obtaining important clues with four dialysis conditions, a more extensive test of lipid mixtures for reconstitution, if sufficient protein is available, will likely give faster results by testing a larger range of lipid combinations. A good starting point can be the composition of the native membranes, possibly simplifying this to the two main lipid components.
3. Temperature: Too high or low temperatures may have an effect on reconstitution, length of successful dialysis, membrane size, as well as structural and functional integrity of the protein. It is mostly recommended to test further temperatures in a subsequent trial if the chosen temperature was very low, e.g., between 4 and 20°C, and (a) membranes are mostly smaller than 100 nm, (b) the time of detergent removal exceeds 7–10 days to an unreasonable timeframe, and/or (c) reconstitution does not occur by any other means such as testing of lipids. High temperatures may be decreased when (a) reconstitution does not occur, (b) membranes possess a "wrinkled" appearance, or (c) activity cannot be maintained. Unless the temperature is significantly above or below room temperature, some of the other parameters listed here may be successfully tested first.
4. Co-purified lipid: An LPR = 0, where no lipid was added to the protein, will give a very good indication of even small amounts of co-purified lipid. Thin-layer chromatography allows for the determination of the identity of co-purified lipid(s). In this case the purification will require further delipidation steps as described above.

5. LPR: Provided that no lipid or very minimal amounts of lipid have been observed in the first 2D crystallization trial and that proteoliposomes were observed at the higher LPRs, a range of LPRs is chosen for the second trial. Presumably LPRs of 0, 3, 15, and 30 were tested. If the high LPRs of 15 and 30 have shown plenty of proteoliposomes, and LPR = 0 showed only precipitation, the next LPRs to be tested will be around 3. If LPR = 3 does not contain any, virtually none, or few, small proteoliposomes, LPRs between 3 and 15 will be tested such as 3, 6, 9, 12 or 3, 7, 11, 15. If LPR = 3 contains plenty of large membranes, LPRs between 0 and 3 will be tested such as 0.3, 0.75, 1.5, 3. With co-purified lipids, molar LPRs of even below 0.1 and as low as 0.25 have resulted in 2D crystals (15). As screening requires time and the first goal is to obtain crystals, it is generally recommended to not duplicate more than one or two conditions unless problems were observed in one dialysis. Whenever possible, we try to include one of the conditions from the previous trials as a control to spot check for reproducibility.

   This approach will work equally well for a larger or smaller number of LPRs within one setup. The LPRs are thus narrowed down until first order is observed. This might be in few to all of the membranes on the grid, can depend on membrane morphology, or might be in small patches or with mosaicity (see Chapter 5).

6. Membrane size and morphology: Generally at this point, unless the membranes possess a morphology that does not clearly resemble a collapsed vesicle, a planar-tubular membrane, a sheet, or a tube, that is larger than 50 nm, the focus should stay on inducing order rather than on increasing the membrane size or changing the morphology. The membrane size will in some cases improve with the crystal size, where a higher LPR can produce smaller and noncrystalline membranes, while a lower LPR under the same conditions induces both order and larger membrane sizes (Schmidt-Krey, unpublished observations). It might be necessary to test salt concentrations, salt combinations of different concentrations, and temperature at this point, if membranes are never larger than 50 nm, before resuming the optimization of the LPR.

7. The amount of protein per proteoliposome: This can rarely be discerned by EM, even for many larger membrane proteins, at a stage before order is induced. Freeze-fracture can yield valuable information and guidance towards a more suitable LPR range. Images of freeze-fractured samples will give a clear indication of whether a membrane contains only few membrane proteins, or is tightly packed, possibly only needing a minimal reduction in LPR. This method is, however, not available in

many laboratories, and thus the relative amount of reconstitution often remains an unknown until order is induced.

Later experiments depend on which areas, such as order, size of 2D crystals, or membrane size, showed promised and are developed from there depending on screening outcomes.

If reconstitution is successful, minimal or no-lipid was co-purified, and membranes fulfill the minimal morphology requirements as outlined under 5 above, the second as well as most subsequent dialysis experiments will focus on the refinement of the LPR. Should the LPR not induce crystal formation, even after extensive testing, chain length, mixture, or charge may achieve order, when examined with appropriate LPRs, which may or may not allow for flexibility. Salt, inhibitors, or temperature gradients (see Chapter 15) as well as the other parameters proposed in Subheading 3 may be tested carefully.

Once the first 2D crystals are identified, these are usually quickly improved in both crystal and membrane size. It may be possible to increase the size by modifying the choice of salt, a combination thereof, glycerol concentration, temperature, lipids, detergent, and/or detergent concentration.

Activity may be tested for each condition, or limited to major changes in parameters as well as upon crystal formation and data collection.

It should be emphasized again here as well as in Chapter 5, that careful screening to identify even low percentages of 2D crystal, as low as 2.0%, will lead to identification of optimal conditions quickly.

4. Notes

1. Spacing between dialysis clips of less than 6–7 cm frequently results in partial or complete loss of sample volume (6). Removal of air from the dialysis tubing will in most cases not be necessary. In fact, the presence of air makes sample recovery easier as the sample will be pooled at one end of the tubing.
2. In addition to detergent removal with traditional tubing, other options for dialysis devices are available. Depending on protein sample availability and other dialysis factors, these alternative devices can be advantageous. Slide-A-Lyzer dialysis cassettes (Pierce) are constructed of two sheets of dialysis membranes sealed on all sides by gaskets inside a plastic frame. Samples volumes of 100 μl or less can be injected into and removed with a Hamilton Syringe from a self-sealing gasket of a 1 ml-cassette, which prevents leaking and sample loss. However,

cutting of the dialysis membrane is preferable over removal by syringe as proteoliposomes may not exit the needle of the syringe or be subjected to sheering forces. So-called Hockey sticks are also convenient for small volumes of 20—50 μl and are constructed of a glass capillaries bent by 90° close to one end (14). The bent end is smoothed and covered with a dialysis membrane, which is fixed with a silicone ring, and the sample can be injected with a Hamilton syringe into the opposite open end, ensuring that it reaches the dialysis membrane. This method is also advantageous because the "elbow" in the 90° bend allows collection and segregation of precipitate from the bulk sample. An important advantage of the hockey sticks is the possibility to monitor the dialysis by easy specimen removal through the open end of the hockey stick. In addition, dialysis buttons are effective for even smaller sample volumes, between 5 and 100 μl, and are also commercially available (Hampton Research Corp.). These devices consist of a small cup-shaped chamber with a diameter of up to several millimeters, into which the sample is injected with a pipette; the chamber is covered with a dialysis membrane and secured on the exterior with an O-ring. Another more cost-effective alternative to commercially purchased buttons is to construct them from microtube tops; the size of the tube determines the dialysis volume (i.e., 35 μl for a PCR tube top and 250 μl for a 1.5 ml tube top). Cut off the top of the tube, deposit the sample into the top, cover with a dialysis membrane, and secure it by reattaching the bottom of the tube. Dialysis cups are similar in concept to buttons, except they enable use of a slightly larger sample volume (~10–100 μl). These are available from Pierce, but they could be constructed in a similar fashion to the buttons described by cutting of the top of a 100 μl microcentrifuge tube, placing the sample in the tube portion, covering it with a membrane, and replacing the cap to seal.

3. With this ratio of protein to buffer, the buffer does not need to be exchanged with fresh buffer during the dialysis period. As the buffer is far in excess of volumes required, lower amounts could easily be sufficient. Beakers should not contain more than one dialysis tube, however, to avoid complicating the system (Schmidt-Krey, unpublished results). In one case, conditions were optimized with two dialysis tubes each containing 100 μl of dialysate under identical conditions. The optimal 2D crystallization parameters necessitated the presence of two tubes of 100 μl and the conditions could not be downscaled in terms of protein without an investment in time and protein.

   The starting point for pH of the buffer should be one in which the protein is optimally stable and enzymatically active, if applicable, which should be close to physiological pH.

Thus, this is a parameter that should be and is likely known prior to crystallization, and the likelihood that it requires adjustment is low. A change of pH might be required if the protein needs to be locked in one conformation for 2D crystallization to occur.

4. Addition of glycerol is believed to disrupt the hydration shell around solubilized protein micelles, which enables merging into extensive 2D arrays (14), and increased crystal sizes have been observed with elevated levels of glycerol (6, 22, 23).
5. While contamination may become problematic during long dialyses (≥2 weeks) at higher temperatures (≥24°C), typically bacteriostatics are not required. For most 2D crystallization experiments we have not needed to add NaN_3 to prevent bacterial contamination. This was the case even for dialysis times of 2–3 weeks above room temperature. It might, however, be advisable to take the precaution of adding NaN_3 especially for prolonged dialysis times at elevated temperatures. Keep in mind that this modification of adding NaN_3 will result in additional salt.
6. The inclusion of enzyme ligands or inhibitors can sometimes aid in crystallization by stabilization of a flexible protein domain or key subunit interface (see also Chapter 15). In addition, the structure determination of an enzyme with a bound ligand can provide useful functional information that would not otherwise be available.
7. The purification should be reproducible to obtain a sufficient yield of protein for many crystallization trials. Although high purity is an important factor, successful crystallization can be accomplished with roughly 95% purity. In special cases in which protein purity is only accomplished with sacrificed yield of an expensive protein, one may benefit by choosing to proceed with preliminary crystallization trials. Bear in mind that crystal size and quality will likely improve with increased purity. One of our proteins, however, routinely yields highly ordered crystals even with 2–3 additional weaker bands of contaminants on the gels after purification.
8. Minimal LPRs have been shown to produce the most highly ordered crystals. Some co-purified lipids can be removed by a sequence of washing steps with buffer or by phospholipase A_2 enzyme treatment. On the other hand, certain lipids may be integral and essential components for the maintenance of structural integrity of the protein monomer and/or contacts between subunits or monomers in complex, which was found true for light-harvesting complex II (26) and photosystem II (29). Complete removal of lipids from these proteins resulted in unfavorable monomerization (26) and prevented growth of 2D crystals (26, 29). Thus, knowledge of the lipid composition

of the solubilized protein will be of benefit before proceeding with crystallization trials. The simplest test, in addition to lipid analysis, is a crystallization trial without the addition of lipid. Resulting proteoliposomes will be a clear indication that excessive amounts of lipids have been co-purified, necessitating further purification/washing steps.

9. In our lab, we have not seen a need to exchange the detergent, but for a transport membrane protein, this was a key step in improving 2D crystals (see Chapter 2).
10. While the presence of detergent can have a significant impact on the protein concentration measured, varying widely with different techniques, use of only one method will allow for consistent determination of the protein concentration and make the LPR a relative value as well as ensure reproducibility. Error by only a small factor in measurement of initial protein concentration will result in addition of an inaccurate fraction of lipids, as well as an inability to reproduce conditions in later experiments. Our preferred method is to compare the image density of the protein band on a gel to densities of protein standards.
11. Solubilizing detergent is only needed if the protein concentration is several mg or more and will be reduced in concentration before dialysis. It is a relatively common mistake that scientists new to membrane proteins will not keep the detergent concentration fixed when reducing the protein concentration.
12. The detergent concentration should be above the critical micelle concentration (CMC). Since the detergent will be removed during the dialysis, the choice of detergent usually weighs far less importance than other mentioned parameters. Most types of non-ionic detergents used for traditional membrane protein purification have been successful in formation of 2D crystals, irrespective of their physical and chemical properties as they will not be, or at the most minimally, a component of the ordered proteoliposome.
13. Specifically, the CMC is the minimum concentration at which micelles can form. For membrane protein purification and maintenance of solubility prior to crystallization, the detergent concentration must remain above the CMC. In addition, the CMC value indicates binding strength. For example, high CMCs indicate weak binding, which require a short dialysis for removal. On the other hand, low CMCs indicate a powerful strength of binding, which require a longer dialysis. Typically, detergents with low CMCs are preferred for crystallization because they enable a slow removal rate, which favors formation and order of crystals.

14. BioBeads are generally washed and blotted with paper before weighing. As very small amounts are used for 2D crystallization, making weighing of the wet beads unreliable, this has prompted a selection of set numbers of beads by size.
15. At www.avantilipids.com a large range of detailed advice, information and protocols can be obtained about lipids.
16. Many lipids are hygroscopic, making handling and weighing of powders problematic. Chloroform can be pipetted immediately after the manufacturers container is opened, which generally results in highly reproducible lipid stock solutions.
17. This value can be roughly estimated and is most easily done by comparing projection maps of similar-sized membrane proteins with comparable hydrophobic and hydrophilic domains. Frequently hydrophobicity and/or mutagenesis data is available for the protein of interest that provides a convenient starting point.
18. The use of lower molar LPRs than 1 might be necessary in cases of co-purified lipid during the protein purification.
19. This type of information will also be useful in interpreting projection maps.
20. If 3D crystallization experiments are to be conducted in parallel, one might decide to work with a detergent known to be amenable to 3D crystallization. This will allow the protein purification protocol and individual protein batches to be used for both 2D and 3D crystallization experiments. Furthermore, insights from one type of crystallization can be used in the other.
21. Excessive amounts of remaining detergent will be noticeable through bubble formation when the sample is pipetted from the dialysis device into a microcentrifuge tube for storage. Often grid preparation will give the first insight into remaining detergent though: minimal detergent will still allow for the formation of a convex meniscus on the grid upon first applying the sample. A specimen with more excessive amounts of remaining detergent though, will rapidly spread without meniscus formation. Crystal stability, longevity, and possibly quality can be dramatically influenced by decreasing a minimal amount of remaining detergent even further by even 1–2 additional days of dialysis (6), resulting in 2D crystals that are stable for months or years versus days or weeks. Also, reconstitution might occur before 2D crystals form. We have observed beautiful, large vesicles ad sheets on day 5 of the dialysis, which were ordered on day 6 (unpublished observations).
22. We found that rapid pipetting with a conventional plastic pipette tip did not have any ill effects on the lipid. The chloroform may, however, affect the pipette itself after multiple uses.

23. Careful visual inspection of each lipid preparation at various stages will be helpful in identifying potential problems or discrepancies in preparation.

24. A large, non-heating sonicator bath should not lead to sufficient heating over minutes to cause any ill effects. The water in some sonicator baths does not heat at all during the solubilization, and these types of baths are ideally used.

25. An intermediate stage of solubilization is usually signified by a clear solution with "fragments" of lipid rather than a milky solution. Complete solubilization of DMPC in 0.5% Na-DOC will involve sonication of roughly 5 min. Prolonged sonication of 10 or more minutes with possibly an intermediate milky phase will likely be due to an older or substandard lipid batch, and we found it safest to discard these types of preparations rather than to use them for crystallization trials.

26. While one should aim to use lipid stocks within 1–2 years of preparation, we found that lipid stocks stored at −80°C over several years were identical in quality and crystallization behavior of the protein compared to freshly prepared lipid.

27. A longer piece of dialysis tubing of ≥9 cm prevents complete or partial sample loss as opposed to a shorter piece of tubing.

28. Volumes can be varied, but 250 ml has been successful for us for a range of proteins with various detergents and buffer conditions. The beaker size is also not critical but should be consistent.

29. The protein may be newly purified or removed from storage at −80°C, depending on the stability of the protein.

30. While we have successfully dialyzed volumes as low as 40 μl, smaller volumes combined with sample loss may be problematic in terms of reproducibility. For smaller volumes than 75 μl, dialysis tubing with a flat width of 4 mm or dialysis buttons are recommended.

31. An easy way to do this is to first hold the dialysis clip between the thumb and fourth fingers of one hand, using the index and third finger to stabilize the longer, open end of the tubing.

32. Unless a protein has specific requirements (e.g., anaerobic conditions), it is not necessary to remove the air from the dialysis tubing. The air makes sample recovery easier. Specimen loss during sampling of dialysis is prevented due to cutting the tubing at the air space.

33. In order to facilitate reproducibility and consistency, each sample should be placed in its individual beaker, rather than placing multiple dialysis conditions in one larger volume of buffer.

34. We found stirring and buffer exchange during dialysis not to be necessary with any of a range of samples and detergent conditions.
35. Observed precipitate may be aggregated protein due to factors such as an unnecessarily speedy detergent removal, protein denaturation/degradation, or buffer-related causes. It is also possible, however, that the precipitate consists of proteoliposomes. Thus, it is recommended to briefly screen this type of sample by EM to verify that it indeed is composed of precipitate.
36. One easy way to do this is during application of the sample onto carbon film for electron microscopy. If the sample shows low surface tension and does not form a convex meniscus on the carbon film surface, then it is very likely that significant detergent remains in the sample (see Chapter 15 also for reference to the drop box). If this is the case, the sample may be placed in a new dialysis membrane and dialyzed against the original buffer.
37. While we found storage at –80°C to be indispensable for samples where the 2D crystals started stacking in register upon storage after several days (15), it needs to be confirmed that storage at –80°C does not affect the high resolution order of any specific 2D crystals.

Acknowledgments

We wish to thank our colleagues in the laboratory as well as in the cryo-EM community for helpful discussions over the years. We are particularly indebted to our collaborators for entrusting us with their valuable samples. Research in the Schmidt-Krey laboratory contributing to some of these observations was funded by NIH grant HL090630 and the Integrative Biosystems Institute.

References

1. Engel A, Hoenger A, Hefti A, Henn C, Ford RC, Kistler J, Zulauf M (1992) Assembly of 2-D membrane-protein crystals—dynamics, crystal order, and fidelity of structure-analysis by electron-microscopy. J Struct Biol 109:219–234
2. Jap BK, Zulauf M, Scheybani T, Hefti A, Baumeister W, Aebi U, Engel A (1992) 2D crystallization—from art to science. Ultramicroscopy 46:45–84
3. Dolder M, Engel A, Zulauf M (1996) The micelle to vesicle transition of lipids and detergents in the presence of a membrane protein: towards a rationale for 2D crystallization. FEBS Lett 382:203–208
4. Hasler L, Heymann JB, Engel A, Kistler J, Walz T (1998) 2D crystallization of membrane proteins: rationales and examples. J Struct Biol 121:162–171
5. Kühlbrandt W (2003) Two-dimensional crystallization of membrane proteins: a practical guide. In: Schägger H, Hunte C (eds) Membrane protein purification and crystallization: a practical approach. Academic Press, San Diego, pp 253–284

6. Schmidt-Krey I (2007) Electron crystallography of membrane proteins: two-dimensional crystallization and screening by electron microscopy. Methods 41:417–426
7. Abeyrathne PD, Chami M, Pantelic RS, Goldie KN, Stahlberg H (2010) Preparation of 2D crystals of membrane proteins for high-resolution electron crystallography data collection. Methods Enzymol 481:25–43
8. Xu C, Rice WJ, He W, Stokes DL (2002) A structural model for the catalytic cycle of Ca(2+)-ATPase. J Mol Biol 8:201–211
9. Ryan C, Stokes DL, Chen M, Zhang Z, Hardwicke PM (2004) Effect of orthophosphate, nucleotide analogues, ADP, and phosphorylation on the cytoplasmic domains of Ca(2+)-ATPase from scallop sarcoplasmic reticulum. J Biol Chem 13:5380–5386
10. Stokes DL, Delavoie F, Rice WJ, Champeil P, McIntosh DB, Lacapère JJ (2005) Structural studies of a stabilized phosphoenzyme intermediate of Ca^{2+}-ATPase. J Biol Chem 280:18063–18072
11. Chen Z, Akin BL, Stokes DL, Jones LR (2006) Cross-linking of C-terminal residues of phospholamban to the Ca^{2+} pump of cardiac sarcoplasmic reticulum to probe spatial and functional interactions within the transmembrane domain. J Biol Chem 281:14163–14172
12. Glaves JP, Trieber CA, Ceholski DK, Stokes DL, Young HS (2011) Phosphorylation and mutation of phospholamban alter physical interactions with the sarcoplasmic reticulum calcium pump. J Mol Biol 405:707–723
13. Rigaud JL, Chami M, Lambert O, Levy D, Ranck JL (2000) Use of detergents in two-dimensional crystallization of membrane proteins. Biochim Biophys Acta Biomembr 1508:112–128
14. Kühlbrandt W (1992) 2-dimensional crystallization of membrane proteins. Q Rev Biophys 25:1–49
15. Zhao G, Johnson MC, Schnell JR, Kanaoka Y, Irikura D, Lam BK, Austen KF, Schmidt-Krey I (2010) Two-dimensional crystallization conditions of human leukotriene C_4 synthase requiring a particularly large combination of specific parameters. J Struct Biol 169:450–454
16. Brisson A, Unwin PNT (1984) Tubular crystals of acetylcholine-receptor. J Cell Biol 99:1202–1211
17. Iacovache I, Biasini M, Kowal J, Kukulski W, Chami M, van der Goot FG, Engel A, Rémigy HW (2009) The 2DX robot: a membrane protein 2D crystallization Swiss Army knife. J Struct Biol 169:370–378
18. Kim C, Vink M, Hu M, Love J, Stokes DL, Ubarretxena-Belandia I (2010) An automated pipeline to screen membrane protein 2D crystallization. J Struct Funct Genomics 11:155–166
19. Hu M, Vink M, Kim C, Derr K, Koss J, D'Amico K, Cheng A, Pulokas J, Ubarretxena-Belandia I, Stokes D (2010) Automated electron microscopy for evaluating two-dimensional crystallization of membrane proteins. J Struct Biol 171(1):102–110
20. Rigaud JL, Mosser G, Lacapere JJ, Olofsson A, Levy D, Ranck JL (1997) Bio-beads: an efficient strategy for two-dimensional crystallization of membrane proteins. J Struct Biol 118:226–235
21. Schmidt-Krey I, Mutucumarana V, Haase W, Stafford DW, Kühlbrandt W (2006) Two-dimensional crystallization of human vitamin K-dependent γ-glutamyl carboxylase. J Struct Biol 157:437–442
22. Schmidt-Krey I, Lundqvist G, Morgenstern R, Hebert H (1998) Parameters for the two-dimensional crystallization of the membrane protein microsomal glutathione transferase. J Struct Biol 123:87–96
23. Schmidt-Krey I, Kanaoka Y, Mills DJ, Irikura D, Haase W, Lam BK, Austen KF, Kühlbrandt W (2004) Human leukotriene C_4 synthase at 4.5 Å resolution in projection. Structure 12:2009–2014
24. Mitra AK, Miercke LJW, Turner GJ, Shand RF, Betlach MC, Stroud RM (1993) 2-dimensional crystallization of *Escherichia-coli*-expressed bacteriorhodopsin and its D96N variant—high-resolution structural studies in projection. Biophys J 65:1295–1306
25. Tsai CJ, Ziegler C (2005) Structure determination of secondary transport proteins by electron crystallography: two-dimensional crystallization of the betaine uptake system BetP. J Mol Microbiol Biotechnol 10:197–207
26. Nussberger S, Dorr K, Wang DN, Kühlbrandt W (1993) Lipid-protein interactions in crystals of plant light-harvesting complex. J Mol Biol 234:347–356
27. Cevc G (1987) How membrane chain melting properties are regulated by the polar surface of the lipid bilayer. Biochemistry 26:6305–6310
28. Mosser G (2001) Two-dimensional crystallogenesis of transmembrane proteins. Micron 32:517–540
29. Kruse O, Hankamer B, Konczak C, Gerle C, Morris E, Radunz A, Schmid GH, Barber J (2000) Phosphatidylglycerol is involved in the dimerization of photosystem II. J Biol Chem 275:6509–6514

Chapter 4

Monolayer Two-Dimensional Crystallization of Membrane Proteins

Luc Lebeau and Catherine Vénien-Bryan

Abstract

This method of two-dimensional crystallization of proteins on a lipid monolayer aims at producing 2D crystals of membrane proteins, which can provide structural information at high resolution by electron crystallography.

A lipid monolayer is spread over the whole air–water interface of a drop, which provides a substrate for protein binding. The protein of interest is then adsorbed onto the lipid monolayer and forms a closely packed layer. The reconstitution step of the membrane protein into a lipid bilayer is realized by elimination of detergent. The combined effect of the elevated protein concentration, the alignment of the protein on the lipid monolayer, and the fluid monolayer film properties are conducive, in some cases, to the formation of 2D crystals. The use of locally fluorinated lipids in avoiding solubilization of the lipid monolayer by detergents is presented and discussed.

Key words: Lipid monolayer, Fluorinated lipid, Membrane protein, Self-assembly, Two-dimensional crystal, Electron crystallography

1. Introduction

This chapter will focus on the formation of two-dimensional (2D) crystals of membrane proteins on lipid monolayers. As first shown by Fromherz in 1971, an ordered 2D arrangement of protein can be generated underneath a lipid monolayer spread at the air–water interface (1). In that seminal work, ferritin molecules were observed to form a closely packed 2D lattice after adsorption to a charged lipid film made of methyl stearate and eicosanoic trimethyl ammonium

Ingeborg Schmidt-Krey and Yifan Cheng (eds.), *Electron Crystallography of Soluble and Membrane Proteins: Methods and Protocols*, Methods in Molecular Biology, vol. 955, DOI 10.1007/978-1-62703-176-9_4,

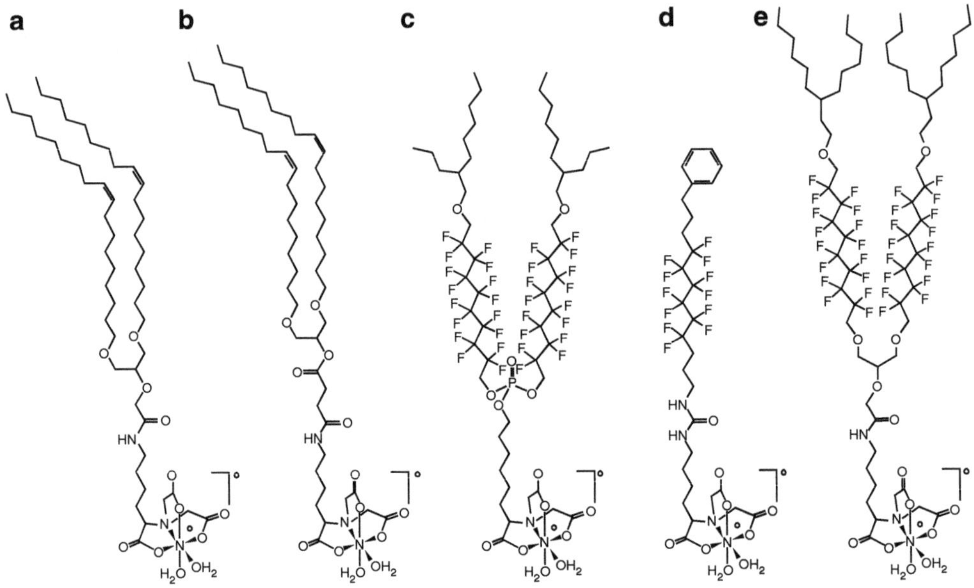

Fig. 1. Structure of Ni^{2+}–NTA lipids used in 2D crystallization experiments with His-tagged water soluble proteins and membrane proteins. The lipids consist of two distinct parts: (1) a hydrocarbon tail to confer stability and fluidity to the subsequent monolayer and (2) a Ni^{2+}-functionalized hydrophilic head group which promotes binding of His-tagged molecules at the air–water interface. Newer examples of these compounds also incorporate fluorocarbon moieties to confer detergent resistance to the monolayers.

bromide. Since then, the method has been developed and has been shown to be useful for the formation of 2D crystals of soluble proteins suitable for structural analysis by electron microscopy (2). Later, this 2D crystallization technique was extended to membrane proteins (3).

The 2D crystallization of a water-soluble protein on a lipid monolayer spread at the air–water interface is a two-phase process. The first step is the adsorption of the protein onto the lipid layer. This leads to (a) the concentration of the protein at the interface and (b) its preorientation along the *z*-axis, i.e., perpendicular to the interface. This second phenomenon depends on the nature of the interaction established between the protein and the lipid film. When the protein orientations are limited, further protein organization within the plane is facilitated possibly leading to 2D crystallization. The polar headgroup of the lipid is responsible for the adsorption of the protein onto the film and operates as a recognition element for the protein. This recognition may take place in two different ways. One involves purely electrostatic and non-specific interactions between a charged lipid within the monolayer and the protein. The other is achieved through the interaction of the protein with a specific ligand attached to the lipid polar head. For example the lipid can be functionalized with a Ni^{2+}–NTA group that can be utilized for protein binding via a His-tag introduced in the protein peptide sequence (Fig. 1a) (3). Following that strategy, His-tagged proteins present in the aqueous subphase are specifically adsorbed at the lipid film, leading to at least a 100-fold increase in protein concentration at the lipid interface (3). The second step of the 2D crystallization process arises from a combination of the elevated protein concentration, the preorientation of the protein (charge- or ligand-mediated) on the lipid monolayer, and the mechanical properties (stability and fluidity) of the lipid film. All three effects may enable the selection of a discrete set of protein–protein interactions via lateral and rotational diffusion, which, in favorable cases, are conducive to the formation of 2D crystals. Efforts have been devoted to the fine-tuning of the mechanical properties of the lipid film and sophisticated lipid architectures with various hydrophobic chains have been evaluated (4–6). Results reveal a marked influence of film fluidity on the rate of protein adsorption and 2D crystallization (6).

In contrast, the general technique of 2D crystallization of membrane proteins is different and involves the reconstitution into a phospholipid bilayer concomitantly with the formation of 2D arrays of protein in the latter (see Chapter 3). To achieve that critical operation, micelles containing a mixture of a (membrane) phospholipid and a detergent are added to the membrane protein of interest (solubilized in detergent). Then detergent is slowly removed (i.e., by dialysis or via adsorption onto Biobeads®) so that the phospholipid self-assembles into a bilayer in which the membrane protein inserts. In some cases, the protein inserts with a unique orientation along the *z*-axis (perpendicular to the bilayer) and the system may evolve towards formation of 2D crystals. But most of the time there is no selection of the *z* orientation of the protein and its random insertion (up and down) within the bilayer produces amorphous patches.

Application of the technique of water-soluble protein 2D crystallization on a lipid monolayer to membrane proteins has the advantage to preferentially orient (*z* orientation) the membrane protein before and during the phospholipid bilayer reconstitution. It is to note that the application of this monolayer 2D crystallization technique to membrane proteins is complicated by the high propensity of detergents to quickly solubilize lipids and, a fortiori, the template lipid used for protein concentration at the air–water interface.

However, unexpected stability of a phospholipid monolayer deposited on a TritonX100 solution was observed (7). Surface pressure measurements and grazing incidence X-ray diffraction (GIXD) have shown that the molecular organization of pure phospholipid domains at the air–water interface is caused by detergent through surface pressure alteration. Importantly, detergent molecules adsorb at the air–water interface, but do not readily solubilize the lipid monolayer. Indeed, GIXD measurements revealed that at the air–water interface, adsorbed detergent and phospholipid molecules segregate in pure phospholipid domains surrounded by a detergent-rich phase. The pressure increase due to the detergent at the interface leads to compression of the lipid domains in the liquid condensed (LC) phase and dissolution of the phospholipid molecules does not occur, even when the concentration of detergent is above its critical micelle concentration (cmc). This resistance to dissolution is likely to be due to the tight molecular packing of the lipid chains and is suppressed upon expansion of the monolayer. It is worth noting that resistance to dissolution by detergents does not exist when the monolayer is spread with a lipid collapsing in the liquid expanded (LE) phase.

For the membrane protein 2D crystallization strategy, some resistance of the lipid domains to dissolution by detergent is necessary so that some tethering molecules (lipid–ligand conjugates) are still available at the interface for promoting the membrane protein concentration at the interface (i.e., lipids-ligands are not totally solubilized as mixed micelles with detergent in the subphase). If there is resistance to solubilization, then, fast concentration of the protein at the interface may induce stabilization of the film at the interface to some extent. However, compression of the lipids into LC domains imposed by the presence of detergent hinders the 2D diffusion of these lipids at the interface. 2D diffusion is necessary for the protein crystallization process because lateral diffusion of the protein molecules attached to the monolayer is crucial for the free selection and accommodation of protein–protein interactions. As a consequence, the use of low surface pressure detergents is to be preferred when classic lipids (i.e., lipids collapsing in the LE phase) are used.

The dependence on a low cmc and low surface pressure detergent is a restriction, which may limit the general application of this approach. The hindrance of lipid diffusion might explain why the membrane proteins crystallized on monolayers made of classic lipids diffract at only medium resolution (in the range of 20–35 Å resolution). Indeed, two His-tagged membrane proteins, FuhA and TF_0F_1, were crystallized on monolayers of a regular lipid bearing a Ni^{2+}-chelating headgroup (Fig. 1b) (8). These experiments were based on the use of low cmc detergents (TritonX100 and *n*-dodecyl-β-d-maltopyranoside) and proteins with high affinity for functionalized lipids in order to quickly stabilize the interface (the equilibrium dissociation constant K_D for the interaction between a His-tag and a Ni^{2+}–NTA complex is in the submicromolar range) (9). The results obtained tend to show that the regular lipid forming the lipid monolayers was not solubilized (or only slowly solubilized) under the experimental conditions but resolution attained was not better than 18.5 Å (FhuA) and 35 Å (TF_0F_1).

To avoid solubilization of the lipid monolayer by detergents (this is especially valuable when working with membrane proteins requiring solubilization with high cmc detergents), locally fluorinated lipids have been developed (Fig. 1c) (10). Due to the non-ideality of hydrocarbon–fluorocarbon mixtures, these lipids are not only hydrophobic but also lipophobic. Consequently, the monolayers made thereof display high resistance to a range of detergents used at high concentration and well above their cmc. Of the utmost interest is that such fluorinated lipids do spread and collapse at the air–water interface in the LE phase. The resulting highly fluid lipid films do allow subsequently formed lipid–protein complexes to freely diffuse at the interface, which may be conducive to protein arrangement into 2D crystals. Indeed these fluorinated Ni^{2+}-lipids were successfully used to produce 2D crystals of the His-tagged plasma membrane proton ATPase from *Arabidopsis thaliana*. Electron microscopy and subsequent image processing yielded a 2D projection map at 9 Å resolution (10). Following this work, other fluorinated lipids were developed for self-assembly and 2D crystallization of membrane proteins (Fig. 1d, e) (11, 12). Locally fluorinated Ni^{2+}–NTA glycerolipid shown in Fig. 1e promoted the 2D crystallization of the His-tagged membrane protein BmrA, an ABC transporter from *Bacillus subtilis* (12). The crystals obtained only diffracted poorly, which would likely be a result of a too-cohesive fluorinated lipid film. Fluorinated lipids designed for the 2D crystallization of membrane proteins have been extensively reviewed recently (13). However, due to the poor availability of such lipids (they are not commercially available), only a few membrane proteins were assayed in 2D crystallization experiments and the results reported up to now probably do not reflect the potential of these tools.

2. Materials

2.1. 2D Crystallization of Membrane Proteins

1. Lipids forming the monolayer: Regular (non-fluorinated) lipids or fluorinated lipids bearing a Ni^{2+}–NTA group for interaction with His-tagged proteins or any other lipid–ligand conjugate for interaction with the corresponding receptor.
2. Lipids for the protein reconstitution (reconstituting lipids): These lipids will form bilayers in which the proteins will insert and possibly organize in 2D crystals. Usually the reconstituting lipids are 1,2-dioleoyl-*sn*-glycero-3-phosphocholine (DOPC), 1-palmitoyl-2-oleoyl-*sn*-glycero-3-phosphocholine (POPC), *Escherichia coli* lipids, or other phospholipids. These lipids should be present in the subphase in the form of mixed micelles with detergent and are different from the lipids forming the film at the air–water interface. The detergent usually is TritonX100, *n*-dodecyl-β-d-maltopyranoside (DDM), decyl

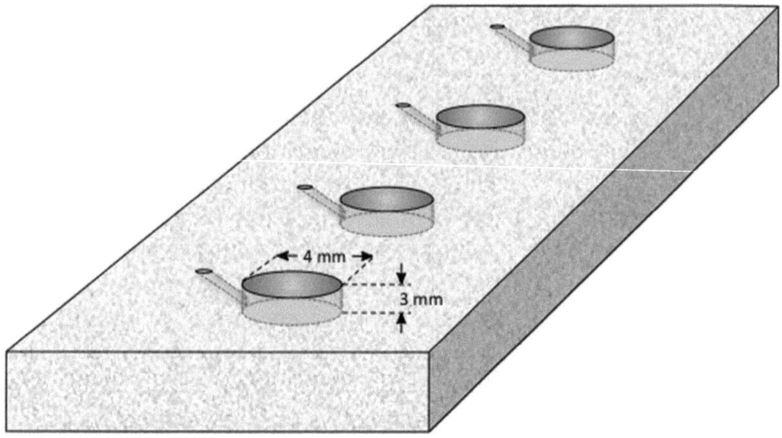

Fig. 2. Teflon® block for lipid layer 2D crystallization trials. The Teflon® block consists of four wells, each 4 mm in diameter and 2–3 mm deep, i.e., with a volume of about 50 μL. Each well has a small, angled, side channel which enables either membrane proteins in detergent with reconstituted lipids or BioBeads® to be injected.

β-d-maltopyranoside (DM), 1,2-diheptanoyl-*sn*-glycero-3-phosphocholine (DHPC), 1-O-*n*-octyl-β-d-glucopyranoside (OBG), or other detergents.

3. A Teflon® block in which wells, 4 mm in diameter and 1–2 mm deep, have been drilled such that each well can contain about 50 μL of aqueous solution (this is because electron microscopy is still the most common way to screen for crystallization conditions and electron microscope grids are 3 mm in diameter). In addition each well has a small, angled, side tube (about 1 mm in diameter) (Fig. 2).

4. A Petri dish as a humidity chamber. Some water saturated paper is placed at the bottom of the dish. The Teflon® block is placed on this water saturated paper and the top of the Petri dish is closed. Water evaporation is prevented by sealing the top of the Petri dish with parafilm®.

5. BioBeads® for detergent removal. Before use, BioBeads® should be washed three times with methanol or ethanol, and three more times with de-ionized water. BioBeads® should be kept in a glass bottle, in de-ionized water with NaN_3 0.01%, at 4°C. The beads should not float at the surface of the water but stay at the bottom of the bottle. Floating beads should not be used further.

6. Various solutions containing the concentrated solubilized membrane protein (with which it is hoped to form 2D crystals) mixed with reconstituting lipids in detergent (see Subheading 2) at various lipid/protein ratios.

2.2. Transfer to an Electron Microscope Grid and Observation

1. Electron microscope grids (400 mesh) covered with a carbon film. It is not the scope of the chapter to describe the deposition of the carbon layer onto an electron microscope grid (see Chapter 5).
2. Optical diffraction bench.
3. Electron Microscope.

3. Methods

3.1. 2D Crystallization of Membrane Proteins

1. Lipids forming the lipid monolayer: The lipids may be prone to oxidation, therefore, are best stored as a dry powder under an argon atmosphere at −20°C in an air-tight glass container. A mother solution (5 mg/mL) is produced by solubilizing the lipids in an organic solvent (chloroform:hexane, 1:1 v/v). This lipid solution also is best stored at −20°C under argon in a glass container with a Teflon® cap. The working solution is made usually of the ligand–lipid and diluting-lipid at a molar ratio between 1:10 and 1:3 at a concentration of 500 μM. The ligand–lipid is the lipid, which may specifically interact with the protein, and the diluting-lipid will act as a fluid matrix. When fluorinated lipids are used, ligand- and diluting-lipids must be fluorinated unless they will segregate.
2. Lipids for the protein reconstitution (reconstituting lipid). These lipids are prepared in detergent. Aliquots of 5 mg of phospholipids DOPC, POPC, *E. coli* lipids, or other lipids should be prepared and stored as a dry powder under an argon atmosphere at −20°C in an air-tight glass container. The detergent can be the same as that used for the solubilization of the membrane protein or a different one such as DDM, DM, DHPC TritonX100, or OBG. The detergent may be supplied as a powder or as concentrated solution in water, or solubilized in a solvent (this is the case for DHPC). When needed, one aliquot of 5 mg lipid is solubilized in 1 mL de-ionized water with detergent, so that the final concentration of lipid is 5 mg/mL. If the detergent is solubilized in a solvent, the solvent should be thoroughly evaporated. Aliquots of dried lipid and detergent can be kept in the same air-tight glass until addition of 1 mL de-ionized water or buffer for complete solubilization of the lipid in detergent. The final concentration of

the detergent should be at least ten times its cmc. A clear indication of the complete solubilization of the lipid in detergent (and this is of utmost importance) is the limpidity of the lipid detergent solution; this preparation should be perfectly clear. Depending on the properties of the reconstituting lipid and detergent, the solubilization of lipid in detergent could be instantaneous or may require stirring overnight at room temperature, or at higher temperature (up to 45°C), or sonication for 15 min (in a sonicator bath taking care that the lipid solution does not heat up too much). If the solution is still opaque after the above treatment it might be useful to add more detergent or apply a more intensive sonication treatment. Solubilised lipid in detergent can be kept for up to a month at 4°C under an argon atmosphere, in an air-tight container.

3. The Teflon® block should be thoroughly cleaned with a detergent, boiled in water for 10 min or sonicated in chloroform for 5 min, followed by extensive rinsing with de-ionized water. The Teflon® surface should be hydrophobic after this treatment.

4. Buffer (30 μL) is placed in the well. It is worth considering adding some azide (0.01%) in this solution to prevent bacteria from growing (see Note 1).

5. The lipid working solution (regular or fluorinated ligand–lipid, 0.5 μL at 500 μM) is deposited at the top of the crystallization drop using a Hamilton syringe. The lipid spreads and leaves a film at the air–water interface while the organic solvent evaporates (14). The amount of lipid is more than what would be required for a single layer, usually five to ten times surplus. The excess lipid forms a reservoir at the edge of the Teflon® well.

6. The solution containing the concentrated membrane protein in detergent and mixed micelles (detergent and reconstituting lipids) is then injected through the side tube into the bottom of the well. A minuscule magnetic stirring bar previously placed at the bottom of the well can be used for a short period to ensure a good homogeneity of the solution. The final concentration of the membrane protein in the well should be between 50 and 150 μg/mL and the reconstituting lipid concentration should be such that the lipid/protein ratio is in the range of 0.2–1.4 (w/w). The final detergent concentration in the well should always be above the cmc (below its cmc, a detergent is fully soluble and does not form micelles). During this step, Fig. 3a, b, which should last up to 24 h, the protein in detergent should bind to the ligand–lipid at the interface. A screen should be done in order to find out which lipid and which lipid/protein ratio gives the best result (the higher protein coverage).

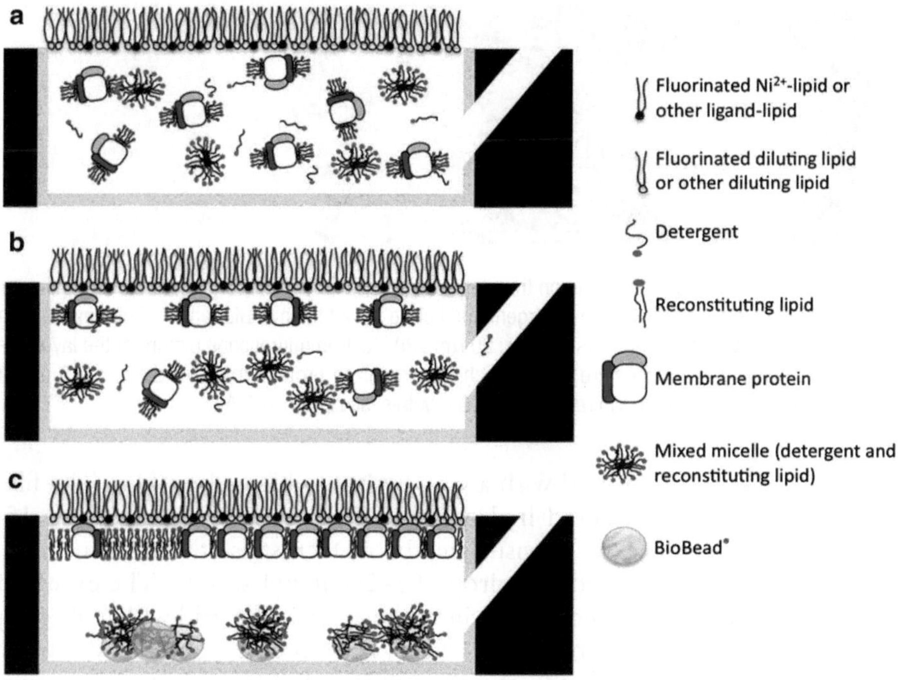

Fig. 3. Method for 2D crystallization of a membrane protein on a lipid monolayer. (**a**) After the lipid working solution (ligand–lipid mixed with diluting lipid) has been deposited at the top of a 30 μL buffer drop and 15–20 μL of a concentrated solution of the membrane protein, mixed micelles have been injected through the side tube to the bottom of the well. (**b**) Over the next 24 h, the membrane protein embedded in detergent binds to the ligand–lipid at the interface. (**c**) BioBeads® are injected through the side tube to the bottom of the well. The detergent binds to the BioBeads® and growth of 2D crystals takes place.

7. The reconstitution step of the membrane protein into the lipid bilayer is realized by elimination of detergent using BioBeads®, which are added through the side tube into the bottom of the well (using forceps). The addition of these BioBeads® (10–20 beads of medium size) should be performed when binding of the membrane protein to the lipid monolayer has reached completion (high protein density).
8. Usually after 24 h, most part of the detergent in solution has been trapped on the BioBeads® and the protein bound to the lipid film at the air–water interface is reconstituted into a lipid bilayer (reconstituting lipid) and hopefully forms 2D crystals (Fig. 3c).

3.2. Transfer to an Electron Microscope Grid and Observation

1. Transfer of 2D crystals onto a microscope grid. The 2D crystals or protein layers are transferred to the electron microscope grid through hydrophobic contacts between the lipid chains of the ligand–lipid and the carbon film on the grids. The grid is deposited on the top of the crystallization drop, left for about 1 min, then withdrawn vertically and prepared for observation under the microscope. The transfer often works best with grids

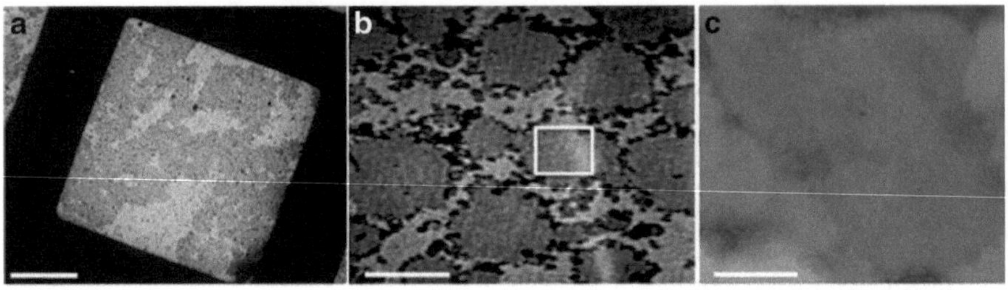

Fig. 4. Observation of the surface of the drop upon transfer to an electron microscope grid. (**a**) Electron micrograph of the layer after injection of the protein solubilized in detergent and before the addition of BioBeads®. Domains of bound proteins (*gray patches*) are seen at low magnification (scale bar: 20 μm). (**b**) Electron microscope picture of the layer after elimination of the detergent. Some *gray patches* are seen, which are organized protein arrays (scale bar: 0.5 μm). (**c**) Protein arrays at higher magnification. These arrays reveal to be 2D crystals (scale bar: 100 Å).

covered with a very hydrophobic carbon film. The film can be rendered hydrophobic through baking for 1 h at 150°C. In order to visualize the 2D crystals, the specimen on grid is stained by a drop of 1–2% uranyl acetate. The excess of stain is blotted by touching the edge of the grid with a filter paper, and the grid is air dried (see Notes 2 and 3). Alternatively, the specimen can be prepared for cryo-electron microscopy observation. In that case, after transfer the grid is blotted for 5–8 s with a filter paper and plunged into liquid ethane. The sample is then preserved in a frozen-hydrated state.

2. Observation of the grid by electron microscopy will show if there is some protein bound to the lipid film (Fig. 4a). Patches of protein can be recognized as extended gray areas. If such patches are observed, higher magnification will show if they are arranged in an ordered way (Fig. 4b, c). When working with a completely hydrophobic membrane protein without any external domain, it is necessary to take images and check them for diffraction at an optical bench. The lack of external domain pointing out of the membrane makes it difficult to locate and identify eventual 2D arrays visually rather than by optical diffraction.

In order to familiarize oneself with the procedure, see Note 4.

3.3. Feed Back Loops

1. During step 6, the protein in detergent should bind to the ligand–lipid. This is easily checked by transferring the layer to an electron microscope grid. At low magnification, some large protein-containing planar grey domains should clearly be seen on the grid (Fig. 4a). If not, modifying and tuning the parameters listed below should lead to a denser protein layer underneath the lipid layer.
   (a) It might be necessary to increase the final protein concentration up to 200 μg/mL.
   (b) Another important factor is the ratio of ligand–lipid to diluting-lipid for the lipid monolayer at the air–water interface. It is worth trying a range from 1:0 to 1:10. This

ratio can have a great influence on the density at which the protein binds to the monolayer.

(c) The incubation period may influence the protein binding to the lipid monolayer. Up to 24–36 h should be allowed for the binding of the protein to the monolayer.

(d) In some cases, changing the buffer conditions may promote the binding of the protein to the film. When using a His-tagged protein in combination with a Ni^{2+}–NTA lipid, a pH value around 8.0 is suitable. EDTA, EGTA, imidazole, or any other metal-chelating species should be avoided or present only at low concentration.

2. When the reconstituted protein tends to form close-packed arrays which do not evolve towards organized protein patches, it is advisable to reduce the kinetics of protein concentration either by increasing the viscosity of the medium by adding glycerol (up to 40%), or by reducing the temperature or the protein concentration, or using a lower ligand–lipid/diluting-lipid ratio.
3. During step 7. The speed of detergent removal during the reconstitution step is directly controlled by the amount of BioBeads® added into the trough (2–5 mg are usually added, i.e., 10–20 beads). This parameter is important and the quality of the 2D crystals often depends on it.
4. Once large crystalline areas are obtained (Fig. 4b, c), electron micrographs are recorded and the extent of order is evaluated by optical diffraction. A large number of parameters such as reconstituting lipid/protein ratio, pH, ionic strength, buffer composition, presence of divalent cations, protein concentration, presence of glycerol, incubation temperature, or incubation time can be modified to improve crystal order.

4. Notes

1. Experiments for 2D crystallization of proteins on a lipid monolayer take between hours and several days. They are usually performed at a temperature around 20°C. This means that the protein has to be sufficiently stable for the course of the experiment. Requirements for buffer conditions are usually flexible enough to be favorable for the stability of the particular protein. If not, experiments may be carried out at 4°C. In that case however make sure that the lipid film still spreads in the fluid phase.
2. Depending on the protein, the contact of the lipid layer with the carbon film on the electron microscope grid during the transfer can distort, bend or even break existing 2D crystals.

Holey grids can be a good answer to that problem as they statistically limit the interactions between the carbon surface and 2D crystals, which may then be observed inside the holes. However, 2D crystals in these holes can sag, which is detrimental when tilted images are necessary for the calculation of the 3D structure.

3. To avoid disruption of the 2D crystals during the transfer, glutaraldehyde (1 μL of a 0.5% aqueous solution), a protein cross-linker, can be added on top of the well (it may diffuse through lipids) or in the subphase before placing the microscope grid. Glutaraldehyde strengthens the crystal (15).

4. In order to familiarize oneself with the technique of crystallization of protein on a lipid monolayer, we advise to practice on a model system using a soluble protein. For instance the 2D crystallization of streptavidin on biotinylated lipids (biotin-LC-1,2-dipalmitoyl-*sn*-glycero-3-phosphoethanolamine) is a very well documented system and easy to be trained with (16). Streptavidin should be used at 10–30 μg/mL in buffer at pH 7.4. The lipid monolayer should be made of biotinylated lipids (commercially available) diluted with DOPC (the diluting lipid) at a molar ratio 1:4. This simple system using soluble protein does neither require the use of detergents nor that of BioBeads®, and upon transfer to an electron microscope grid, very large crystals should be readily observed.

References

1. Fromherz P (1971) Electron microsopic studies of lipid protein films. Nature 231:267–268
2. Lebeau L, Schultz P, Célia H, Mésini P, Nuss S, Klinger C, Olland S, Oudet P, Mioskowski C (1996) Specifically designed lipid assemblies as tools for two-dimensional crystallization of soluble biological macromolecules. In: Barenholz Y, Lasic DD (eds) Handbook of non-medical applications of liposomes. Vol. II. CRC, Boca Raton, pp 153–186
3. Dietrich J, Venien-Bryan C (2005) Strategies for two-dimensional crystallization of proteins using lipid monolayers. Imperial College Press, London
4. Nuss S, Mioskowski C, Lebeau L (1999) Synthesis of new fluidity-enhanced amphiphilic compounds for soluble protein two-dimensional crystallization purpose. Chem Phys Lipids 103:21–35
5. Lebeau L, Nuss S, Schultz P, Oudet P, Mioskowski C (1999) Self-assembly of soluble proteins on functionalized lipid layers: a tentative correlation between the fluidity properties of the lipid film and protein ordering. Chem Phys Lipids 103:37–46
6. Courty S, Lebeau L, Martel L, Lenne PF, Balavoine F, Dischert W, Konovalov O, Mioskowski C, Legrand JF, Venien-Bryan C (2002) Two-dimensional crystallization of a His-tagged protein on monolayers of fluidity-enhanced Ni^{2+}-chelating lipids. Langmuir 18:9502–9512
7. Fontaine P, Faure MC, Muller F, Poujade M, Micha JS, Rieutord F, Goldmann M (2007) Unexpected stability of phospholipid langmuir monolayers deposited on triton X-100 aqueous solutions. Langmuir 23:12959–12965
8. Levy D, Mosser G, Lambert O, Moeck GS, Bald D, Rigaud JL (1999) Two-dimensional crystallization on lipid layer: a successful approach for membrane proteins. J Struct Biol 127:44–52
9. Lata S, Reichel A, Brock R, Tampé R, Piehler J (2005) High-affinity adaptors for switchable recognition of histidine-tagged proteins. J Am Chem Soc 127:10205–10215

10. Lebeau L, Lach F, Venien-Bryan C, Renault A, Dietrich J, Jahn T, Palmgren MG, Kuhlbrandt W, Mioskowski C (2001) Two-dimensional crystallization of a membrane protein on a detergent-resistant lipid monolayer. J Mol Biol 308:639–647
11. Dauvergne J, Polidori A, Vénien-Bryan C, Pucci B (2008) Synthesis of a hemifluorinated amphiphile designed for self-assembly and two-dimensional crystallization of membrane protein. Tetrahedron Lett 49:2247–2250
12. Hussein WM, Ross BP, Landsberg MJ, Levy D, Hankamer B, McGeary RP (2009) Synthesis of nickel-chelating fluorinated lipids for protein monolayer crystallizations. J Org Chem 74: 1473–1479
13. Hussein WM, Ross BP, Landsberg MJ, Hankamer B, McGeary RP (2009) Synthetic approaches to functionalized lipids for protein monolayer crystallizations. Curr Org Chem 13:1378–1405
14. Vénien-Bryan C, Lenne PF, Zakri C, Renault A, Brisson A, Legrand JF, Berge B (1998) Characterization of the growth of 2D protein crystals on a lipid monolayer by ellipsometry and rigidity measurements coupled to electron microscopy. Biophys J 74: 2649–2657
15. Lenne PF, Berge B, Renault A, Zakri C, Venien-Bryan C, Courty S, Balavoine F, Bergsma-Schutter W, Brisson A, Grubel G, Boudet N, Konovalov O, Legrand JF (2000) Synchrotron radiation diffraction from two-dimensional protein crystals at the air/water interface. Biophys J 79:496–500
16. Darst SA, Ahlers M, Meller PH, Kubalek EW, Blankenburg R, Ribi HO, Ringsdorf H, Kornberg RD (1991) Two-dimensional crystals of streptavidin on biotinylated lipid layers and their interactions with biotinylated macromolecules. Biophys J 59: 387–396

Chapter 5

Screening for Two-Dimensional Crystals by Transmission Electron Microscopy of Negatively Stained Samples

Tina M. Dreaden, Maureen Metcalfe, Laura Y. Kim, Matthew C. Johnson, Bridgette A. Barry, and Ingeborg Schmidt-Krey

Abstract

Structural studies of soluble and membrane proteins by electron crystallography include several critical steps. While the two-dimensional (2D) crystallization arguably may be described as the major bottleneck of electron crystallography, the screening by transmission electron microscopy (EM) to identify 2D crystals requires great care as well as practice. Both sample preparation and EM are skills that are relatively easily acquired, compared to the identification of the first ordered arrays. Added to this, membranes may have a variety of morphologies and sizes. Here we describe all steps involved in the screening for 2D crystals as well as the evaluation of samples.

Key words: Negative stain, Grid preparation, EM, Screening, 2D crystals, Proteoliposomes, Liposomes, Fourier transform, Optical diffraction, Membrane morphology

1. Introduction

Screening negatively stained samples for the presence of two-dimensional (2D) crystals by transmission electron microscopy (TEM, EM) is an absolutely critical step in evaluating the outcome of 2D crystallization trials that needs to be undertaken with great care (1). Conditions require thorough evaluation so as not to be dismissed without proper and extensive assessment. Results provide direction for subsequent 2D crystallization trials as well as for the decision to start collecting electron cryo-microscopy (cryo-EM) data. Identifying and evaluating samples necessitates practice and experience, which is unfortunately frequently first

Ingeborg Schmidt-Krey and Yifan Cheng (eds.), *Electron Crystallography of Soluble and Membrane Proteins: Methods and Protocols*, Methods in Molecular Biology, vol. 955, DOI 10.1007/978-1-62703-176-9_5,

obtained in the course of screening for 2D crystallization of a new protein sample.

The first step of evaluation involves identification of proteoliposomes, which will here interchangeably be referred to as membranes and may or may not contain a 2D lattice, that are then assessed for order at higher magnification. Distracting experimental artifacts on the grid can complicate identification of membranes, such as broken carbon film and mica. In some instances, artifacts cannot easily be distinguished from proteoliposomes, but with experience, true membranes can be readily identified in negative stain. Differentiating proteoliposomes from artifacts makes it essential to have familiarity with both the appearance of a large range of possible membrane morphologies as well as artifacts.

Membranes can adopt numerous morphologies, such as vesicles, planar-tubular 2D crystals, sheets, and tubes, and these various appearances may differ further in size and contrast. The membranes can aggregate, and in some cases aggregation of membranes at lower magnifications resembles non-membranous protein precipitate. Often though, aggregated proteoliposomes are also useful indicators of individual membranes that tend to frequently be found in the vicinity of the high-contrast aggregates.

While identification of membranes requires practice that is easily alleviated by some instruction, the main difficulty, however, is the evaluation of early specimens for 2D crystals. This assessment is performed by Fast Fourier Transform (FFT) or optical diffraction. The obstacle primarily lies in the case of initial identification of first, small ordered arrays with an unfavorable signal-to-noise ratio (SNR), where the conclusion may sometimes be hard to reach that the observed order is indeed due to a protein 2D crystal. Detection of even small, poorly ordered arrays, however, gives invaluable information on 2D crystallization conditions and can in subsequent experiments rapidly lead to improvement of the size of crystalline membranes. Thus, detailed examination of images by FFT is essential and requires meticulous assessment when evaluating data at low SNRs. Even though occasionally tricky, the identification of very small ordered arrays also contributes greatly to the power of electron crystallography and 2D crystallization, as progress is often swiftly made upon initial detection. Automation of screening is a critical step towards faster evaluation of 2D crystal samples and is described in Chapters 15–18.

During screening of grids for 2D crystals, all possible membrane morphologies should be evaluated for order by FFTs of digital images or optical diffraction of micrograph negatives, which can be time consuming. Careful and detailed screening of samples should be conducted though to prevent overlooking of a critical 2D crystallization condition. As time consuming as the screening phase can be, the time spent on thoroughly evaluating each grid square will be crucial to identifying 2D crystallization strategies and rapidly inducing order in proteoliposomes.

The use of negative stain in 2D crystal screening is a simple, fast, and inexpensive method for evaluating samples. Negative staining of 2D crystals by electron microscopy provides excellent contrast for weak electron scattering samples such as proteins. Most negative stains consist of a 1–2% heavy metal salt such as uranyl acetate, which is a widely used negative stain for biological electron microscopy, and the protocol will be described here (2). The dried stain superficially encases the protein in a strong electron scattering matrix of heavy metal ions. In essence, the stain provides a topographic cast of the protein surface (3) that is then imaged in projection, providing information that is assessed for crystalline order, size, and membrane morphology. Generally the resolution is limited to roughly 15 Å after image processing of micrographs of ordered arrays.

This chapter outlines the steps from grid preparation of samples in negative stain to screening and identification of 2D crystals by EM. While the main emphasis, in particular to discussions on staining and membrane morphologies, is on 2D crystals grown by reconstitution into phospholipid bilayers, protocols can also be applied or adapted to samples prepared by other means. This could include monolayer 2D crystallization (see Chapter 4) or native membranes.

2. Materials

Clean, moisture- and oil-free supplies and equipment, stored in a dessicator, contribute significantly to a reduced pumping time to achieve the optimal vacuum necessary for carbon evaporation. Clean all forceps with 70% ethanol before use to avoid bacterial and other contaminants. Always wear gloves when handling samples. Longer dialysis times at higher temperatures can more easily promote bacterial growth if care is not taken. Although not necessary, sodium azide can be added as a bacteriostatic agent to buffers if contamination becomes a concern (see Chapter 3).

2.1. Carbon Evaporation

1. A carbon evaporator/coater (such as a Cressington Carbon Coater 208, Ted Pella, Inc.). Alternatively, pre-coated grids can be purchased from any EM supplier should a carbon evaporator not be available.
2. Mica (thickness ~0.1–2 mm, sheet sizes of 2.5 cm × 7 cm), stored in dessicator.
3. A clean straight edge razor blade and/or sandpaper.
4. Forceps (Dumont #5 or similar).
5. Clean sharp iris scissors.

6. Solid carbon rods or wire, depending on available carbon evaporator, stored in dessicator.
7. A carbon or dedicated pencil sharpener may be used to initially sharpen the carbon rods.
8. Washers used to stabilize mica from movement by vibrations of the carbon evaporator.
9. 50 ml Petri dishes (diameter 90 mm) for storage of carbon-coated mica.
10. Filter paper (Whatman #4, 90 mm) stored in dessicator.
11. Triangular pieces of filter paper of ~1.5 cm by 4.5 cm (cut from Whatman #4, 90 mm).
12. Flashlight.
13. Gloves.
14. An airtight container with desiccant for storage of all carbon coating-associated equipment.

2.2. Carbon-Coating of Grids

To avoid exposure from chloroform fumes, all work with chloroform should take place in a hood.

1. 400-mesh Copper grids.
2. 25–50 mL 100% chloroform for cleaning grids.
3. Disposal container to discard excess chloroform. The container should be located in a fumehood.
4. A glass Petri dish (50 mL).
5. Small, water-bath sonicator.
6. Desiccator.
7. 100 mL Glass Beaker.
8. Carbon-coated mica.
9. Forceps (Dumont #5) and anticapillary forceps (Dumont #N5AC).
10. Carbon-floating device that is designed to facilitate carbon coating the grids.
11. Filtered/Deionized water to rinse and fill the carbon-floating device. Deionized and/or filtered water will help prevent artifacts.
12. Clean iris scissors for cutting mica.

2.3. Glow Discharge

1. Glow discharger. The one used here is a carbon coater (Cressington Carbon Coater 208, Ted Pella, Inc.) with a glow discharge attachment. If a standard glow-discharger is not available, it is possible to build one (4).
2. Dry, carbon-coated grids.

3. Clean capillary or regular forceps.
4. Petri dish (50 mL).

2.4. Negative Staining of Specimens on Grids

1. 1–2 μl aliquot of each specimen. A range of volumes up to 5 μl of possible 2D crystal suspensions may be used. Smaller volumes of 1–2 μl, however, save sample, especially valuable specimen potentially used for cryo-EM data collection, and is sufficient for screening.
2. Torn pieces of Whatman #4 filter paper of ~2 cm^2 and/or large enough to thoroughly blot excess liquid.
3. Anti-capillary forceps.

2.5. Preparation of Negatively Stained Grids

1. 1–2% (w/v) Uranyl acetate (UA) solution prepared with water.
2. A uniquely numbered multi-grid storage box with individually indexed grid slots. For long-term grid storage, grid boxes should be labeled in such a way that it is easy to determine contents, marking the box with the start and/or end date, initials, project, or color-coding may be useful.
3. Dessicator cabinet for storage of grid boxes.

2.6. Screening by Transmission Electron Microscopy

1. TEM operated in the range of 60–120 kV is sufficient for screening negatively stained grids. Most commonly a high tension of 100–120 kV is used. A JEM-1400 TEM (JEOL Ltd.), with a 60 μm objective lens aperture, spot size 2, and brightness/intensity reduced to a minimum, is used in the screening described. Low-dose capability is convenient, mainly in easing a transition from screening to cryo-EM data collection for new electron microscopists, but not essential.
2. The TEM may be equipped with one or more CCD cameras and/or image intensifiers for screening and image collection. For the described experiments, a Gatan Orius SC1000 is used for screening of the grid and collecting low and intermediate magnification images, and a Gatan Ultrascan 1000 is used for collecting high magnification images for crystallinity assessment. Tietz cameras (TVIPS) provide another frequently used option for highly reliable cameras applicable for a large range of resolutions and specimens. These types of CCD cameras are equipped with software (e.g., DigitalMicrogaph, Gatan, Inc. and EM-MENU, TVIPS) with Fast Fourier transform capabilities (FFT).
3. Film (i.e., Kodak SO-163 film), if preferred or CCD cameras are not available.
4. Optical diffractometer and light table, in the case of film, are required for screening and evaluation of crystal quality.
5. A darkroom, if specimens are evaluated with the help of film.

3. Methods

All steps involved in grid preparation and screening for 2D crystals are described here. Initial preparation of suitable carbon film is the first step. Secondly, the carbon film is applied to the grid, followed by negatively staining of specimens to be screened. As a third step, the grids are screened at various magnifications by EM and evaluated with CCD cameras or film and an optical diffractometer, depending on the laboratory. Membrane morphologies are described to aid in the characterization of new samples and identification of order.

While uranyl acetate is a common negative stain and will be described here, other types of frequently used heavy metal stains include uranyl formate (UF), phosphotungstic acid (PTA), and ammonium molybdate (5–11).

To new scientists to the field, it cannot be emphasized enough that the objective of screening is *not* to achieve uniform appearance and size of 2D crystals. As long as enough of the favored type of 2D crystal size and morphology is present, one to two optimal membranes on each grid square, or even on every 2–3 grid squares will be sufficient to move on to cryo-EM experiments.

3.1. Carbon Evaporation

Protocols for carbon evaporation will vary with equipment and/or laboratory. Some of the procedures below may be adapted to specific setups. Wear gloves at all times during this procedure to prevent contamination and natural lipid transfer from hands. Frequent and meticulous cleaning to remove evaporated carbon from the viewing glass chamber or bell jar will make the evaluation of the desired carbon thickness significantly easier, should this be done by eye.

1. Polish and sharpen graphite carbon rods to prepare them for evaporation. Use a razor blade to create a perfectly flat surface on the end of one carbon rod, and shine and polish the surface by vigorously rubbing it against a piece of Whatman #4 filter paper on a hard, flat surface or a fine-grained filter paper. Sharpen the second carbon rod to a point with a carbon rod sharpener by insertion and manual rotation to the desired length and again, rub the tip edges against the filter paper to create a smooth, finely pointed tip. Pre-sharpened and pointed carbon rods can also be purchased from various companies.
2. Insert carbon rods into the top of the vacuum chamber, ensuring proper contact between the rods. Use care to avoid breaking the tip of the sharpened rod.
3. Place a piece of filter paper on the stage inside the evaporator.
4. Cut a sheet of mica to an appropriate size using clean iris scissors. The carbon floating device tray will limit the size.

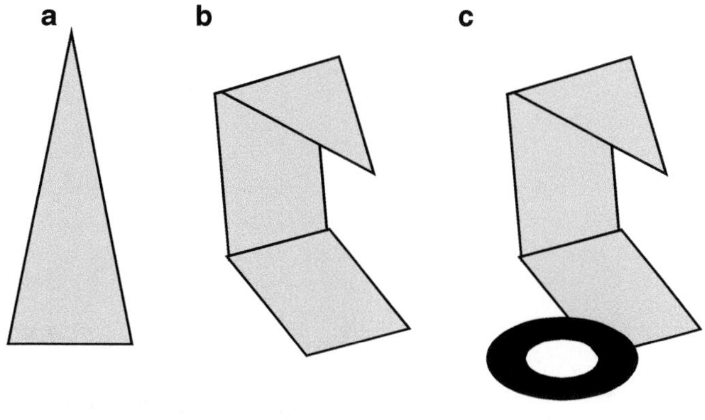

Fig. 1. Schematic representation of folding of filter paper for a reference area for carbon thickness during evaporation. (**a**) Shows the triangular or pie-shaped wedge of filter paper cut from a circular 90 mm filter paper. In (**b**) the filter paper has been folded twice to create a triangular shield at the top, which will produce a carbon-free shadow upon evaporation. (**c**) Shows the folded filter paper stabilized with a washer to prevent displacement through vibrations.

5. Cleave the piece of mica with a pair of fine forceps (e.g., Dumont #5) by parallel insertion of the tip into one end of the mica to separate it into two sheets. As soon as the sheets are separated, gently pull them apart to evenly split. Avoid pushing the tip of the forceps too far inward or rubbing the surface to prevent fracture or scratching of the inner, clean surface. The clean side of the mica will provide a smooth surface for the carbon deposition. Mica used in carbon coating should always be freshly cleaved, so use or discard the unused piece.
6. Place the mica with the freshly cleaved side facing up on the filter paper on the stage inside the evaporator.
7. Fold a piece of triangular wedge of filter paper twice by 90° along its long axis as shown in Fig. 1, and position it alongside the mica in an upright position with the point of the filter paper facing the outside of the evaporator. Use a small washer to weigh down the folded filter paper at one corner (Fig. 1c). This wedge of paper is important because it will be used as an indicator for the carbon coating thickness during the evaporation, as the actual thickness on the mica will be indiscernible.
8. Before replacing the glass chamber, examine the rims for dust or other pieces of debris. Securely situate the chamber on the O-ring and gently close the top of the chamber. Make certain that there is a perfect seal on both ends of the chamber.
9. Turn on the evaporator and evacuate to an appropriate pressure of $\leq 10^{-5}$ mBar, which can take several hours to overnight.

Fig. 2. Contrast level comparable to optimum carbon thickness upon evaporation. A thickness within this range has proven to be appropriate for negatively stained grids.

10. Once the carbon evaporator has reached an appropriate vacuum, slowly apply a current across the carbon rod to around 100 A. If a current is not observed, repeat step 2 above. Avoid increasing the current to the threshold of sparks. Decrease current if sparks occur. Allow evaporation to occur for 5–10 s by holding the current steady. Do not look directly at the brightly glowing rod; protective laser safety goggles can be worn to ensure proper eye safety.
11. Use a flashlight to evaluate the progression of the carbon film thickness by examination of the triangular wedge of filter paper. The folded triangular apex of the paper creates a carbon-protected shadow that can be used for comparison to neighboring carbon-coated areas for assessment of the coating thickness. See Fig. 2 for an example shade of carbon that has the proper thickness for negatively stained grids.
12. Once the desired thickness has been achieved, turn off the evaporator. Carefully remove the mica, place it in a Petri dish on top of a piece of filter paper, write the date on the Petri dish, and place it in a desiccator. The film can also be floated immediately onto copper grids. As a convenient tool to identify the optimal carbon film thickness in future evaporations, save the filter paper used in step 3 with the carbon shadow from the evaporator chamber. A collection of pieces of these filter papers is valuable for reference thicknesses for future experiments.

3.2. Carbon-Coating of Grids

1. Clean the copper grids by placing them in a 50–100 mL glass beaker filled with ~30 ml of 100% chloroform. This step has to be completed in a fume hood.
2. Fill the sonicator with an appropriate amount of water. The glass beaker should not float or be unstable while sonication is occurring. Add more chloroform or clamp in place if beaker is unstable. Sonicate the grids in the beaker for 5 min.

3. Pour or pipette off the excess chloroform from the grids into a designated waste container where it can evaporate. Pour grids into a filter paper-lined glass Petri dish. Again, remove as much as possible of the chloroform with a nonreacting pipette. Allow any residual chloroform to evaporate inside the hood.
4. Clean the carbon-floating device (an example is shown in Fig. 3, which can, however, be substituted by a shallow dish with a platform) with filtered/deionized water.
5. Cut a piece of Whatman #4 filter paper to the size of the metal shelf inside the floating device, and place it on top of the shelf (Fig. 3a).

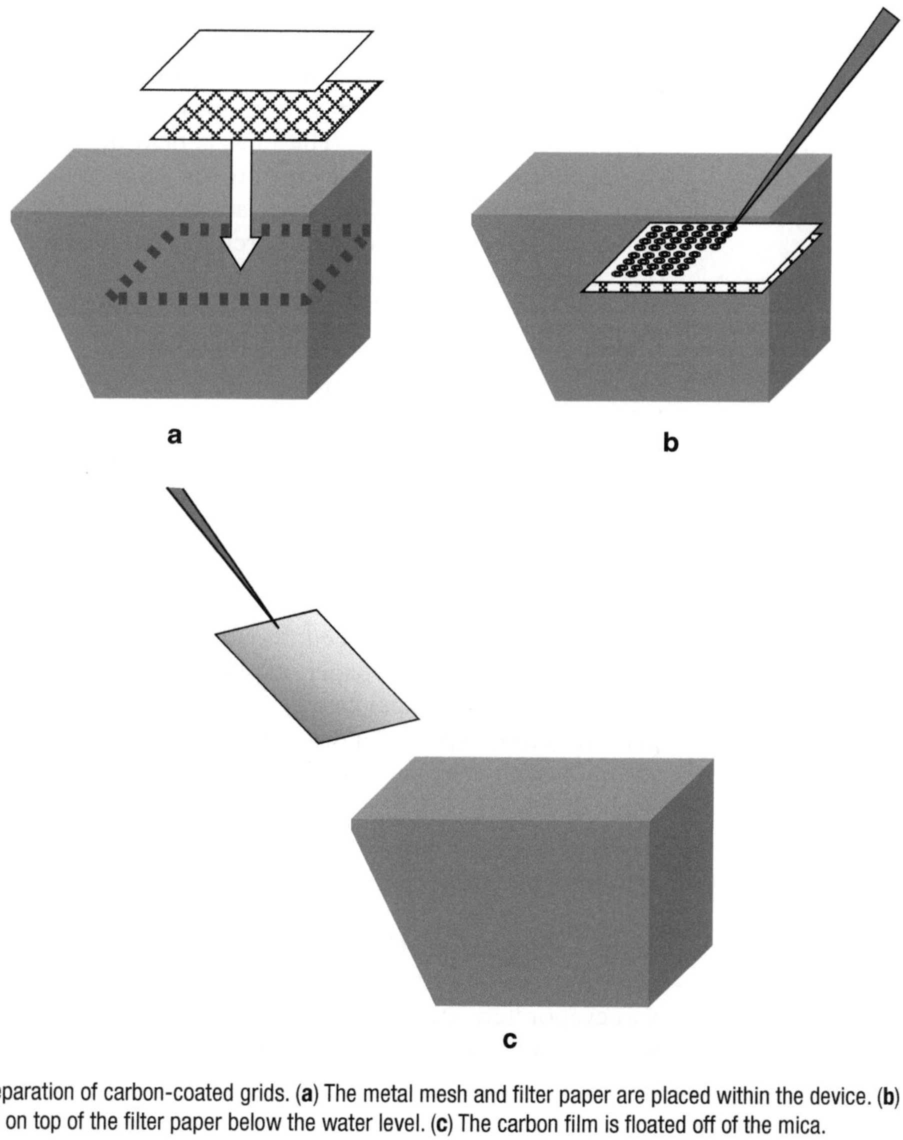

Fig. 3. Preparation of carbon-coated grids. (**a**) The metal mesh and filter paper are placed within the device. (**b**) The grids are placed on top of the filter paper below the water level. (**c**) The carbon film is floated off of the mica.

6. Fill the apparatus with filtered/deionized water by slowly submerging the paper in the water.
7. With fine, clean forceps (Dumont #5), place chloroform-cleaned 400-mesh grids, with the flat side (shiny and darker in appearance) up, on the water-covered filter paper (Fig. 3b). Avoid placing the grids close to the corners and edges, as these areas will later be used to tape the filter paper in place so it can dry without wrinkling. Order the grids as close together as possible, without any edge overlap. This step requires some practice, as the delicate grids are easily displaced in the aqueous environment. Additionally, the grids should be held with the forceps at the outer rim to avoid damaging the mesh.
8. After an appropriate number of grids (6–10 grids by ~6–8 grids) is placed on the filter paper, on an area no larger than the piece of carbon-coated mica, use the angled edge of the carbon floater device to slowly and gently submerge the mica into the water as shown in Fig. 3c. The mica is held by forceps at one of the shorter sides. During this process, the carbon film will float off the mica onto the water surface. Should the carbon film not float off on the water but remain attached to the mica, one or more edges can be cut with iris scissors. Usually removal of a very narrow rim of less than 1 mm is sufficient to ensure that the carbon detaches from the mica easily.
9. Carefully manipulate the carbon with a pair of forceps so that it is oriented to cover the grids.
10. Using the syringe(s) attached to the apparatus, very slowly evacuate water from the chamber, and lower the carbon onto the grids. At the same time, maneuver and adjust the carbon so that it remains positioned above the grids as it is likely to move while the water level is lowered.
11. After the water level has been decreased to at least a few millimeters below the filter paper, carefully lift the metal support shelf with the filter paper carrying the carbon-coated grids, and place it onto two pieces of dry filter paper in a Petri dish. Secure the ends of the wet paper with tape so that they do not curl up while drying.
12. Cover the Petri dish with a lid, but create a small opening with a folded piece of filter paper to enable air circulation and thorough drying. Allow the carbon-coated grids to dry overnight before use. Label the Petri dish with the date and any other information that may be necessary (initials, date carbon was evaporated, etc.).
13. The dried, carbon-covered grids are stored in a dessicator cabinet.

3.3. Glow Discharge of Grids

Glow discharging applies a negative charge to the carbon-coated grid allowing hydrophilic interactions with specimens that normally would not spread well and/or adhere to the carbon surface (1). Glow-discharge is used much more rarely for 2D crystals compared to single particle samples as often even relatively old carbon-coated grids of 1–2 years have sufficient sample adhering to the carbon film. We only use glow-discharge for grid preparation of 2D crystals in very rare instances. Applying glow discharge to the carbon-coated surface in the presence of 10 μL amylamine has helped greatly in cryo-EM sample preparations, however (1, 12–14). A major disadvantage of amylamine is that the glow-discharge unit will be contaminated with amylamine after only one single use and nearly impossible to clean. Thus, if possible, a second dedicated amylamine glow-discharge unit is highly recommended.

1. Setup the glow discharge unit. Check for dust or contaminants around the rims of the glass shield and O-rings.
2. Place grids on the stage of the glow discharge unit.
3. Turn on the glow-discharger and adjust to appropriate pressure.
4. Discharge the grids at 40 mA for 10–15 s. The glow-discharge may require sample- and grid-dependent optimization.
5. After glow-discharge, using forceps, place the grids in a Petri dish.
6. Glow-discharged grids should be used immediately following or within a few hours of the glow discharge to take advantage of the hydrophilicity characteristics.

3.4. Grid Preparation: Negative Staining

Typically, grids are prepared immediately following completion of dialysis since prolonged storage of 2D crystals can result in aggregation and/or deterioration due to incomplete removal of detergent. When samples are solubilized in a high concentration of detergents and/or detergents with low critical micelle concentrations (CMCs), then longer dialysis times may be required. In this case, aliquots may be taken directly from the dialysis tubing or other dialysis device for grid screening during early experimental trials when the required dialysis time is unknown.

Samples dialyzed at a low protein concentration such as below 0.5 mg, due to an expensive or a low yield of protein stock, may need to be concentrated. Concentration of the samples can be accomplished by centrifugation and removal of part of the buffer. This is followed by vortexing of the sample to obtain a concentrated suspension (see Note 1). The concentration step ideally allows for more membranes to adhere to the grid and a sufficient number of membranes to be examined for order more rapidly than a dilute sample. Some samples may settle at the bottom of the storage tube after approximately 1–3 days. Adequate amounts of specimen can then simply be obtained by pipetting from the bottom 1–2 mm of the microcentrifuge tube. Highly concentrated, settled samples may be recognized by visual inspection as slightly milky, white, or colored, depending on the protein.

Allow adequate time for UA to dissolve (overnight). Filter the solution with a 0.2 μm filter to remove any insoluble fragments. The solution should be stored at room temperature in an opaque light sensitive container or an aluminum-wrapped microcentrifuge tube. Although the solution does not require fresh preparation with each experiment, UA should not be kept for longer that approximately 6 months. Higher concentrations provide denser staining, which can obscure fine/delicate features. An optimal staining concentration is dependent on such factors as the protein and the extent by which any soluble domains protrudes out of the membrane.

1. With a pair of anti-capillary forceps, pick up a carbon-coated grid, and carefully place the forceps on a flat surface with the carbon surface facing up.
2. Pipette 2 μl of sample onto the grid, ensuring that the pipette tip does not touch the surface of the grid (see Note 2).
3. After pipetting the sample onto the grid, take note of whether the sample remains a droplet or immediately spreads evenly and flatly across the grid. If the sample remains a small droplet and does not disperse across the grid, then this is an indication that the detergent has been removed by dialysis to a large extent. Should the sample be slightly beaded, simply use the lateral edge of the pipette tip to evenly distribute the drop onto the surface. In this case, the hydrophobic nature of the grid and cohesive forces between the water molecules may prevent an even spread. Use care to prevent disruption of the carbon film with unnecessary pressure or force. On the other hand, instantaneous, unassisted spreading of the droplet to form a flat surface across the grid likely indicates the presence of residual detergent, and one may opt to dialyze the sample for a longer period before screening in negative stain (see Note 3). In instances in which the sample binds to a minimal degree, the grid can be treated with glow-discharge in air or in the presence of amylamine to increase hydrophilic interactions and protein binding (see Subheading 3.3). Please, be aware of the dangers in handling amylamine and take the necessary precautions to not touch or inhale any amylamine.
4. Allow the sample to incubate on the grid for 1 min.
5. Blot away excess liquid by touching the edge of a torn piece of filter paper to the outer rim of the grid. Blot from several locations, usually 2–3, on the grid perimeter to ensure thorough removal of excess 2D crystal suspension (2). More concentrated samples may be blotted by lowering the filter paper on the grid in a parallel manner.

6. Protein samples that are dialyzed in the presence of high concentrations of viscous solutes, such as sucrose, glucose, or glycerol, may result in suboptimal contrast during EM screening and membranes may appear diffuse due to the thick, viscous layer. One or more washing and blotting cycles with 2 μL of water, or ideally buffer, will dilute or remove these background artifacts. After blotting away of excess sample, immediately apply water or buffer, and allow 30–60 s before blotting. Multiple washes may be needed. Alternative methods to this effect applied prior to grid application, include direct dilution of the sample on the grid or in a microcentrifuge tube (i.e., 1:1 or 2:1 ratio of sample to water/buffer) (15) and centrifugation combined with buffer exchange to not alter the sample concentration. Dilution steps can also be used in the case of a highly concentrated protein sample.
7. Immediately apply 2 μl of 1% UA by pipetting onto the grid or touch the grid to 1–3 drops of UA that have been applied on Parafilm. Typical concentrations of UA are 0.5–2% (w/v), and the optimal concentration depends on the experimental conditions (see Note 4 for alternative stains), but 1% UA generally works well for most samples in providing sufficient stain and contrast. The UA container should remain relatively stationary during experiments because movement or agitation can induce unsettling of any precipitate. Additionally, always pipette from the top of the container to avoid introduction of precipitate onto the grid.
8. Incubate the stain on the grid for 30 s.
9. Blot away excess liquid with a piece of torn filter paper, as in step 7.
10. Allow the grid to thoroughly dry for several minutes before storage in a grid box.
11. Store the grid box in a clean, moisture-free environment, such as a desiccator. Under proper storage conditions, UA-stained grids can be preserved for up to several years. Long-term storage of these grids is valuable for demonstrations, teaching applications, or long-term studies.

3.5. Screening at Low Magnification to Assess Grid Quality, Sample Distribution and Morphology

Timeframes for screening will depend very much on the grid preparation, distribution, and number of membranes, and whether sample is present at all. Grids without sample or carbon film or with only protein precipitate can be dismissed within minutes after it has been ascertained that the grid has a uniform appearance. Suboptimal grids, containing either no or little sample, as well as badly prepared grids, will require preparation and examination of additional grids. Ideally a second grid, prepared at the same time as the first grid, is available. Grids containing too little samples might require the preparation of a second grid, where the sample has been concentrated (described in Subheading 3.4), thus saving time in moving from membrane to membrane over long distances and giving a more detailed overall impression of the sample. Grids with an ideal specimen distribution of at least a few membranes per grid square will necessitate systematic assessment of a range of proteoliposomes sizes, morphologies, grid square locations, and a sufficient number as even percentages as low as 2% (16) of small ordered arrays can give critical information on conditions that will rapidly lead to the identification of optimal parameters to grow numerous large and well-ordered 2D crystals. The time spent on each of these types of grids will depend on experience. An absolute minimum of 10 min will be necessary to gain at least a preliminary impression (1) of a grid with a good distribution of proteoliposomes.

Here we will refer to three frequently used magnification ranges as 'low' for magnifications of 300–800×, 'intermediate' at 2,000–10,000×, and 'high' at 30,000–80,000×.

1. Proper alignment of the EM is essential for acquisition of the best images, and thus it is highly recommended to perform an alignment for each EM session (see Note 5).
2. A magnification of approximately 2,000–10,000× will give a first impression of the number, morphology, and distribution of membranes on individual grid squares. If necessary, such as when no or little sample is seen, remove the objective aperture and examine the grid at a lower magnification of ~300–800×. The low magnification will allow for a rapid overview of the entire grid. This will provide information on possible uneven sample distribution, staining, and/or grid quality in terms of percentage of broken carbon film.
3. Ascertain that a reasonable percentage of the carbon film is intact. A quick scan of the entire grid will save time later during the screening at high magnification. Inevitably, the carbon film will break in a several places, resulting in either one or more grid squares completely devoid of carbon film or partially covered grid squares with rolled up pieces or fragments of carbon along the edges of the grid bars (see Note 6). If too many grid squares are broken, devoid of sample, or badly stained, then one will likely choose to prepare another grid before moving forward (see Note 7).
4. Many newer EMs are equipped with software to save and recall grid positions for closer examination later during a screening session. Thus, one can follow two strategies during screening. Either one quickly scans the entire grid and saves positions on the grid that appear to contain proteoliposomes, or alternatively, one can immediately, upon identification at low

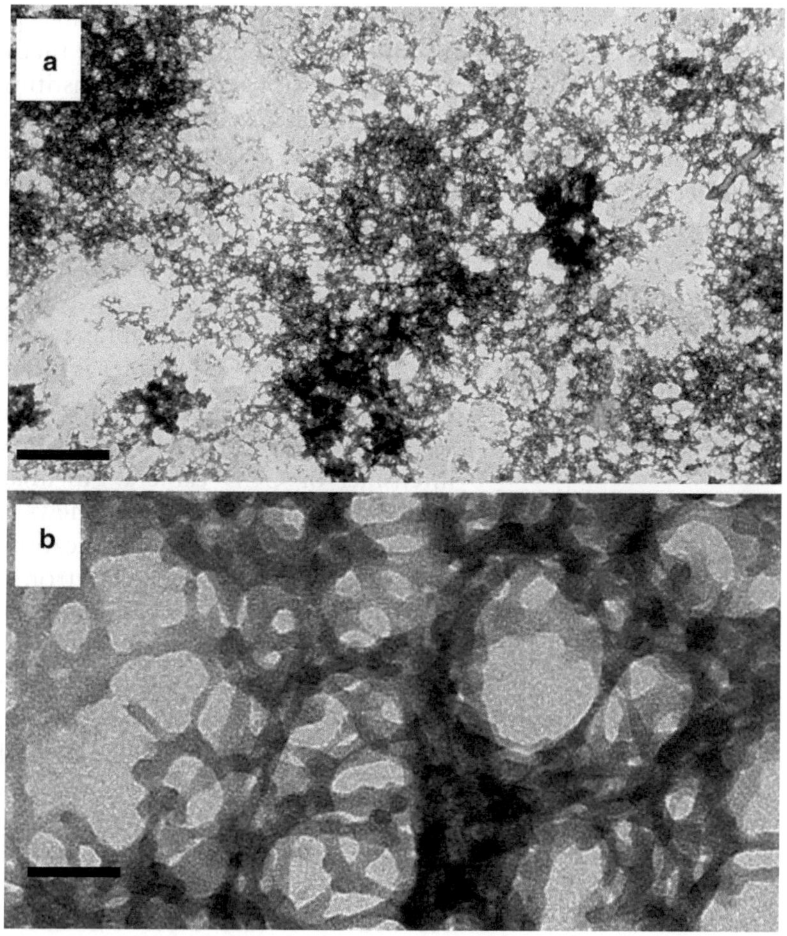

Fig. 4. (**a**) Protein precipitate at low magnification, the scale bar is 1 μm, and (**b**) smaller, web-like membrane structures appear similar at low magnification, but can easily be distinguished at the higher magnification shown here. The scale bar for (**b**) corresponds to 50 nm.

magnification, examine a membrane for order at high magnification as outlined below. As the sample is occasionally unevenly distributed across the grid, one may find that only a few areas actually contain sample. At low magnification, protein or membranous structures (see Fig. 4) will appear as distributed high contrast pin dots against a usually low contrast background. Larger, very dark areas can be an indication of protein precipitation or aggregation, and it is worthwhile to save the positions for closer examination at high magnification as well. If possible, save the positions of at least 10–20 grid squares. The majority of time will be spent at higher magnification on analyzing saved positions.

3.6. Screening at Intermediate Magnification to Assess Proteoliposomes for Distribution, Size, and Morphology

The terminology for membrane and proteoliposome are used interchangeably. It should be noted that reference to a membrane might or might not indicate reconstitution, while a proteoliposome here indicates reconstituted protein within a phospholipid bilayer that might or might not be ordered. The terms ordered array, 2D crystal, and lattice all refer to 2D crystals.

1. Insert the desired objective aperture (60 μm works well for us on the JEM-1400), and change the magnification to an intermediate magnification of roughly 3,000–10,000×. In this magnification range, membrane morphology of even membranes of ~50 nm and smaller is clearly distinguishable.
2. Depending on preference and/or availability, grids can be screened with the EM binoculars or with the use of an image intensifier or CCD camera. An image intensifier or a CCD camera is useful to collect overviews and images of membranes at low and intermediate magnification, but film will easily serve the same purpose. Dim the electron beam to a current density of ~10–15 pA/cm^2 to avoid image oversaturation when using a camera as well as damage to 2D crystal lattice, which will occur even in UA after prolonged viewing.
3. Successful reconstitution into proteoliposomes can result in different types of membrane morphology, and the first step at this stage of screening is to confirm reconstitution of proteins into continuous lipid membranes. If available, freeze-fracture is a highly valuable supplementary technique to confirm reconstitution into proteoliposomes as well as relative amounts of reconstitution (see Note 8). The possible proteoliposome morphologies are vesicles, planar-tubular 2D crystals, sheets, and tubes (see Fig. 5). Round vesicles, resembling a collapsed, soccer-ball-like sphere on the grid, are frequently the first observed morphologies in initial 2D crystallization trials of a new membrane protein (see Fig. 5a). Planar-tubular 2D crystals are essentially elongated vesicles with two parallel, straight edges and rounded ends that have collapsed on the grid, resulting in two 2D crystal layers (1, 14) as shown in Fig. 5b. Sheets (Fig. 5c), unless stacked, are composed of only one bilayer and may have a wide range of shapes from square or rectangular over rounded or round (17, 18) to uneven morphologies (16, 19). Tubes usually have a narrow diameter, do not collapse, and thus retain their circular diameter and a helical arrangement of the membrane protein ((20); Fig. 5d). Note is taken of the morphology, size (see Note 9), and distribution of membranes as well as carbon film and stain quality. It is critical to evaluate all membrane morphologies and sizes at higher magnification, as order may be restricted to only one type of morphology, while several types of membrane morphologies may be present. In many cases, 2D crystal morphology is dominated by one type of membrane and largely protein and

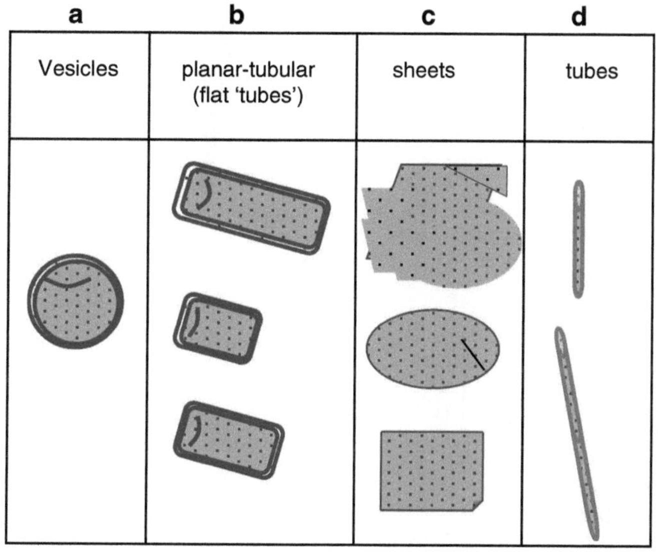

Fig. 5. Schematic representations of different membrane morphologies that may contain 2D crystals. (**a**) represents vesicles that are frequently observed. (**b**) Planar-tubular 2D crystals often have a fairly consistent diameter, but will vary in length, (**c**) While sheets ideally are composed of one membrane, they can have a large range of morphologies. (**d**) Tubes are narrow, long and do not collapse on the EM grid.

condition dependent. On the other hand, some membrane protein 2D crystals are present in multiple morphologies (14). It should be emphasized here, however, that while note is taken of morphology at this point, inducing crystallization is of higher importance.

4. Planar 2D sheets may vary in size and can grow up to several μm in diameter (see Note 10). Figure 6 shows examples of different types of sheets. The contrast of very thin sheets can be low, making them difficult to see to a new user. Occasionally, small broken pieces of carbon film can resemble membrane sheets, even to an experienced user. However, closer examination of the edge of the structure can distinguish carbon versus membrane. At a magnification of ~30,000× or higher, membranes will have an easily visible lighter folded border (Fig. 7a), while carbon film will have a more uniform straighter edge (Fig. 7b). The morphology of membrane edges can vary from straight and smooth to rough and jagged.

5. Planar-tubular 2D crystals appear as flattened cylindrical structures and can vary in length and width; thus, two crystalline lattices are directly superimposed (see Note 11). Examples of planar-tubular membranes are shown in Fig. 7c. Comparable to sheets, optimal diameter for structure determination is at least 1 μm in diameter. Tubes will also have the folded border described above for the sheets. Some proteins appear to preferentially form planar-tubular crystals.

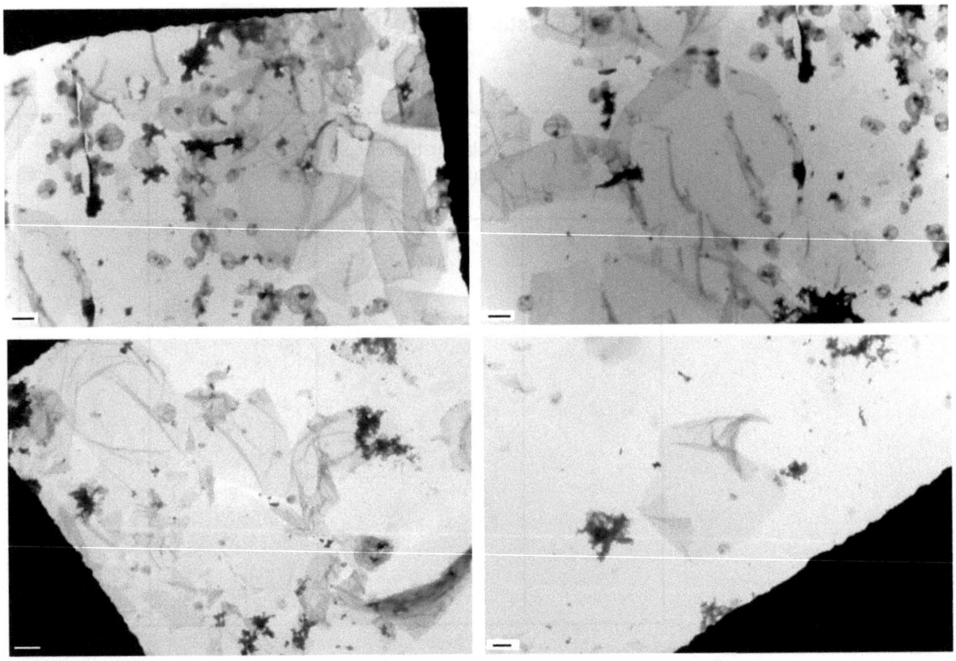

Fig. 6. Images of 2D crystals with sheet morphology. The scale bars are 1 μm (*upper left image*), 1 μm (*upper right*), 2 μm (*lower left*), and 1 μm (*lower right*).

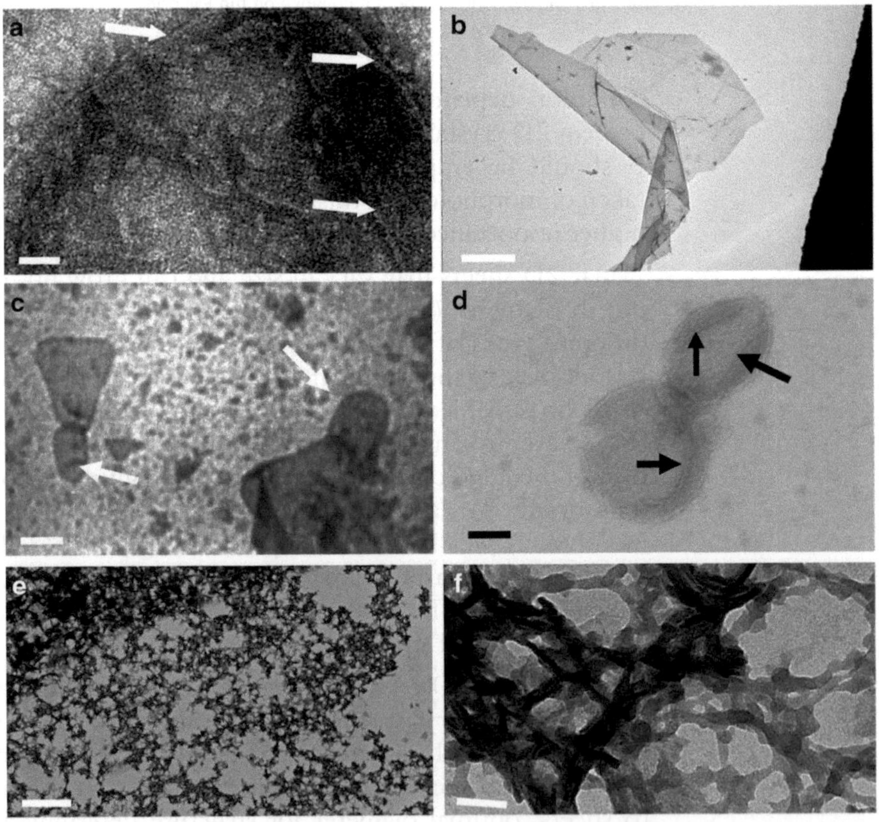

Fig. 7. (**a**) A membrane fold at high magnification, where the *arrows* indicate folds (the scale bar corresponds to 50 nm), (**b**) folded carbon-film (the scale bar corresponds to 5 μm), (**c**) planar-tubular membranes (*arrows*) (the scale bar corresponds to 1 μm), (**d**) a vesicle with the *arrow* pointing to a fold, (**e**) protein precipitation (the scale bar corresponds to 1 μm), (**f**) high-contrast, web-like membrane structures (the scale bar corresponds to 50 nm).

6. Tubes are similar to planar-tubular 2D crystals in that they resemble elongated vesicles. Their diameter is narrow to such an extent though that tubes do not flatten on the grid. Length can be variable.
7. Vesicles are the most common type of 2D crystal, which appear like deflated balloons on a grid. An example of a vesicle is shown in Fig. 7d. Similar to planar-tubular membranes, these crystals usually have two lattices, one from each collapsed membrane bilayer. In addition to the folded membrane edge like the sheets and tubes, vesicles tend to also have other obvious folded areas due to the flattening of a large, round surface area. These folds are indicated in Fig. 7d.
8. Absence of the four crystalline morphologies described above is not an immediate indication of unsuccessful reconstitution. Precipitation of protein can occur (Fig. 7e). However, these areas should be thoroughly evaluated at intermediate and high magnifications since aggregated membrane structures may also resemble protein precipitate. An example of membrane aggregates is shown at low and high magnification in Fig. 7f. Again, one should look for the typical membrane folds along the edges previously described at this magnification for indications of the presence of membranes. While some precipitation and aggregation of membranes is typical and acceptable, a large proportion can indicate that the detergent was removed too quickly during the dialysis or that the lipid and/or buffer conditions require optimization.
9. In addition to proteoliposome 2D crystals, phospholipids may crystallize below their phase transition temperature. Examples are shown in Fig. 8. Analogous to membrane protein crystals, lipid crystals can also form in various morphologies (14). When a dialysis is performed below the lipid phase transition temperature, protein 2D crystals can generally easily be distinguished based on characteristic lipid unit cell sizes as well as by freeze-fracture.
10. Other artifacts such as pieces of mica, carbon film, and bacteria could be mistaken for proteoliposomes and/or 2D crystals. Small broken pieces of mica from the carbon film evaporation can look like thick, layered objects with smooth surface edges. Salt crystals have very sharp and angular edges. Finally, bacterial contamination is an uncommon, yet occasional occur in these experiments. Although the shape of some types of bacteria can resemble planar-tubular crystals, they are in many cases rapidly identified by their flagella and the higher contrast in comparison to the empty proteoliposomes. Flagella may or may not be associated directly with bacteria. Bacterial contamination can be easily avoided with proper measures (see Chapter 3).

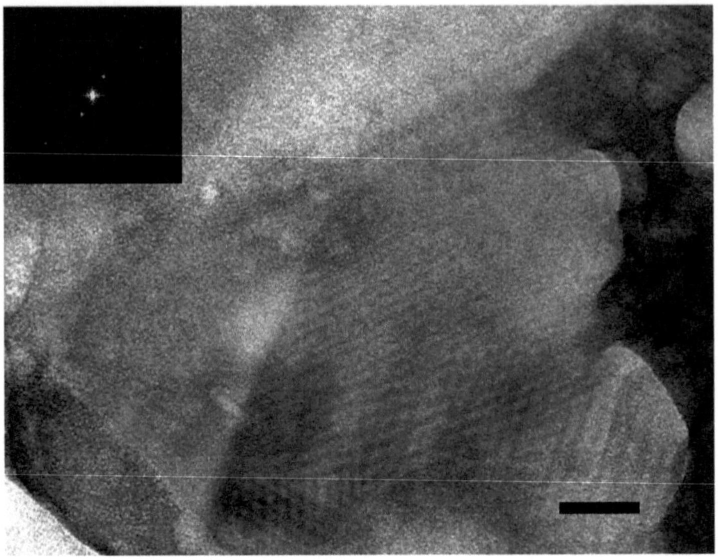

Fig. 8. Appearance of ordered phospholipids bilayers below the lipid phase transition temperature. The inset shows the corresponding FFT. The scale bar corresponds to 50 nm.

11. Save overview images of representative grid squares containing all membrane morphologies and even artifacts. They will be a helpful reference in choosing future 2D crystallization conditions.
12. An important aspect of screening is documentation of results. Since 2D crystallization involves the assessment of numerous conditions, detailed notes on the experimental outcomes for later reference are essential. It can be helpful to create a spreadsheet or table of the conditions used and the results. During screening at intermediate magnification, be sure to comment on the abundance of various sizes and types of membrane morphologies and the presence or absence of precipitation or artifacts. It should be noted that samples do not need to be uniform in appearance, as long as a sufficient number of the largest 2D crystals are available for imaging.

3.7. Screening at High Magnification to Assess for 2D Crystals, Size of Crystalline Areas, and Possible Mosaicity

Once membranes have been identified at low and intermediate magnification, the largest amount of time of the entire EM session will be spent at high magnification, even on samples that are only tens of nanometers in size.

1. Depending on the size of the membrane, screening at a higher magnification of 30,000–80,000× (see Note 12) is used for assessment for crystallinity. The presence of a lattice is determined either with FFTs of images collected with a CCD camera, or alternatively, by optical diffraction of micrographs (see Note 13).

2. Even small membranes of 100 nm or less may and should be examined for order as this can provide critical guidelines on how to proceed with future dialysis experiments to grow large and well-ordered 2D crystals. Thus, all membrane sizes and morphologies will be assessed for order. Very large membranes that exceed the size of the CCD camera chip will require examination for 2D crystals in various locations of the same membrane since the first ordered arrays are not necessarily located in the center of the largest membrane.
3. Unless the morphology clearly indicates the area of interest to be a proteoliposome, which is often the easiest to determine for vesicles and planar-tubular membranes, this will have to be determined at high magnification. Membrane folds are very helpful as they appear as a distinctive lighter rim (Fig. 7d). Curved sides clearly differentiate proteoliposomes from crystal-sized carbon film fragments that tend to have sharp, straight edges (Fig. 7b).
4. The 2D lattice of proteins with a large soluble domain and/or a large protein in general can sometimes be visually discerned with the binocular of the EM on the small screen. Although order may be obvious by visual inspection, identification and assessment by FFT or optical diffraction is essential to evaluate 2D crystal dimensions, possible mosaicity, and first indications of order and symmetry. While the potential resolution of 2D crystals of a new sample will have to be assessed by cryo-EM and image processing, the evaluation by FFT or optical diffraction of negatively stained samples will be the crucial step in the decision to start collecting cryo-EM data as described below.
5. If available, a CCD camera is used to assess crystallinity by an FFT. After focusing on a membrane area of interest at ~50,000×, defocus by minus 1–2 μm. A defocus of −400 nm is typically ideal for screening negatively stained samples on the large screen of the EM as well as for collection of electron micrographs (see Note 14). Figure 9 shows FFTs of the same image area at different defocus values to demonstrate the importance of choosing the appropriate defocus value.
6. Adjust the beam to a current density of ~10–15 pA/cm^2 and acquire an image with the CCD camera. Initial or previous CCD image collection will provide information on optimal current density and defocus with a specific camera. The image might need to be retaken to optimize these parameters (see Note 15).
7. Calculate a live online FFT of the image or subareas of the image with the image analysis software (see Note 16 and Fig. 9).
8. Adjust the γ-value of the displayed FFT to approximately 30. The γ-value to visualize spots most optimally will depend on the

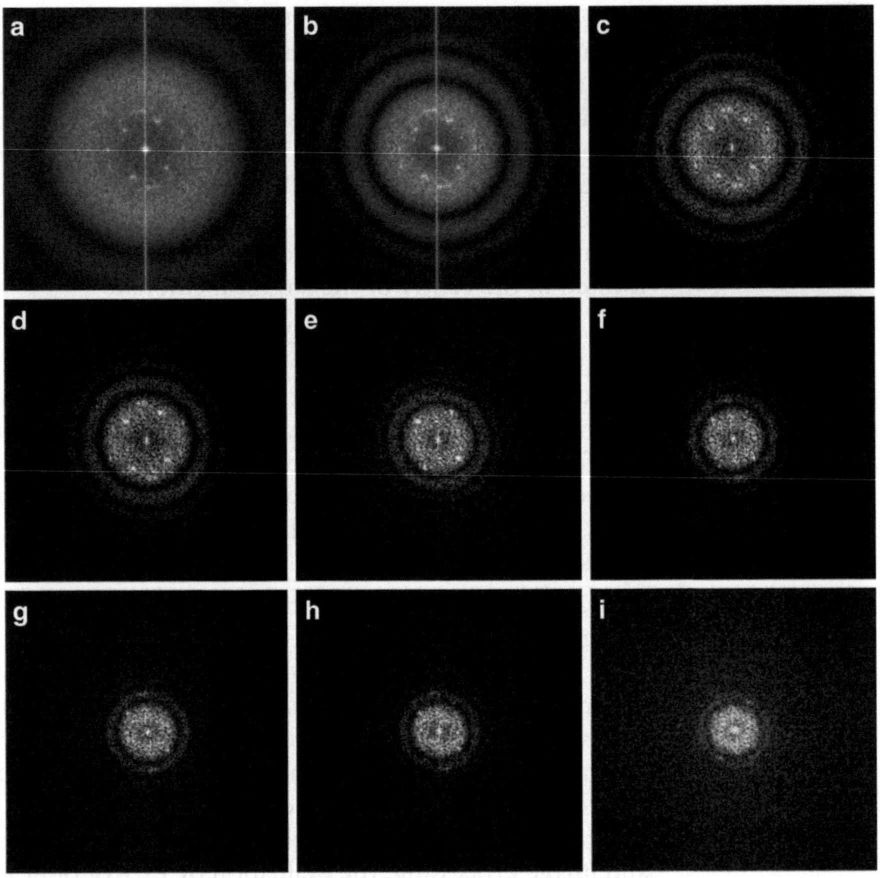

Fig. 9. FFTs at different defocus values (images were selected from a focus series, with (**a**) being the FFT from an image collected closest to focus and (**i**) the FFT from an image at the highest defocus value) collected with a CCD camera showing the effect with increasing defocus of the CTF on the reflections. The real space unit cell dimensions are $a = 83.7$ Å, $b = 76.6$ Å, $\gamma = 91°$.

imaging and sample conditions, such as intensity, specimen thickness, as well as size and quality of 2D crystals (see Note 17).

9. When scanning the image area with a small box to produce an FFT of a partial image area only, slowly move the selected box across all of the image regions containing membrane areas, and observe the changes in the FFT. Spots close to the SNR can be attributed to noise, when they immediately disappear upon incremental movement of the selected box. While looking for ordered arrays by FFT, avoid membrane edges and folds; these obvious lines will result in spots on the FFT. Figure 8 shows an image and an FFT resulting from lipid crystals.

10. If the ordered area is very small, sometimes selection of a smaller box area on the image will provide a much improved SNR and greatly facilitate identification of a lattice. Figure 10 shows the FFTs of a small 2D crystal, covering only part of the image area.

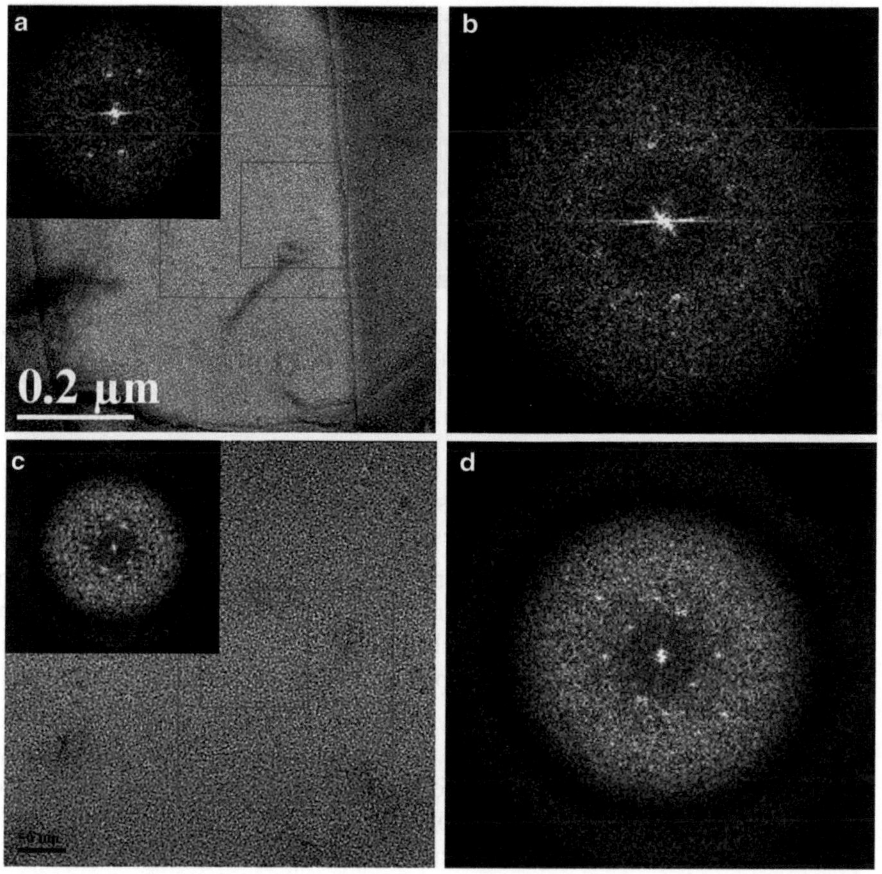

Fig. 10. (**a**) Image of a membrane with small ordered areas and (**b**) the large boxed area was selected for an FFT and (**a**, inset) only a small image area was chosen for the FFT. The FFT of the smaller box area shows a substantially improved FFT due to the reduced SNR. (**c**) The 2D crystal is substantially larger, spanning the entire area of the large box. The inset is the FFT corresponding to the small box, and (**b**) shows the higher order due to the larger number of unit cells included in the bigger box for the FFT. The γ-value of the FFTs was adjusted for easier comparison of the FFTs. The real space unit cell dimensions are a=83.7 Å, b=76.6 Å, γ=91°.

11. For a membrane larger than the image area, either choose a lower magnification or obtain images of several areas and assess them for crystallinity by FFT. In early trials, ordered arrays may be restricted to one or a few small areas in a large membrane. Thus, it is essential to not only collect and evaluate an image from the central area of the largest membrane, but multiple areas of various membranes should be assessed for possible order. Save all images containing a lattice.
12. Depending on the symmetry of the 2D crystal, reflections are observed in an orthorhomibc or hexagonal pattern, as shown in Fig. 11. Typically, negatively stained samples contain only first, or possibly second and third, order spots within the first Thon ring as the negative stain limits the resolution to approximately 15 Å. Thus, a very high defocus might prevent the

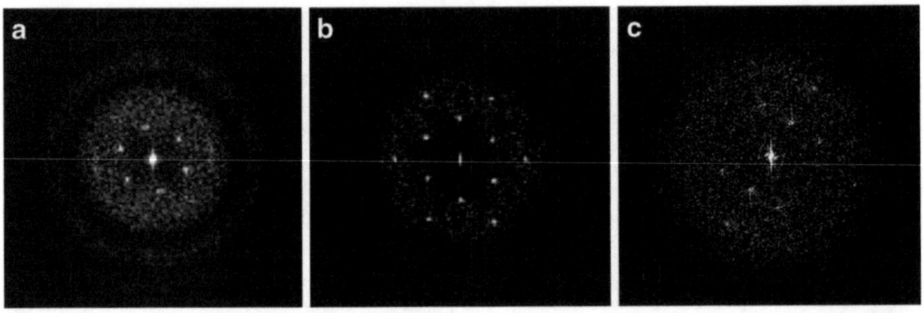

Fig. 11. The FFTs correspond to a hexagonal lattice with (**a**) first order reflections and (**b**) second order reflections. More than first order reflections are not necessarily expected from negatively stained samples, but occasionally more than two orders are observed. (**c**) Shows the FFT of an orthorhombic lattice. The real space unit cell dimensions of the hexagonal and the orthorhombic lattice are $a = b = 73.4$ Å, $\gamma = 120°$ and $a = 83.7$ Å, $b = 76.6$ Å, $\gamma = 91°$, respectively.

identification of spots should these fall on a zero of the CTF. Cryo-EM is necessary to draw any conclusions on resolution.

13. In-register stacking of multiple layers of 2D crystals might not be apparent in negative stain, or even cryo-EM images of untilted crystals, but sometimes different contrast levels of two membranes can provide a good indication.
14. In a large, highly ordered, continuous 2D crystal, the diffraction spots will be sharp and remain in a fixed position when the smaller box area for the FFT is moved over the image. Compared to the smaller boxed area, a large box should then show one set of stronger reflections due to the improved SNR.
15. As mentioned in Subheading 3.6, detailed documentation of results at various conditions is a critical aspect of screening. Notes for high magnification should include information on quality and size of ordered arrays from FFTs for various membrane morphologies. Specifically, one should emphasize the morphology that contains crystalline areas, the size of the areas, locations in the membranes (edges or central), presence of mosaicity, indications towards the symmetry, the number of orders, and the relative intensity and sharpness of spots. The goal is not that ordered proteoliposomes are uniform in size, but that a sufficient number of 2D crystals of an optimal or acceptable size are present.

3.8. Optical Diffraction

Optical diffraction of a micrograph is an alternative as well as important supplementary method to FFTs of images acquired with a CCD camera. One critical advantage of optical diffraction is the quick evaluation of much larger image areas for both 2D crystals and mosaicity. Optical diffraction is a crucial step in the decision to scan specific image areas within a micrograph for image processing, which is particularly critical for images collected by cryo-EM. Currently optical diffraction is routinely used for such cryo-EM images. Access to EMs not equipped with a CCD camera will necessitate image collection and optical diffraction. Within the past few years new EMs used for screening and multiple purposes are, however, often equipped with only one or more CCD cameras to substitute film completely.

1. Collect high magnification images of membranes on electron micrograph film (i.e., Kodak SO-163). Ideally collect ~30–50 images per crystallization condition, or, if less, as many as possible, since the first successful 2D crystallization may yield a very low percentage of ordered arrays (16). As for CCD images, even low percentages of crystals will be immensely valuable in optimizing conditions very quickly (16).
2. Develop the film in a darkroom.
3. Pass the film through the expanded laser beam of the optical bench, taking great care to not damage the eyes by even glancing at the beam, and observe the transmitted diffraction pattern produced at the crossover, either on a card or collected with a CCD camera mounted to the optical bench.
4. Micrographs are evaluated for both image quality and 2D crystal quality and size. Image quality is assessed in terms of the Thon rings indicating the defocus, astigmatism, drift, and charging. The 2D crystals are evaluated for their quality in terms of order, size of a coherent lattice, possible mosaicity, and location. While negative stain is not a suitable method of assessing the quality of 2D crystals in terms of resolution, cryo-EM data collection by electron diffraction or, alternatively, cryo-EM image data collection combined with meticulous selection of the best images and followed by image processing provides a true mean of assessing resolution and is described in Chapters 8 and 9.

3.9. Deciding on Cryo-EM Data Collection

The decision to collect cryo-EM images and test electron diffraction will be simple for large, apparently well-ordered 2D crystals. Smaller 2D crystal of less or far less than 1 μm can require some consideration. A decision will easily be arrived at when the protein is precious and little is available for further experiments, necessitating cryo-EM with whatever 2D crystals are on hand. Extended or even somewhat extended 2D crystallization trials to obtain larger ordered arrays without significant improvement might also be an indication for cryo-EM data collection, even if the results will be limited to a projection map that is possibly only at lower resolution. Furthermore, decisions might simply be based on easy availability of cryo-EMs. In parallel to optimization of crystallization conditions, cryo-EM projection data collection of ordered arrays of even a few hundred nanometers will, however, most frequently reveal valuable information (Chapter 8).

4. Notes

1. Proteoliposomes are remarkably robust and can withstand even sonication.
2. Although a range of sample volumes from 1 to 5 μl can be used, 1–2 μl mostly provides adequate sample and grid coverage at concentrations of ~0.4 mg/ml or higher, and the use of more than this volume will result in a waste of valuable sample.
3. Two-dimensional crystals containing small amounts of detergent can be used for screening and data collection. One should ensure, however, that these 2D crystals do not deteriorate, which may happen after only days of storage (see Chapter 3 on length of dialysis and (1)).
4. The type and concentration (typically 1–2%) of negative stain might require optimization depending on the protein and buffer conditions. Although UA is a frequently used negative stain, solutions have a very low pH that can induce precipitation or aggregation of proteins that have not been absorbed to the carbon film (5). Uranyl salts can also form precipitates with phosphates or other ions in the dialysis buffer (6). In addition, uranyl cations can adversely bind to the negatively charged phosphate and carboxyl groups on proteins, which is called positive staining (7). In contrast, and as a possible alternative, anionic negative stains, such as ammonium molybdate and sodium silicotungstate retain higher pH values and are relatively unreactive with proteins (7). Crystals of catalase stained with 1.24% (w/v) ammonium molybdate (pH 7.0) diffracted to 4.0 Å, while staining with UA did not provide data to such a high resolution (8). Moreover, ammonium molybdate may be helpful in the analysis of osmotically sensitive structures; for example, 2–4% solutions have been beneficial for the study of mitochondria due to the isotonic character of the stain with isolation conditions of these membrane systems (9, 10). In contrast, negative staining of the asymmetric unit membrane (AUM) 2D crystals of the mammalian urinary bladder epithelium with 0.75% (w/v) uranyl formate increased the resolution to 22 Å, relative to the 31 Å resolution obtained with a 2% (w/v) silicotungstate stain (11).
5. For EM users relatively new to EM, frequent practicing of the alignment is of great benefit to easily transition to cryo-EM data collection, where the alignment (and specimen preparation) is key to collecting optimal cryo-EM data.
6. Carbon-film covered areas are distinguished from bare grid squares by a minimal difference in contrast, but at times they

can only be identified from grid squares devoid of carbon film in the presence of smaller structures on the carbon film. Film that is too thin will easily break. Preparation of carbon film of optimal thickness is obtained with practice (see Subheading 3.1).

7. The age of the carbon film has an impact on protein binding. Specifically, the hydrophilic nature of fresh carbon film decreases with age, and hydrophobic film binds protein with less efficiency (21). Glow discharge of carbon film can increase hydrophilicity and protein binding (see Subheading 3.3). This appears to be less of a factor for proteoliposomes as we have effectively prepared 2D crystallization samples on grids that were up to 2 years old.

8. Freeze-fracture can determine successful reconstitution during early 2D crystallization trials (6). A good estimate of relative protein content of the proteoliposomes is achieved, giving valuable insights on necessary adjustments of the lipid-to-protein ratio. While freeze-fracture is a useful technique, it is not essential. Specialized equipment is required and not available in all laboratories, the method is time consuming due to the much lengthier preparation time of each grid compared to negatively stained grids. Furthermore, freeze-fracture cannot substitute for screening in negative stain as generally much fewer membranes are observed (6). Atomic force microscopy (AFM) can provide a wealth of critical information on 2D crystals and their formation (see Chapter 24), but is not a substitute for screening by negative stain either.

9. It is essential to develop a good understanding of sizes observed and noted. CCD camera software with a scale bar provides the most convenient means. Without electronic means of viewing the sample, but with the use of the large and small viewing screens, the user can easily calculate the relative distance of the markings on the large screen to help in assessing size at a specific magnification. Viewing the sample with an image intensifier and monitor will require comparison of distances viewed on the large screen, e.g., a feature of 100 nm, which can then be used to manually mark the monitor or a piece of clear foil with a scale bar of that size corresponding to the 100 nm feature on the monitor.

10. As a general rule, the optimal size of sheets for 3D structure determination is at least 1 μm in diameter for a three-dimensional reconstruction at the maximal resolution. Furthermore, electron diffraction may be possible with sheets (see Chapters 9, 14, and 29).

11. This is rarely a problem as one or the other or both lattices can be used for image processing.

12. The choice of the high magnification roughly in the range of 30,000–80,000× will depend on membrane dimensions, and,

if known, unit cell dimensions and the size of the ordered array.

13. Collection of micrographs for sample evaluation has become less common with the wide distribution of CCD cameras. Once installed, CCD cameras provide a means to inexpensively and rapidly collect a large number of images. Micrographs have the advantage that the often much larger, ideal 2D crystals can be rapidly assessed at the optical bench, and promising images can be used for initial evaluation by image processing of much larger areas than would be possible with a CCD image. Small ordered areas of a few unit cells in a micrograph can sometimes be identified with an eyepiece and a light table. With practice, micrograph data collection and analysis will not take substantially more time than CCD image acquisition and evaluation. Thus, an EM equipped with film but not a CCD should not be a deterrent from proceeding with 2D crystallization experiments and screening.
14. A defocus of −200 nm is typically used for cryo-EM data collection.
15. Usually a large number of images of the same area of a negatively stained 2D crystal will have to be collected before deterioration of the sample is evident by visual inspection.
16. The Thon rings in the FFT should be circular rather than oval (Fig. 9), and, if necessary, an astigmatism of the objective lens is corrected again at this point. Diffuse Thon rings limited to low resolution may indicate the requirement for further alignment. Drift of the sample, as evidenced by interrupted and/or partially diffuse Thon rings, can often be overcome by a short wait while exposing the area to the beam or by returning to the membrane at a later time point.
17. The image area might require inspection at different γ-values if the SNR is low. A γ-value adjustment for specific imaging conditions can usually be identified that can be applied to most FFTs of images collected under the same or similar conditions.

Acknowledgments

We wish to thank our colleagues in the laboratory as well as in the cryo-EM community for helpful discussions over the years. We are particularly indebted to our collaborators for entrusting us with their valuable samples. Research in the Schmidt-Krey laboratory contributing to some of these observations was funded by NIH grant HL090630 and the Integrative Biosystems Institute.

References

1. Schmidt-Krey I (2007) Electron crystallography of membrane proteins: two-dimensional crystallization and screening by electron microscopy. Methods 41:417–426
2. Johnson MC, Rudolph F, Dreaden T, Zhao G, Barry BA, Schmidt-Krey I (2010) Assessing two-dimensional crystallization trials of small membrane proteins for structural biology studies by electron crystallography. J Vis Exp 44:pii: 1846
3. Massover WH, Marsh P (1997) Unconventional negative stains: heavy metals are not required for negative staining. Ultramicroscopy 69: 139–150
4. Aebi U, Pollard TD (1987) A glow discharge unit to render electron microscope grids and other surfaces hydrophilic. J Electron Microsc Tech 7:29–33
5. Kiselev NA, Sherman MB, Tsuprun VL (1990) Negative staining of proteins. Electron Microsc Rev 3:43–72
6. Kühlbrandt W (1992) 2-dimensional crystallization of membrane proteins. Q Rev Biophys 25:1–49
7. Harris JR, Horne RW (1994) Negative staining: a brief assessment of current technical benefits, limitations and future possibilities. Micron 25:5–13
8. Massover WH, Lai PF, Marsh P (2001) Negative staining permits 4.0 angstrom resolution with low-dose electron diffraction of catalase crystals. Ultramicroscopy 90:7–12
9. Munn EA (1968) On structure of mitochondria and the value of ammonium molybdate as a negative stain for osmotically sensitive structures. J Ultrastruct Res 25:362–380
10. Muscatel U, Horne RW (1968) Effect of tonicity of some negative-staining solutions on the elementary structure of membrane-bounded systems. J Ultrastruct Res 25:73–83
11. Walz T, Haner M, Wu XR, Henn C, Engel A, Sun TT, Aebi U (1995) Towards the molecular architecture of the asymmetric unit membrane of the mammalian urinary-bladder epithelium: a closed twisted ribbon structure. J Mol Biol 248:887–900
12. Unger VM, Kumar NM, Gilula NB, Yeager M (1999) Three-dimensional structure of a recombinant gap junction membrane channel. Science 283:1176–1180
13. Aller SG, Unger VM (2006) Projection structure of the human copper transporter CTR1 at 6-A resolution reveals a compact trimer with a novel channel-like architecture. Proc Natl Acad Sci USA 103:3627–3632
14. Schmidt-Krey I, Haase W, Mutucumarana V, Stafford DW, Kühlbrandt W (2007) Two-dimensional crystallization of human vitamin k-dependent gamma-glutamyl carboxylase. J Struct Biol 157:437–442
15. Wang DN, Kühlbrandt W (1991) High-resolution electron crystallography of light-harvesting chlorophyll a/b-protein complex in three different media. J Mol Biol 217: 691–699
16. Zhao G, Johnson MC, Schnell JR, Kanaoka Y, Irikura D, Lam BK, Austen KF, Schmidt-Krey I (2010) Two-dimensional crystallization conditions of human leukotriene C_4 synthase requiring a particularly large combination of specific parameters. J Struct Biol 169:450–454
17. Schmidt-Krey I, Lundqvist G, Morgenstern R, Hebert H (1998) Parameters for the two-dimensional crystallization of the membrane protein microsomal glutathione transferase. J Struct Biol 123:87–96
18. Schmidt-Krey I, Murata K, Hirai T, Mitsuoka K, Cheng Y, Morgenstern R, Fujiyoshi Y, Hebert H (1999) The projection structure of the membrane protein microsomal glutathione transferase at 3 Å resolution as determined from two-dimensional hexagonal crystals. J Mol Biol 288:243–253
19. Schmidt-Krey I, Kanaoka Y, Mills DJ, Irikura D, Haase W, Lam BK, Austen KF, Kühlbrandt W (2004) Human leukotriene C_4 synthase at 4.5 Å resolution in projection. Structure 12:2009–2014
20. Diaz R, Rice WJ, Stokes DL (2010) Fourier-Bessel reconstruction of helical assemblies. Methods Enzymol 482:131–165
21. Henderson R, Baldwin JM, Ceska TA, Zemlin F, Beckmann E, Downing KH (1990) Model for the structure of bacteriorhodopsin based on high-resolution electron cryomicroscopy. J Mol Biol 213:899–929

Chapter 6

Low Dose Techniques and Cryo-Electron Microscopy

Yoshinori Fujiyoshi

Abstract

Our understanding of subcellular structures has been greatly increased owing to electron microscopy, even though radiation damage of biological samples by the electron beam demanded staining techniques. Technological and instrumental advances of electron microscopy have, however, established various highly sophisticated techniques to study biological systems in their native states without staining and thus facilitated comprehension of rather intact structures of biological components. Among these techniques, electron crystallography is a well-established one to analyze membrane protein structures within lipid bilayers, without staining at close-to-physiological conditions. Structures of membrane proteins could be analyzed at resolutions better than 3Å by electron crystallography due to techniques of low dose and cryo-electron microscopy (cryo-EM). Here, recent cryo-EM technological and instrumental advances crucial to optimal data collection in electron crystallography are summarized as well as examples of structures of membrane proteins analyzed with the help of this method.

Key words: Cryo-EM, Low-dose, Multiple dose system, Electron crystallography, Beam damage

1. Introduction

The technological development of electron microscopy (EM) to observe fine structural details of biological components has been a constant struggle against radiation damage to biological samples by the required high-energy electron beam. At the cellular level, the problem of radiation damage was overcome by staining and shadowing methods with many technological variations. Traditional EM with various staining techniques has provided us with rich and important structural knowledge about biological cells and tissues. The achievable resolution by such traditional technology is, however, very much limited to the cellular level. To go beyond the cellular level, that is to understand mechanisms of any functional

Ingeborg Schmidt-Krey and Yifan Cheng (eds.), *Electron Crystallography of Soluble and Membrane Proteins: Methods and Protocols*, Methods in Molecular Biology, vol. 955, DOI 10.1007/978-1-62703-176-9_6,

biological systems at the molecular level, requires structural information at much higher resolution. The need of observing these molecular structures at high resolution required us to combat radiation damage without using staining and thus encouraged us to develop techniques such as low dose method and cryo-electron microscopy (cryo-EM). The modern molecular EM that combines techniques of imaging biological samples with extremely low electron doses and at cryogenic temperature was first developed to study membrane proteins, known as electron crystallography. Electron microscopy has possibilities for wider applications, such as single particle analysis, electron tomography and other analyses. This chapter, however, focuses on electron crystallography, which is the most reliable method to overcome the problem of radiation damage based on the low dose technique as well as averaging techniques utilizing molecular redundancy in periodic crystals.

The low dose technique was used and its usefulness proved for the first time by Henderson and Unwin who determined the three-dimensional (3D) structure of bacteriorhodopsin (bR) to a resolution of 7Å (1). Prior to this, the biological structure was analyzed in negative stain because of two main reasons, i.e., dehydration of the biological sample in a high-vacuum environment of the EM and radiation damage caused by the high-energy electron beam. The column of an electron microscope (EM) is evacuated because the electron beam requires a high vacuum for passing through the column without serious obstruction. Unwin and Henderson developed the glucose embedding technique to overcome the dehydration problem in the EM (2). Glucose embedding prevents dehydration, and thus could preserve high-resolution structures of two-dimensional (2D) crystals even in vacuum. This technique and the low dose technique enabled them to analyze the α-helical arrangement of bR. This structure is historically important in the field of structure analyses of membrane proteins not only by electron crystallography but also X-ray crystallography. Prior to this work, no structure of any membrane protein was analyzed by any technique in 3D. Around the same time, Glaeser discussed theoretically the low dose technique, i.e., using a low exposure to obtain high resolution images by the SNAP-shot method (3).

Through the pioneering work of Henderson and Unwin (1, 2), electron crystallography was proven to be an extremely powerful tool for structural studies of membrane proteins. Later, based on the technology of electron crystallography, Henderson et al. determined the first atomic structure of bR (4). However, even with demonstrated success, the spread of this method in the structural biology community was rather slow presumably because of technological difficulties of this technique itself as well as the difficulty of crystallizing membrane proteins in well-ordered 2D crystals. With the technological development over the years, particularly the development of modern microscopes with extremely stable specimen stages at liquid helium temperature and very efficient cryo-transfer systems, electron crystallography has now become much more accessible. The targets of electron crystallography are primary membrane proteins that were known to play key roles in sensing and regulating signal transduction, regulating energy metabolism as well as many other cellular activities. With the capability of determining atomic resolution structures of membrane proteins in their native functional states within the membrane bilayer close to physiological conditions, electron crystallography could be the core method for establishing structural physiology.

2. Low Dose Technique for an Atomic Resolution Image

It was demonstrated early on that an EM is capable of the recording an atomic image of organic molecules such as chlorinated copper phthalocyanine (Fig. 1). Because of the radiation damage, the illuminating electron dose used was very low, even though the chlorinated copper phthalocyanine is one of the most radiation-tolerant organic molecules. The original image taken by the low dose technique is very noisy (Fig. 1a). By averaging the image utilizing periodic features of the crystal, a noisy image was dramatically improved to a clear image as shown in Fig. 1b. These images resolved for the first time individual atoms, namely 16 chloride atoms and 1 copper atom in the phthalocyanine molecule (5). Obtaining an image of such quality, however, required many trials, mostly because it was extremely difficult, under very low "illumination," to focus a specimen quickly and precisely so that an image could be taken at Scherzer focus. It was well known that different foci change image features dramatically, thus adjustment of focus is very important especially for high-resolution imaging. A minimum dose system (MDS) was therefore developed to collect high-resolution images without excess electron illumination for focusing on the image area, but rather by illuminating a bright beam on an adjacent area for accurate focusing (6). An optimally focused image could then be taken with minimum electron illumination, as shown for an image of more radiation sensitive organic molecule, silver-7, 7′, 8, 8′-tetracyanoquinodimethane complex, Ag-TCNQ (Fig. 2) (7).

Imaging biological molecules is, however, more challenging than imaging ordinary organic molecules. Two important challenges that we have to deal with in order to image a biological sample at high resolution are radiation damage and dehydration. Biological molecules are not only far more sensitive to radiation

Fig. 1. Image of chlorinated copper phthalocyanine (**a**). Averaged image of chlorinated copper phthalocyanine (**b**).

damage but also much more sensitive to dehydration caused by a high vacuum, while organic molecules generally keep their original structure even in high vacuum. While development of MDS resolved a practical problem of how to image a biological sample properly, it does not reduce the radiation damage itself and does not resolve the problem of dehydration.

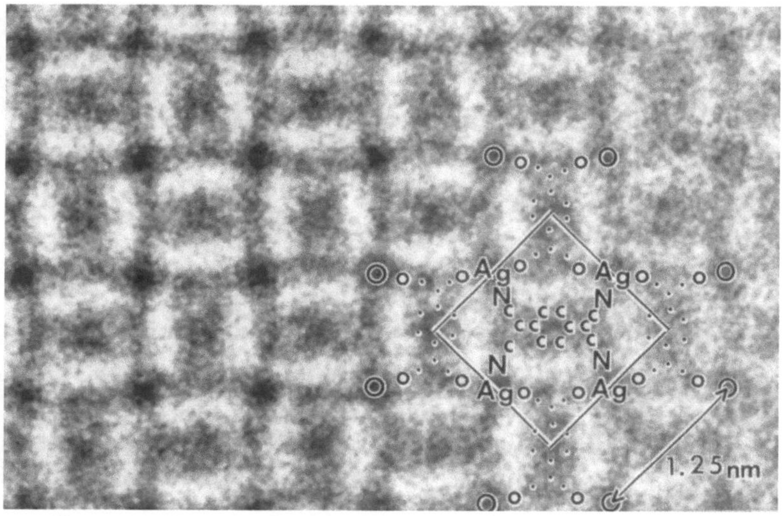

Fig. 2. Image of silver-7, 7′, 8, 8′-tetracyanoquinodimethane complex (Ag-TCNQ) taken by MDS.

3. Dehydration and Radiation Damage

Dehydration frequently deteriorates the quality of protein crystals, making it unsuitable for high-resolution structural study. This is particularly true if the target of resolution is at higher than 3Å. In electron crystallography, embedding 2D crystals in glucose can reduce dehydration to a certain level and enabled us to analyze structures of biological molecules without staining. To keep the specimen in a completely hydrated state requires a different technique in which fully hydrated specimens were kept in vitreous ice generated by rapid freezing (8). To study membrane proteins at a resolution beyond 3Å, however, often requires combining the two techniques, i.e., glucose and ice embedding. In the case of aquaporin 4 (APQ4), trehalose-embedded 2D crystals yielded nice and sharp diffraction spots at a resolution lower than 3Å. Only when specimens were prepared by combining trehalose embedding and ice embedding techniques, the best diffraction patterns of AQP4 2D crystals with diffraction spots to higher than 2Å resolution were recorded (Fig. 3) (9).

Observing a frozen-hydrated sample in an EM is however technically demanding. While transferring glucose-embedded samples does not require a special holder, transferring an ice embedded sample requires an efficient cryo-transfer system. During the transfer, the frozen-hydrated sample must be kept at a temperature lower than 140 K to avoid the phase transition from amorphous ice

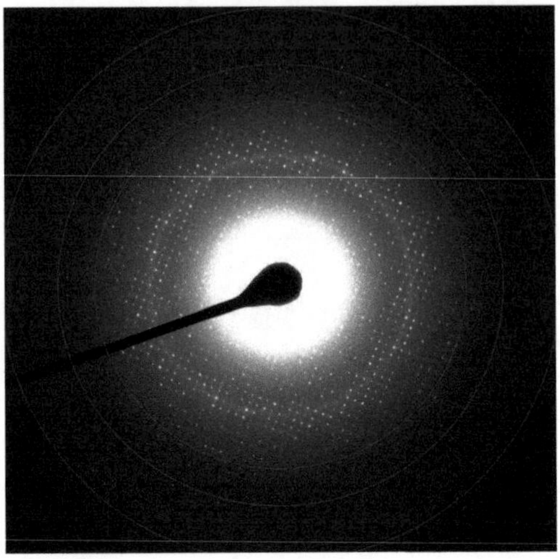

Fig. 3. Typical electron diffraction pattern from a 2D crystal of AQP4 taken by cryo-EM with a helium stage. The three *rings* from the inner (lower resolution) pattern are corresponding with resolutions of 3.0, 2.5, and 2.0Å.

to crystalline ice. Nowadays, with efficient cryo-transfer systems one can exchange ice embedded samples easily below 100 K.

Coincidently, low temperature not only keeps a biological sample in its fully hydrated state in an EM, it also reduces the radiation damage, which is the most serious problem for high resolution structural studies of these biological molecules by EM. The electron beam inevitably causes breakage of chemical bonds of organic and biological materials. Subsequently, the produced radicals may attack circumference bonds and/or molecular fragments diffuse away, leading to collapse of the initial structure. While low temperature does not prevent bond breaking by the high-energy electron beam, it does effectively reduce the movement of molecular fragments. Minimizing the movement of atoms and fragments may also increase re-formation of the same bonds from the broken chemical bonds through emitting energy such as light and/or thermal energy accumulated on the broken bonds. Overall, low temperature helps to maintain the structural integrity of a biological molecule even after chemical bonds are broken. Historically, many attempts were made to find a solution to overcome the difficult problem of radiation damage. Cooling the sample to a low temperature to limit the diffusion of atoms and molecular fragments seemed to be the only effective way of reducing the radiation damage.

Cooling the specimen to liquid helium temperature could therefore be a very effective method of reducing the radiation damage to biological sample. This concept was demonstrated to be true by a number of experiments. First, it was necessary to measure

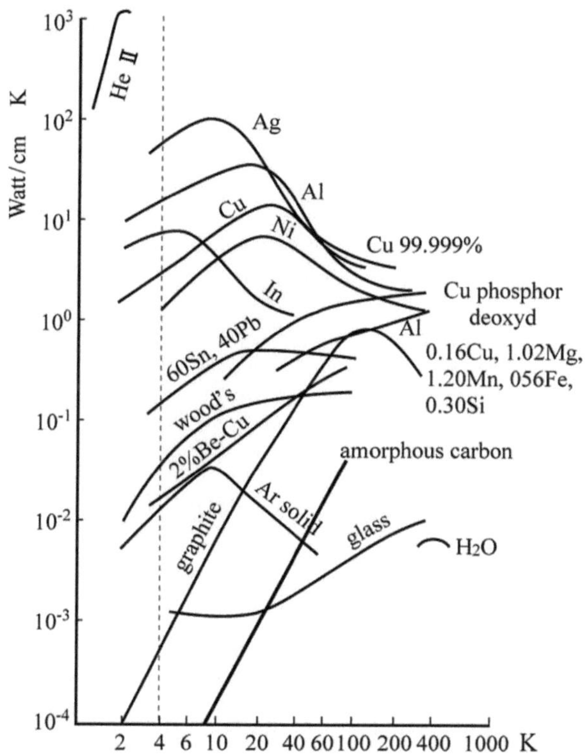

Fig. 4. Heat conductivity of various materials depending on the temperature.

accurately the temperature of a specimen under the illumination of an electron beam. When materials are cooled down to the level of liquid helium temperature, for all the materials the lattice heat capacities (C_l) become very small because a lattice heat capacity is proportional to$(T/100)^3$. The heat conductivity of non-metals is dramatically reduced at liquid helium temperature as shown in Fig. 4. Upon illumination by an electron beam, the specimen temperature can be raised to above the temperature of the specimen stage because the electron beam gives energy to the specimen by inelastic scattering. When the stage is cooled down to 4.2 K by liquid helium, the temperature of the specimen in the beam illuminated area can be significantly higher than the stage temperature. The amount of temperature increase is strongly dependent on the electron beam illuminating conditions as well as the specimen conditions, such as the thickness and electron stopping power of the material. Therefore, measuring the temperature of the beam-illuminated area could be very tricky. To directly detect the temperature of a beam irradiated area, we developed a system in which we could introduce gases, such as Nitrogen and Neon, through a modified objective aperture holder (10). The vapor pressures of Nitrogen and Neon dramatically rise up at 20 K and 7.9 K,

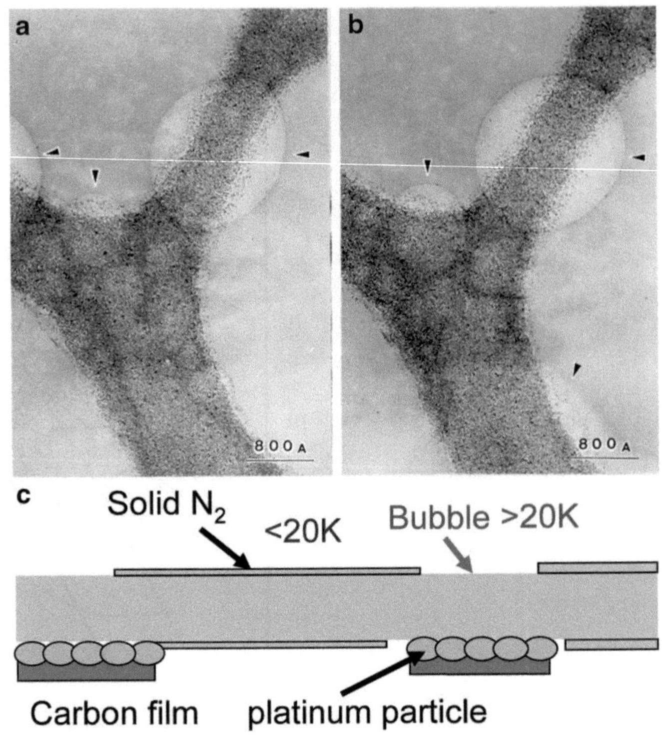

Fig. 5. Solid Nitrogen film and bubbles of the film appearing on the platinum coated frame of a microgrid at serial moments (**a**, **b**). (**c**) Schematic figure explaining the images of (**a**) and (**b**).

respectively. Therefore, the temperature could be confirmed by taking diffraction patterns of solid Nitrogen and Neon on the sample. Although metals have high thermal conductivity, the high electron stopping power of the metal-coated area tends to accumulate beam energy. Thus, the temperature of the metallic area tends to be raised more, and solid nitrogen is sublimated on a platinum-coated area on a holey grid frame (grid bar), but not on the ice layer as shown in Fig. 5. Thus, although the temperature at the metallic grid bars is raised, it is increased much less at the transparent ice layer as shown in the schematic figure in Fig. 5c. The diffraction ring, (1, 1, 1) ring, of the low temperature form of solid Nitrogen proves that the temperature at the beam illuminated area is below 20 K (Fig. 6); at the same time, the temperature of the Nitrogen sublimated area of the platinum coated frame was confirmed to be higher than 20 K. The bubbles, which are formed by the sublimated area on the continuous solid Nitrogen film, change their positions and also their sizes, presumably because sublimating energy cools down parts of the film, and Nitrogen gas moves to the sublimated place (Fig. 5a, b). The type of introduced gas on the specimen can be confirmed by a diffraction pattern as shown in the examples of (1, 1, 1) rings from solid Nitrogen and solid Neon (Fig. 6). With the method described here, we confirmed

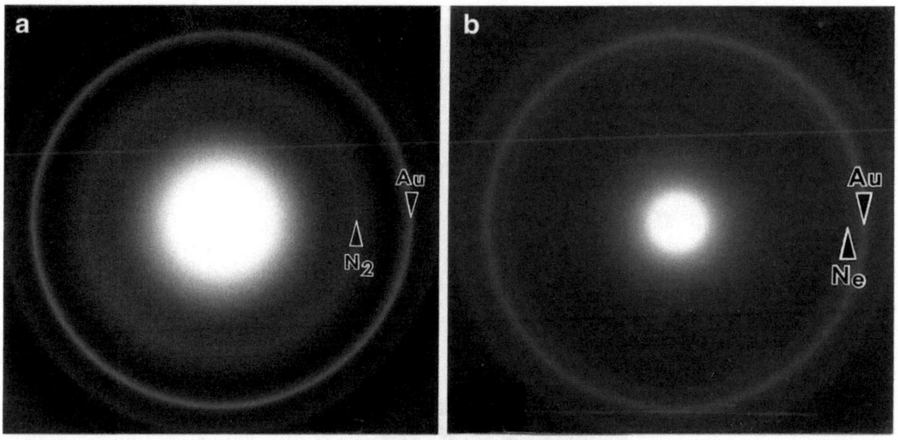

Fig. 6. Electron diffraction patterns of solid Nitrogen (a) and Neon (b) that could be taken at a temperature below 20 K and 8 K, respectively.

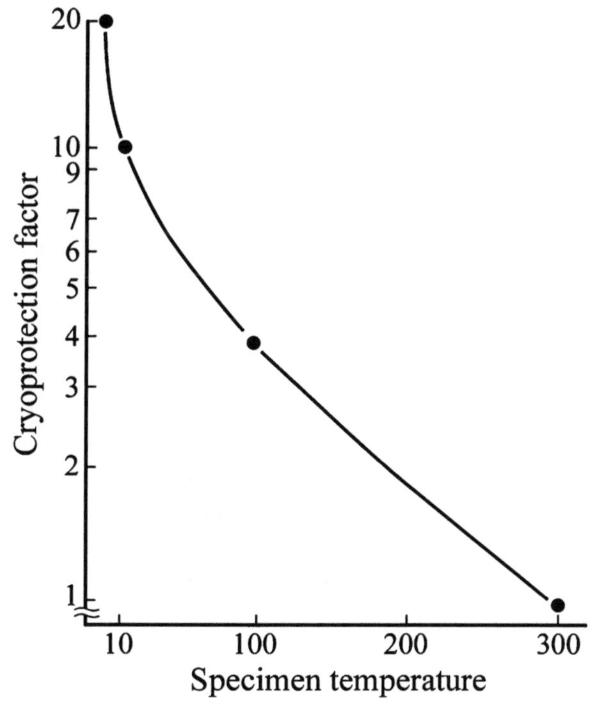

Fig. 7. Cryo-protection factors and specimen temperature. Radiation damage of biological specimens is reduced to 1/10 and 1/20 at specimen temperatures below 20 K and 8 K, respectively, compared with the damage at room temperature.

that the temperature of an area radiated by an electron beam remains under 20 K.

We could further confirm that, when the specimen was cooled below 20 K and 8 K, respectively, the irradiation damage to crystals of biological molecules was found to be reduced to about 1/10 and 1/20 of the damage at room temperature (Fig. 7). Therefore, the radiation damage can be further, and rather, dramatically

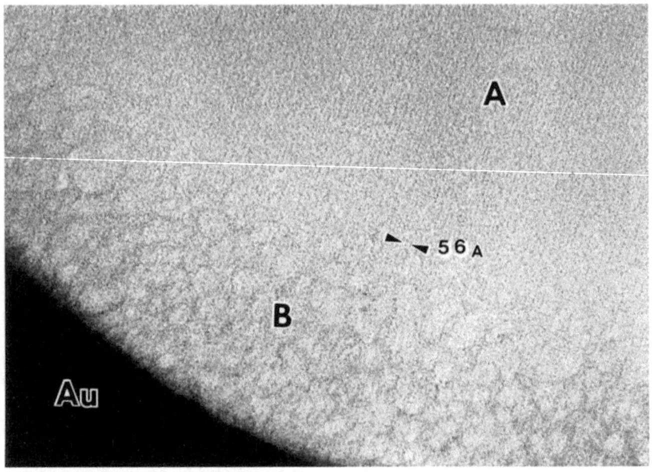

Fig. 8. Ice embedded image of a tRNA crystal on holey carbon film with artificially thick gold coated frame. The lattice lines of the crystal can clearly be seen in area A far from the metallic area, but not in area B close to the gold coated frame of holey carbon film.

reduced below liquid nitrogen temperature (Fig. 8). Even minor variations below or above 20 K make a big difference, as demonstrated by the lattice lines of a tRNA crystal observed far from a metallic area (below 20 K) (area A in Fig. 8) while damage was evident near the metallic area where the temperature was above 20 K (area B in Fig. 8).

Our observations described above suggested that a cryo-EM equipped with a specimen stage that can be cooled by liquid helium would be best for electron crystallography of membrane proteins. Developing such an instrument, however, was quite difficult. Boiling of the coolants, liquid Nitrogen and liquid Helium, causes mechanical vibrations. Temperature changes in the specimen stage introduce specimen drift. To design an ideal cryo-EM, one has to overcome all of these problems.

4. Development of Stable and Effective Cryo-EM

For high-resolution data collection, 2D crystals of membrane proteins are embedded in a thin layer of amorphous ice and/or a sugar solution as shown in Fig. 9. Among different sugars, Trehalose is the best for embedding 2D crystals in partially or fully hydrated conditions (11). Another important factor that one has to consider is the thickness of the ice, which has to be optimized for high-resolution data collection. The ice thickness influences two factors that are critically related to the quality of data collected from 2D

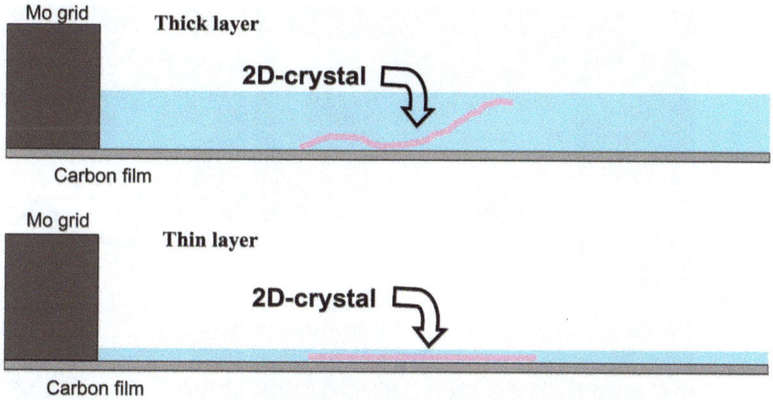

Fig. 9. Schematic drawing of specimen preparation for observation of 2D crystals by cryo-EM. A thinner water layer is better than thicker one for high-resolution data collection, especially for data collection at a highly tilted condition.

crystals of membrane proteins, i.e., flatness and hydration of 2D crystals. Since 2D crystals are not rigid, thinner ice ensures crystal flatness by forcing the ordered membranes down on the atomically flat carbon support film (10) (Fig. 9). On the other hand, such a thin ice layer can cause dehydration of biological samples (Fig. 10). While no high resolution spots were observed in relatively dehydrated condition (Fig. 10a), thick ice causes blurring of high resolution spots at high tilt (Fig. 10b). Only a very narrow range of ice thicknesses can simultaneously satisfy both requirements for flatness and hydration (Fig. 10c). Therefore, it requires many trials to find such optimized conditions for high resolution data collection, especially at a resolution higher than 3Å. Such a trial-and-error approach could be laborious and inefficient, if not entirely impossible, without an efficient cryo-transfer system that enables rapid exchange of frozen hydrated samples.

For the purpose of developing an ideal electron microscope that can satisfy all the factors described here, we devoted a significant effort in the technological development. A prototype cryo-EM with a super fluid helium stage that can achieve a resolution of 2.6Å was developed in 1991 (12). On the second-generation cryo-EM, we invested a significant effort to improve the resolution as well as operational difficulties of this cryo-EM with a top-entry type helium stage. By achieving further improvements of the second generation cryo-EM with a stable helium stage and an effective cryo-transfer system, a third generation cryo-EM with a field emission gun was developed, which makes it possible to collect an image at 2Å resolution of a second sample in less than 10 min after finishing observation of a previous grid (10). An automatic cryo-transfer system was tested in our third generation cryo-EM.

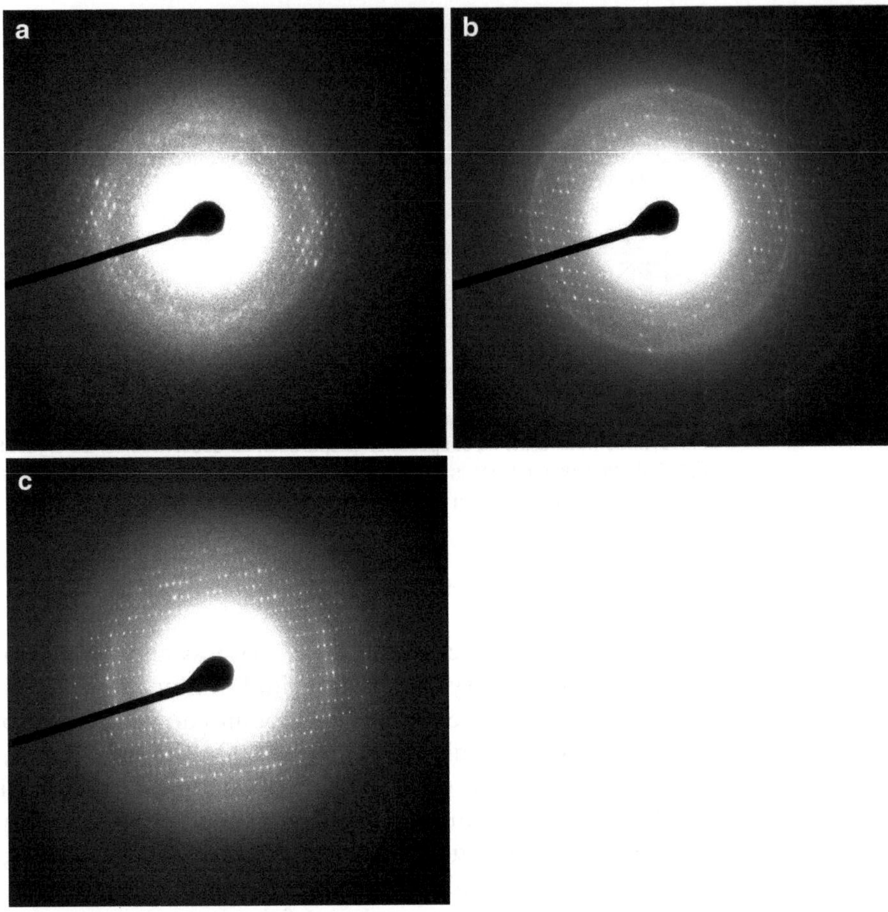

Fig. 10. Electron diffraction patterns of 60° tilted AQP4 crystal at a dehydrated condition (**a**), in too thick water layer condition (**b**), and with a good specimen preparation condition (**c**). (**a**) Diffraction spots are deteriorated in all directions. (**b**) High resolution spots are observed on parallel to the tilting axis, while no high resolution spots are observed perpendicular to the axis. (**c**) High resolution spots are observed in all directions.

A fourth generation cryo-EM equipped with an automatic cryo-transfer system and an omega filter was developed and the fifth generation cryo-EM was also developed to allow for single particle cryo-EM work. Recently we developed yet another new cryo-EM, the sixth generation cryo-EM with an outer control tilting device for electron tomography and additional types of analyses not relevant to electron crystallography. We also improved the sixth generation cryo-EM to a eucentric top entry helium stage and are using the seventh generation cryo-EM for collecting tomographic images (Fig. 11).

The target resolution of electron micrographs is strongly dependent on the specific research project, but it is usually highly desirable to achieve resolutions better than 3Å. At such resolutions, determination of an atomic structure of membrane proteins

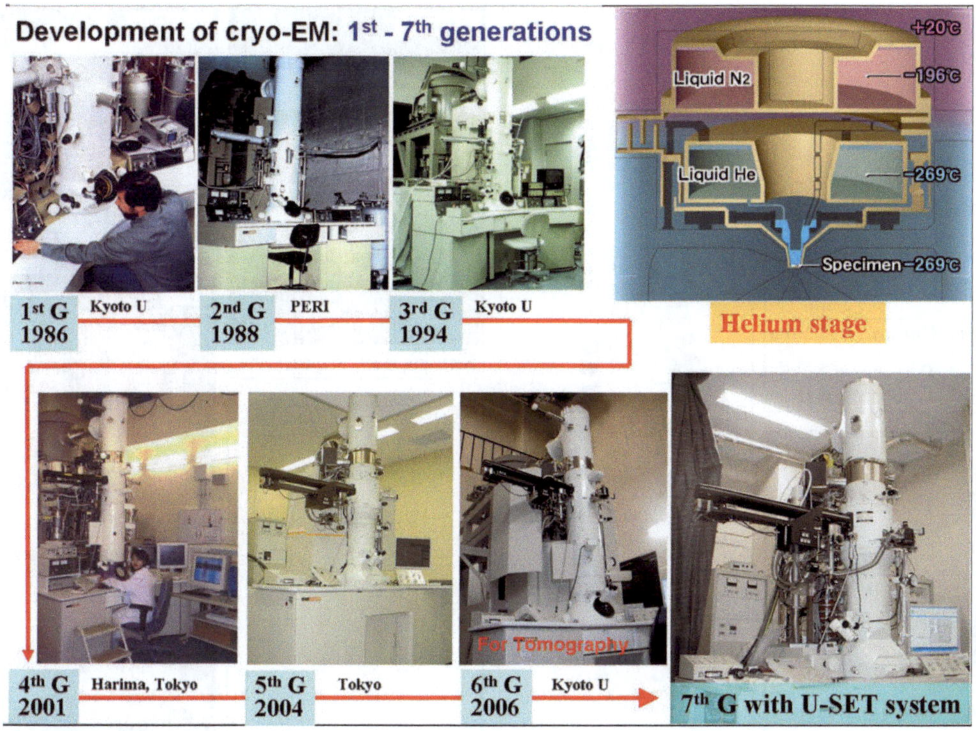

Fig. 11. Advancement of cryo-EM with top-entry helium stage and cryo-transfer system and schematic figure of the stage (*upper right*). The seventh generation cryo-EM was developed in 2009.

becomes possible. An instrument yielding better resolution would be highly beneficial because biological molecules consist mainly of light atoms that exhibit small atomic scattering factors in a high-resolution range. X-ray crystallography has taught us that higher resolution than 2.5Å might be required to identify water molecules. Thus the instrumental resolution was intended to be improved to 2.0Å (10). The third generation cryo-EM was developed as a mature instrument for electron crystallography and has made it possible to collect many data for high resolution structure analyses of membrane proteins as shown in Fig. 12.

5. Conclusions

Electron crystallography requires optimized techniques in the entire process from protein purification over crystallization and data acquisition to computer analysis. For example, specimen preparation is crucial to high-resolution data collection by cryo-EM.

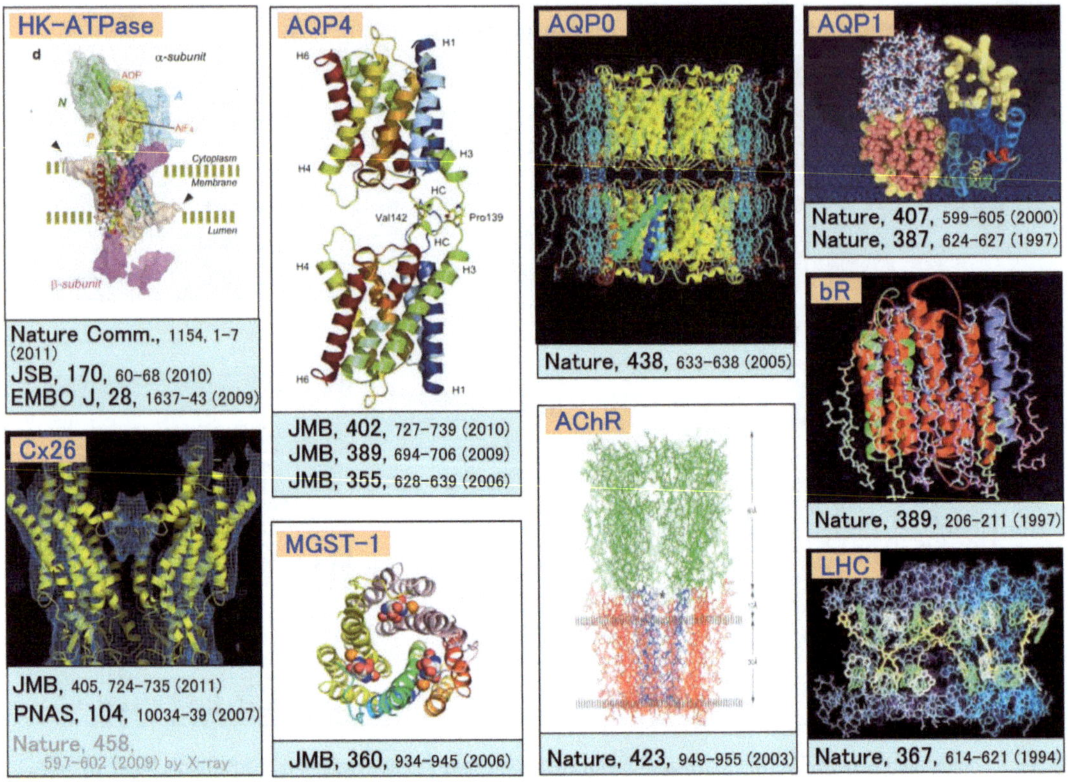

Fig. 12. Structures of membrane proteins analyzed by electron crystallography.

In the process of determining the high-resolution structure of bacteriorhodopsin (13), a serious problem was the charging-induced image shift, especially of highly tilted specimen. The image shift smears diffraction spots perpendicular to the tilt axis even in the medium- or low-resolution range. We obtained only approximately 200 good images out of taking a total of more than 10,000 images. The success rate for acquiring high-resolution images from highly tilted specimens was therefore only roughly 2%. This made image data collection in electron crystallography a tedious and laborious task. Almost all of the images from a tilted specimen prepared on a one-layer carbon support tend to be deteriorated by the image shift caused by beam-induced specimen charging. To overcome this serious problem, the carbon sandwich preparation method, in which crystals are put between two sheets of carbon film, was developed. As discussed in the following chapter (Chapter 7 by Nobuhiko Gyobu), when we used carbon-sandwiched specimens, the number of images which showed clear asymmetric features in the diffraction patterns due to the image shift was significantly decreased as compared to specimens supported by an ordinary single carbon film. Thus, the carbon sandwich preparation method was confirmed to overcome the most difficult problem of charge-induced image shift of highly tilted specimens and contributes greatly to a highly increased rate of structural analysis by electron crystallography (14).

An additional advantage of the carbon sandwich technique is that it minimizes the problem of dehydration, which ultimately leads to an atomic model of AQP0 at 1.9Å resolution (15). Furthermore, the carbon sandwich technique minimizes changes in the salt concentration due to evaporation during the single-layer grid preparation, which tends to deteriorate the quality of crystals. However, in some types of 2D crystals, interactions between the crystal surface and the carbon support film tend to deteriorate the quality of the crystals. In such cases, the carbon sandwich technique cannot be used and therefore many images are needed to be collected for high resolution analysis, particularly at highly tilted angles. This means that a very effective cryo-EM system is crucially important. Therefore, high-resolution analysis by electron crystallography is impossible without a stable cryo-EM with a helium stage, when image data for phase information are required.

The ability of electrons to form images, combined with advances in cryo-technology, enabled us to acquire detailed structural and chemical information about membrane proteins in their physiological lipid and ionic settings. Shown in Fig. 12 are structures of membrane proteins analyzed by electron crystallography. This information complements that obtained by X-ray diffraction of proteins in detergent, where biological relevance of the obtained structure is less certain. The development of cryo-EMs and real-space averaging methods together with instrumental development such as a seventh generation cryo-EM will be extending the possibilities of obtaining high resolution information from increasingly complex protein-lipid arrays, and from cell membrane specialization in situ.

Acknowledgments

These studies were performed in wonderful collaborations with many researchers whose names were recorded as authors in each referenced paper. This work was supported by Grants-in-Aid for Scientific Research (S) and the Japan New Energy and Industrial Technology Development Organization (NEDO).

References

1. Henderson R, Unwin PNT (1975) Three-dimensional model of purple membrane obtained by electron microscopy. Nature 257:28–32
2. Unwin PNT, Henderson R (1975) Molecular structure determination by electron microscopy of unstained crystalline specimens. J Mol Biol 94:425–440
3. Kuo IAM, Glaeser RM (1975) Development of methodology for low-dose exposure, high-resolution electron microscopy of biological specimens. Ultramicroscopy 1:53–66

4. Henderson R, Baldwin JM, Ceska TA, Zemlin F, Beckmann E, Downing KH (1990) Model for the structure of bacteriorhodopsin based on high-resolution electron cryo-microscopy. J Mol Biol 213:899–929
5. Uyeda N, Kobayashi T, Ishizuka K, Fujiyoshi Y (1999) High Voltage Electron Microscopy for Image Discrimination of Constituent Atoms in Crystals and Molecules. Chemia Scripta 14:47–61
6. Fujiyoshi Y, Kobayashi T, Ishizuka K, Uyeda N, Ishida Y, Hrada Y (1980) Digital reconstruction of bright field phase contrast images from high resolution electron micrographs. Ultramicroscopy 5:459–468
7. Uyeda N, Kobayashi T, Ishizuka K, Fujiyoshi Y (1980) Crystal structure of Ag TCNQ. Nature 285:95–97
8. Adrian M, Dubochet J, Lepault J, McDowall AW (1984) Cryo-electron microscopy of viruses. Nature 308:32–37
9. Tani K, Mitsuma T, Hiroaki Y, Kamegawa A, Nishikawa K, Tanimura Y, Fujiyoshi Y (2009) Mechanism for aquaporin's fast and selective water conduction and proton exclusion. J Mol Biol 389:694–706
10. Fujiyoshi Y (1998) The structural study of membrane proteins by electron crystallography. Adv Biophys 35:25–80
11. Hirai T, Murata K, Mitsuoka K, Kimura Y, Fujiyoshi Y (1999) Trehalose embedding technique for high resolution electron crystallography: application to structural study on bacteriorhodopsin. J Electron Microsc 48:653–658
12. Fujiyoshi Y, Mizusaki T, Morikawa K, Yamagishi H, Aoki Y, Kihara H, Harada Y (1991) Development of a superfluid helium stage for high-resolution cryo electron microscopy. Ultramicroscopy 38:241–251
13. Kimura Y, Vassylyev DG, Miyazawa A, Kidera A, Matsushima M, Mitsuoka K, Murata K, Hirai T, Fujiyoshi Y (1997) Surface of bacteriorhodopsin revealed by high-resolution electron crystallography. Nature 389:206–211
14. Gyobu N, Tani K, Hiroaki Y, Kmegawa A, Mitsuoka K, Fujiyoshi Y (2004) Improved specimen preparation for cryo-electron microscopy using a symmetric carbon sandwich technique. J Struct Biol 146:325–333
15. Gonen T, Cheng Y, Sliz P, Hiroaki Y, Fujiyoshi Y, Harrison SC, Walz T (2005) Aquaporin-0 membrane junctions reveal the structure of a closed water pore. Nature 438:633–638

Chapter 7

Grid Preparation for Cryo-Electron Microscopy

Nobuhiko Gyobu

Abstract

Once 2D crystals suitable for electron crystallography have been obtained, grid preparation for cryo-EM is a critical step in obtaining high-resolution structural information. Specimens have to be prepared in a manner that prevents dehydration and disruption of the crystals in the vacuum of the electron microscope. Sugar embedding is an effective way to preserve specimens in the native and hydrated state. Preparation of almost perfectly flat specimens is another prerequisite. Imperfect specimen flatness is a crucial problem in the recording of images and diffraction patterns at higher tilt angles because it causes the blurring of spots perpendicular to the tilt axis.

In this chapter, we describe the protocols of preparing 2D crystal specimen for electron crystallographical data collection. These protocols cover preparation of a flat carbon support film by sparkless carbon evaporation, sugar embedding using back injection, and the recently developed carbon sandwich technique.

Key words: Grid preparation, Specimen preparation, Cryo-electron microscopy, Sugar embedding, Back injection technique, Carbon sandwich technique, Sparkless carbon evaporation, Molybdenum grid

1. Introduction

Grid preparation for cryo-EM is an essential step in the collection of high-resolution data. It is important to establish a reliable and reproducible specimen preparation method that retains the native state and the high degree of order of the original sample.

The flatness of 2D crystals on the support film is a critical factor that strongly affects the attainable resolution in electron crystallography. Imperfect specimen flatness causes diffraction spots perpendicular to the tilt axis to be blurred. The flatness of the carbon support film to which the 2D crystals attach is of particular importance in preserving the planarity of 2D crystals. Carbon films are

Ingeborg Schmidt-Krey and Yifan Cheng (eds.), *Electron Crystallography of Soluble and Membrane Proteins: Methods and Protocols*, Methods in Molecular Biology, vol. 955, DOI 10.1007/978-1-62703-176-9_7,

usually prepared by evaporating carbon onto freshly cleaved mica sheets. The mica-facing side of the carbon film is smoother than the air-facing side, and is used for adsorbing 2D crystal. It is important to use high-quality mica sheets that can be cleaved without serious cracking and high-purity carbon rods. The prerequisite for preparing an atomically flat carbon film is the avoidance of sparks during carbon evaporation. A reliable sparkless carbon evaporation technique using a pre-evaporation method has been developed (1). At the beginning of evaporation, relatively large graphite particles tend to be released from carbon rods, causing sparking, while a carbon chunk is formed between the tips of carbon rods. This carbon chunk tends to reduce sparks. After this pre-evaporation, carbon is evaporated by slowly increasing the heating current and deposited onto freshly cleaved mica sheets. The pressure should be kept below 2×10^{-6} Torr during carbon evaporation.

The material and quality of the EM grid can also affect the flatness of 2D crystals. Cooling of the specimen to liquid nitrogen temperatures leads to cryo-crinkling of the carbon film on a copper grid, which is caused by the mismatch of the linear thermal expansion coefficient for copper and carbon (2). Cryo-crinkling can be significantly reduced by using molybdenum grids, which have an expansion coefficient similar to that of carbon. Molybdenum grids have now become standard in electron crystallography. Some commercially available molybdenum grids, however, have a somewhat rough surface and cause wrinkling of otherwise flat carbon films. Fujiyoshi has designed special molybdenum grids for high-resolution electron crystallography; these grids were manufactured by photochemical etching to preserve a smooth surface (1). A new version of the special molybdenum grids, which is now commercially available from JEOL Ltd. (JAPAN), has a thickness of 10 μm and circular holes of 100 μm in diameter. This relatively large hole size results in a relatively large visible sample area on tilted grids.

The most commonly used method for cryo-EM grid preparation of 2D crystals is sugar embedding using the back injection technique (3, 4). 2D crystals prepared by back injection are adsorbed to carbon film, embedded in a thin layer of sugars and then partially dried in air followed by quick freezing in liquid nitrogen. Nonvolatile sugar molecules are thought to substitute for water molecules and form hydrogen bonds to the polar residues on the protein surface to preserve the specimen. Embedding media that have led to atomic structures include glucose (5, 6), tannin (7), a mixture of glucose and tannin (8), and trehalose (9–15). Although not discussed in detail here, the vitrification of the crystal solution on a holey carbon film grid by plunge-freezing in liquid ethane is an alternative sample preparation method. High-resolution 3D models of helical crystals have been determined using this method (16, 17).

Beam-induced specimen movement and/or image deflection is a severe problem that frequently causes anisotropic diffraction when collecting tilted images and seriously reduces the efficiency of data collection. The carbon sandwich technique has proven to dramatically increase the success rate of obtaining high-quality images at high tilt angles as compared to the conventional back injection technique (18). Another advantage of this method is that the specimen is preserved in its native state because the crystals between two carbon layers are surrounded by buffer and never exposed to the air–water interface. The carbon sandwich technique has been particularly effective in studies of membrane proteins that have relatively large extramembranous domains and are sensitive to drying (19). The carbon sandwich technique can also significantly improve the quality of diffraction patterns at high tilt angles (11, 12, 15), presumably due to the almost complete preservation of the specimen in the buffer.

In this chapter, we describe the optimized protocols for preparing flat carbon films by carbon evaporation, and the two important grid preparation techniques for cryo-EM, back injection, and carbon sandwich.

2. Materials

2.1. Carbon Evaporation

1. Vacuum evaporator with an oil-diffusion pump equipped with a liquid-nitrogen cold trap (e.g., JEE-420, JEOL).
2. Carbon rods with a purity of more than 99.9999% (e.g., Hitachi Spectroscopic Graphite Electrodes, Hitachi Chemical Co., Ltd.).
3. White ceramic tiles (approximately 1.5 × 1.5 cm).
4. Vacuum grease.
5. Mica sheets.
6. Desiccator.

2.2. Back Injection Technique

1. Carbon-coated mica sheets.
2. Crystal solution.
3. Crystallization buffer solution.
4. 7% (w/v) Trehalose solution in crystallization buffer (see Note 1).
5. Molybdenum grids with a thickness of 10 μm and circular holes 100 μm in diameter (JEOL) (see Note 2).
6. Petri dish (approximately 30 mm in diameter).
7. Tweezers.
8. Anti-capillary reverse (self-closing) tweezers (Dumont).

9. Parafilm (Pechiney Plastic Packaging Company, Chicago, IL).
10. Filter paper (e.g., Whatman, No. 4).
11. Liquid nitrogen.

2.3. Carbon Sandwich Technique

1–11 The same materials described in Subheading 2.2.

12. Platinum loop with a diameter of 4 mm (available from EM supplies vendors or home-made).

3. Methods

3.1. Carbon Evaporation

1. Prepare two carbon rods. Sharpen one tip of the carbon rod cylindrically to approximately 1.0–1.5 mm in diameter and at least 7 mm long; grind down one end of the other carbon rod to be flattened and polish to a mirror-like surface.
2. Mount the two carbon rods on the rod holders of the vacuum evaporator. The flat-tipped rod is fixed and the pointed one is held against it. The tips of the two rods are kept in contact, maintained by pressure from springs attached to the holder.
3. Evacuate the bell jar of the evaporator until the pressure in the bell jar becomes lower than 0.1×10^{-4} Pa (0.08×10^{-6} Torr).
4. Evaporate carbon by applying current across the carbon rods until the carbon stops sparking during this pre-evaporation, when a chunk of carbon is formed between the tips of carbon rods (see Note 3).
5. Freshly cleave mica sheets. Make sure that they have no serious cracks.
6. Place the freshly cleaved mica sheets 10–15 cm beneath the carbon rods mounted on the rod holders.
7. Apply a thin layer of vacuum grease on half of the surface of a white ceramic tile and put the tile adjacent to the mica sheets, in order to estimate the thickness of the carbon film (see Note 4).
8. Evacuate the bell jar of the evaporator until the pressure in the bell jar becomes lower than 0.1×10^{-4} Pa (0.08×10^{-6} Torr).
9. Evaporate carbon slowly and stepwise by heating the rods; deposit carbon onto the mica surface until the desired thickness of the carbon film has been obtained (e.g., 5–20 nm thick) (see Note 5). Make sure that no spark is emitted from the tips of the carbon rods throughout the evaporation process, and that the pressure remains lower than 2×10^{-4} Pa (1.5×10^{-6} Torr).
10. Retrieve the carbon-coated mica sheets and store in a desiccator.

3.2. Back Injection Technique (Fig. 7.1a–e)

1. Fill a Petri dish with a crystallization buffer solution (see Note 6).
2. Cut a small piece of carbon-coated mica sheet (approximately 4 × 4 mm square).
3. Float the carbon film off the mica onto the surface of the buffer solution.
4. Check the quality (e.g., dents, cracks) of the floated carbon film by shining a light onto it. Pick up the best-looking part of the carbon film with a molybdenum grid held by anti-capillary reverse tweezers (the anti-capillary, bent leg of the tweezers is facing up for this step).
5. Put three drops of sugar solution on a sheet of Parafilm. Place the grid, with the carbon film facing up, on the three drops of sugar solution.
6. Turn the grid upside down so that the carbon film faces down.
7. Gently scrape the upper surface of the grid mesh using the side of a pipette tip to remove folded edges of the carbon film.
8. Remove excess liquid until a thin layer remains.
9. Add 1–2.5 μl of the crystal solution through the grid bars and mix by gently pipetting approximately ten times (see Note 7).

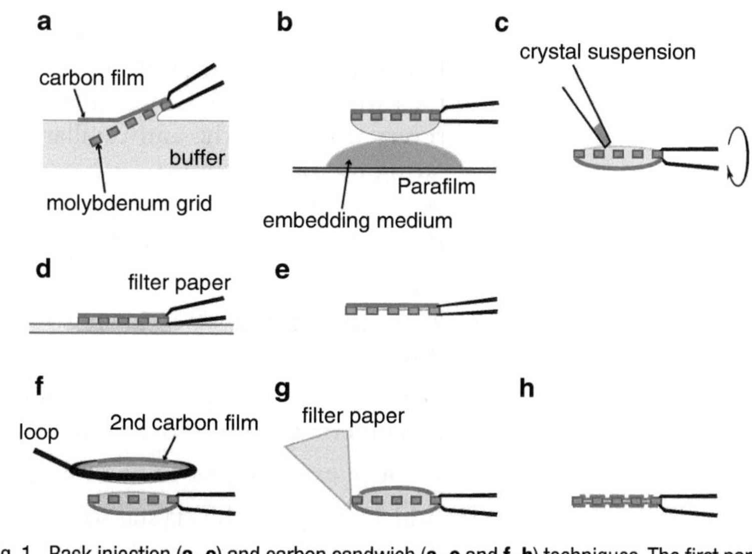

Fig. 1. Back injection (**a**–**e**) and carbon sandwich (**a**–**c** and **f**–**h**) techniques. The first part of the procedure is identical. (**a**) A 3 × 3 mm piece of carbon film is picked up with a grid. (**b**) The buffer is exchanged for a sugar solution by transferring the grid to the surface of a drop of sugar solution with the carbon film facing up. (**c**) The grid is turned over and a crystal suspension is injected and mixed. (**d**) For the back injection technique, the grid is turned over again and placed on a filter paper with the carbon film facing up. (**e**) After air-drying, the grid is quickly frozen in liquid nitrogen. (**f**) For the carbon sandwich technique, a second piece of carbon film (approximately 2 × 2 mm) floated on buffer solution is picked up with a loop and transferred onto the grid. (**g**) Excess liquid is blotted from the side of the grid with filter paper. (**h**) After blotting and air-drying, the grid is quickly frozen in liquid nitrogen.

10. Incubate for approximately 1 min to wait for crystals to be adsorbed to the carbon film (see Note 7).
11. Remove excess liquid with a pipette from the top of the grid.
12. Turn over the grid again and place the grid on two sheets of filter paper, with the carbon film facing up, to blot excess liquid through the grid bars.
13. Pick up the grid with normal tweezers and air-dry for 3–20 s (see Notes 8 and 9). Quickly freeze the grid by manually plunging the grid with the forceps into liquid nitrogen.
14. Transfer the grid still held with the same pair of tweezers into a cryo-transfer specimen holder for TEM observation (see Note 10).

3.3. Carbon Sandwich Technique (Fig. 7.1a–c, f–h)

The carbon sandwich technique is based on the back injection technique, and the first part of the procedure is identical to steps 1–10, above.

1. Fill a Petri dish with a crystallization buffer solution (see Note 6).
2. Cut a small piece of carbon-coated mica sheet (approximately 4 × 4 mm square).
3. Float the carbon film off the mica onto the surface of the buffer solution.
4. Check the quality (e.g., dents, cracks) of the floated carbon film by shining a light onto it. Pick up the best-looking part of the carbon film with a molybdenum grid held by anti-capillary reverse tweezers (the anti-capillary, bent leg of the tweezers is facing up for this step).
5. Put three drops of sugar solution on a sheet of Parafilm. Place the grid, with the carbon film facing up, on the three drops of sugar solution.
6. Turn the grid upside down so that the carbon film faces down.
7. Gently scrape the upper surface of the grid mesh using the side of a pipette tip to remove folded edges of the carbon film.
8. Remove excess liquid until a thin layer remains.
9. Add 1–2.5 μl of the crystal solution through the grid bars and mix by gently pipetting approximately ten times (see Note 7).
10. Incubate for approximately 1 min to wait for crystals to be precipitated and adsorbed to the carbon film (see Note 7).
11. Cut a small piece of carbon-coated mica sheet (approximately 2 × 2 mm square) (see Note 11).
12. Float the second carbon film off the mica onto the surface of the buffer solution (see Note 6).

13. Pick up the second carbon film with a platinum loop.
14. Remove excess liquid with a pipette from the top of the grid.
15. Place the second carbon film on the top of the grid.
16. Carefully blot excess liquid by lightly touching a torn filter paper to the edges of the grid so that the second carbon film stays at the center of the grid. (Otherwise, the second carbon film will easily come off with the filter paper.)
17. Blot extensively around the edge of the grid with a torn filter paper (see Note 12).
18. Air-dry for 3–7 min (see Notes 9 and 12).
19. Pick up the grid off the anti-capillary tweezers, using normal tweezers, and quickly freeze the grid in liquid nitrogen.
20. Transfer the grid to a cryo-transfer specimen holder to observe in the TEM (see Note 10).

4. Notes

1. The type of embedding medium and its concentration should be optimized for each specimen. For preserving specimens, pH-adjusted tannin is used at a final concentration of 0.5–3.0% (w/v). Glucose and trehalose are usually used at a final concentration of 1–10% and 3–7%, respectively. However, higher concentrations of embedding media might work better with specimens sensitive to drying, such as proteins with relatively large extramembranous domains.
2. Molybdenum grids are washed before use as follows: Placed in 100 ml of 1% SDS overnight; rinse with 100 ml H_2O and repeat three times; ultrasonic cleaning in 100 ml H_2O for 5 min; 100 ml of ethanol for 5 min; and finally 100 ml chloroform for 5 min.
3. Alternatively, the freshly cleaved mica sheets can be covered with a shutter until the carbon stops sparking during evaporation. Next, the shutter is removed in order to deposit carbon onto the mica surface.
4. The color of a tile surface without grease turns gray as carbon is deposited, whereas the greased surface stays white. To easily measure the thickness of the carbon film during evaporation, a set of carbon-coated tiles, in which the thickness of coated carbon films ranges from 5 to 30 nm, is commercially available. Filter paper can be used instead of white tiles to estimate the thickness.

5. The heating current is turned off when the vacuum deteriorates to 2×10^{-4} Pa (1.5×10^{-6} Torr) during evaporation. The current is gradually turned up again when the vacuum has recovered. This process is repeated until a desired thickness of carbon film has been obtained.
6. Alternatively, the carbon film can be stripped off on the sugar solution from the start, and picked up using a grid.
7. The number of crystals on a cryo-EM grid is smaller than on glow-discharged negatively stained grids as well as on grids that have not been glow-discharged because the interaction between the specimen and the carbon film is much weaker. This sometimes causes a decrease in the efficiency of data collection. The following strategy to spread a larger number of crystals onto the carbon film should be considered. The population of crystals on a grid can be improved by incubation for several minutes after injection of the sample. A humidity chamber is helpful for preventing sample evaporation during the incubation. In the case of crystals that tend to stack upon each other, extensive agitation (e.g., pipetting some dozen times until the solution is homogeneously mixed) before sampling may help to separate the stacked crystals and allow observation of a larger number of single crystals on the grid. Generally, specimens are a mixture of crystals, small vesicles without ordered arrays, and aggregates. Purification of samples on a sucrose gradient may help remove many of the aggregates and small vesicles (20).
8. Air-drying changes the concentration of protein as well as the chosen solute in the buffer. Therefore, care should be taken when the buffer solution contains a solute at a high concentration. For example, salt is concentrated, or can be precipitated on the grid during specimen drying. Additionally, higher salt concentrations precipitate tannin. We usually keep solutes at a lower concentration in the sugar solution unless deterioration of crystals occurs under such conditions. Removing excess buffer before blotting in step 11 in Subheading 3.2 also helps suppress the concentration of buffer components. Alternatively, precipitation/concentration can be prevented by preparing cryo-EM grids in a cold room, where high humidity is maintained and evaporation is slow. The carbon sandwich (see Subheading 3.3) is also an effective way to prevent concentration of buffer components due to slow evaporation.
9. The time of blotting and air-drying should be optimized for each individual specimen. The optimal time depends on the humidity and temperature of the room and the sensitivity of specimens to drying. The layer of vitreous buffer should be thin enough for the electron beam to penetrate, but thick enough so that the specimen is preserved in the buffer. If preparation is

done in a cold room that is very humid, the time required for blotting and air-drying will increase.

10. After harvesting of crystals, some crystals tend to stack upon each other, aggregate, or deteriorate over time in the storage tube. These problems can be prevented by preparing quick-frozen grids and storing them in a grid box in liquid nitrogen; however, the grid preparation method has to be optimized for the crystals. Alternatively, if the crystal solution tolerates freeze-thawing, it can be stored as flash-frozen aliquots in liquid nitrogen.
11. To effectively reduce beam-induced image movement, the specimen should be symmetrically arranged between two carbon films of the same thickness. The two layers of carbon films for a grid preparation should be from the same source of carbon-coated mica.
12. The grid should be extensively blotted with a torn filter paper and partially air-dried so that the two carbon films eventually stack together, forming a flat carbon-sandwiched specimen. The ice layer between the carbon films needs to be thin enough to observe 2D crystals in cryo-EM.

References

1. Fujiyoshi Y (1998) The structural study of membrane proteins by electron crystallography. Adv Biophys 35:25–80
2. Booy FP, Pawley JB (1993) Cryo-crinkling: what happens to carbon films on copper grids at low temperature. Ultramicroscopy 48:273–280
3. Kuhlbrandt W, Downing KH (1989) Two-dimensional structure of plant light-harvesting complex at 3.7 Å resolution by electron crystallography. J Mol Biol 207:823–828
4. Wang DN, Kuhlbrandt W (1991) High-resolution electron crystallography of light-harvesting chlorophyll a/b-protein complex in three different media. J Mol Biol 217:691–699
5. Henderson R, Baldwin JM, Ceska TA, Zemlin F, Beckmann E, Downing KH (1990) Model for the structure of bacteriorhodopsin based on high-resolution electron cryo-microscopy. J Mol Biol 213:899–929
6. Gonen T, Sliz P, Kistler J, Cheng Y, Walz T (2004) Aquaporin-0 membrane junctions reveal the structure of a closed water pore. Nature 429:193–197
7. Kuhlbrandt W, Wang DN, Fujiyoshi Y (1994) Atomic model of plant light-harvesting complex by electron crystallography. Nature 367:614–621
8. Nogales E, Wolf SG, Downing KH (1998) Structure of the alpha beta tubulin dimer by electron crystallography. Nature 391:199–203
9. Kimura Y, Vassylyev DG, Miyazawa A, Kidera A, Matsushima M, Mitsuoka K, Murata K, Hirai T, Fujiyoshi Y (1997) Surface of bacteriorhodopsin revealed by high-resolution electron crystallography. Nature 389:206–211
10. Murata K, Mitsuoka K, Hirai T, Walz T, Agre P, Heymann JB, Engel A, Fujiyoshi Y (2000) Structural determinants of water permeation through aquaporin-1. Nature 407:599–605
11. Gonen T, Cheng Y, Sliz P, Hiroaki Y, Fujiyoshi Y, Harrison SC, Walz T (2005) Lipid-protein interactions in double-layered two-dimensional AQP0 crystals. Nature 438:633–638
12. Hiroaki Y, Tani K, Kamegawa A, Gyobu N, Nishikawa K, Suzuki H, Walz T, Sasaki S, Mitsuoka K, Kimura K, Mizoguchi A, Fujiyoshi Y (2006) Implications of the aquaporin-4 structure on array formation and cell adhesion. J Mol Biol 355:628–639
13. Holm PJ, Bhakat P, Jegerschold C, Gyobu N, Mitsuoka K, Fujiyoshi Y, Morgenstern R,

Hebert H (2006) Structural basis for detoxification and oxidative stress protection in membranes. J Mol Biol 360:934–945

14. Jegerschold C, Pawelzik SC, Purhonen P, Bhakat P, Gheorghe KR, Gyobu N, Mitsuoka K, Morgenstern R, Jakobsson PJ, Hebert H (2008) Structural basis for induced formation of the inflammatory mediator prostaglandin E2. Proc Natl Acad Sci USA 105: 11110–11115
15. Hite RK, Li Z, Walz T (2010) Principles of membrane protein interactions with annular lipids deduced from aquaporin-0 2D crystals. EMBO J 29:1652–1658
16. Miyazawa A, Fujiyoshi Y, Unwin N (2003) Structure and gating mechanism of the acetylcholine receptor pore. Nature 423:949–955
17. Yonekura K, Maki-Yonekura S, Namba K (2003) Complete atomic model of the bacterial flagellar filament by electron cryomicroscopy. Nature 424:643–650
18. Gyobu N, Tani K, Hiroaki Y, Kamegawa A, Mitsuoka K, Fujiyoshi Y (2004) Improved specimen preparation for cryo-electron microscopy using a symmetric carbon sandwich technique. J Struct Biol 146:325–333
19. Abe K, Tani K, Nishizawa T, Fujiyoshi Y (2009) Inter-subunit interaction of gastric H+, K+-ATPase prevents reverse reaction of the transport cycle. EMBO J 28:1637–1643
20. Krebs A, Villa C, Edwards PC, Schertler GF (1998) Characterisation of an improved two-dimensional $p22_12_1$ crystal from bovine rhodopsin. J Mol Biol 282:991–1003

Chapter 8

Recording High-Resolution Images of Two-Dimensional Crystals of Membrane Proteins

Agustin Avila-Sakar, Xueming Li, Shawn Q. Zheng, and Yifan Cheng

Abstract

Principles underlying the recording of high-quality/resolution images of two-dimensional crystals of membrane proteins are discussed in the context of instrumental conditions and operational procedures. A detailed example of low-dose microscope settings is provided along with an overview of a program that implements a computer-aided data acquisition procedure.

Key words: Electron microscope, Electron crystallography, Image data acquisition, Membrane proteins

1. Introduction

Recording high-resolution electron micrograph and diffraction data from two-dimensional (2D) crystals of membrane proteins has always been a challenging task. The possibility of generating a high-resolution and high-quality three-dimensional (3D) density map may depend not only on the quality of crystals, which is the subject of other chapters in this volume, but also on the instrument and technology incorporated in it. In this chapter we focus on the influence of instrumental conditions and operational procedures on the quality of image data and present a comprehensive strategy for an overall system optimization for high-resolution image acquisition. This strategy encompasses general precautions, maintenance routines, ways of approaching the use of the instrument and data recording techniques that will work synergistically to get the best

Ingeborg Schmidt-Krey and Yifan Cheng (eds.), *Electron Crystallography of Soluble and Membrane Proteins: Methods and Protocols*, Methods in Molecular Biology, vol. 955, DOI 10.1007/978-1-62703-176-9_8,

performance out of any particular instrument. We will loosely refer to these principles as "good practices" and we will explain the rationale of each of these in this chapter.

2. Materials

2.1. Electron Microscope

Electron microscopes from different vendors may differ in appearance, operational procedures, and even in technologies used, but they all share the same physical principles. Chapter 19 of this volume discusses in detail the criteria involved in making a choice for a particular instrument and how to integrate it and its ancillary equipment to an electron microscopy laboratory. In this section we will only discuss the factors that are critical for high-resolution imaging of biological molecules, with special emphasis on 2D crystals of membrane proteins.

2.1.1. Components of an Electron Microscope

Electron Source

To a great extent, the achievable resolution of a microscope depends on the quality of its electron source. Brightness and coherence are the key physical parameters to describe such quality. Compared with a thermal emission source, such as a tungsten filament or LaB_6 crystal, a field emission gun (FEG) generates a much more coherent beam with high brightness (1). Although a microscope with a LaB_6 crystal filament can reach atomic resolution when imaging at near Scherzer focus, its effective resolution decreases steeply as the image condition moves away from near-focus conditions, due to poor spatial coherence. Since defocusing is the main means to introduce contrast in images of unstained biological molecules, it is necessary to use an electron source with better spatial coherence. A 200 keV electron microscope with a FEG can retrieve the 3.4 Å resolution signal from the atomic lattice of graphitized carbon imaged at a 2 μm defocus, something that cannot be achieved with a thermal emission source. Therefore, nowadays, a FEG is an absolute necessity on any electron microscope intended for high-resolution imaging of biological macromolecules (2).

Vacuum System

Modern microscopes rely on one or more ion getter pumps in the microscope column to achieve a column pressure on the order of 1.2×10^{-5} Pa, which is necessary to avoid contamination on the sample. There are two major sources of contamination: build-up of hydrocarbon materials in the specimen area being illuminated by the strong electron beam and accumulation of water molecules absorbed on specimen. Hydrocarbon build-up is a phenomenon often observed during certain electron microscopy applications in materials science, where a highly focused electron probe is shone on the sample. Low-dose illumination with a spread beam and short exposure, as practiced on biological specimens, does not induce such contamination. Water molecules inside the column, however, cannot be pumped out efficiently by ion getter pumps, thus representing a major source of contamination. The cold trap, when cooled to liquid nitrogen temperature, functions as an absorption pump for water molecules and improves the column vacuum. However, water molecules weakly bound to the cold trap, or to the specimen stage when it is also cooled with liquid nitrogen, desorb very slowly and represent a continuous source of contamination. Repeated loading of samples into the microscope, especially cryo-samples, contributes to replenishing the pool of absorbed water molecules. Eventually, the cold trap will be saturated and the column will no longer hold acceptable vacuum levels. In order to clear the column from the excess of water molecules, the inner surface of the microscope column must be warmed up to room temperature and pumped by water-efficient pumps, such as a turbomolecular pump. These maintenance cycles are usually referred to as "cryo-cycles." They should be run regularly according to the manufacturer instructions.

Cryo Protection (Anti-contaminators)

To minimize the radiation damage to proteins caused by the high-energy electron beam, 2D crystals are usually imaged at a low temperature, often at ~100 K or even lower than 10 K (3). However, the cold specimen also acts as an absorption pump for water molecules inside the column. Over time, the accumulated ice on the specimen will increase sample thickness and reduce image contrast. Therefore, in addition to the cold trap, it is necessary to provide some additional means of cryoprotection around the specimen area to reduce such water contamination of the specimen. This is achieved by covering the tip of the specimen holder with a pair of metallic blades or a box with apertures for the beam above and below the grid in the specimen holder. This protective enclosure, called cryo-blades or cryo-box, is cooled to liquid nitrogen temperature via copper braids connected to an external dewar. Their geometry is designed to decrease the probability of water molecules reaching the sample. As with the other internal cold surfaces, the protection is only effective until the cryo-blades or -boxes become saturated, in which case it is necessary to run a cryo-cycle on those devices as well. The apertures on cryo-blades or -boxes are much larger than the objective lens diaphragms, so their presence should exert no influence on the quality of the images. However, we have encountered systems in which such a device affected image shifts and astigmatism and transmitted instabilities to the image. Regarding the former effect, we conjecture that the cause was the presence of magnetic impurities in the cryo-box raw materials and, perhaps, some electric grounding problems. As for the latter effect, it resulted from insufficient mechanical isolation. One should be aware of these possibilities at the time of purchase and acceptance of a new microscope.

Detectors

Traditionally, film has been the detector of choice for recording high-resolution images of biological samples, particularly of 2D crystals of membrane proteins. It has probably the best detective quantum efficiency (DQE) and modulation transfer function (MTF) for electrons, compared with other detectors such as charge coupled device (CCD) cameras and image plates (4). Even to date, many laboratories still use film routinely, despite the tedious and time-consuming chemical processing and digitization required to convert the image to digital format.

The CCD camera is another popular device used for recording electron micrographs. It does not detect electrons directly, but relies on a scintillator to convert electron events into photons. Due to this dual stage process, both DQE and MTF of even the best CCD cameras are not comparable to film (5). This can, in principle, be overcome by using a higher magnification to oversample images. In fact, nowadays, CCD cameras are routinely used to record images of radiation-insensitive inorganic materials at atomic resolution. Such images are recorded at very high magnification with a high electron dose. However, such practice cannot be applied to radiation-sensitive materials. Due to radiation damage, electron exposure of biological macromolecules is limited to a total dose of ~20 $e^{-}/Å^{2}$, which, on a CCD image at high magnification represent much less electron counts per pixel, and therefore a much lower signal-to-noise ratio (SNR) than normally used on radiation-insensitive materials. Furthermore, the higher the magnification, the smaller the imaged area of a 2D crystal, which implies less unit cells per image and therefore even lower SNR for the reflections in the computed Fourier transform. Thus, specimen radiation damage, detector size, and indirect signal detection process determine boundaries for the resolution that can be retrieved from images of 2D membrane protein crystals on CCDs, although the actual limits are not precisely known. Based on single particle cryoEM work on complexes with high symmetry, it is possible to generate 3D reconstructions at near-atomic resolution from images recorded on a CCD camera (6). CCD cameras should also be capable of delivering similar results from images of membrane proteins in a 2D crystal (7). Nevertheless, compared with film, the area covered per image recorded on a commonly used 4 K×4 K CCD camera is too small, which means less unit cells contribute to the reflections in the computed Fourier transform, resulting in a lower SNR for those reflections. This may eventually be overcome with a new generation of large format camera (8 K×8 K pixels, or even 10 K×10 K pixels). For the time being, an alternative approach is to combine a modified form of the spot-scan imaging method (8) with single particle treatment of the molecules in CCD images of 2D crystals (9).

CMOS (complementary metal oxide semiconductor) cameras with direct electron detection capabilities can potentially overcome the problem of the relative low DQE and MTF of CCD cameras (10, 11). They have similar size limitations as CCDs, but, as in that case, proper adaption of imaging conditions and image processing may make it possible to record images of 2D crystals to near atomic resolution.

2.1.2. Ancillary Equipment

Grids

In electron crystallography, 2D crystals are usually supported by a continuous carbon film attached to an EM grid. 2D crystals are not rigid and thus adopt the shape of the supporting carbon film. For that reason, the flatness or lack of flatness of the supporting carbon film plays a key role in determining whether the Fourier averages of large areas of an image reveal high-resolution reflections or not, especially at high tilt angles. The flatness of the support film is influenced by the way in which it is prepared, as well as by the flatness of the grid itself. Furthermore, the large difference in the thermal contraction coefficients of carbon and conventional metals used in the production of EM grids, like copper, causes wrinkling of the carbon support film when the specimen is cooled down to liquid nitrogen temperature. In the case of molybdenum, such difference is small. Therefore, in electron crystallography, molybdenum grids with a polished surface are recommended, instead of copper grids. The following procedure can be used to prepare super-flat carbon film. An extended discussion, including related topics, can be found in refs. 3, 12.

1. Split a carbon rod in two and file one part flat and the other to a pencil point at about 45° angle.
2. Mount the two halves into the holders of a carbon evaporator, pointed half on the spring-loaded position.
3. Secure the vacuum bell jar on the apparatus and start the vacuum pump.
4. Once the pressure has stabilized at a level of less than 1×10^{-4} Pa, turn on the filament.
5. Very slowly increase the current until the carbon begins to evaporate.
6. After the carbon has begun to evaporate, carefully allow for sparking to occur at a low rate until no more sparks are observed.
7. Once the sparking has stopped, turn off the filament and vent the bell jar.
8. Place a piece of freshly cleaved mica inside the apparatus.
9. Put back the bell jar and restart the vacuum.
10. Let the pressure stabilize again at a level of less than 1×10^{-4} Pa and then turn on the filament.

11. Very slowly increase the current until the carbon begins to evaporate. This time no more sparking should occur. If it does, discard mica and go back to item 3 (If sparking persists, carbon source must be replaced).
12. Carefully monitor the bell jar pressure.
13. If the pressure rises, turn off the filament and wait for the vacuum to completely recover.
14. Once the carbon evaporation has been completed, store the carbon film in a desiccator until use.

Cryoholders

Most electron microscopes used for biological macromolecules have side-entry specimen stages. The specimen holder is a long rod that holds the sample at one end and has a dewar for liquid nitrogen at the other end. The volume of liquid nitrogen that it can hold usually lasts for about 3–4 h. Unfortunately, the relatively heavy, large-sized dewar increases the sensitivity of the sample holder to acoustic noise and vibrations, and therefore the attainable resolution is lower than the one achieved with a standard holder of the microscope.

In recent years, manufacturers have produced a new kind of tilting specimen stage that greatly improves the mechanical stability and also increases the useful time of a sample in the column. There are variations, but the basic concept is to mount the EM grid on a carrier of dimensions just slightly larger than the grid itself and load this carrier (also called cartridge) on a sample holder that resides permanently inside the column. This holder is cooled down via long metallic braids connected to a large capacity liquid nitrogen dewar, located outside the column. Examples of this kind of holder are found on the Tecnai TF30 Polara and Titan Krios microscopes, provided by FEI Company, and the JEM3100 and 3200, provided by JEOL. The cooling and stability advantages of this type of holders are also present on microscopes with non-tilting top entry holders, like the JEM3000 from JEOL. On top-entry holders the samples reside either untilted or tilted at a fixed angle, which is suitable for electron crystallography of 2D membrane protein crystals, as long as their azimuthal orientations are random, which is the case for all 2D crystals studied to date.

2.2. Equipment Site Conditions

Environmental conditions can also have a significant effect on the quality of images. Therefore electron microscope sites must meet certain minimal requirements in order for the instrument to perform to its full potential (see also Chapter 19). However, even if the site meets these requirements, users must also keep in mind that their presence and actions affect the environmental conditions, even though, perhaps, only momentarily. The following factors are probably the most relevant in this context.

2.2.1. Environmental Factors

Acoustic Noise and Vibrations

Even if the average acoustic level in the EM room is low and acceptable according to manufacturer specifications, if a user talks or makes some other sound during an exposure, this can be enough to obliterate any high-resolution information present in the image. The same rationale applies to vibrations produced, for example, by leaning one's elbow against the microscope table, rolling carts on a hallway or doors being slammed.

Air Temperature and Circulation

Specimen and image stability require constant temperature of lenses, coils, stage, and sample. Temperature fluctuations of these components result in image drift, which severely compromises the exposure time possibilities. This is especially important in low-dose imaging, since beam intensity cannot be freely increased to compensate for a shorter exposure time. Modern instruments usually have cooling systems that maintain very low lens temperature fluctuations. Also, room ventilation systems can provide fairly constant ambient temperature and laminar airflow toward walls and floors to minimize turbulence in the neighborhood of the column. Nevertheless, the presence of the EM operator in the room is enough to alter an otherwise constant temperature regime, especially if he/she moves frequently. Obviously, the temperature regime is altered even further when there is more than one person in the room or when people come in or out.

2.2.2. Enclosed Environment and Remote Control

A means to minimize the effects of acoustic noise, vibrations, and temperature fluctuations is to create an isolated environment for the electron microscope. One way of creating such an environment is by the use of heavy fabric curtains around the electron microscope. This will reduce acoustic noise and air currents, thus helping to maintain a constant temperature regime around the column. Such a stable system can be achieved more effectively in a small space, rather than over the entire room. In addition, behavioral restrains, such as entering the room without opening the door all the way, walking slowly toward the microscope, and not talking and keeping the body away from the microscope table while taking an exposure are necessary.

However, behavioral restrains can be very stressful both physically and psychologically for an operator confined within a small and enclosed environment for several hours. One way to alleviate such stress and to reduce disturbance to the environment of the microscope is to operate the instrument by remote control, something that is now becoming increasingly feasible thanks to the advances in computer technology and telecommunications. As an example, Fig. 1 shows a Tecnai TF20 microscope (FEI Company) operated remotely.

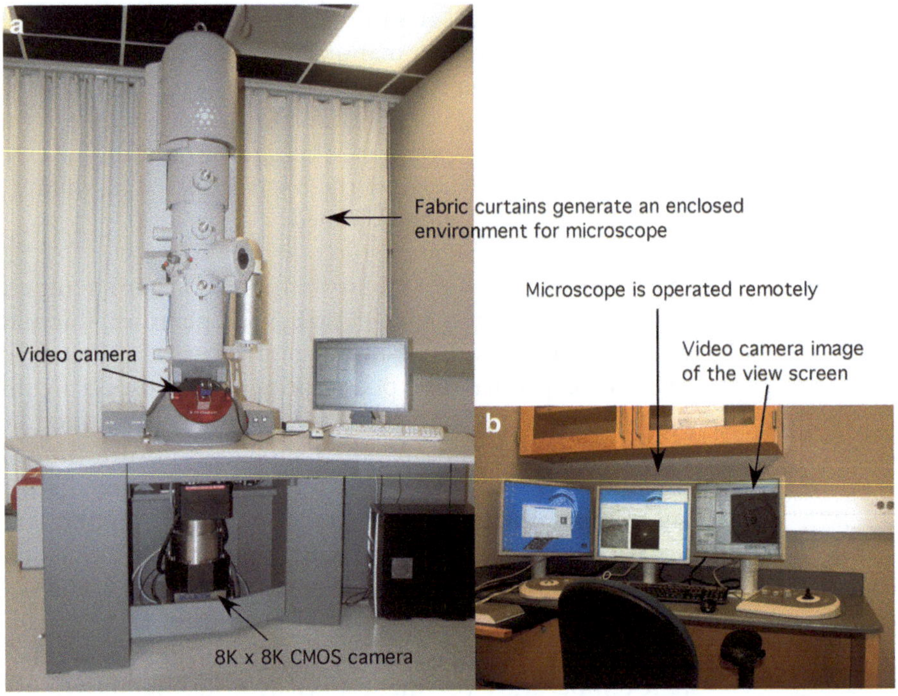

Fig. 1. An example of a simple microscope setup for remote control. (**a**) A Tecnai TF20 (FEI Company) with a video camera mounted in front of the viewing chamber. The microscope is enclosed by a heavy fabric curtain in a small area. (**b**) The operating desk is away from the microscope, outside of the enclosed environment.

3. Methods

3.1. Preparations

3.1.1. Microscope

Vacuum Quality

Instruments used for cryoEM receive a significant load of water with each sample introduced in the column. The vacuum system and anti-contaminators maintain acceptable pressure levels during data recording. However, over time they get saturated. The system's capacity must therefore be regenerated on a regular basis. The process is often referred to as cryo-cycling and is usually included as an item in the user interface of modern microscopes. In essence, two main things take place: (1) the internal cold surfaces, which have become saturated with ice, are warmed up so that water molecules can be released to the column space; and (2) the resulting water vapor is pumped out efficiently. Pumping is usually done with a turbomolecular pump, while the column ion getter pumps are either isolated or turned off during the process. Optimal vacuum conditions are usually restored when the microscope has been put through a cryo-cycle prior to use. Similar to the baking of the cryo-holders, the cryo-cycle should also be carried out for several hours.

Column Alignment

Modern electron microscopes are generally very stable. The column alignment parameters can be stored and reloaded each time a new operator is using the microscope. However, it is always necessary to do some alignments before and even during each data collection session, if the goal is to obtain the highest possible resolution. In particular, it is necessary to check the coma-free alignment to make sure that the beam is aligned precisely to the column optical axis.

3.1.2. Sample Holder

The thermal isolation of the cryo-holder must be such that when cooled with liquid nitrogen, the tip of the cryo-holder reaches a stable temperature of approximately −180°C inside the microscope. In order to achieve this condition the holder has to be baked and pumped prior to use. This process should be done according to the manufacturer's instructions. As a general rule, the minimum baking time to meet the stability requirement is around 2 h. But it is advisable to extend the baking process overnight. In systems with automated controllers, the heating current can be programmed. However, valves opening and closing is still done manually. As a matter of principle one should close the valves while the dewar is still hot, because, once it cools down, the absorbing material in the dewar outcompetes the turbo-molecular pump for the residual water molecules in the lines.

3.2. Collecting Image Data

3.2.1. Low-Dose Settings

To minimize radiation damage, the total dose used for imaging a 2D crystal must be under strict control. The total dose tolerance depends on the intended resolution. As a general rule, in order to preserve near-atomic resolution structural information, the total dose must not exceed 20 $e^-/Å^2$. Some pre-exposure is often necessary to re-center the target or stabilize beam-induced movement of the specimen. However, any exposure of the target area prior to the actual recording of the image must be kept to an absolute minimum. We will refer to the general procedures for controlling the dose in this context as Low Dose Imaging. In modern microscopes, these procedures are included as modules of the computer user interface to the instrument. On FEI instruments this module is called "Low Dose" and on JEOL instruments it is called "LowDose" (one word). In both cases there are three basic modes: Search, Focus, and Exposure (Record, on JEOL). The sequence for recording a low-dose image is, first, to locate a target area (i.e., a 2D crystal) in Search mode and center it with the specimen stage controls, then, set the desired defocus in Focus mode, and finally, record an image in Exposure mode. In setting up the low-dose imaging conditions one should start with Exposure mode. All column alignments should be optimized for this mode, since this is the one in which the data will actually be recorded. The other two modes, Search and Focus, should be defined relative to Exposure mode. The most important aspects of each of these modes are now described as follows.

1. Exposure mode
   (a) *Magnification.* The magnification should be compatible with the target resolution and image recording device. If one uses film to record the image, one should take into account the step size of the scanner used to digitize the data. Based on empirical results, it is generally agreed that the magnification should be such that, when scanned, the pixel size corresponds to ~1/3 of the target resolution. On a CCD camera, this should be 1/4 or less, given the steeper resolution-dependent decay in spectral SNR of current CCD cameras (5, 7).
   (b) *Objective aperture.* The objective aperture is often used to enhance image contrast, similar as a low-pass filter in Fourier space. A relative small aperture, such as 50 μm diameter, is often used when imaging frozen hydrated single particle samples. When collecting images of 2D crystals, one can use a large objective aperture, such as 100 μm diameter, since the image contrast is not critical for image processing (see Chapter 10). The use of the objective aperture has another purpose, to reduce image blurring caused by beam-induced charging. It is believed that an objective aperture can partially neutralize beam-induced charge on specimen, thus reduce image blurring (13).
   (c) *Exposure time.* The exposure should be set according to the limitation imposed by the electron dose tolerance of the crystal, which is usually about 20 $e^-/Å^2$. However, other factors must also be taken into consideration, such as stage drift. Moving the stage always generates some drift. Generally speaking, the shorter the exposure time, the smaller the effect of drift on the recorded image. The most common exposure time is 1 s or shorter. Longer exposure times demand cryo-stages of much higher stability. In the context of the actual image recording sequence, one can evaluate the rate of specimen drift in Focus mode (discussed below) and wait for the drift rate to subside to acceptable levels before taking the picture.
   (d) *Electron dose.* With the total dose at the specimen level fixed, the average number of electron counts per image pixel is related to the image pixel size in an inverse proportion to the square of the magnification. The electron dose per image pixel on the detector should provide a signal that is above the detector noise at the target resolution. Therefore, for a given microscope/detector combination, one must first define the dose rate on the specimen, and then set the magnification in such a way that, while being appropriate for the target resolution, it also generates a sufficient signal on the detector. As mentioned before, the

electron dose that a protein 2D crystal can tolerate before loosing ~3 Å resolution information is about 20 $e^-/\text{Å}^2$. It is worth noting that a target resolution may not be achievable with certain combinations of microscope/detector. For a given dose at the specimen level, the only way to increase the SNR on the detector is to lower the magnification, which ultimately limits the resolution.

(e) *Parallel illumination*. Another factor to be considered for high-resolution imaging is parallel illumination. The convergent angle of the illuminating beam on the specimen influences the envelope of the Contrast Transfer Function (CTF) (1). A large convergent angle leads to a rapid damping of envelope, particularly for highly defocused images. When imaging 2D crystals, defocus values are often smaller than those used in the recording of single particle data. Thus, in general, the requirement for parallel illumination in electron crystallography is less stringent than for single particle cryoEM. However, to achieve atomic or near-atomic resolution of 2D crystals, it is still necessary to take the parallelism of the illuminating beam into consideration, because the high-frequency image components with different tilt angles are transmitted with slightly different phase shifts across the specimen, which results in their incoherent combination.

Modern electron microscopes have a two-condenser lens system and objective lens pre-field (twin) and mini-condenser lens. This provides the means to generate a beam of suitable size on the specimen with a small degree of nonparallelism. Under illumination conditions normally used in low-dose imaging, a diffraction pattern from a sample of gold particles reveals the angular spread of illumination beam as a smear of the diffraction rings of gold particles, as shown in Fig. 2a. To obtain parallel illumination, one must expand the beam until the diffraction rings become as sharp as possible, as in Fig. 2b. However, on a two-condenser system this can only be achieved by strongly exciting the C2 lens, which results in a very weak intensity and an unnecessarily large illuminated area on the specimen, a condition that is not suitable for imaging biological samples. In the absence of a zoom system with a third condenser lens, i.e., Köhler illumination, there is little one can do to alleviate this situation. One way is to reduce the excitation of the mini-condenser lens. Unfortunately, not much is obtained unless one reduces it to the point of actually turning off the lens (or, more precisely, reversing its current). That mode, often called "nano-probe," is normally used to produce a strongly

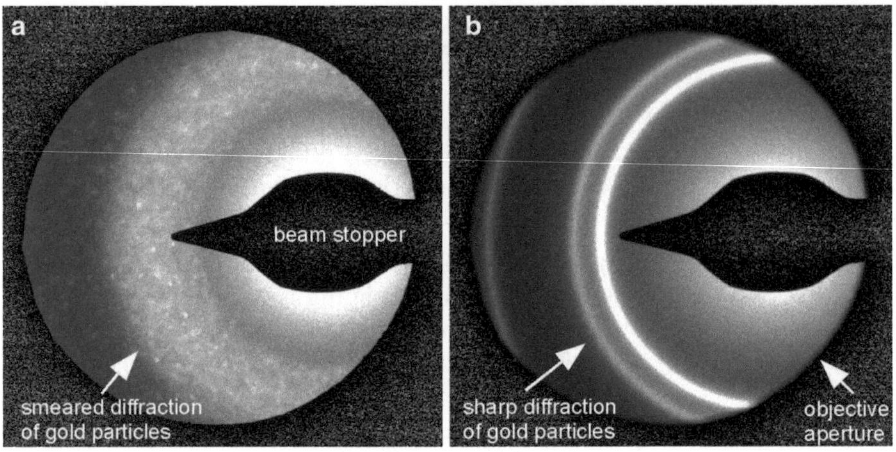

Fig. 2. Setting up parallel illumination in nano-probe. (**a**) A diffraction pattern of a cross-grating grid with random oriented gold particles. The diffraction rings of gold particles appear smeared, indicating a convergent angle in the illuminating beam. This corresponds to the normal illumination condition of micro-probe that is normally used in low-dose imaging. (**b**) A diffraction pattern of gold particles in the presence of an objective aperture, with the aperture focused and the diffraction pattern sharpened. Under these conditions, and based on the assumption that the objective lens aperture and back focal plane are coplanar, the illuminating beam is parallel.

convergent beam for small probes. However, one can also generate parallel illumination in that mode, by moving the crossover to the front focal plane of the objective lens. Thus, on a Tecnai TF20 microscope, one can generate an illumination spot of about 2 μm diameter and 20 $e^-/Å^2/s$ at the specimen with a proper choice of C2 aperture and spot size while keeping the electron beam parallel. The procedure to obtain this condition is as follows:

- Using a specimen that can generate a sharp diffraction pattern, such as a grid with gold particles, switch illumination to "nano-probe" mode;
- Center 50 μm condenser aperture and do all the direct alignments;
- Switch to diffraction mode and focus an objective aperture's limiting disc as sharply as possible by adjusting objective lens current (focus knob). The image on the viewing screen generally appears as in Fig. 2a (i.e., sharp edge of aperture disc, diffused diffraction pattern);
- Change the beam intensity (C2, i.e., intensity knob) to focus the diffraction pattern as sharp as possible, as shown in Fig. 2b (i.e., both aperture disc and diffraction pattern in focus);
- Switch back to image mode. Do not change beam intensity during data acquisition.

The method described here relies on the assumption that the objective aperture is very nearly coplanar with the back focal plane of the objective lens, which is often the case.

(f) *Beam tilt*: Beam tilt introduces coma and must therefore be minimized. This is usually done in conjunction with astigmatism correction. To perform this coma-free alignment, the user should use a CCD camera and image processing software to compute Fourier transforms of the image at acquisition rates fast enough for quick interactive adjustments. Alternatively, the adjustments can be done via automated correction using scripts or packages within existing image acquisition software. Regardless of the actual procedure used, the principle is based on the fact that induced tilt results in defocus change. The coma-free condition is obtained when tilting by the same amount in opposite directions over all orientations results in the same defocus change. This is best evaluated with the aid of a Zemlin tableau (14), which displays the corresponding diffractograms.

Search Mode

The purpose of the search mode is to identify suitable 2D crystals and provide a high-contrast image of the target area and its immediate surroundings in order to accurately center it for image recording. The dose rate at the specimen should be so low (less than 0.01 $e^-/Å^2/s$) that a continuous exposure of a few minutes results in negligible radiation damage. However, at the detector the intensity should be strong enough to allow quick and unambiguous identification of 2D crystals for image recording. Conditions combining high contrast and intensity at the detector with extremely low dose rate at the specimen can be met at the low end of the standard magnification range and in the defocused central image of a diffraction pattern (15). The procedures for setting up these conditions on a Tecnai microscope are now outlined.

1. Search in imaging mode:
   (a) Verify specimen eucentricity.
   (b) Select the lowest magnification in the standard range and the smallest spot size. On a Tecnai TF20 microscope, these are 1,700× and spot size 11, respectively.
   On a Tecnai TF20 microscope, these are 1,700× and spot size 11, respectively.
   (c) Confirm that focus is at eucentric focus by pressing "eucentric focus" button.
   (d) Spread the beam so that the dose rate is sufficiently low.
   (e) Defocus the image by 100–300 μm.
2. Search in diffraction mode:
   (a) In magnification mode select the smallest spot size (11). Spread the beam so that it has a divergent angle and the dose rate is sufficient low.

(b) Switch to diffraction mode and select the largest camera length (6 m).

(c) Focus the central spot with the focus knob (lower objective lens). Because of the divergent illumination, focused central spot appears as a sharp spot within a bright disc. Increase the size of the disc by spreading the intensity, while at the same time focusing the sharp spot in the middle. Note that for an FEG source, the disc has a triangular shape.

(d) Shift the sharp spot to the center of the disc by using beam shift.

(e) Move the disc to the center of the view screen by using diffraction shift.

(f) Spread the sharp central spot by turning the focus knob in the overfocusing direction. One can observe specimen features within the diffraction disc with high contrast. Note that one has to turn the focus knob to overfocus. Otherwise the contrast within the disc is reversed.

(g) The camera length and amount of over focus define the magnification of the image within the central spot. This should be set such that the image within the diffraction disc shows good visibility of specimen features without significant distortion.

Figure 3a is the under-focused image and Fig. 3b is the over-focused diffraction image of the same specimen area with the same

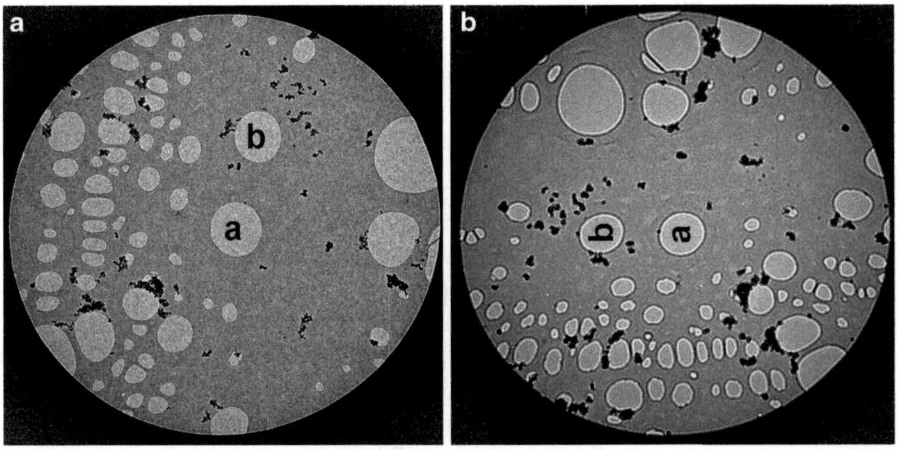

Fig. 3. Search modes in low-mag image and overfocused diffraction. (**a**) An image recorded at the lowest magnification in the normal magnification range, ×1,700, with a defocus of ~200 μm. (**b**) An image recorded from the central spot of an overfocused diffraction pattern, with camera length of 6 m. The dose rates of both images are the same, approximately 0.002 e^{-}/Å^{2}/s. The contrast in both images appears comparable. Note that the image from the overfocused diffraction mode is rotated relative to the image recorded at low magnification. This leads to some inconvenience when moving the specimen by stage movement when searching in the diffraction mode.

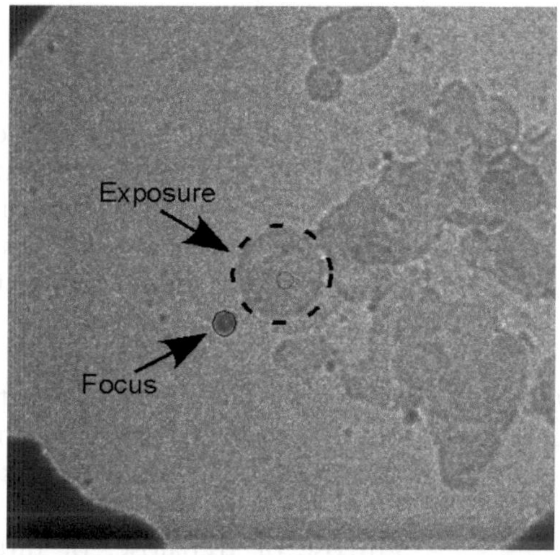

Fig. 4. Illustration of low-dose imaging. The image displayed is the Search mode view. The *large circle* corresponds to the area used in the Exposure mode, and the *small circle* corresponds to that of the Focus mode. The specimen is glucose-embedded 2D crystals of bacteriorhodopsin.

amount of electron doses. Both methods described here are used by many cryoEM laboratories. One may prefer one to the other, but they essentially amount to the same result.

Focus

Focusing is done on a small area adjacent to the selected crystal. This is achieved by inducing an image shift coupled with beam shift. One must carefully adjust the magnitude of this shift so as to avoid overlapping focus and exposure areas. This is best controlled by observing in search mode the imprints left on the sample by the beam in the other two modes (Fig. 4). In addition to focusing, one needs to make sure that the specimen drift is at a tolerable level, given the exposure time set in Exposure mode. For better focusing accuracy and in order to directly detect even very small rates of specimen drift, it is customary to use a magnification higher than that used in the Exposure mode, often the highest magnification in the standard range. Care must be taken to account for the difference in focus between Focus and Exposure magnification. This usually exists as a residual error due to lens hysteresis, which occurs to some extend during both alignment and operation of the microscope.

It is worth noting that modern image acquisition software, as pointed out above, has routines to compute a Fourier transform of the image on the spot and this provides a means to focus accurately without necessarily using very high magnifications. Also one can determine specimen drift rates accurately by cross-correlation at the same magnification used to record the image. Thus, one can set the focus and exposure magnifications to the same value and thereby avoid the need to account for focus differences and hysteresis effects induced by changing from one magnification to the other.

Since the focusing spot is off-axis, one should not correct astigmatism or coma in such state, because there is usually a small difference in beam tilt with respect to the on-axis position. It is better to periodically check astigmatism and coma on-axis on an area of the carbon support film close to the next set of target record areas.

3.2.2. Initial Sample Quality Assessment

A preliminary assessment of the sample quality is convenient and often absolutely necessary. The main goal is to check that the grid has, indeed, absorbed a sufficient amount of 2D crystals and that they are persevered well in the embedding medium, such as glucose or trehalose, or directly in vitrified ice. Only properly embedded 2D crystals can give good quality images. Obtaining properly embedded 2D crystals is often a laborious task that involves screening grids prepared under different conditions, including the percentage of glucose or trehalose, blotting times, and if flash-frozen in liquid ethane.

3.2.3. Data Recording Sequence

The low-dose settings described above can be done on a standard calibration grid or on an area of the specimen that will not be used later for data recording. Once the three modes have been set, only certain adjustments are needed from time to time, for example, objective astigmatism correction. Use Search mode to locate potentially good specimen areas. On each area the recording sequence is: (1) center a suitable crystal in Search mode; (2) focus specimen and set desired defocus in the Focus mode; and (3) record image in Exposure mode. The sequence is repeated on every suitable crystal in that area. Adjust eucentricity on every area that is one or more grid squares away from the preceding one.

Since transitions between modes usually involve large changes in lens excitations one must take steps to minimize hysteresis effects. The following suggestions are recommended:

1. Do a normalization of the relevant lenses after each transition.
2. Execute the transitions between modes always in the same order, i.e., Search → Focus → Exposure → Search. If you want to go, say, from Focus back to Search, do not do it directly, but through Exposure, and blank the beam in the transition to avoid unnecessary exposure.
3. Tune the gun shift/beam shift alignments so as to minimize the beam displacements between the highest and the lowest spot size used.
4. Once the desired beam diameters have been set, avoid using the intensity knob (C2). If adjustments are necessary, make

sure to normalize the illumination system and go one round through the recording sequence (beam blanked, if necessary).

5. Avoid doing image shift adjustments too early after a transition between different magnification ranges. Thermal stabilization after a large change of lens current is slow and the resulting image shifts are high during the first couple of minutes after the change.

3.2.4. Computer Assistance

The precision and reproducibility of image settings is greatly improved by means of computer-aided execution of the recording sequence described above. In modern instruments, there are several levels at which such control can be exercised. (a) The microscope's user interface usually includes a low dose module with a minimal set of options; also, (b) the microscope software accepts external control in the form of scripts; and (c) sophisticated programs can perform lengthy and complicated routines, including image processing tasks to guide the data acquisition process and, in principle, run a fully automated electron microscopy session. In our experience, high-resolution imaging of biological samples still requires direct operator intervention occasionally. An efficient approach combines the best of automated acquisition with careful supervision from the microscopist. In our laboratory, we developed a program, UCSFImage (see Note 1), to assist the user to acquire data efficiently and remotely. UCSFImage is a simple-to-use program that accepts full intervention from users during data acquisition. This sets it apart from programs, such as Leginon, described in Chapter 16, whose goal is to provide a platform for fully automated data acquisition over an extended length of time. Figure 5 shows the user interface of UCSFImage. In Search mode, the user can move the specimen to a new crystal by mouse point-and-click. Only a few short exposures (~0.05 s each) are enough to precisely re-position the specimen, thus controlling the total accumulated dose more stringently than when the operator adjusts the stage position manually based on a live image in Search mode. In Focus mode, the defocus is computed from the power spectrum and the drift rate is computed from the cross-correlation of a sequence of images taken at predefined intervals. Backlash correction is implemented after stage movement. Using multithread programming, such operations can be carried out simultaneously, thus improving data acquisition efficiency.

3.2.5. A Detailed Working Example

Instrument: Tecnai F20 (FEI Company, USA)
Specimen holder: CT3500 (Gatan Inc, USA)
Detectors: F816 (TVIPS, Germany) on-axis and F114 (TVIPS, Germany) off-axis;
Data acquisition software: UCSFimage;
Sample: glucose embedded bacteriorhodopsin 2D crystals.

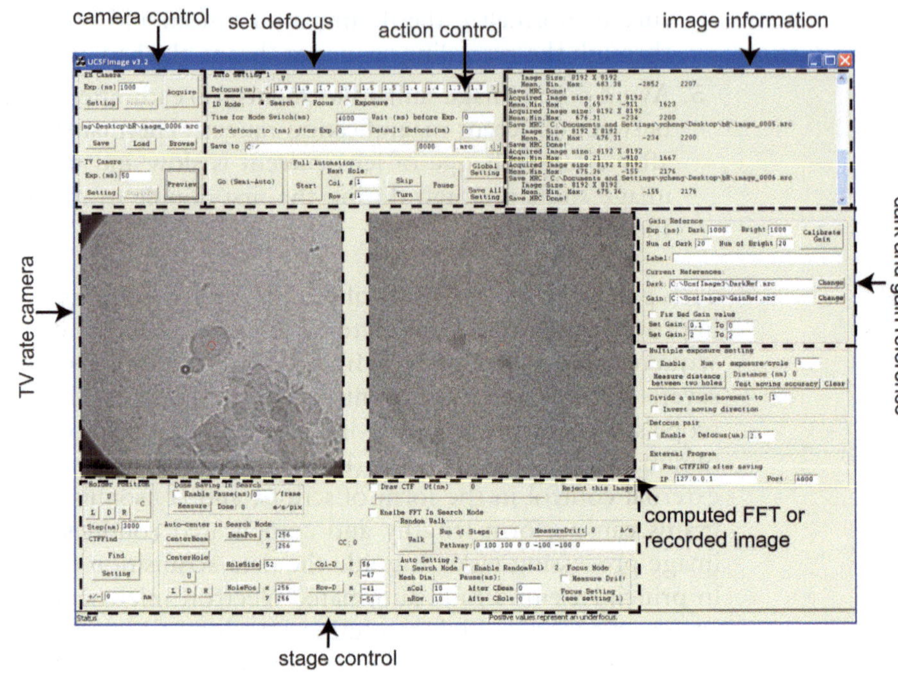

Fig. 5. Illustration of UCSFImage. Function blocks are indicated.

Procedures:

1. Bake cryo-holder at 45°C and cryo-cycle microscope column for 8 h. Verify that the high tension is set to 200 kV, extraction voltage to 4,150 V and gun lens to 3. Note that thermal equilibrium of the lenses can take several hours after changing the high-tension, especially if the change is large.
2. Make sure that the anti-contaminator (cryo-box) is IN and the external dewar has been filled with liquid nitrogen to 3/4 of its capacity.
3. Select and center the condenser aperture (C2) that will be used for data recording, in this example, 50 μm. Leave objective and select area apertures out.
4. Active Low Dose mode on Tecnai user interface menu. Tune the column alignment and set up the low-dose conditions as follows, preferably on a test sample, such as cross-grating or other calibration sample, using a standard, room temperature holder prior to introducing any cryo-sample into the column. Later, on the actual sample, adjust the low-dose setting as needed.
   (a) Go to Exposure and select the working magnification, in this example, 100K×.
   (b) Center the beam using Align Beam shift *x* and *y*. This is actually a direct alignment adjustment. Do not use the default user-dependent Beam Shift.

(c) Expand the beam approximately to the size that will be used for imaging. Go to an empty area of the grid and raise the small focusing screen.

(d) Open Low Dose flap-out window and activate Measure Dose. Select "Read Continuously." Adjust beam diameter until the dose rate is 20 $e^{-}/Å^{2}/s$. If necessary, select another spot size and center it with the Align Beam shift controls.

(e) Perform all necessary alignments here, including coma-free alignment.

(f) Go to Search, select the lowest magnification in the SA range (~5,000×) and select the smallest spot size, 11.

(g) Expand the beam to cover the large viewing screen and switch to Diffraction mode.

(h) Select the largest camera length (6 m).

(i) Focus the central diffraction spot as sharp as possible by using focus knob (objective lens). Increase the size of the disc by spreading the intensity, while at the same time focusing the sharp spot in the middle.

(j) Shift the sharp spot to the center of the disc by using beam shift.

(k) Center the diffraction pattern with the diffraction shift knob.

(l) Spread the sharp central spot by turning the focus knob to overfocus. Observe the specimen features within the diffraction disc. Be sure to use an overfocused diffraction pattern, which generates an image with correct contrast, while an under-focused diffraction pattern generates an image with reversed contrast.

(m) The camera length, and amount of overfocus define the magnification of the image within the central spot. This should be set such that the central beam image shows adequate feature visibility without significant distortion.

(n) Visualizing the dim image in search mode requires enhancement with a photomultiplier or CCD. In the Search Settings page select TV camera and specify its position. If OFF-axis, check the OFF-axis detector shifts for all magnifications and camera lengths to be used and adjust, if necessary.

(o) Select Focus mode and the Focus tab in low dose. Neutralize mode-dependent image/beam deflections.

(p) Set the same magnification and illumination as for Exposure. Set Focus to its standard value by pressing the Eucentric Focus knob on the right-hand side control panel.

(q) Center beam with align-beam controls (direct alignment) and reduce its diameter. Normalize illumination and re-center the beam if necessary. Increase magnification to 280K×.

(r) Select the alternative focus position and neutralize any mode-dependent beam/shift deflections. Set distance and orientation of the spot with either the multi-function knobs or the slide bars of the Focus tab. Orientation is irrelevant for acquisition of untilted views. For images of a tilted sample, place the Focus position on the tilt axis. See Note 2. The Focus position should be as close as possible to the Exposure area, but without any overlaps. On the other hand, since 2D crystals of membrane proteins often extend over several microns, care should be taken to avoid hitting other areas of a given target crystal with the electron beam while focusing. So it might be necessary to keep the distance between Focus and Exposure larger than the minimum required to avoid overlap of the illuminated areas.

(s) Leave the other Focus position on-axis as a resource to use later for the correction of coma and astigmatism.

(t) Go to Exposure mode and re-center beam.

(u) Cycle several times over the three modes and adjust settings until stable. Make transitions always in the same order to minimize hysteresis effects. If hysteresis persist, go to Options menu and try several normalization combinations until transitions are stable and according to the desired conditions. See Note 3.

- The scope is now ready for data recording. The task can be expedited by means of computer software. The following section describes how a program, called UCSFimage, is used for this purpose.

3.2.6. Remote Control and Computer-Assisted Data Acquisition

UCSFImage was originally developed to assist in the collection of large data sets of single particle cryoEM data from frozen-hydrated samples on holey carbon grids, but it can also be used to collect images of 2D crystals. The program uses an off-axis TV rate CCD camera (F114N, TVIPS, Germany) for searching and focusing, and an on-axis large format CMOS camera (F816, TVIPS, Germany) for image recording. With minor modifications, it can also use a single on-axis camera for all low-dose procedures, such as using a K2 camera (Gatan Inc., USA) in a Tecnai F30 Polara microscope (FEI Company, USA). The program has the following capabilities to facilitate image acquisition. In search mode, it usually only needs to record several 50 ms exposures to center the next crystal. This reduces the accumulated electron dose used for repositioning the stage for the next crystal by more than 90%.

In Focus mode, it displays a live image and computes a live FFT from it, thus assists focus and astigmatism correction. It also measures the drift rate of the specimen in Focus mode, and reduces stage drift by means of a build-in backlash correction routine. It saves the recorded image automatically and computes its Fourier transform, which is then available for image quality assessment, thus providing immediate feedback to the user. We are constantly upgrading UCSFImage. Features that will be included in our future version include automatic overall alignment, measurement and correction of objective astigmatism, coma-free alignment, etc. The use of UCSFImage is very simple and does not need extensive calibrations. The procedures are outlined here:

1. Set up Low-Dose as described above.
2. Start UCSFImage program.
3. Acquire dark and gain reference images:
   (a) Either remove the specimen or move the stage to a sufficiently large open area of the specimen, for example, a grid square not covered by carbon film;
   (b) Acquire the dark and gain reference images with the same exposure time that will be used to record images.
   (c) Record an image after acquiring dark and gain references, the image should have a uniform intensity and no patterns or features in the Fourier transform, which should have a smooth Gaussian fall-off.
4. Calibrate the stage movement relative to search view. Move a recognizable fiducial mark to the center of the image. Move the stage for a given distance (for example, 3 μm) along *x*- and *y*-axis, respectively, by clicking the calibration button. Mark the position of the fiducial mark in the image by left clicking the mouse button. A transformation matrix will be calculated that converts any position in the TV image by mouse click to the *x*- and *y*-stage movement.
5. Mark the Focus and Exposure positions on the search view window.
6. Set desired defocus values.
7. Set exposure time in camera control to 50 ms for TV rate camera and 1 s for EM camera.
8. In Search mode, move stage to a specimen area with a sufficiently large number of properly embedded 2D crystals.
9. In Search mode, acquire a 50 ms exposure image with TV rate camera.
10. Select a 2D crystal by mouse click. Stage will be moved to the desired position, and a new TV image (50 ms exposure) will be recorded. If necessary, re-center the target 2D crystal by mouse click again.

11. Click "Go" (semi-auto) button in the action control panel. This will switch to Focus mode. The TV rate camera will record the image in Focus mode and show it in the TV camera widow continuously. A Fourier power spectrum will be computed simultaneously and displayed on the FFT window. A 50 ms exposure offers a TV rate real time image and power spectrum for manual focusing. Focus, astigmatism, and drift values can be measured by clicking corresponding buttons.
12. Once the drift rate has reached the desired level, click on the "GO" button again. This will switch to Exposure mode and record the image. Upon completion of the recording sequence, it will switch the microscope back to Search mode, save the image, and compute the defocus value of the recorded image using CTFFIND (16). All these actions are performed simultaneously by a multi-core CPU.
13. Image quality can be evaluated immediate from the outcome of CTFFIND, displayed on a separate window (not shown here).
14. Continue in the same way with the next image.

4. Notes

1. Obtaining UCSFimage. Currently this program can be downloaded from website http://cryoem.ucsf.edu. Questions and helps can be addressed Yifan Cheng (ycheng@ucsf.edu) and Xueming Li (Xueming.Li@ucsf.edu).
2. Setting focus positions along tilt axis. The tilt axis can be accurately determined from a tilt series prior to any actual data recording. It can be displayed as an overlay on the Search view, on which the imprint of the focused beam is clearly visible (see Fig. 4). The focus position can then be adjusted so that it lies exactly on the displayed axis.
3. Lens normalization options. The user can choose which lens currents to normalize automatically upon transition between any two low-dose modes. Generally speaking, normalizations are required when transitions involve large changes in excitation currents, but the user should determine the best working combinations for a given microscope and low-dose settings. Furthermore, we have noticed that in some cases it is necessary to repeat the normalization. Usually, the following combinations work satisfactorily:

   Search → Focus: Illumination system, Objective Lens, and Projection system

   Focus → Exposure: Illumination system

   Exposure → Search: Illumination system and Projection system

4. General remarks. Procedures for taking low-dose images of 2D crystals are based on protocols and equipment used in our laboratory at this point in time. Obviously, there are many different alternatives to the microscope settings, data acquisition procedures, even the overall philosophy of how to collect best images of 2D crystals. Some of these alternative choices are just personal preferences, such as search in diffraction or image mode. Other choices may reflect the level of technology available for example, in the case of detectors. New advances will keep influencing our choices and the manner in which we collect high-resolution and high-quality images of 2D crystals.

Acknowledgments

Work in the Cheng Laboratory is supported by NIH grants (R01GM082893, R01GM098672, 1S10RR026814-01 and P50 GM082250 (to A. Frankel)) and grants from UCSF Program in Breakthrough Biomedical Research (New Technology Award and Opportunity Award in Basic Science).

References

1. Reimer L (1997) Transmission electron microscopy. Springer, New York
2. Zhou ZH, Chiu W (1993) Prospects for using an IVEM with a FEG for imaging macromolecules towards atomic resolution. Ultramicroscopy 49:407–416
3. Fujiyoshi Y (1998) The structural study of membrane proteins by electron crystallography. Adv Biophys 35:25–80
4. McMullan G, Chen S, Henderson R, Faruqi AR (2009) Detective quantum efficiency of electron area detectors in electron microscopy. Ultramicroscopy 109:1126–1143
5. Booth CR et al (2004) A 9 angstroms single particle reconstruction from CCD captured images on a 200 kV electron cryomicroscope. J Struct Biol 147:116–127
6. Yu X, Jin L, Zhou ZH (2008) 3.88 A structure of cytoplasmic polyhedrosis virus by cryo-electron microscopy. Nature 453:415–419
7. Chen DH, Jakana J, Liu X, Schmid MF, Chiu W (2008) Achievable resolution from images of biological specimens acquired from a 4 k×4 k CCD camera in a 300-kV electron cryomicroscope. J Struct Biol 163:45–52
8. Downing KH (1991) Spot-scan imaging in transmission electron microscopy. Science 251:53–59
9. Morgan DG, Grant RA, Chiu W, Frank J (1992) Patch averaging of electron images of GP3*I crystals with variable thickness. J Struct Biol 108:245–256
10. Faruqi AR, Henderson R, Pryddetch M, Allport P, Evans A (2005) Direct single electron detection with a CMOS detector for electron microscopy. Nucl Instrum Methods Phys Res A 546:170–175
11. McMullan G et al (2009) Experimental observation of the improvement in MTF from backthinning a CMOS direct electron detector. Ultramicroscopy 109:1144–1147
12. Hite RK, Schenk AD, Li Z, Cheng Y, Walz T (2011) Collecting electron crystallographic data of two-dimensional protein crystals. Elsevier
13. Miyazawa A, Fujiyoshi Y, Stowell M, Unwin N (1999) Nicotinic acetylcholine receptor at 4.6 A resolution: transverse tunnels in the channel wall. J Mol Biol 288:765–786

14. Zemlin J, Zemlin F (2002) Diffractogram tableaux by mouse click. Ultramicroscopy 93: 77–82
15. Nakamura N et al (2010) Automated specimen search in cryo-TEM observation with DIFF-defocus imaging. J Electron Microsc (Tokyo) 59:299–310
16. Mindell JA, Grigorieff N (2003) Accurate determination of local defocus and specimen tilt in electron microscopy. J Struct Biol 142:334–347

Chapter 9

The Collection of High-Resolution Electron Diffraction Data

Tamir Gonen

Abstract

A number of atomic-resolution structures of membrane proteins (better than 3Å resolution) have been determined recently by electron crystallography. While this technique was established more than 40 years ago, it is still in its infancy with regard to the two-dimensional (2D) crystallization, data collection, data analysis, and protein structure determination. In terms of data collection, electron crystallography encompasses both image acquisition and electron diffraction data collection. Other chapters in this volume outline protocols for image collection and analysis. This chapter, however, outlines detailed protocols for data collection by electron diffraction. These include microscope setup, electron diffraction data collection, and troubleshooting.

Key words: Electron cryomicroscopy (Cryo-EM), Electron crystallography, Electron diffraction

1. Introduction

In the mid 1970s Henderson and Unwin published their pioneering work using electron crystallography to determine the structure of bacteriorhodopsin. This work established electron crystallography as a *bona fide* technique for membrane protein structure determination (1). In electron crystallography, the membrane protein of interest is crystallized in two dimensions (2D) within a membrane. The environment experienced by the protein closely mimics the native environment of the protein in a cellular membrane (2, 3). Often the protein is functional in these 2D crystals, allowing researchers to investigate both structure and functional rearrangements directly from the very same membranes, as beautifully illustrated in a number of examples (4–6). The methods employed for growing 2D crystals are discussed in detail elsewhere in this volume.

Ingeborg Schmidt-Krey and Yifan Cheng (eds.), *Electron Crystallography of Soluble and Membrane Proteins: Methods and Protocols*, Methods in Molecular Biology, vol. 955, DOI 10.1007/978-1-62703-176-9_9,

Once two-dimensional (2D) crystals are obtained, structure determination ensues either from image analysis of the 2D crystals and/or from analysis of electron diffraction data. Images of 2D crystals contain both phase and amplitude information that are modulated by the contrast transfer function and can be used to determine structures directly, provided the resolution is sufficient. Electron diffraction, however, is not influenced by the CTF nor is it affected by charging effects and drift artifacts, so typically higher resolution data is obtained in diffraction studies.

A number of recent studies used electron diffraction data for the determination of protein structures without the use of image data (7–9), yielding structures whose levels of resolution rival X-ray crystallographic studies. The highest resolution reported to date is 1.7Å anisotropically and 1.9Å isotropically recorded from double-layered 2D crystals of the water channel AQP0 (8). While electron crystallography is capable of delivering atomic resolution information, many of the techniques used to generate such data require years of practice and fine-tuning to master.

In this chapter we outline, in detail, sample preparation protocols for electron diffraction (ED) as well as procedures for collecting high-resolution electron diffraction data. Furthermore, we elaborate upon technical issues including modifications made to the electron microscope itself, the selection and setup of all apertures, and setup for low-dose data collection. Later chapters in this volume concern protocols for phase determination by molecular replacement and structure determination.

2. Materials

2.1. Equipment

An electron microscope (EM) capable of low-dose (minimal dose) protocols (10) and equipped with a cryo box, and a charge-coupled device (CCD) camera. A number of modifications to the standard EM are recommended for electron diffraction studies. Some of these include custom Selected Area (SA) apertures, a slim-line beam stopper, and a free-lens control for selecting the diffraction camera length. The data presented here were recorded on an FEI Tecnai 200 kV EM equipped with a field emission electron source (field emission gun, FEG) and a bottom mount Gatan CCD camera. The setup also includes a bottom mount on axis TV rate camera used for search and focusing functions. If a 200 or 300 kV microscope is not easily available, data can also be collected on a lower kV instrument equipped for cryo. The slim-line beam stopper was manufactured by Fischione (www.fischione.com). Typical SA apertures supplied with FEI EMs are 800, 200, 40, and 10 μm. It is recommended to install different apertures such as 200, 150,

70, and 40 μm. These sizes are usually large enough to record ED from biological 2D crystals and offer the user fine control over data quality and intensity (discussed in Subheading 3.3.4). A cryo holder is also required for these experiments. While the Gatan 626 cryo holder is robust and can tilt to ±70°, the Gatan CT3500 (also known as the Oxford CT3500 cryo holder) is lighter, and thus more stable. The latter can only tilt to ±55°.

2.2. Consumables

The following consumables are required for sample preparation:

- Tweezers (anti-capillary and hooked).
- Glass petri dish.
- Thin layer of carbon evaporated onto a layer of mica.
- MilliQ water, freshly filtered.
- Cryogenic embedding solutions (glucose, trehalose, tannic acid, various PEGs).
- 1–10 μl pipette with tips.
- Filter paper.
- Electron microscopy grids (copper grids are acceptable for untilted specimens, molybdenum grids are recommended for tilt-data collection—see Subheading 3.1).

3. Methods

In this section we outline, in detail, the method used for ED data collection on biological specimens. First we discuss sample preparation, followed by the initial setup of the electron microscope, and finally the actual data-collection process. Examples will be provided, as well as common problems and troubleshooting tips.

3.1. Grid Preparation

Once 2D crystals are grown they must initially be screened by negative staining. Typically 2 μl of the crystalline sample is applied to a freshly glow-discharged electron microscopy grid, washed three times with water and stained with two drops of 0.75% uranyl formate (see Chapters 5 and 15 for additional protocols for negative staining). The grid is then air-dried and viewed in the electron microscope. A sample that is suitable for ED studies should have many crystalline vesicles or sheets containing coherent crystalline areas of ~1 × 1 μm or more in size. The power spectrum of such an area in negative stain should exhibit strong and sharp reflections to at least 20Å resolution (the resolution limit in negative staining). An example is presented in Fig. 1 (5). Once these conditions are met, the sample should be prepared for cryo-EM and ED studies as described below.

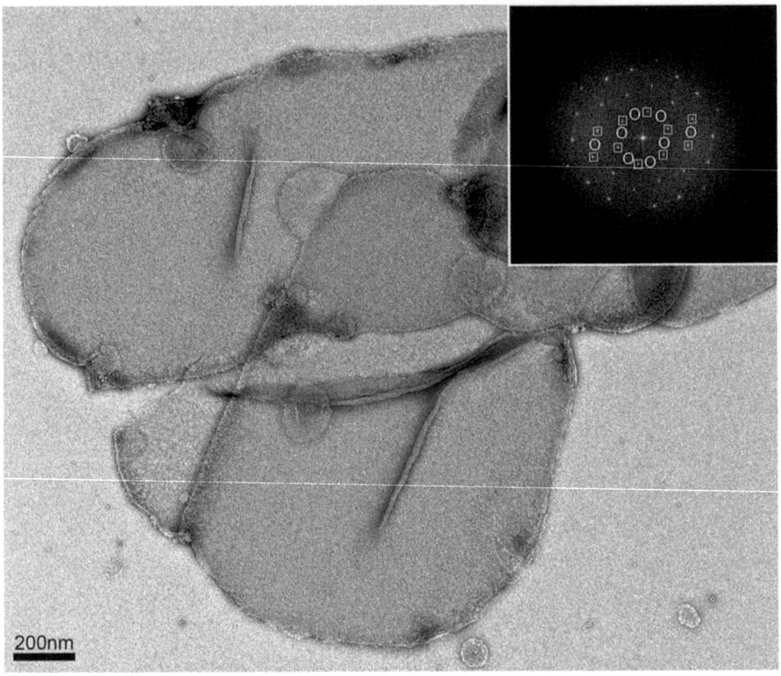

Fig. 1. Negatively stained two-dimensional crystals of the galactose/H^+ transporter GalP. The crystals are large and highly ordered. *Inset*, Fourier transform of the image shows strong and sharp spots to ~18Å resolution. The crystals are vesicular and two sets of spots can be seen in the Fourier transforms (*squares* and *circles*). These crystals are suitable for analysis by cryo-EM and electron crystallography because the crystals are large and the spots in the Fourier transforms are sharp and extend to the resolution limit in negative stain. Figure adapted from (5).

Arguably the most critical step in cryo-EM is the preservation of the sample's finer structure. For electron crystallography, the crystalline vesicles or sheets must be frozen in such a way that preserves the crystalline order within the crystals. Even the very best 2D crystals will not diffract well if the crystals are not frozen properly in a hydrated state (discussed further in Subheading 3.3.1). Two methods are widely used: sugar embedding and vitrification. Our laboratory typically employs sugar embedding, and in the next section we describe the protocol for sample preparation by this method.

In the sugar embedding method, the 2D crystals are mixed with a sugar solution (Note 1) and then applied to an electron microscopy grid. Excess solution is blotted off and the sample is frozen (Fig. 2). The freezing can be done either slowly or very fast. Slow freezing is by far the easiest method. Here the blotted grid is placed into the cryo holder, then inserted into the electron microscope at room temperature, and viewed prior to freezing. An ideal grid should have very few broken grid windows. Each intact grid window should have ideally ten or more usable crystals, and the

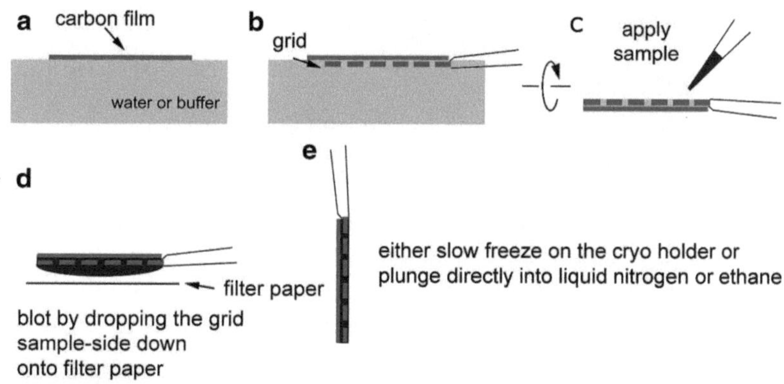

Fig. 2. Preparation of the continuous carbon grid for cryo-EM. (**a**) A thin layer of carbon film evaporated onto mica is floated onto a glass petri dish filled with water or buffer of choice. The buffer can include the cryo protectant of choice such as trehalose, glucose, glycerol, etc. (**b**) The carbon is lifted onto an electron microscopy grid such as that composed of molybdenum or copper. (**c**) The tweezers are turned upside-down and the sample applied via pipette. The grid is then incubated for a few seconds at room temperature to allow the 2D crystals to settle and adhere to the carbon support. (**d**) The preparation is then turned upside-down and blotted directly by dropping the grid onto filter paper. At this stage one can control the eventual embedding by changing the blotting time. (**e**) When blotting is complete, the grid is picked up with tweezers and either placed onto the cryo holder for slow freezing or plunged into liquid nitrogen or liquid ethane for fast freezing.

crystals should appear flat. One can tilt the stage to ~40° to assess whether or not the carbon support is indeed flat. Any wrinkles in the carbon support will affect tilt data (Note 2). If the grid is intended for tilted data collection, it should be discarded if wrinkles are seen. Untilted data are not affected by micro-wrinkles, but grids with very large wrinkles should be discarded. The quality of the embedding is then assessed prior to freezing. A well-embedded crystal will exhibit strong and sharp reflections in power spectra, while a poorly embedded crystal will display smeared reflections, or no spots at all. From good crystals, one should expect to obtain strong reflections to at least ~6Å resolution prior to freezing (Fig. 3). When such a preparation is encountered, liquid nitrogen (LN_2) can be added to the dewar on the cryo holder. This will slowly freeze the grid at the cryoholder tip, a process that usually takes ~30 min.

Alternatively, with the fast-freeze method the blotted grid can be manually plunged into either LN_2 or supercooled ethane immediately after blotting. The grid is then placed into a precooled cryo holder. The temperature of the cryo holder should be lower than −160°C, and preferably stabilized to lower than −180°C for longer than 5 min prior to insertion into the EM (Note 3). The sample preparation can then be viewed and assessed as above. This method is a little more cumbersome because ice buildup on the holder can

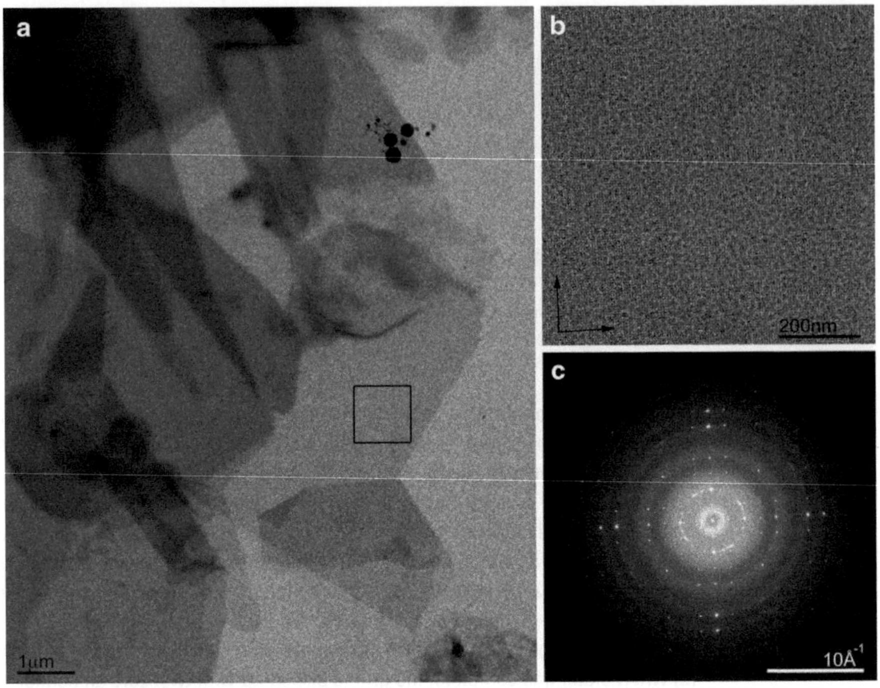

Fig. 3. Cryo EM of two-dimensional crystals. (**a**) Crystals of the water channel aquaporin-0 are large and have sharp edges attesting to the degree of order within. (**b**) High-resolution image of the crystal area highlighted by a *box* in (**a**). (**c**) Fourier transform of the image in (**b**) showing strong and sharp spots to ~6Å resolution. These crystals are ready for analysis by electron diffraction because the crystals appear uniformly *grey* on the grid. The spots in the Fourier transform are sharp and extend to ~6Å resolution without unbending. At this stage the sample should be frozen and the microscope setup should be changed to diffraction and data collected.

be problematic, especially in very humid laboratories. Ice buildup can crash the vacuum on the microscope upon holder insertion.

3.1.1. Protocol for Sample Preparation

A thin layer of carbon is evaporated onto a layer of mica in a Cressington 208°Ccarbon evaporator in a vacuum greater than 2×10^{-6} Torr. An approximately 5 mm × 5 mm square of carbon-coated mica is then cut and floated onto a sugar solution, with the carbon facing up. The mica separates from the carbon leaving behind a thin layer of carbon floating on the water surface (Fig. 2a). An EM grid, grasped on its edge by a pair of anti-capillary tweezers, is then used to pick up the carbon (Fig. 2b). The grid is then turned over and the sample applied directly onto the grid (Fig. 2c). Excess solution is blotted off by dropping the grid face-down onto a piece of filter paper (Fig. 2d). The blotting time can be assayed but a good starting point is 10 s. The grid is then picked up with a pair of tweezers and either cooled in the EM using the slow method or directly plunged into LN_2 or supercooled ethane (Fig. 2e) (Note 4).

3.2. Setup of the Electron Microscope for ED and Data Collection

Most of the data in our laboratory are collected using low-dose electron diffraction on an FEI Tecnai TF20 (200 kV cryo-TEM equipped with an FEG). Low-dose features three settings: search, focus, and exposure (10). The search mode is set as an overfocused diffraction mode allowing the user to examine a relatively large grid area with very little radiation damage to the sample. The beam is set to the smallest possible spot size (on our microscope it is spot size 11). The focus is identified and then overfocused from there to yield a highly contrasted image. Focus mode is an imaging mode (not diffraction) and done at 250,000× magnification at the specimen level. Exposure is in diffraction mode typically using a very small spot size such as 9 or 10. The focus mode is only used once or twice in a session for untilted specimens. For tilted specimens, it is recommended to refocus and realign directly adjacent to the target crystal prior to every ED collected.

When a crystal is identified the Selected Area aperture is inserted in the search mode and diffraction recorded in the exposure mode. The SA aperture is centered over a flat area of the crystal, and data that are recorded will only originate from the area of the crystal that was selected by the SA aperture. Because a small spot size is used, in addition to a small SA aperture, the electron count is quite low. As a result, diffraction data must be recorded over tens of seconds, typically up to 60 s.

Setup of the microscope for diffraction (Note 5):

Setting up the search mode:

1. Choose the smallest spot size.
2. Set magnification in imaging mode to ~15,000×.
3. Turn diffraction on.
4. Change the diffraction length to the largest setting (e.g., 6,000 mm).
5. Focus the beam until you obtain a very sharp spot at the very center of the screen.
6. If the spot is not sharp, use the diffraction astigmators to sharpen the spot.
7. At the end of this procedure (if done properly) the beam should appear as a triangle with a sharp spot at its very center.
8. Overfocus the beam until an image appears on the screen.

Setting up focus mode (Note 6):

1. In search mode use the wobbler to find the eucentric Z-height.
2. Switch to focus mode.
3. Choose spot size 10.
4. Magnification should be set to 250,000×.
5. Set up the distance to ~1.5 μm from the target.

6. Focus the beam crudely at first by using the eucentric focus feature.
7. Focus the beam more finely.
8. Reset focus when the focus is reached (do not defocus at this stage).

Setting up exposure mode:

1. Choose spot size 10.
2. Magnification should be set to ~15,000×.
3. Spread the beam to ~2 cm diameter on the viewing screen.
4. Turn diffraction on.
5. Choose the appropriate diffraction length (see Subheading 3.3.3).
6. Focus the beam until you get a sharp spot at the very center.
7. If the spot is not sharp, use the diffraction astigmators to sharpen the spot.
8. Insert the beam stopper (Subheading 3.3.2).
9. Use diffraction shift to align the diffraction beam with the beam stopper (Subheading 3.3.2).
10. Insert the CCD and record an image to determine the appropriate exposure time (Subheading 3.3.4).

Data collection in ED:

1. Find a crystal that appears flat and not folded in search mode.
2. Place the chosen crystal area at the center of the screen.
3. Insert the SA aperture and center the SA aperture over the chosen crystal area.
4. Insert the beam stopper (see Subheading 3.3.2).
5. Blank the beam.
6. Switch to exposure mode.
7. Raise the screen and cover the camera chamber with the rubber cover that was supplied with the microscope.
8. Unblank the beam.
9. At this stage the CCD is controlling the shutter, so the crystal is not exposed to radiation damage.
10. Record the diffraction using the appropriate exposure time (Subheading 3.3.4).
11. Once the diffraction is recorded, retract the camera.
12. Lower the screen.
13. Realign the beam behind the beam stopper.
14. Remove the beam stopper.

15. Check the beam for focus and diffraction astigmatism.
16. Remove the SA aperture.
17. Return to search mode.
18. Use the diffraction shift to align the exposed area of the crystal with the center of the screen.
19. Go back to step 1 above and repeat the process for collecting ED data of the next crystal.

3.3. Optimization of ED Data Collection

In this section we discuss specific items that need to be optimized for every ED experiment. When conditions for data collection are optimized, the obtainable resolution is far better. We begin by discussing specimen embedding and common problems with crystal flatness, and how to detect these during the ED experiment (Subheading 3.3.1). Next we discuss how the beam stopper is aligned so that maximum information is obtained (Subheading 3.3.2). We discuss the selection of diffraction length as a function of obtainable resolution (Subheading 3.3.3) and the selection of exposure time/saturation of the CCD (Subheading 3.3.4).

3.3.1. Assessment of Crystal Flatness and Embedding by Electron Diffraction

Optimum samples for ED are embedded properly in a cryoprotectant, and the crystals also lay flat against the carbon support. This allows accurate intensities to be recorded and subsequently measured. It is important to identify early in the ED experiment if the crystals are properly embedded, and if the preparation is sufficiently flat. Otherwise hours can be spent on the collection of suboptimal data.

An important objective of sample preparation for analysis by cryo-EM is to ensure the complete preservation of the finer structure of crystals in the sample. This is done by embedding the crystals in a thin layer of a cryoprotectant, as described in Subheading 3.1. Once the grid is prepared and the crystals behave as presented in Fig. 3, the user can set up the microscope for ED and begin recording electron diffraction data according to the protocol in Subheading 3.2.

It is advisable to first record diffraction patterns from untilted crystals to assess the overall quality of the preparation. The best data are always obtained from untilted crystals because these data are not affected by crystal flatness. Indeed, for well-ordered 2D crystals, if quality data are not obtained from untilted crystals the problem is almost certainly from the embedding and cryo preservation of the sample. In such cases, a new grid must be prepared. Once the diffraction from untilted crystals exhibits intense and sharp spots to high resolution, the user can then tilt the stage to 40–60° to assess grid flatness prior to data collection.

There are distinct features one can seek in the electron diffraction patterns of highly tilted crystals that can expedite the identification of flatness and/or embedding issues. Examples are

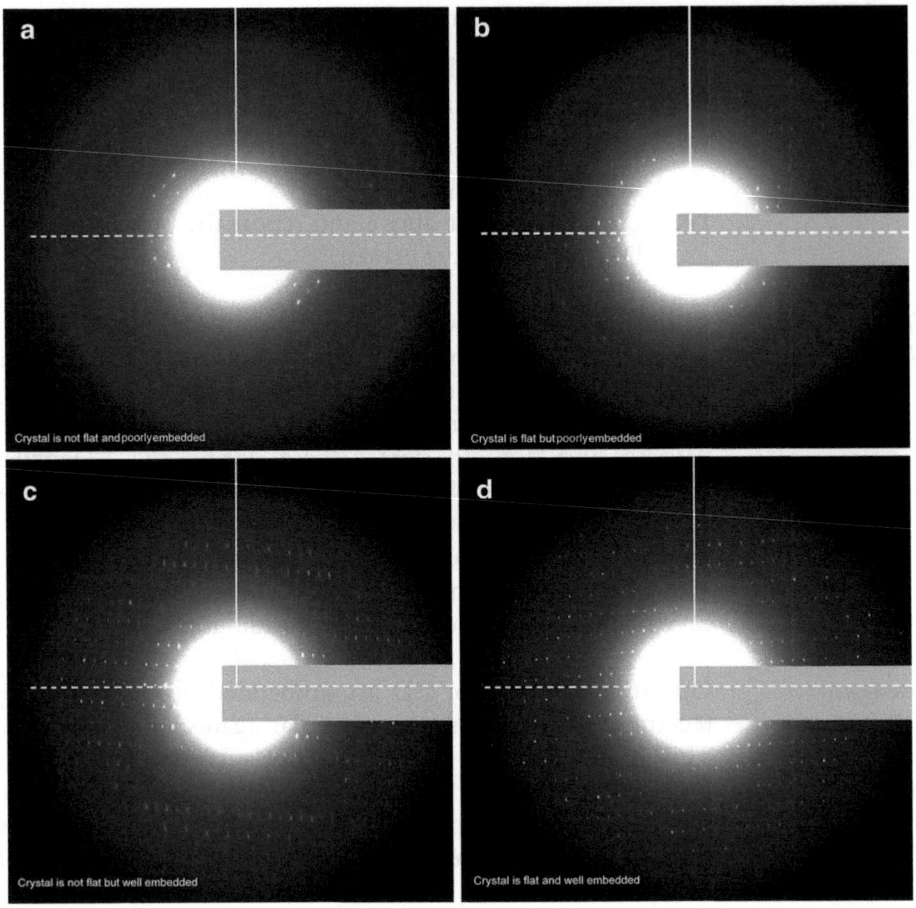

Fig. 4. Assessment of crystal flatness and embedding by electron diffraction. (**a–d**) The flatness of the crystal will affect the attainable resolution. If crystals are not flat the resolution will be cut off in the direction perpendicular to the tilt axis (tilt axis in *dashed line*, perpendicular to the tilt axis in *solid line*). Likewise, a well-embedded crystal will show strong and sharp spots in the diffraction pattern but a poorly embedded crystal will have limited resolution. (**a**) An example of a crystal that is not flat and also poorly embedded. No sharp spots are visible perpendicular to the tilt axis, indicating that this crystal is not flat. Only a limited number of spots are visible on the tilt axis, indicating that the crystal is not embedded properly and was damaged during grid preparation. (**b**) An example of a crystal that is flat but poorly embedded. It is flat because many spots are visible in all directions, even perpendicular to the tilt axis. The crystal, however, is poorly embedded because most of the spots are not sharp and have a weak intensity. (**c**) An example of a crystal that is not flat but is embedded well. The crystal is well embedded because most spots in the diffraction patters appear strong. However, the crystal is not flat because only the spots on the tilt axis are sharp, while the spots perpendicular to the tilt axis are smeared. (**d**) An example of a crystal that is both flat and well embedded. All of the spots in this diffraction pattern are intense and sharp. In all figures the beam stop is masked by the *grey rectangle*.

presented in Fig. 4. The tilt axis is indicated as a dashed line. Any data along this line represent information at low tilt angle. The quality of data along the tilt axis provides information about crystal embedding. When a crystal is optimally embedded, spots along the tilt axis will be strong and sharp. The tilt data can be found perpendicular to the tilt axis, represented by a solid line in Fig. 4. Data along this axis provide information about crystal flatness. When the crystal is perfectly flat, the spots along this axis will be strong and sharp. With such information in hand, one can immediately identify when a crystal is both poorly embedded and not flat (Fig. 4a). If more than 20 consecutive crystals provide this sort of information, it is advisable to prepare a new grid and modify the embedding conditions. Using the guidelines above it is also straightforward to distinguish between crystals that are flat but poorly embedded (Fig. 4b) and crystals that are not flat yet embedded well (Fig. 4c). Crystals that are flat but poorly embedded will have sharp diffraction spots along both axes but the spots will not appear strong (Fig. 4b). Crystals that are not flat but are embedded well will have strong spots in all directions. The spots along the tilt axis will appear sharp but the ones perpendicular to the tilt axis will appear smeared (Fig. 4c). Finally, when the perfect grid is obtained the diffraction spots will appear strong and sharp, both on the tilt axis and in the direction perpendicular to the tilt axis (Fig. 4d).

3.3.2. Alignment of the Beam with the Beam Stop and with the CCD

When an optimal grid is obtained one must ensure that all possible data are recorded, and none are lost to simple alignment issues. There are three things that need to be perfectly aligned with respect to one another: the beam, the beam stop, and the center of the CCD (Note 7). The beam stop must be directly in front of the central beam. In electron diffraction the intensity of the central spot is much larger than all of the diffracted spots, so it is important to protect the CCD from overexposure and saturation that can cause bleeding and damage. The alignment of the central beam with the beam stop therefore ensures that the beam stop will protect the CCD from damage. The beam and the beam stop then need to be placed at the center of the recordable area of the CCD so that all possible information will be recorded evenly in all directions. Common alignment problems are illustrated in Fig. 5. In the first two examples, the beam (indicated by ×) and the beam stop are well aligned with respect to one another, but not with the recordable area of the CCD. In the first example the beam and the beam stop are too far to the right and information is lost on the *x*-axis (Fig. 5a). In the second example the beam and the beam stop are aligned too far to the top and information is lost in the *y*-axis (Fig. 5b). The beam can be centered in respect to the CCD by using diffraction shift, while the beam stop can be centered using the manual controls on the microscope. In Fig. 5c, both the beam and the beam stop are roughly at the center of the recordable area of the CCD but the beam is not aligned with the beam stop. As a result, the CCD is oversaturated and “bleeding” is observed. The diffraction shift can then be used again to center the beam behind the beam stop for perfect alignment, as presented in Fig. 5d.

3.3.3. Selection of Diffraction Length

Another critical factor to consider in an ED experiment is the selection of the diffraction length (Note 8). Once the beam and the

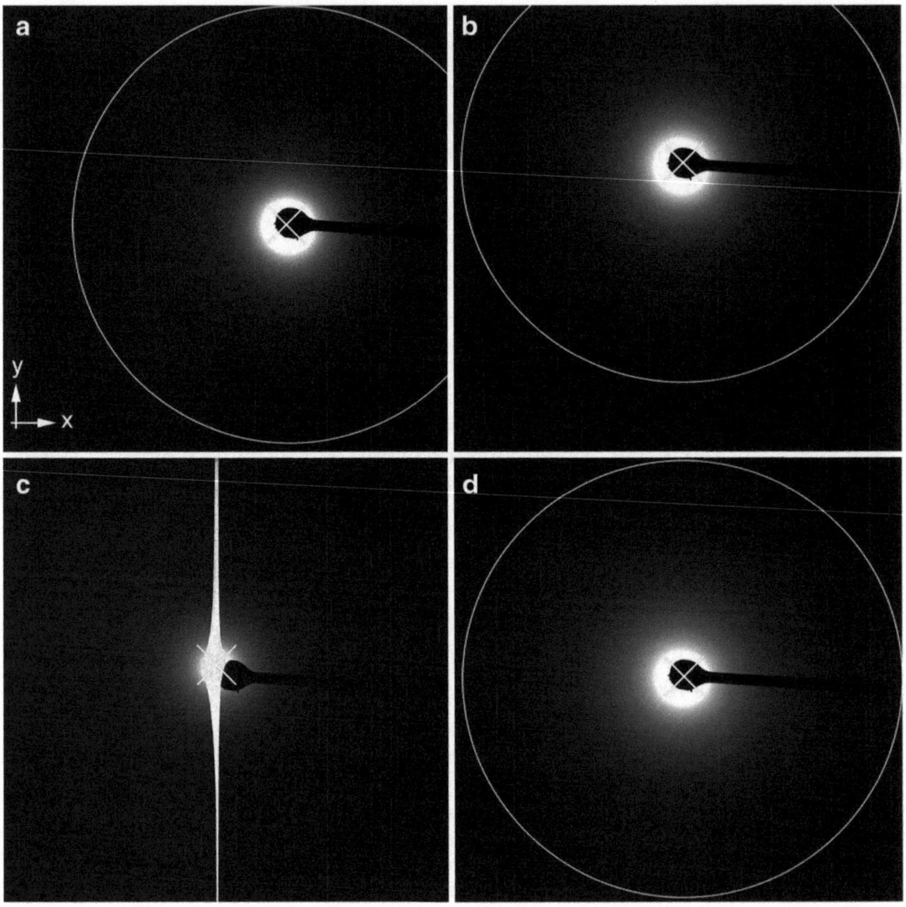

Fig. 5. Alignment of the beam with the beam stop and with the CCD. It is important to align all three elements well; the beam must be at the center of the recordable area of the CCD so that no information will be lost, and the beam stop must be aligned with the beam to prevent damage to the CCD. (**a** and **b**) An example of a beam (indicated by an "*X*") that is well aligned with the beam stop. The beam, however, is not at the center of the recordable area of the CCD and information on the *x*-axis or the *y*-axis is lost (**a** and **b**, respectively). (**c**) An example where the beam is aligned roughly at the center of the recordable area of the CCD but is not aligned with the beam stop. As a result the CCD is oversaturated and bleed-through in the *y*-axis is observed. (**d**) Perfect alignment. The beam is at the center of the recordable area of the CCD and directly behind the beam stop. Data to the same resolution can be recorded in all directions as indicated by the *circle*. In all panels the *white X marks* the center of the beam.

beam stop are aligned with the center of the recordable area of the CCD, one must select the correct diffraction length to optimize the quality of the obtainable data. The larger the diffraction length the lower the recordable resolution; however, the spaces between diffraction spots become larger, which facilitates indexing and background subtraction during data processing. Figure 6 illustrates three examples of diffraction data collected from aquaporin-0 2D crystals with three different camera lengths. The white circle indicates the resolution cutoff. Using the 4,000 mm diffraction length, the attainable resolution is ~10Å (Fig. 6a). The diffraction spots are well separated, and accurate intensities can be extracted.

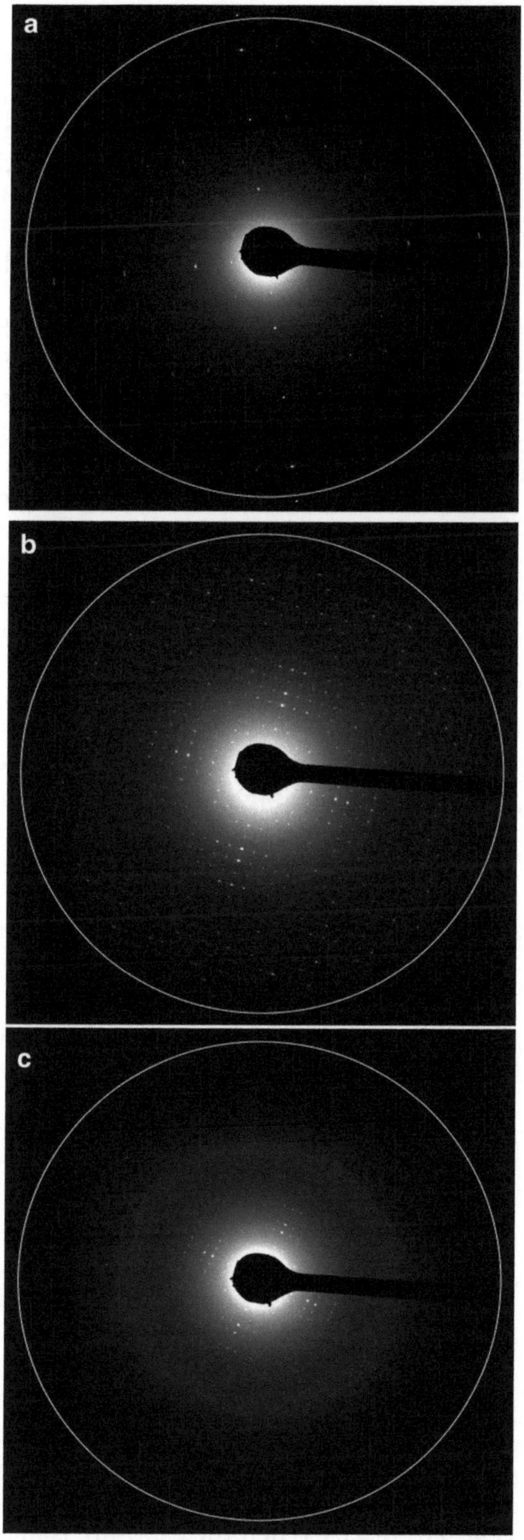

Fig. 6. Selection of diffraction length. As with X-ray crystallography, the diffraction length in electron diffraction influences the attainable results. The longer the length, the lower the recordable resolution is. (**a–c**) Electron diffraction patterns recorded from untilted aquaporin-0 2D crystals with 4,000 mm, 3,000 mm, and 1,750 mm diffraction length lenses, respectively. The attainable resolution is 10Å, 3Å, and 1.5Å, respectively, as indicated by the *white circle*. As the camera length changes the diffraction spots also spread apart or condense. In this particular example the 3,000 mm lens is the ideal one to use because these particular crystals do not appear to diffract significantly better than 3Å under these embedding conditions. Data are therefore lost in (**a**) and no additional data are gained in (**c**).

However, it is clear that the crystals can diffract to even higher resolution because there are some strong and sharp spots that extend all the way to the edge of the recordable area of the CCD.

The resolution cutoff becomes ~3Å when the 3,000 mm diffraction length is used (Fig. 6b). The diffraction spots appear much closer together but are still separated enough to allow accurate determination of the intensities. No spots are visible at the edge of the recordable area of the CCD. The attainable resolution limit is ~1.5Å when the 1,750 mm diffraction length is used (Fig. 6c). The spots appear much closer together, rendering indexing and intensity estimation difficult. Diffraction spots do not extend all the way to the edge of the white circle, so nothing is gained by changing the camera length from 3,000 to 1,750 mm. It is recommended to consider the camera length very carefully so that the best resolution is obtained and the extracted intensities are accurate. In the example presented here, the 3,000 mm diffraction length is ideal because the spots are well separated yet do not extend to the edge of the CCD, so no information is lost (Fig. 6b).

3.3.4. Optimization of Electron Dose and CCD Saturation in Electron Diffraction

A final consideration is electron dose and CCD saturation. When the electron dose is high, the CCD is saturated and accurate determination of spot intensities from diffraction patterns becomes impossible, thus rendering data useless. On the other hand, if the dose is low the beam becomes very weak and background noise increases, so the determination of spot intensity yet again becomes difficult. Moreover, high-resolution data that usually appear as weak spots may vanish in the high noise.

Different CCD cameras have different saturation points that the manufacturer should be able to provide upon request. However, these saturation points can, and should, be determined experimentally during setup of the electron beam prior to ED data collection. Once the beam is set up for diffraction and is aligned with the beam stop and with the CCD, one should record an exposure over 1 s, as illustrated in Fig. 7a. The intensity can then be displayed as a plot of counts versus distance (Fig. 7a, Inset). A dose that is not oversaturated will appear as a Gaussian curve. In this particular example, however, the counts are very low and the background noise is rather high. Without adjusting the beam settings one can increase the dose by increasing the exposure to 30 s or longer (Fig. 7b). Now the background appears more uniform and the counts are quite high, yet the CCD is not oversaturated and the dose plot still appears as a Gaussian curve (Fig. 7b, Inset). When the exposure is increased to 60 s the CCD is clearly oversaturated, as bleeding is observed. The dose plot here no longer appears Gaussian. In these examples the ideal setting was 30 s, as in Fig. 7b.

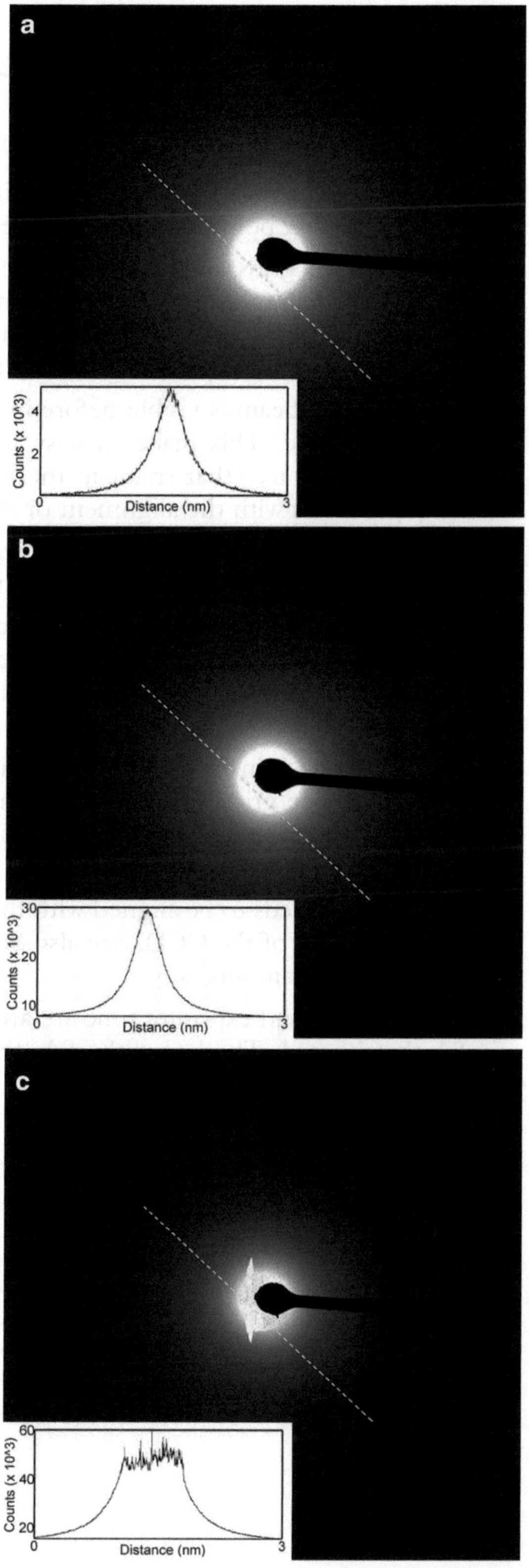

Fig. 7. Optimization of electron dose and CCD saturation in electron diffraction. (**a**–**c**) The intensity over the line drawn is presented in the *inset* as counts versus distance. (**a**) Undersaturation. The beam is weak and as a consequence the background is noisy. The peak intensity *inset* is symmetric but appears noisy and is below the saturation, as indicated by the low counts. (**b**) Optimal dose. The peak intensity *inset* appears symmetric and has high counts. (**c**) Oversaturation. The peak intensity is too high and the CCD is oversaturated. No accurate intensities can be obtained with such settings.

4. Notes

1. Any number of sugars and sugar combinations can be assayed. A good starting point is 3–20% glucose or 3–20% trehalose.
2. Molybdenum grids are recommended for tilted data collection. Molybdenum shrinks less at low temperatures than copper, resulting in fewer micro-wrinkles on the carbon support. Copper grids are acceptable for untilted data collection.
3. Ensure that a beam is visible before inserting a cryo grid into the microscope. This makes it easy to distinguish problems with too thick ice (that result in the beam not being visible) from problems with the alignment or the electron source.
4. The double-carbon, or carbon-sandwich, technique (11) is described in detail elsewhere in this volume.
5. You will need to repeat this entire process three to four times because of hysteresis. Once the optics are stable, there should not be much variation as one changes from search to focus to exposure.
6. For data collection from untilted specimens the focus is only used once per grid. For tilted specimens repeat the entire procedure with focus before performing ED on every single crystal.
7. The SA also needs to be aligned with the beam, the beam stop, and the center of the CCD. See also Subheading 3.3.4—optimization of beam intensity.
8. The intensity and exposure time are affected by the size of the SA that is used. The larger the SA, the higher the dose at a constant exposure time.

Acknowledgments

We thank Noah Shuart (University of Washington) for critically reading this manuscript. Research in the Gonen laboratory is supported by the American Diabetes Association Award # 1-09-CD-05 and by the National Institutes of Health R01GM079233 and U54GM094598 as well as the Howard Hughes Medical Institute.

References

1. Henderson R, Unwin PN (1975) Three-dimensional model of purple membrane obtained by electron microscopy. Nature 257:28–32
2. Andrews S, Reichow SL, Gonen T (2008) Electron crystallography of aquaporins. IUBMB Life 60:430–436
3. Reichow SL, Gonen T (2009) Lipid-protein interactions probed by electron crystallography. Curr Opin Struct Biol 19:560–565
4. Miyazawa A, Fujiyoshi Y, Unwin N (2003) Structure and gating mechanism of the acetylcholine receptor pore. Nature 423:949–955
5. Zheng H, Taraska J, Merz AJ, Gonen T (2010) The prototypical H+/galactose symporter GalP assembles into functional trimers. J Mol Biol 396:593–601
6. Murata K, Mitsuoka K, Hirai T, Walz T, Agre P, Heymann JB, Engel A, Fujiyoshi Y (2000) Structural determinants of water permeation through aquaporin-1. Nature 407:599–605
7. Gonen T, Sliz P, Kistler J, Cheng Y, Walz T (2004) Aquaporin-0 membrane junctions reveal the structure of a closed water pore. Nature 429:193–197
8. Gonen T, Cheng Y, Sliz P, Hiroaki Y, Fujiyoshi Y, Harrison SC, Walz T (2005) Lipid-protein interactions in double-layered two-dimensional AQP0 crystals. Nature 438:633–638
9. Ho JD, Yeh R, Sandstrom A, Chorny I, Harries WE, Robbins RA, Miercke LJ, Stroud RM (2009) Crystal structure of human aquaporin 4 at 1.8 A and its mechanism of conductance. Proc Natl Acad Sci USA 106:7437–7442
10. Fujiyoshi Y (1998) The structural study of membrane proteins by electron crystallography. Adv Biophys 35:25–80
11. Gyobu N, Tani K, Hiroaki Y, Kamegawa A, Mitsuoka K, Fujiyoshi Y (2004) Improved specimen preparation for cryo-electron microscopy using a symmetric carbon sandwich technique. J Struct Biol 146:325–333

Chapter 10

Image Processing of 2D Crystal Images

Marcel Arheit, Daniel Castaño-Díez, Raphaël Thierry, Bryant R. Gipson, Xiangyan Zeng, and Henning Stahlberg

Abstract

Electron crystallography of membrane proteins uses cryo-transmission electron microscopy to image frozen-hydrated 2D crystals. The processing of recorded images exploits the periodic arrangement of the structures in the images to extract the amplitudes and phases of diffraction spots in Fourier space. However, image imperfections require a crystal unbending procedure to be applied to the image before evaluation in Fourier space. We here describe the process of 2D crystal image unbending, using the 2dx software system.

Key words: Membrane proteins, Electron crystallography, 3D structure determination, Electron microscopy, Image processing

1. Introduction

Cryo-transmission electron microscopy (TEM) of biological specimens suffers from a very low signal-to-noise ratio in the recorded images, which makes it difficult to recognize the exact localization and orientation of smaller molecules in the images. Crystalline arrangement of the proteins allows overcoming this problem by exploiting the structural repetition and extracting the common signal from multiple noisy images. The crystallization of the proteins in two dimensions ensures that the proteins lie on a plane and are oriented in the same direction. Tilting this plane in the TEM results in different views of the proteins, which can be combined through image processing into a three-dimensional (3D) reconstruction of the protein.

Ingeborg Schmidt-Krey and Yifan Cheng (eds.), *Electron Crystallography of Soluble and Membrane Proteins: Methods and Protocols*, Methods in Molecular Biology, vol. 955, DOI 10.1007/978-1-62703-176-9_10,

Image processing of two-dimensional (2D) crystal data needs to cope with several problems: deviations from perfect crystallinity have to be recognized and corrected, as well as perturbations introduced by the imaging system. Specific image analysis concepts and algorithms have been developed over the years that accomplish these tasks. In this chapter we give an overview of the methods used to process 2D crystal images. We introduce the algorithms and software systems for image processing of 2D crystals, explain the methods involved, and present a guide for processing an image with the help of 2dx.

2. Software Resources

The field of electron crystallography of membrane proteins was created through the work of Richard Henderson and Nigel Unwin (1, 2). Their developed algorithms were made available to the public in the so-called MRC programs for image processing (3–6). Electron crystallography has produced atomic models for seven transmembrane proteins and tubulin so far: BR (5), LHCII (7), AQP1 (8, 9), nAChR (10), AQP0 (11, 12), AQP4 (13, 14), MGST (15), and Tubulin (16). The 3D map from double-layered 2D membrane crystals of AQP0 from the Walz laboratory at 1.9 Å resolution not only showed the well-resolved lipids in the membrane around the membrane proteins, but even allowed localizing water molecules embedded in the hydrophobic core of the membrane (12).

The data for all these structures were processed with the MRC programs. In the majority of these projects, several years of work were needed to construct a high-resolution map from existing 2D crystals. AQP0 and AQP4 were solved from electron diffraction patterns alone, using molecular replacement methods. Their structure determination could be done significantly faster. It did not require the computer processing of recorded images, since the homologous structural model of AQP1 was used for homology modeling.

The MRC software is composed of a collection of standalone programs, most of them written in Fortran-77, which contain a rich repertoire of algorithms and solutions. The MRC programs are an implementation of brilliant algorithms and theory, so that it comes as no surprise that their usage is not simple. Detailed knowledge about these programs can best be gathered from other advanced users, or by "reading" the Fortran-77 code itself. The MRC programs also include a graphical image display program called Ximdisp (17).

The SPECTRA software from the ICE package was a program that generated Unix Shell scripts that launched an early version of the MRC programs (18, 19). SPECTRA contained its own image visualization program, and facilitated the usage of the MRC programs significantly. To our knowledge, SPECTRA is no longer supported. Similar functions are found in the GRIP system by Wilko Keegstra in the University of Groningen (unpublished), and Sven Hovmoeller's Calidris software package, which reproduces some of the MRC functionality in a commercially distributed PC software package (http://www.calidris-em.com/). A recent effort is the Image Processing Library and Toolbox (IPLT), a new software system that was introduced by Ansgar Philippsen in the laboratory of Andreas Engel in Basel. IPLT is written in C++ and Python (20, 21), and is available at http://iplt.org.

Other program systems that can partly be employed for the image processing of 2D crystal images of membrane proteins are XMIPP (22, 23), Spider (24), bsoft (25), and others. A collection of links to the different systems can be found at http://en.wikibooks.org/wiki/Software_Tools_For_Molecular_Microscopy.

We here describe the processing of 2D crystal images with the 2dx software package (26, 27). 2dx is based on the MRC programs, which are run as background programs behind a graphical user interface (GUI). The installer for 2dx includes an adapted version of the MRC programs, which have been modified to interface with the 2dx GUI, and in some cases additional functions were added to the MRC programs. 2dx features a user-friendly GUI that assists the user in the management of an image processing project, that provides user-guidance, streamlines respectively automates parts of the image processing, and allows detailed monitoring of the processing progress and results. 2dx maintains a data structure in form of simple text files, where image-processing parameters for each image, as well as for the entire processing project are stored. 2dx also includes a single-particle processing module for 2D crystal images, which uses a maximum likelihood (ML)-based algorithm. This method may give a significantly better result, when the crystal quality does not allow high-resolution work through the unbending algorithm. This ML software is fully integrated into the 2dx package, and produces results as Fourier-space amplitude and phase values, to allow merging with other MRC-based processing results in Fourier space. We distribute 2dx as open-source software under the GPL at http://2dx.org.

3. Processing of Individual Images

The processing of 2D crystal images involves several steps, some of which are listed in the simplified block diagram in Fig. 1.

The processing involves the following steps:

1. Defining basic processing parameters
2. Calculating the Fourier transform of the image

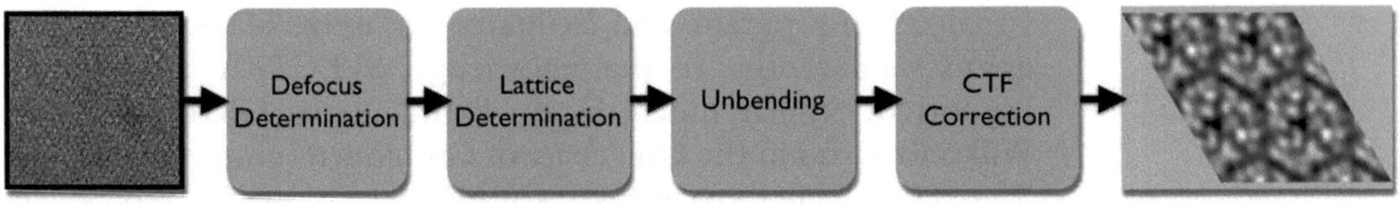

Fig. 1. Block diagram of the processing flow for one image.

3. Measuring the defocus in the image center, and in different locations in the image
4. Calculating potential specimen tilt from the defocus gradient
5. Determining the 2D crystal lattice
6. Refining potential specimen tilt from distortions of the 2D crystal lattice
7. Determining a spotlist of significant Fourier reflections
8. Determine lattice distortion vectors and perform a first image unbending (Unbend I)
9. Iteratively refine the lattice distortion vectors, do refined unbending (Unbend II)
10. Extract amplitude and phase values for each Fourier reflection from the unbent image
11. Correct the list of amplitudes and phases for the instruments CTF
12. Calculate a final projection map from this image

These steps are now discussed in detail, using the 2dx software package as an example.

3.1. Defining Basic Processing Parameters

Chapter 11 describes the management of a 2D crystal image-processing project. We refer to that chapter for the initialization of the processing environment, how to import images into 2dx, and how to open them with *2dx_image*.

Upon opening an image with *2dx_image*, the user needs to define the basic processing parameters that are already known for this project. These concern the magnification under which the image was recorded on the microscope, and the pixel size of the recording medium. *2dx_image* will then calculate the image pixel size from these values. Electron microscope parameters also need to be defined correctly, concerning for example the acceleration voltage, the spherical aberration (Cs) value of the objective lens, and some other parameters. As described in Chapter 11, these values can be saved as project default values, so that for other freshly imported images they are already set correctly.

3.2. Calculating the Fourier Transform of the Image

The technical quality of an image should be verified before starting the image processing. Depending on the method by which the image was recorded, it may be available in byte (8-bit unsigned), short integer (16-bit unsigned), or real (32-bit floating point) modus. Conversion of an image from a recording medium (TIFF or MRC are currently supported by 2dx) to the internal file format (2dx uses the MRC 2000 format) is trivial, once it is implemented, but may pose a problem the first time a new image source is used. Typical errors occur when images are saved as 16-bit signed integer, but are interpreted as 16-bit unsigned integer values, or vice versa. Such images then show the darkest pixels in the displayed maps as white pixels. Another possible source of error is that images recorded on film might be scanned with a film scanner as positive or negative film transmission, instead of the film's optical density (OD). Only the latter is proportional to the protein density on the film so that the logarithm of the pixel values in the transmission files need to be calculated in order to produce OD values.

The first script "Calculate FFT" of `2dx_image` produces among other files a pixel histogram of the input image. The histogram should show a symmetric, Gaussian peak that is centered somewhere in the middle of the histogram and that is not truncated at either end. If the histogram is not symmetric but tilted to either side, then the data are most likely film transmission values that have not yet been transferred into OD values. If the histogram is truncated at either end, then the film digitization or file format transfer was not done correctly, or the image is over or under exposed. This script also calculates down-sampled versions of the input image, and Fourier transforms of the original and the down-sampled images.

3.3. Measuring the Defocus in the Image Center

The recorded image is not a 1:1 projection of the specimen. Instead, it rather represents a collection of measured numbers, which in their entirety contain the information about the high-resolution structure of our sample. The imaging process in the TEM is a complex phenomenon, which can be mathematically described by a contrast transfer function (CTF) (28). This function describes in Fourier space, how different resolution components of the structure were transferred onto the image: The amplitude of the Fourier transformed image underwent strong modulations, and the phases were shifted by 180° where the CTF had a negative value, and may have been affected by additional alterations due to beam-tilt (29–31). The resolution rings where the CTF is of value zero are the so-called Thon rings (32). The CTF is dependent on the defocus under which the image was recorded and therefore strongly varies across an image of a tilted specimen (5, 33).

For 2D membrane protein crystals that are weak phase scatterers to the electron beam, the CTF is an oscillating function that depends among other factors on the defocus and astigmatism of the objective lens, the electron wavelength (and therefore the acceleration voltage), and the spherical aberration constant (Cs) of the objective lens. The CTF is dampened by an envelope function, which depends on the chromatic and spatial coherence of the electron beam, and which is affected by specimen drift and vibration during the electron exposure. In addition, our micrograph is covered by a high amount of noise, which stems from the low number of electrons used to record the image (shot noise), and detector noise. In addition, the recording medium usually modulates the image with a further transfer function, the detector modulation transfer function (MTF), due to the extended point spread function of the electrons on the detector material.

For the processing of a 2D crystal image, correct definition of the acceleration voltage and Cs value are needed, and the defocus and astigmatism for each image have to be measured experimentally from the image. This, together with knowledge about the beam tilt (fitted during the merging process), allows defining the resolution rings in Fourier space (Thon rings), where the contrast reversal occurs.

As described in Subheading 3.9 below, these contrast reversals will be corrected by phase-flipping, which adds 180° to the phase values of the affected zones in Fourier space. The effects of the CTF's amplitude oscillations can be computationally corrected, by multiplication of the amplitudes with a corrective factor. A correction for the dampening envelope function on the CTF and for the detector MTF is usually approximated by employing a negative temperature factor during calculation of the final map, which increases the amplitude of the higher-resolution components.

To determine the defocus and astigmatism one can use the program CTFFIND3 (34). The program's objective is to fit the oscillatory function of the measured power spectrum. It is also integrated in the 2dx program suite and used in its "Determine Defocus" script. CTFFIND3 subtracts from the power spectrum the smooth background estimated from the image. It then fits a two-dimensional CTF to the background-free power spectrum to determine the defocus and astigmatism. In CTFFIND3 the fit is found by maximizing the correlation coefficient of the observed power spectrum and the estimated squared CTF. The maximum correlation is found by doing a grid search of defocus values and astigmatic angle. In 2dx the resulting defocus and astigmatism can be examined and manually adjusted in the power spectrum of the image. The defocus is defined in Ångströms in two orthogonal directions *a* and *b* in the image, while a third parameter defines the angle in degrees between the horizontal *X*-axis and the direction *a*. As always, a positive angle describes a counterclockwise rotation.

3.4. Measuring Potential Specimen Tilt

The defocus is a measure for the distance between the sample and the focal plane of the objective lens. In the case where we record images of tilted samples in the TEM, the defocus varies across the image, due to the varying distance to the objective lens. This defocus variation can be used to determine the tilt geometry of the sample.

The program CTFTILT (34) is used to determine tilt axis and angle by measuring the defocus at several locations of the image. The tilt axis is defined by a line with constant defocus in the image. In practice the tilt axis direction is determined by minimizing the variance of defoci between sub-tiles in the image. Once the tilt axis is estimated, the tiles along the tilt axis are used to determine the defocus and astigmatism for the axis. Having found the tilt axis allows determining the tilt angle. Therefore, CTFTILT performs a search on the tiles of the entire image, which are assumed to lie in a single (tilted) plane. Thus the defocus can be formulated as a function of its image coordinates and the angle between tilt axis and *X*-axis and the tilt angle. The search range for the tilt angle is usually between ±65°. In the end, the estimated values for defocus, astigmatism, tilt axis direction and tilt angle are refined by maximizing the correlation between observed power spectra and the CTF as described earlier with CTFFIND3. An alternative way to determine the tilt geometry is also implemented in `2dx_image`, which cuts the image into 7×7 sub-images, on which the script uses CTFFIND3 to calculate the central defocus on each sub-image. A separate function then fits a tilted plane to those 49 defocus measurements, while excluding outliers from the fitted plane.

The tilt geometry for the specimen plane is defined by two variables, TLTAXIS and TLTANG, in degrees. TLTAXIS defines the angle between the horizontal *X*-axis and the tilt axis in the recorded image, where positive values describe counter-clockwise rotation. (A value of TLTAXIS = +30 defines a tilt axis that is "pointing at 2 o'clock.") TLTAXIS is defined between −89.999° and 90.000°. TLTANG defines the tilt angle around this axis. In the MRC programs and in 2dx, the origin of an image is considered bottom left. The image pixel in the bottom left corner has the coordinates (0,0). Pixels at the bottom of the image starting bottom left and walking right are defined as (1,0), (2,0), (3,0), …, (*x*,0). If the tilt axis is horizontal (TLTAXIS = 0), and the underfocus gets stronger as you go up the image (towards higher *y* values), then TLTANG is positive. TLTANG is defined between −89.999° and 89.999°. Note that TLTAXIS and TLTANG are independent of any crystal lying on the sample plane.

3.5. Determining the 2D Crystal Lattice

Image processing for single particle electron microscopy and electron tomography requires localizing proteins or other image features in the images, which usually is done by cross-correlation of the image with a reference image (Fig. 2). This works for large protein complexes, but becomes difficult for small particles, if the image has a low signal-to-noise ratio (SNR). In electron crystallography, the alignment of the proteins is done biochemically, by inducing 2D crystal formation. Even with a very low SNR in the images, the averaging of the signal from several proteins is then possible, because the image processing can rely on the protein

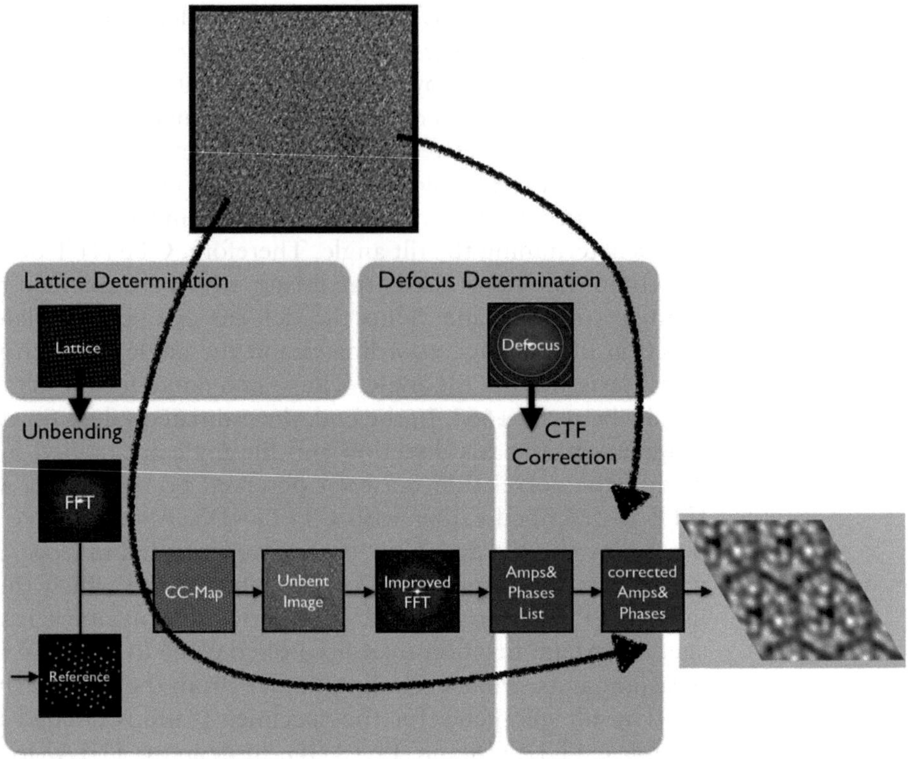

Fig. 2. Flowchart of the image processing steps of a single 2D crystal image.

being approximately at the predicted position. This makes it possible to get high-resolution structural data from small, crystallized proteins, even though the images of the individual proteins have an SNR so low that they would be very difficult to detect individually in the image.

The periodic alignment of the projected proteins in the 2D crystal image is the cause of diffraction peaks in the calculated Fourier transform of the image. The basic unit cell repeat in the image defines the first order diffraction in the Fourier transform. Higher-resolution details of the protein structures cause the lattice in the Fourier transform to show also higher-order diffraction spots. The Fourier transform of a good cryo-EM image of a perfectly ordered 2D crystal will show a lattice that reaches with sharp diffraction spots out to very high resolution.

In practice, however, the 2D crystals are not perfect. Crystals have defects, in form of dislocations in individual unit cells, or distortions from shear stress or other stress on the crystalline lattice. Crystals can be polycrystalline, so that several "2D crystal grains" form a patchwork. In addition, the electron microscope can introduce image distortions, often in form of a pincushion or spiral distortion, which then not only result in displacement of the proteins from the predicted lattice, but also induce slight variations of the image magnification. The Fourier transform of such distorted crystals then only shows low-resolution diffraction peaks, while the higher-resolution peaks are smeared out, and very high-resolution peaks are so strongly smeared out that they become invisible to our detection methods.

Richard Henderson et al. introduced the concept of 2D crystal lattice unbending. The lattice distortions in the image are recognized by software, and computationally corrected, by compiling a corrected output image that is constituted with segments from specific image locations in the input image. The Fourier transform of the resulting "unbent" image would then again show very high-resolution diffraction spots, which can be evaluated for their amplitude and phase information.

For this to be possible, the first step is to find and define the lattice. This is best done in reciprocal Fourier space, by a process called "Lattice Indexing." As can also be done with `Ximdisp` (17), *2dx_image* features a full-screen browser that allows manually indexing a reciprocal lattice. In a displayed Fourier transform in *2dx_image*, the user would manually click on an identifiable diffraction peak, enter its Miller index coordinates, repeat the same for a few more linearly independent peaks, and have the software calculate the best fitting reciprocal 2D crystal lattice. From then on, the user can add more diffraction peaks, and the software will propose the corresponding Miller index automatically.

2dx_image also includes two different algorithms to automatically determine the reciprocal lattice. Both algorithms start by calculating an origin-shifted averaged power spectrum, which shows any present lattice in the Fourier transform with much better clarity and with less absences in the low-resolution spots. This is done by the program *2dx_peaksearch*, which also identifies all peak positions in the resulting averaged power spectrum, as described in (35).

The first lattice identification algorithm is using a program *2dx_getlat*, which calculates difference vectors among all identified peak positions. The two linearly independent shortest difference vectors that occur with a higher frequency between this set of peak positions are assumed to be the base vectors u and v of the reciprocal lattice. *2dx_getlat* doesn't need any prior knowledge about the crystal project, but may fail in case of systematic absences in the reciprocal lattice, or in case of multiple overlapping lattices.

The second lattice identification algorithm implemented in *2dx_image* is using a program *2dx_findlat*. Using prior knowledge about the (project-wide) real-space crystal lattice dimensions, the magnification, pixel size, and the tilt geometry (determined from the defocus gradient), this program calculates a predicted reciprocal lattice. The predicted lattice is then rotated around in the Fourier transform by 360°, taking distortions from the tilt geometry into account and varying the lattice parameters and magnification, while for each orientation the program compares the predicted peak positions with the actually measured diffraction peaks. If a theoretically predicted lattice fits to the experimental peaks, the lattice is refined to the identified peak locations by a least-square algorithm. These test lattices are then evaluated by a score value calculated from the sum of peaks that lie within the listed peak coordinates. The lattice with the best score is identified as `first` *`reciprocal lattice`*, and its diffraction peaks are removed from the list of peaks. The process is then repeated with the remaining diffraction peaks, to identify potential additional lattices, so that more than one reciprocal lattice can be identified. *`2dx_findlat`* requires prior knowledge about the crystal and approximate tilt geometry, but then promises a higher reliability to find lattices also in difficult cases.

Once the lattice is identified, and if the recognized tilt angle is at least 25°, the distortion of the reciprocal lattice from the lattice for a non-distorted lattice is used to refine the tilt geometry. At low tilt angles, this lattice distortion is minimal (proportional to 1/cos(TLTANG)), so that the more reliable source of the tilt geometry is the defocus gradient. At higher tilt angles, an error in magnification for example results in a wrong calculation of the tilt angle, so that the lattice distortion is a more reliable basis to calculate the tilt geometry. In this case, the lattice distortion allows for calculating the tilt geometry, but for geometric reasons in the general case two different tilt geometries can explain any given lattice distortion. In addition, the sign of the tilt angle TLTANG is not defined by the lattice distortion, so that four different tilt geometries could explain any given lattice distortion. For this reason, the script takes from the four possibilities the tilt geometry that is closest to the one determined from the defocus gradient as the refined tilt geometry. This obviously bears a possibility for error, if two possible solutions lie closely together, if the defocus gradient was wrongly measured, or if the lattice was wrongly indexed.

Once the refined tilt geometry for the sample plane is known (in the variables TLTAXIS and TLTANG, as described above), it is possible to describe the orientation of the crystal with respect to that tilted plane. The variable TLTAXA describes how the crystal is seen in the image with respect to the tilt axis, by measuring the angle between the tilt axis and the first lattice vector A* in the recorded image. The variables TAXA and TANGL finally describe the tilt geometry from the point of view of the crystal lattice: TAXA measures the angle between the tilt axis and the vector A* in the 3D space of the crystal (this is similar but not the same as TLTAXA), and TANGL measures the tilt angle for the crystal (the sign may be different than that of TLTANG). Only the latter two variables, TAXA and TANGL, are important for the 3D merging.

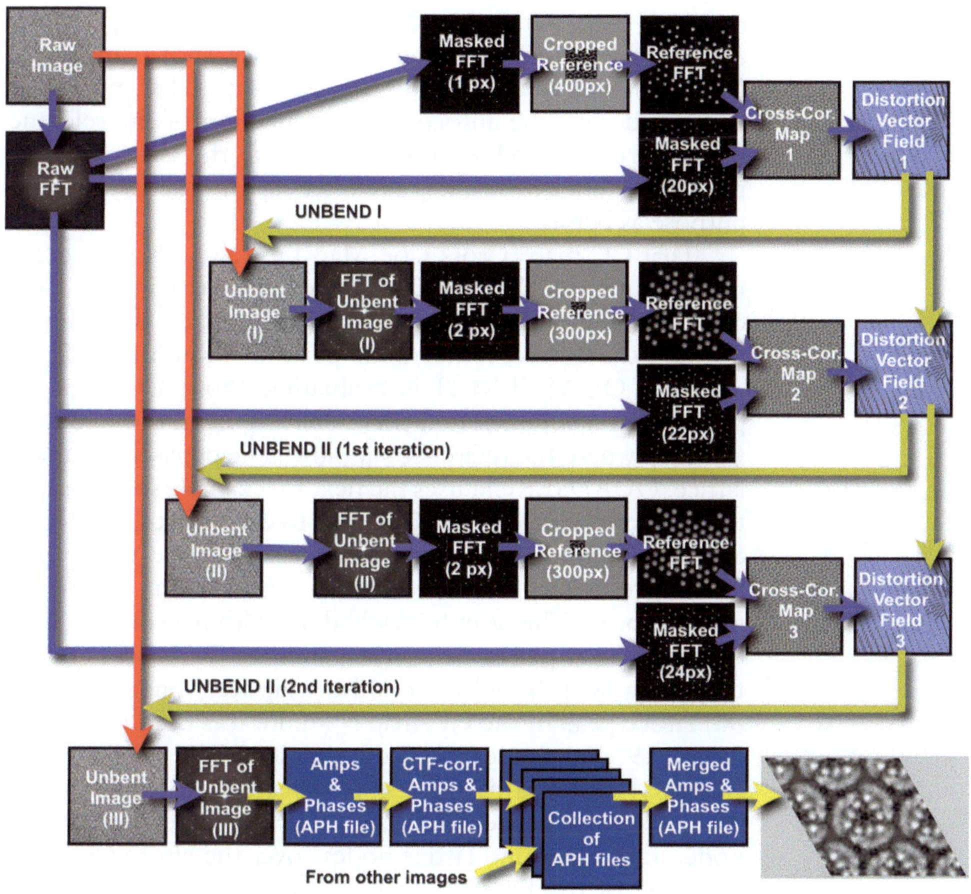

Fig. 3. Detailed flowchart of the unbending process steps for a single 2D crystal image.

3.6. The Unbending Algorithm

Figure 3 introduces the detailed flow diagram of the image processing for one image, as it is implemented in `2dx_image`.

Let us start analyzing Fig. 3. The raw image (top left) is Fourier transformed (Raw FFT), and the resulting Fourier transform is masked on the lattice positions with a very narrow masking radius of 1 pixel, so that only the basic periodicity of the crystal lattice remains. The masked FFT is back-transformed into real space, and a central patch of for example 400 pixels width is masked, while the remainder of the image is grayed out. This gives our first Cropped Reference that shows a very rough approximation of the crystal unit cells. This is again Fourier transformed for later use. In parallel to this, the Raw FFT is also masked with a wider masking radius (e.g., 20 pixels), so that the majority of the noise in the image is removed, without loosing too much of the structure signal. This masked Fourier transform is then multiplied with the complex conjugated Fourier transform of the Cropped Reference. The result is back-transformed into real space, to give a cross-correlation map (CC-map) (Cross-Cor. Map 1) between the Cropped Reference and our raw Fourier filtered Image. Not shown in Fig. 3 is a further refinement step for the CC-map: That map is cross-correlated with the autocorrelation function of the reference, to identify true unit-cell correlation peaks in the CC-map with sub-pixel precision. The result is a refined CC-map that serves the same purpose as before.

That CC-map (Cross-Cor. Map 1) shows peaks for each location of a unit cell. Those peaks can be quite weak and the CC-maps can be quite noisy, so that a simple peak search algorithm would not find the correct lattice node peaks in those maps. The MRC program QUADSERCH is evaluating those CC-maps, and is performing a sophisticated trick: Starting from the middle of the CC-map where the reference should still lie precisely on the crystal lattice, QUADSERCH uses the next for example seven peaks along the predicted lattice in the CC-map, to extrapolate, where the next 8th peak should be, and then only searches the CC-map in the immediate vicinity of the predicted position for the real presence of that 8th peak. This search is only done within a small radius of for example 9 by 9 pixels, where the parameter RADLIM can adjust if the search field should be round or elliptical. Once that predicted 8th lattice peak in the CC-map is found, it is used to predict peak number 9, and the search in the vicinity of the predicted location is started again. This process is repeated, whereby the last for example seven lattice nodes are always used to predict the next node, until the found lattice nodes cover the entire CC-map. The deviation of the found peak positions from the perfect lattice is stored as a vector field (Distortion Vector Field 1).

The MRC program CCUNBEND is then using the determined Distortion Vector Field, to perform an image unbending of the raw image (UNBEND I). This results in an Unbent Image (I), in which each 2D crystal unit cell is moved into a location that corresponds roughly to the predicted lattice position. The Fourier transform of such an unbent image now shows much sharper diffraction peaks (FFT of Unbent Image (I)), so that the narrow masking can be done much more efficiently (now with a mask radius of 2 pixels), to produce a reference map that shows at much better clarity the crystal unit cells. Of this clearer reference, a smaller Cropped Reference can be cut (Cropped Reference (300 px)), so that lattice distortions can be followed more easily. This Cropped Reference is then cross-correlated with the raw image, to produce a refined Cross-Cor. Map 2, which is evaluated by QUADSERCH to refine the Distortion Vector Field 1, giving the Distortion Vector Field 2. This refined distortion vector field is then used by CCUNBEND to unbend the raw image (UNBEND II, 1st iteration). This process can be repeated a few more times (UNBEND II, 2nd (and more) iterations), until a final unbent image is obtained.

The resulting Unbent Image (III) is then Fourier transformed (FFT of Unbent Image III), and this transform now shows sharp diffraction spots. These spots can be evaluated for the Amplitudes and Phases of the diffraction peaks (APH file), which can be CTF-corrected, merged with the APH files from other images, and a resulting final map can be computed from the merged APH files.

3.7. Determining a Spotlist

The above-described unbending process involves masking of Fourier transforms during the calculation of cross-correlation maps. This masking is done on the diffraction peaks in the Fourier transforms such that only diffraction spots remain that are listed in a so-called spotlist. These spots are masked with a circular mask of a given radius (e.g., 20 pixels), while the rest of the Fourier transform is set to zero. The spotlist allows identifying the spots with good signal-to-noise (SNR) ratio, which should be used for the determination of the lattice distortion. Diffraction lattice positions that do not show peaks would only contribute noise to the cross-correlation map and should therefore not be listed in the spotlist. Also, spots that lie not only on the reciprocal lattice of interest, but also lie on or close to a peak from a second interfering lattice in that Fourier transform, should not be part of that spotlist. This can occur, when two 2D crystals are overlaid in the image, so that the Fourier transformation would show two sets of peaks, forming two lattices.

`2dx_image` contains a script to automatically determine a spotlist. This is done by running the MRC program MMBOX on the Fourier transformation of either the raw image (script Get Spotlist for Unbend I) or of the unbent image (script Get Spotlist (complete)), and all diffraction peaks that have an SNR above a certain threshold (above a defined IQ value) are taken as good spots for the spotlist. `2dx_image` also contains a script to do some spotlist arithmetic: Spots that lie within a certain proximity (e.g., of 7 pixels) to peaks from a second lattice, can be deleted from a given spotlist with the script Refine Spotlist.

`2dx_image` also allows manually selecting, viewing, or editing spots in the full-screen browser, by clicking spots in the Fourier transform on or off.

3.8. Unbending

As described above, the 2D crystal image is unbent to remove crystal distortions. This unbending is done with a profile, which is refined in several iterations. However, during this process, the original image is usually only unbent once, but with an iteratively refined profile. Re-unbending of an already unbent image would be dangerous, in the sense that image noise might get aligned to any high-resolution feature in a given reference, after which the statistics of the processed image might appear nice, but the result has little to do with the true protein structure.

If the distortion vector plot resembles a "fur" with smoothly flowing ordered lines, then the unbending process most likely picked up the true distorted lattice of the 2D crystal. However, if the distortion vector plot instead resembles chaotic vectors that point into random directions, then the unbending process might have just followed noise correlation instead of the protein. Given that the vector plot appears coherent, the quality of the unbending procedure can be evaluated by analyzing the signal-to-noise ratio (SNR) of the diffraction spots in the Fourier transform of the newly obtained image. The MRC programs (6) have introduced a quality measure of the spots in form of an Image Quality (IQ) value, which roughly corresponds to the inverse SNR of the diffraction spot: The ratio of a diffraction peak intensity and its local background is translated into an IQ value, such that an IQ=1 means that the peak is at least seven times above background. IQ=7 means that the peak signal is at least as high as the background (the peak height is calculated after background subtraction). A spot of IQ=8 still has a SNR that is one or slightly positive, and a spot of IQ=9 has a negative signal, which means that the Fourier transform shows a dip instead of a peak in that location. In 2dx (26, 27) we introduced a further extension of this concept, by calculating with an empirically defined formula a single quality value QVal for the entire image. The formula to calculate QVal includes among others a weighted sum of the IQ values of the image. This QVal value quantifies the quality gained through unbending, and can be used to judge the quality of an unbent image in general. This can be helpful later in the merging process, where good images can be recognized by their high QVal values, while bad images with low QVal values can be excluded from the merging process. The QVal value can also be useful to automatically refine parameter settings, as done in several "Custom Scripts" for parameter refinement in `2dx_image`.

An alternate reference for cross-correlation and unbending can be calculated from a merged data set. The merged set contains the information of all the individually processed images and therefore is usually capable of producing a better reference than if obtained from a single image. Given that the specimen tilt geometry and defocus were correctly determined, the script Synthetic Unbend can be used, to run the MRC program MAKETRAN (36) to create a synthetic reference from a merged dataset, and use that reference for the lattice unbending. Applying this so called "synthetical unbending" to every individual image will result in an improved merged map. Hence this should be repeated until the unbending of the individual image does not yield any quality improvements.

3.9. CTF Correction

The Fourier transform of a correctly unbent image shows sharp diffraction peaks that extend to high resolution. The amplitude and phase values of these peaks are then evaluated by the MRC program MMBOX, which also records the local background signal around each diffraction peak. The ratio of the signal to the background is translated into an IQ value for each diffraction peak, and can also be used to calculate a Figure of Merit (FOM) for each peak.

This list of Amplitudes and Phases (APH file) is still modulated by the microscope's contrast transfer function (CTF). The effect of the CTF has to be deconvoluted from the image, which can be done in Fourier space with the MRC program CTFAPPLY, by dividing (in form of a Wiener filter to avoid division by zero) the evaluated APH file by the CTF.

This is done in *2dx_image* with the script "Correct CTF." The result is an APH file, where the phases of spots in the negative CTF regions have been flipped by 180°, and where the amplitudes have been corrected through Wiener filtration.

3.10. Calculation of a Projection Map

The result of the processing of one image is a final APH file that lists for each Miller index (H,K) an amplitude and phase value, that is accompanied by a figure of merit. This information can then be used to calculate a corresponding real-space projection map, which also can be symmetrized in case of crystal symmetry. This is done by the script "Generate Map," which will employ MRC and CCP4 programs for this purpose. This script can also run a Spider script (24), to produce larger multi-unit-cell final maps. This, however, requires that Spider be installed on the computer in addition to 2dx and CCP4 (http://www.ccp4.ac.uk, (37)).

4. Application Notes for *2dx_Image*

This section describes the implementation of the presented image processing methods in 2dx (26, 27) for Version 3.3.0.

The processing of an individual image in *2dx* is handled by the program *2dx_image*. In *2dx_image* the whole processing workflow can be automated, which is shown in Chapter 18. Here the focus lies on how *2dx_image* deals with the processing steps of defocus determination, lattice determination, unbending, and CTF correction.

1. At first one would have to choose the image to be processed. Usually in 2dx the images are imported in the program *2dx_merge* (see Chapter 18) and *2dx_image* is launched for that image by double clicking a listed image in the GUI of *2dx_merge*.

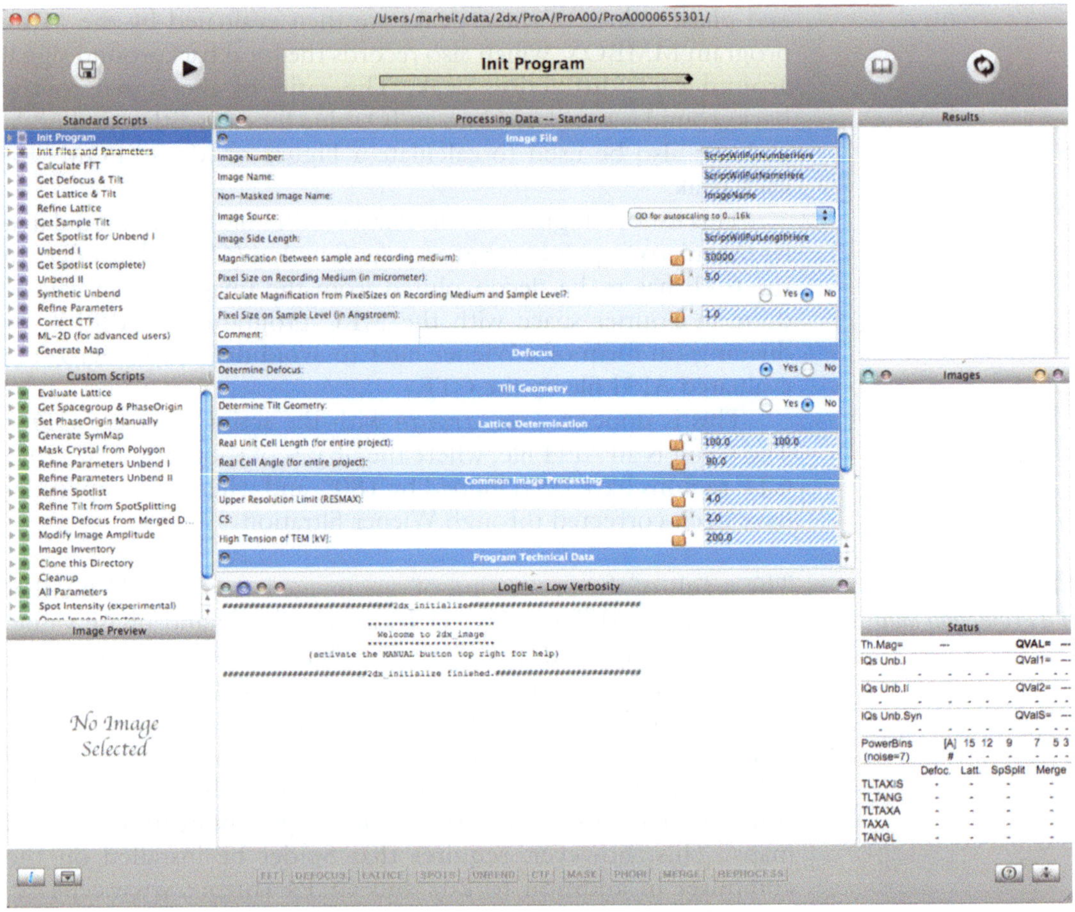

Fig. 4. The GUI of *2dx_image* that holds processing data, parameters, and processing routines.

2. The GUI of *2dx_image* consists of different panels displaying data, processing parameters and image-processing routines, as shown in Fig. 4.
   When *2dx_image* is launched it automatically runs the "Initialization" script as displayed by the progress bar in the top section of the GUI.
3. The script "Init Files and Parameters" fills in the "Image File" parameters in the "Processing Data" panel, e.g., the image side length is taken from the image header displayed in the "Image Header" panel.
4. The image parameters that need the user's attention have striped blue background. It is for example essential to set the correct values for "Magnification" and the "Pixel Size on Recording Medium."
5. If an image of a non-tilted sample is processed the switch "Determine Tilt Geometry" should be set to "No" in the "Tilt Geometry Section" of the "Processing Data" panel. This will

prompt the program to not try to determine any specimen tilt, which would be time consuming and might lead to wrong results.

6. There is a help function for the parameters in the "Processing Data" section. A right click on any parameter will display its description and also give a link to the online documentation. Each script also has a help text, which is shown when the user clicks on the "Manual" button in the top section on the right side of the progress bar.
7. The "Images" panel on the right will show the resulting images and other files that are produced by an executed script. These files will be displayed under their nicknames. The full file names can be seen when placing the mouse pointer over the nickname, or by changing the display mode of that panel. Double-clicking a nickname (e.g., "Non-Masked Image") will show that image in the full screen viewer, where the image display can be adjusted, and additional information through the "Navigator" pull-down menu can be displayed. For example under "Selection based FFT" the user can activate a life FFT of the selected region, which can then be moved over the image to search for a well-diffracting image area. We call this function also a "virtual laser diffractometer." The displayed life FFT can be adjusted in brightness with the keys 1…9, and zoomed with the keys – and + (or =).
8. On the left of the GUI is the "Standard Scripts" panel. As the name states, this holds the scripts that should normally be executed sequentially. Each of these scripts represents an individual processing routine. When double clicking a script entry, the corresponding CShell script template will open in a text editor. When launched, `2dx_image` will incorporate the parameters from the GUI into the script template, create a valid CShell script, save it in the local image directory under the sub-directory "proc," and execute it there. `2dx_image` will then monitor the running of the script, and display the output (logfiles, resulting image files, and determined parameter values) in the GUI.
9. `2dx_image` allows automation in form of consecutive execution of the selected scripts in the "Standard Scripts" Panel. To launch all "Standard Scripts," one can double-click on the top bar of that Standard Scripts panel, and press the "Play" Button. The progress bar then depicts, which script is running momentarily and the "Logfile" panel displays the text output of the running script. The last "Standard Script" is the "Generate Map" script. The result can be examined by double-clicking the "Non-symmetrized Map" in the Images "panel." This will display 2 × 2 unit cells.
10. The automatic processing will only succeed if all the processing parameters are chosen carefully. One can then improve the

processing result by manually refining the parameters of each step.

11. First step to improve the final map would be to look at the determined defocus and astigmatism for the CTF correction. This can be examined by looking at the power spectrum of the image. Therefore, one needs to double click on "FFT of Downsampled Image" in the "Images" panel of the "Calculate FFT" section. This then opens up in the full screen browser. Pressing the C key or selecting "View CTF" in the "Navigator" menu displays the Thon rings as overlays that correspond to the current setting for defocus. The defocus and astigmatism can be altered in the dialog window with the page up/down and cursor up/down buttons. It can help to adjust the contrast and brightness in the "Navigator" menu to see the Thon rings (keys "b" and "n"). The values for defocus and astigmatism are then used in the "Correct CTF" script.
12. The next step would be to examine the lattice that was found. Again this can be done by looking at the power spectrum in the full screen browser. The determined lattice is displayed by pressing the L key. Helpful is the "Display Coordinate Info" (key "i") dialog window that displays amplitude, phase, Miller index, etc. of the pixel that the mouse pointer hovers over.
13. The lattice can be refined manually in "Lattice Refinement Mode" through the "Lattice Refinement" in the "Navigator" menu. This enables the user to add spots to the lattice to refine it. These would ideally be high resolution spots. 2dx helps the user with a smart-mouse feature: A double click in the surroundings of a diffraction spot will select the pixel with the maximum intensity, or it will fit a Gaussian blob to the pixels in the vicinity of the double-clicked position, and take the center of that Gaussian as peak coordinate. The behavior of this "smart mouse" function can be adjusted in the menu "Display Parameters" reached by "d." In the "Lattice Refinement" dialog window an identified spot is added to the lattice refinement table by pressing the Enter key or the "Add Point" button. The Miller index is proposed according to the existing lattice, but can be altered in the dialog window. The "Accept First Lattice" button will apply the lattice refinements according to the picked spots and display the refined lattice directly in the full screen viewer.
14. In many cases, when vesicular 2D crystals were formed and adsorbed to carbon-coated grids, the two surfaces of the vesicles will give two overlapping 2D crystals in the image. The Fourier transform will then always show two lattices. If there are two similarly well ordered 2D crystal lattices in the image, then both lattices should be indexed, and stored as "First

Lattice" and "Second Lattice." Process the image for one lattice first, then copy the image directory to another subdirectory under a different name (e.g., a different sub-frame number), and process it for the other "Second Lattice." In each processing, use the alternative lattice to filter out conflicting overlapping spots from the spotlist, using the script "Refine Spotlist."

15. 2dx uses a minimum of two unbending rounds as depicted in the "Standard Scripts" panel by the "Unbend I" and "Unbend II" scripts. In the unbending scripts one can change the mask radius (holea, holeb) for generating the reference as well as its diameter in real space (boxa, boxb). The fourier filtering mask of the original image is defined by the radius maska and maskb. The suffix depicts for which round the parameters are used. The optimal settings for these box and mask values can be found automatically with the custom script "Refine Parameters Unbend I" and "Refine Parameters Unbend II." As described above, the valitidy of a found set of unbending parameters can be verified by the coherence of the unbending distortion vector plot: It should look like a "well-brushed fur," and not like a chaotic pile of jumping lines.

16. The automatic refinement of box size and mask size bears a trap: If you use an unreasonably small reference area (e.g., a box size of 3 pixels), an extremely large mask that effectively doesn't do any Fourier filtering, and allow the QUADSERCH search to find any peaks in the cross-correlation map with great flexibility (very large quadrada value of for example 20 pixels, and very low number of lattice nodes for predicting the next lattice node, i.e., quadpreda of only three nodes), then QUADSERCH will produce a lattice distortion vector plot that only follows random noise peaks. The unbending routine then will produce an unbent image that shows in the Fourier transform strong diffraction spots out to the highest possible Nyquist frequency, and even your processing statistics look too good to be true, with IQ = 1 spots out to Nyquist. You could get similar results when unbending an image that was already unbent. In such a case, the resulting structure doesn't bear any valuable content, and cannot be merged with results from other images. A situation like this can be easiest recognized by inspecting the unbending distortion vector plot (Does it look like a "fur"?).

17. The progress of the image processing can be monitored in the Status panel in the bottom right corner of the GUI of `2dx_image`. This panel shows a collection of values from the processing, as shown in Fig. 6.

18. The top line of the Status panel shows the Theoretical Magnification, which is calculated from the given pixel size,

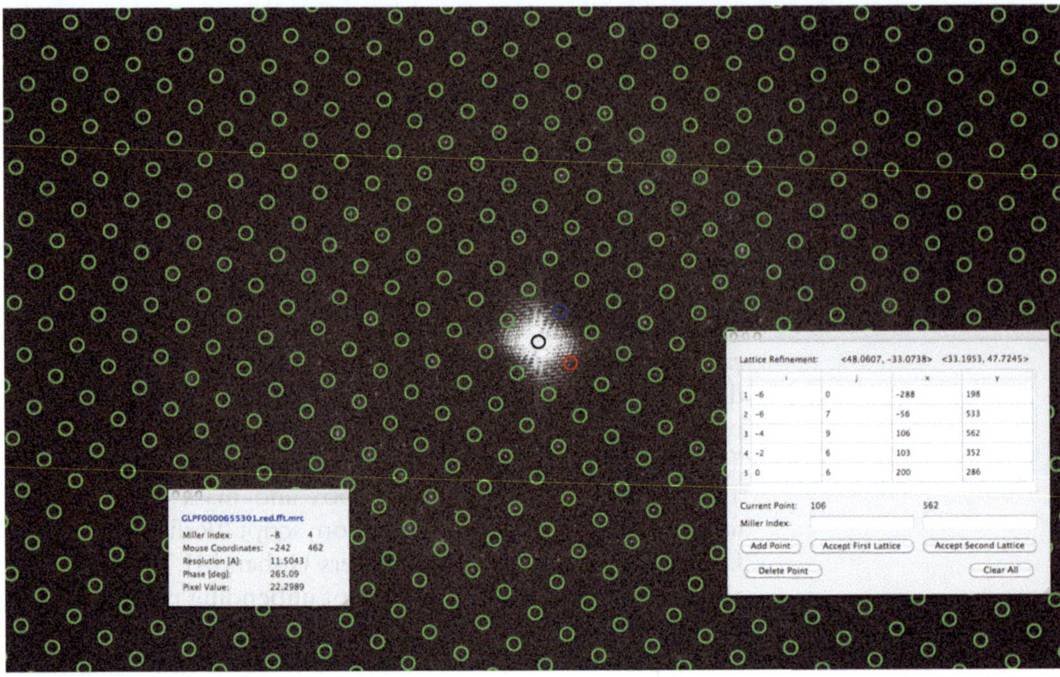

Fig. 5. The "Lattice Refinement Mode" allows manually refining the lattice by picking spots on the power spectrum of the image.

the indexed reciprocal lattice, and the real-space lattice dimensions. That calculated Th.Mag. should correspond to the given Magnification under which the image was recorded. If the Th.Mag. deviates from the nominal magnification, then most likely the lattice indexing is wrong.

19. The Status panel also shows the IQ value statistics for each unbending step: The Unbend I script produces APH files with reflection values, in which each reflection has an IQ value. The number of IQ = 1, IQ = 2, IQ = 3,…, IQ = 9 spots is listed for each Unbending step Unbend I, Unbend II, and Unbend Synthetic in that Status panel. For each unbending script this panel also shows the resulting QVal (QVal1, QVal2, QValS), which summarizes the image quality and the unbending performance into one single number. The last unbending step defines the final QVal, which is shown in the top right corner of this panel. In this example, QValS is lower than QVal2, indicating that the better result was obtained from the conventional Fourier-filter based unbending (Unbend II), rather than using the synthetically generated reference (Synthetic Unbend). This could indicate problems in the available 3D merged dataset, or in the tilt geometry assignment for this image.

20. The Status panel also displays the summed intensity of the diffraction peaks in a certain resolution band, where the

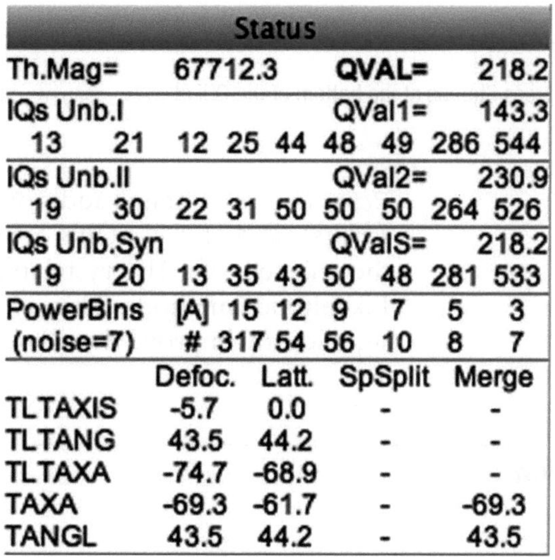

Status									
Th.Mag=	67712.3				QVAL=			218.2	
IQs Unb.I					QVal1=			143.3	
13	21	12	25	44	48	49	286	544	
IQs Unb.II					QVal2=			230.9	
19	30	22	31	50	50	50	264	526	
IQs Unb.Syn					QValS=			218.2	
19	20	13	35	43	50	48	281	533	
PowerBins	[A]	15	12	9	7	5	3		
(noise=7)	#	317	54	56	10	8	7		
	Defoc.	Latt.	SpSplit	Merge					
TLTAXIS	-5.7	0.0	-	-					
TLTANG	43.5	44.2	-	-					
TLTAXA	-74.7	-68.9	-	-					
TAXA	-69.3	-61.7	-	-69.3					
TANGL	43.5	44.2	-	43.5					

Fig. 6. The Status panel of the GUI of `2dx_image`, here for an image of a 45-degree tilted crystal.

background intensity is normalized to the value "7." This is displayed in the lines "PowerBins." The example in Fig. 6 shows strong signal out to 15, 12 and even 9 Å resolution. The range up to 7 Å resolution still has some signal (PowerBin is 10), and this image has almost no signal in the range from 7 Å to 5 Å (PowerBin is 8), and no signal at all beyond 5 Å (PowerBin is 7).

21. The Status panel finally shows the determined tilt geometry parameters (TLTAXIS, TLTANG, TLTAXA, TAXA, and TANGL). These are shown in different columns as determined from the defocus gradient (first column), from the lattice distortions (second column), from the spot-splitting of higher-resolution spots (SpSplit, not calculated in this example since TTF correction wasn't applied yet), and the tilt geometry can also be refined from the merged dataset (last column Merge). The determined tilt geometries should agree with each other, which is the case in the example shown in Fig. 6.

22. The GUI of `2dx_image` shows on the bottom of the window a list of stamps, as shown in Fig. 7. The scripts that have already been executed appear in solid dark, while the steps that still need to be done are pale. These stamps indicate to the user, which steps may need to be repeated: In the above example, the crystal area in the image was not yet masked (MASK is grey). If that were done, then the stamps FFT, SPOTS, UNBEND, CTF would become grey, because those steps would have to be repeated. The steps DEFOCUS determination and LATTICE determination would not have to be repeated and therefore would remain solid.

Fig. 7. The Progress Stamps at the bottom of the GUI of `2dx_image`, here for a partly processed image.

`2dx_image` contains additional functions (e.g., found in custom scripts), and also contains a tool for single particle processing of the unit cells of a 2D crystal image, which is done with a Maximum Likelihood method. This is described in ref. (38), and may significantly improve the resolution for badly ordered 2D crystals.

5. Conclusion

This chapter has focused on the image processing steps to get from a micrograph of a 2D crystal to the projection map. The key concept of extracting the periodicity of the structure by Fourier analysis has been shown. An introduction to the Contrast Transfer Function has been given and methods to correct for it have been presented.

2dx aims at making image processing for 2d crystals user-friendly, while still offering to the expert user the possibility to adjust the algorithms through own scripts, and to fine-tune every detail of the processing. Although the processing is simplified through the software it still needs an experienced and trained operator to get the best results.

The next step in processing would be to merge the results from all the individual processed images together. This is described in the next chapter.

References

1. Henderson R, Unwin PN (1975) Three-dimensional model of purple membrane obtained by electron microscopy. Nature 257:28–32
2. Unwin PN, Henderson R (1975) Molecular structure determination by electron microscopy of unstained crystalline specimens. J Mol Biol 94:425–440
3. Amos LA, Henderson R, Unwin PN (1982) Three-dimensional structure determination by electron microscopy of two-dimensional crystals. Prog Biophys Mol Biol 39:183–231
4. Henderson R, Baldwin JM, Downing KH, Lepault J, Zemlin F (1986) Structure of purple membrane from *Halobacterium halobium*: recording, measurement and evaluation of electron micrographs at 3.5 Å resolution. Ultramicroscopy 19:147–178
5. Henderson R, Baldwin JM, Ceska TA, Zemlin F, Beckmann E, Downing KH (1990) Model for the structure of bacteriorhodopsin based on high-resolution electron cryo-microscopy. J Mol Biol 213:899–929
6. Crowther RA, Henderson R, Smith JM (1996) MRC image processing programs. J Struct Biol 116:9–16
7. Kühlbrandt W, Wang DN, Fujiyoshi Y (1994) Atomic model of plant light-harvesting complex by electron crystallography. Nature 367: 614–621
8. Murata K, Mitsuoka K, Hirai T, Walz T, Agre P, Heymann JB, Engel A, Fujiyoshi Y (2000)

Structural determinants of water permeation through aquaporin-1. Nature 407:599–605
9. Ren G, Reddy VS, Cheng A, Melnyk P, Mitra AK (2001) Visualization of a water-selective pore by electron crystallography in vitreous ice. Proc Natl Acad Sci USA 98:1398–1403
10. Miyazawa A, Fujiyoshi Y, Unwin N (2003) Structure and gating mechanism of the acetylcholine receptor pore. Nature 424:949–955
11. Gonen T, Sliz P, Kistler J, Cheng Y, Walz T (2004) Aquaporin-0 membrane junctions reveal the structure of a closed water pore. Nature 429:193–197
12. Gonen T, Cheng Y, Sliz P, Hiroaki Y, Fujiyoshi Y, Harrison SC, Walz T (2005) Lipid-protein interactions in double-layered two-dimensional AQP0 crystals. Nature 438:633–638
13. Hiroaki Y, Tani K, Kamegawa A, Gyobu N, Nishikawa K, Suzuki H, Walz T, Sasaki S, Mitsuoka K, Kimura K et al (2006) Implications of the aquaporin-4 structure on array formation and cell adhesion. J Mol Biol 355:628–639
14. Tani K, Mitsuma T, Hiroaki Y, Kamegawa A, Nishikawa K, Tanimura Y, Fujiyoshi Y (2009) Mechanism of aquaporin-4's fast and highly selective water conduction and proton exclusion. J Mol Biol 389:694–706
15. Holm PJ, Bhakat P, Jegerschold C, Gyobu N, Mitsuoka K, Fujiyoshi Y, Morgenstern R, Hebert H (2006) Structural basis for detoxification and oxidative stress protection in membranes. J Mol Biol 360:934–945
16. Nogales E, Wolf SG, Downing KH (1998) Structure of the alpha beta tubulin dimer by electron crystallography. Nature 391:199–203
17. Smith JM (1999) Ximdisp – a visualization tool to aid structure determination from electron microscope images. J Struct Biol 125:223–228
18. Schmid MF, Dargahi R, Tam MW (1993) SPECTRA: a system for processing electron images of crystals. Ultramicroscopy 48:251–264
19. Hardt S, Wang B, Schmid MF (1996) A brief description of I.C.E.: the integrated crystallographic environment. J Struct Biol 116:68–70
20. Philippsen A, Schenk AD, Signorell GA, Mariani V, Berneche S, Engel A (2007) Collaborative EM image processing with the IPLT image processing library and toolbox. J Struct Biol 157:28–37
21. Philippsen A, Schenk AD, Stahlberg H, Engel A (2003) Iplt – image processing library and toolkit for the electron microscopy community. J Struct Biol 144:4–12
22. Sorzano CO, Marabini R, Velazquez-Muriel J, Bilbao-Castro JR, Scheres SH, Carazo JM, Pascual-Montano A (2004) XMIPP: a new generation of an open-source image processing package for electron microscopy. J Struct Biol 148:194–204
23. Marabini R, Masegosa IM, San Martin MC, Marco S, Fernandez JJ, de la Fraga LG, Vaquerizo C, Carazo JM (1996) Xmipp an image processing package for electron microscopy. J Struct Biol 116:237–240
24. Frank J, Radermacher M, Penczek P, Zhu J, Li Y, Ladjadj M, Leith A (1996) SPIDER and WEB: processing and visualization of images in 3D electron microscopy and related fields. J Struct Biol 116:190–199
25. Heymann JB (2001) Bsoft: image and molecular processing in electron microscopy. J Struct Biol 133:156–169
26. Gipson B, Zeng X, Zhang Z, Stahlberg H (2007) 2dx – user-friendly image processing for 2D crystals. J Struct Biol 157:64–72
27. Gipson B, Zeng X, Stahlberg H (2007) 2dx_merge: data management and merging for 2D crystal images. J Struct Biol 160:375–384
28. Scherzer O (1948) The theoretical resolution limit of the electron microscope. J Appl Phys 20:20–29
29. Smith DJ, Saxton WO, O'Keefe MA, Wood GJ, Stobbs WM (1983) The importance of beam alignment and crystal tilt in high-resolution electron microscopy. Ultramicroscopy 11:263–281
30. Glaeser R, Downing K, DeRosier D, Chiu W, Frank J (2007) Electron crystallography of biological macromolecules. Oxford University Press, New York
31. Glaeser RM, Typke D, Tiemeijer PC, Pulokas J, Cheng A (2011) Precise beam-tilt alignment and collimation are required to minimize the phase error associated with coma in high-resolution cryo-EM. J Struct Biol 174:1–10
32. Thon F (1971) Phase contrast electron microscopy. Academic, London
33. Philippsen A, Engel HA, Engel A (2007) The contrast-imaging function for tilted specimens. Ultramicroscopy 107:202–212
34. Mindell JA, Grigorieff N (2003) Accurate determination of local defocus and specimen tilt in electron microscopy. J Struct Biol 142:334–347
35. Zeng X, Gipson B, Zheng ZY, Renault L, Stahlberg H (2007) Automatic lattice determination for two-dimensional crystal images. J Struct Biol 160:353–361
36. Kunji ER, von Gronau S, Oesterhelt D, Henderson R (2000) The three-dimensional structure of halorhodopsin to 5 Å by electron crystallography: a new unbending procedure

for two-dimensional crystals by using a global reference structure. Proc Natl Acad Sci USA 97:4637–4642

37. Collaborative Computational Project, Number 4 (1994) The CCP4 suite: programs for protein crystallography. Acta Crystallogr D Biol Crystallogr 50:760–763

38. Zeng X, Stahlberg H, Grigorieff N (2007) A maximum-likelihood approach to two-dimensional crystals. J Struct Biol 160:362–374

Chapter 11

Merging of Image Data in Electron Crystallography

Marcel Arheit, Daniel Castaño-Díez, Raphaël Thierry, Priyanka Abeyrathne, Bryant R. Gipson, and Henning Stahlberg

Abstract

Electron crystallography of membrane proteins uses cryo-transmission electron microscopy to record images and diffraction patterns of frozen-hydrated 2D crystals. Each two-dimensional (2D) crystal is only imaged once, at one specific tilt angle, and the recorded images are then processed with dedicated image processing software packages. Processed image data from non-tilted and tilted 2D crystals can then be merged into a 3D reconstruction of the membrane protein structure. We here describe the process of the 3D merging, using the 2dx software system.

Key words: Membrane proteins, Electron crystallography, 3D Structure determination, Electron microscopy, Data processing, Lattice line interpolation

1. Introduction

In electron crystallography of 2D crystals of membrane proteins, frozen-hydrated 2D crystals are either directly imaged with the transmission electron microscope, or they are subjected to electron diffraction. Due to the beam-sensitive nature of the samples, each crystal is imaged only once, at a given tilt angle, at a typical dose in the range of 10 electrons per square Ångström, which is submitted to the sample in one single exposure. Images and diffraction patterns are then processed individually. Processed images result in a set of Fourier amplitudes and phases from each image, and evaluated diffraction patterns result in a set of amplitudes only. These measurements in Fourier space are accompanied

Ingeborg Schmidt-Krey and Yifan Cheng (eds.), *Electron Crystallography of Soluble and Membrane Proteins: Methods and Protocols*, Methods in Molecular Biology, vol. 955, DOI 10.1007/978-1-62703-176-9_11,

with an approximate knowledge of the tilt geometry under which that 2D crystal was positioned in the microscope during data collection.

Here, we discuss the merging of evaluated amplitude and phase data from images. We first discuss how data from images of non-tilted specimens are merged into a 2D average projection map. We then discuss how image data from tilted specimens are added to that 2D map, to form a 3D reconstruction.

The merging or inclusion of data from electron diffraction is not discussed here.

2. The General Concept

The general goal of electron crystallography data processing is to gain a computer simulated 3D model by integrating sets of 2D data recorded by the electron microscope.

The structure reconstruction typically utilizes Fourier space and exploits the Central Section Theorem for the 3D reconstruction of the protein structures. The Central Section Theorem states that a 2D Fourier transform of a projection image along a certain direction is equivalent to the central section in the reciprocal space of the three-dimensional Fourier transform in the plane perpendicular to the projection direction. This implies that a full sampling of the Fourier transform of the specimen can be gained through the projections along all directions. Through the reversibility of the Fourier transform we can then reconstruct the imaged object.

Application of this theorem to electron crystallography exploits the fact that the images created by a transmission electron microscope correspond (approximately) to projections of the sample. This means that with sufficient projections of samples imaged in different orientation to the electron beam, one can create an approximation to its three-dimensional density distribution. (Effects of nonparallel illumination, multiple electron scattering in the sample, and the curvature of the Ewald sphere are ignored here).

The Fourier transform is a suitable tool for processing of 2D crystal data, since it exploits its repeating structure. The repetitive arrangement of membrane proteins in a 2D crystal results in projection images that can well be studied with a 2D Fourier transform. However, that periodicity only holds true in the X/Y plane of the crystal. In the vertical (Z) direction the crystal is only formed by one layer, so that there are no repetitions in this direction. The Fourier components in the vertical direction are therefore not restricted to discrete diffraction points, but are continuously distributed. The 3D Fourier transform of a three-dimensional 2D crystal structure can therefore be seen as a set of "lattice lines"

(h, k, z^*), where the indices h and k identify a discrete lattice in the Fourier space, while the z^* coordinate defines the (continuously sampled) position along that lattice line. The Central Section Theorem now ensures that all the information gained from projections of non-tilted crystals will contribute to the (h, k, 0) Fourier components. To fill the central section $z^* = 0$ is the task of 2D merging, which is described in more depth in Subheading 3.

For projections of tilted crystals, each diffraction spot in Fourier space corresponds to the value of the lattice line at height z^*. The height depicts where the tilted plane intersects with the lattice line, hence the z^* value can be determined from the tilt angle under which the crystal was oriented during data collection. Each evaluated 2D crystal image produces a set of (h, k, z^*) data, which are used to populate 3D Fourier space. The more tilted 2D crystal data are processed, the denser the lattice lines are sampled. The incorporation of the data gained from tilted 2D crystals is discussed in Subheading 4.

The 2dx software package (1, 2) is available at http://2dx.org. The software package 2dx includes programs for different purposes, such as the program 2dx_image for processing individual images as described in the previous chapter. 2dx also contains the program 2dx_merge, which fulfills two functions: it facilitates the management of a 2D crystal project, and it enables 2D and 3D merging of processed crystal data.

The 2dx package is based on the MRC image processing software (3–8). Most underlying Fortran programs were developed with and for the excellent MRC software. While the original MRC programs can be obtained from Richard Henderson and Jude Smith at the MRC, Cambridge, UK, the 2dx package includes a slightly adapted version of those MRC programs.

The project management with 2dx_merge is described in chapter 18 of this book. In short, you should start by setting up a suitable directory structure with 2dx_merge (e.g., by following the default structure created by 2dx_merge). Import one image into 2dx_merge, process it manually with 2dx_image, save those parameters as defaults for future image processing tasks, import the remaining images from that tilt angle category, and automatically process these newly imported images. If images of tilted specimens (here called "tilted images") are available, the same should be done with those images, until all available images have been processed with 2dx_image.

After successful processing of all images with 2dx_image, each image directory contains so-called "APH files". These are text files that list for each Miller index H and K the evaluated values for *A*mplitudes and *PH*ases. These files also list the background amplitude around each H, K reflection, and may contain information about the contrast transfer function (CTF) value that was valid for that *H*, *K* reflection. These APH-files then have to be merged into a 2D or 3D merged dataset.

2.1. Comparison with Single Particle Image Processing

The merging of electron crystallography data is similar to the merging in single particle electron microscopy: In classical single particle electron microscopy, particle images that show the particles from the same viewing direction and that are all rotated into the same orientation are compared with a reference image by cross-correlation. The *X*, *Y* peak in the cross-correlation map then corresponds to the *X*, *Y* shift, under which a certain particle image would best agree with the current reference. Particle images are then shifted to those locations and are averaged by adding those aligned images up and dividing the result by the number of images.

2.2. Averaging in Electron Crystallography

In electron crystallography, a similar process is implemented to merge different APH files. Data from those APH files in Fourier space are compared with a reference dataset. Instead of shifting images by a certain number of pixels, however, the phase values in the APH files are modified. A phase shift of 360° applied to the reflection with *H*, *K* indices (1, 0) thereby would correspond to shifting an image by one full unit cell in *H* direction. A phase shift of 360° applied to the (0, 1) reflection would correspond to a displacement by one unit cell length in *K* direction. Higher-order reflections require more phase shift, to result in the same amount of image shift: To shift a 10 nm unit cell image by 1 nm in *H* direction would require 36° phase shift applied to the reflection (1,0), a 72° phase shift applied to the reflection (2,0), a 108° phase shift added to reflection (3,0), etc. Arbitrary phase shifts can be applied to the phase variables in the APH files, which bypasses any limitation from discrete pixel steps. Aligned APH files are then averaged. In contrast to single particle microscopy, however, here we have a signal to noise ratio (SNR, encoded by so-called IQ values (4)) measurement for each individual *H*, *K* reflection from each image dataset. That IQ value has been calculated by the MRC programs under 2dx_image from the ratio of the reflection peak amplitude and the amplitude of the background signal around each peak. When averaging APH reflection data from different images, Richard Henderson has devised a method, whereby each *H*, *K* reflection is contributing to the calculated average with its IQ-related weight (7).

When a low-resolution particle image is added to a good average in single particle EM, the average is usually made worse. In contrast, the reflection-based weighting in electron crystallography means that adding data from a mostly bad image does not necessarily make the average worse. Instead, that bad image may still have at least a few good reflection values, which then have a SNR (low IQ value) and benefit the calculated average positively, while the bad reflections from that dataset are down-weighted by their low signal to noise ratios (high IQ values), so that they degrade the average only with minor contributions.

Alignment of data from different image datasets to a current reference is done in iterations, whereby all datasets from non-tilted images are aligned to the reference, and the aligned datasets are then averaged with the above described weighting. The resulting average can be refined through further rounds of alignment against the most recent average structure.

2.3. Extending 2D into 3D

The Fourier space representation of a 2D crystal is a reciprocal space, where the protein structure is encoded in Amplitude and Phase values on Fourier pixels along vertically oriented so-called *lattice lines*. Due to the real-space limits on the 2D crystal (above and below the crystal is emptiness), each lattice line is continuous in the *vertical or z**-direction, as its values are convoluted with a profile that is the Fourier transformation of the vertical structure mask function (e.g., a block function in real space, corresponding to a sinc function in Fourier space). Fourier pixels along one lattice line therefore have neighborhood correlation. In contrast, Fourier pixels on different lattice lines in the general case have no neighborhood correlation. This is also described in (9), and can be exploited to reconstruct the data in the so-called missing cone (10–12).

To create a 3D reconstruction of the 2D crystal protein structure, the available APH files from the images of tilted samples have to be aligned with each other, and with the non-tilted dataset. Once all tilted and non-tilted APH files are correctly aligned, continuous lattice line functions are fitted through the amount of measured values. Those functions (e.g., sinc functions) are then equidistantly interpolated, to create a Cartesian (H, K, L) 3D Fourier space representation for our protein structure.

This Fourier space dataset can then be used to calculate a real-space 3D map for our protein. All this is implemented in the scripts in 2dx_merge. We will now describe the different steps to perform this 2D and 3D merging in 2dx.

3. Merging in 2D

To perform 2D merging, switch the parameter "Modus of Merging" in 2dx_merge to "2D."

3.1. Creating a First Reference

1. The first step is to create a reference, to which the non-tilted images can be aligned. Choose the best non-tilted image in the graphical user interface (GUI) of 2dx_merge. You can usually

recognize this by it having the highest QVal value or the lowest symmetrization phase residual value, as listed in the columns for QVal2 or SymPhaRes in the top central panel of 2dx_merge.

2. Double-click this image, to open it in 2dx_image, and determine its correct phase origin for the determined symmetry. This can be done in 2dx_image either with the Custom Script "Get Spacegroup & PhaseOrigin," or manually with the Custom Script "Set PhaseOrigin Manually."
3. After having correctly centered this image, save the database in 2dx_image, quit 2dx_image, and return to 2dx_merge. In the top central panel of 2dx_merge, select only this one image, by checking the square checkbox in front of that image, and uncheck all other images.
4. Now in 2dx_merge use the Standard Script "Merge Once," to merge the available data from this single image into a reference file. For this, you should specify the resolution range, up to which you believe the data in this image can be reliably used for the alignment (parameter "Resolution of merged dataset for reference"). This script will run several MRC programs to create an amplitude and phase file called merge.aph. The script also eventually creates a MTZ format file with the amplitude and phase values for this image, after phase origin correction, CTF correction, symmetrization, possibly removal of symmetry-forbidden reflections, and correction of symmetry-restricted phase values. This final reference MTZ file is listed as a red entry with the nickname "MTZ: Merged full reciproc. space 2D data" in the "Images" panel bottom right in the GUI of 2dx_merge.
5. If you double-click on that MTZ file, its binary contents will be translated into a temporary HTML text file, which will be displayed in a local web browser. There you can see that the MTZ file contains not only values for amplitudes, phases and FOMs for each reflection, but that it also contains information about the crystal unit cell dimensions, the symmetry, the resolution range, and more.

3.2. Aligning the Non-tilted Image Data onto the Reference

1. After having created a first reference, uncheck that first image, and select the other processed non-tilted images. In a first round you may restrict this to only the reliably and solidly processed images.
2. In 2dx_merge, use the Standard Script "Refine Once" to align the selected images onto the reference file merge.aph. You need to specify the phase origin search space, over which the alignment search should be performed. To perform a search over the entire crystal unit cells, you should scan an area of 360° phase range, which will be searched in the H and K direction by the MRC program origtilt. This search will be

performed in steps of a certain increment size, which you need to specify under the parameter "Stepsize of the phase origin search." In a first round, you could specify a search in 6.0° step size. To cover the entire unit cells, you would need 61 steps of this 6.0° step size, so that the range from (and including) -180° to (and including) 180° is covered. In later rounds, once the rough phase origins have been found for each image, you could perform local refinements of the phase origins by restricting the phase origin search range to a smaller area of for example a 30° range, by specifying 61 steps of 0.5° step size, giving a range of ±15° for the phase origin search.

3. The alignment of all image datasets against the specified reference is done by the MRC program ORIGTILT. This MRC program is a powerful tool that has grown over the years through historic modifications to a complex Fortran program. It requires to be called with a long parameter list, whereby for each image dataset to align, ORIGTILT requires several lines of input parameters. In 2dx_merge we created a separate Fortran program called 2DX_MERGE_COMPILEB, which receives from 2dx_merge the list of image directories to be included in the merging. This program then collects from all 2dx_image configuration files (2dx_image.cfg) in those directories the required information, and creates as output a C-Shell script file, which will later be launched by 2dx_merge to call the actual ORIGTILT Fortran program. This is an awkward, but functional way to utilize the mathematical implementations in origtilt from the graphical user interface of 2dx_merge.

4. Launching the script "Refine Once" in 2dx_merge will therefore create the refinement script, run it, thereby run ORIGTILT, and evaluate the output of ORIGTILT. The refinement script can be inspected in the "Images" panel of the 2dx_merge GUI, where it is listed under "CSH: refinement script." The output of ORIGTILT is listed under the nickname "LOG: origtilt B output." You should inspect that file carefully: In this file, ORIGTILT will document for each aligned image what it did. It will specify the number of H, K values that were used to find the phase origin for a certain image. It will also list for the best identified phase origin position the summed-up phase residual values for the H, K reflections in a certain resolution range, which allows verification of up to which resolution that determined phase origin value is trustworthy for a certain image. This file will also list a larger densely populated field of numbers, whereby each digit corresponds to the normalized cross-correlation value between the reference and the image dataset aligned to this position. This field corresponds to a cross-correlation map, and it should indicate a clear peak in the position identified. In later rounds,

this peak should be centered in the middle of the map, indicating that specific image dataset was already correctly aligned.

5. In 2dx we modified the original MRC program origtilt slightly, so that it writes the determined optimal phase origin values for the images into a separate file, which is later used to update the configuration files 2dx_image.cfg in each image directory.
6. After successful completion of the Standard Script "Refine Once," you should inspect the column for "PhaOri Change" in the top central panel of 2dx_merge, to see by how many degrees the phase origins of each image were changed.
7. Now include the first image in the list of selected image directories, and merge all selected images into a new reference dataset merge.aph, using the "Merge Once" script. If the symmetry is not P1 or P2, you will see in the logfile a table entitled "Phase Residuals in Resolution Ranges," which indicates, up to which resolution the merged data are trust-worthy. Limit the resolution of the reference file with the parameter "Resolution of the merged dataset for reference" to that value.
8. Now refine the phase origin alignment of the selected images, by again using the script "Refine Once," this time with a smaller phase origin search range. You could, for example, perform a search over 61 steps of 0.5° step size (parameters "Number of steps in phase origin search" and "Stepsize of the phase origin search"), to search over a range of ±15°. Run this script, and inspect the output of ORIGTILT in the file "LOG: origtilt B output." You should see clear cross-correlation peaks in the correlation maps for each image.
9. You can use the following script "Merge & Refine (Iterative)" in 2dx_merge to perform three more rounds of merging and refinement. The iteration counter after that script name can be changed from 3 to higher numbers, by double-clicking it.
10. During those iterations, inspect the column "PhaOri Change" in the GUI of 2dx_merge. The indicated values for the phase origin change should be converging towards zero. The image phase origin values under the column "PhaRes" should also converge towards smaller values. If the Phase Origin value for a certain image remains high (90° = random), you could double-click that line in 2dx_merge, to open that image in 2dx_image, to manually see, where the phase origin was placed, and why this image was not correctly aligned.
11. If you activate the parameter "List Reflections into logfile (ILIST)," a separate logfile "LOG: reflections after origtiltk" will be created during the merging step. This file lists for every H, K index the available values from the contributing image datasets. For each H, K reflection, the amplitude, phase, micrograph number, IQ, film weight, background amplitude,

and CTF value are listed. A dashed line appends these values, its length corresponds to the reflection's IQ value (lower IQ value = longer line). For reflections with IQ values 1–4, the values for amplitude and phase are repeated at the end of the lines. Perusal of this file allows you to efficiently recognize suspicious reflections, where despite strong IQ values the measurement for amplitude or phase deviates from the values for the same reflection from other images. In such cases, you might have identified a problem in the defocus assignment for that specific image, so that the Thon rings of the CTF estimation were not correctly placed, and you could manually correct this with 2dx_image.

3.3. Creating Maps

1. After finishing the alignment phase, you can re-create the projection maps for all aligned images with the script "Generate Image Maps" in 2dx_merge. Only after running this script, the maps for all images correspond to the new phase-shift corrected APH files. These maps can be displayed in the GUI of 2dx_merge by "walking" with the cursor line over all image directories in the top central panel. This is facilitated by activating the menu pull-down option "Show Only Selected Directories" (Fig. 1). Misaligned phase origins can often be recognized here, and manually corrected.

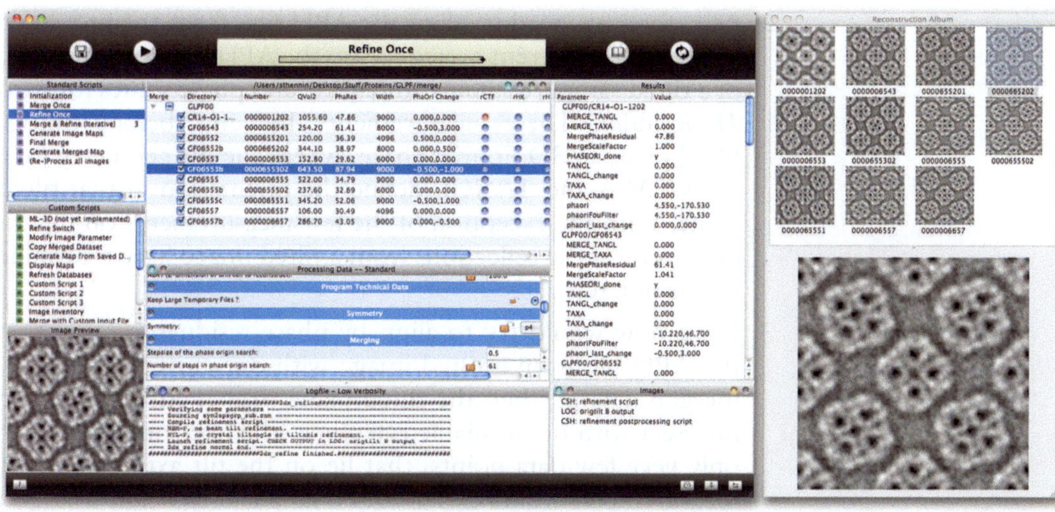

Fig. 1. The GUI of 2dx_merge (*left*), and the Reconstruction Album (*right*). In this example, image number 665202 is *highlighted* in the Reconstruction Album (*right*), it is correctly centered, and displayed as 2 × 2 unit cells. The alignment has not yet converged: PhaOri Change values and phase origins (PhaRes) are too high. Image number 1202 (the *first in the list*) has a phase residual of 47.86°, but the Reconstruction Album shows that its handedness is wrong with respect to the other images. Image number 655302 (highlighted in *blue* in the GUI of 2dx_merge) has an unreasonably high phase residual of 87.94°, and the thumbnail preview in the GUI of 2dx_merge, as well as the Reconstruction Album shows that its phase origin is misplaced by (+180°,+180°).

2. After having manually corrected or excluded all miss-aligned images, you should now merge the aligned images with a reasonable resolution limitation (script "Final Merge"), and create a merged projection map (script "Generate Merged Map"). This map can then be used to create synthetic references to reprocess all images, or it can be used to refine the defocus and astigmatism for all images. This is described in chapter 18.

4. Merging in 3D

To perform 3D merging, switch the parameter "Modus of Merging" in 2dx_merge to "3D."

4.1. Extending from 2D into 3D

1. Extending the current merged dataset from 2D into 3D is not trivial. So far, our calculated 2D average corresponds only to the horizontal plane in *H*, *K* direction in a 3D Fourier space, where all values have a coordinate of zero in the vertical direction (L or z^*). Our 3D Fourier space is mostly empty or undefined, with the exception of this infinitely thin horizontal slab at $z^* = 0$. This Fourier space needs to be filled with continuous reciprocal values for amplitudes and phases along the above introduced lattice lines.
2. As before, we need to align the APH files from the processed images to the current average. Also the APH file from an image of a tilted sample contains amplitude and phase values for each *H*, *K* index. These measurements, however, belong onto a tilted plane in 3D Fourier space, where the tilt geometry variables TAXA and TANGL define the orientation of that tilted plane. With the exception of the few measurements on the tilt axis, most of the reflections from a tilted image have a z^* value that is not zero.
3. If we now would try to align by phase residual optimization the APH file from a tilted image with our currently available merged dataset, then the tilted plane of that APH file would intersect with our available reference (the horizontal plane) in only very few data points, that lie on the tilt axis. Those few measurements would by far not be sufficient to define a correct alignment of that image dataset. Especially in the direction perpendicular to the tilt axis the alignment would most likely fail, see Fig. 2.
4. In order to successfully align the first tilted image dataset to the (non-tilted) average, we need to allow the programs to treat the z^* dimension with greater flexibility, so that a wider

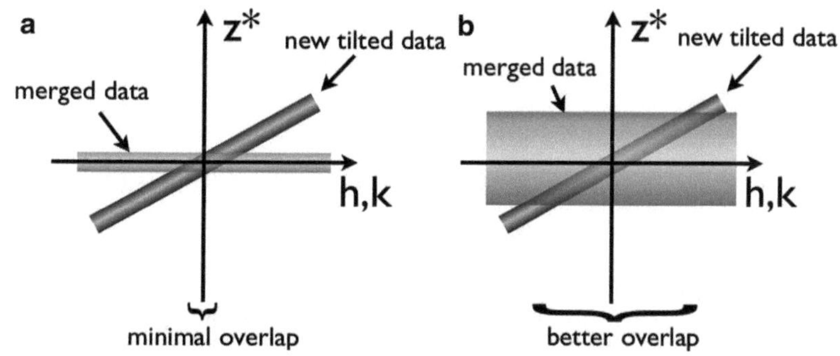

Fig. 2. Phase origin determination for an APH file from a tilted crystal to a reference dataset. (**a**) If only data from non-tilted images are available as a reference ("merged data" slab), then only very few reflections from the "new tilted data" find reference partners in the merged data, resulting in "minimal overlap." The vertical tolerance zone for the merged data is defined by the variable zstarwin. (**b**) If the merged data has already an extension into z^* from successful inclusion of other tilted images, then the phase origin determination for another new tilted dataset is easier.

region of overlap between the tilted image and the non-tilted ($z^* = 0$ plane) reference can be created. In 2dx_merge, the parameter zstarwin defines the vertical z^* tolerance along a lattice line in the search for reference points for the phase origin determination. To start a 3D reconstruction, set this parameter to 0.2 or larger.

5. Once a 3D reconstruction is available, you should set this parameter back to the reciprocal of twice the vertical lattice size, or $1/(2 \times \text{ALAT})$. If for example you reconstruct a unit cell of 200 Å vertical dimension, then you should eventually set this value to 0.0025.

4.2. Aligning Data from Tilted Crystals to the Merged Dataset

1. Start the 3D alignment with backing up your project to a safe location, using the custom script "Synchronize Project with Backup." In case your alignment parameters are lost during the following processing steps, you can open 2dx_merge in your backup location and use that same script to write the saved configuration back to the original data.
2. Also save the current merging result (MTZ and APH files) to one of the ten storage registers in 2dx_merge, using the Custom Script "Copy Merged Dataset."
3. To align tilted APH files to the merged dataset, set 2dx_merge to 3D Modus of Merging.
4. Select the script "Refine Once," and set zstarwin to 0.5 to increase the z^* tolerance during the phase origin search.
5. We already have a non-tilted reference. Select in the GUI of 2dx_merge only a few of the best images of very low tilt angle, and unselect all other images, as well as all non-tilted images.

6. Set the stepsize of the phase origin search back to 6.0° and define the search to run over 61 steps.
7. Limit the resolution of the alignment search to a reliable conservative range.
8. Then run the script "Refine Once."
9. As before, inspect the logfile of the ORIGTILT run "LOG: origtilt B output."
10. Re-create the final maps for all images with the script "Generate Image Maps," and visually verify correct alignment.
11. Uncheck the wrongly aligned tilted images, and include the correctly aligned non-tilted images (use the Save Selection and Load Selection pull-down menu dialogues for this).
12. Merge the non-tilted and tilted datasets into a 3D dataset with the script "Merge Once."
13. Then refine the alignment of the tilted images, by selecting the script "Refine Once" with smaller phase origin search step size, while reducing the z^* tolerance parameter zstarwin (e.g., to 0.1).
14. Repeat the merge and refine cycles with finer step sizes a few more times.
15. Then include successively the datasets from higher tilted crystals. Since you now already have a 3D volume as a reference, there is no further need to set zstarwin to a large value.
16. Use the script "Generate Image Maps" to re-create the aligned final maps for all images.
17. Inspect the Reconstruction Album with the non-tilted and tilted images, to identify wrongly aligned images. As before, these can either be manually corrected with 2dx_image, or removed from the merging process by unchecking their selection boxes.

4.3. Creating a 3D Reconstruction

The phase-origin aligned amplitude and phase datasets from non-tilted and tilted crystals now need to be merged into a 3D reconstruction. This involves transfer of data through different stages. The data flow during this process is illustrated in files on the web server http://2dx.org/documentation/2dx-software/manual/data-flow, where the involved programs are listed together with their input and output data files, and the formats of these files is shown in examples and with explanations.

The program ORIGTILT produces as an output a file called merge.aph, which contains a combined listing of all (h, k) reflections from all images. This file lists in columns the (h, k, z^*) coordinates that ORIGTILT calculated for the reflection, followed by a measurement of the amplitude and the phase of this reflection. The phase values in this file are already phase-flipped (180° added) where needed, and the phase shifts due to phase origin correction have also been applied to these phase values. The file further lists the micrograph number that this reflection came from, and the IQ value, a reflection weight factor, the background signal, and CTF multiplication factor for each reflection. Even though the phases for the reflections have already been corrected, the amplitudes in this file are still not CTF corrected.

The program LATLINPRESCAL uses this file merge.aph to correct the amplitude values. It scales reflections that originate from the same micrograph to equivalent average intensities, and applies the amplitude CTF correction to the amplitude values. The output is a file called latlines.dat. So far, the measured *H*, *K* reflections from non-tilted and tilted datasets were assigned to a z^* coordinate, which specifies its vertical height in 3D Fourier space along the *H*, *K* lattice line. These reflections contribute measurements along one lattice line; however, these measurements are not equidistantly but randomly spaced in z^* (5). The program LATLINE is now fitting sinc functions to the measured data in a least-square approximation. Equidistantly sampled amplitude and phase values along the fitted sinc functions are then written out into the file latfitteds.dat. Precise fine-tuning of parameters for LATLINE is an important part of successful 3D merging. The file "PS: latline.ps" should be inspected, and the lattice line carefully parameters finetuned (13). The produced lattice line data are then subjected to symmetrization by further programs, and finally transferred into the binary file format MTZ. The program PLTILT produces a plot to verify the completeness of the 3D dataset (14). Inspection of its plot "PS: TLTPLOT file" allows to recognize at which tilt angle more data is needed to approach an even sampling of 3D Fourier space.

The final 3D reconstruction map is a volume in CCP4 format, which can be inspected with UCSF Chimera or another suitable display program (15).

4.4. Projective Constraint Optimization

An alternative to lattice line fitting is projective constraint optimization (PCO) (12). This method uses an iterative algorithm to alternately apply known constraints to the dataset between real space and Fourier space, while applying constraints in each space. Gipson also introduced a linear algebra solution to avoid interpolation onto the randomly sampled lattice line measurements in each iteration, by setting up large matrices that directly transform real-space representations of the lattice lines to the measured z^* coordinates of the experimentally determined randomly sampled reciprocal lattice line measurements. After a singular-value decomposition based inversion of these matrices, a method was developed that allows oscillating many times between real space and randomly-sampled Fourier space, without the need to interpolate sinc functions in each iteration.

Fine-tuning of the choice of constraints, their weighting factors, and the way the constraints are applied, allowed Gipson et al. to not only perform a 3D reconstruction of the protein without lattice line interpolation, but allowed for a reconstruction that showed a full coverage of the data in the so-called missing cone. The reconstruction contained correctly determined amplitude and phase values for high z^* values for low-resolution h, k lattice lines, including for the lattice line at ($h=0$, $k=0$). These missing cone data had not been experimentally available from the electron microscopy images and diffraction patterns, but instead were computationally derived from the oversampled data collected at lower tilt angles. An important finding of that work by Gipson et al. is also that this algorithm is capable of eliminating the need for highly tilted image data by strongly oversampling at lower tilt angles: In the specific case of Bacteriorhodopsin (16) analyzed in that work, the data in the tilt angle range up to 45° sample tilt were sufficient to reconstruct the entire protein structure, so that the data collection in the tilt angle range from 45° to 60° tilt would not have been necessary.

As a consequence, focused efforts into the automation of electron crystallography data collection is now needed, so that very large numbers of high-resolution cryo-EM images and diffraction patterns of 2D crystals at lower tilt angles can be recorded automatically. These can then be used to computationally produce the data for higher tilt angles, respectively high z^* coordinates, to fill the missing cone region, and to produce complete 3D reconstructions of the proteins. Cryo-EM 2D crystal data can be collected with much greater success rate at lower sample tilt angles, so that automation of this process appears feasible. This will enable efficient high-throughput structure determination of membrane proteins by electron crystallography.

5. Conclusion

Electron crystallography of membrane proteins entails the collection of cryo-EM 2D crystal data in form of images and diffraction patterns. These are processed individually, and the processed amplitude and phase data from different crystals are merged into a 3D dataset. While the processing of individual images in 2dx is automated, the merging process still requires manual control. The program 2dx_merge assists in the management of an electron crystallography project, and provides user-guidance and user-friendly merging capabilities. Further automation of the data processing for merging is required.

3D reconstruction of membrane proteins from 2D crystal data can strongly benefit from application of projective constraint optimization, to use available oversampled data from 2D crystals imaged at lower tilt angles, to complement the missing data at higher tilt angles, so that eventually the entire so-called missing cone region can be filled. Automation of the cryo-EM data collection of 2D crystal images is therefore now needed to enable automated high-throughput structure determination of membrane proteins by electron crystallography.

Acknowledgments

This work was in part supported by the Swiss initiative for Systems Biology (SystemsX.ch, grant CINA), Swiss National Science Foundation (Grant 205321_126490), the NCCRs TransCure, Nano, and Structural Biology.

References

1. Gipson B, Zeng X, Stahlberg H (2007) 2dx_merge: data management and merging for 2D crystal images. J Struct Biol 160:375–384
2. Gipson B, Zeng X, Zhang Z, Stahlberg H (2007) 2dx—user-friendly image processing for 2D crystals. J Struct Biol 157:64–72
3. Unwin PN, Henderson R (1975) Molecular structure determination by electron microscopy of unstained crystalline specimens. J Mol Biol 94:425–440
4. Henderson R, Unwin PN (1975) Three-dimensional model of purple membrane obtained by electron microscopy. Nature 257:28–32
5. Amos LA, Henderson R, Unwin PN (1982) Three-dimensional structure determination by electron microscopy of two-dimensional crystals. Prog Biophys Mol Biol 39:183–231
6. Henderson R, Baldwin JM, Downing KH, Lepault J, Zemlin F (1986) Structure of purple membrane from *Halobacterium halobium*: recording, measurement and evaluation of electron micrographs at 3.5 Å resolution. Ultramicroscopy 19:147–178
7. Henderson R, Baldwin JM, Ceska TA, Zemlin F, Beckmann E, Downing KH (1990) Model for the structure of bacteriorhodopsin based on high-resolution electron cryo-microscopy. J Mol Biol 213:899–929
8. Crowther RA, Henderson R, Smith JM (1996) MRC image processing programs. J Struct Biol 116:9–16
9. Schenk AD, Castano-Diez D, Gipson B, Arheit M, Zeng X, Stahlberg H (2010) 3D reconstruction from 2D crystal image and diffraction data. Methods Enzymol 482:101–129
10. Agard DA, Stroud RM (1982) Linking regions between helices in bacteriorhodopsin revealed. Biophys J 37:589–602
11. Spence JC, Weierstall U, Fricke TT, Glaeser RM, Downing KH (2003) Three-dimensional diffractive imaging for crystalline monolayers with one-dimensional compact support. J Struct Biol 144:209–218
12. Gipson BR, Masiel DJ, Browning ND, Spence J, Mitsuoka K, Stahlberg H (2011) Automatic recovery of missing amplitudes and phases in tilt-limited electron crystallography of two-dimensional crystals. Phys Rev E Stat Nonlin Soft Matter Phys 84:011916
13. Agard DA (1983) A least-squares method for determining structure factors in three-dimensional tilted-view reconstructions. J Mol Biol 167:849–852
14. Cheng A, Yeager M (2004) A graphical representation of image quality for three-dimensional structure analysis of two-dimensional crystals. Acta Crystallogr A 60:351–354
15. Goddard TD, Huang CC, Ferrin TE (2007) Visualizing density maps with UCSF Chimera. J Struct Biol 157:281–287
16. Mitsuoka K, Hirai T, Murata K, Miyazawa A, Kidera A, Kimura Y, Fujiyoshi Y (1999) The structure of bacteriorhodopsin at 3.0 Å resolution based on electron crystallography: implication of the charge distribution. J Mol Biol 286:861–882

Chapter 12

Evaluation of Electron Crystallographic Data from Images of Two-Dimensional Crystals

Vinzenz M. Unger

Abstract

In the past, various formats have been used to present data fidelity of electron crystallographic data from images of two-dimensional crystals. While each scheme provides some information about the data, not all schemes are equally useful for presenting data reliability. This brief chapter gives guidelines on how to look at datasets that are solely derived from images, and on how to summarize the data structure through graphical outputs that provide unbiased and detailed information about data quality. By the time of publication, the actual procedures described here will have been integrated into the 2dx package and therefore, these evaluation tools will be available at a "push of a button." Because of this, the chapter focuses on brief explanations for why certain ways of presenting data make more sense than others.

Key words: 2D crystal, Projection data, 3D Data, Difference mapping, Phase statistics, Graphical data representation

1. Introduction

Among reconstruction approaches from electron microscopic images, processing, refinement and evaluation of data from two-dimensional (2D) crystals is the most straightforward because of the discrete nature of the individual measurements. Having access to discretely sampled data points allows for unambiguous determination of signal-to-noise ratio for each measurement, which in turn enables proper weighting during data averaging. The discrete nature of the data is also advantageous in determining the resolution of the projection/three-dimensional (3D) reconstruction because the experimentally determined "figure of merit"-values

Ingeborg Schmidt-Krey and Yifan Cheng (eds.), *Electron Crystallography of Soluble and Membrane Proteins: Methods and Protocols*, Methods in Molecular Biology, vol. 955, DOI 10.1007/978-1-62703-176-9_12,

give access to point-spread functions whose dimensions are a direct measure of the effective resolution.

In examining data, the experimentally determined phases are the most useful criterion because phases carry most of the information about the structure. Yet, more often than not phase statistics are reduced to a few global numbers at the time of publication. Such limited disclosure of phase statistics makes it difficult to gauge the reliability of the presented structural data and can leave significant doubts about the merit of a reconstruction. Moreover, there is little consensus—particularly for 3D data—as to what constitutes acceptable data quality, and virtually no consensus exists about how to effectively display the overall sampling of 3D data, or how to go about calculation of difference maps. Adding to this, increasingly more automated and integrated processing workflows minimize exposure to the actual data, which can lead to unintentional errors. It therefore seems timely to implement a generally accepted set of graphical outputs and/or tables that are easy to read and provide an unambiguous record by which to judge data quality. Doing so could prove advantageous much the same way, as the introduction of free R-factors in X-ray crystallography provided a tool for unbiased assessment of data quality.

2. Materials

2.1. Projection Data

1. Merged list of amplitudes and phases for all images that are to be averaged. Image data should have been shifted to the appropriate crystallographic phase origin and reflections should have been folded into the appropriate asymmetric unit. If the dataset contains images of tilted crystals, their data should have been merged with the proper tilt geometry.
2. Programs: AVRGAMPHSC, FOMSTATS, PLOTALLK.

2.2. Difference Mapping (Projection)

1. Separate lists of merged amplitudes and phases for the images from both samples/states. Both datasets should be aligned to a common phase origin. If the symmetry of the sample imposes a centrosymmetric constraint, this constraint should not be enforced (= data should be merged in plane group p1, Note 6), and amplitudes of both datasets should be scaled to a common overall intensity.
2. Programs: AVRGAMPHSC, FOMSTATS, PLOTALLK.

2.3. 3D Data

1. Merged list of amplitudes and phases for all images. Image data should have been shifted to the appropriate crystallographic phase origin and reflections should have been folded into the

appropriate asymmetric unit. Proper tilt geometries should have been applied to the data from images of tilted crystals.

2. Programs: LATLINEK, PLTILTK, IDEAL, PREPNTSPREAD, CCP4.

3. Methods

As mentioned in the introduction, phase statistics along with information about data redundancy are the most useful measures of data reliability in the case of reconstructions from 2D crystals. While this view is not contentious, there is less of a consensus as to how to present this information. The following three sections will deal with assessment of projection data, requirements for "image-only" difference mapping, and the evaluation of 3D data. A detailed discussion of the relevant issues can be found in ref. (1).

3.1. Projection Data

Projection density maps represent the first milestone in each 2D crystallographic project and provide some global information about oligomeric state of the constituent molecules, symmetry and molecular dimensions/shape. In some instances, characteristic density signatures may reveal α-helices that are oriented almost perpendicular to the direction of view. However, interpreting projection maps in terms of the detailed 3D structure is likely to be overreaching and should be avoided unless there are compelling reasons that justify statements other than those related to the global organization of the molecule.

Currently, it is not uncommon to present projection data through an example for the signal-to-noise ratio distribution in the calculated transform of a single contributing image (e.g., CTF-corrected IQ-plot) combined with a table that lists unit cell dimensions, symmetry, number of images, defocus range, number of measurements and unique reflections, overall phase error, and—where applicable—phase deviations of the average phases from their theoretical target values. Of these two elements, the CTF-corrected IQ-plot is not very useful because it has no information about the significance of the final averages, the reliability of CTF-correction, or the degree of sampling. In contrast, data compiled in the "traditional Table 1" (see Table 1, and Notes 1–4) provide a global sense of data quality, which is the reason why such a table should be shown. However, the table fails to communicate sample specific idiosyncrasies and can disguise potential problems with the data such as anisotropy, or problems in CTF-correction. Moreover, "data completeness" is sometimes listed, but is of questionable merit (see Note 2). For these reasons, it is strongly recommended to complement Table 1 with a plot of the combined error

Table 1
Image statistics (projection structure), adapted from ref. (2)

Plane group symmetry[a]	*p*622
Number of images	5
Unit cell parameters	$a = b = 87.9 \pm 0.4$ Å ($n = 10$); $\gamma = 119.9 \pm 0.2$ ($n = 5$)
Range of defocus	4,723–13,626 Å
Total # of observations[b]	578
Total # of unique reflections to 6 Å	57
Overall phase residual compared to 0°/180° (45° is random)[c]:	14.8°

Resolution range (Å)	Number of unique reflections	Phase residual (0°/180°) (45° is random)
88–12	17	7.7
11.9–8.5	14	11.3
8.4–6.9	13	24.0
6.8–6.0	13	18.8

[a]Symmetry was determined using the MRC program ALLSPACE (3)
[b]Reflections with IQ≤6 to cutoff at 6 Å
[c]Amplitude weighted vectorially averaged phase residual which shows the phase deviation from theoretical 0°/180° (where 45° is random). The overall phase error resulting from simple averaging without factoring in the deviation from 0°/180° was 15.7° (where 90° is random)

for each structure factor (Fig. 1), which is the most comprehensive illustration of data structure and quality.

1. Run program AVRGAMPHS. This will produce a list of vectorially averaged structure factors, each with its own figure of merit that is calculated based on the spread of the experimental phase values. The logfile contains detailed information, most of which is intended to provide diagnostic feedback. For instance, a table listing the average error as a function of IQ and resolution is plotted at the end of the logfile. This table is valuable to spot potential errors (e.g., in CTF correction, which will result in resolution bins with high phase errors even for good measurements) and to determine what IQ cutoff to use for averaging. As rule of thumb, the lower the calculated overall error, the better (with 90° being random since AVRGAMPHS does not take into account symmetry-specific constraints). Similarly, the logfile lists how many measurements contributed to each average, how many measurements were rejected based on program settings, and how special cases

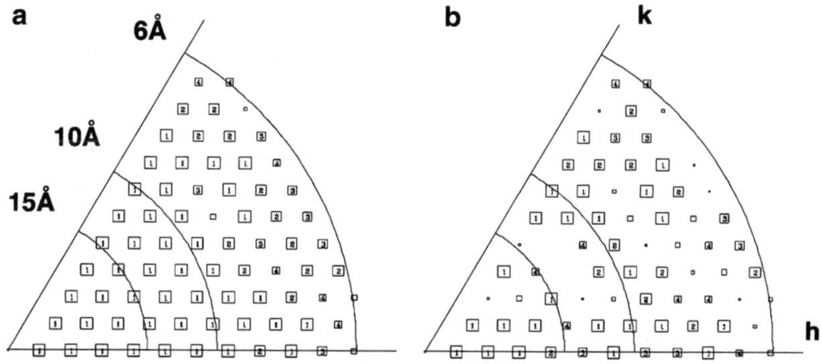

Fig. 1. Comparison of "simple" and "combined" phase errors (adapted from ref. 1). (**a**) Phase error for a gap-junction dataset after simple averaging using the program AVRGAMPHS (3). The following definitions were used to classify the phase errors: 1 <8°, 2 <14°, 3 <20°, 4 <30°, 5 <40°, [6 <50°, 7 <70°, 8 <90° encoded by decreasing box size); (**b**) Using the same dataset, but displaying the combined phase error after averaging and rounding to the closer of the two theoretical values (0/180°). For this purpose the combined phase error is defined as: (°) = $\cos^{-1}(\text{FOM}^*\cos\alpha_{av})$ where FOM is the figure of merit ($0 \le \text{FOM} \le 1$) and α_{av} is the average phase of the structure factor after simple vectorial averaging. Both FOM and α_{av} are calculated by AVRGAMPHS. That is: plot (**b**) equals (**a**) compounded by the amount by which the phase needs to be adjusted to conform to the closer of the two centrosymmetric values (0/180°). In both diagrams, an error of 90° indicates a random distribution of the contributing data. For plane groups with centrosymmetric constraints, this plot is the most useful and comprehensive representation of the data since it completely reflects how faithfully each reflection obeys the symmetry constraint, and how much weight was given to each reflection during calculation of the density map. If no centrosymmetric constraint applies, plot (**a**) would be the relevant representation since the statistical significance of the vectorial average for each structure factor is the only criterion that allows objective evaluation.

were treated (|CTF| < 0.2, single measurements, absence of measurements with IQ of better than 3). All of this information is useful in assessing whether gaps exist in data coverage. For instance, the number of structure factors obtained from only a single measurement should, for obvious reasons, be kept to a minimum.

2. Run program FOMSTATS. This will bin structure factors into user-defined resolution bands, calculate the average phase error for each bin, and produce a listing of the binned errors (to be used in "Table 1"). In plane groups with centrosymmetric constraints, additional operations adjust phases to the closer of the two theoretical values, adjust the figure of merit proportional to the amount of rounding (see Note 5) and produce a listing of the centrosymmetric phase error (to be used in Table 1 instead of the binned errors after simple averaging). Moreover, forbidden reflections are removed from the datalist and their contributions are ignored in the phase error calculation. This process yields a properly symmetrized list of data for further use in map calculations, and for generating a plot that reflects the quality and weight for each structure factor.
3. Run program PLOTALL to produce a plot showing the quality/weight of each unique structure factor. In plane groups without centrosymmetric constraint, this plot will reflect the

phase error associated with the calculated average phase. If centrosymmetric constraints apply, this plot will reflect the combined phase error after averaging and rounding to the closer of the two target values (0/180°). For the purpose of display, errors are categorized/binned as follows: 1 <8°, 2 <14°, 3 <20°, 4 <30°, 5 <40°, 6 <50°, 7 <70°, 8 <90°. Categories 1–4 are indicated by numerals inside the box representing each structure factor, grades 5–8 are encoded by decreasing box size.

3.2. Image-Based Difference Mapping

Studies on the light driven proton pump bacteriorhodopsin have shown that 2D crystals can serve as useful substrates to visualize conformational changes in membrane proteins through difference mapping (e.g., (4–8)). Possibly motivated by the studies on bacteriorhodopsin, a few other examples have appeared in the literature over the years (e.g., (9)). However, in contrast to the studies on bacteriorhodopsin, the more recent examples were based on image data only. While opportunities to study conformational changes of membrane proteins by 2D_crystallography may be tempting, image-based difference mapping is prone to visualizing experimental artifacts rather than real structural changes. Therefore, extreme rigor is required in assuring comparable quality of the component datasets if direct electron diffraction data are not available.

1. If difference mapping is the goal, evaluation begins at the time of data collection. Boundary conditions that must be met (see Note 7):
   (a) Size of the crystals used for each of the samples must be identical (can be achieved by boxing equivalent areas from 2D_lattices exposed to different conditions),
   (b) The number of contributing images in each dataset must be the same
   (c) The spread of defocus values for the images in each dataset should be matched as closely as possible (this is particularly critical if a small number of images is being used because image-derived amplitudes are impacted by the CTF. Hence differences in the spread of defocus values will result in artificial differences in mean amplitude)
   (d) All images, regardless of which dataset they belong to, must have identical B-factors (gauged by the resolution-dependent fall-off of image derived amplitudes)
2. After image processing and merging are complete, run programs, AVRGAMPHS, FOMSTATS, and PLOTALL for each dataset (as for simple projection datasets). When running FOMSTATS, disable the "twofold" option to avoid rounding of the phases.
3. Evaluate phase data at two levels: (1) the number of unique reflections sampled and the phase errors in equivalent resolution

bins in both samples (output from FOMSTATS) should be the same. (2) Side-by-side comparison of PLOTALL outputs. The phase error of the majority of reflections should be defined to within 8° (plotsymbol: 1) (see Note 8). Because of the intricate interplay between factors that affect image-derived amplitudes, quantitative evaluation in terms of changes of amplitudes is not really feasible.

4. Self-difference maps should be calculated for each component dataset by splitting each dataset into two. These difference maps should be almost featureless, and should be shown in addition to the real difference maps.

3.3. 3D Data

Calculation of the final 3D density map often represents the culmination of much effort. In more ways than one, the map itself will provide feedback on the quality of the dataset because it either shows structural details clearly or it does not. It is beyond the scope of this chapter to discuss why the outcomes may fall short of expectations, however the following brief instructions will allow one to obtain a quantitative assessment of the structure's merit

1. In contrast to projection data, whose quality can easily be displayed in a single plot (see Subheading 3.1), displaying the overall data structure and quality of 3D data is less straightforward. The most creative solution to this problem has been presented by Cheng and Yeager (10). For each image in the dataset, their program PLTILTK evaluates a ±45° wedge perpendicular to the tilt axis and summarizes completeness of the data within this wedge through a quality factor IMQ (see Fig. 2). In addition, the program calculates the maximal permissible difference in the orientation of the tilt axis between two crystals that will still result in 50% data overlap (parameter: κ_{50}). This plot should be a mandatory display item upon publication (either in the main-text, or as supplemental information). It is useful because it unambiguously identifies gaps in the data and how well the 3D-transform has been sampled overall. Specifically: IMQ values should be as small as possible (= good completeness for data perpendicular to the tilt axis = good contribution to the vertical resolution of the map), and images should evenly sample the distribution of the tilt axis orientations within the sector that is unique for a given symmetry (for further details, see ref. (10)). Moreover, display of the κ_{50}-value unambiguously shows how much the different images contribute to redundancy of data for fitting of the lattice lines. While not all images need to be within κ_{50} of each other, coverage of the different geometries should show a high degree of data overlap, and no gaps should exist that are close to, equal to, or even larger than what Cheng and Yeager call κ_0 (= maximal angular difference in tilt axis for which there

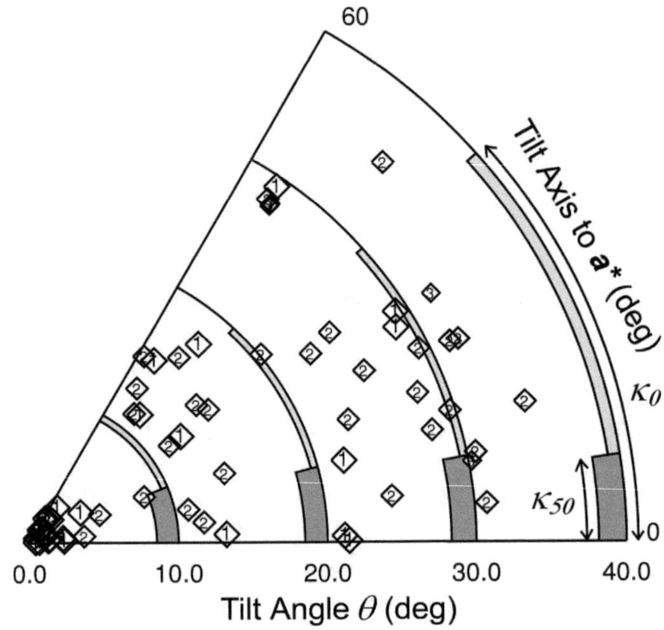

Fig. 2. PLTILTK plot for a connexin 43 dataset (taken from ref. (10)). The 60° wedge of the plot covers the spread of unique tilt geometries for the p6-plane group of the sample. The *x*-axis of the plot refers to the crystal tilt in degree. Each diamond corresponds to one specific image in the dataset. Numbers inside the diamonds reflect the following quality criteria: 1 ≥50% complete in bin between (4/3)*r* and *r* with *r*=resolution limit; 2 ≥50% complete in bin between 2*r* and (4/3)*r*; 3 ≥50% complete in bin between 4*r* and 2*r*; 4 ≥50% complete in bin between ∞ and 4*r*; 5 does not meet quality criterion in any resolution bin perpendicular to tilt axis. Images with an IMQ of 5 should be excluded from the dataset. *Diamonds* separated by more than κ_{50} represent images that have less than 50% overlap in their data perpendicular to the tilt axis. Diamonds separated by more than κ_0 refer to images that have no data overlap perpendicular to the tilt axis.

is any overlap in the data between two images). If such gaps become apparent in the plot, then data fidelity in the corresponding *z**-bin will be low, resulting in significant shortcomings in data/map quality.

2. The second level of assessment for 3D data comes from the actual fitting of the lattice lines. To this end, run program LATLINEK to obtain the fitted structure factors and some global statistical information (the input data should be properly formatted and individual measurements should be properly weighted based on their signal-to-noise ratio—program LATLINPRESCAL). The most important information to be gained is the plots for each lattice line and the overall statistics that are given at the end of the log-file.

   (a) The lattice line plots complement the information obtained from PLTILTK by giving specific visual feedback about the sampling along the various lattice lines. More importantly, however, lattice line plots also facilitate identification of errors in the CTF-correction for individual images that may otherwise be hard to notice in the actual merged list of data points (= phases of good measurements (e.g., IQ1-3) being ~180° off the fitted phase value). Furthermore, the plots will give an impression of the vertical resolution of the data because this measurement will be closely tied to the z^*-values that are reliably sampled in the fitted dataset. Regarding presentation of data, showing a few select lattice lines as supplemental information is both time-honored and courteous. However, this information is insightful only if the selected lattice lines are close to the chosen resolution cutoff because it will be those data that are the most affected by application of B-factors during map calculation. In cases where crystallographic symmetry constrains phase values along entire lattice lines, showing a few select lines close to the chosen resolution cutoff can also be useful because it demonstrates that applicable symmetry constraints are met at the highest resolution.

(b) The overall statistics given at the end of the log-file contain information that should be included in a table summarizing the crystallographic data. Specifically, LATLINEK lists the total number of measurements used, and the overall weighted phase residual. As a rule of thumb: the number of measurements should be 5–10× the number of expected unique reflections (which can be obtained from the output of the program IDEAL), and the overall weighted phase residual for an intermediate resolution map (5–8 Å range) should be ~20° or less. Datasets missing these targets are likely to result in maps where the structure is poorly resolved.

3. The final element in assessing the 3D-reconstruction is the calculation of the effective resolution cutoff. While the in-plane resolution will almost always be at the nominal resolution cutoff, vertical resolution will always be lower because of the missing cone of data (see Note 9). The most objective way of estimating the actual resolution limits of a reconstruction is to calculate the data's pointspread-function in all directions (see Note 10, Fig. 3). To this end two steps need to be carried out:

(a) An ideal synthetic dataset is generated based on the in-plane resolution cutoff and maximum tilt angle (program IDEAL, Note 11). CCP4 is then used to calculate the inverse transform of this dataset, in which all structure factors have unity amplitude, a phase angle of 0°, and a figure of merit of 1. The *xy*-, *xz*-, and *yz*-sections through the resulting ellipsoid are contoured at half-height (= if the volume were scaled to range from 0 to 250, the peak

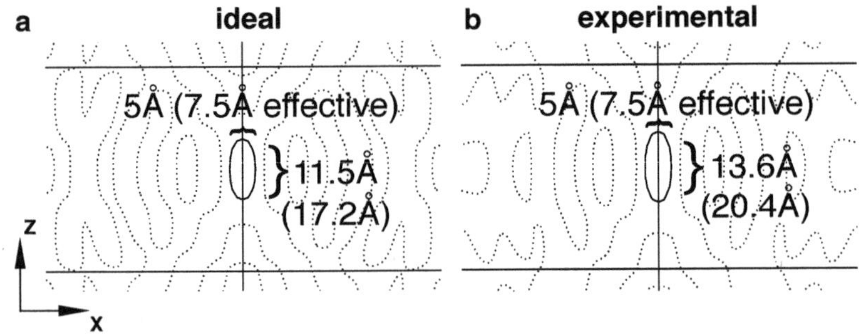

Fig. 3. Example for pointspread function of 3D-dataset (taken from ref. 1). In this example, only the *zx*-sections are shown for the simulated, ideal dataset (**a**) and the actual, experimental dataset (**b**). Measures obtained for the ideal dataset are 5 Å along a-axis at half height. The nominal cutoff for map calculation was 7.5 Å and therefore, the conversion factor equals 7.5 Å/5 Å = 1.5. Based on this calibration, the best possible vertical resolution would be expected to be 11.5 Å (half-height), translating to 17.2 Å effective resolution for this dataset which had a nominal resolution cutoff of 7.5 Å and assuming a maximum crystal tilt of 30°. Applied to the experimental data the effective resolutions are 7.5 Å in plane and 20.4 Å vertical.

would be contoured at 125; if the volume were scaled to range from 0 to 1, the peak would be scaled at 0.5) and plotted for evaluation. It is important to print a background grid for internal calibration purposes because post-calculation scaling may affect the spacings in the plot. Next, the diameter/principal directions are measured and converted into Å. For the *xy*-section the nominal resolution is known because it was used as cutoff for generating the ideal reference dataset. Correlating the width of the pointspread peak at half height with the nominal resolution of the dataset gives a conversion factor (typically around 1.4) that will allow the resolution(s) of the real dataset to be estimated.

(b) The pointspread function of the real data is calculated by first generating a copy of the dataset in which all experimental phases are replaced by 0° while the experimental amplitudes and FOM values are retained. This dataset is then used for calculating the inverse transform as was done for the ideal data. However, in the case of the experimental dataset, the same empirical B-factor should be used that was applied during the calculation of the actual 3D-density map. The volume is contoured at half height and the *xy*-, *xz*-, and *yz*-sections through the resulting ellipsoid are plotted. The dimensions along the principal axes are measured, and converted into resolution cutoffs by multiplying with the conversion factor (see Note 12).

4. In addition to the summary plot obtained by PLTILTK, a brief table should be shown that summarizes the overall data

Table 2
Example of summary table for 3D-data (modified from ref. (11))

Parameter	Value
Number of images	58
Unit cell parameters	$a=b=89$ Å; $\gamma=120°$
Two sided space group	p622
Range of crystal tilts[a]	1.2° to ~40°
Range of underfocus	~4,000–15,000 Å
Total number of measurements	13,764
Total number of fitted measurements	1,147
Effective resolution Cutoffs[b]	In plane: 7 Å Vertical:15 Å
% Complete to 6 Å in plane with 40° of tilt	79%
Overall weighted phase residual[c]	19.0°

[a]Nominal tilt range if PLTILTK plot is provided; effectively sampled tilt range (based on %-completeness) if PLTLTK plot is not available
[b]Measured from the pointspread function of the fitted dataset
[c]For fitted lattice lines, including data to IQ8 and to a nominal in-plane resolution of 6.0 Å

properties. At a minimum, this table should list the total number of images, unit cell parameter, two-sided plane group, range of crystal tilts (see Note 13), range of defocus, total number of observed measurements, total number of fitted measurements, effective resolution cutoffs, % completeness, and overall weighted phase residual. An example is shown as Table 2 (taken from ref. (11)).

4. Notes

1. "Table 1"; concerning the number of images—how many are needed for a "good" dataset? There is no general answer to this question and much depends on the unit cell size, symmetry, the actual structure of the molecules, and the ultimate use for the projection data (= simple rendering of the projection vs. difference mapping). In many cases, statistically significant averages can be obtained from just two to three images. A larger number is likely to be required for samples with large unit cell dimensions because in these cases the average intensity of the diffraction maxima is lower (=noisier measurements). Similarly, thick specimen ($z \geq 150$ Å) may require more than three images

since even small tilts can result in reciprocal space sampling outside the $z^* = 0$ plane. Notably, not all phase errors are created equal. Errors that are obtained after simple vector averaging of the data behave as expected in that the error decreases as the number of measurements increases (provided that the CTF-correction of the phase values is correct in each contributing image). In stark contrast, centrosymmetric phase errors in plane groups with twofold symmetry along the *z*-axis do not markedly improve through increasing the number of images that are averaged. This behavior is caused by specific features of the 3D-molecular transform, and cannot be sensibly presented in table format, which is the reason why showing a plot of the combined phase error (averaging + rounding) is the most comprehensive and objective representation of the data (also see Note 5).

2. "Table 1"; concerning "completeness"—sometimes, investigators provide a measure of completeness. This information is of limited use, especially in plane groups that in projection call for systematic absences of reflections (e.g., c222, $p22_12_1$). Specifically, it is rare that forbidden reflections are truly absent from the calculated transforms of the images (= IQ9). With that in mind, how does one deal with this issue in terms of "completeness," where "complete" would have to mean that one has reliably measured the absence of these reflections? This question is particularly relevant for centric plane groups where 50% of the projection terms are forbidden. That is, if "completeness" were to refer to the percentage of terms for which structure factors with amplitudes > 0 have been determined, then completeness in for example c222 could never exceed 50%. For these reasons, including information about completeness in "Table 1" is not particularly helpful (and also is not necessary if a plot of the combined phase errors (Fig. 2) is shown either as main figure or in the supplement).
3. "Table 1"; redundancy (= # of observations/ # of unique reflections) should be 5–10.
4. "Table 1"; phase errors—two different evaluation criteria apply depending on whether the errors refer to simple averaging (90° error is random) or deviation of the average from the theoretical value (45° error is random). The generally accepted criterion is that in the centrosymmetric case, data should not be used if their overall centric error approaches or exceeds 45° for a given resolution bin. This treatment does not imply that each measurement in such a bin is invalid/troubled, but merely states that the measured phases on average do not comply with the centrosymmetric constraint and that inclusion of these data will add a significant amount of random noise to the map.
5. Figure 1 illustrates the difference between significance of simple average and their compliance with the centrosymmetric

constraint that is applicable for the plane group in this example (p6). Specifically, all structure factors are well determined after averaging (Fig. 1a), yet many of them have averages that are significantly different from 0/180°. The latter cannot be due to errors in the CTF-assignments because such errors would result in a very low FOM during simple averaging. In this particular case, the 3D-transform shows that "non-conformity" is caused by rapid changes of the phases in the vicinity of the $z^*=0$ projection term. In such cases, small sample tilts or a lack of sample flatness can cause sampling of the phase far away from 0/180° and hence give a solid average that is significantly off the theoretical target. Unfortunately, information from the 3D-transform will not be available in the case of genuinely new samples for which no tilt data have been previously analyzed. Therefore, deviations from the expected target phase values cannot initially be put into proper perspective. For instance, noncompliance could also indicate that the sample does not have the assumed symmetry, even if ALLSPACE suggests it does. The latter is true because ALLSPACE only tests for overall compliance with plane group specific phase constraints. This global analysis can return an incorrect answer if the number of comparisons made close to the resolution cutoff is small compared to the number of comparisons that are made at low resolutions. To avoid potential errors/problems, it is therefore advisable to use data that have been properly weighted for the uncertainty of noncompliant structure factors. FOMSTATS takes care of this aspect by adjusting the FOM obtained from averaging by the "offset" from the expected target phase value. The adjusted FOM represents the "overall" quality of a structure factor, and is used in CCP4 to weigh the structure factor amplitude during summation. Accordingly, a plot of the adjusted FOMs, representing the combined phase errors of each structure factor, is the singly most useful way of presenting the merit of the dataset. If there is insufficient space to present this information in the main text, it should be provided as a supplemental figure.

6. Image based difference mapping should take into account changes in both amplitude and phase data because without a priori knowledge of the scale of structural changes it is safest to assume that changes will occur throughout the unit cell. In this case, using variations in just the amplitudes or phases is not sufficient (see ref. (12)). The need for vectorial difference mapping also requires the entire procedure to be carried out in a symmetry that does not impose centrosymmetric constraints (best: p1). This is necessary because changes in phases are typically small (12), and in many cases will be smaller than the "rounding error" made by adjusting the phase to their "target" value.

There are two additional arguments in favor of omitting symmetry in image based difference mapping. First: if centrosymmetric constraints are enforced then 180° phase changes are the most likely to occur for the least determined and noisiest structure factors in both datasets. This is unlikely to be meaningful. Another argument in favor of p1 is that omitting symmetrization will provide an internal consistency check. That is, if the observed changes are real, they will (1) be more significant than the observed differences between symmetry related subunits/molecules, and (2) will affect symmetry related molecules in similar ways (assuming that the conformational changes do not break the symmetry within oligomers or between different molecules on the lattice).

7. Meeting the set conditions is important for minimizing the risk of errors because in contrast to electron diffraction amplitudes, image-derived amplitudes are modulated by the contrast transfer function, and thus, image-derived amplitudes are very sensitive to the specific imaging conditions used. For instance, if the images in the two datasets are not recorded at very similar defocus values, then differences of amplitudes will become a "random" function of where the CTF-zero values happened to be located. The only way to prevent this from happening would be to record a large number of images with a large spread of defocus values for each of the datasets. But even in this case, all these images would need to be of crystals roughly the same size and order because otherwise, changes in amplitudes between samples may reflect different B-factors, which in turn would result in rendering the differences in data quality rather than a representation of genuine structural differences.

8. Since changes in phase are expected to be small, any structure factors with poorly defined phase values are not likely to make meaningful contributions to the difference map. Therefore, data redundancy should be large enough to yield highly significant averages during the initial combination of data (i.e., small phase errors for the majority of structure factors). If structure factors show noticeable quality differences in the two datasets, then the cause for these differences should be analyzed. The only acceptable differences are those where changes in amplitudes occurred independently of any issues related to the CTF (see Note 7) and B-factor of the image. For this analysis (as well as for calculation of the difference map) both datasets should be scaled to each other (using the sum of intensities for one dataset as reference). The scale factor should be close to 1 as any significant deviation from this value suggests that the conditions outlined in Subheading 3.2 step 1 were not satisfied.

9. Missing cone: in 2D-crystallographic applications, the physical dimensions of the holder limit data collection to a wedge of at most ±70°. Consequently, no data can be collected along the (0,0,l) lattice line. The absence of data from the (0,0,l) lattice line and the cone shaped region around it lowers the vertical resolution and causes the structure to be elongated/stretched along the z-axis.
10. Concept of pointspread function: structure factors describe complex waveforms for all phase angles except 0° and 180°. For the latter two, only the real component of the structure factor exists and the corresponding waveform is a simple cosine. Consequently, if all phases are 0° then Fourier summation of a 100% complete 3D-dataset generates a "solid" sphere that is centered about the real space origin at $z = 0$ (if all phases were set to 180°, the sphere would be "hollow"). Lack of data in any direction causes a distortion of the sphere into an ellipsoid. This effect, which is commonly referred to as "pointspread function," can be exploited to estimate the effective resolution of a final reconstruction because the diameter of the sphere and its distortion by lacking data depend on what frequencies contributed to the sphere's (ellipsoid's) generation. In 2D-crystallography, the in-plane resolution (x,y-plane) usually equals the highest resolution that was observed in projection. Customarily, this resolution is chosen as the "nominal resolution cutoff." However, the missing cone, and any additional limitations in tilt, always cause incomplete data along the z^* direction of the transform. The impact of such data imperfections can readily be evaluated by setting all experimental phase values to 0° and summing of the Fourier terms. Using the Raleigh definition of resolution, the vertical resolution can be obtained from the calculated ellipsoid as described in the main text.
11. In calculating the ideal dataset, the maximum tilt angle should be adjusted such that the resulting number of unique reflections is close to the number of actually fitted, experimental structure factors. A fairly good estimate will be available from scrutinizing the plot generated by PLTILTK, which will show what tilt range was effectively sampled. As a rule of thumb: when compared to the ideal dataset, the completeness of the experimental dataset (= # of fitted unique reflections/# of theoretically possible unique reflections) should be ~80% or higher.
12. In plane groups where symmetry constrains phases along certain lattice lines, the point-spread function likely will overestimate the actual resolution. This is because LATLINEK assigns FOM = 1 to the fitted structure factors of constrained reflections. Close to the resolution cutoff this will almost certainly not be the case. At the same time, however, seemingly

perfect data close to the in-plane resolution cutoff significantly contribute to the fraction of data with high z^*-values. Depending on how many of the data points are constrained, this can lead to distortions of the pointspread function that greatly overestimates the actual resolution of the data. The only fix for this problem is to isolate the experimental data for the respective lattice lines, fit them in p1 and substitute the now unconstrained, true FOM back into the dataset where phase constraints were enforced. Furthermore, in analyzing/stating resolutions one should always keep in mind that the quality of the actual density map should match/support the claimed resolution. For instance—at vertical resolutions below ~6 Å the helical repeat distance should be clearly visible. Similarly, density maps claimed to have a resolution of ~7 Å (in plane) and reasonable tilt data (e.g., ~40° of crystal tilt) should clearly resolve all transmembrane segments and throughout the entire bilayer region (assuming helix tilts are all below 40°). Disjointed helical segments, or significant "fusion" of adjacent helical spans almost always are hallmarks of poor sampling of data at low crystal tilts, incomplete data at higher tilts or inclusion and over-sharpening of large amounts of noisy "high" resolution data. Notably, the "precise" resolution is often irrelevant for the biological information that can be obtained from a structure. That is, if a reconstruction falls short of atomic resolution where the backbone and side chains can be unambiguously fitted, it often does not really matter whether the resolution is 6 or 8 Å. Similarly, if resolution falls short of 10 Å, it likely will not matter whether the density reflects the structure as seen at 11 Å or 15 Å. In cases like this, it may be preferable to choose a resolution cutoff that avoids inclusion of noisy data and therefore will give a clean and well-defined volume at the lower, yet sufficient resolution cutoff.

13. If output of PLTILTK is not shown, then the information about tilt should refer to the effective tilt sampled, not the highest nominal tilt angle based on goniometer reading.

References

1. Unger VM (2000) Assessment of electron crystallographic data obtained from two-dimensional crystals of biological specimens. Acta Crystallogr D Biol Crystallogr D56:1259–1269
2. Aller SG, Unger VM (2006) Projection structure of the human copper transporter CTR1 at 6-Å resolution reveals a compact trimer with a novel channel-like architecture. Proc Natl Acad Sci USA 103:3627–3632
3. Valpuesta JM, Carrascosa JL, Henderson R (1994) Analysis of electron microscope images and electron diffraction patterns of thin crystals of Θ29 connectors in ice. J Mol Biol 240: 281–287
4. Subramaniam S, Gerstein M, Oesterhelt D, Henderson R (1993) Electron diffraction analysis of structural changes in the photocycle of bacteriorhodopsin. EMBO J 12(1):1–8

5. Han BG, Vonck J, Glaeser RM (1994) The bacteriorhodopsin photocycle: direct structural study of two substrates of the M-intermediate. Biophys J 67(3):1179–1186
6. Vonck J (1996) A three-dimensional difference map of the N intermediate in the bacteriorhodopsin photocycle: part of the F-helix tilts in the M to N transition. Biochemistry 35:5870–5878
7. Hendrickson FM, Burkard F, Gleaser RM (1998) Structural characterization of the L-to-M transition of the bacteriorhodopsin photocycle. Biophys J 75:1446–1454
8. Subramaniam S, Henderson R (1999) Electron crystallography of bacteriorhodpsin with millisecond time resolution. J Struct Biol 128: 19–25
9. Appel M, Hizlan D, Vinothkumar KR, Ziegler C, Kühlbrandt W (2009) Conformation of NhaA, the Na/H Exchanger from Escherichia coli, in the pH-activated and ion-translocating states. J Mol Biol 386(2):351–365
10. Cheng A, Yeager M (2004) A graphical representation of image quality for three-dimensional structure analysis of two-dimensional crystals. Acta Crystallogr D Biol Crystallogr A60:351–354
11. De Feo CJ, Aller SG, Siluvai GS, Blackburn NJ, Unger VM (2009) Three-dimensional structure of the human copper transporter hCTR1. Proc Natl Acad Sci USA 106(11): 4237–4242
12. Lindahl M, Henderson R (1997) Structure of the bacteriorhodopsin D85N/D96N double mutant showing substantial structural changes and a highly twinned, disordered lattice. Ultramicroscopy 70:95–106

Chapter 13

Modeling, Docking, and Fitting of Atomic Structures to 3D Maps from Cryo-Electron Microscopy

Gregory S. Allen and David L. Stokes

Abstract

Electron microscopy (EM) and image analysis offer an effective approach for determining the three-dimensional structure of macromolecular complexes. The versatility of these methods means that molecular species not normally amenable to other structural methods, e.g., X-ray crystallography and NMR spectroscopy, can be analyzed. However, the resolution of EM structures is often too low to provide an atomic model directly by chain tracing. Instead, a combination of modeling and fitting can be an effective way to analyze the EM structure at an atomic level, thus allowing localization of subunits or evaluation of conformational changes. Here we describe the steps involved in this process: building a homology model, fitting this model to an EM map, and using computational methods for docking of additional domains to the model. As an example, we illustrate the methods using an integral membrane protein, CopA, which functions to pump copper across the membrane in an ATP-dependent manner. In this example, we build a homology model based on the published atomic coordinates for a related calcium pump from sarcoplasmic reticulum (SERCA). After fitting this homology model to a 17 Å resolution EM map, computational software is used to dock a metal-binding domain (MBD) that is unique to the copper pump. Although this software identifies a number of plausible interfaces for docking, the constraints of the EM map steer us to select a unique solution. Thus, the synergy of these two methods allows us to describe both the location of the unknown MBD relative to the other cytoplasmic domains and the atomic details of the domain interface.

Key words: Electron microscopy, Structure modeling, Protein–protein docking, P-type ATPases, Computational biology

1. Introduction

Electron microscopy can be used to generate 3D structures of proteins using several different reconstruction strategies: electron crystallography, helical reconstruction, single particle averaging, or tomography. Due to the limited resolution (8–30 Å), however, it is

Ingeborg Schmidt-Krey and Yifan Cheng (eds.), *Electron Crystallography of Soluble and Membrane Proteins: Methods and Protocols*, Methods in Molecular Biology, vol. 955, DOI 10.1007/978-1-62703-176-9_13,

often difficult to directly evaluate the conformation of the polypeptide chain. This limitation can be overcome by building a model based on related X-ray crystallographic or NMR structures and then fitting this model into the lower resolution EM map. Such a model is frequently useful for evaluating conformational changes due to different conditions for EM sample preparation or for localizing accessory domains or subunits that were not present in the X-ray or NMR structure.

As an example of this modeling procedure, we have used an ATP-dependent copper pump from *A. fulgidus* called CopA. CopA belongs to the large family of P-type ATPases that couple the energy of ATP hydrolysis to the transport of ions across the membrane, thus generating ion gradients that are essential for the homeostasis of cells. X-ray crystallographic structures exist for Ca^{2+}-, Na^{+}/K^{+}-, and H^{+}-ATPases from this family. However, CopA inhabits the P1b subclass of P-type ATPases that contains large insertions and deletions with respect to these existing structures. Of particular interest are the metal-binding domains (MBD) on the N- and C-termini of CopA, which are homologous to a large family of soluble metal-binding proteins. These MBDs are connected to the main body of CopA by flexible linkers, which allow the MBDs to interact with the other cytoplasmic domains responsible for binding and hydrolysis of ATP (Fig. 1). Previous work with CopA suggested that the N terminal MBD was involved in protein–protein interactions with one or more of the cytoplasmic domains (1).

For our modeling studies, we used X-ray crystallographic structures of related P-type ATPases, related metal-binding proteins, and isolated cytoplasmic domains of CopA. The shape of this model was constrained by a 12-Å resolution map of CopA determined by helical reconstruction of tubular crystals that were imaged by cryo-electron microscopy. We will describe our efforts first to model the structure of CopA with truncated N- and C-termini (ΔNΔC-CopA) and then to fit it to our EM map (1). Additionally, we will describe docking of an MBD to this ΔNΔC-CopA model in order to identify the orientation of the bound MBD that is consistent with our EM map and which provides the lowest energy domain interface.

2. Materials

2.1. EM Maps

The EM maps were obtained by helical reconstruction of tubular crystals of CopA, as described elsewhere in this book and by Wu et al. (1). Alternative reconstruction strategies are possible depending on the nature of the sample (e.g., crystalline or a homogeneous

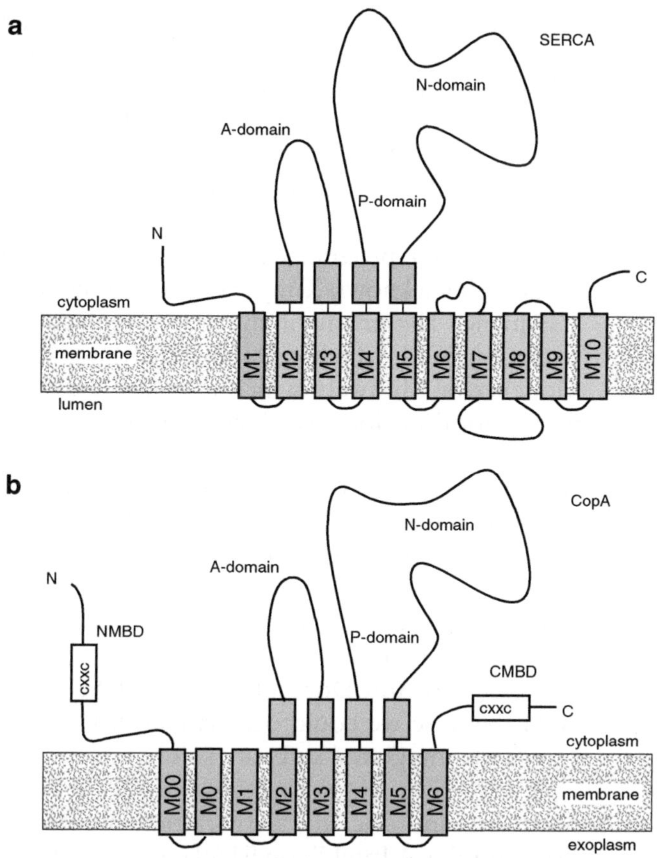

Fig. 1. Topology of the proteins used for modeling. (**a**) The Ca^{2+}-ATPase from sarcoplasmic reticulum (SERCA) provided the template for homology modeling. SERCA consists of ten transmembrane domains and two large cytoplasmic loops. In the 3D structure of SERCA, these cytoplasmic loops form three domains, which are denoted A-, N-, and P-domains. (**b**) The Cu-ATPase from *Archaeoglobus fulgidis* (CopA) has eight transmembrane helices. Six of these helices and the two cytoplasmic loops are analogous to the catalytic core of the SERCA molecule. Specialized metal-binding domains with the conserved CxxC motif are present on both the N-terminus (NMBD) and C-terminus (CMBD). Two additional transmembrane helices are also associated with the N-terminus of CopA, denoted M0 and M00.

preparation of isolated macromolecules). Depending on the software suite used for reconstruction, the user will obtain maps in a variety of different formats (e.g., SPIDER (2) or MRC (3)). These formats may be interconverted using em2em, which is a free component of the IMAGIC software suite (4). We have used the MRC format throughout and also encourage this practice.

2.2. Crystal/NMR Structures

Crystallographic or NMR structures of relevant proteins or domains may be obtained from the Protein Data Bank (http://www.rcsb.org), which is maintained by the Research Consortium for Structural Bioinformatics (RCSB). Coordinates are generally downloaded in PDB format.

2.3. Sequence Analysis Software

There is a plethora of sequence analysis programs available. Most of these programs will be adequate for this application since we are using structures having high homology with our target. The following is a small sampling of available programs. For multiple alignments: ICM (5), ClustalW (6), MULTALIN; for pairwise alignments: PyMol (Schrödinger, LLC).

2.4. Modeling Software

The following programs can automatically produce a structural model using a known X-ray crystallographic or NMR structure and a sequence alignment of the target molecule with this template structure: ICM, which is commercially available from Molsoft (5), Modeller or Modweb, which is available either as a Web application or for download to a local workstation (7).

2.5. Visualization Software

In order to display the EM maps as well as the models as they are developed, we have used Chimera (8), but PyMol (Schrödinger, LLC), Coot (9), or O (10) may also be used for this purpose.

2.6. Docking Software

There are many docking programs available, but the following performed well in an analysis by Cross (11): ICM (5), GLIDE (12), Surflex (13), AutoDock (14), and UCSF DOCK (15). We used ICM, because we could perform all of the necessary tasks within this single integrated software suite.

2.7. Computer Workstations

Modeling and docking were carried out on SGI-Linux workstations running 4 Intel Xeon 5160 cpus (3 GHz) with 2 Gbyte RAM. Using this hardware, ICM modeling of CopA (664 residues) took ~2 h; ICM docking of NMBD to cytoplasmic domains of CopA took ~12 h.

3. Methods

The methods described below illustrate how to build a protein model into a density map obtained by electron microscopy, how to validate the models obtained, and how to use the models in a protein–protein docking experiment. We assume that at least one structure exists with significant homology to the target protein. The steps involved in this process are as follows. First an appropriate template structure must be chosen as a basis for building a model for the target protein. Next, the sequences of the template structure and the target protein must be aligned. A structural model for the target protein can then be built based on this sequence alignment. This model should be carefully validated based on common sense and with regard to the existing data. Most importantly, the model

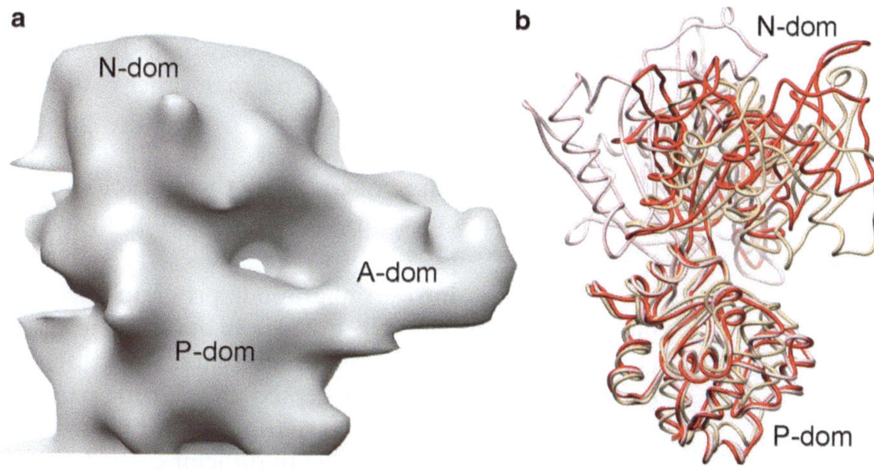

Fig. 2. Selecting the appropriate conformation for homology modeling. In the case of SERCA, the various different X-ray crystal structures show that the linker between the N- and P-domains is flexible. (**a**) The cytoplasmic domains from the EM density map of CopA at 17 Å resolution (1). The locations of the A-, N-, and P-domains are indicated. (**b**) After superimposing the P-domains from three atomic structures from SERCA, their N-domains are seen to adopt different positions. The position of the N-domain in the red model, which corresponds to 1IWO, is most consistent with the EM density map. PDB codes for the three structures are 1IWO (*red*), 1SUH (*pink*), and 1VFP (*olive*). The corresponding A-domains have been omitted for clarity.

should fit the EM density as closely as possible and we describe several means to optimize this fit. Finally, this model can be used to explore its interaction with known binding partners in silico by performing a protein–protein docking experiment.

3.1. Choose Structure as a Template for Constructing a Homology Model

1. The first step is to use the target protein sequence to search for homologs using Psi-Blast (blastp) on Pubmed with default parameters (16). From the resulting set of potential homologs, focus only on those that have exiting structures, because a homologous structure from the Protein Database (PDB) can be used as a template for building a model of the protein of interest. If structures of the homologous protein exist in multiple conformations, as was the case for Ca^{2+}-ATPase from sarcoplasmic reticulum (SERCA), it is necessary to determine the most appropriate conformation for use as a homology model. Criteria for this choice include the conformation that is most consistent with the shape of the EM map and/or the conformation that is most consistent with the biochemical conditions used for crystallization (Fig. 2).
2. For building a model of CopA, we chose the E2 conformation of SERCA, which is represented by the PDB entry 1IWO (17). This conformation is induced by chelation of the primary transport ion (EGTA for chelating Ca^{2+}) and is thus consistent with the conditions used for crystallization of CopA

(BCDS for chelating Cu^{+}). The resolution of the SERCA structure was acceptable (3.1 Å) and the overall shape of this structure matched our EM map better than other conformations of SERCA in the PDB (see Note 1).

3.2. Align Protein Sequences of Template Structure with Target Protein

1. Obtain the protein sequences of the homologous template protein and the target protein by searching a public database such as UniProt (13).
2. Feed sequences into an alignment program such as Clustal, GCG, and ICM (see Note 2). It is beyond the scope of this chapter to discuss ideal alignment parameters for the various programs. We generally accepted the default parameters and were guided by the positive alignment of the known signature sequences of P-type ATPases. In difficult cases the related publications for the software should be consulted.

3.3. Model Building

1. To make a homology model, an automated building routine can be used within ICM directly after sequence alignment. Under the "Homology" tab, select "Build Model." The resulting model will have the sequence of the target protein that has been folded into a three-dimensional protein based on the structure of the template molecule. ICM then allows the user to choose among several possible conformations of variable loops between secondary structure elements (see Note 3). Alternatively, the Modeller Web server (MODWEB) only requires that the user pastes in the target protein sequence, after which the alignment and model building are performed non-interactively.
2. Inspect the homology model (Fig. 3). After automatic model building, the researcher must verify that the new model makes sense and that it also fits the target map. To do this, view the homology model using visualization software (e.g., Chimera) and specifically evaluate regions of the molecule that have low homology to the template molecule (see Notes 4 and 5).

3.4. Fitting the Model to the EM Map

1. Manual fitting of the homology model to the EM density map: Manual fitting is a straightforward procedure that involves using the visualization software to move the model as a rigid body until it resides within the envelope defined by the EM density map. This process generally involves selecting the model through the graphical user interface and then using the computer mouse to manipulate its position relative to the density map. It will also be necessary to rotate the view in order to evaluate the fit from different angles. Consult the software documentation for specific instructions for these processes.

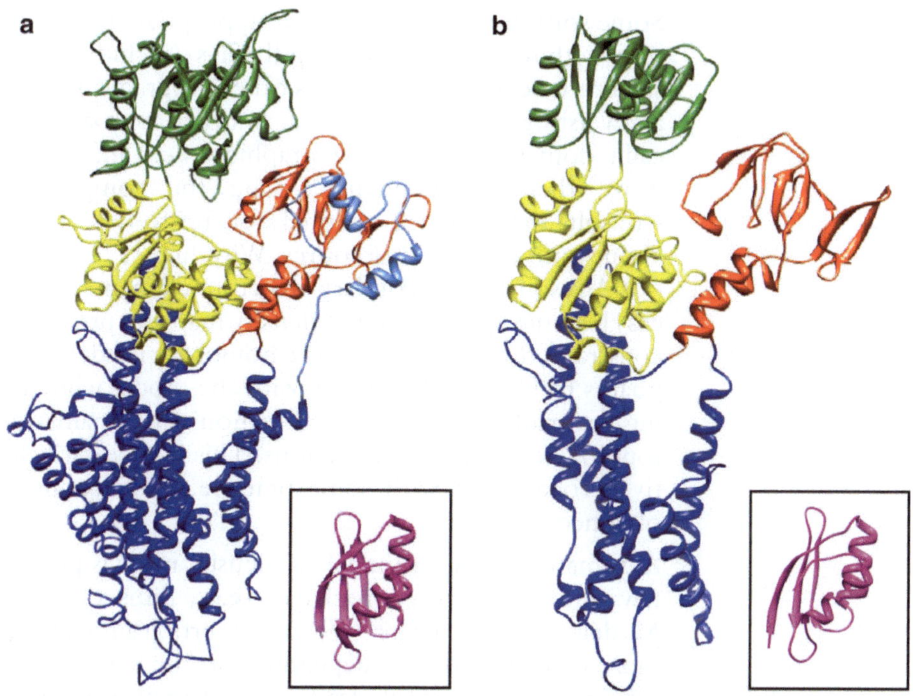

Fig. 3. Homology model for CopA. (**a**) The crystal structure of SERCA in the E2 conformation (1IWO) provided the template for building the homology model for CopA. SERCA does not have an NMBD, so a separate template (*inset*) was used, namely, a copper metallochaperone (2QIF) which has high sequence homology with NMBD. (**b**) The resulting homology models were produced by ICM. Note that CopA lacks a number of loops in the three cytoplasmic domains but that the overall configuration of these domains is preserved. Some of the transmembrane helices appear to have been misfolded and will need manual rebuilding for optimal fitting to the map. Domains have been colored as follows: N-domain *green*, P-domain *yellow*, A-domain *red*, transmembrane domain *blue*, metal-binding domain *purple*. SERCA also has two N-terminal helices associated with its A-domain (*light blue*).

Shape complementarity is the main criterion that guides the placement of the model into the map, though properties of the macromolecule may also guide this process (e.g., the membrane domain should be placed within the corresponding region of the map).

2. Optimized manual fitting: Once a reasonable overall fit is achieved, the homology model can be broken into individual domains and each domain can be moved independently in order to best match the density map. Prior knowledge of domain boundaries in the map and flexible joints in the structure are helpful in this regard. One strategy for separating the domains is to edit the pdb file for the model and to save the coordinates for each individual domain into a separate file.

Some software programs require appropriate header information within each individual pdb files (including secondary structure definitions), whereas other programs will work with a raw listing of the coordinates. Another approach is to select each domain through the graphical user interface and to move them independently using the computer mouse. Consult the software manual for up-to-date information about both of these procedures. With higher resolution maps (<8 Å resolution) secondary structure elements (e.g., α-helices) become apparent in the density map and provide a strong constraint for manipulating the homology model. The overall goal is to fill the EM density with the model while minimizing protruding, poorly fitted regions and maintaining the topology of domain connections. This is necessarily a subjective process, but with some practice, good results may be obtained.

3. Automated fitting of models to density maps is provided by a few different software packages (e.g., Situs (18), Chimera, Modeller, and Sculptor (19)). The procedures are generally more complicated and, depending on the nature of the project, may not necessarily lead to great improvements. In particular, although these procedures may be successful for rigid-body docking, the introduction of conformational changes in the homology model during fitting is an especially difficult problem (20, 21). Nevertheless, these programs generally quantify the goodness of fit, e.g., with a cross-correlation coefficient, which provides a useful parameter for comparing several slightly different fits.
4. A good compromise between manual and fully automated fitting is provided by the "Fit in Map" function in Chimera, which can be used to facilitate or to refine manual fitting. This function transforms the model, or some portion of it, into a map with a specified resolution. This model map is fit to the EM map using cross-correlation and quantifies the fit with the corresponding cross-correlation coefficient. Using "Fit in Map" at various resolutions together during the process of manual fitting can help find a global minimum.
5. Once the optimal location of the model within the EM map has been established, Crystallography and NMR System (CNS) may be used to fine-tune the model fitting by restraining geometry and symmetry during simulated annealing (22). In the "Utilities" section of CNS, an input script called "em_map_to_hkl.inp" can be used to transform an EM map into a reflection file used in all crystallographic refinement routines. Then, for example, symmetry could be imposed

during the refinement in CNS. With a 17 Å map of CopA, we found that CNS only marginally improved the manual fitting of our homology model. At higher resolution, one could use rigid body fitting of the various domains to improve their fits.

6. Iterate these procedures: Analysis and refinement are necessarily iterative, especially if CNS is used. The researcher should monitor the outcome of each round to determine if changes are appropriate or if the results have converged to a stable answer. This process necessarily entails visual inspection of the model and evaluating its fit to the EM map.

3.5. Docking of Additional Components to the Homology Model

1. Obtain model for the element to be docked. For our work, we built a model for the N-terminal metal-binding domain of CopA (NMBD), which has many homologous structures in the Protein Data Bank. We chose a copper metallochaperone (2QIF), which has an X-ray crystallographic structure with high resolution and which has high homology to NMBD. We then followed the procedure outlined in Subheadings 3.2 and 3.3 to produce a homology model of the NMBD of CopA.
2. Docking of the two elements using ICM: In the protein–protein docking module of ICM, select one element as the receptor (ΔNΔC-CopA in our case) and the other element as the ligand (NMBD). Next select a region of the receptor for docking of the ligand. Based on this selection ICM calculates various potentials (electrostatic, hydrophobic, etc.) on a grid around the receptor (23). Solvent is accounted for by calculating its shielding effect on the solvent-exposed charges and including this in the grid potentials (24). The size of the grid may be varied, but the defaults generally produce satisfactory results. The output from our docking experiments is shown in Fig. 4. The table lists a number of alternative docking solutions sorted according to their energy (see Note 6). In our case, we found that the lowest energy solution gave a plausible location of NMBD that was consistent with extra density within the EM map.
3. Check validity of docking result relative to the EM map. Because the EM map was not considered during the docking experiment, the correspondence between the extra EM density and the location of the NMBD provides a powerful validation of the result (Fig. 5). Thus, the combination of docking and fitting acts synergistically to cross-validate the model and to suggest not only a location but also a specific binding interface that neither technique could independently produce.

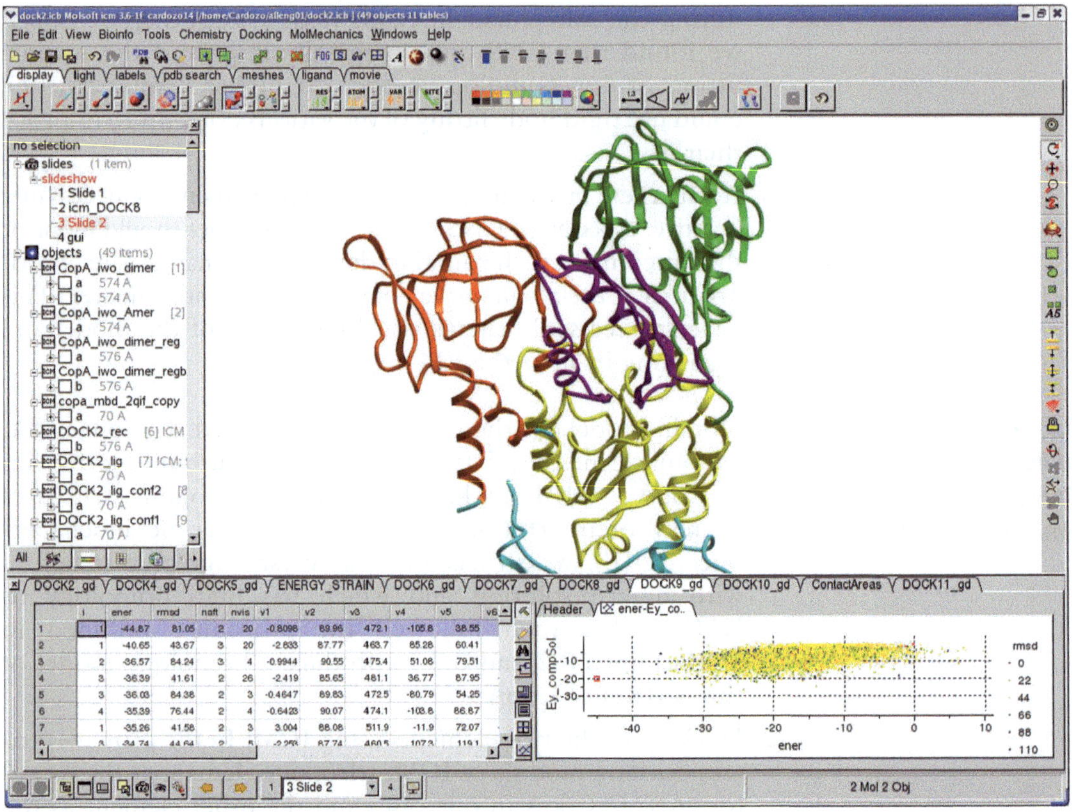

Fig. 4. Docking Results in ICM. The graphical user interface for ICM has several windows for viewing relevant information. The main display window allows interactive visualization of the model. The *top-left* window allows selection of objects for viewing in the main display window. The *bottom-left* window shows a table of docking results sorted by energy value; clicking an individual row displays that solution in the visualization window. The *bottom-right* window displays a plot of each docking solution versus energy (on the abscissa). Low-energy solutions (e.g., *red square*) that are well separated from the remainder indicate a successful docking experiment.

4. Notes

1. When choosing between equivalent structures to use as a template, the highest resolution structure is preferable. Higher resolution structures are, by definition, better determined statistically and thus more reliable as a template. α-helices are recognizable in X-ray maps with resolutions better than 5.5 Å while β-sheets require <3 Å resolution (25). Thus, if the target contains significant amounts of β-sheet, the template model should have a resolution better than 3 Å to be confident of correct placement of modeled β-sheets; an all α-helical model could be based on a homologous structure of lower resolution. Nevertheless, higher resolution is valuable for providing information about the

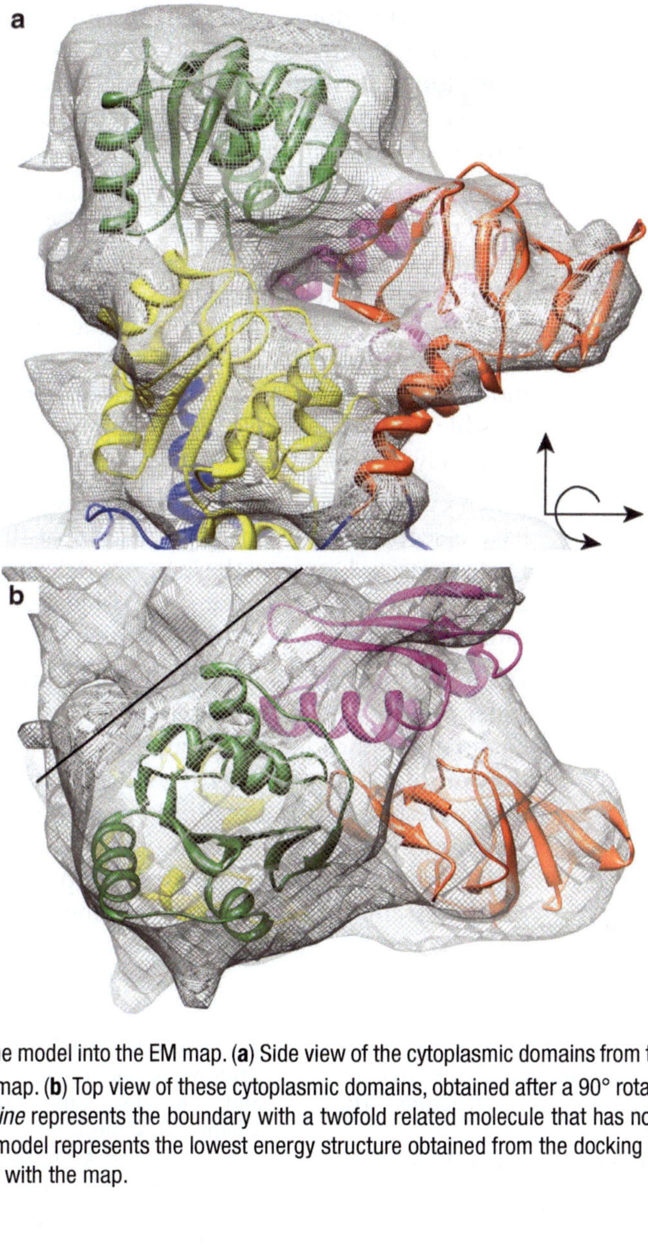

Fig. 5. Result of fitting the model into the EM map. (**a**) Side view of the cytoplasmic domains from the final model fitted into the 17 Å resolution EM map. (**b**) Top view of these cytoplasmic domains, obtained after a 90° rotation about the horizontal axis. The *oblique black line* represents the boundary with a twofold related molecule that has not been fitted with a corresponding model. This model represents the lowest energy structure obtained from the docking experiment and appears to produce an optimal fit with the map.

side chain configurations, which can prove valuable for docking experiments.

2. In an alignment of the CopA and SERCA sequences, it was immediately apparent that CopA contains many common features but that there were discrepancies in the size of cytoplasmic domains and the number of transmembrane helices. These discrepancies arise from the fact that CopA belongs to a distinct subclass of P-type ATPases and indicate areas of potential problems in model building.

3. After model building with ICM the user is given the possibility to choose among several different loop conformations for many loops throughout the model. Unless the EM density map has sufficient resolution to discriminate between the alternatives, we generally accept the default.
4. In our case we had the opportunity to compare our new model with crystal structures of the three isolated cytoplasmic domains of CopA (2HC8 and 3A1D). This comparison indicated that the model building routine of ICM failed to make a reasonable N-domain for CopA, presumably due to the numerous deletions relative to SERCA. Therefore we opted to replace this modeled N-domain with the crystal structure 2HC8. Another possible approach would be to align and build each domain separately.
5. When evaluating a model built by software it can be helpful to have a secondary structure prediction of the target sequence. Discrepancies between the new model and the prediction in either secondary structure or residue numbering should be carefully analyzed to determine which is likely correct.
6. We used default settings for all our protein–protein docking experiments. The top solution or lowest energy pose for the ligand was reproducibly located in EM density that we had previously assigned to NMBD. This energy of interaction does not correspond to the real binding energy due to the various assumptions and approximations made for the calculation. Rather, these calculated energies only provide a measure of the relative strength of the various binding poses.

References

1. Wu CC, Rice WJ, Stokes DL (2008) Structure of a copper pump suggests a regulatory role for its metal-binding domain. Structure 16: 976–985
2. Frank J, Radermacher M, Penczek P, Zhu J, Li Y, Ladjadj M, Leith A (1996) SPIDER and WEB: processing and visualization of images in 3D electron microscopy and related fields. J Struct Biol 116:190–199
3. Crowther RA, Henderson R, Smith JM (1996) MRC image processing programs. J Struct Biol 116:9–16
4. van Heel M, Harauz G, Orlova EV, Schmidt R, Schatz M (1996) A new generation of the IMAGIC image processing system. J Struct Biol 116:17–24
5. Cardozo T, Totrov M, Abagyan R (1995) Homology modeling by the ICM method. Proteins 23:403–414
6. Thompson JD, Gibson TJ, Higgins DG (2002) Multiple sequence alignment using ClustalW and ClustalX. Curr Protoc Bioinformatics, Chapter 2, Unit 2.3
7. Sanchez R, Sali A (2000) Comparative protein structure modeling. Introduction and practical examples with modeller. Methods Mol Biol 143:97–129
8. Pettersen EF, Goddard TD, Huang CC, Couch GS, Greenblatt DM, Meng EC, Ferrin TE (2004) UCSF Chimera–a visualization system for exploratory research and analysis. J Comput Chem 25:1605–1612
9. Emsley P, Lohkamp B, Scott W, Cowtan K (2010) Features and development of Coot. Acta Crystallogr D Biol Crystallogr 66:486–501
10. Jones TA, Zou JY, Cowan SW, Kjeldgaard M (1991) Improved methods for building protein models in electron density maps and the location

of errors in these models. Acta Crystallogr A 47:110–119

11. Cross JB, Thompson DC, Rai BK, Baber JC, Fan KY, Hu Y, Humblet C (2009) Comparison of several molecular docking programs: pose prediction and virtual screening accuracy. J Chem Inf Model 49:1455–1474
12. Halgren TA, Murphy RB, Friesner RA, Beard HS, Frye LL, Pollard WT, Banks JL (2004) Glide: a new approach for rapid, accurate docking and scoring. 2. Enrichment factors in database screening. J Med Chem 47:1750–1759
13. Jain AN (2003) Surflex: fully automatic flexible molecular docking using a molecular similarity-based search engine. J Med Chem 46:499–511
14. Goodsell DS, Morris GM, Olson AJ (1996) Automated docking of flexible ligands: applications of AutoDock. J Mol Recognit 9:1–5
15. Moustakas DT, Lang PT, Pegg S, Pettersen E, Kuntz ID, Brooijmans N, Rizzo RC (2006) Development and validation of a modular, extensible docking program: DOCK 5. J Comput Aided Mol Des 20:601–619
16. Altschul SF, Madden TL, Schaffer AA, Zhang J, Zhang Z, Miller W, Lipman DJ (1997) Gapped BLAST and PSI-BLAST: a new generation of protein database search programs. Nucleic Acids Res 25:3389–3402
17. Toyoshima C, Nomura H (2002) Structural changes in the calcium pump accompanying the dissociation of calcium. Nature 418:605–611
18. Wriggers W, Milligan RA, McCammon JA (1999) Situs: a package for docking crystal structures into low-resolution maps from electron microscopy. J Struct Biol 125: 185–195
19. Rusu M, Birmanns S (2010) Evolutionary tabu search strategies for the simultaneous registration of multiple atomic structures in cryo-EM reconstructions. J Struct Biol 170:164–171
20. Trabuco LG, Villa E, Mitra K, Frank J, Schulten K (2008) Flexible fitting of atomic structures into electron microscopy maps using molecular dynamics. Structure 16:673–683
21. Hinsen K, Reuter N, Navaza J, Stokes DL, Lacapere JJ (2005) Normal mode-based fitting of atomic structure into electron density maps: application to sarcoplasmic reticulum Ca-ATPase. Biophys J 88:818–827
22. Brunger AT (2007) Version 1.2 of the crystallography and NMR system. Nat Protoc 2:2728–2733
23. Fernandez-Recio J, Totrov M, Abagyan R (2002) Soft protein-protein docking in internal coordinates. Protein Sci 11:280–291
24. Fernandez-Recio J, Totrov M, Abagyan R (2002) Screened charge electrostatic model in protein-protein docking simulations. Pac Symp Biocomput. 552–563.
25. McRee DE (1999) Practical protein crystallography. Academic, San Diego

Chapter 14

Phasing Electron Diffraction Data by Molecular Replacement: Strategy for Structure Determination and Refinement

Goragot Wisedchaisri and Tamir Gonen

Abstract

Electron crystallography is arguably the only electron cryomicroscopy (cryo EM) technique able to deliver atomic resolution data (better then 3 Å) for membrane proteins embedded in a membrane. The progress in hardware improvements and sample preparation for diffraction analysis resulted in a number of recent examples where increasingly higher resolutions were achieved. Other chapters in this book detail the improvements in hardware and delve into the intricate art of sample preparation for microscopy and electron diffraction data collection and processing. In this chapter, we describe in detail the protocols for molecular replacement for electron diffraction studies. The use of a search model for phasing electron diffraction data essentially eliminates the need of acquiring image data rendering it immune to aberrations from drift and charging effects that effectively lower the attainable resolution.

Key words: Electron cryomicroscopy (Cryo-EM), Electron crystallography, Electron diffraction, Molecular replacement, Structure refinement

1. Introduction

Electron crystallography became a bona fide method for determining the structures of membrane proteins following the pioneering work by Henderson and Unwin in the mid 1970s (1). Since then, the field of electron crystallography has been steadily evolving with recent advancements in methodology and technology that led to a number of success stories of membrane protein structures that have been determined to resolutions that rival those achieved by X-ray crystallography. Some of these developments are described in detail elsewhere in this book, including advances in grid preparation and

Ingeborg Schmidt-Krey and Yifan Cheng (eds.), *Electron Crystallography of Soluble and Membrane Proteins: Methods and Protocols*, Methods in Molecular Biology, vol. 955, DOI 10.1007/978-1-62703-176-9_14,

sample embedding, improved hardware for data collection, the development of the field emission electron source as well as the development of a highly sophisticated helium-cooled stage (2).

In electron crystallography the membrane protein of interest is crystallized in two-dimensions within a lipid-bilayer where its structure and function can be assayed directly. The environment experienced by the protein closely mimics the native state of the protein in the cell. Lipids and membrane proteins coevolve to form biological membranes and lipids intimately influence the structure and function of membrane proteins. Extensive lipid–protein interactions occur within the lipid bilayer as lipids fit into crevices and irregularities on the protein surface to maintain an electrochemical seal across the membrane. One of the strengths of electron crystallography is that both protein and lipid structure can be determined and lipid–protein interactions studied directly if high enough resolution is achieved as illustrated by a number of examples (3–6). The methods used for growing two-dimensional (2D) crystals are discussed elsewhere in this book.

Once 2D crystals are obtained, structure determination follows either from images of the 2D crystals and/or from electron diffraction data. Images of 2D crystals contain both phase and amplitude information, and can be used directly for structure determination if the resolution is sufficient. The protocols used for image acquisition in electron crystallography are discussed in another chapter in this book (Chapter 8). Once images are collected, they are scanned and Fourier filtered in order to generate a 3D density map using the MRC suite of programs (7), 2DX (8), and/or IPLT (9). In practice it is very difficult to collect high-resolution images of 2D crystals, particularly if these are tilted. The main reasons are instability of the cryo-stage (mechanical and thermal drift) and various charging effects that "smear out" the high-resolution features in the images perpendicular to the tilt axis. The result is that high-resolution structure determination by this technique can take years, depending also on factors such as cryo-EM access, experience, and time that can be dedicated to the image collection.

In contrast, electron diffraction can deliver atomic-resolution information because it is not affected by drift and charging effects and is independent of the contrast transfer function (CTF). A number of recent studies used electron diffraction to determine the structures of membrane proteins to resolutions where water molecules become visible and the density for the protein approaches atomicity (3, 4, 6), with the highest resolution recorded at 1.7 Å anisotropically and 1.9 Å isotropically (3). We outline detailed protocols and strategies for the collection of high-resolution electron diffraction data elsewhere in this book. One drawback of electron diffraction is that it provides only intensities (amplitude information) but not phase information. To solve the structure, phases need to be determined by other means. In X-ray crystallography, heavy-atom labeling can be used to determine phases but this is not possible in electron crystallography because the phasing power is too small, and the phases thus obtained are not accurate enough to reliably interpret the map (10). In the case of AQP0, phases were determined by molecular replacement (MR)—a common procedure in X-ray crystallography but one that was implemented only recently for electron crystallography (11). This procedure rapidly generated an atomic-resolution electron density map into which it was possible to build an atomic model using protocols analogous to X-ray crystallography. In this chapter we outline in detail protocols for molecular replacement in electron crystallography as well as strategies for structure determination and refinement.

2. Materials

2.1. Equipment

A computer running any Linux platforms or Macintosh OS-X, or Microsoft Windows, supported by the CCP4 software for macromolecular crystallography.

2.2. Program

Among the most widely used and well-supported MR programs are Amore (12, 13), Molrep (14, 15), and Phaser (16, 17) which are distributed as part of the Collaborative Computational Project No. 4 (CCP4) software for macromolecular crystallography (18) and MR implementation in CNS (19, 20). All programs mentioned here are freely available for academia. However, we recommend inexperienced readers to use the Molrep program because of the ease of use and its several automated MR features. Molrep is installed as part of the CCP4 program suite and can be run from the CCP4i graphical user interface (21), command lines, or shell scripts. A link for software download and instructions for installation can be found at the CCP4 Web site (www.ccp4.ac.uk). For other MR programs, readers are referred to appropriated program manuals for more information.

3. Methods

3.1. Patterson Function

MR is not a homology modeling technique, in which the amino acid sequence of the unknown structure is simply threaded computationally onto the known structure of a homologous protein. MR is a method to estimate initial phases of an unknown crystal structure using a structure of a related known molecule (search model). MR comprises complex calculations for comparing and matching mathematical information derived from the structure of the search model with data derived from the diffraction intensities of the unknown structure in order to arrive at approximated phases. The mathematics involved is known as the Patterson function—the Fourier transform of the squared amplitude ($|F|^2$) with phases set to zero. In simple terms, the Patterson function corresponds to a map of inter-atomic vectors that can be calculated directly from experimental diffraction intensities ($I = |F|^2$) of the unknown structure without phase information. Likewise, the Patterson function can be calculated from amplitude parts of the search model structure factors without using the phase components. When the structure of the search model is similar to the unknown structure, their inter-atomic vectors resemble each other, resulting in similar Patterson functions.

The Patterson function for a protein structure in a crystal lattice is complicated by having multiple sets of inter-atomic vectors within the same protein molecules (self-vectors) from different orientations in the lattice related by crystallographic symmetry, and inter-atomic vectors across to neighboring molecules (cross-vectors). These vectors provide valuable information on how protein molecules are arranged in the crystal lattice. However, if we draw a sphere centered at the origin with a small enough radius, the vectors included in the sphere are mostly self-vectors. The use of the vector maps (Patterson function) is the basis for MR.

3.2. Data Preparation

Prior to starting MR calculations it is important to check the quality and completeness of the experimental diffraction data, as well as any anisotropy, intensity statistics, and possible twinning (the latter not usually being a problem in electron crystallography). These factors may impact the success of finding MR solutions and MR parameters can be adjusted appropriately in some difficult cases to increase the signal.

Even though there are only 17 possible plane groups for 2D crystals (Table 1), compared to 230 space groups for 3D crystals (of which 65 are for biological macromolecules), finding the right space group still requires some effort. For example, indexing and merging programs cannot distinguish space groups within the same Laue group such as P222, $P222_1$, and $P2_12_12$, without knowledge of systematic absences. Systematic absences may not be reliable when data are not complete because certain wedges containing reflections of systematic absence information are missing or when low-resolution intensities are overloaded and eliminated by a merging program. Also noncrystallographic translation in the asymmetric unit often introduces awkward intensity profiles that appear as if systematic absences are possible. Nevertheless, molecular replacement may be able to pick out the true space group based on the success of finding a correct solution with high correlation. With additional information on the native Patterson function (also

Table 1
17 space groups in 2D crystals and their 3D equivalents for biological macromolecules

Cell geometry	Plane group	2D space group	2D space group number	IUCr space group	IUCr space group number	Bravais lattice	Systematic absence (*h, k, l*)
Oblique	*p1*	P1	1	P1	1	Primitive triclinic	–
Oblique	*p2*	P2	2	P2	3*		–
Rectangular	*pm*	P12	3	P2	3*	Primitive monoclinic	–
Rectangular	*pg*	$P12_1$	4	$P2_1$	4*		(0, 2*n*+1, 0)
Rectangular	*cm*	C12	5	C2	5	C centered monoclinic	–
	p2mm	P222	6	P222	16		–
Rectangular	*p2mg*	$P222_1$	7	$P222_1$	17	Primitive orthorhombic	(0, 0, 2*n*+1)
	p2gg	$P22_12_1$	8	$P2_12_12$	18		(2*n*+1, 2*n*+1, 0)
Rectangular	*c2mm*	C222	9	C222	21	C centered orthorhombic	–
	p4	P4	10	P4	75*		–
Square	*p4mm*	P422	11	P422	89	Primitive tetragonal	–
	p4gm	$P42_12$	12	$P42_12$	90		(0, 2*n*+1, 0)
	p3	P3	13	P3	143*		–
	p3m1	P321	14	P321	150		–
Hexagonal	*p31m*	P312	15	P312	149	Primitive hexagonal	–
	p6	P6	16	P6	168*		–
	p6mm	P622	17	P622	177		–

*Denotes polar space groups

known as self-Patterson), one might be able to confirm noncrystallographic translation in the crystal. An advantage of electron crystallography over X-ray crystallography is its ability to capture images of 2D crystals directly where the crystal lattice is visualized and the correct space group can be deduced based on the packing pattern in the lattice.

Most crystallographic programs for structure determination were developed for X-ray crystallography and crystallographic symmetry is dictated by space group information that complies with the International Union of Crystallography (IUCr) (and preferably using the Hermann–Mauguin notation). In electron crystallography, the unit cell of the 2D crystals may be assigned by indexing programs as *a* and *b* on the plane of the crystal. Therefore, when a plane group is converted to a space group for structure determination, crystallographic axes in some space groups that do not follow the convention may need to be specified or swapped (Table 1). For example, in plane group *p2* (P2) the symmetry axis is perpendicular to the plane of the membrane while in plane group *pm* (P12) or *pg* ($P12_1$), the symmetry axis is parallel to the membrane plane. By convention, the symmetry axis in space groups P2 and $P2_1$ is parallel to the *b* axis where $\beta \neq 90°$. The orthorhombic plane group *pgg* ($P22_12_1$) is equivalent to space group $P2_12_12$ for which the twofold axis runs parallel to the *c* axis (perpendicular to the membrane plane). However, in the plane group *p2mg* ($P222_1$), the 2_1-screw axis is parallel to the *b* axis on the membrane plane and should be swapped to the *c* axis.

3.3. Model Preparation

The success of a molecular replacement search often depends on the quality of the search model. The more similar the model to the unknown structure, the higher the correlation of their Patterson functions. The search model is generally identified by amino acid sequence alignment. Structures with sequence identity higher than 25% often present structural similarity. However, some structures can appear to be similar (same fold) even with less than 25% sequence identity but their pairwise superposition could yield large root-mean-square (r.m.s.) deviations. Since Patterson functions are distance based (inter-atomic vectors), models with lower r.m.s. deviation from the unknown structure are generally more suitable than models with larger r.m.s. deviations. If the choice of model is limited, having a decent model is still better than having no model. If multiple unique structures can be identified form homology search, it does not hurt to try them all for MR. However, the model with higher sequence identity is expected to have smaller r.m.s deviation from the unknown structure and that can make MR easier. Alternatively, multiple homologous structures (of the same part in the unknown structure) can be used for MR by superimposing them into one ensemble and using the ensemble as a search model. The ensemble model may work better than a single model in some cases.

After a known homologous structure is selected, the model should be modified to remove hydrogen atoms, alternative conformations, water molecules, and heteroatoms (such as ligands, and metals). Any parts of the models that may pose major differences to the unknown structure should also be removed. If the model has high sequence identity (>50%) to the unknown structure, the model side chains may be left unchanged. If the sequence alignment can be obtained with high confidence, identical residues may be kept intact and only different residues should be modified to alanine. Alternatively, all side chain C_γ may be kept in addition to C_β (as alanine) since removing too many atoms that are otherwise similar to those in the unknown structure will decrease the signal and correct solutions may become less clear in some difficult cases. The program Chainsaw (22) offers a convenient way to generate models from PDB files with choices of model modification from sequence alignment. If the sequence identity is low (<25%), all side chains are often removed by converting the model to poly-alanine (with glycine intact) as sequence alignment becomes less reliable. Protein termini and flexible loops connecting secondary structure elements should be examined and removed if these regions are highly variable (by length and by structure flexibility). This can be guided by checking atomic *B*-factors. Flexible regions in the structure usually have higher *B*-factors than the average for the entire molecule and they can be removed. After the coordinates are modified, atomic *B*-factors of the model should be set to a low level (15–20 $Å^2$). Alternatively, the model *B*-factors can be computed based on surface accessibility. In this approach, atoms on the surface will be given higher *B*-factors than buried atoms to smear out electron density at the surface to account for conformational flexibility of surface residues.

If the model and the unknown structure contain multiple domains connected by loops, it is possible that their domain arrangement may be different due to flexibility. It is recommended that multiple search models each containing only one domain should be prepared if no clear solutions can be found using the complete multi-domain model. No clear solution is likely caused by the difference in position of each domain relative to another in the search model compared to the domains in the unknown structure. After each domain is separated into different search models, MR should be performed by first searching for the largest domain with highest homology (accounting for largest structure amplitudes of the asymmetric unit content with most similar Patterson function). The multiple copy search strategy is given in detail in Subheading 3.7.

Before performing the molecular replacement search, the model structure is placed in a large P1 unit cell to simplify the Patterson function. In this way, the inter-atomic vectors of the model are only self-vectors clustered near the origin within a certain spherical radius far separated from cross-vectors. In most cases, the model should also be shifted to the origin. Many MR programs have automated features to prepare the model in a large P1 cell and shift the model to the origin. By default, users only need to provide the model coordinates.

3.4. Self-Rotation Function

The self-rotation function is a useful tool for determining how many molecules are present in the asymmetric unit and how they are related by noncrystallographic rotational symmetry. Essentially, the self-rotation function is a product of rotating and imposing the Patterson function onto itself and does not require a search model. The self-rotation function always produces the largest peak at the origin corresponding to its un-rotated self-imposing. If there is more than one molecule in the asymmetric unit that is not related by translation, rotated self-vectors for each molecule within a certain spherical radius can be matched to un-rotated self-vectors from other molecules which then produce additional peaks at the angles corresponding to the relative rotation between these molecules.

3.5. Cross-Rotation Function

The typical MR method comprises two essential steps: (1) determination of the orientation of the unknown structure by rotational searches of the model, and (2) determination of the position of the unknown structure in the asymmetric unit by translational searches of the oriented model from the first step. The first step involves the calculation of the cross-rotation function (RF) which is essentially a correlation between the model and the crystal Patterson functions within a limited spherical radius. By rotating the model, the Patterson function is also rotated by the same angles. Therefore, small increments of rotation angles of the model Patterson function can be sampled to cover the entire angular possibility, and the match to the un-rotated crystal Patterson function can be calculated by RF. At any rotation angle, a large score (peak) appears when the rotated model Patterson vectors and the crystal Patterson vectors coincide. The higher the scores, the more likely the solutions are correct. In practice, the correct solutions are usually separated from the incorrect ones by a large drop in the score. The number of correct solutions also depends on how many molecules are in the asymmetric unit and how they are oriented relative to each other.

3.6. Translation Function

The second step involves the calculation of the translation function (TF). In space groups with a symmetry higher than P1, the Patterson function also includes cross-vectors generated from atoms that belong to different molecules that are related by crystallographic symmetry. When the search model is translated, the cross vectors change as the positions of symmetry-related molecules calculated from the model for that particular space group also change according to the symmetry operators. The correlation between the Patterson functions of the translated model and the crystal data are calculated by the TF. The translation search is relative to the space group symmetry. In polar space groups such as P2 (Table 14.1 indicated by asterisk), the translational shift is necessary only in the plane perpendicular to the rotational symmetry axis as the Patterson function does not change in the direction parallel to the symmetry axis.

3.7. Molecular Replacement Search Strategies

Biochemical knowledge of the proteins such as their oligomeric states and symmetry can often help with the interpretation of MR results such as noncrystallographic symmetry and relative orientations of the molecules in the crystal. Most membrane proteins have simple oligomeric states (related by rotational symmetry instead of complicated point group symmetry) due to the constraint of the membrane plane both in the native membrane bilayer in cells and in 2D crystals. However, one has to be open minded that crystal packing and crystallographic symmetry may border on the multimeric proteins in such a way that the contents of the asymmetric unit are not necessarily the entire biological ensemble. For example, if a dimeric protein related by a twofold rotational noncrystallographic symmetry crystallizes in the space group containing a crystallographic twofold that coincides with the dimer twofold axis, the asymmetric unit content may only contain one subunit.

MR gives a clearer solution when the search model approximates the totality of the asymmetric unit content. When there are two copies or more of the molecule in the asymmetric unit, multiple approaches can be used to find correct solutions. The first approach is to perform a standard MR search for one copy at a time from a search model containing one copy of the molecule. In this approach, the first copy is searched for and the best rotation function can be selected. Note that in polar space groups in which the origin is not defined along the symmetry axis, the first copy is placed in an arbitrary origin and TF is only calculated in the direction perpendicular to the symmetry axis. By fixing the first copy, the model is then used to search for a second copy. If a solution is found, RF and TF are usually further improved. Additional copies can be investigated by fixing the position of initial units until all copies are found. This approach works well in most cases but becomes problematic when the asymmetric unit contains too many copies and the search model only accounts for less than 20% of the asymmetric unit content.

The second approach is to perform searches with two copies simultaneously. This approach has been implemented in the program Molrep (23) as a dyad search (for two identical copies) to find top orientations from RF for a monomer and construct dimer search models (dyads) based on monomer orientations identified from RF. From the properly oriented dyads, the program calculates a special translation function (STF) that gives the intermolecular vector between properly oriented monomers. This information is then used to calculate standard TF and correlation coefficients from the dyad search model. This approach may be useful when solutions cannot be found by searching with only a single copy.

The third approach is to use the locked cross rotation function (LRF) and locked translation function (LTF) (24–26). This approach takes advantage of point group NCS operators if present in the assembly. The point group NCS is first identified using the self-rotation function. Rotational search is then carried out for the first and second copies. Only orientations consistent with the NCS found by self-rotation functions are selected. The NCS is then expanded for additional copies to define their orientations in the entire assembly and to produce LRF. Next, the translational search is performed for the first copy, followed by the NCS expansion for additional copies in the assembly. The LTF is calculated based on vectors among NCS related molecules to position them correctly relative to the center of the NCS in the assembly before performing standard TF to locate the position of the assembly in the unit cell. This approach is only applicable to proteins that have a point group symmetry and proves particularly beneficial to large protein assemblies.

Alternatively, if the homologous protein has a known functional oligomeric assembly, sometimes it can be assumed that the unknown structure may also form a similar assembly. Therefore, the search model containing a complete assembly (dimer, trimer, etc.) should be tried if searching by a monomer does not find clear solutions. However, the assembly of the model can be slightly different from that of the unknown structure by a combination of a small rotation and translation from the center of the NCS for each subunit, resulting in failure of MR. This problem may be overcome but requires some effort to vary the orientations and positions of each monomer in the model assembly and perform several MR runs with different modified models.

3.8. Protocol for Molecular Replacement Using CCP4i

The advancement in computer technologies, software development, and automation has been tremendously helpful in speeding up or eliminating tedious procedures dealing with data and model preparation for MR. Programs such as Molrep (15) and Phaser (16) can finish automated model preparation and MR in all possible space groups in a short period of time. Program pipelines such as Balbes (27) and MrBUMP (28) also offer a convenient way to solve structures by MR based on only the input data and amino acid sequence of the unknown structure.

Here we will focus on the use of Molrep together with additional CCP4 programs (18) and, as a test case, for phasing the electron crystallographic structure of lens aquaporin-0 (AQP0, protein data bank (PDB) accession code 2B6O) (3) using the X-ray structure of bovine aquaporin-1 (AQP1, protein data bank accession code 1J4N) (29). We will use the CCP4i graphic user interface (21) to run the relevant programs. The program names are given at the beginning for each step for users who are interested in running the programs from command lines or shell scripts.

3.8.1. Downloading Files from Protein Data Bank

1. Create a directory where you store files and perform MR. This directory will be referred to as the working directory.
2. Go to the PDB Web site (www.pdb.org).
3. In the search box for "PDB ID or text," type "2B6O" and click search to retrieve a PDB entry for the electron crystallographic structure of AQP0. On the right side of the PDB code, click on the drop-down menu "Download files," and select by clicking "Structure Factor (text)." You will be asked whether to open or save file "2b6o-sf.cif." Save this file in the working directory. This file contains amplitudes ($|F|$) in mmCIF format from diffraction data of aquaporin-0 crystal which will be treated as an unknown structure. The first part of this file contains header information describing in each line each column in the second part. The second part contains reflection data in column format. Columns 4–6 contain $h\ k\ l$ indices for each reflection, and columns 5 and 6 contain measured amplitudes ($|F|$) and their signal-to-noise ratio expressed as standard deviation ($\sigma|F|$), respectively. The last column is the reflection status which tells each reflection whether to be used for refinement or as a test set for cross-validation.
4. In the same PDB entry as above, click on "Sequence" tab. Under "Chain Display," click on "[fasta]" to download file "2B6O_A.fasta.txt" containing the aquaporin-0 sequence in fasta format. Rename this file to "2B6O_A.fas" and save it in the working directory where you want to perform MR. Next, use a text-editing program to remove the first line (see below) and keep only the amino acid sequence.

   >2B6O:A|PDBID|CHAIN|SEQUENCE

   Save the file as "2B6O.seq" in the working directory.
5. In the search box on the PDB Web site, type "1J4N," then click search to retrieve a PDB entry for the crystal structure of the AQP1 water channel. On the right side, click on the drop-down menu "Download files," and select by clicking "PDB File (text)." You will be asked to open or save file "1J4N.pdb." Save this file in the working directory. This file contains coordinates of AQP1 which will be used as a search model for molecular replacement.

3.8.2. Setting up CCP4

1. For Linux or Macintosh OS-X, open a terminal window and change directory into working directory. Launch CCP4i by typing "ccp4i" and hit enter. For Windows, double click on CCP4i icon.

2. If CCP4i is started for the first time, users will be asked to create a "CCP4i project." Fill in a one word alias such as "AQP0" for the project name and the full directory paths for the project (where the working directory is located) and temporary directory (for temporary files created by the program to be stored), then click "Apply & Exit." If the directories do not exist, users will be asked to allow the program to create them by clicking on "create directory."

3.8.3. Converting Data to mtz File Format

The CCP4 software suite uses mtz binary files as a format for reflection data. Users are encouraged to visit the CCP4 Web site for more information on the mtz file format. If the reflection data file contains intensities (*I*) (after indexing and merging) instead of amplitudes (|*F*|), the file should be converted using the "data reduction" module. Appropriate tasks may be chosen depending on the data file format. Here the file "2b6o-sf.cif" already contains |*F*| and we will use the program cif2mtz to convert mmCIF format to mtz.

1. To convert reflection data in mmCIF file format to mtz format, open the "Reflection Data Utilities" module on the left side of the CCPi window and select "Convert to/modify/extend MTZ" task. A task interface window appears. Fields colored orange are required and fields in gray can be left blank for default parameters.
2. From the top in the "Job title" field, type in a job title that can be easily recognized for back-tracking such as "convert from mmCIF to mtz." In the next line "Import reflection file in," select from the drop-down menu "mmCIF." The box in front of "create full unique set of reflections" should be checked and "keep existing FreeR data" is selected from the drop-down menu (default). (Optional) If the data does not have FreeR reflections flagged, "generate FreeR data" may be selected to flag FreeR reflections at this step. To enter input files, click on "Browse" on the right of the "To" field and select the file name "2b6o-sf.cif" from a list in the file window. The file name "2b6o-sf.mtz" automatically appears in the "Out" field. This file will be the output reflection file in the mtz format. Type in the fields "AQP0" for crystal name and "HighRes" for data set name. This is useful if multiple data sets from the same crystal or data from multiple crystals will be stored in the same file (in different columns) as their names can be established from the identifiers. However, these fields may be left blank. In the cell space group name or number, type in "P422" for space group. This should be a space group in which diffraction intensity data are indexed and merged. In the cell dimensions field, type a "65.5," b "65.5," c "160.0," alpha "90," beta "90," gamma "90." The numbers are in Ångstrom unit for unit cell dimensions and degrees for angles. In the line "FreeR column label," type "FREE" (default).

3. Click on "Run" drop-down menu and select "Run now." A new job will appear in the job database in the central window. When the run is finished, the job status in the job database window changes from "Running" to "Finished." Output and Log files can be viewed by clicking on (which will highlight) the line corresponding to the desired job and click on "View files from job" drop-down menu on the right hand side and select appropriate files. The log file is in the upper section of the drop-down menu. File names "2b6o-sf.mtz" should appear on the list of output files in the middle section.

3.8.4. Checking Data Quality

Data quality can be checked using different validation tools. In this protocol, we will focus on program Sfcheck (30). Program Truncate (31) (also available in CCP4i) can be used in addition to Sfcheck. Both programs analyze and report statistics for intensities and amplitudes such as data completeness, anisotropy, Wilson *B*-factor, twinning, and pseudo-translation (only in Sfcheck). The output log file should be examined for possible pathological cases.

1. Select "Validation & Deposition" module on the left side of the CCP4i window and select "Validate model and/or data" task. A task interface appears in a new window.
2. In the task interface, type in a job title such as "Sfcheck data analysis" in the "Job title" field. Uncheck boxes in front of "Run Rampage to calculate structure geometry" and "Run Procheck to calculate structure geometry." Check the box in front of "Run Sfcheck to analyse" and select "experimental data only" from the drop-down menu. Next line "Run Sfcheck against," select "native SF" data from the drop-down menu. Uncheck the box in front of "Generate anisothermally corrected SF amplitude." In the "MTZ in" field click on "Browse" on the right and select the file "2b6o-sf.mtz" from the file window. FP, SIGFP, and FREE should appear in the F, Sigma, and FreeR drop-down menus, respectively. In the "Sfcheck Output PS" field, the file "2b6o-sf_sfcheck1.ps" should appear. This postscript format file summarizes the analysis result. Click on "Run" drop-down menu and select "Run now."
3. Examine the file 2b6o-sf_sfcheck1.ps and/or Log file. In the output file, important values to pay attention to are numbers of strong reflections ($I > 1\sigma$ and $I > 3\sigma$), completeness (53.4% in this case), *B*-factor (34 Å^2 by Patterson, and 32.8 Å^2 by Wilson plot), pseudo-translation (not detected in this case), anisotropic distribution of Structure Factors.

3.8.5. Calculating Cell Content and Matthews' Coefficient

Analysis of cell content gives an idea of how many molecules could be present in the asymmetric unit. The correct number is usually consistent with the self-rotation function and the native Patterson analysis. Knowing the number of molecules can be helpful with the interpretation of rotation and translation functions. Here we will use the program Matthews_coeff to analyze the cell content.

1. In the module menu, select "Molecular replacement" module and click on "Analysis" folder. Select "Cell content analysis" task. A task interface appears in a new window.
2. In the "Job title" field, type "content analysis" as a job title. In the next line "Calculate Matthews coefficient for," select "Protein only" from the drop-down menu. Check the box in front of "Read crystal parameters from mtz file." In the MTZ file field, click on "Browse" on the right and select file "2b6o-sf.mtz" from the file window. Space group "P422" and high resolution limit "1.8" should automatically appear and the box in front of this is checked. In the line "Use molecular weight," select "enter in Daltons" from the drop-down menu. Type "28000" in the field on the right of "Molecular weight of protein or nucleic acid." Alternatively, the molecular weight can be entered by selecting "estimated from number of residues" and typing the number of residues per molecule (the program assumes 112.5 Da per amino acid residue), or by selecting "estimated from sequence file" and enter the sequence file in the "Sequence file" field. Click on "Run" drop-down menu and select "Run now."
3. The result will appear in the same task window (and also in the Log file) as follows:

N_{mol}/asu	Matthews coeff	% solvent	*P*(reso)	*P*(tot)
1	3.06	59.89	0.99	0.99
2	1.53	19.77	0.01	0.01

N_{mol}/asu shows possible number of molecules per asymmetric unit. Matthews coeff and % solvent are Matthews' coefficient (32) and estimated solvent content, respectively, at a given number of protein molecules. *P*(reso) is the normalized probability by high resolution limit based on a recent survey of crystallographic PDB entries (33). The highest *P*(total) is a strong indicator of the preferred solution. Based on cell content analysis, AQP0 crystals most likely contains one molecule per asymmetric unit. On average, protein crystals usually have % solvent content in the range of 40–60%. However, extreme cases have been observed for crystals with solvent contents below 30% and above 70%.

3.8.6. Calculating the Native Patterson MAP

Analysis of the native Patterson map can help identify the presence of noncrystallographic translation in the asymmetric unit. The program Sfcheck analyzes the native Patterson map but currently does not output a Patterson map and peak search results. Here we will use the program FFT (Fast Fourier Transform) (34) to calculate a Patterson map. This step is not essential for AQP0 in this case as there is only one molecule in the asymmetric unit. However, this analysis is recommended for data of unknown structure containing more than one molecule per asymmetric unit and is therefore included here.

1. Select the "Map & Mask Utilities" module and "Generate Patterson map" task. Type in a job title such as "native Patterson" in the "Job title" field. Check the box in front of "Run FFT to generate" and select "Patterson" map in "CCP4" format from drop-down menus. Check the box in front of "List peaks to file" and uncheck the box in front of "Plot default Harker" map sections. In the "MTZ" in field, click "Browse" and select file "2b6o-sf.mtz" from the file window. FP and SIGFP should automatically appear in the drop-down menus for F1 and SigmaF1, respectively. In the "Map" field, the file "2b6o-sf_patterson1.map" should appear. Select "AQP0" from drop-down menu for the map file to be output in the working directory (default in "temporary" directory). The rest on this task window can be left as defaults. Click on "Run" drop-down menu and select "Run now."
2. Examine the output peak files ("2b6o-sf_peaks1.pdb" in orthogonal coordinates and "peaks.ha" in fractional coordinates). In this case, there is only one large peak (peak height 114σ) at 0 0 0, corresponding to the origin peak indicating no noncrystallographic translation. If a noncrystallographic translation is present, additional peak(s) of at least 25% the size of the origin peak should be found. Alternatively, the Patterson map (2b6o-sf_patterson1.map) can be viewed using the program Mapslicer in CCP4i or through a command line.

Sometimes when the input resolution range is too low for calculating a native Patterson map, an extra peak at similar peak height as the origin peak may be observed. This is usually due to the crystallographic translation because the Patterson function is more prone to overlap with a neighboring origin at low resolution. By adjusting the resolution to a higher resolution range, the second peak should disappear. Native Patterson analysis is also implemented into the "MR data analysis" task under the "Molecule Replacement" module. This task runs the programs FFT, Peakmax, Wilson, and Baverage to analyze the data and the search model.

3.8.7. Calculating the Self-Rotation Function

The self-rotation function can indicate the presence of noncrystallographic rotational symmetry in the asymmetric unit. It can often tell the number of molecules in the asymmetric unit and their relative orientation and/or symmetry without the use of a model. Parameters such as resolution range can be adjusted to enhance peak signals. The radius of integration should approximate the diameter of the monomer (twice the radius of gyration) and this value can be estimated from search models. Applying a negative

B-factor to enhance amplitudes can sharpen the data. With automation features in Molrep, the program can calculate its own default setting that often works well. The default value for radius of integration in Molrep is 30 Å. This value should be corrected if a better estimation is available.

1. In the module menu, select the "Molecular replacement" module and select the "Run Molrep – auto MR" task. A task interface appears in a new window. Alternatively, a task "self RT with molrep" under "Analysis" will also lead to the same task window.
2. In the "Job title" field, type "self rotation" as a job title. In the next line, "Do," select "self rotation function" from the drop-down menu (if it has not been selected). Select "MTZ" from the drop-down menu for "Get input structure factors from." In the "MTZ in" field, click on "Browse" on the right and select file "2b6o-sf.mtz" from the file window. Leave the box in front of "Use intensities" unchecked (unless you are using intensities from the input file for the calculation). "FP" and "SIGFP" should automatically appear from the drop-down menus. The rest on this task window can be left blank in order to use default parameters chosen by the program. Click on "Run" drop-down menu and select "Run now."
3. To examine the output files, click on "View Files from Job" on the right side of the user window. The file containing the "srf.molrep_rf" suffix lists all the peaks identified from the self-rotation function. In this case, there is a large peak of 11.20σ at theta = 0, phi = 0, chi = 0 corresponding to the origin peak. Additional peaks are insignificant because their peak heights are much smaller than the origin peak (<2σ). This indicates no NCS rotation in the asymmetric unit. The postscript output file (containing the "rf.ps" suffix in the file name) should also be examined. The plot at chi = 180° can be viewed to identify any twofold axes.

3.8.8. Preparing the Search Model

Many modern MR programs such as Molrep offer an option of inputting the amino acid sequence of the unknown structure together with a homologous structure for the program to generate its own sequence alignment and modify the model to greatly improve the initial model quality. Some automated MR pipelines (Subheading 3.8.12) offer automatic model searches of the Protein Data Bank from the amino acid sequence of the unknown structure. In this section, protocols for manual model modification will be explained in detail.

1. Open the file "1J4N.pdb" using a text editing program of your choice. (For Linux, in a terminal window, go to the working directory and type "nedit 1J4N.pdb" to use the nedit program, or "gedit 1J4N.pdb" to use the gedit program.)

2. Scroll down on the text-editing window, the card "REMARK" in the first column contains experimental information. Under the *B* values section, the mean *B* value (overall) for this structure is 53.20 Å^2, similar to the *B* value from the Wilson plot (42.70 Å^2), indicating that the refined atomic *B*-factors approximate the *B*-factors calculated from the diffraction data well. Noted that for structures determined at lower resolution than 3 Å, the Wilson *B*-factors are not reliable and a large discrepancy is normal. Residues containing atomic *B*-factors in their main chain atoms much larger than the average are highly flexible (as their electron densities are less well-defined) and should be removed from the model.
3. Lines starting with the "HELIX" (for helices and "SHEET" in other PDB files for β-sheets) card contain secondary structure information derived from the coordinates. Since secondary structure is usually conserved in homologous proteins, residues outside defined secondary structure may be removed.
4. The actual coordinates of the AQP1 structure start when the lines start with the "ATOM" card in the first column. In this PDB file, the structure contains only one subunit (chain A) from residue Met 1 to Ser 249 (columns 4 = residue name, column 5 = chain name, column 6 = residue number).
5. If the structure contains hydrogen atoms (usually from NMR spectroscopy or ultra-high resolution X-ray crystallography), the lines containing hydrogen atoms should be removed. This can easily be done with the CCP4 program pdbcur in the "Coordinate Utilities" module and "Edit PDB file" task. This program can also be used to remove alternate conformations, atoms with low/zero occupancy, and anisotropic *U*'s (for high-resolution structures determined by X-ray crystallography where each atom has an additional line in the PDB coordinates for anisotropic *B*-factors).
6. Lines starting with the "HETATM" (for heteroatom) card contain coordinates of nonprotein entities. In this case, the AQP1structure contains ordered BNG (β-nonylglucoside) detergent and ordered water molecules bound to the protein. Since it is unlikely that BNG and water molecules will be present at exactly the same location in AQP0 structure, lines containing HETATM should be removed from the model.
7. Any other lines that do not start with an "ATOM" card are not important for the model preparation purpose and can be deleted. Lines containing CRYST, ORIGX, and SCALE cards may be kept since some programs may require this information to properly interpret the PDB file.
8. In this protocol, we will modify "1J4N.pdb" to create 3 new PDB files for MR comparison. The first file contains the

complete protein coordinates from residues 1 to 249, but everything else is removed. This PDB file is saved as "1J4N-simple.pdb."

9. The second file has flexible loops/termini removed. Residues 1–4 which are not part of the helices and contain high *B*-factors in their main chain atoms (>2 times mean *B*-factor), are deleted. Residues Pro38 to Thr45 are also deleted as these residues are likely to form a flexible loop (high *B*-factor). The PDB file is saved as "1J4N-edited.pdb."
10. Molrep has automated model preparation features for users to have the model modified to poly-alanine or to set *B*-factors to target values (currently fixed at 20 $Å^2$ or to values related to surface accessibility). Therefore, the model does not need to be edited further from step 8. However, if other MR programs lacking model preparation features are to be used, Gerald Kleywegt's program moleman from the Uppsala Software Factory (URL http://xray.bmc.uu.se/usf/moleman_man.html) is a quick and convenient way to edit the model. In this protocol, the third PDB file containing the poly-alanine model is created using the program Chainsaw in CCP4i. Select the "Molecular Replacement" module in the main CCP4i window, and click on the "Model Generation" box to expand this group for additional tasks and then click on "Create search model." In the "Job title" field, type "Create poly alanine model." In the line "Create search model" select "as polyA model" from the drop-down menu. In the "PDB in" field, click "Browse" on the right and select file "1J4N-edited.pdb" from the file window. The file name "1J4N-edited_chainsaw1.pdb" should automatically appear in the "PDB out" field. Rename this file to "1J4N-edited-polyA.pdb." Click on the "Run" drop-down menu and select "Run now."
11. Alternatively, if a good sequence alignment between the model and the unknown structure (in PIR format) is available, the program Chainsaw (22) can be used for better model preparation by pruning the model atoms based on sequence conservation. To use the program Chainsaw, in the line "Create search model" select "using Chaisaw" from the drop-down menu. Selective modification of nonconserved residues can be performed by selecting "gamma atom," "beta atom," or "last common atom" in the drop-down menu of "prune nonconserved residues to." Chainsaw outputs a coordinate file containing identical residues in the alignment left unchanged and nonidentical residues trimmed to specified atoms in their side chains. Residues that do not align to the sequence are deleted.

3.8.9. Running Molecular Replacement Using Molrep

Since Molrep is an automated MR program, most parameters can be left blank or unchanged from the default values. In some cases, however, default parameters may not be optimal or correct and users are required to set them to more appropriate values in order to get better signal. Normally, users may start with an MR run using the default setting. After inspecting MR results, appropriate parameters should be adjusted or varied for subsequent runs. In this protocol, we will focus on how to run Molrep using the default parameters. Choice of parameters will be discussed when MR results are explained later in the text. In addition, to demonstrate the importance of using a good search model, we will use the different model options below and compare the results:

(a) Unedited model (1J4N-simple.pdb)

(b) Model edited by removing flexible loop and termini (1J4N-edited.pdb)

(c) Poly-alanine model of (b)

(d) Automated model preparation using sequence and unedited model in (a)

1. Select "Molecular Replacement" module from the left side of the CCP4i window and click on "Run Molrep – auto MR" task. A new task interface window will appear.

2. In the "Job title" field, type "MR from 1J4N-simple" as a job title. In the next line, "Do," select "molecular replacement" from the drop-down menu (if it has not been selected) and "performing" select "rotation and translation function." In the line "Get input structure factors from," select "MTZ" from the drop-down menu. Leave the three boxes below unchecked for now. In the "MTZ in" field, click on "Browse" on the right and select file "2b6o-sf.mtz" from the file window. Leave the box in front of "Use intensities" unchecked (unless you are using intensities from the input file for the calculation). "FP" and "SIGFP" should automatically appear from the drop-down menus for FP and SIGFP, respectively.

   (a) In the "Model in" field, click on "Browse" on the right and select file "1J4N-simple.pdb." File name "1J4N-simple_molrep1.pdb" will automatically appear in the "Coords out" field below. The rest on this task window can be left blank for now in order to use default parameters chosen by the program. As default, the *B*-factors of the model are set related to accessibility and the model is shifted to the origin. Click on the "Run" drop-down menu and select "Run now."

   (b) Run Molrep with the file "1J4N-edited.pdb" as the model and change the job title to "MR from 1J4N-edited." The file name "1J4N-edited_molrep1.pdb" will automatically appear in the "Coords out" field. Click on the "Run" drop-down menu and select "Run now."

(c) Run Molrep with the file "1J4N-edited-polyA.pdb" as the model and change the job title to "MR from 1J4N-polyA," respectively. The file name "1J4N-edited-polyA_molrep1.pdb" will automatically appear in the "Coords out" field. Click on the "Run" drop-down menu and select "Run now."

(d) Run Molrep using the automated model preparation from the sequence (35). In the same task interface window, type "MR from sequence" in the "Job title" field. Underneath, check the box in front of "Use sequence" and a new section for "Parameter for SEQ" will appear near the bottom of the window. In the "Model in" field, click on "Browse" on the right and select file "1J4N-simple.pdb." The file name "1J4N-simple_molrep1.pdb" will automatically appear in "Coords out" field below. Rename this file to "1J4N-sequence_molrep1.pdb." Under the group "Parameter for SEQ," click "Browse" on the right of the "Seq in" field and select file "2B6O_A.seq." Click on the "Run" drop-down menu and select "Run now."

3. By default, Molrep runs MR in a space group defined in the input mtz file. To run MR with all possible space groups, click on the "Infrequently used Parameters" group and in the line "change space group," select "check all" from the drop-down menu. To test a particular space group, select "define space group" and select a test space group to run from the drop-down menu in the line below.

4. Depending on the CPU, Molrep may take several minutes to run. When the run is finished, the job status in the job database window changes from "Running" to "Finished." Results from the MR run can be checked by viewing the Log files. Output and Log files can be viewed by clicking on the line corresponding to the desired job and then clicking on the "View files from job" drop-down menu on the right hand side and selecting the appropriate files. Three files should appear on the list of output files: an output coordinates (.pdb) file, a complete search (.doc) file, and a rotation function (.molrep_rf) file. The output coordinate file will be needed for future steps (to check for packing, rigid body, and restrained refinement).

3.8.10. Interpreting Molecular Replacement Solutions from Molrep

1. Begin by checking the Log file from an MR run from step 2a, Subheading 3.8.9. In this run, the program estimated one monomer from the number of atoms in the model corresponding to a volume of the asymmetric unit V_{mol} of 47.5% (important for calculating COMPL). The resolution range used here

is between 22.9 and 2.25 Å. By default, the program detects data anisotropy and corrects it for proper scaling. In this case, the data is highly anisotropic (as shown by Ratio of Eigen values). In this run, the program calculates the "COMPL" and "SIM" parameters as 0.475 and 0.350, respectively. These numbers correspond to model completeness based on asymmetric unit volume (from 0.1 to 1.0) and model similarity to the unknown structure (also from 0.1 to 1.0), which are used to compute B_{off} and B_{add}, respectively. B_{off} and B_{add} control low and high resolution cut off, respectively, for MR and also for scaling of F_{model}. In this case, the model is quite complete (47.5% of the asymmetric unit volume) and has high similarity to the unknown structure (~50%). The values for COMPL and SIM may be entered to override default values. In the Molrep task window, under "The model" group, enter Expect "0.475" fraction completeness of model with "0.5" fraction similarity to input the structure. The program calculates a radius of integration (RAD) of 32.75 Å from the model. If the model is an oligomer (not in this case), the correct radius of integration (twice the radius of gyration) corresponds to the monomer being entered, otherwise the default value will be the radius calculated for the oligomer. To input a search radius, click on the "Infrequently Used Parameters" group and enter the correct number for "Search radius." The next section includes peaks from cross RF. In addition to plain RF, Molrep uses Rf/sigma to enhance signals and to use for peak ranking. The RF peaks show a big contrast after the first four peaks when the Rf/sigma drops from 5.82–5.65 to 4.57. The next section includes peaks from TF. Molrep calculates Tf/sig and TFcnt (multiplications of different Tf) to enhance peak contrast. In addition, Molrep calculates the packing function (PF) that is an overlapping function of the models (1 = no overlap) which is then used to calculate scor (scoring function = correlation coefficient × packing function). Finally, the program calculates Contrast, the ratio of the top Scor to the mean Scor. Usually Contrast >2.5 indicates a solution. A contrast <1.5 is probably not a solution because it is not significantly different than the mean. The final section summarizes allowed TF results after results containing overlapping models are removed. In this case, Contrast is 4.47 indicating a clear solution.

Rf	TF	theta	phi	chi	tx	ty	tz	TFcnt	Rfac	Scor
1	1	178.44	−144.90	95.53	0.277	0.732	0.356	4.58	0.529	0.543

2. Check the result from step 2b, Subheading 3.8.9 using edited model "1J4N-edited.pdb." In this case, the contrast is 3.69,

indicating a correct solution. Because the model is improved over the simple model by the removal of flexible loops and termini, the improvement in results is shown by the lower wRfac and higher Scor over the use of the simple model.

Rf	TF	theta	phi	chi	tx	ty	tz	TFcnt	Rfac	Scor
1	1	178.76	-144.90	95.55	0.778	0.235	0.356	4.50	0.523	0.562

However, when the COMPL and SIM parameters are entered in a new run as 0.45 and 0.5, respectively, the result is further improved by the lower wRfac and higher scor with the contrast increased to 4.44.

Rf	TF	theta	phi	chi	tx	ty	tz	TFcnt	wRfac	Scor
1	1	0.76	4.57	174.66	0.235	0.221	0.356	4.29	0.517	0.573

3. Check the result from step 2c, Subheading 3.8.9 using polyalanine model "1J4N-edited-polyA.pdb." In this case, the completeness of the search model suffers from the loss of atoms by conversion of residues to polyalanine. The program calculates a V_{mol} of 33% and expects two monomers, which is incorrect. An incorrect number of monomers may result in an unreliable scoring function. However, the program uses the self-rotation function to determine the actual number of monomers to be used for scoring. In this case, the program found the correct number of a single monomer. The program calculates COMPL and SIM as 0.330 and 0.350, respectively. The RF from this run does not show a big contrast until after 11 peaks, compared to 4 and 3 peaks in (a) and (b), respectively. The TF of the first molecule gives statistics as shown below. Strikingly, the best TF score in this case comes from the 11th and 7th RF peaks (instead of the first peak in the previous two cases) and the second best score is consistent with other MR runs.

Rf	TF	theta	phi	chi	tx	ty	tz	TFcnt	wRfac	Scor
11	1	0.00	0.00	83.34	0.210	0.765	0.360	3.19	0.563	0.498
7	1	178.01	-148.95	96.11	0.210	0.765	0.360	3.89	0.564	0.498

The program continues to search for the second molecule (unless overridden by entering NMON parameter=1 under "Search Parameters" group and searches for "1" monomers in the asymmetric unit). The program uses the previously calculated RF done for the first monomer to calculate the TF of the second monomer.

Since there is no second monomer in this case, the program cannot find a solution as indicated by the low contrast of 1.42.

4. Check the result from step 2d, Subheading 3.8.9 using the automated model prepared from the sequence and a simple model "1J4N-simple.pdb." In this option, the program calculates a structure-guided sequence alignment. This is essentially a sequence alignment but gaps are not allowed within secondary structure segments and buried residues contribute to the total alignment score more than residues at the surface. The program detects 47.7% identity between the sequence and the model with a gap of nine residues. The model is then modified as follows. First, residues that align with gaps in the sequence are deleted. Second, side chain atoms in the aligned residues of the search model that have no counterpart in the sequence are deleted and only atoms the sequences have in common are kept. Lastly, the residues and atoms of the modified model are renamed and renumbered. The program calculates COMPL and SIM to be 0.429 and 0.477, respectively. The result is much improved over using the simple model in (a) and the polyalanine model in (c) and on par with the edited model in (b). The rotation functions show a big contrast after three peaks from 6.23–6.10 to 4.15. TF scoring gives a contrast of 2.76, indicating a correct solution. The automated model preparation seems to be the best option when a good sequence alignment can be obtained.

Rf	TF	theta	phi	chi	tx	ty	tz	TFcnt	wRfac	Scor
1	1	178.86	−136.33	95.92	0.784	0.241	0.352	3.61	0.521	0.573

5. After the solution is found, a new Molrep run can be performed using the slow MODE parameter. The program default is in the "fast" mode that includes standard RF and TF without rigid body refinement. The "slow" mode uses advanced RF and TF with rigid body refinement to improve MR results and to pave the way for restrained refinement. However, the slow mode generally takes much longer computational time. To change the MODE parameter in the Molrep task window, click on "Infrequently Used Parameters" and in the line "Use," select "advanced RF and TF with rigid body refinement ('slow' mode)" from the drop-down menu. Compared to the fast mode in step 2d, Subheading 3.8.9 the slow mode MR results in improved solution statistics.

Rf	TF	theta	phi	chi	tx	ty	tz	TFcnt	wRfac	Scor
1	1	178.88	-134.54	95.95	0.784	0.241	0.352	3.18	0.451	0.632

3.8.11. Verifying a Probable Solution

After a probable solution has been found, the next step is to check that the solution is indeed correct. In many cases, when Contrast from Molrep is borderline it is possible that the solution may be incorrect. Correct solutions may be hidden among top results but their signal-to-noise ratio may not be high enough to stand out and give good contrast. This usually happens when the search model is poor or MR parameters are not optimal. In this case, additional runs with model modification or different parameters (such as radius of integration) should be tried. In addition, different MR programs such as Phaser (16) (also in CCP4i) should be used. Phaser uses maximum likelihood for scoring instead of using standard RF and TF peak heights and this may help for the correct solution to stand out better. Also, when searching for more than one monomer, Phaser recalculates rotation functions after fixing the first monomer. This process can be slow but often useful to increase the contrast because Phaser modifies F_{obs} by taking the amplitudes of the fixed monomer into account. Furthermore, Phaser can automatically check all space groups, perform rigid body refinement, and output phase information from which density maps can be calculated and examined.

1. A correct solution is likely to give reasonable packing. An easy way to check for packing is to open the output PDB file in molecular rendering programs such as Pymol (www.pymol.org) or Coot (36, 37) that can calculate (and display) neighboring molecules related by crystallographic symmetry. Good packing should not have serious main chain clashes with neighboring molecules while maintaining protein contacts and a clear separation for protein and solvent regions. However, when the model is not perfect, there might be regions in the model that are not present in the unknown structure (due to a different conformation, shorter loops, disordered regions, etc.) and the clashes between molecules are artifacts that can be eliminated by model corrections. It is recommended that several peaks be checked in this way. Sometimes the correct solution can be found in a much lower peak.
2. The correct solution is usually found consistently in different MR runs using different model preparations or parameters and is consistent with the self-rotation function. Check and compare orientations (theta, phi, chi angles) and positions (tx, ty, tz fraction coordinates) of several peaks from different runs.
3. Perform rigid body and restrained refinements and check the refinement *R*-factors (Subheading 4). Molrep outputs results as PDB coordinates that then can be used for refinement using the program Refmac (38) (also part of CCP4i) which automatically calculates phases from the model. However, some other refinement programs may require input phases prior to refinement. In that case, phases can be calculated from the

PDB file using the program SFall (in CCP4i). After restrained refinement, the correct solution will result in a large drop of R-factors (especially R-free) to below 40%. If R-factors stay at high levels, then there is something wrong and could point out that a wrong MR solution has been obtained. However, incomplete models (missing domains or subunits) and model error (different conformation) could also result in high R-factors and so the density maps ($2F_{obs} - F_{cal}$ and $F_{obs} - F_{calc}$) should be examined carefully.

4. Check different space groups. If the space group is wrong, no correct solution should be found. Any probable solutions are basically incorrect and will not refine further. To check, perform MR with different space groups and try basic refinements. The correct space group will give the best R-factor following refinement.

3.8.12. Running Molecular Replacement Using Automated MR Pipelines

The success of MR often depends on the choice of search models and how they are prepared for MR searches. The essence of automated program pipelines for MR such as MrBump (28) and Balbes (27) is to take advantage of an increasing number of available structures in the Protein Data Bank to find search models based on sequence similarity to an unknown structure and use all of them for MR searches. The main advantages of MR pipelines are their automation and ease of use. Users only need to provide the sequence of an unknown structure and its amplitude data and the pipelines perform the following three major steps automatically:

1. Database searches from the input amino acid sequence to find as many potential homologous structures (domains, multimers, etc.) as possible. Each program uses several different databases, for example, MrBUMP uses the SCOP database to identify additional search models based on fold similarity.

2. Model preparation. In MrBUMP, a multiple sequence alignment of the potential homologous structures identified in (1) is generated and the structures are ranked and edited. MrBump can generate four different search models for each sequence according to (i) the PDBclip method (removal of water, hydrogen atoms, and alternative conformations); (ii) the Polyalanine method (all side chain truncated to the C_β atom); (iii) the Molrep method (35); and (iv) the Chainsaw method (22). Balbes uses Molrep for model preparation. In addition, both pipelines create additional search models at the domain level. MrBUMP uses the SCOP database to generate additional models truncated to the domain boundary according to domain definitions. Balbes, on the other hand, defines domains based on compactness and truncates the models accordingly. Ensembles of models are also generated as search models.

3. Molecular replacement and restrained refinement. Both pipelines use the programs Molrep (MrBUMP also uses Phaser) for MR, and Refmac (38) for refinement. Each pipeline uses its own definition for solution scoring based on refinement results and suggests the best solution. The pipelines output partially refined models and mtz files containing phases and sigmaA-weighted amplitudes for map calculations.

Running MrBump

1. In the "Molecular Replacement" module, select the "Run MrBUMP" task. A new task interface window will appear.
2. In the "Job title" field, type "Automated MR." In the line "Program Mode," select "Model search and Molecular Replacement" from the drop-down menu. In the "SEQ in" field, click "Browse" on the right and select file "2B6O_A.seq" from the file window. In the "MTZ in" field, click "Browse" and select file "2b6o-sf.mtz." File names "2b6o-sf_mrbump_soln1.mtz" and "2b6o-sf_mrbump_soln1.pdb" should automatically appear in the "MTZ out" and "PDB out," respectively. The rest of the parameters can be left unchanged to use default settings. If desired, the program Phaser can be selected to run MR in parallel or as an alternative to Molrep in some difficult cases.
3. Click on the "Run" drop-down menu and select "Run now."

Running Balbes

1. In the "Molecular Replacement" module, select the "Run Balbes" task. A new task interface window will appear.
2. In the "Job title" field, type "Automated MR." In the line "Do," select "Standard MR" from the drop-down menu. In the "Structure factor file (MTZ or CIF) in" field, click "Browse" on the right and select the file "2b6o-sf.mtz" from the file window. In the "SEQ in" field, click "Browse" and select file "2B6O_A.fas" (Balbes takes input sequence in fasta file format). File names "2b6o-sf_balbes_out1.pdb" and "2b6o-sf_balbes_out1.mtz" should automatically appear in the "Solution PDB" and "Solution HKL," respectively.
3. Click on the "Run" drop-down menu and select "Run now."

4. Methods for Refinement

After a probable MR solution is found, the output coordinates should be refined to check whether or not the solution is correct. Program Refmac (38) in CCP4i can be used to check the refinement after MR by running rigid body refinement followed by restrained refinement cycles. If rigid body refinement has already been performed using MR programs (slow mode in Molrep or Phaser), users can start with restrained refinement directly. Refmac can be used further after manual model building (such as in the program Coot), an automatic model building/rebuilding program such as ARP/wARP (39, 40), or Resolve (41–43) to complete the structure determination process.

4.1. Rigid Body Refinement

1. In the "Refinement" module, select "Run Refmac5" task. A new task interface window will appear.
2. In the "job title" field, type "Rigid body refinement." In the next line "Do," select "rigid body refinement" using "no prior phase information" from the drop-down menus.
3. In the "MTZ in" field, click "Browse" on the right and select the file "2b6o-sf.mtz" from the file window. "FP" and "SIGFP" should automatically appear from the drop-down menus, and the file name "2b6o-sf_refmac1.mtz" appears in the "MTZ out" field. This file name can be changed to a recognizable new name. In the "PDB in" field, click "Browse" on the right and select a Molrep coordinate file such as "1J4N-sequence_molrep1.pdb" from the file window. The file "1J4N-sequence_molrep1_refmac1.pdb" should automatically appear in the "PDB out" field. This file name can be changed to a recognizable new name. By default, Refmac in CCP4i runs 20 cycles of rigid body refinement using all data. The number of cycles and the output resolution can be changed in the "Refinement Parameters" group.
4. If there is more than one identical domain or monomer in the asymmetric unit, it is important to allow them to move independently by treating each domain and/or monomer as a rigid body. To enter domain information, under the "Rigid Domains Definition" group, click on "Add domain Definition" and enter chain id and residue numbers for each domain (or monomer).
5. The rest can be left unchanged to use default parameters. Click on the "Run" drop-down menu and select "Run now" to run the refinement.

4.2. Restrained Refinement

1. In the same task window as Subheading 4.1, type "Restrained refinement" in the "job title" field. In the line "Do," select "restrained refinement" using "no prior phase information" from the drop-down menus.
2. Enter input files as described in Subheading 4.1. By default, Refmac in CCP4i runs 10 cycles of rigid body refinement using all data. To change the number of refinement cycles and the resolution range, click on the "Refinement Parameters" group and enter the desired parameters.
3. By default, Refmac automatically determines a weight matrix (experimental data vs. ideal geometry) for every refinement cycle. However, a defined number can be entered to override

the default. Under the "Refinement parameters" group, uncheck the box in front of "Use automatic weighting," and in the field "use weighting term" below, enter a number (usually between 0.01 and 10). A small number should be used for low-resolution data since the refinement uses less weight for experimental data and restrains the model to tighter geometry. An optimal weight matrix can be determined by trying different weights until a suitable r.m.s.d. for bond lengths and bond angles from ideal geometry is found.

4. If there are multiple monomers (or domains) related by NCS, it is useful to restrain them. To set up NCS restraints, under "Setup Non-Crystallographic Symmetry (NCS) Restraints" group, click "Add NCS restraint." In the drop-down menus, select chains that need to be restrained together and enter the residue numbers to define range. Select a level of restraints for the main chain and side chain from the drop-down menu. "Tight," "medium," and "loose" restraints allow for up to 0.05, 0.5, and 5 Å deviation, respectively. Tight to medium NCS restraint should be selected at the beginning of the refinement. Later when it becomes clear that each monomer should be allowed to deviate more from each other, appropriate NCS restraints should be used and the refinement R-free should drop further.
5. Click on the "Run" drop-down menu and select "Run now."
6. A drop in the R_{work} and R_{free} value often indicates good progress in the refinement and R values below 40% can be expected. If the R values do not decrease during the refinement, one needs to check whether or not the solution is indeed correct. One possibility is that the chosen space group is incorrect and MR should be tried with different space groups. Another possibility is that there are serious clashes between protein subunits. In addition, factors such as model incompleteness (missing parts of the model) and coordinate errors (large r.m.s.d between the model and the target structure) also result in large R values. Therefore after a macro cycle of the geometry restrained refinement, the σ_A-weighted $2F_{obs} - F_{cal}$ and $F_{obs} - F_{cal}$ density maps (output by the Refmac program) should be examined for positive and negative densities (in a graphic program such as Coot) and the model should be edited according to the density maps that are obtained from the refinement. New atoms should be added in the positive density regions as additional residues or modified amino acid side chains, while residues in the negative density regions should be deleted or revised. Several cycles of model building followed by the geometry restrained refinement should be carried out until the protein model is complete.
7. If the data resolution is better than 3 Å, additional density for water, lipid, and ligand molecules may show up in the later stage of the refinement when the protein model is nearly

complete and the R values are low. Cycles of water and ligand adding to the model followed by the geometry restrained refinement should be performed to complete the structure determination.

Acknowledgments

Research in the Gonen laboratory is supported by the American Diabetes Association Award # 1-09-CD-05 and by the National Institutes of Health R01GM079233 and U54GM094598 as well as the Howard Hughes Medical Institute.

References

1. Henderson R, Unwin PN (1975) Three-dimensional model of purple membrane obtained by electron microscopy. Nature 257:28–32
2. Fujiyoshi Y (1998) The structural study of membrane proteins by electron crystallography. Adv Biophys 35:25–80
3. Gonen T, Cheng Y, Sliz P, Hiroaki Y, Fujiyoshi Y, Harrison SC, Walz T (2005) Lipid-protein interactions in double-layered two-dimensional AQP0 crystals. Nature 438:633–638
4. Hite RK, Li Z, Walz T (2010) Principles of membrane protein interactions with annular lipids deduced from aquaporin-0 2D crystals. EMBO J 29:1652–1658
5. Mitsuoka K, Hirai T, Murata K, Miyazawa A, Kidera A, Kimura Y, Fujiyoshi Y (1999) The structure of bacteriorhodopsin at 3.0 A resolution based on electron crystallography: implication of the charge distribution. J Mol Biol 286:861–882
6. Tani K, Mitsuma T, Hiroaki Y, Kamegawa A, Nishikawa K, Tanimura Y, Fujiyoshi Y (2009) Mechanism of aquaporin-4's fast and highly selective water conduction and proton exclusion. J Mol Biol 389:694–706
7. Crowther RA, Henderson R, Smith JM (1996) MRC image processing programs. J Struct Biol 116:9–16
8. Gipson B, Zeng X, Stahlberg H (2007) 2dx_merge: data management and merging for 2D crystal images. J Struct Biol 160:375–384
9. Philippsen A, Schenk AD, Signorell GA, Mariani V, Berneche S, Engel A (2007) Collaborative EM image processing with the IPLT image processing library and toolbox. J Struct Biol 157:28–37
10. Ceska TA, Henderson R (1990) Analysis of high-resolution electron diffraction patterns from purple membrane labelled with heavy-atoms. J Mol Biol 213:539–560
11. Gonen T, Sliz P, Kistler J, Cheng Y, Walz T (2004) Aquaporin-0 membrane junctions reveal the structure of a closed water pore. Nature 429:193–197
12. Navaza J (1994) Amore—an automated package for molecular replacement. Acta Crystallogr A50:157–163
13. Trapani S, Navaza J (2008) AMoRe: classical and modern. Acta Crystallogr D64:11–16
14. Vagin A, Teplyakov A (1997) MOLREP: an automated program for molecular replacement. J Appl Crystallogr 30:1022–1025
15. Vagin A, Teplyakov A (2010) Molecular replacement with MOLREP. Acta Crystallogr D66:22–25
16. McCoy AJ (2007) Solving structures of protein complexes by molecular replacement with Phaser. Acta Crystallogr D63:32–41
17. McCoy AJ, Grosse-Kunstleve RW, Adams PD, Winn MD, Storoni LC, Read RJ (2007) Phaser crystallographic software. J Appl Crystallogr 40:658–674
18. Collaborative Computational Project, Number 4 (1994) The CCP4 suite: programs for protein crystallography. Acta Crystallogr D50:760–763
19. Brunger AT (2007) Version 1.2 of the crystallography and NMR system. Nat Protoc 2:2728–2733
20. Brunger AT, Adams PD, Clore GM, DeLano WL, Gros P, Grosse-Kunstleve RW, Jiang JS, Kuszewski J, Nilges M, Pannu NS, Read RJ, Rice LM, Simonson T, Warren GL (1998) Crystallography & NMR system: a new software suite for macromolecular structure determination. Acta Crystallogr D54:905–921

21. Potterton E, Briggs P, Turkenburg M, Dodson E (2003) A graphical user interface to the CCP4 program suite. Acta Crystallogr D59:1131–1137
22. Stein N (2008) CHAINSAW: a program for mutating pdb files used as templates in molecular replacement. J Appl Crystallogr 41:641–643
23. Vagin A, Teplyakov A (2000) An approach to multi-copy search in molecular replacement. Acta Crystallogr D56:1622–1624
24. Tong L (2001) How to take advantage of non-crystallographic symmetry in molecular replacement: 'locked' rotation and translation functions. Acta Crystallogr D57:1383–1389
25. Tong L, Rossmann MG (1990) The locked rotation function. Acta Crystallogr A46:783–792
26. Tong LA (1996) The locked translation function and other applications of a Patterson correlation function. Acta Crystallogr A52:476–479
27. Long F, Vagin AA, Young P, Murshudov GN (2008) BALBES: a molecular-replacement pipeline. Acta Crystallogr D64:125–132
28. Keegan RM, Winn MD (2008) MrBUMP: an automated pipeline for molecular replacement. Acta Crystallogr D64:119–124
29. Sui HX, Han BG, Lee JK, Walian P, Jap BK (2001) Structural basis of water-specific transport through the AQP1 water channel. Nature 414:872–878
30. Vaguine AA, Richelle J, Wodak SJ (1999) SFCHECK: a unified set of procedures for evaluating the quality of macromolecular structure-factor data and their agreement with the atomic model. Acta Crystallogr D55:191–205
31. French S, Wilson K (1978) Treatment of negative intensity observations. Acta Crystallogr A34:517–525
32. Matthews BW (1968) Solvent content of protein crystals. J Mol Biol 33:491–497
33. Kantardjieff KA, Rupp B (2003) Matthews coefficient probabilities: improved estimates for unit cell contents of proteins, DNA, and protein-nucleic acid complex crystals. Protein Sci 12:1865–1871
34. Teneyck LF (1973) Crystallographic fast Fourier-transforms. Acta Crystallogr A29:183–191
35. Lebedev AA, Vagin AA, Murshudov GN (2008) Model preparation in MOLREP and examples of model improvement using X-ray data. Acta Crystallogr D64:33–39
36. Emsley P, Cowtan K (2004) Coot: model-building tools for molecular graphics. Acta Crystallogr D60:2126–2132
37. Emsley P, Lohkamp B, Scott WG, Cowtan K (2010) Features and development of Coot. Acta Crystallogr D66:486–501
38. Murshudov GN, Vagin AA, Dodson EJ (1997) Refinement of macromolecular structures by the maximum-likelihood method. Acta Crystallogr D53:240–255
39. Cohen SX, Ben Jelloul M, Long F, Vagin A, Knipscheer P, Lebbink J, Sixma TK, Lamzin VS, Murshudov GN, Perrakis A (2008) ARP/wARP and molecular replacement: the next generation. Acta Crystallogr D64:49–60
40. Perrakis A, Harkiolaki M, Wilson KS, Lamzin VS (2001) ARP/wARP and molecular replacement. Acta Crystallogr D57:1445–1450
41. Terwilliger TC (2003) Improving macromolecular atomic models at moderate resolution by automated iterative model building, statistical density modification and refinement. Acta Crystallogr D59:1174–1182
42. Terwilliger TC (2003) Automated main-chain model building by template matching and iterative fragment extension. Acta Crystallogr D59:38–44
43. Terwilliger TC (2003) Automated side-chain model building and sequence assignment by template matching. Acta Crystallogr D59:45–49

Chapter 15

High-Throughput Methods for Electron Crystallography

David L. Stokes, Iban Ubarretxena-Belandia, Tamir Gonen, and Andreas Engel

Abstract

Membrane proteins play a tremendously important role in cell physiology and serve as a target for an increasing number of drugs. Structural information is key to understanding their function and for developing new strategies for combating disease. However, the complex physical chemistry associated with membrane proteins has made them more difficult to study than their soluble cousins. Electron crystallography has historically been a successful method for solving membrane protein structures and has the advantage of providing a native lipid environment for these proteins. Specifically, when membrane proteins form two-dimensional arrays within a lipid bilayer, electron microscopy can be used to collect images and diffraction and the corresponding data can be combined to produce a three-dimensional reconstruction, which under favorable conditions can extend to atomic resolution. Like X-ray crystallography, the quality of the structures are very much dependent on the order and size of the crystals. However, unlike X-ray crystallography, high-throughput methods for screening crystallization trials for electron crystallography are not in general use. In this chapter, we describe two alternative methods for high-throughput screening of membrane protein crystallization within the lipid bilayer. The first method relies on the conventional use of dialysis for removing detergent and thus reconstituting the bilayer; an array of dialysis wells in the standard 96-well format allows the use of a liquid-handling robot and greatly increases throughput. The second method relies on titration of cyclodextrin as a chelating agent for detergent; a specialized pipetting robot has been designed not only to add cyclodextrin in a systematic way, but to use light scattering to monitor the reconstitution process. In addition, the use of liquid-handling robots for making negatively stained grids and methods for automatically imaging samples in the electron microscope are described.

Key words: Electron crystallography, Electron microscopy, Membrane proteins, Protein structure, High-throughput, Crystallization, Dialysis, Cyclodextrin, Negative stain

Ingeborg Schmidt-Krey and Yifan Cheng (eds.), *Electron Crystallography of Soluble and Membrane Proteins: Methods and Protocols*, Methods in Molecular Biology, vol. 955, DOI 10.1007/978-1-62703-176-9_15,

1. Introduction

Electron microscopy (EM) has made a significant contribution to our understanding of membrane protein structure through the application of electron crystallography (1–3). As with X-ray crystallography, the formation of suitable crystals is the first and often biggest hurdle to overcome (4). Crystallographic methods cannot be applied without crystals and their quality is primarily responsible for the resolution of the final structure. The number of structures solved by X-ray crystallography has experienced exponential growth in the last two decades. Although membrane proteins have lagged behind their soluble counterparts, recent successes show a marked acceleration not only in numbers of structures, but also in their biological impact (5). Much of this success is attributable to automation, which allows X-ray crystallographers to implement high-throughput approaches at various stages of the crystallization pipeline. Specifically, automation is employed for screening genetic constructs for expression, screening of detergents for protein stability and, of course, screening tens of thousands of conditions for producing well-ordered crystals (6, 7). With regard to EM, methods for automation are routinely employed for collecting image tilt series for tomographic reconstruction (8–12) and for collecting images to include in single particle reconstructions (13–16). In addition, prototypical robotic systems have been reported for exchanging samples in the electron microscope (17–19). However, relatively little attention has been paid to automating the process of forming two-dimensional (2D) arrays of membrane proteins within the lipid bilayer, so-called 2D crystals. Such regular assemblies are amenable to atomic scale resolution assessment by electron crystallography and they yield the structure of membrane proteins in their native environment and thus in a functional state. High-throughput automation of 2D crystallization is critical to more widespread application of electron crystallography. In this review, we will describe recent developments in high-throughput 2D crystallization employing both detergent dialysis and detergent complexation with cyclodextrin. In addition, we describe facilities necessary for imaging these large-scale 2D crystallization screens in the electron microscope.

2. Materials

2.1. Sample Characterization

1. Purified, detergent-solubilized protein sample with concentration 1–2 mg/ml. Unlike X-ray crystallography, it is not necessary to have a highly concentrated protein solution and

fractions collected from the last step in the purification (e.g., size-exclusion column) may be sufficient to use directly for crystallization trials (see Note 1).

2. SDS polyacrylamide gel electrophoresis system, including glass plates, combs, tank, casting frame, power supply (e.g., Mini-Protean Electrophoresis System, BioRad Laboratories, Hercules, CA).
3. Bovine serum albumin (BSA) stock solution of 1 mg/ml in pH 7 buffer (Sigma-Aldrich, St. Louis, MO).
4. Densitometer for acrylamide gels (e.g., Molecular Imager from BioRad Laboratories or an inexpensive document scanner).
5. Device for measuring contact angles from small drops of detergent solutions. A self-contained drop-box as described by Kaufmann et al. (20) can be built according to plans available at http://temimps.nysbc.org/ or can be purchased commercially. A homemade device can be built by mounting a digital camera on a tripod and by using an optical rail to support an *x–y–z* translation stage (Fig. 1). Software specifically developed for the drop analysis can be downloaded at http://temimps.nysbc.org/.
6. Parafilm M, laboratory sealing film (Pechiney Plastic Packaging Company, Chicago, IL).
7. Bench-top centrifuge (e.g., Eppendorf).
8. Thin layer chromatography (TLC) setup for measuring lipid concentration. This consists of 20×20 cm TLC plates with silica gel $60F_{254}$ coating (Merck & Co., Whitehouse Station, NJ), a Glass chromatography tank (25 cm×27 cm×10 cm), a Kontes Chromatography TLC Reagent Sprayer with Standard Ground Joint (Fisher Scientific, Pittsburgh PA).
9. Chloroform, methanol, aqueous ammonia, and 8-anilino-1-naphthalene sulfonic acid (ANSA) (Sigma, St. Louis, MO) solubilized in water at 0.1% (w/w).
10. UV lamp and digital camera or UV imaging densitometer.
11. Spectrophotometer (e.g., NanoDrop 2000c, Thermo Scientific, Wilmington DE).

2.2. Preparation of Lipid

1. Selection of detergents that are compatible with the protein(s) of interest (Anatrace, Inc. Maumee OH). Commonly used long-chain detergents are *n*-dodecyl-β-D-maltopyranoside (DDM), *n*-decyl-β-D-maltopyranoside (DM), dodecyl octaethylene glycol ether ($C_{12}E_8$) Triton X-100 (TX-100), all of which have a very low critical micelle concentration (cmc, see Note 2) in the range of 0.1–1 mM. Commonly used short-chain detergents are *n*-octyl-β-D-glucopyranoside (OG), *n*-octyl-β-D-thioglucopyranoside (OTG), pentaethylene glycol

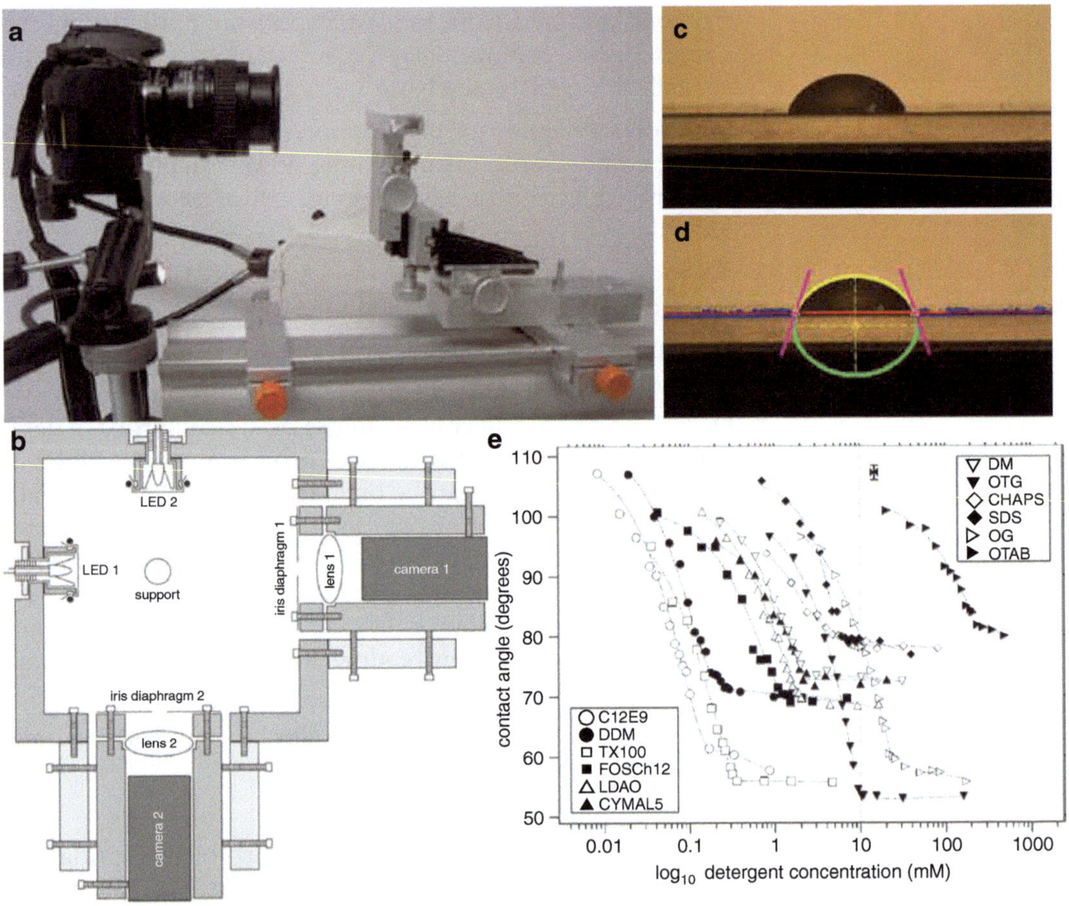

Fig. 1. Using drop shape as a measurement of detergent concentration. (**a**) Simple apparatus for recording an image of a drop. A digital camera is mounted on a tripod and drops are placed on a horizontal plate (*arrow*). The use of optical components to support the plate allows convenient translation of the drops along *x–y–z* axes. Diffuse backlighting of the drops provides a suitable image for analysis. (**b**) Diagram of the "drop box" described by Kaufmann et al. (20) (more detailed planes available at http://temimps.nysbc.org). In this device, the drop is imaged from two orthogonal angles by CCD cameras connected to a frame grabber. (**c**) Typical image of a drop containing 15 mM OG. (**d**) Fitting of the drop shape by the drop box program. *Yellow line* indicates the points used for fitting an ellipse to the drop. *Red line* corresponds to the surface of the susbstrate and the *pink lines* are the tangents to the drop at this surface, thus defining the contact angle. (**e**) Calibration curves for several detergents; adapted from Kaufmann et al. (20).

monooctyl ether (C_8E_5), which have relatively high cmc's in the range of 10–30 mM. Aqueous stock solutions should be made at 10 mg/ml.

2. Selection of lipids that are compatible with the protein(s) of interest (Avanti Polar Lipids, Alabaster, AL). These should include different chain lengths with differing degrees of saturation as well as head groups with different sizes, shapes, and charge distribution. A good initial selection includes 1,2-Dimyristoyl-sn-glycero-3-phosphocholine (DMPC), 1-Palmitoyl-2-oleoyl-sn-glycero-3-phosphocholine (POPC), 1,2-Dioleoyl-sn-

glycero-3-phosphocholine (DOPC), 1,2-Dioleoyl-sn-glycero-3-phospho-(1'-rac-glycerol) (DOPG), 1-Palmitoyl-2-oleoyl-sn-glycero-3-phosphoethanolamine (POPE), 1-Palmitoyl-2-oleoyl-sn-glycero-3-phosphatidic acid (POPA) (see Note 3).

3. Glass test tubes (7 cm × 1 cm) or 25 ml round-bottom flasks.
4. Hamilton syringes.
5. Lyophilizer or vacuum desiccator.
6. Dry N_2 gas supply.

2.3. Crystallization by Dialysis

1. 96-Well dialysis block capable of dialyzing 96 unique protein samples against 96 unique dialysis buffers. One design is described by Vink et al. (21), another is available commercially (XZ-HT-96-8k, GN Biosystems, Santa Clara CA).
2. 8–12 Channel micropipettor (e.g., Hamilton Co., Reno, NV) or liquid handling robot (e.g., Biomek FX or Biomek NXP, Beckman Coulter, Fullerton, CA).
3. Dialysis membrane in sheets with 12 kDa MW cutoff (Spectrum Laboratories, Inc. Rancho Dominguez, CA).
4. Small crystallization incubator.
5. Stock solutions of buffers for the crystallization matrix. Buffers with p*K*a of 6, 7, 8 are appropriate for initial screens, e.g., 2-(*N*-morpholino)ethanesulfonic acid (MES), *N*-Tris(hydroxymethyl)methyl-2-aminoethanesulfonic acid (TES), and 3-[4-(2-hydroxyethyl)-1-piperazinyl] propanesulfonic acid (EPPS). Stock solutions should be 0.5–1 M with their pH adjusted to the desired value.
6. Stock solutions of NaCl and $MgCl_2$ with a concentration of 0.5 or 1 M.
7. Dialysis buttons with volumes of 5, 10, 20, and 50 μl for optimization of crystal conditions (Hampton Research, Aliso Viejo, CA).

2.4. Crystallization with Cyclodextrin

1. Methyl-β-cyclodextrin and/or 2-hydroxylpropyl-β-cyclodextrin (Aldrich). Stock solutions should be made in deionized water at 50 mg/ml and filtered through a 0.2 μm filter.
2. 96-Well microtiter plate.
3. Cyclodextrin dispensing robot as described by Iacovache et al. (22) (Seyonic PCNC-0034-00-V00 pipettor) or 8–12 channel micropipettor (e.g., Hamilton Co., Reno, NV) or liquid handling robot (e.g., Biomek FX or Biomek NXP, Beckman Coulter, Fullerton, CA).
4. Stock solutions of buffers for the crystallization matrix as specified in Subheading 2.3.
5. Stock solutions of NaCl and $MgCl_2$ with a concentration of 0.5 or 1 M.

2.5. Negative Staining

1. Uranyl acetate (Electron Microscopy Science, Hatfield, PA) dissolved in water at 2 mg/ml.
2. 300 Mesh Ni electron microscopy grids (Electron Microscopy Sciences).
3. 1% Solution of collodion (parlodion) dissolved in amyl acetate (Electron Microscopy Sciences, Hatfield, PA).
4. Crystallizing dish (10 cm diameter).
5. Vacuum evaporator equipped with carbon rod or carbon thread (e.g., Auto 306, BOC Edwards, Crawley, UK).
6. Glow discharge apparatus, typically an accessory for the vacuum evaporator. Because a large number of grids are prepared simultaneously, the small chamber offered by plasma cleaners is not convenient.
7. 96-Position magnetic grid support platform (SPRI plate 384 Post Magnet Plate, Agencourt, Beverly MA) or magnetic grid tray described by Coudray et al. (19).

2.6. Electron Microscopy

1. Electron microscope suitable for screening negatively stained grids (e.g., Morgagni M268 or Tecnai T12, FEI Corp., Hillsboro OR, or JEM-1400 JEOL Ltd. Tokyo).
2. Digital camera for electron microscope (e.g., ES500W Erlangshen, Gatan Inc, Pleasanton CA).
3. Software for automated imaging of samples (where available).

3. Methods

As with 3D crystallization of detergent-solubilized protein for X-ray crystallography, 2D crystallization within the lipid bilayer is favored by a homogeneous starting solution, in which the protein adopts a single oligomeric state and the mixed micelles of protein, detergent, and lipid are monodisperse. Detergents for promoting monodispersity and stability can be screened using an FPLC system equipped with a size exclusion chromatography column. A protein solution that is well behaved will form a single Gaussian-shaped peak on the elution profile. The presence of peaks in the void volume, or the presence of multiple or asymmetric peaks in the elution volume indicates that the protein micelles are not monodisperse, either due to the presence of multiple oligomeric states or aggregation. After using chromatography to optimize the biochemical parameters of a preparation, it is also useful to examine the sample by negative stain EM to further verify its monodispersity and stability over time.

Prior to screening a membrane protein for crystallization, it is essential to characterize the relative amounts of protein, detergent, and lipid in the starting ternary mixture. This information allows one to accurately and reproducibly control important parameters such as the lipid-to-protein ratio and the rate of detergent removal. Crystallization itself relies on the controlled removal of the detergent in the presence of a defined mixture of lipids. We present two alternative approaches for removing the detergent, namely by dialysis and by complexation with cyclodextrin. Both of these techniques are amenable to parallelization, though some special facilities are required. Specifically, a 96-well dialysis chamber has been devised, which is compatible with the use of a liquid-handling robot or multichannel pipettor. For cyclodextrins, a pipetting robot has been designed and built to systematically add nanoliter quantities of cyclodextrin stock solutions to commercially available 96-well microtiter plates. This so-called 2DX robot is also equipped with a level sensor, to allow compensation for evaporation, a shaker to ensure mixing, a temperature controller, and a laser for measuring light scattering to follow the reconstitution process. All of these parameters are displayed in real time by the control software. Alternate approaches for crystallization are dilution of the detergent below its cmc (23) or addition of polystyrene beads (BioBeads) to adsorb the detergent (24). Although the former is suitable for parallelization, it's disadvantage is the inevitable dilution of the protein as well. Whereas good results have been obtained with BioBeads, this method is not amenable to automation, given the difficulty in handling the beads.

3.1. Sample Characterization

3.1.1. Protein

In this section, we present methods for determining protein concentration using SDS-PAGE, detergent concentration using contact angle measurements, and lipid concentration using TLC. Although there are easier, faster ways to determine protein concentration (i.e., absorbance at 280 nm, Lowry assay, Bradford Assay), these are often inaccurate and are sometimes influenced by the detergent present in the solution. We therefore recommend SDS-PAGE to compare the staining of the purified protein relative to known amounts of BSA. This method removes the variability due to detergent, requires only minimal amounts of material and also verifies the purity of the sample. Furthermore, the Coomassie-stained protein band can be excised from the gel and sent to a protein sequencing facility to confirm identity, which is especially useful if the sample is provided by another laboratory. It should be noted that the Coomassie stain used for SDS-PAGE does not stain all proteins equally, so the comparison of staining intensity relative to BSA cannot be assumed to represent an exact measure of concentration. Moreover, SDS-PAGE is time consuming and may not be the best choice for unstable proteins. In these cases it may be necessary to use one of the aforementioned spectroscopic techniques, perhaps after calibrating them relative to SDS-PAGE. The NanoDrop 2000c spectrophotometer is an excellent alternative given its extremely small sample size.

1. Prepare acrylamide gel with an appropriate concentration (e.g., 8%) to resolve your protein of interest as well as BSA (66 kDa).
2. Load a series of lanes with 0.25, 0.5, 1.0, and 1.5 μg of BSA as well as the protein of interest.
3. Run gel according to standard protocols and stain with Coomassie blue.
4. Scan gel with a densitometer and use either associated software or ImageJ to calibrate the integrated intensities of each protein sample. Plot a standard curve for the BSA samples and use this to determine the concentration of the protein of interest.

3.1.2. Detergent

Purified membrane proteins may have a poorly defined detergent concentration due to the use of a concentrator, which often retains detergent and increases its concentration in an unpredictable way. Although a concentrator with a molecular weight cutoff of 50 kDa will remove short- and medium-chain detergents (e.g., OG and DM) together with the filtrate, long-chain detergents are generally retained with the protein with molecular weight cutoffs up to 100 kDa, which may be larger than the expected micelle size. In this case, detergent concentration can be determined from the shape of droplets. This shape is governed by surface tension. Detergent molecules that line the air/water interface decrease the surface tension and cause the drop to spread. The contact angle between the drop and the supporting surface decreases monotonically with increasing detergent concentration up to the cmc. This behavior is characteristic for all detergents and needs to be calibrated (20). Above the cmc, the concentration of free detergent molecules remains unchanged, so there are no further changes in drop shape. Thus, solutions with higher detergent concentrations, such as those used to purify proteins, must be diluted to bring the detergent below the cmc. This may cause the protein to precipitate, but this precipitate can be removed by centrifugation.

1. Cover the substrate (e.g., glass slide) with a fresh piece of parafilm or Teflon tape (see Note 4).
2. Collect data for a standard curve. Prepare a series of detergent solutions with concentrations ranging from 0 to twice the cmc (e.g., 0, 2, 4, 8, 16, 24, 30, 36, 48 mM OG). Place two drops on the substrate for each concentration. Equilibrate for 30 s before recording an image.
3. Collect data for unknown sample. Dilute protein solution based on estimated detergent concentration to achieve half the cmc.

Centrifuge this solution for 30 s using a benchtop centrifuge. Place two drops on the substrate and, after 30 s incubation, record images.

4. The substrate may be washed and reused several times, but should be changed after a maximum of five drops. Thus, using a very long plate with the ability to translate the plate along its length can save a lot of time.
5. Use a computer program (e.g., ImageJ, xdroptrace, dropbox) to measure contact angles (see Note 5) and to plot a standard curve from which the concentration of detergent in the protein sample can be determined.

3.1.3. Lipids

It has frequently been observed that lipids co-purify with membrane proteins. Also, addition of exogenous lipid is a common way to optimize membrane protein stability during purification. TLC is a well-established method both to quantify the total amount of lipid and to identify different lipid species. Specifically, if the appropriate standard lipids are run together with the protein sample, it is possible to fully characterize the lipid composition in the solution.

1. Combine 65 ml chloroform, 25 ml methanol and 5 ml 25% aqueous ammonia in a glass graduated cylinder. Add this solution to a chromatography tank lined with filter paper and allow it to equilibrate for 10 min.
2. Cut TLC plates into 5 × 10 cm pieces using a glass cutter. Draw a light pencil line across the plates ~1.5 cm from the bottom. Use gloves to handle the plates to prevent contamination.
3. Prepare standard samples of lipid with a concentration of 1 mg/ml. These standards can be either in water or in chloroform/methanol.
4. Spot 2–4 μl of sample at 5 mm intervals along the pencil line. Include 1–5 μg of the standard lipids (see Note 6).
5. Dry TLC plates under a stream of N_2 gas.
6. Place plates into glass tank such that samples lay above the level of the solvent. Cover the tank and allow solvent front to ascend to the top of the plate (30–45 min).
7. Remove plates and dry under a stream of N_2 gas in a chemical fume hood.
8. Spray plates with 0.1% ANSA solution.
9. Image plates with either transmitted or reflected UV illumination using a digital camera or a specialized imaging device.
10. Quantify strength of the bands using ImageJ or similar software and determine the lipid composition and concentration by comparison with the standard solutions (Fig. 2).

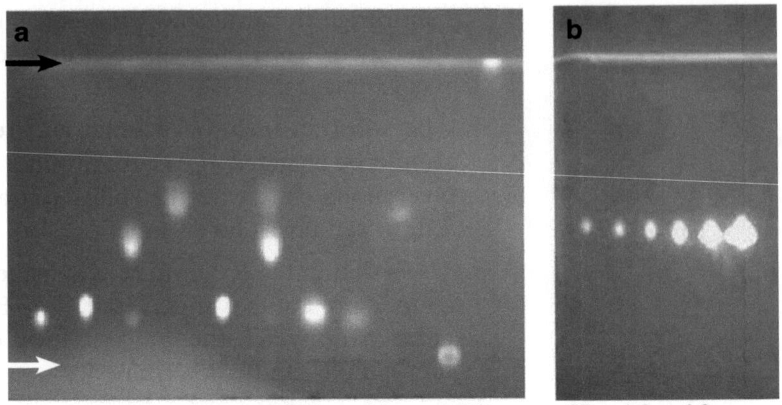

Fig. 2. Use of thin layer chromatography to characterize lipid. (**a**) Various species migrate differently on the TLC plate and can thus be distinguished. DMPC, DOPC, DOPE, DOPG, POPG, and *E. coli* correspond to the lipids mentioned in the text. DDM, DM, and OG are detergents mentioned in the text. h-CD and m-CD are 2-hydroxypropyl-β-cyclodextrin and methyl-β-cyclodextrin, respectively. The *white arrow* indicates the origin, where samples were all initially applied, and the *black arrow* indicates the solvent front at the end of the run. (**b**) Ladder of lipid concentrations show how TLC can be used to determine lipid concentration as well as separate different species. This amount of lipid added to each lane ranges from 0.25 to 8 μg, as indicated along the bottom.

3.2. Preparation of Lipid Solutions for 2D Crystallization Screens

1. Prepare stock solution of lipids in chloroform at 25–50 mg/ml. Lipid is susceptible to oxidation and care must be taken during storage. For long-term storage, it is safest to keep lipid in the powdered form at −80°C. When storing chloroform stock solutions at −80°C, vials with Teflon seals should be used and should be purged with N_2 gas before sealing. Chloroform solutions are also susceptible to evaporation, which will alter the lipid concentration, so care should be taken to keep these solutions cold (e.g., on ice) and capped when not in use.
2. Make a thin film of dried lipid. Use a Hamilton syringe to pipette the desired amount of lipid (e.g., 500 μl) into a glass test tube or 25 ml round-bottom flask (see Note 7). Swirl the test tube while allowing the chloroform to evaporate in a chemical fume hood. A thin stream of N_2 gas will accelerate this process, though the rate of evaporation should be relatively slow in order to obtain a thin, even film of lipid around the walls of the test tube. This film should appear homogeneous and white. If thick, clear areas are present, the film should be redissolved with a small amount of chloroform and the process repeated.
3. If making multiple lipid stock solutions, purge the test tube with N_2 gas and store capped on ice until the next step. This

thin film is particularly susceptible to oxidation given its high surface area.

4. Remove traces of chloroform by placing lipid film in a lyophilizer or vacuum desiccator. A cold trap is highly recommended in order to prevent contamination of lipid samples with back-streaming oil vapor from the vacuum pump. Pump for at least 1 h.
5. Prepare detergent stock solutions at 100 mg/ml. All of the detergents should dissolve readily in water. These can be aliquoted and stored briefly at 4 °C, but should be frozen at −20 or −80 °C for longer periods.
6. Prepare aqueous lipid stock solutions by adding detergent solutions to the dried lipid films (see Note 8). The amount of detergent required for solubilization depends both on the lipid species and on the detergent (Fig. 3). Generally speaking, the choice of both detergent and lipid will be governed by the preference of the protein. For particularly difficult lipid/detergent combinations, one should consider using a different detergent for solubilization and thus attempting crystallization from a solution with a mixture of detergents. After adding detergent to the lipid film, the solution should be stirred on ice for 1 h to ensure a clear, homogeneous solution of solubilized lipid.

3.3. Crystallization by Dialysis

1. Define conditions for the crystallization screen. General parameters that influence crystallization are pH, lipid composition, and lipid-to-protein ratio (LPR) of the bilayer, and ionic composition of the aqueous phase. Therefore, a basic screen for a novel protein with unknown behavior, includes pH 6, 7, and 8, a range of lipids such as DMPC, DOPC, POPC, DOPG, DOPE, or *Escherichia coli* lipid extract for bacterial proteins, LPR's from 0.25 to 1.5, presence of either 100 mM NaCl or 5–10 mM $MgCl_2$. By using a 96-well dialysis block, it is possible to obtain a linear sampling for all of these parameters: e.g., a matrix of conditions with three values of pH, five species of lipid, three LPR's, in the presence of either NaCl or $MgCl_2$.
2. Information about the protein of interest may guide the selection of parameters used for the screen. Specific ligands should be tested as an additive, either in native state (e.g., transported ions) or as an enzymatically inactive state (e.g., non-hydrolyzable nucleotides or transition state analogues). Mammalian proteins call for testing of cholesterol or sphingolipids as a component of the bilayer, or a species-specific lipid extract may be used (e.g., from egg-yolk, brain, wheat germ, or *E. coli*).
3. Prepare the protein/detergent/lipid solutions in 96-well microtiter plates and incubate at 20 °C for an hour to ensure complete mixing of these amphipathic components. The goal is to start with a completely homogeneous preparation of

a

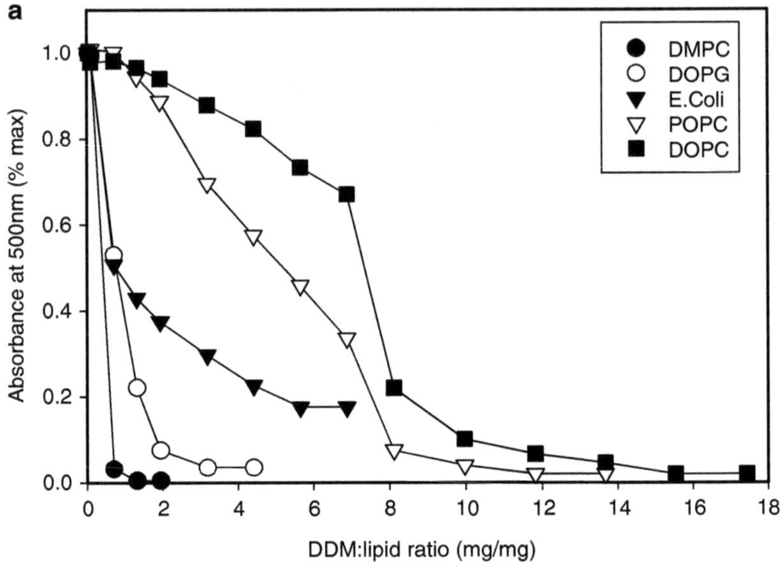

b

Solubilization of lipids by detergent*

	DMPC	DOPC	DOPG	POPC	E. coli
OG	4.4	6.8	2.5	6.8	5.6
OTG	5.0	5.0	2.5	3.7	3.1
C8E5	2.5	5.0	1.9	3.7	3.7
DM	1.0	15.7	n.d.	15.7	2.0
DDM	1.3	10.0	1.9	10.0	4.4
C12E8	1.0	1.4	1.0	1.4	1.0
TX100	1.0	1.9	1.3	1.9	1.9

* values correspond to the weight ratio of detergent required to reduce Abs^{500} to baseline for each lipid species at a concentration of 1.5 mg/ml

Fig. 3. Solubilization of lipid by detergent as followed by light scattering. (**a**) Light scattering (absorbance at 500 nm) is plotted against the weight ratio of DDM relative to a variety of lipids as indicated on the upper right. Solubilization is complete once the light scattering signal reaches a baseline close to zero. Note that some lipids are considerably easier to solubilize than others, and this effect depends on the detergent. (**b**) Concentrations of detergent required to solubilize 1.5 mg/ml solution of various lipids. This table summarizes the results of plots like that shown in panel (**a**). These results will depend on the state of lipids and will be most reproducible if starting with unilamellar vesicles. For these data, dried lipid films were resuspended in small amounts of each detergent by vortexing. Additional aliquots of detergent were then added, the solution stirred and Abs^{500} was measured.

mixed micelles. Shaking the plates during this incubation helps with this mixing. Both the protein and the lipid solutions should contain sufficient detergent for solubilization, so no additional detergent is required at this point. However, it may be desirable to adjust the detergent concentrations to be equal in all of the crystallization trials, thus ensuring a consistent rate of detergent removal across all the conditions. This strategy requires addition of detergent to conditions with lower LPR. For long-chain detergents, this will unnecessarily prolong an already long crystallization period (1–2 weeks) and it may be better to consider the variable reconstitution rates as another parameter in the crystallization.

4. Prepare dialysis solutions in a way that is compatible with the liquid handling device. The geometry of the 96-well format restricts the volume of the dialysis buffer relative to the protein solution, but this restriction can be overcome by frequently changing the dialysis buffer over the course of the crystallization. Typically, this buffer is changed twice per day and a programmable liquid-handling robot is therefore extremely useful. To facilitate this process, the dialysis buffers should be arrayed in a 96-well format (e.g., deep-well plates). Dialysis buffers should contain any soluble components that distinguish the crystallization trial, e.g., pH, salts, nucleotides, and other soluble ligands. Membrane bound additives are likely to stay inside the dialysis chamber in association with either the micelles or the bilayer.

5. Assemble the dialysis block (Fig. 4) in a way that minimizes bubbles in the individual dialysis chamber. Specifically, excess protein solution should be added to each well (e.g., 55 μl to a 50 μl well) to ensure complete whetting of the dialysis membrane prior to sealing the device. Bubbles will potentially block the dialysis membrane and slow the rate of detergent removal. To minimize this problem, the dialysis block can be stored on its side so that any bubbles will rise to the side of the well rather than block the dialysis membrane.

6. Place the dialysis block in a temperature-controlled environment at the selected temperature. Typical temperatures to test are 4, 20, and 37 °C and the initial choice should depend on the stability of the protein sample, as assayed by size-exclusion chromatography. Temperature cycling has also been effective in promoting crystal formation, which involves shuttling samples between high and low temperature environments. Higher temperatures promote bilayer fluidity, mixing of hydrophobic components, and perhaps formation of crystal seeds, whereas the lower temperatures promote annealing of larger crystalline domains. Prolonged periods at high temperature may also lead to denaturation of protein, especially in the presence of high amounts of detergent. Therefore, a useful strategy might be to remove detergent initially at low temperature and then to switch to a period of thermal cycling as the solution transitions from the micellar phase to the bilayer phase.

7. The time required to complete the dialysis should be studied prior to a crystallization screen and can be estimated by visual

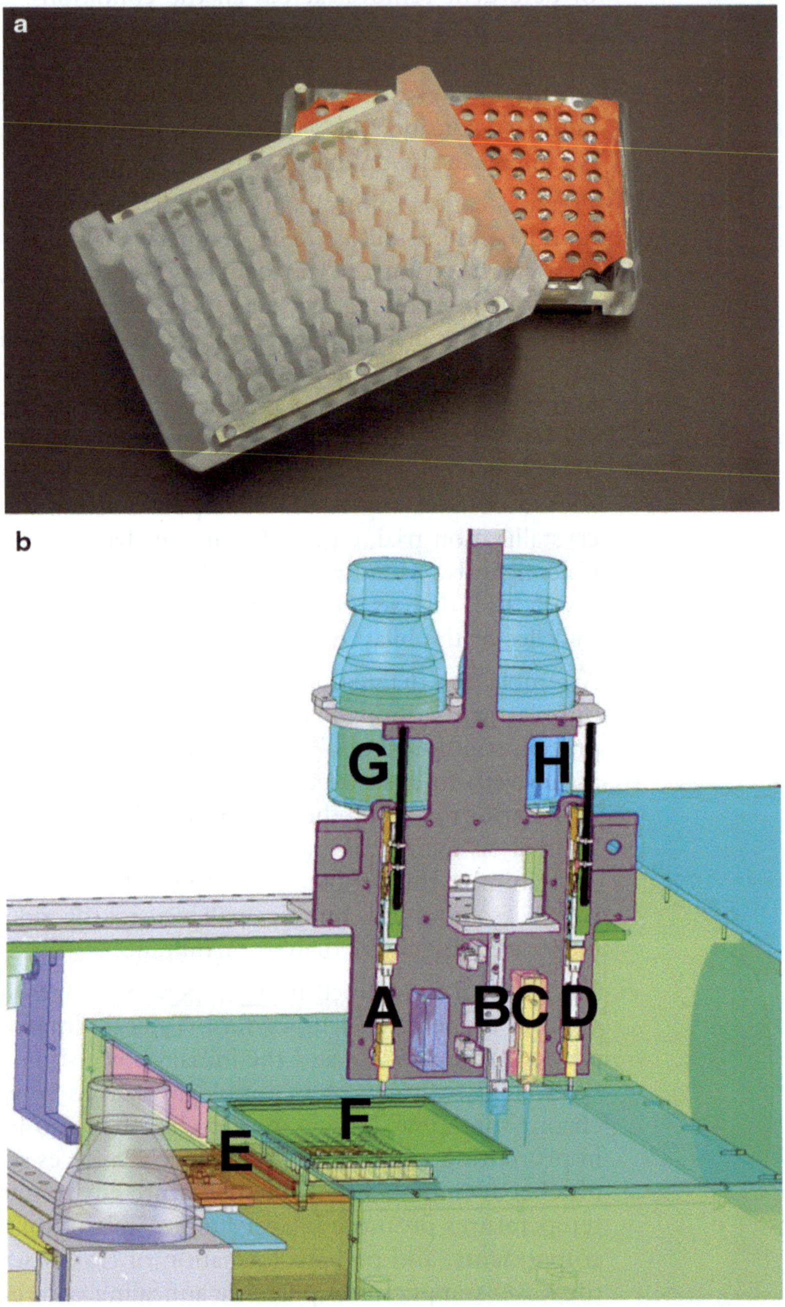

Fig. 4. Devices for high-throughput crystallization trials. (**a**) Dialysis block accommodating 96 independent samples, each with a unique dialysis buffer, as described by Vink, et al. (21). The samples containing protein and lipid are placed in the lower chamber, fitted with a sheet of dialysis membrane and a gasket (orange), and secured to the upper chamber, which holds the dialysis buffers. The wells are arrayed with standard 96-well geometry such that loading and dialysis buffer exchanges can be done with a commercial liquid-handling robot. (**b**) Cyclodextrin crystallization robot described by Iacovache et al. (22) with plans available at http://temimps.nysbc.org. This robot uses nanoliter scale pipettors (Seyonic) to add either a cyclodexrin stock solution (A/G) or water (D/H) to each of 96 well in a commercial microtiter plate. It is equipped with a volume sensor (B) and additions of water are used to compensate for evaporation during the period of crystallization. In addition, light scattering is measured (C) and the plates are shaken to ensure mixing of the solutions (F).

inspection of the turbidity of the samples during crystallization. Certain additives, such as glycerol, have a tendency to slow dialysis significantly, whereas elevated temperatures may speed dialysis; such parameters need to be taken into consideration. Therefore, a mock crystallization trial should be set up in advance, omitting the protein from the sample but otherwise maintaining concentrations of detergent and lipid. After each exchange of dialysis buffer, the detergent concentration of these buffers can be analyzed, using either drop shape measurements or TLC, and the removal of detergent can thus be plotted as a function of time.

8. The time required to reach the transition point can be inferred from the rate of detergent removal and knowledge of the starting concentrations of lipid and detergent. A good rule of thumb is that the transition to lipid bilayers will occur when the detergent-to-lipid ratio is 1:1. In reality, this transition varies depending on the specific lipid and detergent species and can be studied by measuring light scattering of detergent–lipid solutions as the detergent concentration is increased (Fig. 3).

3.4. Crystallization Using Cyclodextrin

1. Define conditions of the screen as above.
2. Prepare the individual detergent/lipid/protein samples in a 96-well microtiter plate as above, but include all cofactors to be tested, because unlike with dialysis, these cofactors cannot be introduced through the dialysis buffer. This fact means that a more concentrated protein solution is desirable for the cyclodextrin method, which can be diluted into a buffer with defined salt composition and pH. Alternatively, the protein solution may be dialyzed prior to crystallization setup.
3. Set up the 2DX robot (Fig. 4). The rate of adding the cyclodextrin solution (μl/h) depends on the detergent concentration in the well and the specific detergent adsorption capacity of the cyclodextrin. One should plan to reach the cmc within 6–12 h. For example, if the ternary mixture containing protein, lipids, and detergent contains 5 mM DDM, a total of 10 mM cyclodextrin will need to be added, because it takes 2 cyclodextrin molecules to chelate 1 molecule of DDM (25). Given a well volume of 20 μl, 0.286 mg of cyclodextrin will be required for full complexation of the detergent, which corresponds to 5.72 μl of a 5% stock solution of cyclodextrin. In this case, a rate of 0.5–1 μl/h would be suitable for adding the cyclodextrin, typically at intervals of one drop every 15 min (~200 nl/drop).
4. Set the initial temperature. Ideally, the initial temperature should be as low as possible during early stages of detergent removal, while remaining above the phase transition temperature of the

lipid. In particular, the transition temperature is 25 °C for DMPC and <0 °C for the lipids with unsaturated acyl chains.

5. Monitor the reconstitution/crystallization process by light scattering. The light scattering signal will increase as the detergent concentration reaches the cmc, indicating the aggregation of membrane proteins and lipids, hopefully forming bilayers.
6. Increase the temperature over the next 6–12 h, typically to 37 °C. Cyclodextrin may be added over another 12 h to ensure complete complexation of the detergent.

3.5. Negative Staining

1. Prepare a solid plastic film by pipetting 25 μl of collodion solution onto the surface of water-filled crystallizing dish. Prior to adding the collodion, the surface of the water should be swept free of dust with a glass rod or pipette. After allowing the film to dry briefly, EM grids are placed on its surface in a defined orientation (e.g., polished side contacting the film). Plastic-coated EM grids are picked up from the water surface using a flat piece of lint-free, unprinted newspaper (see Note 9). Allow the grids and newspaper to dry in a clean environment.
2. Coat the grids with carbon using a vacuum evaporator (see Note 10).
3. Make the grids hydrophilic by glow discharge (see Note 11).
4. Place the grids onto the magnetic posts of 96-position magnetic grid support platform or tray (Fig. 5).
5. Use liquid handling robot or multichannel pipette to carry out pipetting steps for negative staining. The sequence of steps depends on the particular application. Specifically, Kim et al. (26) added 2 μl individual crystallization trials to individual EM grids and allowed them to incubate for 30 s. Thereafter, three sequential drops containing 8 μl of 0.25% uranyl acetate were added, using the pipette to remove each drop prior to adding the next. Alternatively, a home built staining robot pipettes 4 μl of each protein sample on the grids, incubates for 60 s, and then washes grids three times with water to remove cyclodextrin. After removing the final water wash, 3 μl of 2% uranyl acetate are added and removed after 15 s by aspiration (19).
6. Remove final drop of stain either by aspiration with the pipettor or by blotting the grids with filter paper. For the latter, a strip of filter paper can be used to blot an entire row of EM grids simultaneously (see Note 12).

3.6. Electron Microscopy

1. Insert negatively stained samples into an electron microscope. This can be done manually by following the manufacturers procedures, or in an automated fashion using a robotic sample changer. The latter is preferable for screening large numbers of samples and a number of alternative systems have been

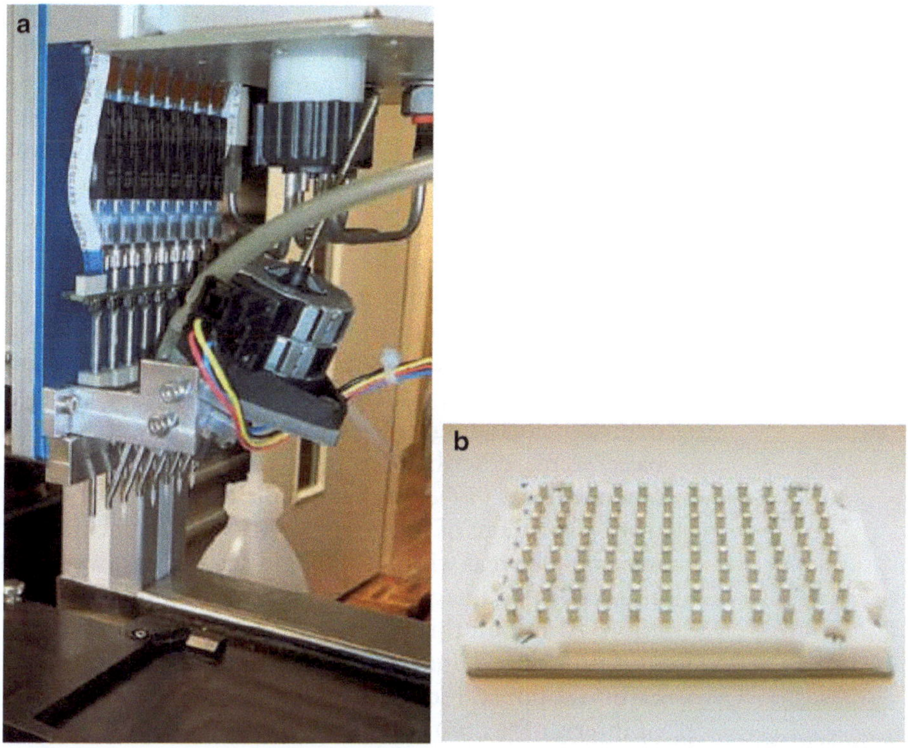

Fig. 5. Apparatus for high-throughput negative staining. (**a**) A negative staining robot described by Coudray et al. (19) for applying negative stain to grids placed on a magnetic platform. This device stains four grids at a time and is capable of working through a set of 96 grids in ~2 h. A computer control interface has been developed that allows the user to specify all the relevant parameters of staining, such as sample incubation time, number of washes, number of stain cycles. (**b**) A commercial magnetic platform that can be used for negative staining. The posts contain rare earth magnets and firmly hold Ni grids, such that they are not lifted by capillary action during the multiple pipetting steps. The posts are arrayed according to standard 96-well dimensions and can be accessed either with commercial liquid-handling systems or with the negative staining robot in panel (**a**).

developed. Most recently, Coudray et al. (19) have described the retrofit of a Tecnai T12 electron microscope with the "Autoloader" from FEI Corp coupled with a carousel for holding up to 96 samples. For this system, EM grids are loaded into 12-grid Autoloader cassettes and 8 of these cassettes are mounted on the carousel within the vacuum of the microscope. A similar strategy was used for the Gatling Gun designed by Lefman et al. (27), in collaboration with Gatan Inc. This device consisted of a spiral drum that accommodates 100 grid cartridges from the FEI Polara microscope, also within the vacuum of the microscope. An alternative approach for loading samples was originally reported by Cheng et al. (17) and more recently elaborated by Hu et al. (18). Both of these groups built free-standing robots to load grids from a tray onto the standard EM sample holder and then to transfer the holder through the air-lock of the microscope. The latter systems take slightly longer for loading each grids and require considerable care in maintaining the mechanical alignment between the robot and the microscope (specifically the axis of sample insertion). However, they require minimal modifications to the electron microscope and grids do not have to be individually mounted into specialized cartridges.

2. Record a series of representative images from each sample. A number of different software packages have been developed for automated imaging, though they have varying abilities to target suitable objects on the grid. Both of the external robots described in the previous section are controlled by the Leginon software program, which also includes applications for automatically collecting images from each grid after loading. The basic steps, described in detail by both Hu et al. (18) and Cheng et al. (17), involve identifying suitable grid squares from a montage of images at very low magnification, identifying suitable areas within these grid squares at low magnification and recording images at medium magnification for later evaluation. Although this software is able to identify simple objects in the intermediate images, like regular holes in a perforated carbon support, it currently lacks the ability to identify objects of interest for crystallization screens, though this is an area of active development. Similarly, the Gatling Gun included a script for directing Gatan's Digital Micrograph software to record a series of images at designated magnifications, but no effective algorithms for targeting (27). Commercial software called PASys from JEOL has been developed in collaboration with the Fujiyoshi labortory that does include a shape analysis of objects in images, but this software appears to only be compatible with JEOL microscopes. A more satisfactory solution is provided by the AnimatedEM program that was developed in conjunction with the Autoloader retrofit of the Tecnai T12 microscope (19). This program collects the usual series of images at increasing magnification, but performs a sophisticated shape analysis of intermediate images to select uniform, isolated sheets that might contain 2D arrays. Higher magnification images are then recorded from these areas and Fourier transforms are inspected for the presence of Bragg reflections, which would reflect not only the presence of 2D arrays, but also their relative order.

3. Store images of crystallization trials in a database for evaluation and cross-referencing. Although a laboratory notebook is the conventional way of recording crystallization results, as the number of conditions and the number of protein targets increase, accessibility of images and correlation with their respective crystallization conditions becomes extremely important. A laboratory information management system (LIMS)

such as those used by high-throughput X-ray crystallography centers is ideal, given their existing facility for storing information about sequence, expression, purification, crystallization. In particular, the Sesame LIMS (28, 29) has been modified to import images and associate them with their respective crystallization conditions (18). This import can be done with any generic set of images, or it can be done automatically from the Leginon database using a shared database identifier. The Sesame user interface then allows users to scan through images and assign corresponding scores reflecting the success of the crystallization trial.

3.7. Conclusion

The dialysis and cyclodextrin methods represent two complimentary approaches to crystallization. On the one hand, the ability to dictate the rate of cyclodextrin addition can be used to precisely control the rate of detergent removal and consequent reconstitution. It is reasonable to assume that the transition between the micellar phase and the bilayer phase is critical to the crystallization process, and manipulation of cyclodextrin addition offers the opportunity to explore this transition. On the other hand, dialysis has a longer track record in reconstitution and membrane protein crystallization. Although the level of control is somewhat limited, the resulting solutions do not have high levels of cyclodextrin present, simplifying the preparation of samples for electron microscopy. With either approach, differences in the physical chemistry of detergent removal can be expected to produce different results and it is impossible to know a priori which will yield the best crystals for a given protein target. In the spirit of assessing the effect of all possible parameters, it may be useful to attempt both methods of crystallization. This is analogous to empirical testing of different 3D crystallization methods for X-ray crystallography (e.g., hanging drop, sitting drop, batch, lipidic cubic phase) in order to obtain the best possible result.

A combination of these two methods may also be a productive avenue to explore. In particular, the addition of cyclodextrin to the dialysis buffer has the potential to accelerate and to provide finer control over dialysis rates. Dialysis rates are fundamentally limited by the relatively low concentration of monomeric detergent molecules, which are the only species that can equilibrate across the dialysis membrane. Detergent micelles typically contain ~100 molecules and are therefore too large to move across this membrane. The limitation is particularly acute for long-chain detergents, which have a very low cmc. However, both cyclodextrin and the cyclodextrin-detergent complexes are able to migrate across the dialysis membrane, thus effectively increasing the pool of monomeric detergent molecules that can be equilibrated and thus removed by dialysis. Preliminary analysis has shown cyclodextrin to accelerate the removal of DDM greatly, achieving complete detergent removal in only 4 days, compared with the typical 2 weeks in the absence of cyclodextrin. The overall cyclodextrin concentration remains relatively low throughout the process and can be completely eliminated by simply omitting cyclodextrin in the last change of dialysis buffer. Because this buffer is frequently changed throughout the procedure, one can maintain a certain control over the rate of detergent removal, for example, pausing when the protein solution reaches the transition from micellar to bilayer phases. Also, a bolus of cyclodextran can be added to the starting protein solution to establish a starting condition that is relatively close to the transition, thus minimizing the unproductive period that a protein spends in the micellar state. This may be particularly important for unstable membrane proteins.

4. Notes

1. Ideally, protein is used immediately after purification without freezing, storage, or concentration. By freezing or otherwise storing the protein, the possibility of denaturation or aggregation increases. Concentrators typically increase the detergent concentration in an unpredictable way, making it difficult to precisely define the starting point for crystallization.
2. The cmc is defined as the concentration at which micelles start to form. Below this concentration, all detergent molecules are monomeric in solution. Above this concentration, there is a mixture of monomeric detergent molecules and micelles, which are in dynamic equilibrium. Above the cmc, the concentration of monomers remains constant at the cmc. The cmc is characteristic for each detergent, depending on its chemical composition, and is affected by physical/chemical parameters such as the presence of solutes and temperature.
3. This selection of lipids has hydrocarbon chain lengths of 14 (myristol), 16 (palmitoyl) and 18 (oleoyl). The latter has one unsaturated bond, whereas the former are both fully saturated. The head groups range from the zwitterionic phosphatidylcholine and phosphatidyl ethanolamine, to the negatively charged phosphatidylglycerol and phosphatidic acid. Although this is a good starting set of lipids, in many cases it is also worth using other membrane components, such as cardiolipin, cholesterol, and sphingolipids, or using lipid extracts from the relevant organism or tissue, such as liver, heart, brain *E. coli*, or yeast.
4. Parafilm is the standard substrate reported by Kaufmann et al. (20) that generally works well. Teflon tape (e.g., for pipe fittings) is an alternate choice of substrate and may be suitable for short-chain detergents. Parafilm has the advantage of

maximizing changes in the contact angle as a function of concentration. Teflon is more hydrophobic and thus limits the extreme flatness of drops produced by short-chain detergents (e.g., OG) as their concentration approaches the cmc. With some software, it is difficult to determine the contact angle for very flat drops and measurements may therefore be more reproducible using Teflon tape.

5. Although contact angle is the conventional and most physically rigorous measure of surface tension, the ratio of drop width and height also produces reliable results, at least for small drops (e.g., 20 μl where the shape is well approximated by an ellipse). The programs Xdroptrace and DropBox both fit the drop shape with an ellipse and calculate either the width/height ratio (based on the major and minor axes) or the contact angle (based on the intersection with the planar substrate). The measurement of contact angle is very sensitive to the precise position of the substrate and the axial ratio may therefore be more robust in practice. Both Xdroptrace can be downloaded from http://temimps.nysbc.org.
6. If the concentration of lipid in the protein solution is low, then the lipids may be readily extracted from the solution with chloroform. After separating the organic phase from the aqueous phase, the chloroform can be evaporated and the resulting lipid film resolubilized in a smaller amount of chloroform prior to running the TLC. Extraction into an organic phase is also a method for separating the lipid from the detergent, which will be abundant and perhaps interfere with the signal from the lipid. Although nonionic detergents partition between the aqueous and organic phases, multiple washings of the organic phase with water will greatly reduce the amount of detergent present; lipid has a negligible solubility in water and will remain almost completely in the organic phase.
7. It is judicious to include a small amount of butylated hydroxytoluene (BHT) as a scavenger of free radicals and therefore as a mechanism to minimize lipid oxidation. Prepare a stock solution of 20% BHT in ethanol and add a sufficient amount to obtain 0.2% BHT in the final aqueous stock solution of lipid.
8. It is possible to bring lipid into the aqueous phase in the absence of detergent, but the lipid will form very heterogeneous multivesicular structures that may prove difficult to solubilize at a later stage. A more homogeneous, pure lipid solution can be obtained by sonicating the solution using a probe sonicator (e.g., Branson Sonifier S-250 fitted with a microtip). The effectiveness at producing unilamellar vesicles depends on the nature of the lipid; lipids with a net charge more readily form unilamellar structures, whereas neutral lipids are more difficult to disperse. Sonication puts a significant

amount of energy into the solution and can cause chemical changes (oxidation, cleavage of headgroups, creation of lysolipids); the solution should therefore be kept cold (on ice and in a cold room) and sonication periods should be minimal (1–3 minutes with 50% duty cycle and with breaks to allow the solution to remain cold). Since a detergent-solubilized, homogeneous solution is required to start crystallization, it therefore makes sense to directly solubilize the lipid film with detergent and avoid these potential problems, but the ability of a given detergent to solubilize the lipid also depends on the physical state of the lipid.

9. To remove the EM grids from the water surface, layer the newspaper on top of the EM grids, allow the paper to absorb a bit of water and then peel the paper off of the surface, thus removing the EM grids as a sandwich between the newspaper and the plastic film. Place the sandwich into a petri dish with the plastic facing upwards and allow to air dry (30).

10. Depending on the apparatus, either a carbon rod (graphite or amorphous carbon) or a carbon thread can be used for evaporation. For high resolution work, the flatness and conductivity of the carbon support is critical (31). For evaluating negatively stained samples, however, one simply requires a uniform carbon film that provides a reasonable proportion of unbroken grid squares (the broken squares can be identified by automated imaging programs and easily avoided). The underlying plastic film is helpful in this regard. It is also important to control the current during carbon evaporation in order to avoid sparking, which produces inhomogeneities in the resulting film. Also, many investigators evaporate several thin layers of carbon in order to improve strength.

11. An optional step in EM grid preparation is to remove the plastic film prior to glow discharge. This improves the clarity of high magnification images and also reduces residual stickiness that results from the ragged edges of the plastic film at the periphery of the grid. However, this step is not easily automated and represents an interruption in the workflow. To remove the plastic, place several pieces of filter paper into a glass petri dish. Saturate the filter paper with amyl amine. Place EM grids onto the filter paper with carbon side facing upwards. Cover the petri dish and allow grids to incubate for 15–30 min in a chemical fume hood. Remove the grids and place on a fresh piece of filter paper for glow discharge. An alternative method for producing carbon film is evaporate carbon onto a freshly cleaved mica surface and then to float this carbon onto a water surface. This method, also less amenable to high-throughput methods due to fragility in the resulting film, is described by Stokes and Ubarretxena (30).

12. The choice between blotting and aspiration will depend on the quality of the samples as assessed by electron microscopy. The goal is to obtain an even, thin layer of solution across the entire EM grid, which after drying encases all of the samples in an even layer of negative stain.

Acknowledgments

The authors would like to acknowledge a number of individuals who have developed the methodologies discussed in this chapter. In particular, Martin Vink, Changki Kim, and Minghui Hu, Thomas Kaufmann, Ioan Iacovache, Nicolas Coudray Hervé Rémigy and many others were responsible for designing and implementing the devices and protocols described in this chapter. The authors are indebted to them for their efforts. The authors belong to the Transcontinental EM Initiative for Membrane Protein Structure, which is a center for membrane protein structure determination funded by the NIH Protein Structure Initiative under grant U54GM094598. More information about this center can be found at http://temimps.nysbc.org. Additional research support was provided by NIH grants R01GM081817, R01GM095747, and R01GM079233.

References

1. Ubarretxena-Belandia I, Stokes DL (2010) Present and future of membrane protein structure determination by electron crystallography. Adv Protein Chem Struct Biol 81:33–60
2. Schenk AD, Castano-Diez D, Gipson B, Arheit M, Zeng X, Stahlberg H (2010) 3D reconstruction from 2D crystal image and diffraction data. Methods Enzymol 482:101–129
3. Hite RK, Raunser S, Walz T (2007) Revival of electron crystallography. Curr Opin Struct Biol 17:389–395
4. Abeyrathne PD, Chami M, Pantelic RS, Goldie KN, Stahlberg H (2010) Preparation of 2D crystals of membrane proteins for high-resolution electron crystallography data collection. Methods Enzymol 481:25–43
5. White SH (2009) Biophysical dissection of membrane proteins. Nature 459:344–346
6. Luft JR, Collins RJ, Fehrman NA, Lauricella AM, Veatch CK, DeTitta GT (2003) A deliberate approach to screening for initial crystallization conditions of biological macromolecules. J Struct Biol 142:170–179
7. Rees DC (2001) Crystallographic analyses of hyperthermophilic proteins. Methods Enzymol 334:423–437
8. Mastronarde DN (2005) Automated electron microscope tomography using robust prediction of specimen movements. J Struct Biol 152: 36–51
9. Nickell S, Forster F, Linaroudis A, Net WD, Beck F, Hegerl R, Baumeister W, Plitzko JM (2005) TOM software toolbox: acquisition and analysis for electron tomography. J Struct Biol 149:227–234
10. Suloway C, Shi J, Cheng A, Pulokas J, Carragher B, Potter CS, Zheng SQ, Agard DA, Jensen GJ (2009) Fully automated, sequential tilt-series acquisition with Leginon. J Struct Biol 167:11–18
11. Zheng SQ, Keszthelyi B, Branlund E, Lyle JM, Braunfeld MB, Sedat JW, Agard DA (2007) UCSF tomography: an integrated software suite for real-time electron microscopic tomographic data collection, alignment, and reconstruction. J Struct Biol 157:138–147
12. Koster AJ, Chen H, Sedat JW, Agard DA (1992) Automated microscopy for electron tomography. Ultramicroscopy 46:207–227
13. Suloway C, Pulokas J, Fellmann D, Cheng A, Guerra F, Quispe J, Stagg S, Potter CS, Carragher B (2005) Automated molecular

microscopy: the new Leginon system. J Struct Biol 151:41–60

14. Zhang J, Nakamura N, Shimizu Y, Liang N, Liu X, Jakana J, Marsh MP, Booth CR, Shinkawa T, Nakata M, Chiu W (2009) JADAS: a customizable automated data acquisition system and its application to ice-embedded single particles. J Struct Biol 165:1–9
15. Zhang P, Beatty A, Milne JL, Subramaniam S (2001) Automated data collection with a Tecnai 12 electron microscope: applications for molecular imaging by cryomicroscopy. J Struct Biol 135:251–261
16. Oostergetel GT, Keegstra W, Brisson A (1998) Automation of specimen selection and data acquisition for protein crystallography. Ultramicroscopy 74:47–50
17. Cheng A, Leung A, Fellmann D, Quispe J, Suloway C, Pulokas J, Abeyrathne PD, Lam JS, Carragher B, Potter CS (2007) Towards automated screening of two-dimensional crystals. J Struct Biol 160:324–331
18. Hu M, Vink M, Kim C, Derr K, Koss J, D'Amico K, Ubarretxena-Belandia I, Stokes DL (2010) Automated electron microscopy for evaluating two-dimensional crystallization of membrane proteins. J Struct Biol 171(1): 102–110
19. Coudray N, Hermann G, Caujolle-Bert D, Karathanou A, Erne-Brand F, Buessler JL, Daum P, Plitzko JM, Chami M, Mueller U, Kihl H, Urban JP, Engel A, Remigy HW (2010) Automated screening of 2D crystallization trials using transmission electron microscopy: a high-throughput tool-chain for sample preparation and microscopic analysis. J Struct Biol 173(2):365–374
20. Kaufmann TC, Engel A, Remigy HW (2006) A novel method for detergent concentration determination. Biophys J 90:310–317
21. Vink M, Derr K, Love J, Stokes DL, Ubarretxena-Belandia I (2007) A high-throughput strategy to screen 2D crystallization trials of membrane proteins. J Struct Biol 160:295–304
22. Iacovache I, Biasini M, Kowal J, Kukulski W, Chami M, Van der Goot FG, Engel A, Remigy HW (2009) The 2DX robot: a membrane protein 2D crystallization Swiss Army knife. J Struct Biol 169(3):370–378
23. Remigy HW, Caujolle-Bert D, Suda K, Schenk A, Chami M, Engel A (2003) Membrane protein reconstitution and crystallization by controlled dilution. FEBS Lett 555:160–169
24. Rigaud J-L, Mosser G, Lacapere J-J, Olofsson A, Levy D, Ranck J-L (1997) Bio-beads: an efficient strategy for two-dimensional crystallization of membrane proteins. J Struct Biol 118:226–235
25. Signorell GA, Kaufmann TC, Kukulski W, Engel A, Remigy HW (2007) Controlled 2D crystallization of membrane proteins using methyl-beta-cyclodextrin. J Struct Biol 157: 321–328
26. Kim C, Vink M, Hu M, Stokes DL, Ubarretxena-Belandia I (2010) An automated pipeline to screen membrane protein 2D crystallization. J Struct Funct Genomics 11(2): 155–166
27. Lefman J, Morrison R, Subramaniam S (2007) Automated 100-position specimen loader and image acquisition system for transmission electron microscopy. J Struct Biol 158:318–326
28. Zolnai Z, Lee PT, Li J, Chapman MR, Newman CS, Phillips GN Jr, Rayment I, Ulrich EL, Volkman BF, Markley JL (2003) Project management system for structural and functional proteomics: Sesame. J Struct Funct Genomics 4:11–23
29. Haquin S, Oeuillet E, Pajon A, Harris M, Jones AT, van Tilbeurgh H, Markley JL, Zolnai Z, Poupon A (2008) Data management in structural genomics: an overview. Methods Mol Biol 426:49–79
30. Stokes DL, Rice WJ, Hu M, Kim C, Ubarretxena-Belandia I (2010) Two-dimensional crystallization of integral membrane proteins for electron crystallography. Methods Mol Biol 654:187–205
31. Gyobu N, Tani K, Hiroaki Y, Kamegawa A, Mitsuoka K, Fujiyoshi Y (2004) Improved specimen preparation for cryo-electron microscopy using a symmetric carbon sandwich technique. J Struct Biol 146:325–333

Chapter 16

Automated Grid Handling and Image Acquisition for Two-Dimensional Crystal Screening

Anchi Cheng

Abstract

Routine large-scale two-dimensional (2D) crystallization trials are good candidates for automation. For imaging the large number of grids prepared from 2D crystallization trials, a two-pass imaging protocol using Leginon and a grid handling robot is described. A manual target selection bridges the two imaging passes. The two passes can be combined into one if objects of interests at different stages of trials can be reliably found using automated methods.

Key words: Automation, Screening, Two-dimensional crystals, Robotics, Electron crystallography, Electron microscopy

1. Introduction

Screening crystallization conditions require preparation of negatively stained transmission electron microscope (TEM) grids and imaging them systematically in the microscope. This is a time-consuming process as soon as the scale of the screening trial expands to any reasonable size. A number of efforts have been made to improve the throughput of these operations. Liquid-handling robots coupled with multi-well plates have been tested both for dispensing the potential crystal suspension onto grids as well as for staining in various ways (1–3). Methods of attaching the grids to the staining plate have been explored (2, 3). Finally, software-controlled grid exchange instrumentation has been developed in both academic and commercial environments (3–6), and applications written to automatically acquire images suitable for evaluating the crystallization results (1, 3, 7, 8). In at least three cases, automated

Ingeborg Schmidt-Krey and Yifan Cheng (eds.), *Electron Crystallography of Soluble and Membrane Proteins: Methods and Protocols*, Methods in Molecular Biology, vol. 955, DOI 10.1007/978-1-62703-176-9_16,

analysis of the images has been performed to locate membranes and to analyze the crystal quality (3, 7, 8).

From my experience, target selection criteria are not the same during initial novel protein crystallization screening as during refinement of crystallization conditions. At the initial trial, broad sampling of different objects is more important than the focused sampling of common objects that becomes necessary during refinement. Algorithms need to be tuned at different stages of the crystallization trials to minimize false negative results at the initial trials and minimize false positives at the refinement stage. In addition, few of the sampled objects found in the initial screening need to be imaged at high magnification where the time overhead is high due to the need for focusing. Therefore, progressive tuning of the sampling algorithm is necessary if it is used at different stages of the crystallization trials. Until a uniform automated solution is reached, it may be more practical and efficient to use a semi-automated approach so that specific objectives can be identified by a human operator. However, the recent publication by Coudray et. al. (3) showed very promising results by fully automated solution on protein with both known and unknown crystallation condition. Such success suggests that perfect tuning between the two extremes may be feasible in the near future.

The protocol described here is for automation of grid insertion and imaging. The imaging strategy consists of two rounds of automated imaging, using Leginon for image acquisition as well as communication with the grid-handling robot. Between the two passes, an operator spends a minimal amount of time evaluating the images acquired in the first automated pass and choosing potential crystalline targets for a second automated pass. If an efficient and accurate automated crystal-finding algorithm is available, a single-pass screening can be achieved by extending the first-pass application to acquire images at higher magnification with precise focusing. As is the case for most calibration-based automation, calibration, alignment, and setup are the most time-consuming and critical steps in the overall process. Details of the calibration, alignment, and setup required for Leginon multi-scaled imaging (MSI) applications are not covered here since continuously updated information is available in the online manual at http://leginon.org/.

2. Materials

2.1. Transmission Electron Microscope

The TEM used for automated crystal screening should be capable of a resolution high enough to enable assessing the quality of a 2D crystal (see Chapter 19 for further details). In addition, the TEM used for automation needs to have an interface to enable external software control. Currently, Leginon runs on FEI Tecnai series and above, some JEOL microscopes such as JEM3100FSC (9) and 1230 (4), and some Zeiss microscopes.

2.2. Grid-Handling Robot for Screening

The grid-handling robot transfers the selected grid from its grid reservoir (defined below) to the imaging position in the TEM after which it can be manipulated by the specimen stage of the TEM. There are only a handful of examples of grid-handling robots: (a) A 6-axis articulate robot arm mounted separately from the microscope to mimic the human operation of inserting the loaded side-entry specimen rod through the air lock (1, 6). The grid reservoir is a 96-grid tray formatted after a microtiter plate. The grids are placed in the open wells of the tray by the operator and can be stored for a long term after attaching a cover. (b) A pair of Cartesian and SCARA robots insert the loaded specimen rod through the air lock (4). The grid reservoir is again an open 96-grid tray. (c) A Gatling gun design, with a cylindrical grid reservoir and linear drive under vacuum, that is mounted on the microscope on the opposite port from that of the side-entry specimen stage (5). The grids in the 100-grid reservoir are pre-mounted in individual cartridges and attached to the cylinder. (d) A modified FEI "autoloader" that can accept, from a cylindrical grid reservoir, one of 8 Titan autoloader cassettes with preloaded grids (3), making it a 96-grid reservoir.

The first two robots have been integrated into Leginon using a database table that registers the handshake signals between the robot's grid insertion/extraction control and requests from Leginon (4). Robot server programs capable of communication to the database table could be written to integrate the latter two robots with Leginon.

3. Methods

3.1. Calibration of the Grid-Loading Robot and Its Alignment with the TEM Grid Positioning Stage

Whether a custom-built or commercial grid-loading robot is used, its calibration and alignment with the TEM grid positioning stage are crucial for reproducibility and safe use of the system.

For robots that insert individual grids through the TEM air lock, the alignment is especially important since the TEM is often supported on a pneumatic air cushion which means that the column is not in a rigidly fixed position. In addition, service or rebuilding of the specimen stage can result in repositioning of the stage relative to the column. During grid insertion, each of the intermediate positions must be known by the robot and the timing must be dictated by the air lock pumping. In contrast, realignment of the robot to the grid reservoir or the loading station where the

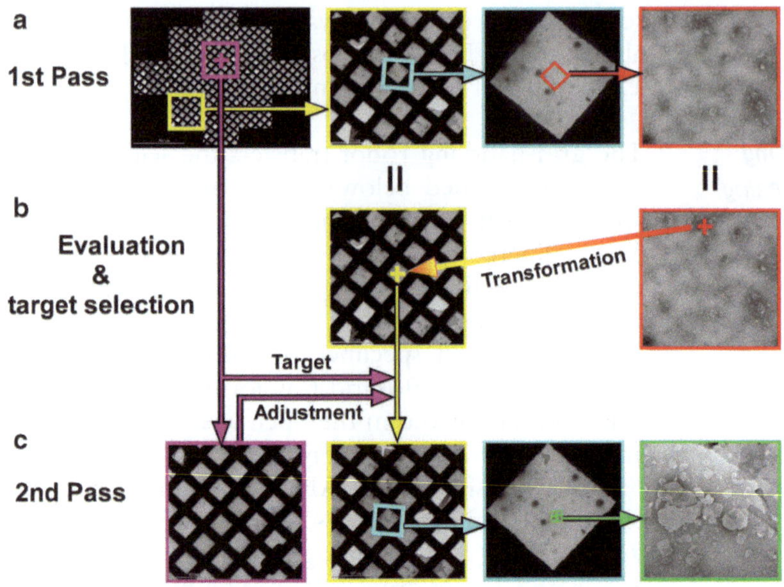

Fig. 1. Leginon Robot-MSI Screen applications. (**a**) First-pass multi-scaled imaging up to a magnification where the membranes can be recognized; (**b**) evaluation and transformation of targets selected manually; (**c**) second-pass multi-scaled imaging to obtain high-magnification images of the transformed targets with adjustments.

grid is placed and clamped down on the specimen rod is rarely needed since it is in a fixed format and there are no moving parts.

3.2. First of the Two-Pass Screening Using the Grid-Loading Robot

During the first step of the two-pass screening application, the grid is surveyed up to an intermediate magnification where membrane, vesicles, or other interesting objects can be identified without requiring accurate focusing (Fig. 1a). This screening application requires only a subset of the Leginon MSI calibration and alignment setup steps to be performed since no autofocusing is required during the operation. The stage is simply set near the eucentric height and the defocus set to eucentric focus at the start of image acquisition.

1. Use the same specimen rod as the robot and a test grid to reset the microscope defocus to zero at the eucentric focus and write down the specimen stage z position when the specimen is at the eucentric point (see Note 1).
2. Place the grids to be screened in the grid reservoir of the robot and register the tray and grids in Leginon's database (Fig. 2).
3. For robots using a modified specimen rod to transfer the grid through the stage air lock, clean and re-grease the o-ring on the specimen rod if a large-scale screening (more than 50 grids) is to be performed. This is required because multiple insertions/

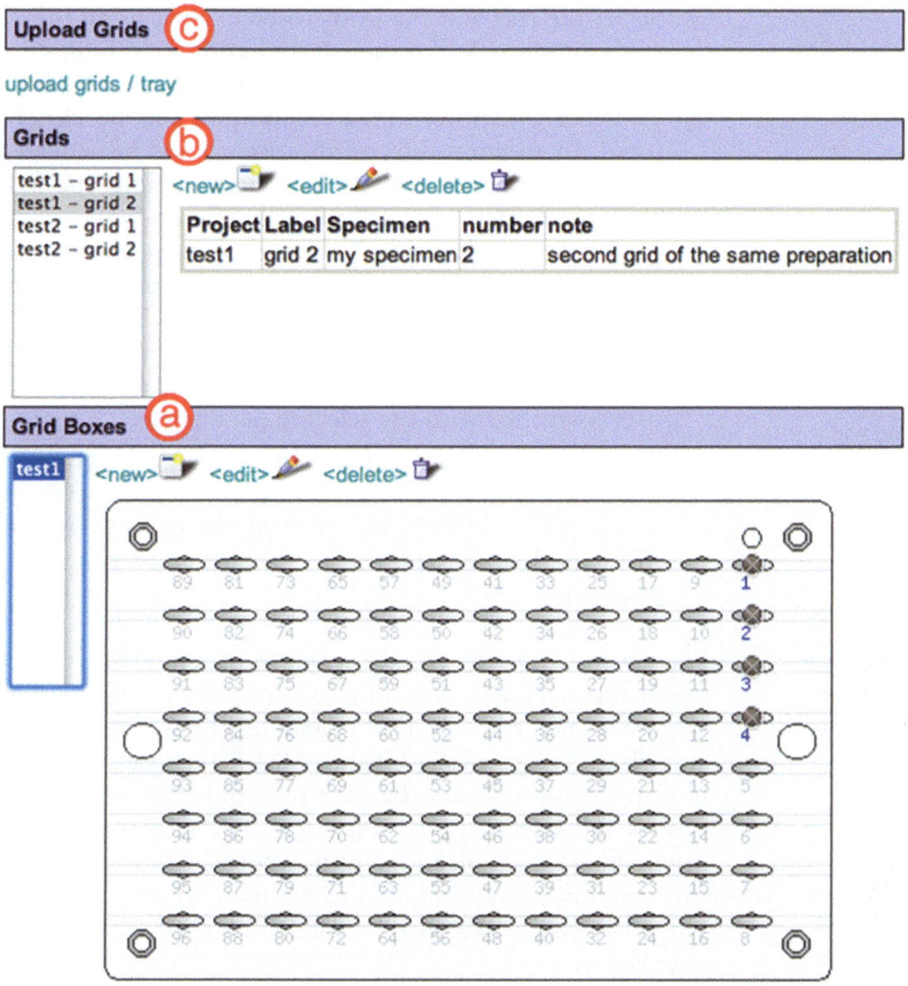

Fig. 2. Grid registration interface in the Web-based Leginon database tools: (*a*) The robot grid tray is registered in the Grid box section. (*b*) Registered grids in a grid box contain information about the grid. (*c*) Multiple grids can be registered at the same time by using an uploaded file.

extractions dry out the o-ring and possibly degrade the seal in the air lock.

4. Establish Leginon/robot/microscope communication: Start Leginon and the robot server, and the Leginon client program on the microscope.
5. Start the "MSI-Robot 1st Pass" application in Leginon: This Leginon application allows the user to systematically screen multiple areas on different grids (see Note 1).
6. In the "Robot" node (see Note 2) within the Leginon interface, enter the predetermined stage z position that puts the grid at the eucentric height as determined in step 1.

7. Select the tray and the grids to be imaged using the "Robot" node interface. When the selected grids are submitted, the rest of the operations will be performed automatically. For the first grid, the grid insertion/extraction should be closely monitored. In addition, user verification should be activated at the "Square Targeting" and "Mid Mag Targeting" node in order to fine-tune the target selection parameters before proceeding. In the "Square Targeting" node, the parameters are selected to choose squares with intact carbon and reasonable staining. In the "Mid Mag Targeting" node, the spacing and number of the sampling positions chosen should efficiently cover the grid square. The user verification option is then turned off once the user is satisfied with the selected parameters.
8. The grids should not be handled by a human operator between the two passes of the screening. If an open tray system is used as the grid reservoir, the tray can be covered and removed from the loading station for storage until required for the next pass.

3.3. Manual Target Selection for the Second of the Two-Pass Screening

For setups where the robot is capable of reinserting the grid with reasonable accuracy, second-pass screening can be performed to acquire higher magnification images of a limited number of manually selected targets. The "Robot MSI-Screen Evaluation" application in Leginon allows the user to browse through the acquired images and select targets of interest. The targets identified at the intermediate magnification are transformed to the coordinates on the very-low-magnification images used to survey the grid (see Note 3) and registered in the database for the second-pass screening (Fig. 1b).

1. Start "Robot MSI-Screen Evaluation" application under the same Leginon session as used for the first-pass screening. The microscope and robot do not need to be available during this step.
2. In the settings dialog of the node "2nd Pass Targeting" (Fig. 3), select the preset used to acquire images at the "Square" or "Mid Mag Survey" in the first-pass screening application as the "Transform From" preset (Fig. 3b) and the preset used in the "Grid Surveying" node to acquire grid atlas (Fig. 3c) as the "Transform To" preset. The images limited by the "Transform From" preset will be displayed as the "Child" image while the ones selected by the "Transform To" preset as the "Ancestor" image (see Note 3).
3. View the first image pair by clicking "Next" tool (Fig. 3d).
4. Select the intended targets on the "Child" image and click the "Transform" tool (Fig. 3e) to register the target as a target on the "Ancestor" image (see Note 4).

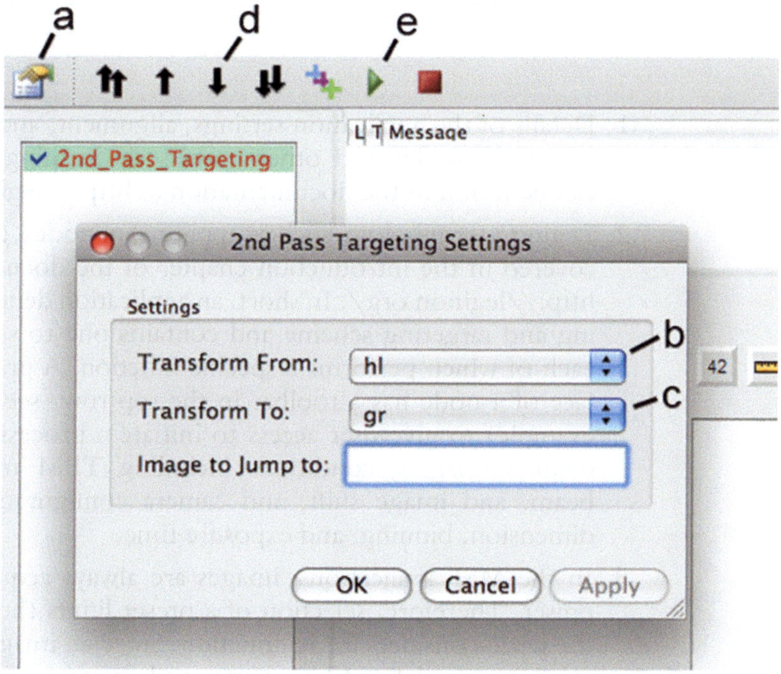

Fig. 3. Tools and settings of "2nd Pass Targeting" node in Leginon "Robot MSI-Screen Evaluation" application. (*a*) Tool on the toolbar of the node for opening settings dialog shown floating at the front of the main Leginon panel. (*b*, *c*) Preset selection boxes for choosing a pair of related images collected with the presets for the target selection (*b*) and transformation (*c*). (*d*) Tool to continue target selection/transformation on the next pair of images. (*e*) Tool to transform the selected target(s).

5. Repeat steps 3 and 4 on all images taken at the "Square" and "Mid Mag Survey."

3.4. Second of the Two-Pass Screening

The second pass of the screening requires the typical MSI calibration and microscope alignments to be performed. The targets selected in the evaluation application are adjusted to compensate for any reinsertion error, followed by focusing and image acquisition at a magnification suitable for crystal quality analysis and assessment (Fig. 1c).

1. Start "Robot MSI-Screening 2nd Pass" application under the same Leginon session used for the first pass and establish communication to the robot, the microscope, and the camera (see Note 1).
2. All registered targets will appear on the corresponding grid atlas in "2nd Pass Targeting" node (see Note 5). More targets may be added if desired.
3. Click on the "Submit" tool in "2nd Pass Targeting" node to start the acquisition series (see Note 2). All subsequent steps are completely automated.

4. Notes

1. Details of the application settings, alignment, and calibrations required for this and other multi-scale imaging applications can be found in the documentation at http://leginon.org/.
2. Leginon terminology such as application, nodes, and presets is covered in the introduction chapter of the documentation at http://leginon.org/. In short, an application defines the imaging and targeting scheme and contains one to several nodes, each of which performs a specific function. A graphical interface of a node has a toolbar in the top row (see Fig. 3 as an example) to give user access to initiate a process. Presets are defined imaging conditions including TEM magnification, beam, and image shift, and camera configurations such as dimension, binning, and exposure time.
3. In the MSI applications, images are always acquired using a preset. Therefore, selection of a preset limits the images that the node considers for its function. Since an image is acquired at a location defined by a target and the target is selected on another image, we call the lower magnification image on which the target of the higher magnification image is selected the parent image. All images on the same grid can then be related to each other like on a family tree.
4. The transformation is required because the very-low-magnification grid atlas images are the only ones accessible to the second-pass screening application.
5. Although this node shares the same name as the node in the evaluation application, it is derived from a different python class and behaves very differently.

Acknowledgments

This work is supported by National Institutes of Health (NIH) through the National Center for Research Resources (NCRR) P41 program (Grants RR17573 and RR023093). I would like to thank Bridget Carragher for valuable comments and editing of the manuscript.

References

1. Cheng A, Leung A, Fellmann D, Quispe J, Suloway C, Pulokas J, Abeyrathne PD, Lam JS, Carragher B, Potter CS (2007) Towards automated screening of two-dimensional crystals. J Struct Biol 160:263–404
2. Vink M, Derr K, Love J, Stokes DL, Ubarretxena-Belandia I (2007) A high-throughput strategy to screen 2D crystallization trials of membrane proteins. J Struct Biol 160:295–304
3. Coudray N, Hermann G, Caujolle-Bert D, Karathanou A, Erne-Brand F, Bruessler J-L, Daum P, Plitzko JM, Chami M, Mueller U, Kihl H, Urban J-P, Engel A, Remigy W (2010) Automated screening of 2D crystallization trials using transmission electron microscopy: a high-throughput tool-chain for sample preparation and microscopic analysis. J Struct Biol 173:365–374
4. Hu M, Vink M, Kim C, Derr K, Koss J, D'Amico K, Cheng A, Pulokas J, Ubarretxena-Belandia I, Stokes D (2010) Automated electron microscopy for evaluating two-dimensional crystallization of membrane proteins. J Struct Biol 171:102–110
5. Lefman J, Morrison R, Subramaniam S (2007) Automated 100-position specimen loader and image acquisition system for transmission electron microscopy. J Struct Biol 158:318–326
6. Potter CS, Pulokas J, Smith P, Suloway C, Carragher B (2004) Robotic grid loading system for a transmission electron microscope. J Struct Biol 146:431–440
7. Coudray N, Beck F, Buessler J, Karathanou A, Rémigy H, Kihl H, Engel A, Plitzko JM, Urban J (2008) Automatic acquisition and image analysis of 2D crystals. Microscopy Today 16:48–49
8. Nakamura N, Fujiyoshi Y, Mitsuoka K, Murata K, Shinkawa T (2003) An image evaluation system coupled with Auto-TEM. Microsc Microanal 9(Suppl 2):1038CD–1039CD
9. Glaeser RM, Lee J, Typke D (2006) Advantages and objectives of high-throughput data collection in single-particle cryo-EM. Microscopy and Microanalysis, Chicago

Chapter 17

Automation of Data Acquisition in Electron Crystallography

Anchi Cheng

Abstract

General considerations for using automation software for acquiring high-resolution images of 2D crystals under low-dose conditions are presented. Protocol modifications specific to this application in Leginon are provided.

Key words: Automation, Electron crystallography, Electron microscopy, Leginon

1. Introduction

Automated image acquisition software has gained popularity in the last few years both for single particle averaging (1–10) and for electron tomography (11–13). A few examples of semi-automated screening of 2D crystals have also been reported (14–17). While few of these packages have focused on the acquisition of high-resolution 2D crystal images, it is worth pointing out that several of the generally available automation software packages can already be used for this purpose without further modification. For example, Leginon (6) can be used with the goniometer set at a fixed-angle in a semi-automated mode which is ideal for acquiring images of 2D crystal preserved in vitreous ice.

To achieve the same quality of imaging, current generation of automated image acquisition software requires additional microscope alignment and setup time. While this disadvantage may become less obvious as software improves and excels in the area of automated microscope alignment at near-atomic resolution, setting up for fully automated acquisition of cryo 2D crystal images with present technology is often not worth the overhead of extra

Ingeborg Schmidt-Krey and Yifan Cheng (eds.), *Electron Crystallography of Soluble and Membrane Proteins: Methods and Protocols*, Methods in Molecular Biology, vol. 955, DOI 10.1007/978-1-62703-176-9_17,

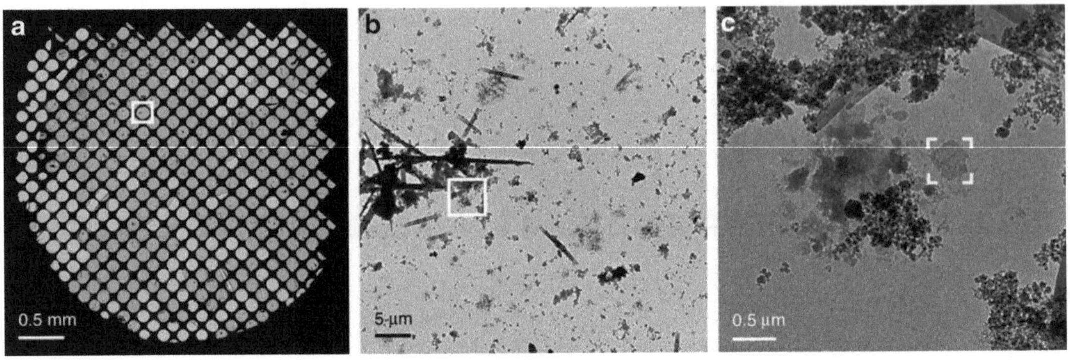

Fig. 1. An example of 2D crystal data acquisition that benefited from automation. (**a**) Very low magnification image (60×) of an EM grid containing gap junction 2D crystals. (**b**) Enlarged view (660×) of the grid square in the *white box* in (**a**). The vast majority of the material on the grid represents non-junctional membrane aggregates and non-crystalline precipitates. (**c**) Enlarged view of the region in the *white box* in (**b**). The *white brackets* identify a gap junction crystal.

setup time since crystals from purified proteins are frequently very abundant, the number of total images required is generally less than 100, and data from different experimental sessions can easily be merged. There are, however, cases in which semi-automated procedures have been shown to be more efficient than using manual procedures. The first example is for acquisition of rare crystals as in the case of imaging gap junction crystals made of connexin 43 protein (18–20). The gap junction-enriched membrane preparation contains approximately one crystal of a size <1 μm per EM grid square, often hidden among massive non-crystalline membrane fragments (Fig. 1). Because these crystals could be more easily identified at low magnification on a CCD image than on the viewing screen, Leginon provided an efficient tool for data collection. Automated image acquisition was also used in the systematic study of monolayer crystallization (21). The lipid monolayer used in the crystallization stretched over almost all micron-sized holes in these frozen-hydrated grid preparations. However, not all patches of the monolayer contained crystallized protein. Subtle differences in the ice thickness gradient across the holes were found to provide a crucial signal for the presence of crystals. Recognition and categorization of these areas was again easier using an automated Leginon application than by manual inspection.

Since the automation software I describe in this chapter is constantly changing and being improved and expanded, and the latest manuals are available online (see below) no formal protocols will be provided here. Instead, a few general considerations are discussed to provide an introduction to the use of automation in 2D crystal image acquisition. Readers who would like more details, or who wish to use the software in a practical application, should consult the detailed manuals of the particular software available to them. For example, Leginon users can find an online manual at http://leginon.org.

2. General Considerations

2.1. Microscope Alignment and Specimen Eucentricity Are Critical for Automation

While automated image acquisition offers many benefits, it can be time-consuming to set up the instrument so as to ensure stable and consistent performance. The cause of the differences lies in the alignment requirements. It is important for users of automation software to understand that software that relies on calibrations for navigation will generally require excellent microscope alignments; frequently better than those that a casual user typically performs. For example, setting the beam-tilt pivot point has minimal impact when acquiring images manually using a low-dose kit but is critical for ensuring accurate image shift targeting in Leginon. Eucentricity of the 2D crystal has also traditionally had only a small impact on the data processing since the crystal lattice provides an internal standard for magnification. However, when using automation software, if the microscope alignments are not performed at the eucentric focus, beam-tilt based auto-focusing will not be accurate. Overall, manual acquisition tolerates more alignment error. However, good alignment practice as enforced by automation requirements will promote better practices as the user aims for high-resolution data acquisition.

2.2. Targeting Used in Typical Automation Software Uses Much Less Dose Than Manual Targeting

Manual methods of target selection on the viewing screen followed by moving the target to the center for the final exposure use a feedback loop in which the operator refines the movement by observation of each previous movement. Even for the best and most efficient users, this targeting process can consume 1 s from the time the target appears in the viewing area to the time it is perfectly centered. During this time, the target crystal is continuously exposed to the electron beam. In contrast, automation software typically selects the target on an acquired binned image and relies on calibrations to bring the target to the center of view. Since a good signal-to-noise ratio can be obtained using very short exposure times for a binned image (going as the square of the binning factor), the total pre-exposure to the beam can be reduced to the millisecond range. In addition, CCD images provide better resolution and contrast than can be achieved using the viewing binoculars and the florescence screen, making it possible to characterize a potential crystal at a lower microscope magnification. For example, using the standard Leginon multi-scale targeting application (images for targeting are acquired at 120×, 550×, and then 5,000× before the final exposure is acquired at 50,000×), the accumulated dose from the three images at lower magnification (binned by 8) is 0.00016% of the dose used to acquire the final image. In practice, this accumulated dose may be somewhat higher mainly due to the lack of accurate specimen and camera shutter control below ~5 ms time scale, but still much less than manual low-dose targeting procedure. Therefore, based on the above argument and our experience, the simple targeting method in TEM imaging mode employed by most automation software does not cause significant damage to the cryo-specimen. In the electron crystallography community, defocused diffraction images covering a large area are traditionally used in searching for crystals. Although strong distortion need to be corrected, there is some indication that searching in diffraction mode also provides better optical stability than using a low magnification imaging mode (22). JADAS (8, 22) is the only current generation of automation software that offers the ability of using defocused diffraction images for targeting purpose.

3. Equipment and Software

3.1. Transmission Electron Microscope

The Transmission Electron Microscope (TEM) used for automation needs to have an interface for external software control. This option is known to be available on FEI Tecnai series and above, and newer types of JEOL JEM series microscopes, and some Zeiss microscopes.

3.2. Automated Image Acquisition Software

User-defined target selection at low magnification followed by imaging at progressively higher magnifications is the basic automation function required for 2D crystal imaging. Accurate low-dose focusing by the software when the grid is tilted is an advantage. Generalized automation software such as Leginon (6) and JADAS (8), as well as packages originally focused on tomography such as TomToolbox (13) and SerialEM (12), all appear to offer these capabilities.

4. Methods

4.1. Using the Leginon MSI Application for Semi-Automated Image Collection at Fixed Tilt Angles

In this operation, the target selection is done manually while centering of the target and low-dose focusing are performed by the software.

1. Start Leginon as per detailed instructions in the documentation at http://leginon.org.
2. Launch either the MSI-T or MSI-Edge application. These differ only in the automated target finding routine that is not required.
3. Change the settings in both "Hole Targeting" and "Exposure Targeting" nodes to skip automated target finding and to allow the user to verify the targets before submission.
4. Align the microscope.

5. After tilting the goniometer to the angle at which data will be collected, acquire a grid atlas, and perform all other setups except that related to automated hole-finding.
6. Select the focus and acquisition targets (see Notes 1 and 2 below) and acquire images as described in the online documentation. If film will be used for the final high magnification image, select this as an option in the high magnification preset (see Note 3).
7. If desired, a second fixed stage tilt can be applied and more images taken as in the previous step. The grid atlas can be used for targeting at the new tilt angle and does not need to be reacquired, and no adjustments are required to any other calibrations, microscope alignment, or node settings (see Note 4).

5. Notes

1. One focus and any number of acquisition targets should be selected per parent image.
2. The design of side-entry specimen stage is such that the stage *xy* plane tilts with the alpha angle change. As a result, objects brought to the tilt axis by movement in the stage *xy* plane always share similar height in the microscope. Only the height of the objects located away from the tilt axis is affected by the tilting. General practice in acquiring tilted images recommends that the low-dose focus positions be set along the tilt axis direction relative to the center of the search area to which the crystal target is moved. By the same analogy, the focus target is best chosen along a line parallel to the tilt axis and through the crystal target so that upon target centering, the two types of the targets are at the same height in the microscope.
3. Because exchanging film magazines increases the contamination of the specimen, experimental sessions that involve more than two film magazine exchanges may not be advisable (23).
4. Leginon adjusts the targeting and auto-focusing calibrations automatically based on the stage tilt and thus no changes are required between tilts.

Acknowledgments

This work is supported by National Institutes of Health (NIH) through the National Center for Research Resources (NCRR) P41 program (Grants RR17573 and RR023093). I would like to thank Bridget Carragher for valuable comments and editing of the manuscript.

References

1. Carragher B, Kisseberth N, Kriegman D, Milligan RA, Potter CS, Pulokas J, Reilein A (2000) Leginon: an automated system for acquisition of images from vitreous ice specimens. J Struct Biol 132:33–45
2. Lei J, Frank J (2005) Automated acquisition of cryo-electron micrographs for single particle reconstruction on an FEI Tecnai electron microscope. J Struct Biol 150:69–80
3. Marsh MP, Chang JT, Booth CR, Liang NL, Schmid MF, Chiu W (2007) Modular software platform for low-dose electron microsopy and tomography. J Microsc 228:384–389
4. Nickell S, Beck F, Korinek A, Mihalache O, Baumeister W, Plitzko JM (2007) Automated cryoelectron microscopy of "single particles" applied to the 26S proteasome. FEBS Lett 581:2751–2756
5. Shi J, Williams DR, Stewart PL (2008) A Script-assisted microscopy (SAM) package to improve data acquisition rates on FEI Tecnai electron microscopes equipped with Gatan CCD cameras. J Struct Biol 164:166–169
6. Suloway C, Pulokas J, Fellmann D, Cheng A, Guerra F, Quispe J, Stagg S, Potter CS, Carragher B (2005) Automated molecular microscopy: the new Leginon system. J Struct Biol 151:41–60
7. Yoshioka C, Pulokas J, Fellmann D, Potter CS, Milligan RA, Carragher B (2007) Automation of random conical tilt and orthogonal tilt data collection using feature-based correlation. J Struct Biol 159:335–346
8. Zhang J, Nakamura N, Shimizu Y, Liang N, Liu X, Jakana J, Marsh MP, Booth CR, Shinkawa T, Nakta M, Chiu W (2009) JADAS: a customizable automated data acquisition system and its application to ice-embedded single particles. J Struct Biol 165:1–9
9. Zhang P, Beatty A, Milne JLS, Subramaniam S (2001) Automated data collection with a Tecnai 12 electron microscope: applications for molecular imaging by cryomicroscopy. J Struct Biol 135:251–261
10. Zhang P, Borgnia MJ, Mooney P, Shi D, Pan M, O'Herron P, Mao A, Brogan D, Milne JLS, Subramaniam S (2003) Automated image acquisition and processing using a new generation of 4K x 4K CCD cameras for cryo electron microscopic studies of macromolecular assemblies. J Struct Biol 143:135–144
11. FEI (2005) Advanced Tecnai TEM software for easy acquisition, reconstruction and visualization, Technical Note, FEI software
12. Mastronarde DN (2005) Automated electron microscope tomography using robust prediction of specimen movements. J Struct Biol 152: 36–51
13. Nickell S, Förster F, Linaroudis A, Net WD, Beck F, Hegeri R, Baumeister W, Plitzko JM (2005) TOM software toolbox: acquisition and analysis for electron tomography. J Struct Biol 149:227–234
14. Cheng A, Leung A, Fellmann D, Quispe J, Suloway C, Pulokas J, Abeyrathne PD, Lam JS, Carragher B, Potter CS (2007) Towards automated screening of two-dimensional crystals. J Struct Biol 160:263–404
15. Hu M, Vink M, Kim C, Derr K, Koss J, D'Amico K, Cheng A, Pulokas J, Ubarretxena-Belandia I, Stokes D (2010) Automated electron microscopy for evaluating two-dimensional crystallization of membrane proteins. J Struct Biol 171:102–110
16. Nakamura N, Fujiyoshi Y, Mitsuoka K, Murata K, Shinkawa T (2003) An image evaluation system coupled with Auto-TEM. Microsc Microanal 9(Suppl 2):1038CD–1039CD
17. Coudray N, Hermann G, Caujolle-Bert D, Karathanou A, Erne-Brand F, Bruessler J-L, Daum P, Plitzko JM, Chami M, Mueller U, Kihl H, Urban J-P, Engel A, Remigy W (2011) Automated screening of 2D crystallization trials using transmission electron microscopy: a high-throughput tool-chain for sample preparation and microscopic analysis. J Struct Biol 173:365–374
18. Cheng A, Pulokas J, Suloway C, Zhu Y, Yeager M, Potter CS, Carragher B (2003) Adaptation of Leginon software for semi-automated recording of electron microscopic images of two-dimensional crystals. Microsc Microanal 9(suppl 02):1518–1519
19. Cheng A, Schweissinger D, Dawood F, Kumar NM, Yeager M (2003) Projection structure of full length connexin 43 by electron cryo-crystallography. Cell Commun Adhes 10:187–191
20. Cheng A, Yeager M (2007) Bootstrap resampling for voxel-wise variance analysis of three-dimensional density maps derived by image analysis of two dimensional crystals. J Struct Biol 158:19–32
21. Taylor DW, Kelly DF, Cheng A, Taylor KA (2007) On the freezing and identification of lipid monolayer 2-D arrays for cryoelectron microscopy. J Struct Biol 160:305–312
22. Nakamura N, Shimizu Y, Shinkawa T, Nakata M, Bammes B, Zhang J, Chiu W (2010) Automated specimen search in cryo-TEM observation with DIFF-defocus imaging. J Electron Microsc 59:299–310
23. Cheng A, Fellmann D, Pulokas J, Potter CS, Carragher B (2006) Does contamination buildup limit throughput for automated cryoEM? J Struct Biol 154:303–311

Chapter 18

Automation of Image Processing in Electron Crystallography

Marcel Arheit, Daniel Castaño-Díez, Raphaël Thierry, Bryant R. Gipson, Xiangyan Zeng, and Henning Stahlberg

Abstract

Electron crystallography of membrane proteins records images and diffraction patterns of frozen-hydrated two-dimensional (2D) crystals. To reconstruct the high-resolution three-dimensional (3D) structure of a membrane protein, a multitude of images of 2D crystals have to be processed. Certain processing steps are thereby similar for batches of images that were recorded under similar conditions. Here we describe how the 2dx software package can be used to automate the processing of 2D crystal images, and how the 2D and 3D merging results can be used to iteratively reprocess the images. While the processing of 2D crystal images has been fully automated, the merging process is still semi-manual.

Key words: Membrane proteins, Electron crystallography, 3D structure determination, Electron microscopy, Image processing

1. Introduction

Electron crystallography of two-dimensional (2D) membrane protein crystals involves several steps. Sufficient quantities of pure, homogeneous, and stable membrane protein in a suitable detergent have to be produced, which then have to be two-dimensionally crystallized. Grid preparation for cryo-electron microscopy (cryo-EM) is a difficult task that requires persistence and experience. Data in form of images and diffraction patterns on the cryo-electron microscope has to be collected. Lastly, the data have to be processed and merged into a high-resolution 3D structure.

Significant progress has been made in all these steps, not only in terms of quality but also in speed. With the exception of sample

Ingeborg Schmidt-Krey and Yifan Cheng (eds.), *Electron Crystallography of Soluble and Membrane Proteins: Methods and Protocols*, Methods in Molecular Biology, vol. 955, DOI 10.1007/978-1-62703-176-9_18,

preparation for cryo-EM, all steps have been streamlined, made more user-friendly, and are partly automated. Here, we discuss the automation of the processing of a larger number of 2D crystal images with the 2dx software package, version 3.3.0.

2. The General Concept

The 2dx software package (1, 2) is available at http://2dx.org. It is based on the MRC programs for image processing (3), and includes additional programs for automation (4) and other functions (5). The 2dx software suite contains a module *2dx_image*, which allows the processing of one single image of a 2D crystal, and *2dx_logbrowser*, which facilitates the analysis of the log files. 2dx also includes the module *2dx_merge*, which is used for the management of a 2D crystal structure determination project, and to perform 2D or 3D merging of the data from different 2D crystal images.

For a new project, the user should first setup a suitable file structure via *2dx_merge*, then process one image of a non-tilted 2D crystal with *2dx_image* manually, then save the determined optimal processing parameters as project default values, followed by launching the automatic processing for the remainder of the images of non-tilted specimens via *2dx_merge*.

For including images of tilted specimens into the processing pipeline, the user should again process the first image of a certain tilt angle category manually with *2dx_image*, save the optimal processing parameters as default values for that tilt angle category, and launch the automated processing of the remainder of the images in that category with *2dx_merge*.

The 2D or 3D merging of the processed image data still has to be done manually, but *2dx_merge* is designed to provide the user with guidance and feedback.

Automatic reprocessing of all non-tilted and tilted images can then again be launched from *2dx_merge*, after which the reprocessed data again have to be merged manually.

These steps will now be explained in detail.

2.1. The Project File Structure

2dx is designed for a certain arrangement of the files on the hard drive. It is strongly recommended to the users to adhere to this arrangement, so that the automated processing finds the files and data where it expects them.

A new membrane protein project is to be initiated in a certain directory. We here call that protein `ProA`, and assume it was initiated by user "John", who placed it into a project base directory called `/usr/people/john/proteins/ProA`. For convenience we abbreviate this directory with `~/proteins/ProA`.

If images were recorded on film and digitized, or if images were recorded on a CMOS or CCD camera, the most suitable image file format for the 2dx software is TIFF, and image names should follow a certain convention. By default, 2dx assumes a file naming in the form of `ProA0012345601.tif`, where the first four letters are the protein name (here "`ProA`"), the next two digits are the tilt angle (here "`00`"), the next six digits are the image number (here "`123456`"), and the last two digits are a possible subframe number from that image (here "`01`"). Subframes can, for example, be used, if several images were cropped from one large digitized film, or if one image should be processed for different crystal layers separately. The image import tool of *2dx_merge* will later extract this information (e.g., the nominal tilt angle) from the file name. If a different file name convention is used, one can adjust the filter setting in *2dx_merge* by defining a different regular expression (see below).

If 2dx is launched for the first time with a base directory `~/proteins/ProA`, *2dx_merge* will initiate a file and directory structure under this base directory. An example is shown in Fig. 1.

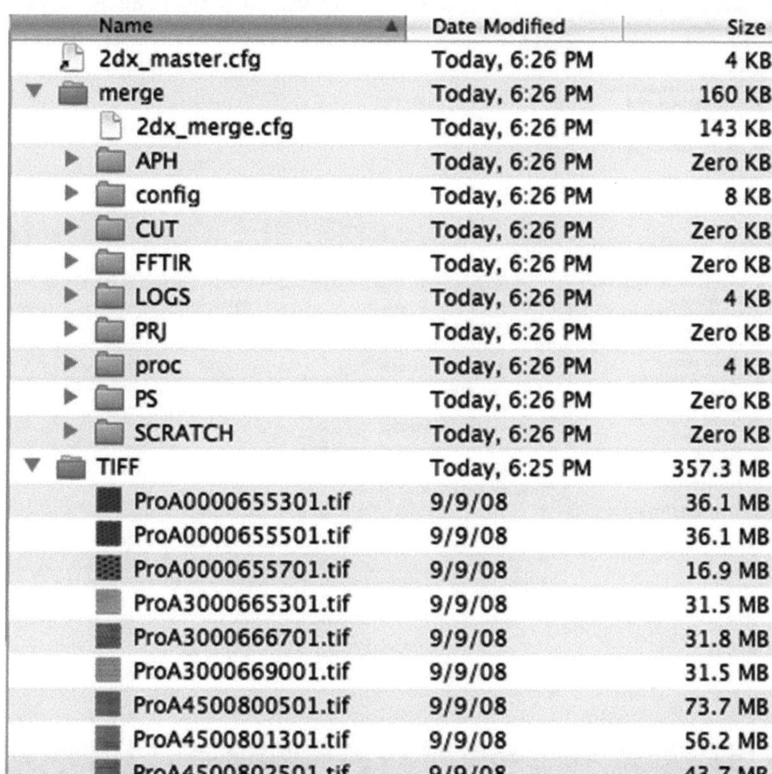

Fig. 1. The file structure created under the 2dx project base directory upon project initialization.

2.2. `2dx_merge` The Project Manager

1. `2dx_merge` will open its graphical user interface (GUI) with the Initialization screen (Fig. 2). This GUI is composed of several panels. Each panel has a certain function, indicated by the title on its top bar. The top left panel contains a list of the available "Standard Scripts," which should be executed sequentially for the merging of processed images. Below it is a panel for "Custom Scripts," which may be executed if needed. The central top panel will list the processed or non-processed images, once they are available. Below it is a panel to define the "Processing Data," which by default displays only the "Standard" set of parameters. Additional "Advanced" parameters can be accessed with the buttons in the top left corner of that panel. The other panels will be explained below.
2. The user should now look through the "Processing Data," and see if any of those image processing parameters can already be defined for this project. For example, the user may have an

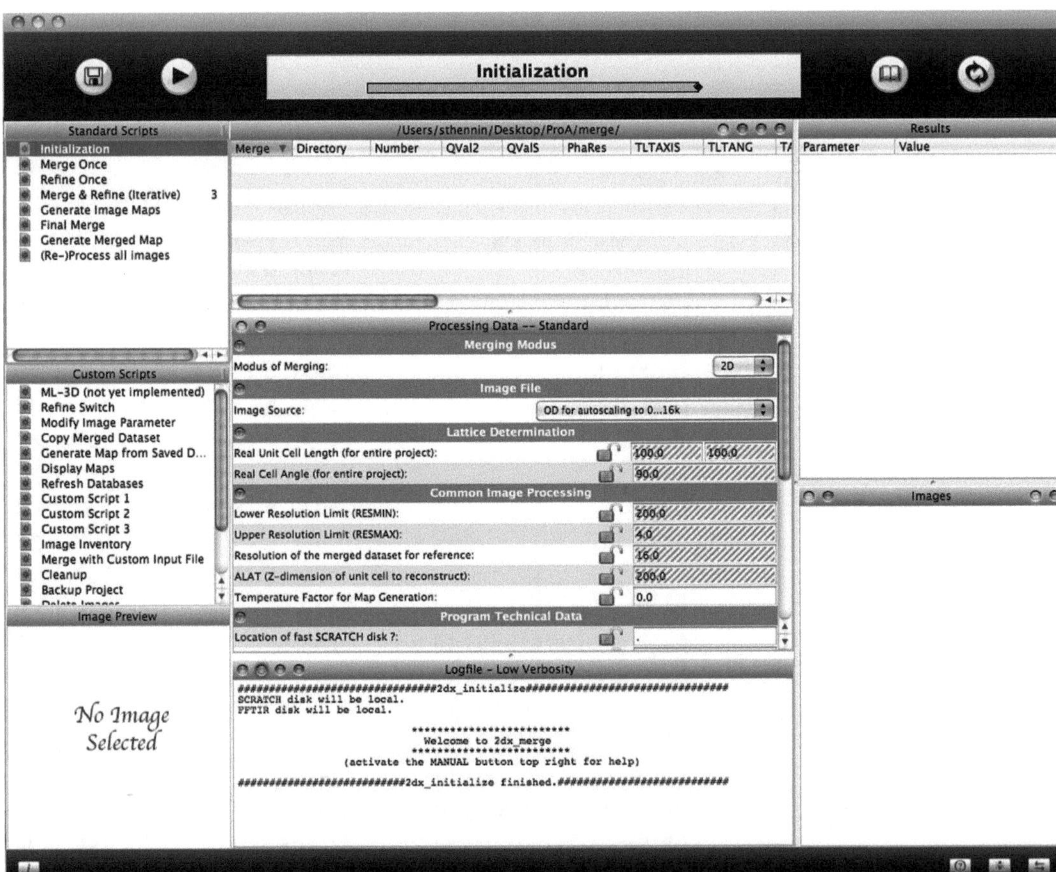

Fig. 2. The GUI of 2dx_merge upon project initialization.

estimate for the upper and lower boundaries for the resolution range to be considered. For a typical negative stain project, these would be set as RESMIN to 200.0 Å, and RESMAX to 16.0 Å. For a beginning cryo-EM project, a RESMIN of 12.0 Å may be adequate, depending on the quality of crystal images.

3. The "SAVE" button on the top left in the GUI of *2dx_merge* allows to save the current set of parameters in a file called 2dx_merge.cfg. This file is a text file, which can be opened with a conventional text editor for inspection. It contains the definition of various parameters, and the current settings for those parameters. The entry for the upper resolution limit RESMAX, for example, reads:

```
# LABEL: Upper Resolution Limit (RESMAX)
# LEGEND: The upper limitation in resolution, in
   Angstroem. This one should be high resolution
   (e.g. 4.0 A), while the other RESMIN should be
   the lower resolution cutoff (e.g. 200A).
# EXAMPLE: RESMAX = "4.0"
# HELP: http://2dx.org/documentation/2dx-soft-
   ware/parameters/Resolution
# TYPE: Float "MIN=0.1;MAX=10000000.0;DEFAULT=16.0"
# LOCKED: NO
# INHERITABLE_UPON_INIT: YES
# SYNC_WITH_UPPER_LEVEL: NO
# ISWRONG: YES
set RESMAX = "4.0"
```

4. The meaning of the lines trailed by a "#" is as follows: LABEL defines the name under which this parameter will appear in the GUI. LEGEND, EXAMPLE, and HELP define the contents of the right-mouse-click help panel for that parameter in the GUI. TYPE defines the parameter type. LOCKED defines if that parameter value can be changed manually or through the programs, or if it is currently protected. INHERITABLE_UPON_INIT defines if this parameter setting should be inherited, when initializing a new image for processing, see below. SYNC_WITH_UPPER_LEVEL defines those parameters that should be synchronized throughout the project. The crystallographic space group symmetry for a crystal project, for example, should be the same for all images in this project: If a crystal has a fourfold symmetry P4, then all images should be processed under this assumption. ISWRONG defines if the current setting for this parameter is still possibly wrong. If so, it will show up in the GUI with a blue-white striped underlay, to remind the user that this parameter still might have to be modified.

And finally, "`set RESMAX`" defines the actual current value for this parameter. This configuration parameter file `2dx_merge.cfg` is used by *2dx_merge* to store a set of processing parameters within the "`~/proteins/ProA/merge`" directory under the project base directory.

3. Workflow for Automated Image Processing

3.1. Importing a First Image into 2dx_merge

1. Make sure the local database in *2dx_merge* is saved, by clicking the top left "Save" icon. Now use the pull-down menu "`Import Images...`" under the "`File`" menu, to select one TIFF image of a good non-tilted 2D crystal. Navigate into the TIFF folder, and select only one image file, which you open. *2dx_merge* will now display the "Import Images" window, displaying on the left side the filename of your TIFF file, and on the right side the interpretation of parameter strings for the protein "Name," the nominal tilt "Angle," the image "Number," and the "Sub-image" number (Fig. 3). If the TIFF files are named following the above described convention, then *2dx_merge*'s default regular expression "^(\w{3,4})(\d{2})(\d{6})(\d{2})?$" will correctly populate the fields for those values. Regular expressions are defined at http://www.regular-expressions.info and are a versatile way to define a parameter lookup scheme. The regular expression in this example will search for a *w*ord of *3 or 4* letters, then a number of *2 d*igits, then a number of *6 d*igits, and finally a number of *2 d*igits. Clicking on "OK" will cause *2dx_merge* to create a directory for the tilt angle (`~/ProA/ProA00`), create an image directory for the image processing of this image under this tilt angle directory (`~/ProA/ProA00/ProA00006553101`), copy the TIFF file into this directory, create an image processing parameter file *`2dx_image.cfg`* in this directory (see below), and

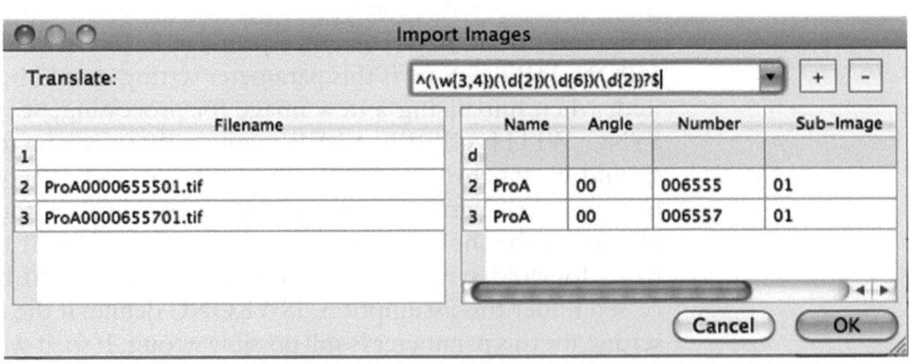

Fig. 3. The file import dialogue of *2dx_merge*.

launch *2dx_image* with its first "`Init Program`" script for this image.

2. The here-created image processing configuration file *2dx_image.cfg* results from a hierarchical processing of several configuration files that *2dx_merge* traverses sequentially:
   (a) For this, *2dx_merge* starts with the system-wide default configuration file, which is usually stored in `/usr/local/2dx/kernel/config/2dx_master.cfg`.
   (b) It then updates all parameters with settings from the personal choice of the users, usually stored in `~/.2dx/2dx_master.cfg`.
   (c) It then updates all parameters with settings from the project database, here in `~/proteins/ProA/2dx_master.cfg`. (Note that by default this file is a symbolic link to the configuration file in the merge directory, here in `~/proteins/ProA/merge/2dx_merge.cfg`, which then is the actual master configuration file for this project.)
   (d) It then updates all parameters with the tilt-angle-specific databases, here in `~/proteins/ProA/ProA00/2dx_master.cfg`, but for each parameter it only does this last update, if the parameter "INHERIT_UPON_INIT" in the last `2dx_master.cfg` file was set to "YES." Note that by default this file is another symbolic link, pointing the `2dx_master.cfg` file one level higher up (`~/proteins/ProA/2dx_master.cfg`), which by itself is a link to the `2dx_merge.cfg` file in the merge directory. By default, all reference `2dx_master.cfg` files are therefore symbolically linked to the `2dx_merge.cfg` file in the merge directory. This assures that initially only one definition of the project-wide parameters is available.
3. *2dx_merge* then stores the resulting database in a configuration file for the newly imported image, here under `~/proteins/ProA/ProA00/ProA00006553101/2dx_image.cfg`.
4. *2dx_merge* also creates several sub-directories in this image directory, which will be needed later for the processing of this image. This concludes the import of this image.

3.2. Processing the First Non-tilted Image with 2dx_image

1. This first imported image of a non-tilted 2D crystal is processed with *2dx_image* by double-clicking in *2dx_merge* onto the line entry for this image in the central top panel of the GUI of *2dx_merge*. This will open *2dx_image* with this image. Proceed to process this image carefully, as described in Chapter 10. Make sure you correctly define the acceleration voltage, Cs value, magnification, pixel step size, etc. Perform the processing, and fine-tune the unbending parameters. Determine the real lattice unit cell with the specific script

"Evaluate Lattice," and save the newly determined real-space lattice as project-wide real-space lattice.

2. Once these parameters are all well known for this image, save the local configuration file (this will be stored in `~/proteins/ProA/ProA00/ProA00006553101/2dx_image.cfg`).
3. This configuration should now also be saved as "Project Default" database, by using the pull-down menu option "`Save as Project Default`" under the "`File`" menu. This will overwrite the file `~/proteins/ProA/2dx_master.cfg`, thereby overwriting the merge database `~/proteins/ProA/merge/2dx_merge.cfg`. You can now close this instance of *2dx_image*.
4. Now it is important to close also the program *2dx_merge*, *without saving its local configuration*. This way, the local database for *2dx_merge* in `~/proteins/ProA/merge/2dx_merge.cfg` remains as *2dx_image* has just saved it. Now reopen *2dx_merge*, so that it has a fresh working copy of the current database. This now represents the default values for this project.

3.3. Processing the Remaining Non-tilted Images

1. You can now proceed to import the remaining non-tilted images. In *2dx_merge* launch the pull-down menu option "Import Images." Specify the other non-tilted images in the file section dialogue. With suitable naming convention, the import dialogue will have already determined the nominal tilt category (00), and the image numbers, see Fig. 3.
2. The first line in the import dialogue headed by "d" can be used to manually define default values, which are then populating the fields for all images to import. For the image number, the given default value will be incremented for each of the following images.
3. This import function will copy the TIFF files into newly created sub-directories, and establish local `2dx_image.cfg` configuration files for each image in those sub-directories.
4. To automatically process these imported images, select them in the central top panel in the GUI of *2dx_merge*, by checking the box in front of each line for those images. Then launch the script "(Re)process all images." This will prompt *2dx_merge* to launch *2dx_image* with all standard scripts on all selected images. For this, *2dx_merge* launches a GUI-less instance of *2dx_image*, and gives it as parameters the directory of the image to work on, and a "*" as symbol to process all standard scripts in sequential order, see also ref. (1). For traps and pitfalls in this process, see Note 4.1.
5. After all images have finished processing, update the database in *2dx_merge* by clicking the top right update icon, and

examine the result in the central top panel of the GUI of *2dx_merge*.

3.4. Verifying the Automatic Processing of the Non-tilted Images

1. The result of the processing of all selected images can be inspected in the top central panel of the GUI of *2dx_merge*. As described in Chapter 10 and also in (1, 2), the 2dx software calculates a singular value termed "QVal" to quantify the quality of an image and the performance of the image processing. A higher QVal value means a better-processed image. The available images in the top central panel in *2dx_merge* can be sorted by the column QVal2, which is the QVal after the second unbending script UnbendII in *2dx_image*. This will list the best images on the top of the image list, while the weaker images move to the bottom of the list.
2. The weaker images, where the automatic processing didn't produce satisfactory results, could simply be excluded from further processing by unclicking their selection boxes. This could be an option, if a very large number of 2D crystal images were available through automatic data collection on an automated cryo-EM system (e.g., driven by Leginon (6) or other systems). In most cases, however, it is better to double-click on the images in that central panel of *2dx_merge*, to open them in *2dx_image*, and inspect or correct the processing steps.
3. To process different image batches with different default settings, see Note 4.2.
4. To implement the project-wide synchronization of parameters, see Note 4.3.

3.5. Processing of Images of Tilted Specimens

The processing of images of tilted crystals is done analogously to the method described above. Import one good image of a tilted crystal with *2dx_merge*, which for a 30° tilted crystal would in the above example create a sub-directory `ProA30`. Process that image manually with *2dx_image*, and save the determined optimal processing parameters as (temporary) project default values. Then continue importing the remaining images from this (tilt) batch and try the automatic processing via the *2dx_merge* script "`(Re)Process all images`". See also Notes 4.4.

3.6. Merging

The image processing of the individual images results in text files with amplitude and phase values for the individual Miller indexed diffraction spots of each image. For images of non-tilted crystals, these can be merged in 2D into a merged projection map. If images of tilted specimens of sufficient quality are available, these can be merged into a 3D merged dataset. *2dx_merge* will produce a final 2D projection map, or respectively a 3D volume. The amplitudes and phases of these reconstructions are available in so-called APH and MTZ files in *2dx_merge*, as described in Chapter 11.

3.7. Re-unbending with a Synthetic Reference

The merged reconstructions can be used to create references to reprocess all available images. As described in ref. (1), the MRC program MAKETRAN (7) can be used by `2dx_image`, to create a synthetic reference for the unbending of the individual images. For this, `2dx_image` would not use the conventional unbending scripts UnbendI and UnbendII, but rather use the script Synthetic Unbend. The parameter SYN_UNBENDING (label "Use Synthetical Reference if possible?") determines, if the UnbendI and II scripts, or rather the Synthetic Unbend scripts should be called. Also the remaining standard scripts in `2dx_image` change their behavior depending on the setting of this switch. For example, if the Synthetical Reference is to be used, then the scripts to determine the defocus or lattice or spotlist are not running, in order to not loose previously determined settings.

The unbending algorithms for the scripts Unbend I and II are different than for the script Synthetic Unbend: Unbend I and Unbend II use a Fourier-filtered version of the original image, from which a box of a certain width (e.g., 500 × 500 pixels, corresponding to, e.g., 5 × 5 crystal unit cells) is cropped. While Unbend I uses a very narrow Fourier filter mask radius (e.g., 1 pixel) from the raw image, Unbend II uses multiple iterations of Fourier masking with larger masks on Fourier transforms of already unbent images, to produce increasingly improved references from this one image. In Unbend I and Unbend II, the raw image is always unbent only once, but each time with an increasingly improved reference structure. Optimization processes here concern two parameters: the Fourier masks for the filtering and the diameter of the boxed reference structure.

In contrast, the script Synthetic Unbend does not require iterative reference refinement, because its synthetically created reference is based on the entire dataset and therefore is likely a near-perfect reference. The optimization parameter here is now only one parameter: The diameter of Gaussian-shaped blobs placed into a synthetic Fourier transform, which is then used directly as a reference structure.

Even though we assume that the first synthetically generated reference structure is already near-perfect, the implementation in `2dx_image` for the Synthetic Unbend nevertheless creates two different reference structures. These are created with blob diameters SYN_maska and SYN_maskb, where typically the first is a smaller value than the second. The reason for this is that the first reference with the more narrow Fourier blobs corresponds to a larger-area reference pattern in real-space, which is used in a first round to determine the coarse lattice distortion parameters. These are stored by the MRC program QUADSERCH in an ERROR field. The second synthetic reference, created with the larger masking value `SYN_maskb`, shows a smaller-area reference pattern in real space. This second synthetic reference is used to refine the ERROR field, while only allowing minor deviations from the originally determined lattice distortions (ERROR field).

The image unbending with the synthetic reference should result in an improved reconstruction over the conventional unbending. However, the availability of a correct synthetic reference depends on the presence of a true and sufficiently sampled 2D or 3D reference dataset, and for tilted images sensibly depends on a correct assignment of the tilt geometry for this image.

3.8. Automating Custom Scripts and Procedures

2dx_merge also offers the possibility to apply the user's own custom scripts or procedures to a larger number of images. This can be done by adapting and launching one of the Custom Scripts "`Custom Script 1`" to "`Custom Script 3`". Double-clicking such a script name will open it with a text editor. These scripts contain code to apply specific (custom) commands to all selected image directories. The commands to adapt appear after the text section "Modify only the section below." You can use such a script for example to set a specific variable to a certain value for all selected image directories. This can be done by un-commenting and adapting a command like

```
echo "set TANGL = 0.0" >> LOGS/${scriptname}.results
```

This would reset the tilt angle variables TANGL to zero for all selected image directories. Another useful command is

```
echo "set SYN_Unbending = 1" >> LOGS/${scriptname}.results
```

which would switch all selected images to processing with the synthetically generated reference.

These custom scripts can also be used to execute one or several standard or custom scripts of *2dx_image* for all selected images. This makes use of the GUI-free command-line modus of *2dx_image*. If *2dx_image* is called from the command line with the image directory and the script identifier as parameters, it will execute the script as background process. This could for example be done with `${app_2dx_image} ${newdir}"+2dx_refinedefocus"` to launch the script `2dx_refinedefocus`. You could for example use a Custom Script in *2dx_merge* to apply the *2dx_image* custom script "`Refine Defocus from Merged Data`" to all selected images. In this example, this script would use the MRC program CTFSEARCH to refine the defocus and astigmatism for all selected images, using the comparison between the image amplitude and phases and the merged reference dataset.

3.9. Optimizing the Performance of 2dx on Your Computer

The 2dx software package launches MRC programs, which were originally written in the 1970 in Fortran-77, when computers had little amounts of RAM and were still using tape drives. Consequently, these programs rarely store images in the RAM, but permanently read and write files from and to the local hard drive. Some of these operations read and write the image files in stripe sections of the images, which further intensifies the disk access. The performance of the 2dx image processing workflow on a modern multi-core computer therefore depends mostly on the speed of your hard drive. To allow acceleration of disk access, 2dx arranges the majority of the larger image files in sub-directories called FFTIR and SCRATCH.

The disk access can be significantly accelerated, if those FFTIR and SCRATCH sub-directories are located on a separate disk with lower latency, e.g., a Solid State Drive (SSD). Alternatively, a smaller performance improvement can already be achieved by placing those sub-directories on separate hard drives, so that the data input and output flow are spread over three drives: The main data drive, another drive for the FFTIR directory, and a third drive for the SCRATCH directory. Big speed increases are also gained when those FFTIR and SCRATCH directories are located on a solid state drive (SSD).

1. To allow comfortable management of such sub-directories on separate disk, use the 2dx_merge parameters "Location of fast FFTIR disk" and "Location of fast SCRATCH disk" to define a path to a fast disk. If a fast SSD drive is mounted under /Volumes/SSD, then these two parameters could be set to "/Volumes/SSD." This would then prompt 2dx to replace all FFTIR and SCRATCH sub-directories in all image directories with links to appropriate sub-directories on that SSD drive. Especially when processing several images simultaneously on one multi-core computer, speed gains by a factor of 10 or more can be gained with this setting.

3.10. The Overall Workflow for a 3D Structure Determination

The 3D structure of a 2D crystal membrane protein project can be determined as described above: A first non-tilted image is processed with `2dx_image` (semi-) manually, and all image processing parameters are refined and saved as project default values. These are then used to automatically process the remaining non-tilted images. Results can be manually inspected and manually optimized, if desired.

Additional tilted images are then added, by processing the first image of each tilt angle category manually, followed by automatic processing of the remaining images from that tilt angle category.

After successful 2D and/or 3D merging of the data, all available images can be reprocessed from `2dx_merge`, using the merged dataset to produce synthetic reference data.

Iterative 2D/3D merging and reprocessing of the entire dataset can be used to improve the final resolution.

4. Notes

4.1. Comments on the Automatic Processing via 2dx_image

2dx_image optionally allows fully automatic processing of one 2D crystal image, by double-clicking in the GUI of *2dx_image* on the top bar of the Standard Scripts panel. This will select all standard scripts, which are then sequentially executed by clicking on the "Run" button. *2dx_image* includes programs and scripts to automatically determine the processing parameters, and then perform the processing. For this, we extended the MRC programs (3) with additional programs and functions and also combined it with other programs, e.g., CTFFIND (8). The following steps are performed in the fully automatic mode in *2dx_image*.

- The defocus in the center of the image is determined via CTFFIND.
- The approximate tilt geometry is determined via CTFFIND or CTFTILT.
- One or several lattices are determined, taking a potential tilt geometry into account (4).
- If the project-wide crystal lattice dimensions are known, they are used to refine the tilt geometry, based on the distortion of the identified lattice in this image, and provided that the tilt angle is higher than 25°.
- The list of spots in the Fourier transform is determined via MMBOX and refined.
- Unbending in step I and step II is done, and the FFT of the unbent image is evaluated to give measurements for amplitudes, phases, background values, and IQ values.
- These measurements are then subjected to correction for the CTF.
- A final map is calculated.

This process can go wrong at several points.

If on an image of a tilted crystal the defocus determination or the tilt geometry determination is failing, e.g., because the image was recorded very close to focus so that no Thon rings are discernible, then a wrong tilt geometry determination will make it unlikely that the correct lattice is found. Without lattice identification, all following processing steps fail.

If the lattice identification fails, for example because a SpotScan image has interfering low-resolution patterns in the Fourier transform, then an adjustment of the lattice search parameters to ignore low-resolution diffraction spots, may help. If the lattice identification fails due to the presence of multiple closely overlapping lattices, or due to other reasons, then manual lattice identification at this single step may rescue this image, and the subsequent image processing steps may still be executed automatically.

2dx_image allows locking specific parameters, e.g., the lattice parameters, so that their values will be protected from overwriting. This would be advisable in the above example: The manually determined lattice parameters should be locked, to protect them from accidental deletion or changes through other (automatic) processing steps.

4.2. Processing Different Image Batches with Different Default Settings

The here described automatic processing assumes that all imported images can be processed with the same default parameter settings. This would be the case, if the images stem from the same microscope, operated under the same acceleration voltage, recorded from specimens on the same grid or prepared by the same grid preparation method, and imaged under the same magnification and defocus settings.

If, however, images from different sessions on different microscopes are present, the above-described approach would not work. In order to enable automatic processing of different image batches, the user can create more than one super-directory for this tilt angle image class. In the above example, we called the protein ProA, and the non-tilted images were imported into a sub-directory called ProA00. The user can manually rename that directory name ProA00 to ProA00_batch1. (Please note that no spaces are allowed in file or directory names.)

Additionally imported non-tilted images will now be imported into a new sub-directory ProA00, so that they can be processed and handled in the *2dx_merge* GUI separately from the previously imported images. At a later stage, the new directory ProA00 can be renamed to ProA00_batch2.

In the above example, the master configuration files `2dx_master.cfg` are still all symbolically linked to the configuration file in the merge directory, called `2dx_merge.cfg`. Because the initialization of the *2dx_image*.cfg files happens during import of the new images via *2dx_merge*, the default settings for acceleration voltage and magnification, etc., are inherited from `2dx_merge.cfg` at the time of import. This feature can be exploited to import a larger number of images recorded at one magnification, then rename the super-directory ProA00 to ProA00_batchX, then import one additional image from the new batch, which creates a fresh ProA00, define in it the new magnification, save that as new project default, and then go on to import the remaining new images, which then will be all imported with the new (other) magnification setting into the new ProA00.

If the user wants to permanently uncouple the default settings for one batch of images from another batch of images, those symbolic links `2dx_master.cfg` can be deleted and replaced with local copies of the file `2dx_merge.cfg`. This allows for processing and reprocessing of large numbers of images with different default parameters. For this, see also the comments on parameter synchronization.

4.3. Project-Wide Synchronization of Parameters

Certain image processing parameters need to be synchronized throughout the entire 2D crystal processing project. These concern, for example, the crystal symmetry space group (e.g., P4), and the real-space dimensions of the crystal lattice (e.g., 103.0 Å × 103.0 Å with an included angle of 90°). Such parameters should be valid for all images that belong to one project.

The configuration files are `2dx_image.cfg` for images and `2dx_merge.cfg` for the merging. They all have identical structure and syntax, as described above.

If *2dx_image* is launched for a certain image directory, it will read the local configuration file `2dx_image.cfg` for this image. If in this configuration file the parameter

```
# SYNC_WITH_UPPER_LEVEL: YES
```

were set to YES, *2dx_image* will ignore the local value setting for this parameter, and instead keep the setting from the upper-level configuration file. If for example the configuration file `2dx_merge.cfg` has a symmetry definition of P4, and the upper-level files `2dx_master.cfg` are all symbolic links pointing to this one `2dx_merge.cfg` file, and if for the symmetry the parameter SYNC_WITH_UPPER_LEVEL is set to YES, then that P4 setting will overwrite the local setting in each image as soon as *2dx_image* is opened on them, so that the entire processing for all images assumes the same symmetry.

This allows the user to change a central parameter only once, and all images in this project will then be (re)processed with this coherent setting. If instead you wish to locally try different settings for a certain parameter, i.e., not to have a parameter sync with the upper level configuration file, you can change that SYNC_WITH_UPPER_LEVEL parameter in the GUI of *2dx_image*, via right-clicking on its parameter, and then clicking on the current value (yes or no) of the SYNC_WITH_UPPER_LEVEL setting.

4.4. Processing Tilted Images

In general, processing up to 30° tilt can be done with CTF correction algorithms, while the processing of high-resolution images at tilt angles of 40° or higher require TTF (titled transfer function) correction algorithms.

Since the processing of images of specimens at higher tilt angles becomes increasingly more complicated due to the anisotropic distortions in the lattice and other image parameters, we recommend to first start with lower tilt angle images, and then include the higher tilt angles later in the project.

4.5. Available Sources of Information

The 2dx image processing software is based on the so-called MRC image processing programs (3), which are in their original form available from Richard Henderson and Jude Short at the MRC, Cambridge. These MRC programs are also included in the distribution of 2dx, though in a modified form, allowing interaction with the 2dx environment, and we added in few cases additional functions to the MRC programs, mostly related to automation. The detailed flow of data from one MRC program to the next and the involved file formats are documented at http://2dx.org/documentation/2dxsoftware/manual/data-flow. 2dx also includes additional programs for a Maximum Likelihood (ML)-based single particle processing, which for non-perfect crystals may significantly improve the resulting resolution (5). The resulting map of the ML processing is thereby translated into amplitudes and phases, so that ML results for individual images can be used for 3D merging with lattice line interpolation via the remainder of the MRC-based 2dx suite.

2dx is available as open-source under the GPL software at http://2dx.org. This website also contains a documentation of the 2dx software, the included MRC programs, the parameters, and the general algorithms. A general introduction to electron crystallography on this web server, as well as detailed instructions to specific MRC programs were contributed by Anchi Cheng and Vinzenz Unger. The usage of 2dx and also the image processing packages IPLT (9) and MRC was lectured on at several workshops, including the EMBO Cryo-EM workshops, and the biennial workshops on Electron Crystallography of Membrane Proteins, described at http://2dx.org/workshop.

4.6. Other Image Processing Systems to Facilitate Processing of 2D Crystal Data

For the processing of electron diffraction data, the MRC software package offers robust programs that automatically index diffraction patterns and evaluate the data from those patterns. The XDP software tool (10) can be used to facilitate the evaluation of diffraction pattern.

A new development for electron crystallography data processing for images and diffraction patterns is IPLT (9, 11), a package that does not make use of the MRC software suite, but provides a reimplementation of existing algorithms as well as newly developed algorithms for 2D crystal processing in a modular way.

These software systems were recently comparatively evaluated be Schenk et al. (12).

Another image processing system for 2D crystal data is the Calidris system available at http://www.calidris-em.com (13–15).

5. Conclusion

Automation in the processing of images of 2D crystals of membrane proteins is implemented in the 2dx software package for the processing of individual non-tilted and tilted images. Successful automation of the image processing depends on the correct setting of suitable default parameters and on the availability of images of comparable image and crystal quality.

While 2dx offers user guidance and streamlines the 2D and 3D merging process, further software development is needed to also automate this merging process. A first step into this direction is the projective constraint optimization (PCO) method that was recently developed for 2D crystal data (16). As is described in more detail in Chapter 11 of this volume, with sufficiently oversampled data at lower tilt angles PCO was shown to be able to fully automatically calculate a 3D reconstruction from the available dataset from non-tilted and tilted 2D crystals, while bypassing lattice line interpolation, and on the same creating correct amplitude and phase values also in the so-called missing cone. 2dx_merge is also prepared for the processing of large numbers of images of 2D crystals, so that future efforts can now be extended to the automation of the collection of high-resolution electron crystallography data.

Acknowledgment

This work was in part supported by the Swiss initiative for Systems Biology (SystemsX.ch, grant CINA), Swiss National Science Foundation (Grant 205321_126490), the NCCR TransCure, and the NCCR Structural Biology.

References

1. Gipson B, Zeng X, Stahlberg H (2007) 2dx_merge: data management and merging for 2D crystal images. J Struct Biol 160:375–384
2. Gipson B, Zeng X, Zhang Z, Stahlberg H (2007) 2dx—user-friendly image processing for 2D crystals. J Struct Biol 157:64–72
3. Crowther RA, Henderson R, Smith JM (1996) MRC image processing programs. J Struct Biol 116:9–16
4. Zeng X, Gipson B, Zheng ZY, Renault L, Stahlberg H (2007) Automatic lattice determination for two-dimensional crystal images. J Struct Biol 160:353–361
5. Zeng X, Stahlberg H, Grigorieff N (2007) A maximum-likelihood approach to two-dimensional crystals. J Struct Biol 160:362–374
6. Carragher B, Kisseberth N, Kriegman D, Milligan RA, Potter CS, Pulokas J, Reilein A (2000) Leginon: an automated system for acquisition of images from vitreous ice specimens. J Struct Biol 132:33–45
7. Kunji ER, von Gronau S, Oesterhelt D, Henderson R (2000) The three-dimensional structure of halorhodopsin to 5 Å by electron crystallography: a new unbending procedure for two-dimensional crystals by using a global reference structure. Proc Natl Acad Sci USA 97:4637–4642
8. Mindell JA, Grigorieff N (2003) Accurate determination of local defocus and specimen tilt in electron microscopy. J Struct Biol 142:334–347

9. Philippsen A, Schenk AD, Signorell GA, Mariani V, Berneche S, Engel A (2007) Collaborative EM image processing with the IPLT image processing library and toolbox. J Struct Biol 157:28–37
10. Mitsuoka K, Hirai T, Murata K, Miyazawa A, Kidera A, Kimura Y, Fujiyoshi Y (1999) The structure of bacteriorhodopsin at 3.0 Å resolution based on electron crystallography: implication of the charge distribution. J Mol Biol 286:861–882
11. Philippsen A, Schenk AD, Stahlberg H, Engel A (2003) Iplt – image processing library and toolkit for the electron microscopy community. J Struct Biol 144:4–12
12. Schenk AD, Castano-Diez D, Gipson B, Arheit M, Zeng X, Stahlberg H (2010) 3D reconstruction from 2D crystal image and diffraction data. Methods Enzymol 482:101–129
13. Hovmöller S (1992) CRISP: crystallographic image processing on a personal computer. Ultramicroscopy 41:121–135
14. Zou XD, Sukharev Y, Hovmöller S (1993) Quantitative electron diffraction – new features in the program system ELD. Ultramicroscopy 52:436–444
15. Zou XD, Hovmöller A, Hovmöller S (2004) TRICE – a program for reconstructing 3D reciprocal space and determining unit-cell parameters. Ultramicroscopy 98:187–193
16. Gipson BR, Masiel DJ, Browning ND, Spence J, Mitsuoka K, Stahlberg H (2011) Automatic recovery of missing amplitudes and phases in tilt-limited electron crystallography of two-dimensional crystals. Phys Rev E Stat Nonlin Soft Matter Phys 84:011916

Chapter 19

Choice and Maintenance of Equipment for Electron Crystallography

Deryck J. Mills and Janet Vonck

Abstract

The choice of equipment for an electron crystallography laboratory will ultimately be determined by the available budget; nevertheless, the ideal lab will have two electron microscopes: a dedicated 300 kV cryo-EM with a field emission gun and a smaller LaB_6 machine for screening. The high-end machine should be equipped with photographic film or a very large CCD or CMOS camera for 2D crystal data collection; the screening microscope needs a mid-size CCD for rapid evaluation of crystal samples. The microscope room installations should provide adequate space and a special environment that puts no restrictions on the collection of high-resolution data. Equipment for specimen preparation includes a carbon coater, glow discharge unit, light microscope, plunge freezer, and liquid nitrogen containers and storage dewars. When photographic film is to be used, additional requirements are a film desiccator, dark room, optical diffractometer, and a film scanner. Having the electron microscopes and ancillary equipment well maintained and always in optimum condition facilitates the production of high-quality data.

Key words: Cryo-electron microscopy, Electron diffraction, High-resolution imaging, Photographic film, CCD camera, Direct electron detector, Specimen preparation, 2D crystals

1. Setting Up an EM Lab

The first consideration when setting up a laboratory for electron crystallography is the type of electron microscope that will be purchased and the type of cryo stage and image detection system this transmission electron microscope (TEM; EM) will be fitted with, as this will dictate the laboratory setup and the additional ancillary equipment that will be required (Fig. 1). Obviously other EM applications (conventional EM, single-particle cryo-EM, cryo-electron tomography) that the EM unit will be used for have to be

Ingeborg Schmidt-Krey and Yifan Cheng (eds.), *Electron Crystallography of Soluble and Membrane Proteins: Methods and Protocols*, Methods in Molecular Biology, vol. 955, DOI 10.1007/978-1-62703-176-9_19,

Fig. 1. A large cryo-electron microscope like this JEOL 3000 SFF cooled with liquid helium has very special room and maintenance requirements.

taken into consideration when deciding on the equipment to be purchased. Very important considerations are the available budget and space (Fig. 2).

The ideal microscope for collecting 2D crystal data is a high-resolution TEM with a field emission gun (FEG) and 300 kV accelerating voltage. It has an in-built cryo stage that allows the sample to be imaged at or below 77K for a long period of time (hours) with no specimen contamination or drift. Such electron microscopes exist, but are very expensive and require special room installations. In addition

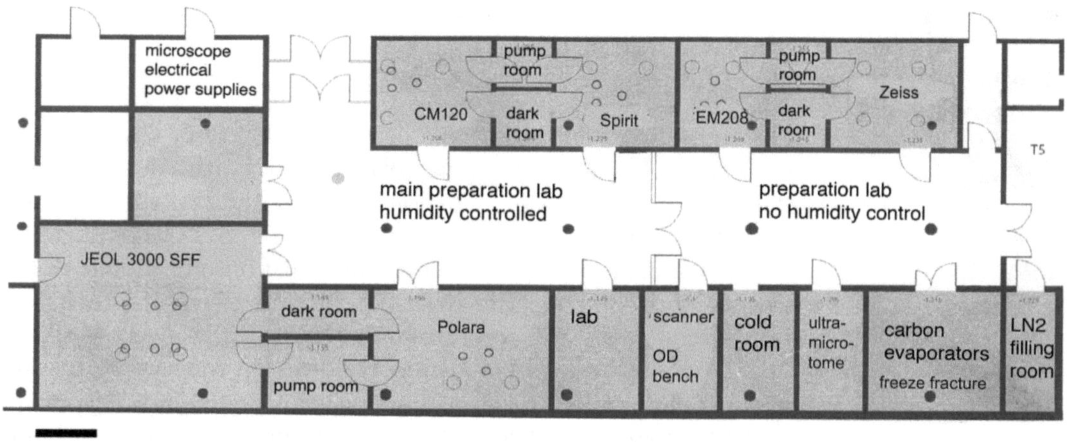

Fig. 2. EM unit floor plan. The state-of-the-art EM unit at the Max-Planck-Institute of Biophysics in Frankfurt houses six electron microscopes and rooms for ancillary equipment around a central specimen preparation area. This type of high-end EM facility with a range of TEMs is designed for a large number of users and multiple applications (electron crystallography, single particle EM, electron tomography, and conventional EM). The scale bar represents 2 m.

to this major TEM it is useful to have a smaller conventional TEM for rapid screening of samples at room temperature, such as a 120 kV TEM fitted with a LaB_6 filament. The type of image detection system(s) fitted to the electron microscopes is the next important consideration. Ideally, a CCD camera should be installed. The larger the CCD the better, and if it is going to be used for high-resolution data collection a CCD of 8K×8K or even 10K×10K would be essential. However, sheet film can always provide a format for high-resolution data collection and combination with a medium range CCD (2K×2K) for screening purposes is a good compromise. Cost and available space are usually the deciding factors on the latter point. When using film, additional equipment, and room space is required: a darkroom for film development, an optical diffractometer for film evaluation, a film scanner for digitizing, and a film archiving cabinet. In addition, film and film processing chemicals will have to be budgeted for.

Other ancillary equipment as well as lab space is required for specimen preparation: a good quality carbon evaporator for producing clean, high-quality carbon film as a support for the 2D crystals, equipment for carbon-coating electron microscopy support grids, a glow discharge unit for changing the adhesive properties of the carbon film in order to allow good spreading and adhesion of the 2D crystals to the support film.

A specimen preparation laboratory should be set up adjacent to the EM room for clean preparation of samples. There should be a safe area with a fume hood for housing a fast freezing device (e.g., Vitrobot) where samples can be rapidly frozen using liquid ethane. Additionally the laboratory should have a source of fresh triple distilled water and also clean contamination-free liquid nitrogen. The lab should have adequate space for one or more long-term liquid nitrogen dewars (50 l) where frozen samples can be stored until time for viewing in the EM.

2. The Cryo-Electron Microscope

For high-resolution cryo-data collection of 2D crystals the ideal EM has a 300 kV field emission gun and is equipped with a dedicated cold stage and good anticontamination facilities. The EM should have a resolution of better than 2 Å when the specimen is cold with a drift rate of less than 1 nm/min. The contamination rate should be less than 0.5 nm/h. The latter is very important, as acquiring data can take several hours and cryo-images are recorded at a very low dose (10–20 $e^-/Å^2$), so contamination build-up on the sample (usually water in the form of ice) will suppress valuable signal and decrease the signal-to-noise ratio. The performance of the EM cryo stage should be equally good when the sample is tilted to at least ±60°, especially in terms of drift rate, resolution, and contamination rate. The cryo stage should allow rapid transfer of cryo specimens into the EM such that they remain contamination free. The time the specimen takes to stabilize in the EM before data recording can start is also an important factor; ideally this should be less than 20 min, and in the latest generations of dedicated cryo-EMs it can be much faster.

The microscope should be equipped with a "low dose image acquisition system" to prevent pre-irradiation of the areas of interest (see also Chapters 6 and 8). The low dose system has a "search" mode that allows screening of the sample at a very low electron dose to select the areas (crystals) to be imaged. The "focus" mode is used to set the precise defocus, adjust objective astigmatism, and judge specimen stability at a magnification above or the same as the image-recording magnification. In this mode the beam is shifted away from the area of interest so as not to pre-irradiate and damage the sample. The image shift can be adjusted to achieve a minimum shift from the image area, depending on the magnification and size of the area of interest (see Note 1). The third mode is the "image" or "exposure" mode where the image is taken. This mode is set up in the imaging magnification with the correct electron dose and beam size. The system should blank the beam above the specimen when switching between the various modes.

Another useful accessory on an instrument to be used for electron crystallography is a spot-scan imaging mode (1). Irradiating the sample with a very small spot of illumination reduces beam-induced movement of the specimen, and ensures a much higher success rate of imaging, especially at high tilt angles (2–4). It is also possible to use dynamic focusing to reduce the range of defocus in images of highly tilted specimens (5).

Finally the cryogenic components of the EM cold stage and specimen airlock should have a long hold time when cold (a minimum of 8 h), as refilling cold stage dewars causes mechanical instability that can take a long time to stabilize.

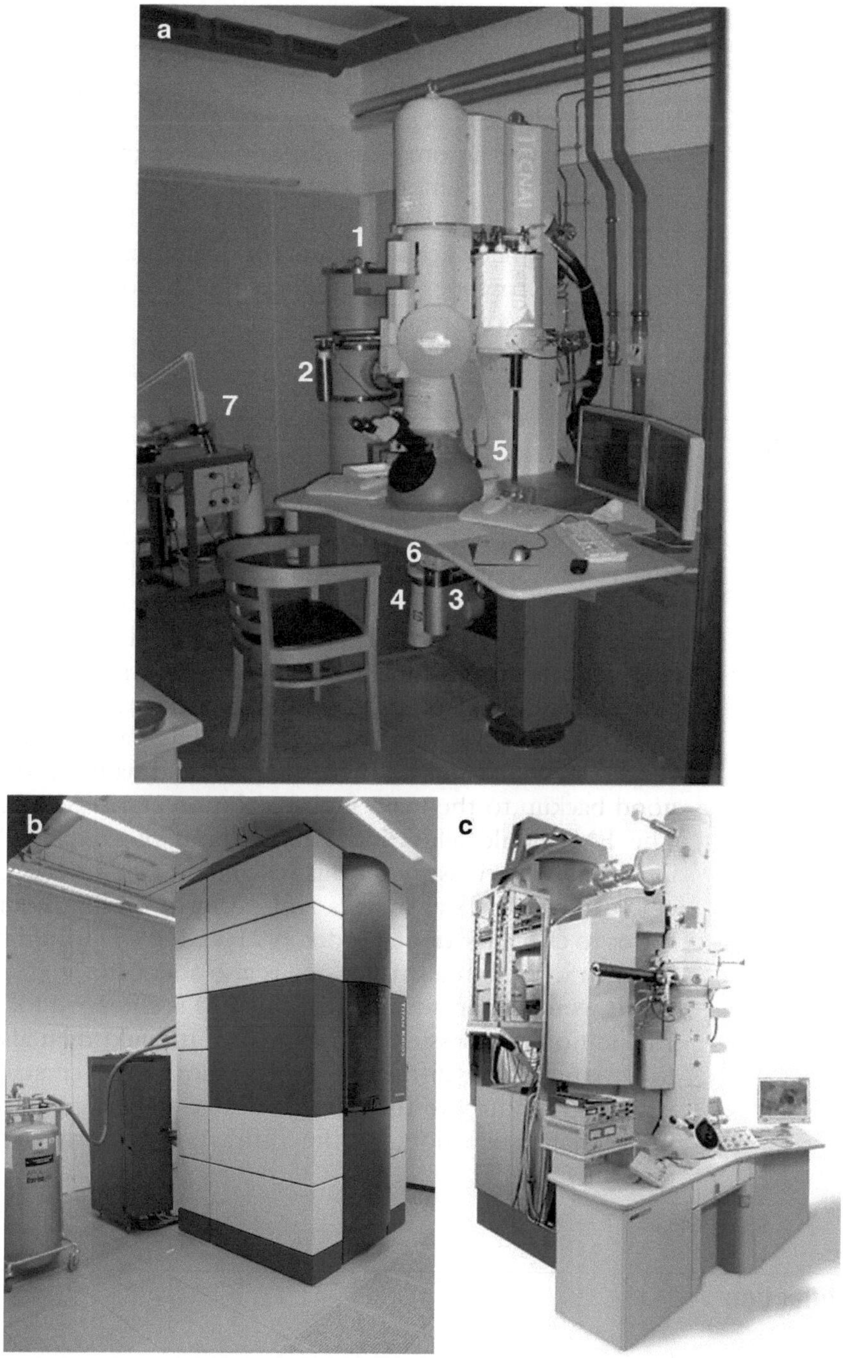

Fig. 3. Modern 300 kV FEG cryo-electron microscopes. (**a**) Tecnai Polara. (1) high tension tank; (2) specimen transfer device; (3) energy filter; (4) off-axis TV camera; (5) plate camera; (6) CCD camera; (7) cryo workstation. (**b**) The Titan Krios enclosed in its environmental chamber (image courtesy of Henning Stahlberg and Ken Goldie). (**c**) JEOL JEM 3200 FSC.

Currently available microscopes that fulfill all of the above are the FEI Tecnai G2 Polara, the JEOL JEM 3200 FSC, and the FEI Titan Krios (Fig. 3). All of these microscopes can be used at liquid nitrogen temperature, but have the option of cooling with liquid helium for operation at lower temperature if desired. They have specimen stages decoupled from the outside, and the option of loading multiple specimens into the specimen airlock for higher throughput and automation of specimen loading. Also, all have the option of being remotely controlled, which helps providing an even more stable, isolated environment. The Titan Krios is housed in an environmental chamber and the operating controls are remote from the microscope, which reduces the environmental constraints on the microscope room.

3. The Screening Electron Microscope

Before optimal 2D crystallization conditions are identified, much time is spent screening crystallization trials by EM (see also Chapter 5). This work does not require a high-end microscope with an FEG, and ideally a second, smaller EM is available for screening of negatively stained samples. This microscope should have reasonable high-resolution capabilities and a good stable tilting stage. An EM that has a 120 kV accelerating voltage and a LaB_6 filament (Fig. 4) would be more than adequate and would provide a very good backup to the main high-resolution cryo-EM (see Note 2). The EM should allow rapid throughput of samples. An almost indispensable accessory on this machine is a mid-sized CCD camera to record crystal images and perform on-line Fourier transforms to check for crystal order. A 2K×2K CCD would be optimal for such work, but a smaller one is also adequate.

It would be an advantage if this EM is equipped for cryo-EM. In addition to assessing crystal quality, it could then also be used for cryo-electron diffraction work (should the crystals permit), which does not have the stringent stage stability requirements of image data collection, as the diffraction pattern is translation insensitive (6, 7).

4. Image Detection Systems

The image detection systems are of crucial importance when setting up a cryo-EM laboratory. They will constitute a large part of the cost of the installation, and if photographic film is to be used, also a considerable investment in infrastructure. The radiation sensitivity of the specimens means that the electron dose has to be kept at a minimum at all times. For convenient specimen screening

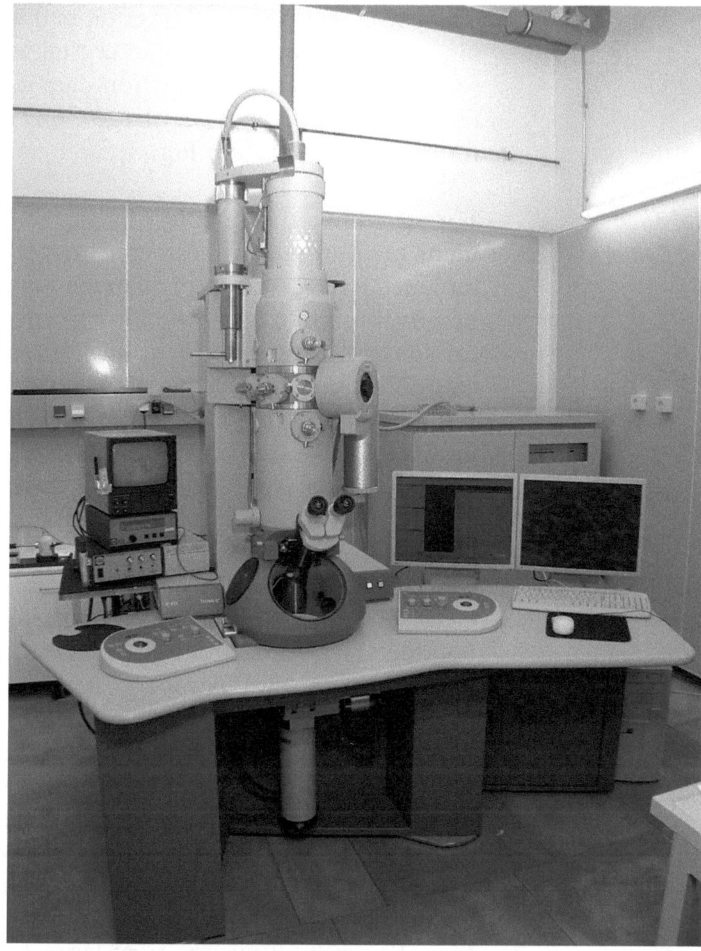

Fig. 4. Tecnai Spirit 120 kV electron microscope with LaB_6 filament, equipped with a 2K × 2K CCD camera and an on-axis TV camera.

at low magnification and very low electron dose, a real-time detection system is needed, and for data collection an image recording system with an adequately large size that will achieve the desired resolution. For electron crystallography, the area of the detector is an especially important consideration. Images of large, well-ordered crystals require a large number of pixels: to achieve 3 Å resolution on a crystal with a size of 1 μm, the pixel size should be no larger than 1 Å × 1 Å, which means that 10,000 × 10,000 pixels are necessary. Until very recently this could only be achieved by photographic film, but now there are CCD cameras as well as CMOS direct electron detectors that achieve such a size. When electron diffraction patterns are to be collected, film is not the medium of choice because of its low dynamic range, but a CCD camera of 2,000 × 2,000 pixels would be optimal.

The ideal EM would be equipped with a TV camera for screening, a CCD camera for rapid crystal assessment, astigmatism correction and electron diffraction data collection, and photographic film (or alternatively a very large CCD camera or CMOS detector) for data collection (see Fig. 3).

4.1. Real-Time Imaging Cameras

For screening the samples for suitable crystals, the most convenient medium is a TV camera (image intensifier). Samples can be screened in real time, using a much lower electron dose than would be possible when using the fluorescent screen. In addition, the TV camera can be used for focusing and astigmatism correction at high magnification. The TV camera is normally mounted directly below the column, but depending on other devices present (e.g., a non-retractable CCD camera) it can also be mounted off-axis (see Note 3). As an alternative, small CCD cameras are now on the market that allow real-time imaging (see Note 4).

4.2. Slow-Scan CCD Cameras

CCD cameras have a high dynamic range and a linear response to electron dose (8). They are particularly well suited for electron diffraction data collection, where these qualities give them a great advantage over film. For image data collection on 2D crystals, their use is more limited, because of their generally small size and their large point spread function, which reduces the effective resolution. CCD cameras of 2K×2K or even 4K×4K pixels are not large enough to collect a high-resolution image of a large crystal and the fact that crystals are often not easily recognizable and can be occupying only parts of membranes means that the success rate of data collection is highly increased on a large detection medium, i.e., film. Recently, however, very large CCD cameras of 8K×8K to 10K×10K pixels have come on the market which would start to compete with film (see Note 5).

Nevertheless, a smaller CCD camera is a very useful accessory on an electron microscope used for electron crystallography. A great advantage of CCD cameras is the immediate availability in digital format of the image. This makes it possible to do on-line image processing, e.g., calculation of Fourier transforms, so that the alignment of the microscope can be assessed and optimized (see also Chapter 20). In a rapid readout mode a "live" Fourier transform gives the possibility to accurately focus and adjust astigmatism. In addition this feature is extremely useful for screening crystal setups, where each crystal candidate can be immediately checked for order.

The CCD camera is usually purchased together with the microscope and its operation should ideally be completely integrated in the microscope control software.

4.3. Photographic Film

Because of its large area, film is still the medium of choice for recording images of 2D crystals, although this will change in the near future (see Subheading 4.4). Sheet films measure 8 × 10 cm, which translates to about 11,000 × 14,000 7-μm pixels and have excellent imaging qualities for low-dose cryo-EM (see Note 6). Film also has the advantage of providing a permanent raw data record. Use of film has several practical disadvantages, including the need to develop the film and to load each fresh film into an individual cassette in a dark room and to dry the film before introduction in the microscope vacuum. Also the image quality is not known until the films are developed, and the optical quality of the image and crystallinity of the sample have to be judged on an optical diffractometer. There are also environmental disadvantages, considering the water consumed and chemical waste generated. Finally the films have to be digitized on a high-quality scanner, which can be time consuming. The following items are necessary when film is to be used.

1. The EM has to be equipped with a film exposure mechanism. Depending on the make of microscope, each unexposed film sheet is held in its own cassette in a light-tight box (see Note 7). Each box holds typically about 50 films.
2. Film desiccator for unexposed film (Fig. 5). The film emulsion contains moisture and should not be introduced in the EM vacuum without drying (see Note 8).
3. A darkroom for film processing (Fig. 6). The darkroom can be equipped with red safelights and should include a dry bench for film handling, a tank for developer and one for fixer, and one or two tanks of flowing, clean water for rinsing. The developer temperature should be kept at a constant temperature (20°C) for reproducible results and the chemical tanks should be purged with nitrogen gas. Alternatively, the film racks can be agitated by hand instead of with nitrogen gas during developing and fixing. Further darkroom necessities include film racks and timers (see Note 9).
4. Film drying closet. This is not an absolute necessity in a humidity-controlled laboratory, but will reduce the drying time of the film to about 45 min.
5. Optical diffractometer (Fig. 7). This is used for checking films for correct focusing and lack of drift and astigmatism, but in electron crystallography especially for assessing the crystal quality and finding the best crystalline area prior to scanning. An optical diffractometer consists of a laser with a series of lenses, a device for mounting the negative in the front focal plane of the imaging lens and a CCD camera in the diffraction plane. A beam stop is essential to block the direct beam and a series of apertures of different sizes should be present to select the area of interest on the negative. Finally a diaphragm should make it possible to attenuate the laser beam to an appropriate

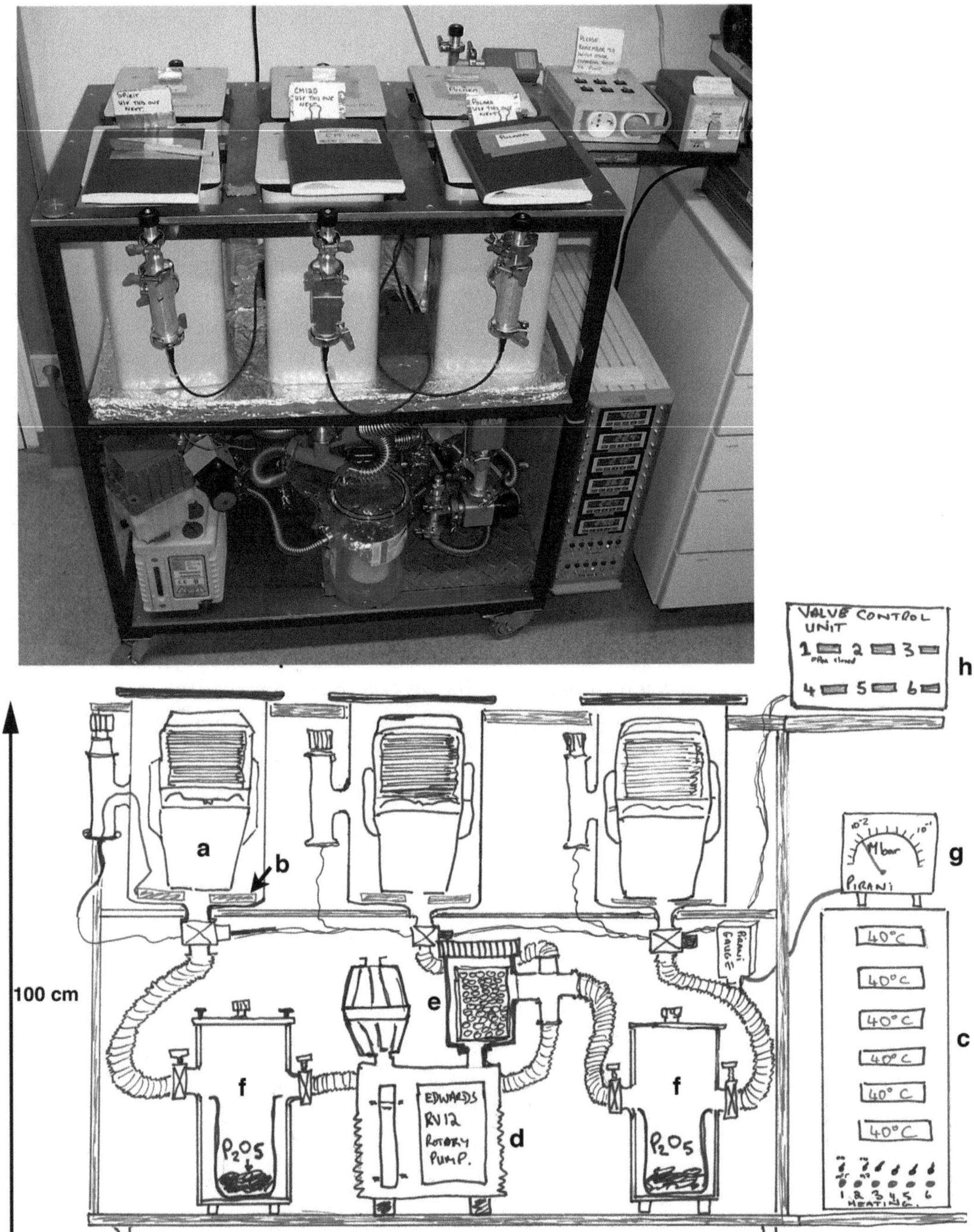

Fig. 5. Desiccator for unexposed film. Loaded film magazines are stored in vacuum chambers (a) equipped with a heater pad (b). The temperature is controlled to 40°C with the controller (c). Chambers are pumped with vacuum pump (d) through an oil mist filter (e) and a water sorb filled with phosphorus pentoxide (f). The vacuum is measured by gauge (g). Six vacuum chambers can be individually pumped using valve controller (h).

Fig. 6. Darkroom for film developing.

intensity, depending on the film density. The setup can be mounted horizontally or vertically, but the vertical version is more economical with space and easier to use.

6. Film scanner. A high-quality flatbed scanner is needed to digitize the micrographs. Instruments that have been routinely used in recent years are the Zeiss SCAI scanner (see Note 10), and the much more economical Nikon CoolScan 9000ED. These scanners have pixel sizes of around 7 μm, which makes it possible to achieve high resolution (3 Å) from images taken at an electron optical magnification around 70,000×.
7. A film storage cabinet for archiving.

4.4. CMOS Detectors

Recent technological advances have lead to the development of CMOS direct electron detection cameras for electron microscopy (9, 10). These cameras have several advantages over other forms of pixel detectors and have a sensitivity comparable, if not superior, to that of film, but have the convenience of current CCD cameras (see also Chapter 20). The instant readout of the CMOS camera and its extreme sensitivity make it very useful for low dose biological applications. Although the image quality of these new cameras should be better than that of CCD cameras, there still remain some questions about the durability of the CMOS detectors.

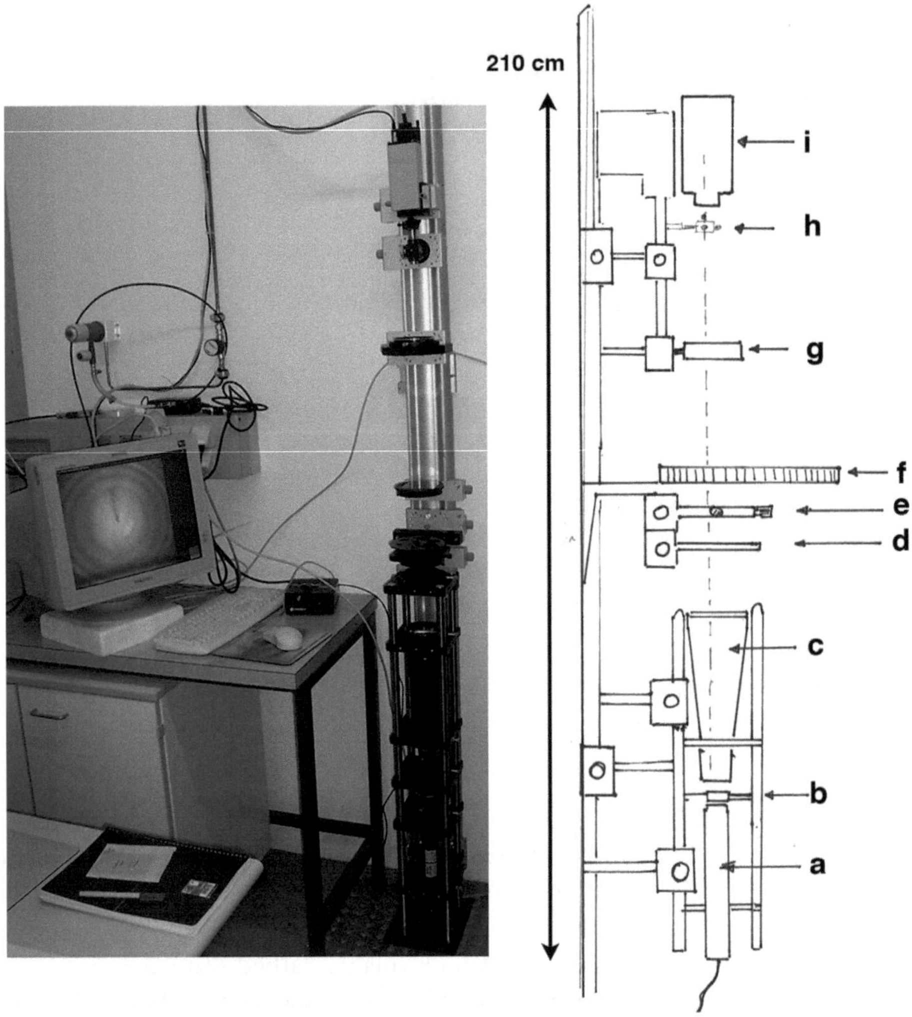

Fig. 7. Vertical optical diffractometer for assessing photographic film: (a) red helium–neon laser; (b) polarizer for attenuating the beam intensity; (c) beam expander; (d) adjustable aperture; (e) optional slit; (f) rotating film holder; (g) condenser lens (76 mm lens with *f* = 300 mm); (h) beam stop; (i) 1 MP KAF-1001 Kodak CCD camera. The diffraction pattern is viewed on a monitor via software supplied with the CCD camera. The laser was bought from Coherent (www.coherent.com) and the optical components from Spindler & Hoyer UK Ltd.

5. Equipment for Cryo-Specimen Preparation

1. Carbon evaporator to produce reproducible, ultra-clean, flat carbon support films on mica under high vacuum (Fig. 8). Flat, high-quality carbon films are essential for preparing grids of 2D crystals that will give high-resolution images (11, 12). Because of noise and vibrations produced by the vacuum pumps, this machine should be housed in a separate room away from the microscopes (Fig. 2) (see Notes 11 and 12).

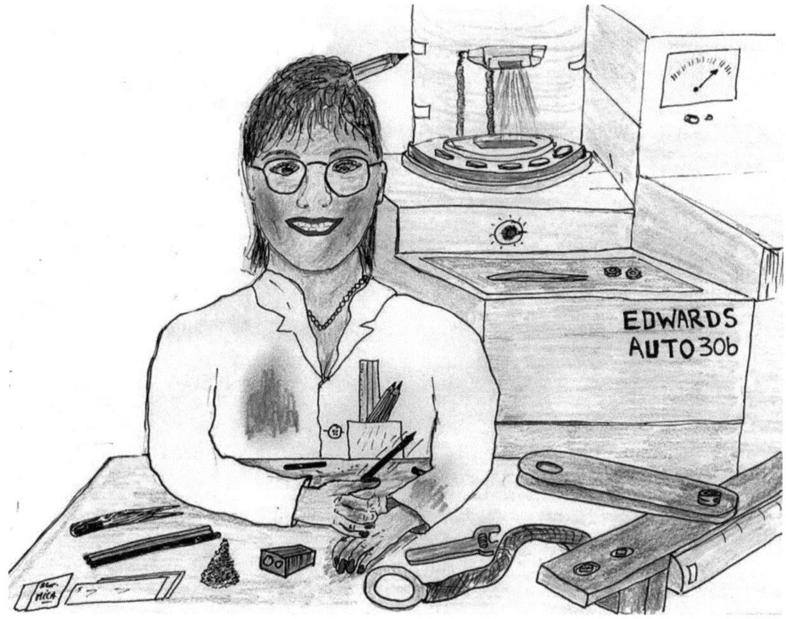

Fig. 8. Carbon evaporator.

2. Glow discharge unit for making carbon-coated grids hydrophilic. With adaptions, these units can also be used for film evaporation, sputter-coating and aperture cleaning (see Note 13).
3. Light microscope for assessing the quality of carbon-coated grids or holey carbon film grids. For this purpose, the light microscope should be equipped for dark field and phase contrast.
4. Plunge freezer for the vitrification of samples (13) (Fig. 9). With this unit grids can be frozen in liquid ethane, at such a fast rate that vitreous water rather than crystalline ice is formed. The unit can also provide a humidity- and temperature-controlled environment to prevent drying of the samples prior to freezing. For safety reasons the plunger needs to be housed in a fume hood or special safety regulated area due to the danger of explosions from the liquid ethane (see Note 14).
5. Cryo-specimen holder with a dedicated specimen loading station (for microscopes without an in-built cold stage). This holder would usually be ordered together with the electron microscope. For stable performance, the dewars and tips of these holders have to be regularly pumped. A dedicated pumping station is required for this purpose. They are commercially available or can be home-built (Fig. 10).
6. Dewar for long-term storage of specimens. 2D crystals are difficult to produce and it may be advantageous to prepare

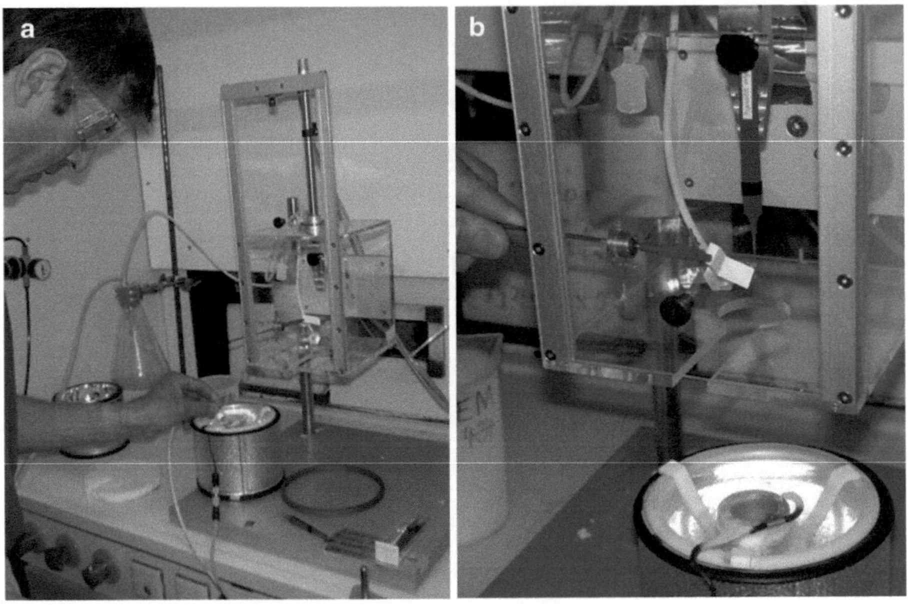

Fig. 9. Home-built cryo-plunger. The device has an environmental chamber with a sliding door at the front for loading the tweezers (**a**) with a port for blotting the grid (**b**). The ethane pot is temperature-controlled to prevent the ethane from solidifying (**b**). The guillotine is operated pneumatically by a foot control (not shown) at 5 m/s. For safety purposes, the unit is contained in a fume hood.

Fig. 10. Multi-purpose pumping station. With this station, the tips and dewars of up to four side-entry cryo-holders can be pumped simultaneously. Such a pumping station can also be used for pumping other vacuum components, for example, liquid gas transfer rods and multi-specimen transfer holders of the Tecnai Polara.

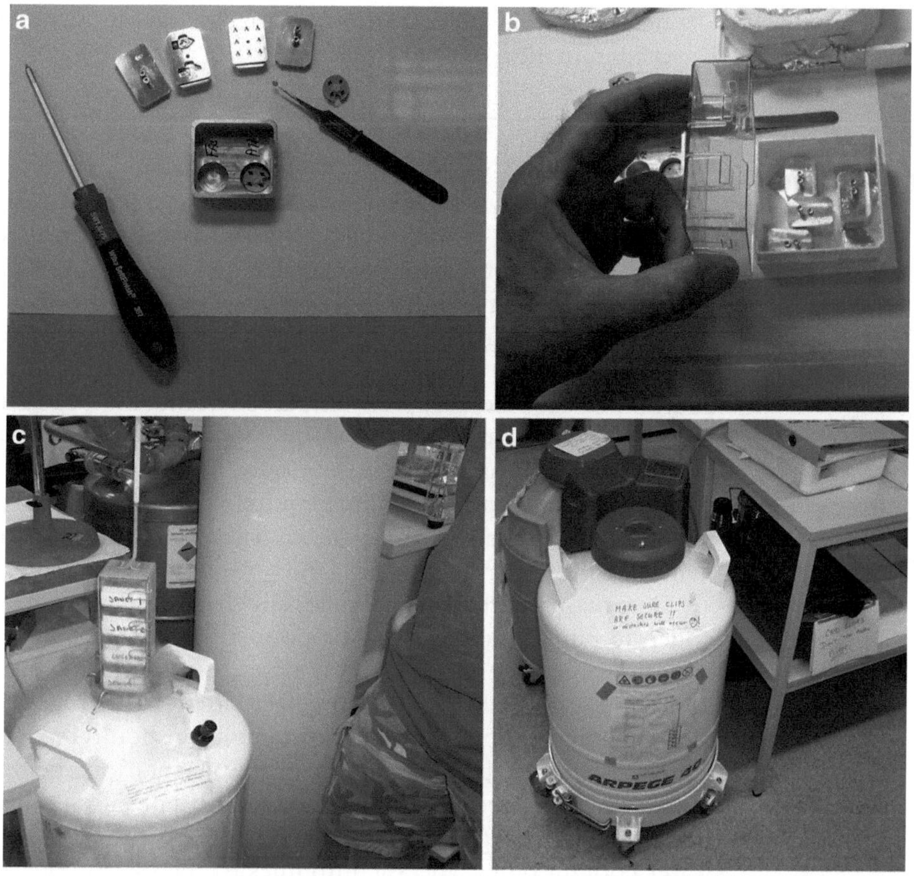

Fig. 11. Storage of cryo-specimens. (**a**) a selection of grid containers and Polara cartridge containers. (**b**) Cryo-boxes holding several grid or EM cartridge containers. (**c**) Removable racks hold five cryo-boxes. (**d**) long-term liquid nitrogen storage dewar (50 l) with liquid nitrogen level alarm.

cryo-samples and store the frozen grids until viewing time is available. Grids can be stored in special containers (see Note 15) which should keep the grids contamination-free. Grids can also be stored mounted in the cartridges for microscopes like the Polara or JEOL for repeated data collection sessions (Fig. 11).

7. A source of fresh triple-distilled water for making stains and buffers.

8. Adequate supply of clean liquid nitrogen. A large amount of liquid nitrogen is used for cryo-EM for sample preparation and for cooling anticontaminators and storage containers. A 300 kV FEG cryo-EM can use up to 300 l of liquid nitrogen per week. The ideal setup would have one or two 100 l self-pressurizing containers and a 10 l dewar per microscope (see Note 16). Care should be taken to keep the containers ice-free and the

dewars should be emptied and dried regularly to prevent ice contamination of the samples.

6. The Electron Microscope Room

The room for the main FEG cryo-EM has very special requirements. The size of the room will be dictated by the type of EM that it will house. Most 300 kV cryo-EMs will require a room at least 6 × 7 m and 5 m high (see Note 17). A room of this size will accommodate the most advanced cryo-EMs on the market today. In addition to the microscope room, an adjacent room measuring at least 2 × 3 m will be required for the microscope water coolers and ancillary vacuum pumps. It is important to liaison with the microscope supplier regarding room requirements at an early stage, as microscope sizes and room specifications can vary for different manufacturers.

The main microscope room ideally should have a central heavy concrete plinth for the microscope to sit on to provide a stable, vibration-free base (Fig. 12). A concrete plinth measuring 3 × 3 m and 1 m high would provide a perfect base for the largest 300 kV electron microscopes. The surrounding floor should be decoupled from this base to prevent any mechanical vibrations being transmitted to the microscope. Having a floating or false floor around the EM plinth (Fig. 12b) allows the running of electrical cables, water cooling pipes, and vacuum lines to the microscope without causing awkward obstructions around the EM, thus allowing easier access for maintenance and the easy passage of liquid gas dewars.

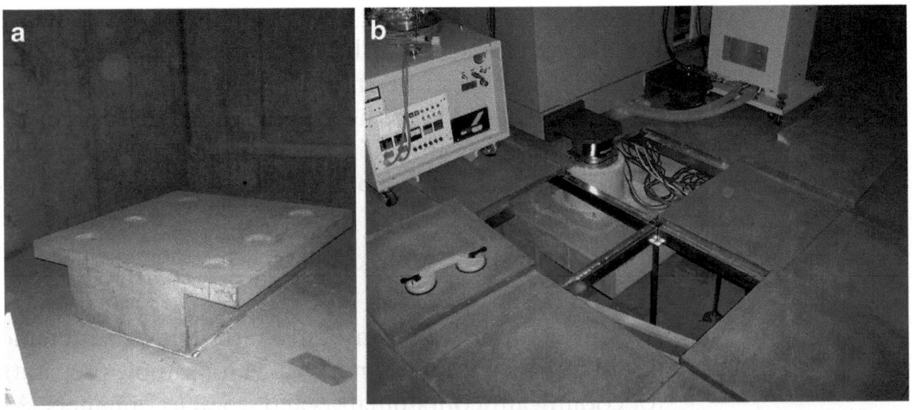

Fig. 12. Microscope foundations and EM room floor. A foundation block suitable for a 300 kV EM (**a**). The block measures about 3 × 3 m and is 85 cm high. The microscope is mounted on three to six aluminum blocks which fit the indentations on the concrete plinth (**b**). Also in (**b**) the decoupled floor is visible, which consists of 60 × 60 cm ceramic tiles on a 1 m high metal support frame. The tiles are resistant to liquid nitrogen and have a conductive coating to prevent electrostatic discharges in the dry atmosphere.

Ancillary power supplies and the high voltage tanks can be positioned around the EM at distances recommended by the EM manufacturer. Normally, most power supplies for the ion pumps, the lenses and the high-voltage tank need to be at least 1.5 m from the EM column. Adequate space should be left at the front of the EM for a cryo workstation where loading of the cryo specimens can take place.

The temperature of the room should be stable and controlled such that no large temperature fluctuations take place. The air delivery to the room is best done with large wind socks which prevent drafts and distribute the air evenly around the room (Fig. 13). Most important of all, the ventilation system should be very quiet.

If possible, the humidity of the room should be controlled to 10–20%. This will reduce ice contamination during sample loading and it will ultimately help maintaining low contamination rates in the EM. It is very important to have adequate ventilation, also in regards to the large volumes of liquid nitrogen used for cryo-EM. The installation of oxygen sensors in these rooms is essential for safety.

Acoustic paneling on the walls of the EM room (Fig. 13b) is also critical to suppress noise generated by the EM itself or any ancillary equipment, as noise can have a dramatic effect on high-resolution data. In general, the room noise should not exceed 40 dB.

Lastly, careful consideration should be taken with regard to any electrical wiring to the room. Large power cables should be run into the room at ground level and directly to the power units where they are required. Lighting circuits should be terminated at a room power switch. Adequate lighting should be provided for routine maintenance. For certain electron microscopes (e.g., the JEOL 3000 series), an overhead gantry crane is extremely useful for the maintenance of the EM column.

The requirements for the screening microscope are less stringent. A room measuring 4×4 m with a 1.5×2 m adjoining room for the water cooler will accommodate most 120 kV EMs. The room must fulfill the minimum specifications by the EM manufacturer: it should have a stable foundation and be free from excessive noise and electrical fields and have a stable temperature.

7. Maintenance and Quality Control

In order for a microscopy lab to function optimally so that high-quality EM data can be consistently obtained, it is essential that routine maintenance is carried out on both the electron microscope and ancillary equipment.

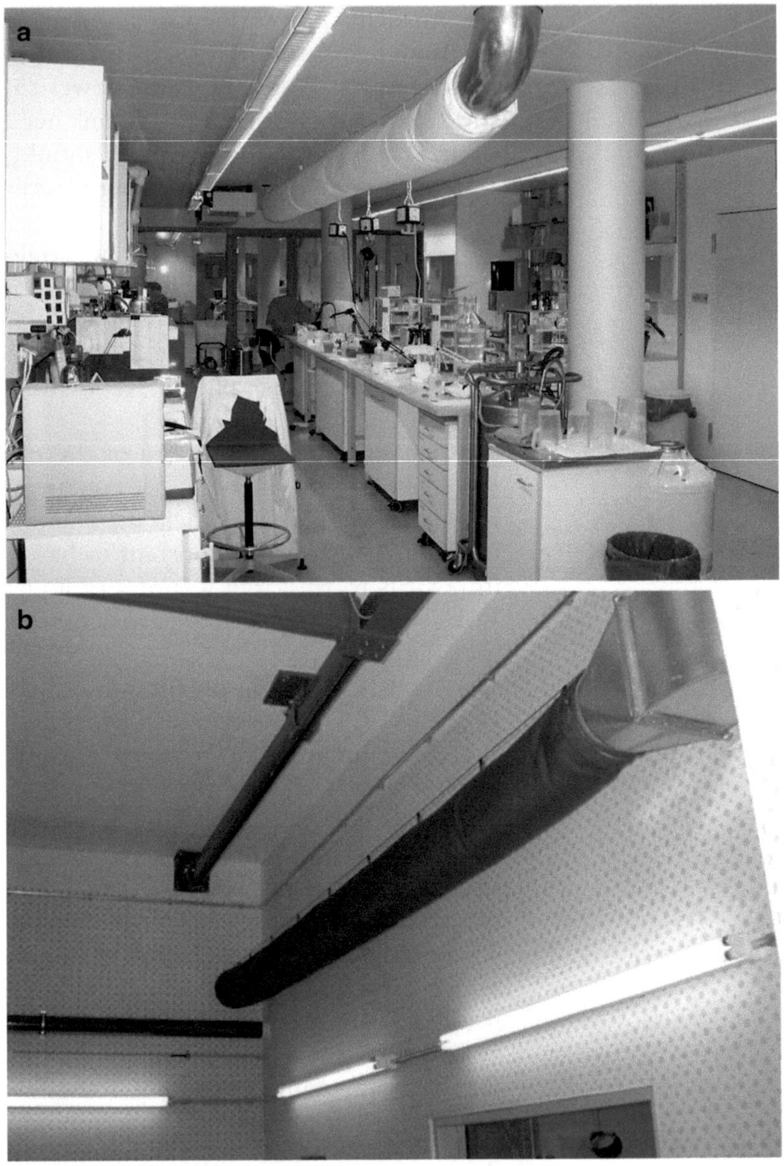

Fig. 13. Room climatization. The main specimen preparation lab (**a**) has a constant temperature and humidity; wind socks provide draft-free, filtered air. The microscope rooms have the same climate control and in addition acoustic paneling (**b**) to suppress room noise.

For the electron microscopes extra careful alignments should be performed regularly, with special attention to illumination alignment for FEG cryo-EMs to correct for beam tilt or coma. It is good practice with high-resolution electron microscopes to regularly check the performance with a test specimen (see Note 18). By recording images of these test specimens the image quality can be quickly assessed in terms of resolution and information limit.

Problems with mechanical instability of the stage, which can manifest themselves as drift or vibrations, can easily be identified (a CCD camera is useful for this). It must be remembered that high-resolution electron microscopes are superb detectors of electrical or magnetic fields, and regular checks of image quality can show if these have become a problem. Disturbances in close proximity to the EM, for example by the installation of a large instrument like a centrifuge or, worse, an NMR machine, can cause vibrations or magnetic fields which can result in an information cut-off in one or more directions.

The cooling of the electron microscope, especially of the lenses, is of the utmost importance and it is essential to check for constant temperature and flow rate of the microscope cooling water on a daily basis. Any variation in the aforementioned can have detrimental effects on the quality of the data.

These very large, specialized cryo-electron microscopes are very sophisticated pieces of equipment, which unfortunately require occasional specialized service and repairs (e.g., the exchange of a field emission gun on average once every 2 years). It is worth mentioning here that on purchase of these machines a service contract should be negotiated to avoid long down times when problems occur. These contracts should include routine maintenance and spare parts and can also cover the electron detectors bought as part of the microscope purchase.

The maintenance of ancillary equipment can be just as specialized. Equipment like scanners, high vacuum coating units and side entry cold stages also require routine quality control tests, so that one can quickly realize a deterioration of performance. They may also require regular service and maintenance by the supply company.

If the lab is fortunate to have an engineer or technician it is vital to get this personnel trained to the highest standard with the equipment when it is installed; this will save money and time in the future and helps provide a steady flow of high-quality data from the electron microscope laboratory (14–16).

8. Notes

1. The image shift direction can be selected too. When collecting data from tilted specimens, it is important to set the image shift along the tilt axis to prevent a focus difference between focus and image mode.
2. A lower kV instrument and/or a tungsten filament will do for screening if that is the only microscope available.

3. Unfortunately, the Gatan 622 TV camera is not produced anymore, but it may be available second hand.
4. A CCD camera that allows real-time imaging is the TVIPS FastScan 1K or 2K camera (www.tvips.com).
5. CCD cameras for electron microscopy are supplied by TVIPS (www.tvips.com) and Gatan (www.gatan.com), and FEI supplies its own CCD camera, the Eagle (www.fei.com).
6. Currently the only available electron imaging film is Kodak SO-163.
7. The cassettes in common use have a solid back for stiffness. When used at 300 kV or above, some electrons will pass through the film and cause fogging by backscattering off the cassette backing. When use at 300 kV is desired, special, unbacked cassettes should be ordered from the EM manufacturer.
8. Film needs to be dried at least 24 h before introduction into the EM to keep the vacuum clean. It is therefore desirable to have at least three sets of cassettes to make a high throughput possible.
9. Optimal results for low-dose images on SO-163 films are 12 min developing in full-strength D-19 developer, 1 min rinse in water, 8 min fixing in Kodak fixer, followed by 20 min rinsing.
10. Production of the SCAI scanner was discontinued in 2008, but it is still used in many laboratories and service and parts are still available.
11. Suitable carbon evaporators include the BOC-Edwards High Vacuum 306 and the Bal-Tec CED 030.
12. Other support films should be considered, such as holey carbon films, which can be used for freezing samples in vitreous ice. These can be useful for tubular crystals or other crystals that cannot be imaged directly on continuous carbon film. Commercially available holey films include Quantifoil (15) (www.quantifoil.com) and C-flat (14) (www.protochips.com).
13. Suitable units include the EMS 100 Glow Discharge Unit (www.emsdiasum.com) and Gatan Solarus (www.gatan.com).
14. Commercially made units for reproducible vitrification of samples include the Vitrobot (www.fei.com) and the Cryoplunge (www.gatan.com).
15. Grid containers are usually modified grid boxes with four grid slots and a rotating lid held in place by a screw. These can be stored in the dewar in 50 ml plastic Falcon tubes with a string attached to the lid for easy removal and several holes in the sides of the tubes. More sophisticated storage containers can be manufactured by a good mechanical workshop (see Fig. 11).

16. Manufacturers provide exact specifications for each EM and will assess potential EM rooms for suitability before purchase.
17. Pressurized liquid nitrogen containers include the Cryotherm Apollo 50, 100, and 150.
18. Combined test specimens (holey carbon film with gold islands and graphitized carbon) for high-resolution performance are commercially available. A good specimen for cryo-performance would be a well-diffracting 2D crystal like purple membrane (16).

Acknowledgments

We thank Werner Kühlbrandt and all past and present members of the Department of Structural Biology for their support and contributions to the EM facility.

References

1. Downing KH (1991) Spot-scan imaging in transmission electron microscopy. Science 251:53–59
2. Williams KA (2000) Three-dimensional structure of the ion-coupled transport protein NhaA. Nature 403:112–115
3. Breyton C, Haase W, Rapoport TA, Kühlbrandt W, Collinson I (2002) Three-dimensional structure of the bacterial protein-translocation complex SecYEG. Nature 418:662–664
4. Vonck J, Krug von Nidda T, Meier T, Matthey U, Mills DJ, Kühlbrandt W, Dimroth P (2002) Molecular architecture of the undecameric rotor of a bacterial Na^+-ATP synthase. J Mol Biol 321:307–316
5. Downing KH (1992) Automatic focus correction for spot-scan imaging of tilted specimens. Ultramicroscopy 46:199–206
6. Wang DN, Kühlbrandt W (1992) Three-dimensional electron diffraction of plant light-harvesting complex. Biophys J 61:287–297
7. Kühlbrandt W, Wang DN, Fujiyoshi Y (1994) Atomic model of plant light-harvesting complex by electron crystallography. Nature 367:614–621
8. Krivanek OL, Mooney PE (1993) Applications of slow-scan CCD cameras in transmission electron microscopy. Ultramicroscopy 49:95–108
9. McMullan G, Chen S, Henderson R, Faruqi AR (2009) Detective quantum efficiency of electron area detectors in electron microscopy. Ultramicroscopy 109:1126–1143
10. McMullan G, Faruqi AR, Henderson R, Guerrini N, Turchetta R, Jacobs A, van Hoften G (2009) Experimental observation of the improvement in MTF from backthinning a CMOS direct electron detector. Ultramicroscopy 109:1144–1147
11. Han B-G, Wolf SG, Vonck J, Glaeser RM (1994) Specimen flatness in electron crystallography of glucose-embedded biological materials is significantly affected by the choice of carbon evaporation stock. Ultramicroscopy 55:1–5
12. Vonck J (2000) Parameters affecting specimen flatness of two-dimensional crystals for electron crystallography. Ultramicroscopy 85:123–129
13. Dubochet J, Adrian M, Chang J-J, Homo J-C, Lepault J, McDowall AW, Schultz P (1988) Cryo-electron microscopy of vitrified specimens. Q Rev Biophys 21:129–228
14. Stagg SM, Lander GC, Pulokas J, Fellmann D, Cheng A, Quispe JD, Mallick SP, Avila RM, Carragher B, Potter CS (2006) Automated cryoEM data acquisition and analysis of 284742 particles of GroEL. J Struct Biol 155: 470–481
15. Ermantraut E, Wohlfart K, Tichelaar W (1998) Perforated support foils with pre-defined hole size, shape and arrangement. Ultramicroscopy 74:75–81
16. Henderson R, Baldwin JM, Ceska TA, Zemlin F, Beckmann E, Downing KH (1990) A model for the structure of bacteriorhodopsin based on high resolution electron cryomicroscopy. J Mol Biol 213:899–929

Chapter 20

Future Developments in Instrumentation for Electron Crystallography

Kenneth H. Downing

Abstract

Advances in instrumentation have proceeded at an impressive rate since the invention of the electron microscope. These advances have produced a continuous expansion of the capabilities and range of application of electron microscopy. In order to provide some insights on how continuing advances may enhance cryo-electron microscopy and electron crystallography, we review some of the active areas of instrumentation development. There is strong momentum in areas including detectors, phase contrast devices, and aberration correctors that may have substantial impact on the productivity and expectations of electron crystallographers.

Key words: Cryo-electron microscopy, Electron crystallography, Charge coupled device, CMOS detector, Phase plate, Aberration corrector, DTEM

1. Introduction

A look ahead at some of the instrumentation that will impact electron microscopy (EM) and, in particular, electron crystallography may provide some inspiration on how new technologies will change the way we work, the types of samples with which we work and the scientific questions we address. This speculation about how the field will develop must be tempered by our limited knowledge of the future, of course. There is a saying, attributed to quite a few different sources, that "it is dangerous to make predictions, especially about the future." Certainly, few people a few generations ago would have predicted the ability to see the details of protein structures that electron microscopists now take for granted. Guessing what technology might offer in terms of molecular and

Ingeborg Schmidt-Krey and Yifan Cheng (eds.), *Electron Crystallography of Soluble and Membrane Proteins: Methods and Protocols*, Methods in Molecular Biology, vol. 955, DOI 10.1007/978-1-62703-176-9_20,

cellular microscopy, along with electron crystallographic studies, within the next generations is likely to have a large degree of error. Still, understanding the trends in technology and instrumentation that are likely to develop over the next years can be helpful in planning for the future and allocating resources efficiently. In this chapter we will look at some of the emerging instrumentation developments in electron microscopy and how they may impact macromolecular studies and, particularly, electron crystallography. A look at how development of some of the instrumentation and methods that are now in common use has influenced EM, along with the factors that have driven and enabled these advances, is useful for setting the stage for predicting how future developments will impact our field. The factors that drive and factors that enable these developments are often tightly coupled but sometimes quite independent.

For example, the concept for low-dose imaging was developed quite independently of the technology that would be required to make it easy to implement. In the mid-1970s a rapidly developing appreciation of the effects of radiation damage on biological samples led Robley Williams Sr. to devise the first "low dose" imaging protocol (1). Not having an image shift coil in his microscope, he simply physically shifted the projector lens pole piece in order to allow focusing on an area adjacent to the target for imaging, and then shifted the lens back before taking an image. At about the same time microscope manufacturers were beginning to add image shift coils, presumably mainly to enhance alignment, and it was easy to incorporate them into the low-dose protocol. Image shift capability is now a standard feature of electron microscopes, and is an essential component of the routine low-dose imaging protocols that have enabled the development of molecular microscopy.

The development of cryo-microscopy capabilities in the biological context was driven by the two goals of retaining specimen structure in the microscope vacuum and reducing effects of radiation damage. Experiments with wet-cell stages had shown a remarkable enhancement of specimen preservation when a crystal was kept hydrated (2). Cold stages had been developed for materials research, but their use with frozen biological samples required development of cryo-transfer specimen holders (3, 4). As soon as these were available, the prospects for specimen preservation were rapidly appreciated, along with the enhanced radiation tolerance that comes with cooling the specimen (5). Once high-resolution cryo-stages became available with cryo-transfer capabilities, refinement of freezing methods (6) opened the door to the vast array of applications we are now familiar with in areas ranging from macromolecular structure to whole cell studies.

The application of electronic imaging systems was a goal of many electron microscopists, both for improving visualization in EM and for direct data recording to avoid the tedium and time delay of developing photographic film. Development of most of this technology was driven by military needs. Electron microscopists and, perhaps to an even greater extent, astronomers were quick to adapt the latest advances and even help to push the field forward. Application to EM of TV-rate video systems using sensitive camera tubes and image intensifiers brought a great relief from the physical strain of low-dose work and improved the efficiency of data collection by helping to align and focus the image as well as to more easily identify good specimen areas and targets, e. g. crystals of a good size with the right amount of embedding medium. Particularly in the context of electron crystallography, being able to preview and evaluate electron diffraction from a protein crystal produced a quantum leap in efficiency of preparing and identifying good crystals. Slow scan CCD (charge-coupled device) cameras brought a similar enhancement of efficiency and capabilities to EM. Their impact in electron crystallography was particularly strong in collection of diffraction data, where CCDs quickly became the preferred data recording medium. They overcame some of the serious limitations of photographic film such as limited linearity and dynamic range as well as the tedious process of developing and accurately digitizing films.

Computer control of electron microscopes has enabled another round of efficiency enhancements. From the manufacturers' standpoint, computer control allows more complete and economic control of the complex and interacting components of the microscope. From the users' standpoint, it can take over some of the highly repetitive operations that one must perform throughout a session and carry them out faster and more accurately than a human. Early implementations that involved connecting a computer to the microscope simply took over some of the analog functions of the microscope, but as soon as even simple microprocessors were used to control microscope functions the availability of scripting languages allowed more and more complicated routines to be developed. Simple tasks such as spot scan imaging (7), which were straightforward to implement with an external computer, are now frequently included within the microscope's control software. Communication between microscope control software and a digital imaging system to provide feedback on microscope performance and specimen characteristics has allowed development of software packages such as Leginon (8), the TOM Toolbox (9) and SerialEM (10) which can assume full control of the microscope session and carry on extended imaging sessions in the absence of a human operator.

Other instrumentation developments have improved the quality of recorded data in a more incremental fashion and have been driven by the universal goal of improving performance of electron microscopes in a broad range of applications. Field emission guns provide an increase in brightness that leads to better resolution through improved coherence and shorter exposure times.

Energy filters can be used to increase contrast and signal-to-noise ratio (SNR) by removing inelastically scattered electrons from either diffraction patterns or images. Intermediate voltage microscopes, operating at 300–400 kV or even 200 kV provide a number of advantages that translate to improved resolution and data fidelity. Methodological advances in data treatment (see Chapter 18), perhaps most notably application of crystal unbending software, are somewhat beyond the scope of this chapter but have contributed substantially to the current capabilities of EM in macromolecular structure determination. These have most often been driven by the particular problem at hand. All together, these developments have brought us to the point that one might consider cryo-electron microscopy and electron crystallography to be mature and sufficiently productive fields. This view could be supported by the flow of publications on structures determined by EM.

When one surveys the advances in instrumentation for electron microscopy that have been made over the past few decades and the capabilities that they have brought, it might be easy to slip into the unconstructive attitude once (incorrectly) attributed to a US Patent Department head that everything that could be developed has already been done. However, as has been pointed out (11) both of the Patent Office heads with whom this idea seems to have been associated actually had the much more enlightened belief that the pace of discovery would continue to accelerate. Indeed when we look at the present momentum of EM developments we can easily anticipate that a continuing series of instrumentation advances will have substantial impact on macromolecular electron microscopy and electron crystallography and that there are still revolutionary developments to come.

In spite of the current publication stream on EM-derived structures, electron crystallography has not produced the number of high-resolution structures that were predicted by some as the field was developing through the 1990s. To a large extent, of course, this has much to do with the rate of development of suitable two-dimensional protein crystals and the success of X-ray crystallographers with some membrane proteins. But one might well ask what still needs to be done in EM instrumentation to make the technology even more efficient, ultimately even competitive with X-ray methods in terms of speed and resolution. It is not hard to identify several areas where improvements are critically needed and where they may also be now within reach. In the following sections we discuss some of the trends in instrumentation and how they may impact cryo-EM, and electron crystallography in particular. We focus on three main areas where there is rapid progress: devices that produce Zernike-type phase contrast, detectors, and aberration correctors. Several other upcoming developments are also briefly discussed.

2. In-Focus Phase Contrast Devices

The development that is likely to have the greatest impact on macromolecular cryo-EM is the advent of devices that provide in-focus phase contrast, analogous to Zernike phase plates in light microscopy. It has long been recognized that thin, unstained protein samples (including typical protein crystals) behave primarily as phase objects in the electron microscope. Contrast is produced by applying a phase shift between the scattered and unscattered electrons, usually by introducing some defocus. Thus microscopists have had to deal with a tradeoff of generating sufficient contrast in images by the use of defocus against the loss of resolution that results from high defocus. This is a much more significant problem in single particle work than in crystallography, since the low frequency components of the image are needed for finding particles within an image, while optical or computed diffraction patterns from crystal images can be used not only to identify the crystals but also to delineate their boundaries and to derive an initial estimate of the image quality. Thus a defocus of 1 μm might be considered rather low for single particle work but quite high for crystallography. For images of both types of specimens, within the resolution range of interest, the typical defocus will produce a number of oscillations and gaps in the contrast transfer function (CTF). Some typical transfer functions are shown in Fig. 1, illustrating some of

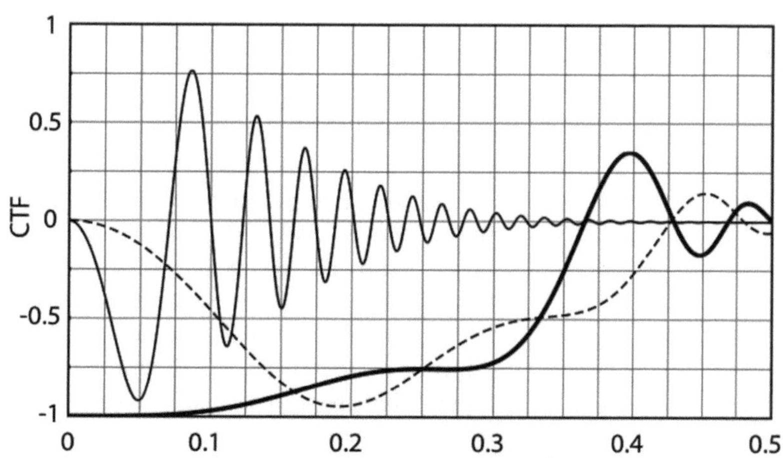

Fig. 1. Examples of phase contrast transfer functions under several fairly typical conditions. Parameters for calculations: $C_s = 2.2$ mm; $C_c = 2.5$ mm; accelerating voltage = 200 kV; illumination angle = 0.009 rad; energy spread = 0.8×10^{-5}. Defocus: dashed curve, 800 Å; thin solid curve, 1 μm. The thick solid curve represents the CTF for an ideal phase plate at defocus = 400 Å. The illumination angle and defocus spread can generally be made better than in these calculations, so the envelope functions (see Fig. 6) are often not as severe.

the problems associated with the usual CTFs encountered in macromolecular studies. For example, as discussed below, spatial coherence limits the resolution when high defocus is used. The purpose of a phase plate is to introduce a $\pi/2$ phase shift between the scattered and unscattered electrons, in addition to the phase shift that occurs upon scattering. The sum of these two shifts gives optimal contrast when the specimen is in focus. By providing contrast across a wide range of spatial frequencies without gaps in the CTF, a well functioning phase plate would improve image quality and the resultant structure.

One factor that has not yet received much attention is determination of defocus when using a phase plate. On the one hand, with no oscillations of the CTF it would not even be necessary to know the defocus to have reliable phases. On the other hand, if one wanted to determine amplitudes from the images rather than from electron diffraction patterns, it would be necessary to determine the amplitude of the CTF, which would be given analytically if one knew the defocus. Under the conditions used for the curves in Fig. 1, changes of 100 Å in defocus produce very large changes in the CTF amplitude, even in the 5–10 Å resolution range, without affecting the phase. Thus one would require an accuracy of substantially better than 100 Å in defocus determination in order to obtain reliable amplitudes. While this accuracy is not difficult to achieve using the Thon rings or phase reversals in Fourier transforms of typical images currently used in electron crystallography, a different approach would be needed with the much more uniform CTF that a phase plate should produce.

The theoretical possibility of a phase plate for electrons was described by Boersch over 60 years ago (12), where he discussed several possible implementations of such a device. His ideas are summarized in Fig. 2. All of the devices would be placed in a diffraction plane following the specimen, where the scattered and unscattered components of the electron beam are spatially separated. Conceptually, the simplest device would use the inner potential of a thin foil to apply a phase shift to the scattered electrons, while the unscattered electrons would pass through a hole in the film. Even for such a simple device, though, there was no suitable technology to fabricate a practical device at the time of Boersch's paper. His proposals to use a small electrostatic lens that would affect either the unscattered or scattered electrons was even farther from available fabrication technologies of the time. The problem with all of these devices was mainly that elements would have to be made at what then was an impossibly fine scale.

The size scale is determined by a critical factor in phase plate performance termed the "cut-on" frequency, i.e., the lowest spatial frequency for which the device produces good contrast. For the thin film device, this frequency is set by the size of the hole through which the focused, unscattered beam passes. For other devices it is

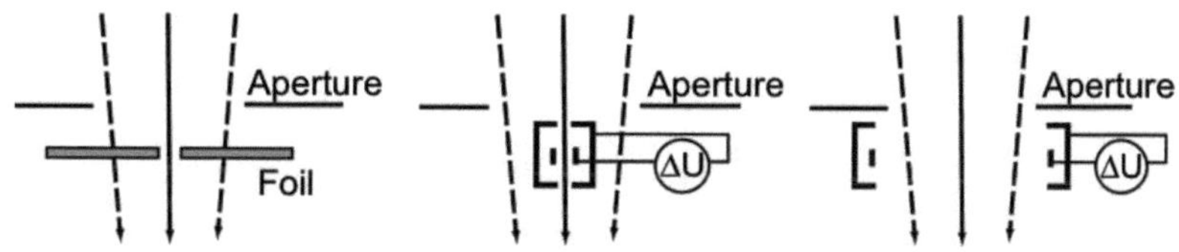

Fig. 2. Boersch's ideas for generating phase contrast. *Vertical solid line* represents unscattered electrons, *dashed lines* are scattered electrons. *Left*: passage through a thin foil shifts phase of scattered electrons. *Middle*: potential from a small einzel lens changes phase of unscattered electrons. *Right*: a larger einzel lens shifts phase of unscattered electrons. While this would be easier to fabricate, it would not provide a distinct shift of electrons scattered at low angle with respect to unscattered beam. Adapted from ref. 12.

determined by the size of the electrodes which surround the beam and block some of the low frequencies. For pure phase objects of a size D, it has been shown that the cut-on frequency should be no higher than 1/2D in order to obtain a reasonable amount of contrast of the full object (13). In practice, a useful target would be a device that would allow imaging macromolecular complexes on the order of 100–300 Å in diameter. For a cut-on frequency of 1/500 Å in a 200 kV microscope with an objective lens having a 2 mm focal length, the device would have to have a radius of less than 0.1 μm. Since this is impractically small for current fabrication technologies, a transfer lens can be introduced into the EM column to magnify the diffraction pattern. Unfortunately this also increases the effective lens aberrations and may compromise the resolution, although this effect can be taken care of with aberration correctors (see below). Even with a magnified diffraction pattern it would still be impractical to fabricate a phase plate with a cut-on frequency appropriate for larger objects such as intact cells. Fortunately, phase contrast is not such an essential consideration for such large objects since amplitude contrast provides a sufficient amount of contrast.

Aside from the fact that the technology for fabrication did not exist at the time, when experiments were attempted with devices that would function similarly to Boersch's suggestions, instability and signal loss due to charging were common problems. Single sideband, or Hilbert, imaging was moderately successful at generating contrast, but charging prevented simple implementation of practical devices (14). Similarly, introduction of a charge on a thin fiber centered near the unscattered beam in the diffraction plane showed some promise but there was no way to effectively control the charge with sufficient precision to make this approach practical (15). Thus Boersch's ideas lay quite dormant for close to half a century. As microfabrication technology developed, the inspiration to develop phase plates was reawakened. The demonstration of successful imaging with the simplest type of phase plate, made of a carbon film with a hole in the center (16), has stimulated several active efforts to develop more sophisticated devices. Some of these

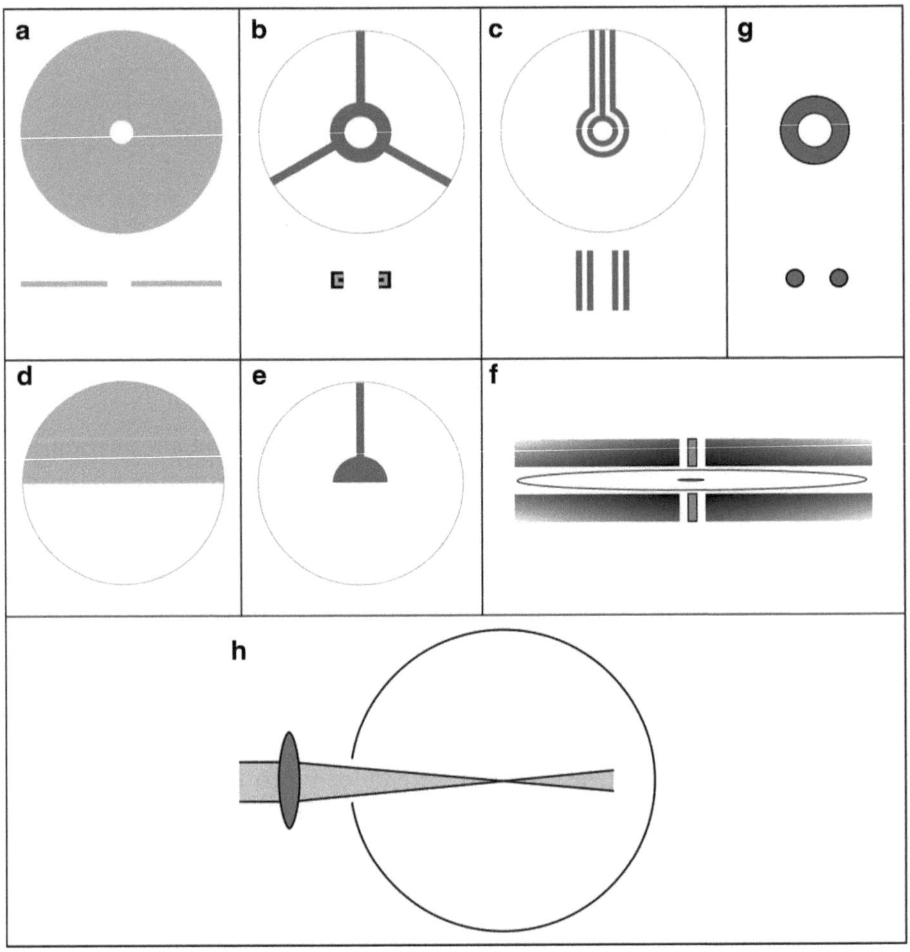

Fig. 3. Diagrams of several of the devices proposed for shifting the phase of electrons in the diffraction plane. (**a**) Top and central section views of the thin foil device. The foil thickness is chosen to produce a $\pi/2$ phase shift. A small hole in the center allows passage of the unscattered beam without effect of the foil (12, 16). (**b**) Top and central views of the einzel lens, suspended by three legs. Voltage is brought to the central electrode through the legs (12, 18, 19). (**c**) The drift tube design uses a central cylinder biased with respect to the outer tube at ground potential (20). (**d**) A foil covering half the plane produces strong Hilbert contrast if its thickness is set to produce a $\pi/2$ phase shift (24), or a single sideband (Foucalt) image if it is made opaque (7). (**e**) An aperture covering a small region of the half plane produces single sideband contrast out to the frequency where the usual CTF reaches a value of 0.5 (56). (**f**) An anamorphotic system produces a highly elongated diffraction pattern (shown by the ellipses); electrodes provide a potential that shifts the phase of the unscattered electrons near the center (22). (**g**) A magnet confined within a shell produces a vector potential which differs in sign between the regions inside and outside the toroid, effecting a phase shift between unscattered and scattered electrons (23). (**h**) A laser beam shining into a highly reflective, resonant cavity produces an intense field at the center which causes a phase shift of the unscattered electrons (25).

are based directly on Boersch's concepts and several on more advanced physical phenomena. Figure 3 shows schematic diagrams of several of the devices that have been fabricated or proposed.

To date the most impressive results have been obtained with the carbon film device (Fig. 3a). Careful calibration of the film thickness is needed to obtain the correct phase shift, of course, and

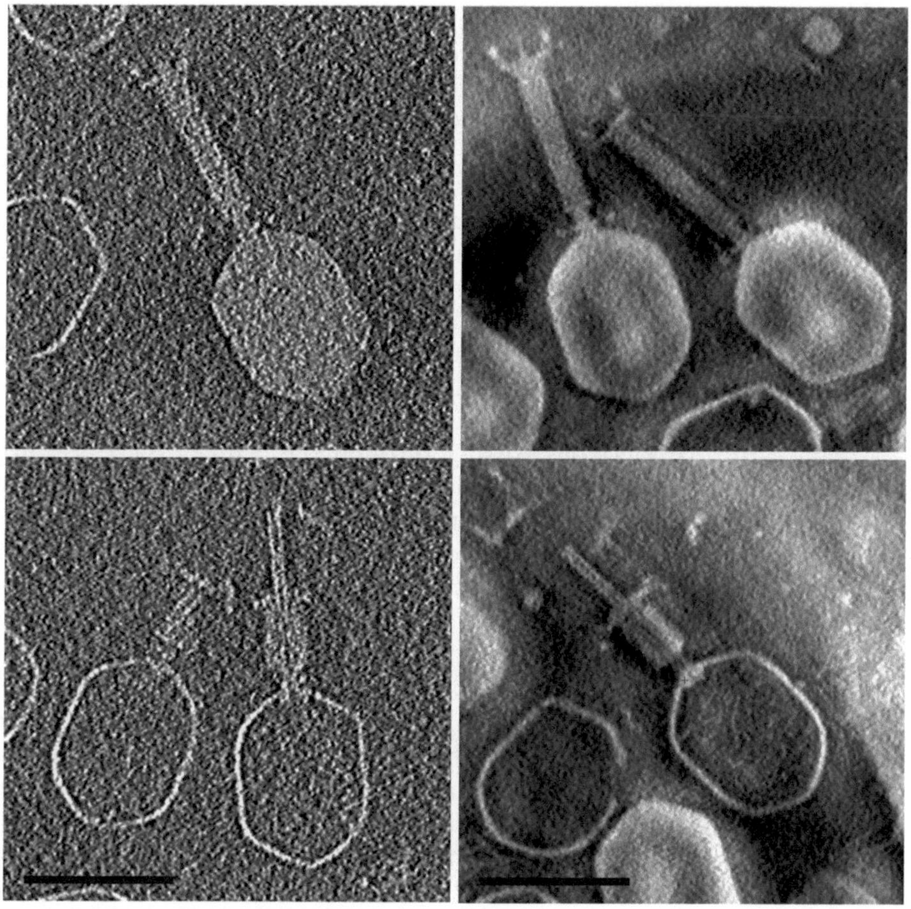

Fig. 4. Example of the improvement in low frequency contrast obtained with a thin film phase plate. *Left*: 5 nm-thick slice through a tomogram of T4 phage obtained under conventional conditions with a defocus of 3 μm. *Right*: Similar slice through tomogram obtained in-focus with the phase plate. Scale bars: 100 nm. Reprinted from ref. 17, with permission.

charging must be prevented to avoid introduction of spurious phase shifts. Heating the device appears to greatly reduce the rate of contamination and extend the lifetime. While the devices do have a lifetime that is still shorter than one would like, it has been possible to collect full data sets for single particle structure determination and even tomographic tilt series (17). Figure 4 shows one example of the difference that is seen in image contrast with conventional phase contrast and Zernike phase contrast using a thin film phase plate. The most obvious difference is the complete lack of contrast in the defocused image for the lowest spatial frequencies, which are the most immediately noticeable features in the phase plate image. The improvement in visibility and SNR at these frequencies should be a tremendous benefit when visual interpretation of the data or other operations such as particle identification are highly dependent on their presence.

Conceptually, the einzel-lens design for a phase plate (Fig. 3b) would appear quite ideal, but this design pushes the limits of microfabrication technologies. A central electrode is biased with respect to upper and lower electrodes providing an electrical potential that affects an electron's velocity and thus its phase. Voltages are brought in through the supporting legs, and the electrodes must be separated by insulating material. Several groups have produced functioning devices and shown that the phase shift can be accurately controlled to provide the complement of a typical phase contrast transfer function (18, 19). However, the stability of these devices (or of the environment in which they are used) has not been sufficient for extended applications to biological samples.

Another concept uses a coaxial pair of tubes that are biased so that electrons passing through the center experience an altered potential (Fig. 3c). This drift tube design has been used to produce functioning phase plates (20) but again stability and charging issues have hampered applications. The main problem with all three of these types of devices appears to be beam-induced charging of insulating material that is either present on the device as a residue from fabrication or that is deposited during operation, and possibly changes in material properties resulting from irradiation. As with the thin film devices, heating to 150–300 C greatly reduces (and may completely eliminate) the problem of contamination deposition. Thus it would seem that the only problem is to avoid the presence of any charging material near the unscattered beam, although it has been difficult to eliminate all non-conductive material left over from the fabrication process. Charging may be a continuing problem with the einzel lens design, which requires an insulating layer between the electrodes, or it may be possible to make this part with some material of sufficient conductivity to provide the necessary resistance but still eliminate charging.

As of this writing, the carbon film device is the only one commercially available, but it is reasonable to expect that the production problems with at least one of the other devices will soon be overcome well enough to make them commercially viable as well.

Several more advanced concepts for generating in-focus phase contrast have been described recently, spurred by the success of the first phase plate applications and the fabrication and electron optics capabilities now available. An extension of the carbon film device, which introduces a 90° phase shift to the scattered electrons, would be to use a film that covers just half of the diffraction plane and introduces a phase shift of 180° (Fig. 3d) (21). While the resulting "Hilbert contrast" produces an image that is strongly asymmetric, computational methods can be used to restore a directly interpretable image. This method combines the advantages of single sideband imaging, which eliminates the zeros in the CTF and provides half of the maximum phase contrast for all frequencies, with those of the symmetric phase plate, which optimizes the contrast.

The demonstration that careful fabrication and heating the devices can eliminate charging actually makes single sideband imaging itself appealing. A recent suggestion of a modified SSB device (56) would block electrons in half of the diffraction plane only out to the frequency where the normal defocus-based phase contrast would provide good signal. This would provide all the contrast normally achieved at high resolution with phase contrast plus half the theoretically available contrast in the low-frequency region where defocus does not provide much contrast, down to the cut-on frequency.

One proposed device, the anamorphotic system, would take advantage of some of the elements incorporated into aberration correctors that are now available for electron microscopes (22). This device would use an element roughly equivalent to a cylindrical lens to produce a highly elongated diffraction pattern, as illustrated in Fig. 3f. Electrostatic elements would then introduce a phase shift to just part of the pattern. i.e., to the unscattered electrons. In one implementation, a pair of such elements orthogonal to each other would each produce a $\pi/4$ phase shift, providing symmetry to the image. In another possible implementation, a single element would be used to produce Hilbert contrast. This device has the great advantage that it does not have to be placed in such close proximity to the unscattered beam as the others, and should thus suffer less from charging problems. The multipole elements used in aberration correctors provide the essential facility to produce the anamorphotic, or elongated, image. Implementing a phase plate requires designing the system so that the focused diffraction pattern lies on the appropriate plane of the element.

The Aharonov-Bohm effect is an interaction best described by quantum mechanics in which the phase of an electron is shifted by the vector potential that is associated with a magnetic field. A phase shift, which depends on the relation between the electron path and the adjacent magnetic field, occurs in a field-free region, providing another approach to achieving phase contrast. In a proof-of-principle experiment, Tonomura et al. constructed a device containing a circular magnet inside a superconducting shell, which they showed to produce this type of phase shift (Fig. 3g) (23). Such a device could, in principle, function ideally as a phase plate, but it is still well beyond the range of current fabrication methods. Nagayama suggested an alternative arrangement in which a thin bar magnet spanning the diffraction plane would be used to produce Hilbert contrast (24). Unfortunately, both of these implementations require placing material adjacent to the unscattered beam and thus suffer the same issues as the others. In addition, it may not be possible to heat the magnetic materials sufficiently to avoid contamination.

Another approach based on quantum theory would use the interaction of electrons with the electric field from an intense, focused light beam. The so-called "ponderomotive force" arises from interactions between electrons and photons which can be described by quantum electrodynamics theory. While the intensity required to achieve a quarter wave phase shift is high, it can be reached at the focus from commercially available lasers, either when operating in a pulsed mode or by use of low-loss reflecting optics. A suggestion to place a resonant, reflecting sphere around the diffraction pattern has been put forward (Fig. 3h) (25). This design would provide a 50–100-fold enhancement of the electric field in a small spot at the center and thus reduce the required laser power. Other configurations with reflecting optics were also suggested that would produce somewhat less amplification but would be simpler to fabricate. This approach has the very large advantage that it would not place any materials that could cause charging near the beam, and while it may seem highly futuristic at the moment this advantage makes it very appealing. The use of a pulsed laser in conjunction with a pulsed electron source (see below) may have some advantages in terms of laser power. The pulsed electron source also opens the door to using time varying fields to generate arbitrary phase shifts at the diffraction plane. This approach is still far enough from any practical implementation that it will not be discussed further.

3. Detectors

As noted above, the development of slow scan CCD cameras has had a notable impact on electron microscopists by giving the ability to obtain immediate feedback on specimen and image quality and to record many types of data without the tedium of developing and digitizing films. In some cases the quality of the data is substantially higher with a CCD than with film. This is particularly true in the case of electron diffraction data, where the spatial resolution of the detector is less of an issue than dynamic range, linearity, and background noise. On the other hand, for recording images of protein crystals the number of pixels available in most detectors is still not sufficient to compete effectively with film. For a typical crystal area 1–2 μm on edge, recording an image with 1.5 Å pixel size in order to reach 3 Å resolution would require an image array of about 7,000–14,000 pixels on edge. While this is routinely achievable with film, there are only a few prototype devices currently available that would be able to provide such performance. The problem is compounded by the fact that the resolution of most current detectors used in electron microscopy is lower than one would calculate based on the pixel size, so that it is often necessary to oversample the image by as much as a factor of two, requiring an even larger array.

The tremendous success of the consumer market for digital cameras, such as in video cameras and cell phones, as well as scientific and military applications, sparked rapid development in CCD cameras. More recently, an alternative imaging technology based on CMOS (complentary metal-oxide-semiconductor) fabrication has become much more popular in consumer devices and is making inroads into the scientific market as well.

Development of large format CCD chips led to a fairly rapid increase in the array size available to electron microscopists from the first useful devices with typically 512 × 512 pixel arrays to 4,096 × 4,096 (4 k) chips well within two decades. Further developments have been much slower, although there are presently chips with up to at least 10 k arrays. Along with increases in the pixel count the pixel sizes have been shrinking from the standard of about 24 μm in the mid 1980s to 14–15 μm common on 4 k chips. The newer, very large array chips have pixels in the range of 9 μm. All of these devices use large and expensive silicon chips and fabrication technologies. At the same time, all of the available CCDs employ a scintillator and either lens or fiber optic coupling to the CCD. Both bonding the fiber optics to the CCD chip and the coupling lens substantially increase the cost of the system. Decreasing the pixel size would decrease the cost with either coupling method, but would seriously compromise performance, particularly with the higher voltage electron microscopes that are becoming more popular for most applications in biological studies. One of the basic problems with fiber optic coupling is that electron scattering within the scintillator and supporting fiber optics produces a high level of noise and a broad point spread function (see Fig. 5a). These problems become significantly worse as the electron energy is increased, say from 100 to 300 keV. To some extent they can be mitigated with a lens coupled design, but the construction and alignment of the lens become major limitations with such large field, high-resolution designs and the requirement for efficient light capture. Thus, while the development of a 10 k CCD camera for EM will bring distinct advantages, the trend in detectors appears to be shifting rapidly to other technologies. Most promising among these are devices that can be classified as hybrid pixel devices or directly exposed CMOS devices.

CMOS fabrication technology is favored for production of large format devices and was the basis for development of the Tietz 8 k camera (26). This camera uses a scintillator and fiber optics similar to those of most CCD cameras, so in some regards the performance is similar to CCDs. The CMOS design allows more flexible readout, though, with pixels being directly addressable rather than the signals having to be shifted along columns and then along a row into the output amplifier. It is thus easier to implement much faster readout with a CMOS design. CCD chips that can provide readouts of several hundred frames per second are available, but not in nearly as large formats as CMOS chips at such high frame rates.

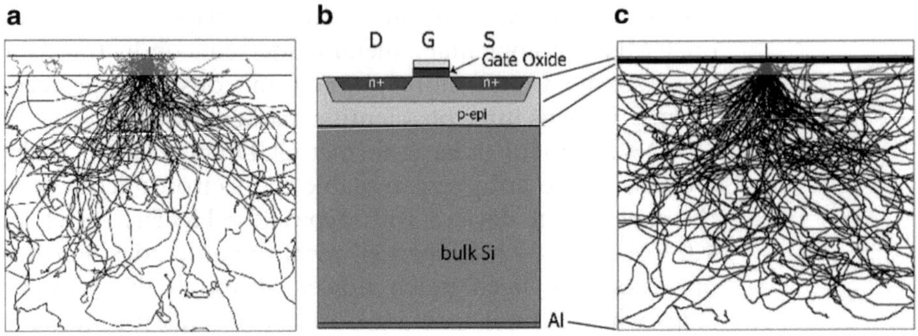

Fig. 5. Monte Carlo electron scattering calculations and CMOS device structure. (**a**) Electron tracks in a 40-μm thick scintillator mounted on glass fiber optics, as in a conventional CCD detector. Lighter colors indicate regions of tracks which produce signal. (**b**) Structure of a typical CMOS detector. The electronics are built on the top surface, above the thin active layer from which electron–hole pairs are collected (thickness not to scale). *D* drain, *G* gate, *S* source. (**c**) Simulation of electron tracks in the CMSO device. Lighter areas indicate signal-producing parts of tracks.

The CMOS design also opens the possibility for direct exposure of the device to the electron beam, avoiding the problems associated with the scintillator and coupling to the chip. Direct exposure of CCDs had been tested years ago, but it quickly became clear that radiation damage in the most sensitive areas of the on-chip electronics would be very difficult to eliminate (27). Damage to CMOS devices is also a potential problem, but over the last decades significant progress has been made in radiation hardening of devices designed for high-energy physics applications, and the principles are now being implemented in electron detectors.

Figure 5b illustrates the geometry of the typical CMOS chip design and signal production. Electron–hole pairs are produced in a very thin layer, giving the potential for very high resolution, as long as they can be collected before diffusion spreads them out too much. Even though the active layer in which the signal is generated is very thin, a sufficient number of electron–hole pairs are generated that the response to a single electron is well above the noise level from thermally generated electron–hole pairs. Several of these devices have been evaluated and more are actively being developed. Generally the performance meets the expectations, with very high signal-to-noise ratios that allow single-electron detection (28–30).

Backscattering from the substrate is a potential problem with the CMOS devices, which could introduce significant noise and degrade the DQE (detective quantum efficiency; see Fig. 5c). In fact, as is the case with CCDs, electrons may produce higher signals when they re-enter the active layer than on their primary transit through it. On the other hand, the SNR and spatial separation of these two signal components are generally sufficient that they can be discriminated and only the initial track counted (31). A more attractive approach to eliminating backscattering may be to thin the device to 30–50 μm. This has been shown to improve the DQE (30), although at the expense of making the device rather fragile.

The pixel size that is compatible with the expected point spread function and efficient signal collection also allows additional electronics to be fabricated to provide simple functions such as thresholding within each pixel. This would, in principle, allow readout in an electron counting mode. However, as discussed below, methods have now been demonstrated that produce substantial improvements in resolution based on the normal analogue readout, albeit with very low illumination rates.

When the individual electron events are sufficiently separated, it is possible to determine the centroid of each, as is done with super-resolution light microscopy. Because the active area of the device is so thin, electron tracks through this layer are quite straight and most of the spread of signal arises from diffusion. The signal is thus reasonably symmetric around each electron's point of incidence, unlike the case where significant scattering can take place within the active (scintillator) layer of a CCD-based detector. Centroiding gives a significantly better measure for the location of each electron impact and can thus greatly improve the MTF (modulation transfer function), even to the point where the effective array size of the device is increased by a factor that may be 2 or more (32). Thus even a 4 k×4 k device could give the spatial definition of an 8 k×8 k device, which would be extremely useful for a context such as electron crystallography. On the other hand, the tracks through the active layer are not perfectly straight, so a more sophisticated model for asymmetric generation of the signal may be required to fully realize this improvement. An alternate approach that gives a similar improvement in MTF is to normalize the signal for each electron. The pixel values around each identified single-electron event are scaled so that their sum is a constant, and these values then taken as representing a probability distribution for the impact location (31).

The tradeoff for the ability to centroid the signal is that the intensity must be low enough that electron events essentially never overlap, which requires that the exposure per frame be on the order of 1 electron per 100 pixels or less. If the target pixel size is 1 Å and the expected total exposure is 10 e/Å^2, this would then require integrating 1,000 frames. Devices have been fabricated that easily can be read at several hundred frames per second, and with a device that could be read out at 1,000 frames per second this would then require a 1 s exposure—not at all a limitation compared to current exposures on film.

Directly exposed CMOS devices for electron microscopes are currently available from several manufacturers, and more are under development. It can be expected that this type of device will be the main system for data recording within just a few years.

Another type of directly exposed device is known as a hybrid pixel device (31, 33), the best developed of which is the Medipix2. These devices generally use CMOS electronics on a silicon wafer that is bump bonded to a solid semiconductor wafer which serves as the detector element. Electron–hole pairs generated within this wafer, which is thick enough to stop electrons of up to about 120 keV, are separated by a bias voltage and the signal at each pixel is collected through the bump bond. The very large signal produced by each electron allows essentially noise-free operation, which is perhaps the principal advantage of these devices. The large pixel size, typically at least 50 μm, is a requirement of the fabrication process, but also allows even more electronics, for example a simple counter, to be built into each pixel. The device can thus operate in a true single electron detection mode without the need for very low beam intensity. Disadvantages include the necessity for large pixels and degradation of the MTF with electrons with energies much above 100 keV. Unfortunately it may be impractical to produce a device with such large pixels and an array size large enough (i.e., at least 4 k) to compete with other readout systems.

4. Aberration Correctors

One of the most significant developments in the quest to extend resolution in electron microscopy is the successful implementation of aberration correctors. Following Scherzer's proof that lenses with cylindrical symmetry and time-invariant fields would always have positive spherical and chromatic aberration coefficients (34), various schemes were proposed to employ non-round elements such as quadrupoles and hexapoles to reduce aberrations. Rose has presented an interesting review of the development of these systems (35). In spite of some ambitious efforts, it was not until power supply stabilities were improved and computer controls could effectively manage the alignment of these highly complex systems that it became possible to build and use correctors efficiently. Correctors for spherical and chromatic aberration, the main electron optical limits in modern microscopes, are now commercially available and have been critical in pushing the point resolution of electron microscopes to the half angstrom level. The value of correctors can be seen mathematically from the influence of the aberration coefficients in the envelope functions that damp the contrast transfer function (36, 37). The spatial coherence envelope depends on spherical aberration C_s, as well as the defocus ΔZ and source size θ:

$$E_{\mathrm{sp}} = \exp[-\pi^2(\theta/\lambda)(C_s\lambda^3 s^3 - \Delta Z s)^2]$$

(where λ is the electron wavelength) while the temporal coherence envelope depends on the chromatic aberration constant C_c and

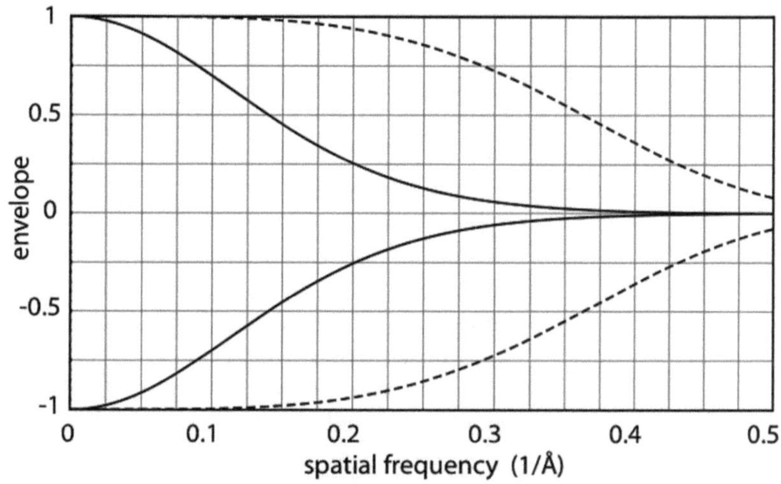

Fig. 6. Spatial and temporal coherence envelope functions corresponding to the CTF curves shown in Fig. 1. *Solid line*, spatial coherence for defocus = 1 μm. With a smaller illumination angle this curve would fall more slowly. At the lower defocus values of Fig. 1 the envelope retains a value very near unity out to the limits of the graph. *Dashed line*, temporal coherence envelope arising mainly from assumed energy width and instability of 1.6 V ($\delta E/E = 0.8 \times 10^{-5}$).

energy spread of the beam $\delta E/E$ along with other factors, including lens instability and vertical specimen motion, that introduce equivalent variation in focus:

$$E_{\text{sp}} = \exp[-\pi^2 (C_c (\delta E / E)\lambda s^2)^2 / 4]$$

The envelope functions for the conditions of Fig. 2 are shown in Fig. 6.

In principle, the spatial coherence envelope can be made arbitrarily close to unity by simply reducing the illumination angle. In practice, of course, the reduction that can be achieved is limited by the brightness of the electron source, and partial spatial coherence can produce a significant limitation of resolution and contrast. Field emission sources have made a substantial improvement because their high brightness allows one to improve both the spatial coherence and intensity of the illumination. Thus one is generally not limited by the spatial coherence, or spherical aberration in particular, in recording data for protein structures. The brightness-induced limitations are actually of even less significance in the case of electron crystallography than in single particle work. In the latter case, the need to identify particles within the image often requires the use of a defocus higher than is common when recording crystal data, and the coherence envelope falls faster at high defocus.

Another factor associated with the need to use significant defocus, which is not so obvious from the mathematics but will be familiar to anyone who has spent much time studying graphitized carbon, is the delocalization of the image signal that occurs with defocus. This can be seen in the sweeping of lattice fringes from a small crystallite across the image as the focus is varied or the buildup of Fresnel fringes around the edge of a particle with increasing defocus. Delocalization becomes increasingly important at high resolution and with small particles such as macromolecules, but would generally not be a significant problem with typical protein crystals containing dozens of unit cells in each direction. On the other hand, in materials science work at very high resolution the delocalization caused by spherical aberration does cause difficulties in interpreting images, so C_s correction is a significant help in this case.

One can expect that almost all of the proteins studied by electron crystallography can be considered to behave as weak phase objects to a good approximation, since they are sufficiently thin and composed of light atoms. In many materials science applications, this approximation is not so relevant and there is a need to determine both the amplitude and phase of the scattered wave. In this case reduction of C_s provides the ability to obtain in- or near-focus data which extends to higher resolution and is free from delocalization effects. In the case of protein studies, though, reducing C_s to zero actually has the disadvantage in that one cannot achieve the interplay between defocus and spherical aberration that gives rise to the broad contrast transfer band of Scherzer focus. In particular, the CTF for an in-focus image would have a value of zero within the range of interest for most biological work. On the other hand, the ability to adjust C_s provides the possibility to tune it to optimize contrast within some desired range of spatial frequencies. For example, the use of a larger value could move the relatively broad Scherzer band toward lower spatial frequencies that are more relevant in biological work. Unfortunately, this also causes the CTF to oscillate even faster at higher frequencies and so probably has a limited usefulness.

Correction of chromatic aberration has much more potential relevance to biological studies, although the C_c correctors are such a recent development that there is so far not a great deal of experimental confirmation of this potential. First, reduction of C_c improves the temporal coherence envelope function that generally sets the information limit in the microscope, assuming that energy spread and not lens or specimen instability is the main contributor to this function. Second, some incremental benefit may arise from the ability to use electrons which have been both elastically and inelastically scattered. It has been known for some time that the inelastic scattering can be used to produce a normal phase contrast image, although this image is at a different focus which depends on the amount of energy lost. Correction of chromatic aberration ensures that the focus change is negligible, so that all of the electrons which have been elastically scattered once contribute properly to the image, even if they have also been inelastically

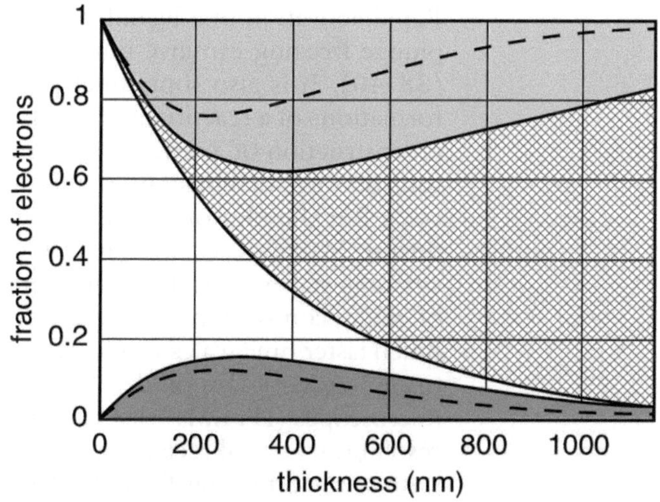

Fig. 7. Calculations of the specimen (ice) thickness range for different scattering models. *Upper and lower gray areas* represent fractions of electrons which undergo pure inelastic or elastic scattering, respectively. *Dashed lines* within these regions indicate fractions that undergo single scattering. *Hatched area* indicates fraction with mixed scattering. For thicknesses up to around 100 nm the single scattering approximation is fairly well satisfied. Thus for most applications of electron crystallography the approximation is valid. For thickness over 200 nm, better treatment of the scattering model in the process of image interpretation would be beneficial. Adapted from ref. 57.

scattered. However, as seen in Fig. 7, this represents a small fraction of the electrons for any thickness, and it may simply be more effective to remove all inelastic scattering to reduce the noise level.

Chromatic correction is expected to be most important in biological studies with thicker samples where essentially all electrons have been scattered inelastically. As seen from Fig. 7, in this regime multiple scattering dominates over the single scattering approximations. In this case the mixture of phase and amplitude contrast mechanisms makes quantitative analysis of images difficult at best. On the other hand, for applications such as whole cell tomography, empirical interpretation of the images and resulting density maps is generally sufficient, so the prospects for substantial improvements with C_c correction are high.

5. Dynamic TEM Studies

Many electron microscopists, particularly those focused on cryo-EM, may be resigned to the fact that our specimens are basically static and we have little access to information about dynamic processes. Various techniques for trapping structural intermediates have been worked out to provide some time resolution.

Rapid activation of a ligand or mixing of components just before plunge freezing can give time resolution in the millisecond range (38–40). It is also sometimes possible to sort out various conformations of a reaction pathway from a collection of images. The reconstruction of a series of steps in ribosome activity represents a tour-de-force demonstration of how far this approach can be taken (41). But the idea of studying proteins in motion is still mostly a dream. Materials scientists have for years used in-situ heating and straining stages to study dynamic effects, but on a time scale of tens of milliseconds at best. For both materials science and biology much faster time scales would be of great interest. Progress is being made in several labs on developing dynamic transmission electron microscopes (DTEMs) that can provide this improvement in time resolution. Several recent reviews give a good sense of the current state of the art in this field (42–44).

For many studies, the time scale of interest is on the order of 100 fs. This corresponds to the time for an atom to move 1 Å at the speed of sound. This time resolution range has now been reached using novel designs for pulsed electron sources. These ultra-fast electron optics systems employ a photoemission electron source which can be triggered by an intense laser flash. A second laser flash is generally used to stimulate some activity within the specimen at a predetermined time before the electron pulse reaches it. Current technology makes it fairly easy to obtain photon pulse lengths and timing precisions well under 1 fs. However, it appears that it will be difficult to take full advantage of such high time resolution with electrons. The Boersch effect, i.e., repulsion of electrons from each other within a small volume, severely restricts the intensity and pulse sharpness that one can achieve. Attempts to pack more electrons into a pulse cause extension of the pulse length and broadening of its energy distribution. Shortening the column to reduce the time during which electrons can interact reduces pulse broadening to some degree but also results in restriction to operating at fairly low accelerating voltages (43). For samples that can be stimulated repeatedly on a rapid time scale, stroboscopic imaging can give a way to overcome pulse broadening. In this mode each electron pulse can contain a single electron, eliminating electron–electron interactions. Summing many images to obtain sufficient statistics is then the rate-limiting step. There are many more active directions of this continuing instrumentation development, including optimization of the electron source and detector, pulse compression to reverse broadening effects and improving sample-beam interactions.

With the systems available now, there have already been many reports on rapid phenomena in materials, mainly relating to order–disorder transitions. Thermal and non-thermal melting have been characterized in different materials, along with the propagation of laser-induced melting fronts. The time scales successfully investigated have ranged down to the 100 fs region, providing access to a wide range of physical phenomena that can now be studied by imaging and diffraction.

There are certainly many dynamic processes that one might like to study by electron crystallography. For example the photocycle intermediates of bacteriorhodopsin received great attention by electron diffraction of thermally trapped states (45, 46). These could be good candidates for study by DTEM. Of even greater potential interest would be conformational changes related to biochemical processes. Generally these would occur at or near room temperature and in solution, but not within a crystal held at low temperature in the microscope. Thus the idea of combining a DTEM with an environmental stage, where the sample might even be in liquid, has some appeal. One could then use a laser pulse, for example, to release caged ATP or other ligand and then follow the time course of subsequent activity. Whether this could be applied with sufficient efficiency to protein crystals is a matter more of specimen preparation and the ability to find good crystals than of engineering the microscope. It may be that this approach would be more productive in single particle studies or in using single particle approaches on 2D crystals. Liquid stages that would be appropriate for this work have recently become commercially available, so there is likely to be significant activity in this direction over the next few years. However, it should be kept in mind that enzymatic activity, and thus most of the effects one might hope to view in real time, stop with an exposure on the order of 1 electron/nm^2, about a thousand times less than is generally used for high-resolution studies (47). Radiation damage is a very limiting problem for dynamic visualization of biological processes. Meanwhile, the development of technology for collecting X-ray diffraction from single molecules has many people thinking well beyond the limits of current technology.

The fundamental physics of radiation damage places the ultimate limit on what we can see by electron microscopy of macromolecules and is a similar problem in X-ray crystallography. The development of free electron lasers and other sources of extremely short and intense pulses of both X-rays and electrons has inspired proposals to overcome these fundamental limits (48). The basic idea is to illuminate the specimen with a pulse that is so short that all of the scattering would take place before the sample atoms can respond to irradiation and move away from their initial positions. The timescale required is on the order of 10 fs, and this is now within the range for X-ray sources. Efforts to develop such extremely short electron pulses, at least for diffraction if not for imaging, have been considered, but face the problem of obtaining short enough pulses with a sufficient number of electrons to produce adequate statistical definition.

6. STEM

Scanning Transmission Electron Microscopy (STEM) has many proponents, some of whom would suggest that it has great advantages over conventional TEM for macromolecular imaging. Indeed, in some applications there is no question that STEM is the preferred technique. These applications generally take advantage of the dependence of scattering on atomic number, particularly the stronger dependence of inelastic scattering or high angle scattering. The ability to capture electrons out to high scattering angles enhances the utility of STEM, but so far there are few examples where more is learned about proteins by STEM than TEM. A large factor in this limitation has been the loss in STEM of information that is captured in TEM by interference of the scattered and unscattered electrons, since the angular range covered by the incident beam in STEM overlaps so much with the elastic scattering, particularly in the case of protein crystals. Presently essentially all STEMs use simple annular detectors to collect the signals. The development of segmented detectors (49, 50) should make it possible to recover the interference information and may make STEM vastly more useful for protein studies than it currently is. The amount of data involved, which corresponds to the equivalent of a small image for every pixel, may seem overwhelming, but one can expect that data processing capabilities will keep up with the development of these detectors.

7. System Automation

When one considers the rate limiting steps in electron crystallography, the first roadblock is usually identified as obtaining suitable specimens. A vast parameter space generally needs to be searched to find optimal crystallization conditions for each new protein. For example, varying lipid, detergent and buffer type and concentration, along with temperature and pH alone can generate a large array of samples, not to mention testing various protein sources and constructs as is routinely done in much of X-ray crystallography. High-throughput systems are available or not so difficult to construct for testing crystallization conditions (51), but the EM processes for screening the results then represent a very time consuming and tedious step. Various approaches to automating this part of the work have been implemented and demonstrate that, in principle, even this bottleneck can be overcome. Robot driven grid preparation, mounting in a specimen holder, and insertion into the microscope can provide access to hundreds of samples in an automated series (52). Then intelligent software for microscope control and candidate crystal identification, followed by evaluation of crystal quality, relieves the human operator of most of the work of specimen development (53). Even for data collection, be it images or diffraction patterns, replacing the operator by a computer is really only a matter of implementation of sufficient analytical algorithms in the control software. While many microscopists might at this point consider that there are too many subjective factors in deciding whether, for example, a particular feature visible when searching the specimen represents a promising crystal, objectifying decisions related to the choice or areas is certainly possible. Eventually it should be possible to teach computers to make these decisions better than a human can.

8. Conclusion

Beyond the elements of specimen preparation, it is not hard to identify a number of technological and methodological advances that could substantially improve efficiency, throughput, and capability in electron microscopy and crystallography. Having everything in a microscope session run faster would be an obvious advantage. Being able to search the specimen with a real-time readout rather than a slow scan CCD, for example, is a tremendous help. This capability can already be achieved on a microscope equipped with two different cameras that are separately optimized for high- and low-speed readout, and to a limited extent with a single CCD. It is likely that the much higher readout speed of the new generation of CMOS cameras will greatly improve this implementation. Having a stage that would stop immediately without perceptible drift after moving would eliminate time that is now often spent either waiting for drift to subside or adjusting a drift compensator—which can be a significant amount of time. Substantial progress has been made in stage stability, but there is still much room for improvement. The next generation of fully piezo-driven stages may bring substantial improvements in this regard, but again there is not yet a lot of data to judge, especially under cryo conditions. Faster detectors will make a great difference in areas such as tomography, where one can imagine recording a series of 100–200 images within a very few minutes, rather than the current times which can range to an hour or more. In electron crystallography, the need is more for larger pixel arrays than for high speed, so the impact will be somewhat less. Still, we can expect something like an order of magnitude increase in throughput with implementation of even a few of these developments.

While these instrument developments will change the rate at which work proceeds, they will not fundamentally alter what we do. There are other issues that could be addressed to have broader impacts. The weak phase approximation has been very productive in electron crystallography, although it is widely recognized that it does not hold perfectly even for typical single particle work where one generally finds about 5–7% amplitude contrast. Figure 7 gives a sense of the scope of this problem and the limited range of the weak scattering approximations. A more complete treatment of amplitude contrast and, in particular, multiple (including inelastic) scattering would allow more quantitative treatment of thicker specimens. In materials science, where mixed contrast modes are more the norm, a number of approaches have been developed for determining the complex wave produced by the specimen. For example, exit wave reconstruction, based on a focal series that can include tens of images (54), is now commonly used in very high resolution work. In principle, a small set of images obtained under different conditions is sufficient to fully determine the exit wave. An intriguing option that makes use of an adjustable phase plate to produce varying phase shifts has been proposed (55). Using three images with phase shifts of $-\pi/2$, 0, and $\pi/2$ radians makes the mathematical treatment of this method particularly simple, but the phase shifts could be rather arbitrary—just as they are currently generated by varying the defocus. Again, the impact of such developments might be felt more in tomography than crystallography.

Higher voltage microscopes significantly improved the accuracy of the single scattering and weak phase approximations, giving more flexibility in specimen thickness with the 300 kV microscopes that are nearly ubiquitous today. The current trend in materials science, though, is toward lower voltage. The ability to remove the lens aberrations that become worse at lower voltages has led to the prospect of achieving very high resolution at low voltages. Working below the knock-on damage threshold—i.e., the energy that produces atom displacements, not just ionization and bond breakage—allows studies that would otherwise be impossible due to radiation damage. There is thus great interest in the possibility of reaching atomic resolution in the 20–80 kV range. Of course, it is ionization damage that limits work with organic materials much more than knock-on damage. One can expect that issues of specimen thickness will dominate biologists' choices for microscopes and lead them toward 300 kV even as the manufacturers may focus more on lower voltage. Prospects for a 400 or even 500 kV FEG microscope, which a few years ago seemed reasonable, now seem largely to have faded. The incremental advantages that they would bring to biology, as well as materials science, are unlikely to be sufficient driving force to influence the manufacturers. In fact it does seem that for the near term much of the excitement in instrumentation development will be driven by materials science applications. The resulting microscopes have become very expensive. While some of the upcoming developments, such as a phase plate and a CMOS camera, will likely be seen as almost essential for all high-end microscopes within a few years, few groups working in the biological sciences will be able to afford a fully corrected, cryo-capable microscope with a top-level detector, at least until the benefits are more thoroughly understood and documented. One can certainly be skeptical that the high-priced innovations driven by materials scientists will benefit biologists, but some microscopists will certainly take advantage of such systems to characterize the improvements they bring. It would be a surprise if we do not see innovative ways both to use the new technologies in biological studies and to push their further development more to our own needs.

References

1. Williams RC, Fisher HW (1970) Electron microscopy of tobacco mosaic virus under conditions of minimal beam exposure. J Mol Biol 52:121–123
2. Matricardi VR, Moretz RC, Parsons DF (1972) Electron diffraction of wet proteins: catalase. Science 177:268–270
3. Heide HG, Grund S (1974) Deep-freeze link for transport of water-containing biological objects to the electron microscope. J Ultrastruct Res 48:259–268
4. Taylor KA, Glaeser RM (1975) Modified air-lock door for the introduction of frozen specimens into the JEM 100B electron microscope. Rev Sci Instrum 46:985–986
5. Taylor KA, Glaeser RM (1974) Electron diffraction of frozen, hydrated protein crystals. Science 186:1036–1037
6. Adrian M, Dubochet J, Lepault J, McDowall AW (1984) Cryo-electron microscopy of viruses. Nature 308:32–36
7. Downing KH (1991) Spot-scan imaging in transmission electron microscopy. Science 251:53–59
8. Carragher B, Kisseberth N, Kriegman D, Milligan RA, Potter CS, Pulokas J, Reilein A (2000) Leginon: an automated system for acquisition of images from vitreous ice specimens. J Struct Biol 132:33–45
9. Nickell S, Forster F, Linaroudis A, Net WD, Beck F, Hegerl R, Baumeister W, Plitzko JM (2005) TOM software toolbox: acquisition and analysis for electron tomography. J Struct Biol 149:227–234
10. Mastronarde DN (2005) Automated electron microscope tomography using robust prediction of specimen movements. J Struct Biol 152:36–51
11. Sass S (1989) A patently false myth. Skeptical Inquirer 13:310–313
12. Boersch H (1947) Uber die Kontraste von Atomen im Elektronenmikroskop. Z Naturforsch A – J Phys Sci 2:615–633
13. Danev R, Glaeser RM, Nagayama K (2009) Practical factors affecting the performance of a thin-film phase plate for transmission electron microscopy. Ultramicroscopy 109:312–325
14. Downing KH (1979) Possibilities of heavy atom discrimination using single-sideband techniques. Ultramicroscopy 4:13–31
15. Unwin PNT (1972) Electron microscopy of biological specimens by means of an electrostatic phase plate. Poc Roy Phil Soc A Math Phys Sci 329:327–359
16. Danev R, Nagayama K (2001) Transmission electron microscopy with Zernike phase plate. Ultramicroscopy 88:243–252
17. Danev R, Kanamaru S, Marko M, Nagayama K (2010) Zernike phase contrast cryo-electron tomography. J Struct Biol 171:174–181
18. Majorovits E, Barton B, Schultheiss K, Perez-Willard F, Gerthsen D, Schroder RR (2007) Optimizing phase contrast in transmission electron microscopy with an electrostatic (Boersch) phase plate. Ultramicroscopy 107: 213–226
19. Alloyeau D, Hsieh WK, Anderson EH, Hilken L, Benner G, Meng X, Chen FR, Kisielowski C (2010) Imaging of soft and hard materials using a Boersch phase plate in a transmission electron microscope. Ultramicroscopy 110:563–570
20. Cambie R, Downing KH, Typke D, Glaeser RM, Jin J (2007) Design of a microfabricated, two-electrode phase-contrast element suitable for electron microscopy. Ultramicroscopy 107:329–339
21. Danev R, Nagayama K (2004) Complex observation in electron microscopy: IV. Reconstruction of complex object wave from

conventional and half plane phase plate image pair. J Phys Soc Jpn 73:2718–2724
22. Schröder RR, Barton B, Rose H, Benner G (2007) Contrast enhancement by anamorphotic phase plates in an aberration corrected TEM. Microsc Microanal 13(suppl 3):136–137
23. Tonomura A, Osakabe N, Matsuda T, Kawasaki T, Endo J, Yano S, Yamada H (1986) Evidence for Aharonov-Bohm effect with magnetic-field completely shielded from electron wave. Phys Rev Lett 56:792–795
24. Nagayama K (2008) Development of phase plates for electron microscopes and their biological application. Eur Biophys J 37:345–358
25. Muller H, Jin JA, Danev R, Spence J, Padmore H, Glaeser RM (2010) Design of an electron microscope phase plate using a focused continuous-wave laser. New J Phys 12
26. Tietz H (2008) Design and characterization of 64 megapixel fiber optic coupled CMOS detector for transmission electron microscopy. Microsc Microanal 14(suppl S2): 804–805
27. Roberts PTE, Chapman JN, Macleod AM (1982) A CCD-based image recording-system for the CTEM. Ultramicroscopy 8:385–396
28. McMullan G, Chen S, Henderson R, Faruqi AR (2009) Detective quantum efficiency of electron area detectors in electron microscopy. Ultramicroscopy 109:1126–1143
29. Battaglia M, Contarato D, Denes P, Doering D, Giubilato P, Kim TS, Mattiazzo S, Radmilovic V, Zalusky S (2009) A rad-hard CMOS active pixel sensor for electron microscopy. Nucl Instrum Methods Phys Res A 598:642–649
30. McMullan G, Faruqi AR, Henderson R, Guerrini N, Turchetta R, Jacobs A, van Hoften G (2009) Experimental observation of the improvement in MTF from backthinning a CMOS direct electron detector. Ultramicroscopy 109:1144–1147
31. McMullan G, Clark AT, Turchetta R, Faruqi AR (2009) Enhanced imaging in low dose electron microscopy using electron counting. Ultramicroscopy 109:1411–1416
32. Battaglia M, Contarato D, Denes P, Giubilato P (2009) Cluster imaging with a direct detection CMOS pixel sensor in Transmission Electron Microscopy. Nucl Instrum Methods Phys Res A 608:363–365
33. Fan GY, Datte P, Beuville E, Beche JF, Millaud J, Downing KH, Burkard FT, Ellisman MH, Xuong NH (1998) ASIC-based event-driven 2D digital electron counter for TEM imaging. Ultramicroscopy 70:107–113
34. Scherzer O (1949) The theoretical resolution limit of the electron microscope. J Appl Phys 20:20–29
35. Rose HH (2009) Historical aspects of aberration correction. J Electron Microsc 58:77–85
36. Frank J (1973) The envelope of electron microscopic transfer functions for partially coherent illumination. Optik 38:519–536
37. Glaeser R, Downing K, DeRosier D, Chiu W, Frank J (2007) Electron microscopy of biological macromolecules. Oxford, New York
38. Berriman J, Unwin N (1994) Analysis of transient structures by cryomicroscopy combined with rapid mixing of spray droplets. Ultramicroscopy 56:241–252
39. Lu ZH, Shaikh TR, Barnard D, Meng X, Mohamed H, Yassin A, Mannella CA, Agrawal RK, Lu TM, Wagenknecht T (2009) Monolithic microfluidic mixing-spraying devices for time-resolved cryo-electron microscopy. J Struct Biol 168:388–395
40. Shaikh TR, Barnard D, Meng X, Wagenknecht T (2009) Implementation of a flash-photolysis system for time-resolved cryo-electron microscopy. J Struct Biol 165:184–189
41. Fischer N, Konevega AL, Wintermeyer W, Rodnina MV, Stark H (2010) Ribosome dynamics and tRNA movement by time-resolved electron cryomicroscopy. Nature 466:329–333
42. Reed BW, Armstrong MR, Browning ND, Campbell GH, Evans JE, LaGrange T, Masiel DJ (2009) The evolution of ultrafast electron microscope instrumentation. Microsc Microanal 15:272–281
43. Miller RJD, Ernstorfer R, Harb M, Gao M, Hebeisen CT, Jean-Ruel H, Lu C, Moriena G, Sciaini G (2010) 'Making the molecular movie': first frames. Acta Cryst A 66:137–156
44. Zewail AH (2010) Four-dimensional electron microscopy. Science 328:187–193
45. Vonck J, Han BG, Burkard F, Perkins GA, Glaeser RM (1994) Two progressive substates of the M-inermediate can be identified in glucose-embedded, wild-type bacteriorhodopsin. Biophys J 67:1173–1178
46. Subramaniam S, Lindahl I, Bullough P, Faruqi AR, Tittor J, Oesterhelt D, Brown L, Lanyi J, Henderson R (1999) Protein conformational changes in the bacteriorhodopsin photocycle. J Mol Biol 287:145–161
47. Hahn M, Seredynski J, Baumeister W (1976) Inactivation of catalase monolayers by irradiation with 100 keV electrons. Proc Natl Acad Sci USA 73:823–827
48. Neutze R, Wouts R, van der Spoel D, Weckert E, Hajdu J (2000) Potential for biomolecular imaging with femtosecond X-ray pulses. Nature 406:752–757
49. Daberkow I, Herrmann KH, Lenz F (1993) A configurable angle-resolving detector system in STEM. Ultramicroscopy 50:75–82
50. Caswell TA, Ercius P, Tate MW, Ercan A, Gruner SM, Muller DA (2009) A high-speed

area detector for novel imaging techniques in a scanning transmission electron microscope. Ultramicroscopy 109:304–311

51. Vink M, Derr K, Love J, Stokes DL, Ubarretxena-Belandia T (2007) A high-throughput strategy to screen 2D crystallization trials of membrane proteins. J Struct Biol 160:295–304
52. Lefman J, Morrison R, Subramaniam S (2007) Automated 100-position specimen loader and image acquisition system for transmission electron microscopy. J Struct Biol 158:318–326
53. Cheng A, Leung A, Fellmann D, Quispe J, Suloway C, Pulokas J, Abeyrathne PD, Lam JS, Carragher B, Potter CS (2007) Towards automated screening of two-dimensional crystals. J Struct Biol 160:324–331
54. van Dyck D, de Beeck MO, Coene W (1993) A new approach to object wave-function reconstruction in electron-microscopy. Optik 93:103–107
55. Gamm B, Dries M, Schultheiss K, Blank H, Rosenauer A, Schröder RR, Gerthsen D (2010) Object wave reconstruction by phase-plate transmission electron microscopy. Ultramicroscopy 110:807–814
56. Buijsse B, van Laarhoven FM, Schmid AK, Cambie R, Cabrini S, Jin J, Glaeser RM (2011) Design of a hybrid double-sideband/single-sideband (schlieren) objective aperture suitable for electron microscopy. Ultramicroscopy 111:1688–1695
57. Koster AJ, Grimm R, Typke D, Hegerl R, Stoschek A, Walz J, Baumeister W (1997) Perspectives of molecular and cellular electron tomography. J Struct Biol 120:276–308

Chapter 21

Tubular Crystals and Helical Arrays: Structural Determination of HIV-1 Capsid Assemblies Using Iterative Helical Real-Space Reconstruction

Peijun Zhang, Xin Meng, and Gongpu Zhao

Abstract

Helical structures are important in many different life forms and are well-suited for structural studies by cryo-EM. A unique feature of helical objects is that a single projection image contains all the views needed to perform a three-dimensional (3D) crystallographic reconstruction. Here, we use HIV-1 capsid assemblies to illustrate the detailed approaches to obtain 3D density maps from helical objects. Mature HIV-1 particles contain a conical- or tubular-shaped capsid that encloses the viral RNA genome and performs essential functions in the virus life cycle. The capsid is composed of capsid protein (CA) oligomers which are helically arranged on the surface. The N-terminal domain (NTD) of CA is connected to its C-terminal domain (CTD) through a flexible hinge. Structural analysis of two- and three-dimensional crystals provided molecular models of the capsid protein (CA) and its oligomer forms. We determined the 3D density map of helically assembled HIV-1 CA hexamers at 16 Å resolution using an iterative helical real-space reconstruction method. Docking of atomic models of CA-NTD and CA-CTD dimer into the electron density map indicated that the CTD dimer interface is retained in the assembled CA. Furthermore, molecular docking revealed an additional, novel CTD trimer interface.

Key words: Cryo-EM, Tubular crystals, Helical reconstruction, Real space, Docking, HIV-1 capsid

1. Introduction

Helical structures are often observed in cells and play important functions in different life forms. Examples include actin filaments and microtubules, as part of the cell cytoskeleton, flagella, functioning as a propeller in bacterial locomotion, and the contractile sheath in tailed phages, responsible for injecting genetic material into host cells. Helical structures, such as amyloid fibrils, can also be involved in diseases (1–4). In addition, many proteins, including the integral

Ingeborg Schmidt-Krey and Yifan Cheng (eds.), *Electron Crystallography of Soluble and Membrane Proteins: Methods and Protocols*, Methods in Molecular Biology, vol. 955, DOI 10.1007/978-1-62703-176-9_21,

membrane proteins acetylcholine receptor (5) and Ca^{2+}-ATPase (6), and soluble protein HIV-1 viral capsid protein (7), can be induced to assemble into helical structures, which are favorable for structural determination. Helical structures can be thought of as a special case of two-dimensional (2D) crystals, in which the crystalline sheet has been deformed into a cylinder. Such structures have particular properties that make it relatively straightforward to obtain a three-dimensional (3D) view of the subunit. A single projection image of the object provides all the views of the subunit for 3D reconstruction, thus, eliminating the need for tilting and the attendant missing cone as in the case of 2D crystals. As such, several helical structures have been solved to near-atomic resolution by cryo-electron microscopy (cryo-EM) (5, 8, 9).

The mature HIV-1 capsid is composed of ~1,500 copies of the capsid protein (CA), generally forming conical or tubular assemblies. Based on early EM analysis of native cores and in vitro assemblies, a fullerene cone model of HIV-1 capsid was proposed (10) in which the capsid comprises a hexameric lattice containing a variable number of CA pentamers that allow for closure of tubular or conical structures. HIV-1 CA consists of two domains, a ~150 amino acid N-terminal domain (NTD) and an ~80 amino acid C-terminal domain (CTD), connected by a hinge. The full-length recombinant CA assembles into highly ordered helical structures in vitro. These in vitro assemblies are likely to contain subunit interactions present in the native HIV-1 capsid, since they exhibit features similar to those of native cores (11). A number of atomic structures of individual CA domains (12–18), as well as full-length hexameric CA have been determined by X-ray crystallography (19). In this chapter, we describe the methods used to obtain a cryo-EM structure of HIV-1 CA in the assembled state. With molecular docking, the reconstructed pseudo-atomic model of HIV-1 CA tube indicates that the inter-hexamer interactions in the surface lattice of the capsid are mediated by two distinct CTD–CTD interfaces, a CTD dimer interface that was previously described and a novel CTD trimer interface. We further validate the structural model and demonstrate the significance of the new CTD interface by mutational studies (20). The protocol described here can be easily adapted to other helical specimens for structural determination using cryo-EM. The programs used in this protocol are freely available.

2. Materials

2.1. Equipment

1. FPLC system: AKTA Explorer with Unicorn software, 5 ml Hi-Trap QP column and 5 ml Hi-Trap SP column, and Hi-Load Superdex 75 26/60 column (GE Healthcare, Piscataway, NJ).
2. Vertical electrophoresis system: XCell SureLock Mini-Cell (Invitrogen, Carlsbad, CA).

3. Glow-discharge device model 100X (EMS, Hatfield, PA).
4. FEI transmission electron microscopes (FEI Corp., OR): Tecnai T12 with LaB_6 filament, a single-tilt room-temperature specimen holder and a Gatan 2K×2K CCD camera; Tecnai Polara F30 microscope with a Field Emission Gun, equipped with a Gatan 4 K×4 K CCD camera.
5. Plunge-freezing devices: Vitrobot Mark III Automated Plunger (FEI Corp., OR) and home-made MRC-type gravity plunger.
6. Nikon super coolscan 9000 ED scanner: resolution of4,000 dpi (Nikon, Japan).
7. Computers running Linux operating system and image processing and visualization software packages: EMAN (21), IHRSR (22), Spider (23), MRC based helical processing software (24), Real-space helical refinement software (9), Situs (25), Chimera (26).

2.2. Reagents

1. The cDNA encoding gag polyprotein, $pr55^{gag}$ can be obtained from the NIH AIDS Research and Reference Reagent Program, Division of AIDS, NIAID, NIH (27).
2. pET21 and Rosetta 2 (DE3) (EMD chemicals, Inc. San Diego, CA).
3. Electron microscope carbon-coated copper grids (SPI Supplies, West Chester, PA), Quantifoil R2/1 200 mesh holely,.carbon copper grids (Quantifoil Micro Tools, Jena, Germany).
4. KODAK Electron Image Film SO-163, KODAK D-19 developer, fixer, and PHOTO-FLO solution (Eastman Kodak Co., Rochester, NY).
5. Filter paper type #1: 90 mm diameter (Whatman, Clifton, NJ).
6. DUMONT biology anti-capillary reverse tweezers (Ted Pella, Redding, CA).
7. All other reagents except those listed below are from Sigma (St. Louis, MO) and are of analytical grade or higher.
8. Pre-cast 4–12% SDS PAGE gel, 4× SDS sample buffer and 20× running buffer (Invitrogen, Carlsbad, CA), Amicon concentrators (Millipore, Billerica, MA). Luria-Bertani media (Bacto, NJ),

2.3. Buffer Systems

1. Lysis buffer: 25 mM sodium phosphate, pH 7.0, 0 mM NaCl.
2. Hi-Trap QP column buffer: 25 mM sodium phosphate, pH 7.0, 1 mM DTT, and 0.02% sodium azide, 0 mM NaCl (Low Salt)—1,500 mM NaCl (High Salt).
3. Hi-Trap SP column buffer: 25 mM sodium phosphate, pH 5.8, 1 mM DTT, and 0.02% sodium azide, 0 mM NaCl (Low Salt)—1,000 mM NaCl (High Salt).

4. Hi-Load Superdex 75 26/60 column buffer: 25 mM sodium phosphate, pH 5.8, 100 mM NaCl, 1 mM DTT, and 0.02% sodium azide.
5. CA assembly buffer: 50 mM Tris–HCl, pH 8.0, 1 M NaCl.
6. Dilution buffer: 50 mM Tris–HCl, pH 8.0, 100 mM NaCl.

3. Methods

3.1. Expression and Purification of HIV-1 CA and CA Mutants

1. Amplify and subclone wild-type full-length CA (Pro1-Leu231) into pET21 using *Nde*I and *Xho*I sites.
2. Express full-length CA in *E. coli*, Rosetta 2 (DE3), cultured in Luria-Bertani media and induce with 0.4 mM IPTG. Culture at 23°C for 16 h.
3. Harvest cells 16 h post-induction and lyse by sonication in the lysis buffer.
4. Clear the cell lysate by ultracentrifugation at 100,000 × *g* for 1 h at 4°C and apply it to a Hi-Trap QP column equilibrated with the lysis buffer.
5. Collect flow through fractions.
6. Adjust the pH of the sample to 5.8 with diluted acetic acid. Add Milli-Q water to lower conductivity to below 2.5 ms/cm. Centrifuge the sample at 25,000 × *g* with a SLA-1500 rotor for 1 h at 4°C.
7. Apply the resulting supernatant to a Hi-Trap SP column and collect peak fractions. CA should elute at ~10 ms/cm.
8. Concentrate the sample to 15–18 ml and apply it to a Hi-Load Superdex 75 26/60 gel-filtration column.
9. Collect peak fractions and concentrate with Amicon concentrators to 10 mg/ml, based on the extinction coefficient at UV 280.
10. Analyze protein samples by SDS-PAGE and stain with Coomassie Blue to assess protein purity (Fig. 1, lanes marked t).
11. For storage, add 5% glycerol to the sample. Store at −80°C in 25μl aliquots that have been flash-frozen with liquid N_2.

3.2. HIV-1 CA Assembly

1. Add 2× assembly buffer (2 M NaCl and 100 mM Tris–HCl pH 8.0) to freshly thawed stock protein to a final CA concentration of 2 mg/ml (see Note 1).
2. Incubate the mixture at 37°C for 1 h. At the end of incubation, transfer the sample to ice for further analysis.
3. For SDS-PAGE analysis of CA assemblies, centrifuge the samples with an Eppendorf centrifuge 5417R (4°C) at 20,000 × *g*

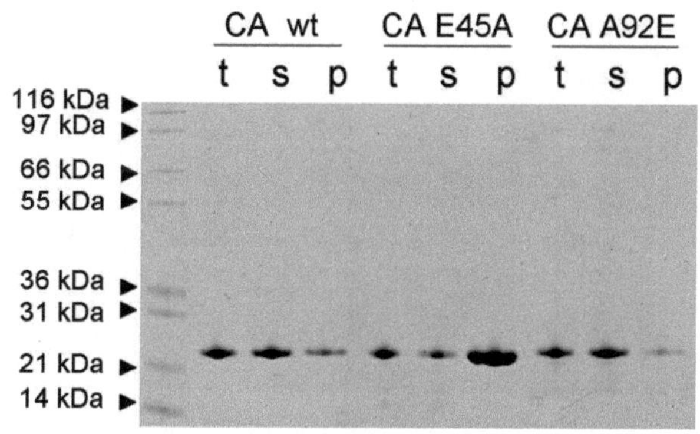

Fig. 1. Coomassie Blue-stained SDS-PAGE of purified HIV-1 CA and CA mutants before assembly (t) and following assembly and centrifugation (s and p). Assembled CA tubes are present in the pellet (p) after high-speed centrifugation.

for 15 min. Load the resulting supernatant and pellet samples onto 4–12% gradient SDS-PAGE gel and stain with Coomassie Blue (Fig. 1).

3.3. EM Specimen Preparation

3.3.1. Negatively Stained EM Specimen Preparation

CA assembly conditions have been optimized by screening an array of parameters, such as CA protein concentration, pH value, salt concentration, assembly temperature, etc. The screening is carried out by checking the integrity of negatively stained tubes, assembled under various conditions, on an FEI T12 microscope. The optimized assembly conditions should produce well-ordered, long, straight, separated tubes in reasonable quantities.

1. Glow discharge 200 mesh solid carbon grids under 25 mA for 25 s using a glow discharge device with the HT polarity set to be negative.
2. Lay two droplets (~40 μl) of 1% uranyl acetate on a piece of parafilm for each specimen.
3. Apply a 3–5 μl sample to the freshly glow-discharged grid, which is held by DUMONT anti-capillary reverse tweezers, and allow absorption of the sample for 30 s.
4. Blot the solution with a piece of filter paper and quickly place the carbon side of the grid on top of the first uranyl acetate droplet surface and gently move the grid back and forth for 5 s.
5. Blot away the uranyl acetate, quickly transfer the grid to the second uranyl acetate droplet, and leave it for 30 s. Then, completely blot off the uranyl acetate solution and let the grid air-dry.
6. Examine the grids with a Tecnai 12 electron microscope (Fig. 2).

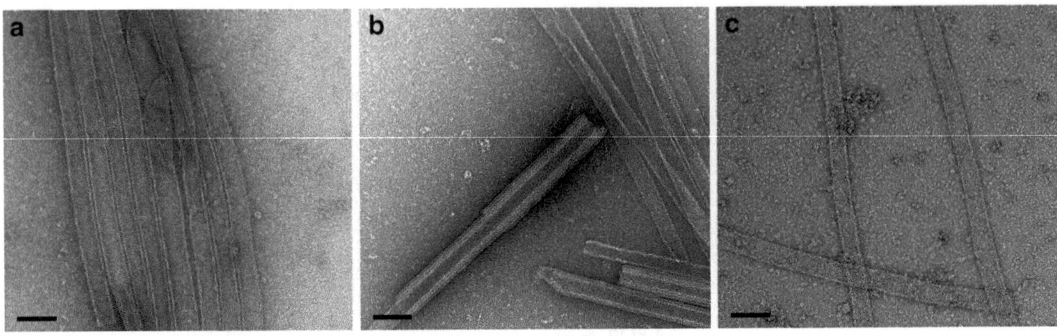

Fig. 2. EM of negatively stained CA wt and CA mutant assemblies. (**a**) CA wt assemblies show bundled tubes. (**b**) CA E45A assembles more efficiently, but with multiple layers. (**c**) CA A92E assembles single, long, and well-ordered tubes. Scale bars, 100 nm.

3.3.2. Frozen-Hydrated EM Specimen Preparation

Embedding biological macromolecules in vitreous ice is the best way to preserve their structure in the hydrated state. During rapid freezing, the molecules can be locked in the state they were just prior to freezing. Thus, high-resolution structural information can be well preserved using this technique (28). These advantages make cryo-EM a powerful method to study protein structure and to study short-lived transitional structures that are hard to catch (29).

Because the HIV-1 CA assemblies are stable only in high salt (1 M NaCl) buffer, which contributes strong background noise in cryo-EM images, we use a rapid dilution and back-side blotting method to transiently reduce salt concentrations when preparing the frozen-hydrated grid.

1. Glow discharge 200 mesh R2/1 Quantifoil grids under 25 mA for 25 s.
2. Turn on the humidifier for the home-made gravity plunger until the humidity reaches approximately 90% in the humidity chamber.
3. Prepare cold liquid ethane cooled by liquid nitrogen in the FEI plunge freezing dewar (see Note 2).
4. Apply 2.5 μl of CA assembly solution on the carbon side of the grid, which is mounted on forceps, and load the forceps onto the plunger with the carbon side (front side) of the grid facing away from you. Add 3 μl of dilution buffer to the back side of the grid.
5. Immediately blot the grid with a piece of filter paper, placed against the back side of the grid. The grid back surface should be in good contact with the filter paper. After ~6 s, remove the filter paper and release the plunger at the same time (see Note 3).
6. Remove the forceps from the plunger and quickly transfer the grid into a grid storage box.

3.4. Cryo-EM of CA Tubular Assemblies

For flexible HIV-1 CA tubular assemblies, the most common problem encountered during data collection is deformation caused by surface tension, which can flatten or bend the tubes. Such deformation can make the sample unsuitable for 3D reconstruction. Thus, finding an ice area suitable for imaging is critical, not only to maximize the signal to noise ratio under low-dose conditions but also to best preserve the helical symmetry of HIV-1 tubular assemblies.

1. Under low-dose search mode with a dose <0.01 e/Å^2, screen the whole grid at a magnification of ~200 and note the areas with suitable ice in a position file.
2. Further screen these areas at a magnification of 3,900 and select the best sample area for data collection. The best sample area should have a reasonable number of tubes, not too sparse or dense. The ice should be thin. Tubes should be straight and not squashed. A few high-resolution CCD images can be collected to confirm the tube quality. Save all the best areas in a different position file.
3. Switch to exposure mode, insert a 100 μm objective aperture, and adjust objective stigmatism.
4. Identify a good tube at 3,900 magnification and position the tube at the center of the camera's field of view. Switch to the focus mode and adjust the Z-height until the image is completely in focus, meaning no contrast. Switch to the exposure mode, at a nominal magnification of 59,000, set a defocus, normally between 0.5 and 2.5 μm, and an exposure time of 0.3–0.5 s with a dose of 10–15 e/Å^2, and collect an image (Fig. 3). When images are collected on a plate camera, the film should be allowed to settle 10 s before an exposure is taken.
5. Move to the next tube and repeat the steps to collect an image.

3.5. Image Processing and 3D Reconstruction

3.5.1. Digitization

1. We digitize the films using a Nikon super coolscan 9000 ED scanner connected to a Mac computer and interfaced with Adobe Photoshop.
2. Place each film onto a clean film holder with emulsion side down, such that it is flush with the stops, and insert the film holder slowly into the holder slot with the arrow facing up and toward the scanner.
3. Launch Nikon scanning software and select settings: Neg (Mono) in the film-type menu, 6×9 in the format size menu, and Grayscale in the color space menu. In the scanner extras palette, choose 16 bit for data size, 215 for manual focusing, and superfine for the CCD scan mode (6.35 μm/pixel).
4. After adjusting the contrast and density range and choosing the area to be scanned in the preview window, click the scan button to start the high-resolution scan.
5. Save the scanned image in a "TIFF Uncompressed" format.

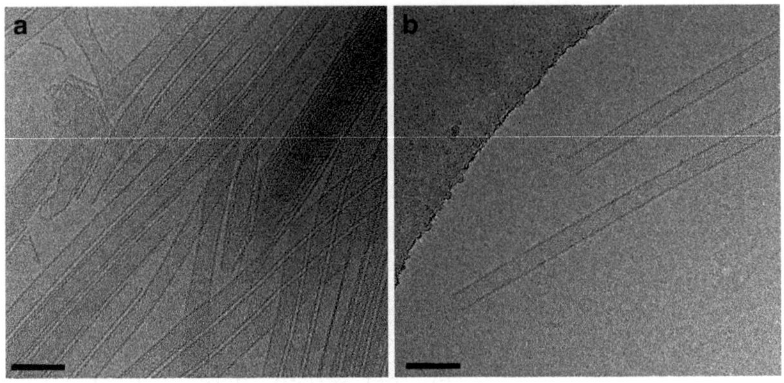

Fig. 3. CryoEM micrographs of recombinant HIV-1 CA tubular assemblies. (**a**) CA wt tubes. (**b**) CA A92E tubes. Scale bars, 100 nm. (Reproduced from ref. 20 with permission from Elsevier Science).

3.5.2. Helical Indexing

Helical Parameters

A helical array may be defined with respect to three parameters, the radius (r), the pitch (P), and the axial distance between neighboring units (p). As illustrated in Fig. 4a, the radius of a helix is simply the distance from the helix axis to all points on the helix. The pitch is the distance of one complete helix turn in the helix axial direction.

Diffraction patterns of a helical array (Fig. 4b) are composed of a set of layer lines, which follow the "helical selection rule", $l = tn + um$. Here l and n are layer line indices. m and n are integers, and t and u are constants of the helix. For any given helix, there may be exactly u units in exactly (or very closely) t complete turns. The helical repeat distance (c) is defined as $c = Pt = pu$.

Indexing Helical Parameters in a Fourier Transform

For helical indexing, we have borrowed the notation from two-dimensional (2D) lattices, since the surface of a helical object is basically a 2D lattice. Thus, the layer line indices l and n can be associated with the 2D lattice vectors, noted as $n_{(1,0)}$ and $l_{(1,0)}$ and $n_{(0,1)}$ and $l_{(0,1)}$ (Fig. 4c, d). Since the intensity of the nth layer line is determined by the Bessel function, $J_n(2\pi Rr)$, the first maximum, R_{max}, measured from the meridian can be used to calculate n, as approximated by the following Eq. 1 (30), in which,

$$2\pi R_{max} r \approx J_n \approx 1.1|n| + 0.9 \quad (1)$$

1. Using the program "helixboxer" in EMAN, frame a straight and long tube with uniform diameter and save the boxed area in MRC format.
2. Determine the helical repeat distance, c, using a cross-correlation based program, "imgccf", in the MRC package.
3. Calculate the Fourier transform with a new box length that is an integral of the repeat distance c (see Note 4).

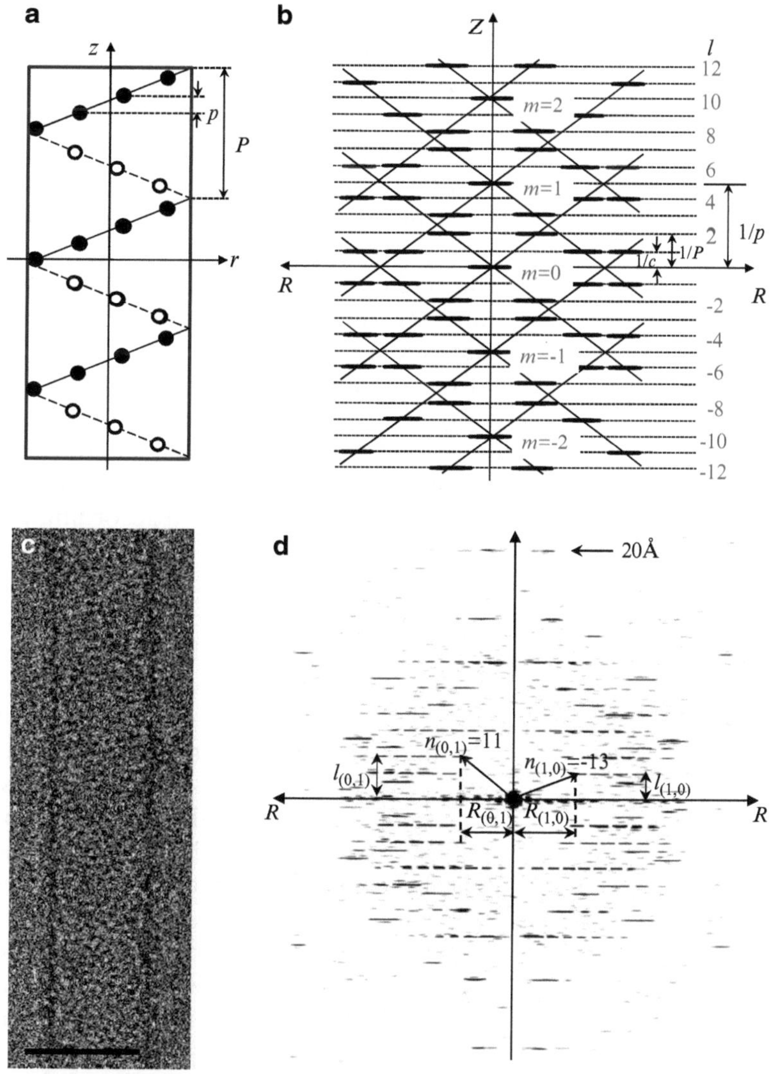

Fig. 4. Indexing of HIV-1 CA helical tubes. (**a**) The geometry of a discontinuous helical structure. A motif (*circle*) is repeated with a vertical spacing *p* along a helix of pitch *P*. (**b**) The diffraction pattern of the helical structure shown in (**a**). Sets of cross-like intensities repeated with a vertical spacing of 1/*p*. There are two principal spacings: 1/*P* within any single *X* pattern and 1/*p* between the centers of separate *X* patterns. All layer lines are separated by 1/*c*. (**c**) An HIV-1 CA A92E tube image. Scale bars, 40nm. (**d**) The Fourier transform of (**c**) with helical indices. The *arrow* at the top points to the layer lane at 20 Å resolution.

4. Measure the radius of the tube, *r*, and the height and radius of the two principal layer lines in Fourier transform, $l_{(1,0)}$, $R_{\max(1,0)}$, $l_{(0,1)}$, $R_{\max(0,1)}$, respectively.
5. Calculate $n_{(1,0)}$ and $n_{(0,1)}$ according to Eq. 1 (see Note 5). The handedness of the helix needs to be determined separately (see Note 6). A right-handed helix assumes a positive *n* and a left-handed helix has a negative *n*. For instance, the estimated helical symmetry for the tube in Fig. 4c is $n_{(1,0)} = -12$ and $n_{(0,1)} = 10$.

6. Test some pairs of n numbers to find the correct helical symmetry using Eq. 1 and the programs for 3D reconstruction, IHRSR (Subheading "Segmentation"). For example, we tested $n_{(1,0)} = -12$ and $n_{(0,1)} = 10$, $n_{(1,0)} = -11$ and $n_{(0,1)} = 10$ and $n_{(1,0)} = -13$ and $n_{(0,1)} = 11$ and found that only $n_{(1,0)} = -13$ and $n_{(0,1)} = 11$ gave us a stable converged reconstruction. Thus, the correct helical symmetry for the tube in Fig. 4c is $n_{(1,0)} = -13$ and $n_{(0,1)} = 11$.

3.5.3. Determining the Screw Symmetry

The screw symmetry of a helical object is described by two real numbers: the rotation between subunits ($\Delta\varphi$) and the axial rise (Δz) as well as a possible point group symmetry, which would be an n-fold rotation about helix axis (22). If the helical object can be completely described by a rotation and axial rise of an asymmetric unit along a 1-start helix ($n = 1$), the point group symmetry does not need to be set. Since the helical arrays obey the selection rule, Δz and $\Delta\varphi$ can be calculated according to the selection rule.

In practice, we use a simple script to calculate the screw symmetry by using the $l_{(1,0)}$, $l_{(0,1)}$, $n_{(1,0)}$, and $n_{(0,1)}$ values and by testing a range of m values. For instance, given $l_{(1,0)} = 27$, $l_{(0,1)} = 45$, $n_{(1,0)} = -13$, $n_{(0,1)} = 11$, $-50 < m < 50$, and a repeat distance of 5,721.15 Å, we have $\Delta z = 6.4854$ Å, $\Delta\varphi = 165.306°$.

3.5.4. Reconstruction Using IHRSR Programs

The iterative real-space helical reconstruction (IHRSR) method (22) treats helical objects as a string of single particles to resolve their three-dimensional (3D) structure using cryo-EM. It is especially useful for partially disordered helical objects when Fourier-Bessel helical reconstruction falls short. The IHRSR algorithm involves determining the helical screw symmetry, the rotation between subunits ($\Delta\varphi$) and axial rise (Δz), that best fits the reconstructed volume at each cycle, as well as a point group symmetry. The procedure is not sensitive to the choice of an initial reference volume, and even a solid cylinder can be used quite effectively as an initial reference as long as a reasonable starting symmetry is used.

Segmentation

One important parameter for segmentation is the size of the segments, which can only be determined empirically. Segments should be extensively overlapped (~90%) when cutting these from the helical images. Thus, 10% of the box size along a helical axis is typically used as a shift for each box.

1. Use the boxer program in the EMAN package to open a graphical user interface and to select helical objects from micrographs. In the control panel of boxer, choose "Helix" and set the parameters for boxing: the size of the box should be larger than the diameter of tube and a value for "Olap" should be ~90% of box size.
2. After selecting two boxes on either end of the helical object, click on the image to automatically generate a series of particle boxes along the tube length.
3. Save boxed particles as well as their coordinates.

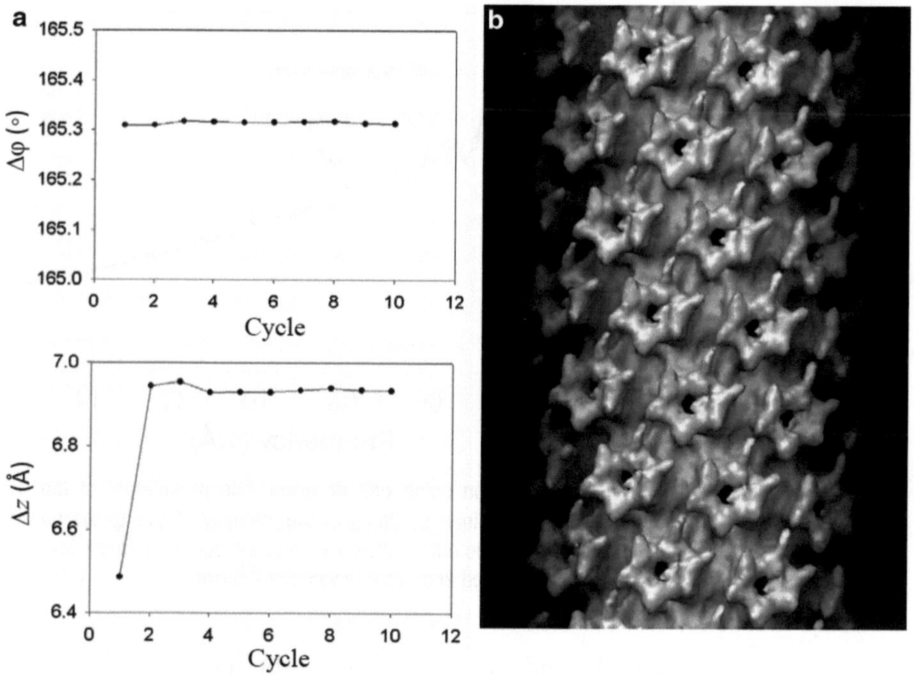

Fig. 5. Initial reconstruction using IHRSR. (**a**) Helical symmetry determination for each iterative cycle. $\Delta\varphi$ and Δz, starting from the initial values, converge to the stable values after ten iterative refinement cycles. (**b**) The initial map using IHRSR.

Generating the First Reconstruction

1. Invert the contrast of the cryo-EM images and apply low pass filtering (optional) prior to processing with the iterative helical real space reconstruction method (IHRSR).
2. Use the program "Generator" to open the graphical interface, and provide it with the name and path of the raw images that will be used by IHRSR to generate the reconstruction script, b25.spi. Use a solid cylinder as an initial reference and allow the procedure to cycle until there are no changes in the helical symmetry. The helical symmetry usually converges after a few cycles (Fig. 5a).
3. When $\Delta\varphi$ and Δz values are stabilized after several iterations, an initial 3D reconstruction is generated in the last cycle (Fig. 5b).

3.5.5. CTF Correction

The images generated by cryo-electron microscopes are not true projections of the specimen. They suffer from a set of artifacts, including the contrast transfer function (CTF) and the envelope function of the microscope. In addition, noise is present from a variety of sources. The CTF in particular can cause serious artifacts in a 3D reconstruction. The severity and type of these effects, of

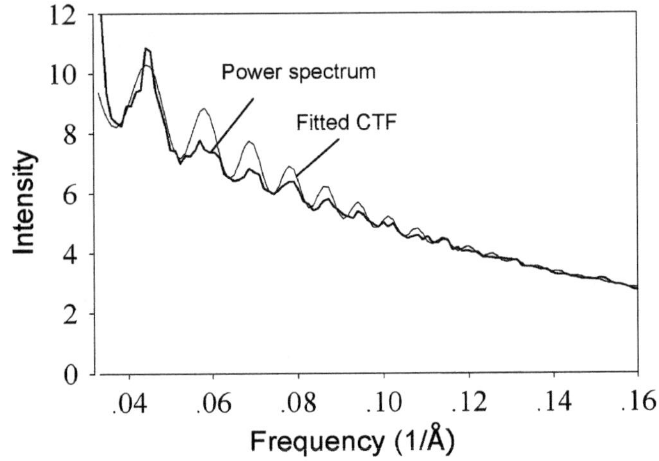

Fig. 6. CTF determination using ctfit program. The parameters of the CTF in each micrograph were determined by fitting of the intensity of CTF to the power spectrum calculated from segmented tubes. Shown are a power spectrum (*thick line*) and fitted CTF (*thin line*) of tubes recorded at a defocus value of 2.5 μm.

course, depends on many factors, the most significant of which is the defocus of the data used in the reconstruction. Therefore, defocus determination and CTF correction is crucial for reconstructing a reliable 3D density map.

1. Use the segments from one micrograph to calculate the power spectrum and determine an initial defocus value using ctfit in the EMAN package (Fig. 6).
2. Further refine the defocus values using programs CTFFIND3 and CTFTILT and determine an average defocus value.
3. The CTF correction involves the use of a Wiener filter (see Note 7). This operation corrects both phases and amplitudes. With a Wiener filter, the Fourier transform of an image, F_i, is first multiplied by the CTF, which serves as phase reversals, then divided by the sum of the squares of the CTF plus a noise-to-signal term, shown in Eq. 2. The f term effectively serves as a trap to avoid boosting noises.

$$S_{\mathrm{WF}} = \frac{F_1 * \mathrm{CTF}_1 + F_2 * \mathrm{CTF}_2 + \ldots + F_i * \mathrm{CTF}_i}{\sum_{i=1}^{n} \mathrm{CTF}_i^2 + f} \quad (2)$$

3.5.6. Reconstruction with Iterative Refinement

1. Refinement cycles need to be carried out to determine the relative orientation of each segment, namely Euler angles. This can be done using a multi-reference alignment procedure, as illustrated in the flow chart (Fig. 7). We use the image processing package SPIDER (23) for many of the steps of refinement cycles. A series of operations can be controlled with

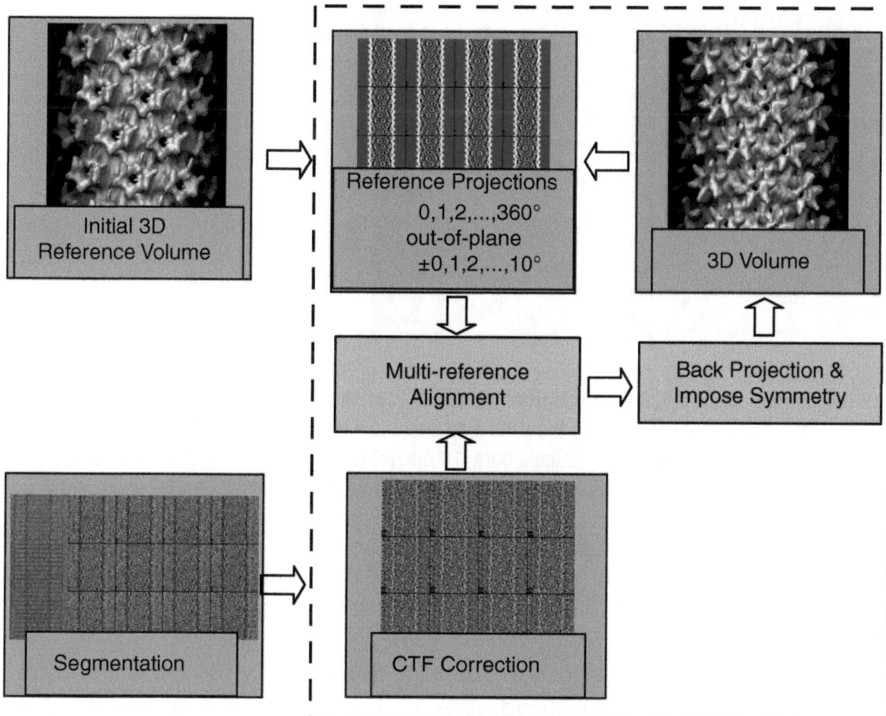

Fig. 7. Flow chart for the refinement of the 3D map. The CTF-corrected segments are aligned against the reference projections and back-projected to generate a 3D volume. The helical symmetry was then imposed onto the 3D volume, which is then used as the new reference for the next refinement cycle. The cycle (*dashed box*) is iterated until a stable reconstruction is obtained.

SPIDER scripts (9), which are user-created batch control files containing sequences of operations and parameter values.

2. The 3D map generated by the IHRSR programs (Subheading "Generating the First Reconstruction") can be used as an initial reference. During refinement, the helical symmetry is fixed with the values of $\Delta\varphi$ and Δz, determined from IHRSR (Subheading "Generating the First Reconstruction").
3. Prior to 3D reconstruction, CTF phase and amplitude correction are applied to the segments with Eq. 2. Projection matching is then performed by comparing projections of the reference volume with the CTF-corrected images using multi-reference alignment. Projections of the reference are generated in 1° increments around the helical axis. The variation of the out-of-plane tilt angle is limited and sampled in 1° steps. Constraints are imposed to the alignment parameters of each segment. Only those segments that satisfy constraints such as high correlation coefficients, near 0° or 180° in-plane angles, and limited *x*-shifts, are kept.
4. After each iterative refinement cycle, a 3D density map is generated by back projection followed by imposing the helical symmetry. The iterative refinement is stopped when the

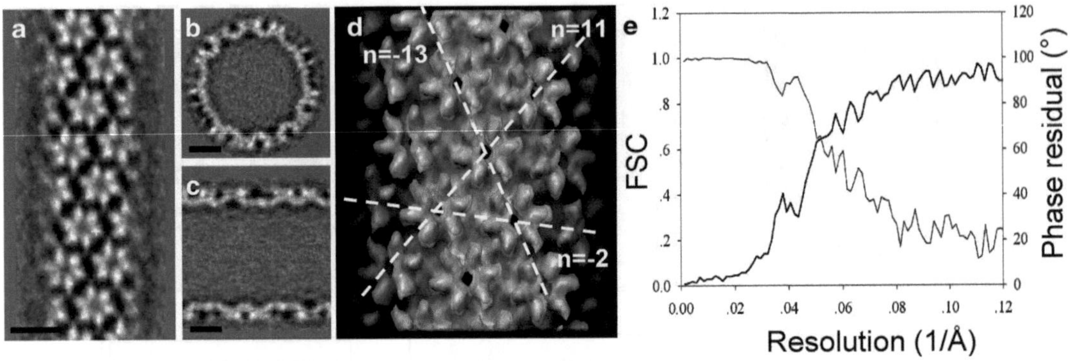

Fig. 8. The 3D density map of the A92E CA tubes from the (−13, 11) helical family. (**a**–**c**) The density map of CA tubes is displayed as three orthogonal slices: parallel to the tube axis and close to the surface (**a**), perpendicular to the tube axis (**b**), and parallel to and through the tube axis (**c**). Scale bars, 10 nm. (**d**) Surface rendering of the 3D density map contoured at 2.3σ (*solid*) and 1.3σ (*transparent*) enclosing 65% and 100% volume, respectively. *Dashed lines* connect hexamers in the three distinct helical arrangements denoted as $n = -2$, 11, and −13 helices; n is the Bessel order and the sign indicates the handedness of the helix (− sign as left hand, and + sign as right hand). (**e**) Fourier shell correlation and phase residual plots of the 3D density map reconstructed from 197 tubular segments. The resolution of the map is 16 Å at FSC = 0.5, or at phase residue of 65°. (Reproduced from ref. 20 with permission from Elsevier Science).

resolution of the new 3D map doesn't have any improvement (see Subheading 3.5.7 for resolution estimation). The final 3D map is shown in Fig. 8a–d.

3.5.7. Resolution Estimation

The resolution measures used in single particle reconstruction are based on comparisons of averages calculated for subsets of the data. These measures are the differential phase residual and the Fourier ring correlation (FRC). "RF 3" in SPIDER can be used to calculate the resolution of our 3D density map. The resolution of the map is 16 Å at FSC = 0.5, or at a differential phase residual of 65° (Fig. 8e).

3.6. Molecular Docking

We reconstructed pseudoatomic models of CA hexamers by docking the following models (PDB codes 1gwp for CA-NTD and 2kod for CA-CTD) into the tubular EM density map using an automated map-fitting feature implemented in Chimera (Fig. 9). A new CTD–CTD trimer interface was identified in the reconstructed pseudoatomic model (Fig. 10).

1. Open pdb models and a 3D density map with Chimera (26).
2. Align the model and the map manually. Move and rotate the model so it is similarly oriented and superimposed on the map.
3. In the Chimera Window, click Tools > Volume Data > Color Zone, select surface to be colored and coloring radius. Then click "Color".
4. Optimize the fit of the model in the map using the tool, "Fit in Map".
5. The docking was further refined using the program "colacor" in Situs package (25). Two parameters can be adjusted to give

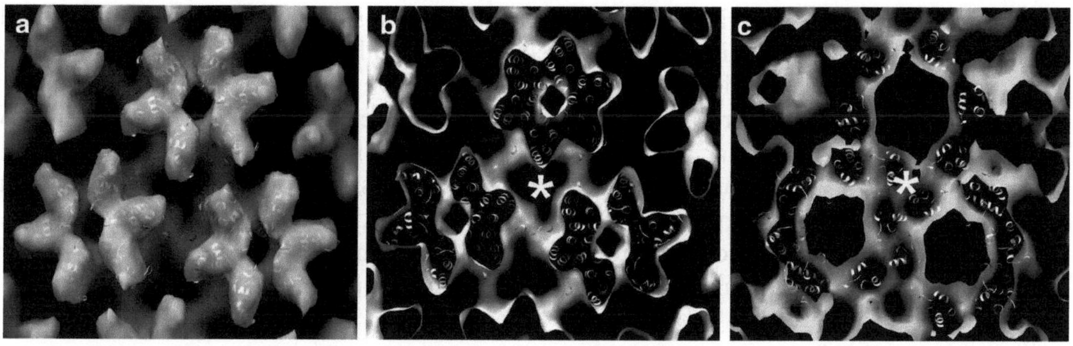

Fig. 9. Molecular Docking of NTD and CTD domain models independently into the tubular density map (contour enclosing 90% volume). (**a**) An overlay of docked pseudo-atomic model and 3D density map viewed from the tube surface. (**b**, **c**) Slab views to show the model fitting at NTD region (**b**) and the CTD region (**c**). Three docked CA-hexamers are displayed. The local threefold axis is marked by *asterisk*. (Reproduced from ref. 20 with permission from Elsevier Science).

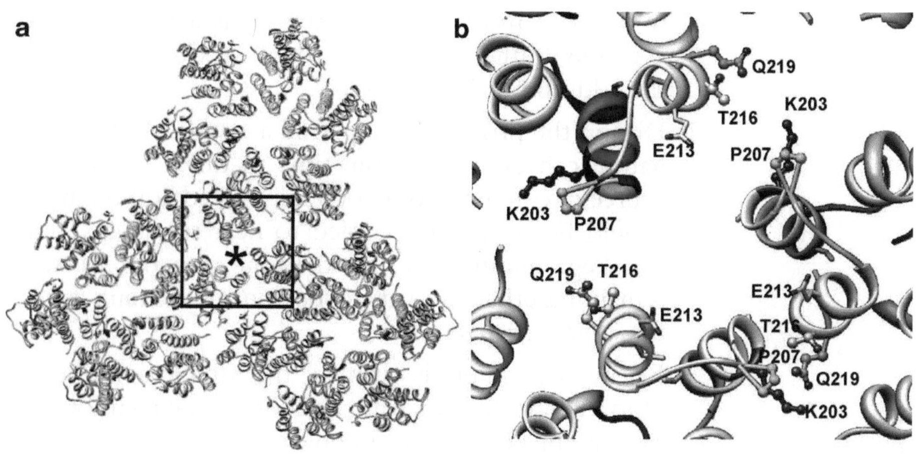

Fig. 10. Novel CTD–CTD interactions at the local threefold axis in CA tubular assemblies. (**a**) A pseudo-atomic model of three CA-hexamers in the assembled tube. *Boxed region* encloses the new CTD–CTD interface at the local threefold axis (indicated by *asterisk*). (**b**) A detailed view of the boxed region illustrating the interactions at the interface. Side chains of K203, P207, E213, T216, and Q219 are shown in ball and stick representation. Mutations of these residues are known to affect in vitro assembly and capsid stability. A pair of spatially close residues, P207/T216, were tested by cysteine cross-linking (20). (Reproduced from ref. 20 with permission from Elsevier Science).

the highest correlation. These parameters are density cutoff and Laplacian filter (on or off).

3.7. Conclusion

We used cryo-EM and iterative helical real-space reconstruction methods described above and obtained a 3D density map of helically assembled HIV-1 CA hexamers at 16 Å resolution. Docking of the atomic models of CA domains into the 3D density map allowed us to construct a pseudo-atomic model of assembled HIV-1 CA. Our cryo-EM structure model of assembled CA revealed a novel CTD–CTD interface at the pseudo-threefold axis that has not been characterized previously. This new CTD trimer interface plays a critical role in capsid assembly, disassembly, and stability, as evidenced by the available mutational data and by additional novel mutations that we have generated based on our structural data. We further validated the trimer interfaces in CA tubular assemblies and in HIV-1 virions by engineering intermolecular cross-linking of cysteine residues, P207C and T216C, at the trimer interface. The methods demonstrated here are applicable not only to HIV-1 capsid assemblies but also to many other helical objects for structure determination at medium resolution.

4. Notes

1. For the CA assembly assay, the sequence of mixing reaction components has an impact on the final number of assembled tubes. The most efficient assembly is achieved by adding 2× assembly buffer to the CA protein stock solution followed by the addition of Milli-Q water.
2. During the process of preparing a cryo-EM grid, ice contamination can be visible from time to time. The most common source of contamination is contaminated liquid ethane. Ethane should be completely cooled to near solidifying temperature and kept minimally exposed to air to avoid frost. Also, the remaining ethane gas at the bottom of the tank is not suitable for sample preparation.
3. Before the sample is vitrified, the sample solution should be blotted away and left with a uniform, thin solution layer that is slightly thicker than the sample size. There are several different ways to blot the sample, depending on the freezing apparatus. FEI mark IV vitrobot and Gatan cryo-plunger both use two-side blotting, whereas the Leica EM GP and most home-made plunge freezing devices deploy single-side blotting. Two-side blotting produces an ice layer with a gradient of thickness, resulting in a band of suitable ice. However, the sample itself sometimes dictates the blotting method. For bacterial cells and tubular crystals such as HIV-1 CA assembly, one-side blotting, particularly from the back-side, is more appropriate, since the contact with the filter paper can result in sample loss or damage. One-side blotting is also useful for rapid buffer exchange before freezing, as many complexes are stable under buffer conditions that are not suitable for cryo-EM. For these samples, a drop (3–5 μl) of dilution buffer can be added to the back side of the grid, and, then, a filter paper can be used to quickly blot away excess solution from the back side. Due to the quick exchange between sample solution and dilution buffer, structural effects on the sample can be kept to a minimum.

4. When box length is an integral of the repeat distance (c), the layer line signals fall on the exact pixels instead of between pixels, thus producing sharp layer lines.
5. As with any two-dimensional lattice, all the lattice vectors are linear combinations of two basic vectors $a_{(1,0)}$ and $b_{(0,1)}$ with $c_{(h,k)} = h^* a_{(1,0)} + k^* b_{(0,1)}$. We may express $n_{(h,k)}$ of any other layer lines in terms of $n_{(1,0)}$ and $n_{(0,1)}$: $n_{(h,k)} = h^* n_{(1,0)} + k^* n_{(0,1)}$. R_{max}, combined with an estimate of the radius r of a helix, gives the correct n according to Eq. 1. In practice, however, R_{max} value may be smaller due to out-plane tilt (31) or the effects of thickness (32), leading to an underestimation of r. Typically, there will be a range of r values estimated from each set of helices, so the best estimate of n will be from a range of possible values.
6. The handedness of a helix can be determined prior to image reconstruction using freeze-etching, followed by rotary shadowing method (33). It can also be verified post-reconstruction, as 3D atomic models of individual components should fit well into the density map when a correct handedness is assumed. Otherwise, the opposite handedness should be chosen.
7. Wiener filtering is used in image processing for both phase and amplitude correction in such a way as to reduce the level of noise amplification. It is more powerful when multiple projection sets are available, each with a different value of the contrast transfer function. In this case, inverse filtering can be used to recover some of the frequencies for which at least one of the CTF values is not zero.

Acknowledgments

The authors would like to thank Dr. Jinwoo Ahn and Danxia Ke for technical support, and Dr. Teresa Brosenitsch for critical reading of the manuscript. This work was supported by GM082251 and GM085043.

References

1. Wetzel R (2002) Ideas of order for amyloid fibril structure. Structure 10:1031–1036
2. True HL, Lindquist SL (2000) A yeast prion provides a mechanism for genetic variation and phenotypic diversity. Nature 407: 477–483
3. Jimenez JL, Guijarro JI, Orlova E, Zurdo J, Dobson CM, Sunde M, Saibil HR (1999) Cryo-electron microscopy structure of an SH3 amyloid fibril and model of the molecular packing. EMBO J 18:815–821
4. Sachse C, Fandrich M, Grigorieff N (2008) Paired beta-sheet structure of an Abeta(1–40) amyloid fibril revealed by electron microscopy. Proc Natl Acad Sci USA 105:7462–7466
5. Miyazawa A, Fujiyoshi Y, Unwin N (2003) Structure and gating mechanism of the acetylcholine receptor pore. Nature 423:949–955

6. Zhang P, Toyoshima C, Yonekura K, Green NM, Stokes DL (1998) Structure of the calcium pump from sarcoplasmic reticulum at 8-A resolution. Nature 392:835–839
7. Li S, Hill CP, Sundquist WI, Finch JT (2000) Image reconstructions of helical assemblies of the HIV-1 CA protein. Nature 407:409–413
8. Maki-Yonekura S, Yonekura K, Namba K (2010) Conformational change of flagellin for polymorphic supercoiling of the flagellar filament. Nat Struct Mol Biol 17:417–422
9. Sachse C, Chen JZ, Coureux PD, Stroupe ME, Fandrich M, Grigorieff N (2007) High-resolution electron microscopy of helical specimens: a fresh look at tobacco mosaic virus. J Mol Biol 371:812–835
10. Ganser BK, Li S, Klishko VY, Finch JT, Sundquist WI (1999) Assembly and analysis of conical models for the HIV-1 core. Science 283:80–83
11. Briggs JA, Wilk T, Welker R, Krausslich HG, Fuller SD (2003) Structural organization of authentic, mature HIV-1 virions and cores. EMBO J 22:1707–1715
12. Gamble TR, Vajdos FF, Yoo S, Worthylake DK, Houseweart M, Sundquist WI, Hill CP (1996) Crystal structure of human cyclophilin A bound to the amino-terminal domain of HIV-1 capsid. Cell 87:1285–1294
13. Gamble TR, Yoo S, Vajdos FF, von Schwedler UK, Worthylake DK, Wang H, McCutcheon JP, Sundquist WI, Hill CP (1997) Structure of the carboxyl-terminal dimerization domain of the HIV-1 capsid protein. Science 278:849–853
14. Kelly BN, Howard BR, Wang H, Robinson H, Sundquist WI, Hill CP (2006) Implications for viral capsid assembly from crystal structures of HIV-1 Gag(1–278) and CA(N)(133–278). Biochemistry 45:11257–11266
15. Momany C, Kovari LC, Prongay AJ, Keller W, Gitti RK, Lee BM, Gorbalenya AE, Tong L, McClure J, Ehrlich LS, Summers MF, Carter C, Rossmann MG (1996) Crystal structure of dimeric HIV-1 capsid protein. Nat Struct Biol 3:763–770
16. Worthylake DK, Wang H, Yoo S, Sundquist WI, Hill CP (1999) Structures of the HIV-1 capsid protein dimerization domain at 2.6 A resolution. Acta Crystallogr D Biol Crystallogr 55:85–92
17. Ternois F, Sticht J, Duquerroy S, Krausslich HG, Rey FA (2005) The HIV-1 capsid protein C-terminal domain in complex with a virus assembly inhibitor. Nat Struct Mol Biol 12:678–682
18. Ivanov D, Tsodikov OV, Kasanov J, Ellenberger T, Wagner G, Collins T (2007) Domain-swapped dimerization of the HIV-1 capsid C-terminal domain. Proc Natl Acad Sci USA 104:4353–4358
19. Pornillos O, Ganser-Pornillos BK, Kelly BN, Hua Y, Whitby FG, Stout CD, Sundquist WI, Hill CP, Yeager M (2009) X-ray structures of the hexameric building block of the HIV capsid. Cell 137:1282–1292
20. Byeon IJ, Meng X, Jung J, Zhao G, Yang R, Ahn J, Shi J, Concel J, Aiken C, Zhang P, Gronenborn AM (2009) Structural convergence between Cryo-EM and NMR reveals intersubunit interactions critical for HIV-1 capsid function. Cell 139:780–790
21. Ludtke SJ, Baldwin PR, Chiu W (1999) EMAN: semiautomated software for high-resolution single-particle reconstructions. J Struct Biol 128:82–97
22. Egelman EH (2007) The iterative helical real space reconstruction method: surmounting the problems posed by real polymers. J Struct Biol 157:83–94
23. Frank J, Radermacher M, Penczek P, Zhu J, Li Y, Ladjadj M, Leith A (1996) SPIDER and WEB: processing and visualization of images in 3D electron microscopy and related fields. J Struct Biol 116:190–199
24. Yonekura K, Toyoshima C (2007) Structure determination of tubular crystals of membrane proteins. IV. Distortion correction and its combined application with real-space averaging and solvent flattening. Ultramicroscopy 107:1141–1158
25. Wriggers W, Milligan RA, McCammon JA (1999) Situs: a package for docking crystal structures into low-resolution maps from electron microscopy. J Struct Biol 125:185–195
26. Pettersen EF, Goddard TD, Huang CC, Couch GS, Greenblatt DM, Meng EC, Ferrin TE (2004) UCSF Chimera – a visualization system for exploratory research and analysis. J Comput Chem 25:1605–1612
27. Erickson-Viitanen S, Manfredi J, Viitanen P, Tribe DE, Tritch R, Hutchison CA 3rd, Loeb DD, Swanstrom R (1989) Cleavage of HIV-1 gag polyprotein synthesized in vitro: sequential cleavage by the viral protease. AIDS Res Hum Retroviruses 5:577–591
28. Taylor KA, Glaeser RM (1974) Electron diffraction of frozen, hydrated protein crystals. Science 186:1036–1037
29. Conway JF, Cheng N, Ross PD, Hendrix RW, Duda RL, Steven AC (2007) A thermally induced phase transition in a viral capsid transforms the hexamers, leaving the pentamers unchanged. J Struct Biol 158:224–232
30. Toyoshima C, Unwin N (1990) Three-dimensional structure of the acetylcholine receptor by cryoelectron microscopy and

helical image reconstruction. J Cell Biol 111: 2623–2635

31. Toyoshima C (2000) Structure determination of tubular crystals of membrane proteins. I. Indexing of diffraction patterns. Ultramicroscopy 84:1–14
32. Moody MF (1967) Structure of the sheath of bacteriophage T4. I. Structure of the contracted sheath and polysheath. J Mol Biol 25:167–200
33. Zhang P, Hinshaw JE (2001) Three-dimensional reconstruction of dynamin in the constricted state. Nat Cell Biol 3:922–926

Chapter 22

Single Particle Electron Microscopy

Wilson C.Y. Lau and John L. Rubinstein

Abstract

Single particle electron microscopy is a versatile technique for the structural analysis of protein complexes in near-native conditions. While tremendous progress has been made during the past few decades in techniques for specimen preparation, imaging, and image analysis, the field is still in development. In the context of this volume on electron crystallography, the following chapter gives practical guidelines on how to begin single particle EM studies, including preparing specimens, selecting imaging conditions, and choosing which of the many approaches to image analysis are appropriate for a specific sample.

Key words: Single particle cryo-EM, Specimen preparation, Image acquisition, Image analysis, Three-dimensional, Two-dimensional

1. Introduction

Single particle electron microscopy (EM) allows for the study of macromolecular assemblies in near-native conditions. Macromolecular assemblies, such as protein complexes, are often too large, scarce, or heterogeneous to be studied by X-ray crystallography or nuclear magnetic resonance (NMR) spectroscopy. However, these attributes do not prevent the use of EM for their analysis. In fact, large protein complexes provide a distinct advantage for structural studies by single particle electron cryomicroscopy (cryo-EM) because the signal in images is proportional to the mass of the complex studied. Recent improvements in cryo-EM instrumentation, map refinement algorithms, and dramatic increases in available computational power have enabled structures of symmetric complexes to be determined to near-atomic resolution, allowing protein backbones to be traced (1, 2) and even the construction of an

Ingeborg Schmidt-Krey and Yifan Cheng (eds.), *Electron Crystallography of Soluble and Membrane Proteins: Methods and Protocols*, Methods in Molecular Biology, vol. 955, DOI 10.1007/978-1-62703-176-9_22,

all atom model from a 3.3 Å resolution map (3). For asymmetric particles, such as ribosomes, recent maps have allowed for visualization of secondary structure elements (4).

In cryo-EM of single particles, as with many electron crystallography experiments, the specimen of interest is embedded in vitreous ice and is maintained near liquid nitrogen temperature (77 K) or colder and imaged with an electron microscope. Limited exposures of electrons must be used when acquiring images to minimize radiation damage. These low exposures result in images that have low signal-to-noise ratios (SNRs). The SNR at low resolution must be sufficient for the orientation of the particle to be determined. Once particle orientations are known, images can be coherently averaged to retrieve the higher-resolution signal obscured by noise in an individual image, and different views can be combined to generate a three-dimensional (3D) map. The amount of averaging required to reveal high-resolution features in a map depends on image SNR at high resolution. Electron microscopes do not transmit signals at all spatial frequencies equally. Radiation-sensitive specimens are susceptible to beam-induced motion during imaging (5, 6). Movement of the specimen during imaging causes blurring that reduces the contrast, and thus the SNR, of high-resolution features. Movement can even suppress the signal from high-resolution features completely. Beam-induced motion appears to be of varying direction and magnitude, reducing contrast far below what is expected from theory (5, 6) and requiring many independent images to be averaged to increase the SNR of single particle EM maps at high resolution. The theoretical lower size limit for cryo-EM of single particles when using ideal images is about 100 kDa for protein, but this limit increases quickly as image quality decreases (7). Currently, due to a variety of mechanisms of contrast loss (5), single particle EM usually becomes practical for protein particles of ~200 kDa and larger when imaged in negative stain, and even larger, when imaged in amorphous ice. The size of biological specimens studied to date ranges between ~50 kDa for an RNA folding intermediate (8) and ~50 MDa for an icosahedral virus (9). Trends in the literature suggest that it is easier to build high-resolution maps for large particles because the increased signal from these complexes facilitates more accurate determination of particle orientations.

In its most simple application, single particle EM assumes that the specimen studied exists as many identical copies of the same structure. In reality, assemblies often have intrinsic conformational flexibility and compositional heterogeneity. Computational analysis of this heterogeneity offers the possibility of gaining important insight into biological structures (10, 11). Examples have demonstrated that cryo-EM can be used to study dynamic protein complexes and heterogeneous samples, such as the 70 S ribosome–EF-G complex (11), ribosomes with tRNA in a hybrid state (12), GroEL–ES-substrate complexes (13), and the transcription factor TFIID (14). As these studies illustrate, the scope of biological systems that can be studied by single particle EM is diverse, and many different tools have been developed to address different types of questions. The objective of this chapter is to introduce the basic methodology of single particle cryo-EM and provide some practical guidelines for carrying out each of the essential steps.

2. Methods for Preparing Specimens

2.1. Background

One of the major advantages of single particle cryo-EM when compared to other structural analysis techniques is that it requires only small quantities of sample. The protein concentrations used for experiments are sometimes as little as a few tens or hundreds of μg/ml, as opposed to the mg/ml concentrations often needed for X-ray crystallography and NMR spectroscopy. These low concentrations, in combination with the small volumes needed for each specimen grid (~2–4 μl), mean that the total mass of protein used for single particle cryo-EM can be orders of magnitude less than that required for structural analysis with other methods.

While it is always desirable to work with the most pure samples possible, purity is not an absolute requirement for single particle EM. Instead, it may be possible to select the complex of interest from images while ignoring contaminants, in a sort of in silico purification of the protein. However, when working with less pure samples, the number of particles that can be selected from each field of view will be reduced. Unless explicitly addressed during image analysis, structural homogeneity is extremely important for single particle analysis because, assuming that a reliable map of a heterogeneous complex can be built at all, the final map will be an average of all of the structures imaged during the experiment. Therefore, all possible efforts should be made to ensure the structural homogeneity of the specimen, such as inhibiting enzymes to reduce their conformational heterogeneity. For specimens where conventional biochemical steps cannot ensure homogeneity, it may even be advantageous to use mild chemical cross-linking (15).

Samples for single particle cryo-EM studies are often screened by negative stain EM. Some electron microscopists prefer to perform preliminary image analysis with stain-embedded samples because they provide greater contrast than frozen-hydrated samples. One area of particular utility for negative stain EM is the study of complexes that are either too small (16) or too flexible (17) for study in ice. There are a variety of negative staining protocols currently in use, each with advantages and disadvantages. For example, the carbon sandwich method (18), originally developed for ribosome particles by the Joachim Frank group, is effective at preventing partial staining, while cryo-negative staining appears to reduce specimen flattening and anisotropic orientation of particles on the grid (19). In our research group, we usually use a simple protocol for negative staining (20), which we describe below in Subheading 2.1.1. More sophisticated protocols can be found elsewhere (18). Because negative stain EM produces an image of the stain envelope around the molecule of interest, the technique has a greater capacity to generate image artifacts than cryo-EM of frozen-hydrated samples. Therefore, many research groups tend to screen samples by negative stain EM but start the process of 3D map building directly from images of ice-embedded specimens.

For cryo-EM of single particles, the buffer conditions of the solution to be vitrified require consideration. Organic molecules such as glycerol and sucrose can decrease image contrast both by decreasing the density difference between protein and ice and by causing bubbling of the specimen upon irradiation. Any buffer reagents that have higher densities than vitrified water (~0.95 g/ml) (21) should be kept to low concentrations. Perforated carbon-coated grids are often used for single particle cryo-EM, although some groups prepare specimens on a thin carbon film supported by a more robust perforated carbon film (22). Perforated carbon film-coated grids can either be made in house (23) or prefabricated grids, such as Quantifoil (http://www.quantifoil.com) or C-flat (http://www.protochips.com) grids (24), can be purchased. While homemade grids are significantly less expensive than prefabricated grids if the standard EM laboratory equipment needed to make them is available, prefabricated grids are convenient, conducive to automated data collection, and can aid in making specimen preparation reproducible. The final resolution of a 3D map depends strongly on the quality of images used to construct the map. Therefore, map resolution depends on preparing grids with ice that is neither too thin to accommodate the specimen without distorting it nor so thick that it reduces image contrast unnecessarily (25, 26). Since ice thickness can also vary between grid squares, it is important to locate "good" areas by visual inspection of the grid before images are acquired.

2.1.1. Specimen Preparation: Conventional Negative Staining

Continuous Carbon Film-Coated Grid preparation

Notes

- Floating carbon from mica can take some practice. If the carbon does not separate from the mica as the mica is slid under water, try allowing the carbon to age for 24 h prior to floating.
- Oil contamination of the carbon surface is a common problem with poorly maintained carbon evaporators. If the carbon film

collapses on itself after floating from the mica, try gently rinsing the carbon while still on the mica by putting the mica between two pieces of filter paper and wetting the filter paper with ethanol.

1. Carefully cleave a mica sheet (e.g., 2.5 × 7.5 cm, V5 grade mica is sufficient) into halves along the plane of the sheet in order to expose a clean, flat mica surface. Place the mica on a piece of filter paper in a Petri dish with the freshly cleaved surface facing up.
2. Using a carbon evaporator, deposit a thin layer of carbon onto the mica. The carbon layer should be kept relatively thin (~5 nm) and its thickness can be judged by monitoring the color of the filter paper adjacent to the mica, which should have a yellow/brown tone rather than the dark grey color that indicates thick carbon.
3. Fill a shallow container with deionized water. Place a flat metal mesh (e.g., 5 cm × 5 cm) onto a test tube stand and lay a 9 cm diameter disc of filter paper on the mesh. Water should be able to flow through the mesh as it is drained from the container.
4. Arrange EM grids on the filter paper in an area no bigger than the piece of mica coated with carbon. We prefer 400 mesh copper/rhodium grids for their rigidity. If you are using copper/rhodium grids, place the grids with the rhodium side up.
5. Remove any debris from the surface of the water by dragging a pipette across the surface. Gently add more water to the container so that the surface meniscus bulges from the container, maximizing the surface tension.
6. Float the thin, continuous carbon film from the mica by slowly pushing the mica, carbon side up, under the water at an angle of ~45°.
7. Without disturbing the floating carbon film, siphon water from the dish using a rubber or silicone tube (the siphon effect can be started with a syringe). As water drains from the dish, use forceps to position the carbon film so that it lowers directly onto the grids when the water level falls below the supporting mesh.
8. After the grids are covered with the carbon film, allow them to dry completely overnight, covered with the lid of a Petri dish to keep them dust free. Grids will become hydrophobic over time but can be used for approximately 1 month.

Staining

Note

- The buffer in which a protein is stored is not as critical for negative stain experiments as it is for cryo-EM. Buffer may be rinsed away and replaced by water after the protein particles have adsorbed to the carbon film.

Safety Note

- The depleted uranium used to make uranyl acetate is mildly radioactive and is toxic by ingestion, inhalation, or contact with abraded skin. Caution should be used when handling uranyl acetate and all waste should be disposed of appropriately.

1. Prepare a small volume (~500 μl) of fresh 2 % (w/v) uranyl acetate solution. Other stains, including uranyl formate, phosphotungstic acid, methylamine tungstate, and ammonium molybdate, can also be used.
2. Render grids hydrophilic by placing them carbon-side-up in a glow discharge device and subjecting them to plasma-discharge in a partially evacuated air environment (2×10^{-2} mbar) for ~10–15 s at 30 mA. We find this procedure necessary for most proteins to adsorb to the carbon surface.
3. Immediately after glow-discharge, secure a grid in a pair of self-closing anti-capillary forceps and pipette 2–4 μl of sample solution onto the carbon-coated side of the grid. Allow the sample to adsorb to the grid for 2 min. The typical concentration of sample needed is ~20 nM (0.01 mg/ml for a 500 kDa complex) but can be adjusted depending on the adsorption properties of the specimen. If the sample concentration is low, the incubation time can be extended to allow more adsorption to the grid.
4. Arrange three drops of Milli-Q water (50 μl) and one or two drops of the uranyl acetate solution (50 μl) on a clean piece of parafilm.
5. Rinse the sample solution from the grid by inverting the forceps and touching the carbon surface of the grid to the first drop of wash water for ~10 s. Blot the water from the edge of the grid with filter paper (Whatman #1) and repeat the rinsing step with the remaining two drops of water. Do not allow the grid to dry completely between rinsing steps.
6. To stain the grid, blot the wash water from the grid and touch the carbon surface of the grid to a drop of uranyl acetate solution. At this point the grid can be blotted and dried in air, or blotted and rinsed with a second drop of uranyl acetate. It is important not to blot the final drop of stain excessively. Before drying in air, the grid should have a slight sheen, still appearing somewhat wet.

2.1.2. Specimen Preparation: Frozen-Hydrated Grids for Cryo-EM

The following protocol assumes the use of a Vitrobot grid freezing robot (http://www.fei.com/vitrobot). Other plungers are commercially available or can be made in a local workshop.

Notes

- While commercially available perforated grids have several desirable qualities, as described above, their surface properties may depend on the age of the grid and can vary from batch to batch.
- Prefabricated grids should be treated to produce a consistent surface before use. Some groups will deposit a fresh layer of carbon onto the grid surface immediately before use to guarantee a consistent surface. In our research group, we generally rinse grids in acetone prior to use to remove residual or adsorbed organic molecules.
- Using more than one piece of filter paper on each blotting pad of a Vitrobot may reduce the required blotting time.
- *Safety note: Always wear safety glasses when handling liquid ethane as it can freeze tissue much more quickly than liquid nitrogen.*

1. Place grids onto several pieces of filter paper in a Pyrex Petri dish in a fume hood. Gently wet the filter paper with acetone without causing the grids to float and move. Cover the Petri dish and wait approximately 30 min for the acetone to evaporate. Repeat this procedure twice more.
2. When using a Vitrobot for grid freezing, the instrument must be precooled to the desired temperature and equilibrated at the desired relative humidity (RH). We generally set the temperature and humidity to 4 °C and 100 % RH, respectively, to avoid evaporation of the sample and consequent concentrating of the buffer reagents (e.g., salts and detergents).
3. Change the blotting paper in the Vitrobot if necessary.
4. Cool the freezing container of the Vitrobot according to the manufacturer's instructions, and fill the central brass chamber with liquid ethane. Avoid mixing liquid ethane and liquid nitrogen. Liquid nitrogen falling into the liquid ethane can cause splashing. Ethane falling into the nitrogen will freeze immediately. However, even if the frozen ethane is removed, the nitrogen surface will remain covered by bubbles that make it hard to see beneath the surface.
5. Glow-discharge the grids for 10 s to 2 min at 30 mA in a 2×10^{-2} mbar air atmosphere. Clamp a grid with the forceps from the freezing device and secure the forceps in place.
6. Move the forceps into the temperature- and humidity-controlled chamber. Apply 2–4 μl of sample solution onto the grid and wait for 30 s.

7. Begin the blotting process. The necessary blotting time varies between experiments, depending on the specimen, buffer composition, and properties of the blotting device. Ice thickness should be kept as thin as possible to maximize the image contrast. With a Vitrobot grid-freezing robot, the offset of the blotting pads can also be adjusted to vary the pressure with which they blot the grid. This parameter can dramatically affect the thickness and uniformity of ice across grid squares. In our work with the V-ATPase membrane protein complex from *Thermus thermophilus* (27), we used ~20 s blot time and −5 mm offset with a single piece of filter paper on each blotting pad.
8. After blotting is completed and the grid plunged into the ethane bath, release the forceps from the freezing device. By hand, raise the grid near the surface of the liquid ethane to allow excess ethane to drain from the forceps. Rapidly transfer the grid from the ethane bath into liquid nitrogen and store the grid in a grid box in liquid nitrogen until it is needed.

2.1.3. Specimen Preparation for Membrane Protein Complexes

In principle, it should be possible to use single particle EM to study detergent-solubilized membrane protein complexes with precisely the same techniques that are used for soluble protein complexes. In practice, slight modifications of the protocol appear to be necessary and interpretation of maps requires some extra consideration.

Negative stain EM is minimally different for membrane protein complexes. The micelle of detergent that keeps the complex soluble excludes stain and therefore appears indistinguishable from protein in images and 3D maps. However, the rinsing steps described in step 5 of the second protocol of Subheading 2.1.1 almost certainly remove some of the detergent monomers from the micelle of detergent surrounding the protein. However, it is unlikely that the rinsing removes all of the detergent and therefore the size of the micelle may vary from particle to particle.

In our experience (27–29), for cryo-EM of membrane protein complexes using perforated carbon film-coated EM grids, it is often necessary to use long glow-discharge times. With insufficient glow discharge particles do not seem to populate the ice layer, and instead can only be seen adsorbed to the carbon at the edges of holes. These glow-discharge times can be up to ~2 min with the conditions described in the third protocol of Subheading 2.1.1 and long blotting times are usually necessary (e.g., around 20 s). With these conditions, particles do not seem to concentrate at the air–water interfaces when preparing grids, as they can with soluble protein complexes. Therefore, for particles of under ~ 1 MDa, 2–3 mg/ml samples are necessary. With detergents, a detergent monolayer will form at the air–water interfaces at both surfaces of the grid. The presence of this monolayer may mean that specimens of detergent-solubilized complexes must be somewhat thicker than specimens of soluble protein complexes in order to accommodate similarly sized particles. Free micelles of detergent may add additional features to the ice; however, in our experience these have been difficult to detect. The concentration of detergent micelles may be decreased by reducing the overall detergent concentration. However, working with a detergent below its critical micelle concentration (CMC) may compromise the behavior of the protein sample and has not been necessary in our experiments. Detergents may have densities that are higher than, equal to, or lower than the density of ice (20, 30). With dodecylmaltoside, which has a higher density than ice, the detergent micelle that keeps the protein soluble is easily visible in the final 3D map (27).

2.2. Choice of Electron Source and Specimen Temperature

Instrument requirements for single particle electron microscopy are similar to those for imaging 2D crystals. One significant difference is that the image must not only contain signal at high resolution but must also have sufficient contrast at low spatial frequencies to allow for identification of protein particles in images and determination of the relative orientations of the particles. Near focus, a conventional bright-field phase contrast electron microscope has almost no contrast at low spatial frequencies (31). The necessary contrast is achieved by underfocusing the microscope to produce phase contrast from the interference of the scattered and unscattered beams, which is augmented by the amplitude contrast present in the image. This mechanism of contrast generation relies on a spatially coherent electron source. Therefore, a field emission gun (FEG) provides a substantial advantage for single particle cryo-EM because the high spatial coherence allows images to be acquired far from focus while preserving signal at high resolution. At increased electron accelerating voltages, the cross section of interaction of electrons with the specimen decreases. Therefore, in order to produce the same number of scattering events and the same image contrast, a higher electron exposure must be used as microscope voltage is increased. The accelerating voltage of the microscope has almost no impact on the amount of radiation damage caused once exposures are adjusted to obtain the same contrast with different voltages (7, 32). However, a higher-voltage microscope, such as one that operates a 300 kV, may be necessary for imaging of thick specimens (e.g., over 800 Å). Because increased electron energy decreases the number of multiple scattering events that occur within a sample and distortion of the image by specimen charging, in general, any microscope used for cryo-EM should be used at the highest available accelerating voltage. The decreased number of scattering events at higher voltages should be accounted for as described in Subheading 2.4.

Recent studies that have sought to determine the optimal temperature for cryo-EM experiments (26, 33) have concluded that there may be some additional advantage for working below liquid nitrogen temperature. However, the advantage may not be sufficient to warrant the additional expense and complications required to work at these temperatures (34).

2.3. Electron Detectors

Images may be recorded either on photographic film that is subsequently digitized or directly on a digital detector. Useful measurements for making quantitative comparisons between detectors are the modulation transfer function (MTF) and detective quantum efficiency (DQE). Digital detectors, such as CCDs that are coupled to a scintillator via optical fibers, have the advantage of real-time inspection of images and the possibility of automated image acquisition. Automated image acquisition has been developed for both of the major cryo-EM instrument manufacturers in the Leginon package for FEI microscopes (35 and Chapters 16 and 17) and the JADAS system for JEOL instruments (36). Analysis of the spectral SNR of images collected at 300 kV from specimens in vitreous ice using a CCD (37) suggested that up to 0.2 times the Nyquist frequency CCDs have better SNRs than film (38), which may help with alignment of particle images during map refinement (39). Film is as good or better than CCDs out to 0.5 times the Nyquist frequency and superior at higher frequencies. As with an earlier study (38), this work demonstrated that CCDs are an effective way to capture images for low-resolution map generation. In comparison, film is able to preserve high-resolution information better than current electronic detectors. It is important to note that when film is used for acquiring images, the combined properties of the film and film scanner will affect the final properties of the images (40). Direct electron detectors, which can have superior MTFs and DQEs relative to CCDs and film, promise to supercede both CCDs and film in the near future for EM imaging of radiation-sensitive specimens (41). There is still some debate about the benefits of energy filtering to remove inelastically scattered electrons before the image is formed. Some studies have concluded that energy filtering can improve the quality of 3D maps from single particles (26, 42), while others have argued that, for thin specimens, deterioration of the image from the contribution of inelastic electrons is negligible (43).

2.4. Imaging Conditions

Unlike with 2D crystals where the orientation of each molecule is known from its position within the crystal lattice, the orientation of each complex in single particle EM must be determined from the images themselves. Therefore, it is often necessary to sacrifice some of the SNR at high spatial frequencies in order to obtain higher SNRs at low spatial frequencies so that the overall shape of the particle can be seen and its relative orientation determined. The emphasis of low spatial frequencies is currently obtained by defocusing the microscope more during single particle experiments than for 2D crystal experiments and occasionally by using a higher electron exposure. The choices of defocus and electron exposure are both affected by the accelerating voltage of the microscope.

2.4.1. Defocus

The amount of defocus required for imaging depends on whether the single particle experiment is being performed on stained or frozen hydrated particles, size of the particles, and thickness of the ice or the stain layer. For imaging of large protein particles in negative stain at 80 or 100 kV, setting the objective lens to a few hundred nanometers under focus is usually sufficient, while imaging small protein complexes in ice at 300 kV will necessitate setting the lens several micrometers under focus. Note that underfocusing the objective lens is always preferable to overfocusing so that the amplitude contrast and the phase contrast of the microscope reinforce each other at low resolution.

At higher voltage, a greater degree of underfocus will be required for the same contrast with an equivalent exposure. The equivalent defocus between voltages can be easily calculated by keeping the product of defocus and electron wavelength constant. By keeping this product constant, microscopes operating at different voltages will produce contrast transfer functions (CTFs) that are similar except for the highest resolution oscillations of the functions. Electron wavelengths are λ=0.0370 Å at 100 kV, 0.0251 Å at 200 kV, and 0.0197 Å at 300 kV. Therefore, the ratio of equivalent electron defocuses are 0.68:1.00:1.22 at 100 kV:200 kV:300 kV. Higher voltage microscopes also provide the advantage of a reduced curvature of the Ewald sphere at high resolution (44, 45), a problem that may require computational correction beyond 5 Å resolution at 200 kV.

2.4.2. Electron Exposure

A detailed analysis of the effects of electron exposure at different resolutions, as determined from the fading of calculated diffraction spots from images of thin crystals of catalase, has been described elsewhere (39). Briefly, the SNR at low spatial frequencies is optimized at high exposures while high-resolution information can be obtained with the best possible SNR when low exposures are used to form an image. The precise exposure at which the SNR at different spatial frequencies are optimized depends on the accelerating voltage of the microscope because higher energy electrons undergo fewer interactions with the specimen than lower energy electrons. Equivalent exposures are proportional to the reciprocal of the linear energy transfer of electrons with protein, which are approximately 4.1, 2.8, and 2.3 MeC cm^2/g at 100, 200, and 300 keV, respectively (43). Electron exposures that optimize the SNR at selected resolutions are given in Table 1. Specimen holders must not have significant drift during the course of the exposure because resolution in an image will be worse than the distance moved by the specimen during the course of the exposure. This effect may influence the choice of exposure time used.

2.5. Cryo-EM Data Collection

The details of "low-dose" electron microscope operation are nearly identical between single particle EM and the electron crystallography experiments described elsewhere in this volume. In many ways, data collection for single particles is more straightforward than

Table 1
Electron exposures at 200 kV that optimize the signal-to-noise ratio (SNR) at different resolutions were taken from ref. (39) and used to calculate exposures for other voltages based on the linear energy transfers quoted above (43)

Resolution (Å)	100 kV (e^-/Å^2)	200 kV (e^-/Å^2)	300 kV (e^-/Å^2)
3	7 e^-/Å^2	10 e^-/Å^2	13 e^-/Å^2
8	8 e^-/Å^2	12 e^-/Å^2	15 e^-/Å^2
16	12 e^-/Å^2	17 e^-/Å^2	21 e^-/Å^2
27	14 e^-/Å^2	21 e^-/Å^2	26 e^-/Å^2
50	21 e^-/Å^2	30 e^-/Å^2	37 e^-/Å^2
80	31 e^-/Å^2	45 e^-/Å^2	55 e^-/Å^2

data collection for 2D crystals because good crystals do not need to be located. Furthermore, most single particle EM data can be collected without tilting the microscope's specimen stage. However, for single particle EM specimen grids, there is usually variation between different grid squares, and regions of the grid where the ice thickness and particle distribution are optimal must be found. Once a suitable region has been identified, one tries to obtain as many good-quality exposures as possible from these regions, a process that is compatible with the use of specimen grids with a regular array of holes and automated data collection, as described in Subheading 2.1.

Images, whether obtained directly with a digital detector or using film that has subsequently been digitized, must be sufficiently highly sampled for single particle EM. The Shannon–Nyquist information limit, given by twice the sampling frequency, dictates the highest-resolution information that can be contained in a digital image (46). Information should be somewhat oversampled to avoid loss of resolution due to interpolation errors when images are rotated or shifted. Oversampling followed by averaging adjacent pixels will also help to avoid signal loss due to the electron detector, which usually has a decreased DQE and MTF at high spatial frequencies. On the other hand, dramatic oversampling of images will significantly increase the computational costs associated with image analysis while not contributing accuracy or improved resolution. Attempts to oversample by increasing the magnification of the microscope may increase electron exposure and consequently the amount of radiation damage. Selecting a sampling frequency so that the highest-resolution features are found at ~2/3 the Shannon–Nyquist limit is usually considered sufficient (47).

2.6. Selection of Micrographs

Before particle images are extracted from micrographs, one may first wish to discard micrographs that show specimen drift or obvious astigmatism, which may reduce the quality of particle images included in image analysis. These image defects can be identified by using optical diffraction or computed power spectra to identify anisotropic fading of Thon rings, which indicates specimen drifts, or ellipsoidal tone rings, which indicates objective lens astigmatism or beam tilt. Most image analysis suites provide a program for inspecting power spectra from micrographs (48), and stand-alone programs also exist (49). Some image analysis programs correct for any measurable astigmatism (50) while others assume that perfectly nonastigmatic images have been obtained (51).

2.7. Particle Selection and CTF Determination

In the particle selection step of image processing, individual protein complexes are identified in micrographs and their approximate centers marked. This process is usually followed by extraction of many small images from the larger micrograph image, each containing only one particle image. With many types of computational image analysis, it can be beneficial to extract images with edge lengths of 2^N pixels in order to allow for optimal use of Fourier transform algorithms. The challenge in particle selection is to identify as many particles as possible in each micrograph while avoiding features that are not the particle of interest (such as ice contamination and contaminating protein complexes), and protein particles that are too close together to be distinguished during subsequent image analysis. Individual particles can be selected from the digitized micrographs either interactively or automatically. Interactive particle picking can be time-consuming, tedious, and prone to user bias, but usually has the advantage of low rates of false positive selection (features of the micrograph that are not protein particles or protein particles that are too close together being selected) and false negative selection (protein particles in the micrograph not being selected). Automatic particle picking is faster and less tedious, but can suffer from significantly higher false positive and/or false negative selection rates than manual selection. Automated particle selection tends to be more robust for symmetric or nearly spherical protein particles, and less robust for asymmetric, non-globular particles. Different approaches can generally be divided into template based and feature identifying methods, and have been reviewed recently (52).

It should also be recognized that features are blurred out around their true locations in defocused images. Higher-resolution features are displaced more than low-resolution information, as shown by the equation $R = \frac{\lambda \Delta F}{d}$, where R is the distance that information is displaced (in Å), λ is the electron wavelength (in Å, with λ=0.0335 Å at 120 kV, 0.0251 Å at 200 kV, and 0.0197 Å at 300 kV), ΔF is the defocus (in Å), and d is the resolution of the information being considered (in $Å^{-1}$) (53). Therefore, the box used to extract particle images from a micrograph should be larger than simply the radius of the particle. If one aims to recapture as much displaced information as possible, the edge length of the box should be $D+2R$ where D is the particle diameter and R is the distance that information of the highest resolution of interest is displaced (53).

Nominal values for the defocus used to obtain a micrograph should be logged at the time of image acquisition. However, if CTF correction is to be applied to a 3D map, it is necessary to determine defocus parameters precisely by inspection of the micrograph. Several software packages include tools for CTF estimation that fit the power spectrum of the micrograph. As described above, stand-alone programs also exist (49).

2.8. Image Analysis

The topic of image analysis for single particle electron microscopy encompasses a number of subjects, including how the orientations of individual particle images are determined, how the effects of the CTF are corrected, and how images are combined to build a 3D map. There are many different approaches to solving each of these problems. Therefore, the topic is far too broad to be covered fully in this book chapter and the interested reader is pointed towards a number of review articles, each one describing the general approaches used by several of the most popular image analysis software packages, including SPIDER (54), EMAN(51, 55), Imagic (56), Frealign (50), and Xmipp (57). In the text below we indicate a few possibilities for several of these steps.

Building a 3D map usually starts with 2D analysis of images (Subheading 2.8.1), where individual projections that represent the same view are grouped together and averaged to generate high SNR projections of the assembly being studied. An initial 3D map is then calculated (Subheading 2.8.2) which is subsequently refined (Subheading 2.8.3) to generate the best possible 3D map. The process of CTF correction is usually performed simultaneously with 3D map refinement. Once a final map is calculated, the resolution of the map is measured and its accuracy validated (Subheading 2.8.4). Finally, the biological information from a 3D map is interpreted, which may include processes such as segmentation of the map and fitting of available crystal structures (Subheading 2.8.5).

2.8.1. 2D Image Analysis

Analyzing images in 2D prior to 3D map construction can provide valuable information about the quality of the dataset and whether structural or conformational heterogeneity is present. In addition, classification and generation of class averages are needed for initial map construction by random conical tilting (58), orthogonal tilt reconstruction (59), and most common lines methods (60, 61) as described below. 2D analysis does not make assumptions about the existence of a single, underlying 3D structure that has given rise to the experimental projections. Instead, each projection may be rotated in plane (one angle) and shifted (*x*- and *y*-translations) to align it with other images so that similar projections may be grouped and averaged. The identification of equivalent views of a particle, once aligned, is usually accomplished by multivariate data analysis followed by the use of clustering algorithms (54, 62).

Invariably, not all class averages represent true projections of the 3D structure of the specimen. Distinguishing real projections from artifactual class averages is a challenging problem. This separation can sometimes be accomplished if the symmetry of the 3D structure is known so that the symmetry of the true 2D projections can be predicted. For asymmetric particles, if the particles are free to adopt any orientation on the specimen grid, one would expect to see a mirror image pair for each true projection that corresponds to the particle rotated by 180° about an axis that lies on the plane of the grid (29). Therefore, class averages may be eliminated where a corresponding mirror image class average cannot be derived from the dataset. However, this criterion for selecting class averages does not guarantee that every pair of mirror image class averages are true projections of the 3D structure (29). Class averages are often incoherent averages of particle images and it is therefore essential to confirm that the individual images included in a class do indeed resemble the final average. Statistics from each class can be inspected to help infer whether a class average represents a true projection of the 3D structure or an incoherent average (63). More sophisticated 2D image analysis strategies can be performed to sort out conformational and structural heterogeneity or partial ligand occupancy, with some approaches based on the multivariate data analysis described above (10) and other techniques relying on approaches such as maximum likelihood methods (64).

2.8.2. Initial Map Generation

Several approaches have been developed to use single particle EM images to create an initial 3D map that can subsequently be refined to high resolution. The most obvious approach to generating a 3D map of protein particles is to perform electron tomography with individual protein complexes. By collecting a tilt series of images from an area of the grid containing many identical copies of the complex, 3D maps may be created for each particle. The low electron exposure used to obtain each image necessarily means that the SNR will be low for each of these tomograms. Therefore, to further improve the initial map, tomogram averaging can be performed to combine the information obtained from individual tomograms. Six degrees of freedom (three Euler angles and *x*-, *y*-, and *z*-translations) must be explored for each tomogram in order to align them in 3D. There are technical challenges associated with this averaging due to the missing wedge of data that arises due to the inability to tilt the microscope stage beyond ~70° (65). While this approach has been used successfully (66), its application is still relatively rare due to the technical difficulties inherent in performing the precise tilting experiments required.

In comparison to performing electron tomography, the simplest approach to generating a 3D map is to combine a dataset of images by assigning random Euler angles to each image, or similarly, taking a 3D structure that is known to be incorrect as the starting map. The structure is then refined using one of the methods described below until convergence has been achieved. This technique has proven successful with several protein structures, particularly those that have high symmetry and when using a refinement approach specifically designed to deviate from incorrect starting maps (51, 55). However, there is no guarantee that convergence will occur and, unless some external information can be used to determine whether the resulting structure is correct, it is difficult or impossible to know when it has succeeded or failed. Consequently, use of random or incorrect starting maps for refinement is best left to situations where the map that is produced can be validated by independent means.

Two conceptually related methods for determining initial maps are the random conical tilt (RCT) and orthogonal tilt reconstruction (OTR) techniques (58, 59). In the RCT method, a first image is acquired from the specimen with the microscope stage tilted to between 50 and 70° and a second image is acquired with the stage untilted. The images of the particles acquired without tilt are subjected to alignment and classification and the resulting alignment parameters are used to determine the orientations of the particles in the tilted image. The RCT method works best when the particles take on one or a limited number of orientations on the specimen grid. If particles are flattened towards the plane of the grid, each RCT map will be flattened, an attribute that may prevent merging of independent RCT maps. In comparison, the OTR method only works when the particles take on random orientations on the specimen grid with no flattening of particles. In the OTR method, images are obtained with the stage tilted to +45 and −45°. Class averages are calculated for both tilt angles and matching class averages are located in the two datasets. Once identified, the relative orientation of the individual particle images in the two class averages is defined by the rotation required to align each particle image with each class average (59). The RCT and OTR methods are implemented, via scripts, in the SPIDER software package (54, 67).

The common lines algorithm was developed in the early 1970s for use with icosahedral virus particles (60). It, and its real-space equivalent, angular reconstitution (61), makes use of the projection theorem (46), which states that the Fourier transform of each

2D projection is a central plane through the 3D Fourier transform of the map. Therefore, any two 2D projections will cross along a 1D common line, and by identifying three different common lines shared by three different projections, the relative orientations of the three projections can be determined. While the method can work well for symmetric protein particles and is particularly robust for icosahedral viruses, it is highly susceptible to noise for asymmetric particles and complexes with low symmetry. New methods based on the common lines method may improve its applicability to low-symmetry complexes but have not yet been used widely (68, 69). The original implementation of the common lines algorithm, specific for application to icosahedral viruses can be found in the MRC image analysis suite (48). The common lines method is implemented in Fourier space in the SPIDER (54, 67) and EMAN (51, 55) software packages, and in real space in the Imagic software package (56).

In addition to these general approaches, some specific situations lend themselves to the development of specialized algorithms for generating initial maps. For example, with particles that tend to adopt a preferred orientation on the EM grid with their long axis parallel to the surface of the grid, initial maps can be constructed using the rotational analysis algorithm (70). As can be seen from the above discussion, there is no rule for how initial maps should be created and the experimentalist must use experience and judgment in deciding how to start the process of 3D map building from single particle EM data. Therefore, this aspect of the image analysis problem is one that requires significant future investment and development. Table 2 summarizes a variety of initial map construction approaches.

2.8.3. Structure Refinement

Once an initial map has been calculated, one usually attempts to refine the map to improve its resolution or quality. While some software packages, such as the Imagic system (56), use an algorithm that searches for common lines between images and reference projections, the most common approach for map refinement is to use an algorithm based on projection matching (71). Experimental images are compared in real space to projections of the initial map or, equivalently, Fourier transforms of experimental images are compared to central planes through the 3D Fourier transform of the initial map. Once a similar projection or a Fourier plane has been identified, the orientation of that matching projection or plane defines the new Euler angles of the experimental image. These new parameters associated with each image are used to calculate the next iteration of the reference map and the process is repeated until refinement statistics cease to improve. The details of the algorithm can differ significantly from program to program, including differences in the figure of merit used to calculate similarity.

Table 2
Applicability, advantages, and disadvantages of initial map construction algorithms

Method	Advantages	Disadvantages	When to use
Tomography	Generally applicable, robust	Can be technically challenging to obtain and merge tomograms	Nothing is known about the structure under study and facilities are available to perform electron tomography
Random map or incorrect map	Easy to do	Can be difficult or impossible to tell if the final map is correct. Map handedness must be determined separately	When the specimen has high symmetry and/or convergence to a correct result can be established by independent means
Random conical tilting	A robust, general method	Can produce flattened maps from stained specimens and can be challenging if there are a large number of particle orientations on the grid. Requires tilting the microscope stage (one image at 0° and one at a high tilt angle)	Particles take on one or a few orientations on the grid, with little or no specimen flattening
Orthogonal tilt	A robust, general method	Requires specimens to adopt random orientations on the grid and will fail if particles are flattened due to staining. Requires tilting the microscope stage (one image collected at 45° and one image at –45°)	Particles take on random orientations on the grid with no specimen flattening
Common lines/angular reconstitution	Does not require tilting the microscope stage	Algorithm can fail and produce incorrect maps with low symmetry specimens. Map handedness must be determined separately	High-symmetry particles
Rotational analysis	A robust method when side views of the particle predominate	Only applicable in a special case. Map handedness must be determined separately	Particles present a single axis rotation series with an unknown rotation angle

Exhaustive projection matching with very fine angular increments can be an extremely slow process, so some programs will incorporate a refinement step where angles and shifts, and even other parameters (such as defocus), are adjusted using a local minimizer algorithm, such as Powell's method (72), which is employed in the Frealign package (50). Other programs, such as EMAN (51), will refine image shifts only by using a downhill simplex method based on the one developed by Nelder and Mead (72).

Correction for the CTF of the microscope is essential for high-resolution map construction. The periodic oscillation in the sign of the CTF leads to 180° changes in the phases of Fourier components in images at frequencies where the CTF is negative. An approximate CTF correction can be applied before image analysis simply by reversing the signs of Fourier components (phase flipping) where necessary, as determined from the known parameters of the CTF associated with each image. A full correction for the effects of the CTF can only be applied during construction of a map after projection matching, or during generation of class averages, because complete correction requires adjusting the amplitude of each averaged Fourier component based on the value of the CTF at that frequency in all of the images that contributed to the component.

Since the SNR in individual images is low, alignment of the images to projections can be affected by noise in the images, which can lead to model bias in averaged images. One strategy to reduce model bias is by iterative 2D class averaging, as employed in the Imagic, Xmipp, or EMAN software packages. Another approach is to use a figure of merit for alignment that is designed to resist noise (73). A third approach is to make use of algorithms, such as maximum likelihood methods, explicitly designed to work in the presence of noise (64, 74).

Because the SNR in images decreases at high resolution, it is usually better to depend on low-resolution information for refinement, particularly in the early stages of the process. One refinement strategy is to gradually allow higher-resolution frequencies to contribute to the image similarity measure as refinement proceeds. Alternatively, the choice of frequencies to include in the alignment process can be optimized by testing the accuracy of particle orientation determination throughout refinement (53). As refinement proceeds, convergence of the process can be monitored by computing the Fourier Shell Correlation between maps produced at each iteration and monitoring the estimated resolution of the map, as described in Subheading 2.8.4. However, it is essential to remember that convergence does not guarantee map accuracy.

2.8.4. Map Validation, Hand Determination, and Resolution Estimation

After refinement of a map has been completed, the resolution of the map can be estimated. With crystalline specimens, the resolution of the raw data can be determined easily by the presence of reflections on a lattice that show some level of significance above the background noise. For single particle EM, the tools for assessing final map resolution are less robust. Usually, resolution is estimated by making use of a Fourier shell correlation plot (75). To obtain this plot, the data used to create the map is divided in half and two maps are created, each from half of the data. The correlation between the maps as a function of spatial frequency is then calculated and the resolution at which the correlation drops below a specified level is taken as the resolution limit of the map. FSC plots can be misleading because the two maps, each made from half of the available data, are not usually completely independent of each other, as each half of the data was aligned to the same reference. Therefore, if the map refinement process has been dominated by alignment of noise to a reference, these maps can be strongly correlated at high resolution even when the final map bears little resemblance to reality. With this caveat in mind, the FSC curve has become the standard tool for measuring EM map quality.

The subject of an appropriate cutoff for significance of the FSC curve has been controversial in the single particle EM community. The arbitrarily chosen but conservative 0.5 threshold remains as a popular choice (76). The 0.143 threshold relates single particle EM resolution to X-ray crystal structure resolution, and identifies the highest resolution frequency bin that will improve the quality of the map if included (77). However, being lower than the 0.5 threshold, it is easier to abuse this criteria to report overly high resolutions if a map has been refined improperly. Non-fixed-value criteria for assessing resolutions from FSC curves have also been proposed (78). The presence of localized disordered regions in a structure will also not be revealed in the FSC curve. Similarly, a disordered detergent micelle in a map of a membrane protein complex may lower the apparent resolution of a map even though its presence does not affect interpretation of the protein structure (79). In addition to reporting a resolution, the shape of the FSC curve may be useful in revealing the occurrence of noise fitting in the refinement process (73). Considering the limitations of the FSC curve, the resolution determined from the curve should only be treated as an estimate, rather than an absolute measure of map quality. More recently, an ab initio map resolution determining algorithm has been proposed based on correlation between neighboring voxels in Fourier space (80). This new criterion appears to have several advantages over the FSC curve, but its behavior has not yet been fully characterized.

For a high-resolution map (<10 Å) where secondary structures can be reliably observed, the resolution of a map may be estimated by inspection of the map features (81). For example, one would expect to see α-helices as rod-shaped densities at ~10–6 Å resolution, beta-sheets and the handedness of helices should be visible at ~5-4 Å resolution, and bulky side chains should become visible at 4 Å resolution. At around 3 Å resolution, it should be possible to build an all-atom model directly into the singe particle EM map (1).

2.8.5. Map Handedness, Sharpening, and Interpretation

Maps of proteins determined from images of untilted specimen grids will have an ambiguous handedness. At sufficiently high resolution, handedness can be determined by the identification of the handedness of α-helices in the structure. At lower resolution it may be possible to identify the hand of a map by comparison to a crystal structure. In all cases, the hand of a map can be determined by using tilt pairs of images with a technique such as the Free hand test (53). In addition to being used to determine map handedness, the Free hand test may be used as an independent test of the validity of a map (53). Maps that have been refined from RCT, OTR, or electron tomography should be of the correct hand, assuming that care was taken throughout image analysis to preserve this information. However, it is easy to accidentally reverse the hand of a map, such as when converting from one image format to another, and it is therefore advisable to determine the handedness even for maps that have been calculated by these methods.

Because the electron microscope systematically decreases the true amplitudes of high-frequency Fourier components, the amplitudes of these components can be increased in a process known as map sharpening (53). Map sharpening, if done appropriately, will reveal high-resolution features in a map, but if done to excess will produce artifacts. The degree of attenuation of the high-frequency signal in 3D maps can be determined by estimation of a B factor from theory (53) or comparison of the power spectrum of the map to an X-ray scattering curve (82). Sharpening is sometimes carried out during the refinement cycle (e.g., in the EMAN program) or can be performed after refinement (e.g., with SPIDER and Frealign). In the latter case, sharpening can be performed with a stand-alone program such as EM-BFACTOR (53, 83).

Various software tools exist for gleaning structural information from maps at different resolution. For maps with resolutions coarser than ~10 Å, molecular envelopes and some subunit boundaries are visible and interpretation is largely based on rigid-body docking or flexible fitting of independent atomic models into the map (84–87). It is also possible to perform volume segmentation on low-resolution maps (27, 29), a process that becomes more reliable with high-resolution maps. Segmentation can be preformed manually or automatically, with tools existing in the SPIDER, EMAN, and UCSF Chimera, and Amira (http://www.amira.com) software packages (51, 54, 88). At sufficiently high resolution, atomic models can be built into cryo-EM density maps with the same tools used to build atomic models into electron density maps from X-ray crystallography (3). Finally, rendering of 3D maps, with and without relevant atomic models, is an essential step for interpreting the biological lessons that can be learned from the map. While there are many packages available for graphical representation of data, the UCSF Chimera package has undergone significant development to optimize its use with cryo-EM maps (88).

2.9. Single Particle Refinement of Ordered Arrays

In the study of ordered arrays of proteins the availability of a real-space image from the electron microscope allows for computational correction of imperfect crystal lattices (89–91). This ability provides an appealing advantage to using images in processing of crystallographic data rather than just diffraction patterns. Traditionally, this "unbending" procedure has been done by identifying displacements of the unit cells by comparison with an ideal lattice, as estimated from a reference array in the center of the crystal (90), or a reference derived from the best current 3D map of the structure (91). Similar approaches have also been used to "unbend" images of helical specimens (92).

By reducing the reference region of the 2D crystal or helical array to a single asymmetric unit, these refinement methods become a single particle refinement of the crystallographic data. This single particle refinement of 2D crystal data has been used to correct for differing orientations of reaction centers in 2D crystals of complexes of reaction centers with light harvesting complex I (93). Further refinements of the technique include using maximum likelihood methods to improve the determination of the single particle orientation in 2D crystal analysis (94). These single particle tools have also proven effective for determining 3D structures from images of helical specimens (95).

Acknowledgements

We thank Lindsay Baker for a critical reading of this chapter. J.L.R. was supported by a New Investigator award from the Canadian Institutes of Health Research (CIHR). This work was funded by operating grant MOP 81294 from the CIHR.

References

1. Zhang J, Baker ML, Schroder GF, Douglas NR, Reissmann S, Jakana J, Dougherty M, Fu CJ, Levitt M, Ludtke SJ, Frydman J, Chiu W (2010) Mechanism of folding chamber closure in a group II chaperonin. Nature 463: 379–383
2. Zhang X, Settembre E, Xu C, Dormitzer PR, Bellamy R, Harrison SC, Grigorieff N (2008) Near-atomic resolution using electron cryomicroscopy and single-particle reconstruction. Proc Natl Acad Sci USA 105:1867–1872
3. Zhang X, Jin L, Fang Q, Hui WH, Zhou ZH (2010) 3.3 A cryo-EM structure of a nonenveloped virus reveals a priming mechanism for cell entry. Cell 141:472–482
4. Frank J (2009) Single-particle reconstruction of biological macromolecules in electron microscopy – 30 years. Q Rev Biophys 42:139–158
5. Henderson R (1992) Image contrast in high-resolution electron microscopy of biological macromolecules: TMV in ice. Ultramicroscopy 46:1–18

6. Glaeser RM, McMullan G, Faruqi AR, Henderson R (2010) Images of paraffin monolayer crystals with perfect contrast: minimization of beam-induced specimen motion. Ultramicroscopy 111:90–100
7. Henderson R (1995) The potential and limitations of neutrons, electrons and X-rays for atomic resolution microscopy of unstained biological molecules. Q Rev Biophys 28:171–193
8. Baird NJ, Ludtke SJ, Khant H, Chiu W, Pan T, Sosnick TR (2010) Discrete structure of an RNA folding intermediate revealed by cryo-electron microscopy. J Am Chem Soc 132:16352–16353
9. Jiang W, Baker ML, Jakana J, Weigele PR, King J, Chiu W (2008) Backbone structure of the infectious epsilon15 virus capsid revealed by electron cryomicroscopy. Nature 451: 1130–1134
10. Elad N, Clare DK, Saibil HR, Orlova EV (2008) Detection and separation of heterogeneity in molecular complexes by statistical analysis of their two-dimensional projections. J Struct Biol 162:108–120
11. Zhang W, Kimmel M, Spahn CM, Penczek PA (2008) Heterogeneity of large macromolecular complexes revealed by 3D cryo-EM variance analysis. Structure 16:1770–1776
12. Julian P, Konevega AL, Scheres SH, Lazaro M, Gil D, Wintermeyer W, Rodnina MV, Valle M (2008) Structure of ratcheted ribosomes with tRNAs in hybrid states. Proc Natl Acad Sci USA 105:16924–16927
13. Clare DK, Bakkes PJ, van Heerikhuizen H, van der Vies SM, Saibil HR (2009) Chaperonin complex with a newly folded protein encapsulated in the folding chamber. Nature 457:107–110
14. Elmlund H, Baraznenok V, Linder T, Szilagyi Z, Rofougaran R, Hofer A, Hebert H, Lindahl M, Gustafsson CM (2009) Cryo-EM reveals promoter DNA binding and conformational flexibility of the general transcription factor TFIID. Structure 17:1442–1452
15. Kastner B, Fischer N, Golas MM, Sander B, Dube P, Boehringer D, Hartmuth K, Deckert J, Hauer F, Wolf E, Uchtenhagen H, Urlaub H, Herzog F, Peters JM, Poerschke D, Luhrmann R, Stark H (2008) GraFix: sample preparation for single-particle electron cryomicroscopy. Nat Methods 5:53–55
16. Cheng Y, Wolf E, Larvie M, Zak O, Aisen P, Grigorieff N, Harrison SC, Walz T (2006) Single particle reconstructions of the transferrin-transferrin receptor complex obtained with different specimen preparation techniques. J Mol Biol 355:1048–1065
17. Brignole EJ, Smith S, Asturias FJ (2009) Conformational flexibility of metazoan fatty acid synthase enables catalysis. Nat Struct Mol Biol 16:190–197
18. Ohi M, Li Y, Cheng Y, Walz T (2004) Negative staining and image classification – powerful tools in modern electron microscopy. Biol Proced Online 6:23–34
19. Adrian M, Dubochet J, Fuller SD, Harris JR (1998) Cryo-negative staining. Micron 29:145–160
20. Rubinstein JL (2007) Structural analysis of membrane protein complexes by single particle electron microscopy. Methods 41:409–416
21. Mishima O, Calvert LD, Whalley E (1985) An apparent first-order transition between two amorphous phases of ice induced by pressure. Nature 314:76–78
22. Frank J, Penczek P, Agrawal RK, Grassucci RA, Heagle AB (2000) Three-dimensional cryoelectron microscopy of ribosomes. Methods Enzymol 317:276–291
23. Harris WJ (1962) Holey films for electron microscopy. Nature 196:499–500
24. Quispe J, Damiano J, Mick SE, Nackashi DP, Fellmann D, Ajero TG, Carragher B, Potter CS (2007) An improved holey carbon film for cryo-electron microscopy. Microsc Microanal 13:365–371
25. Stagg SM, Lander GC, Pulokas J, Fellmann D, Cheng A, Quispe JD, Mallick SP, Avila RM, Carragher B, Potter CS (2006) Automated cryoEM data acquisition and analysis of 284742 particles of GroEL. J Struct Biol 155:470–481
26. Fujii T, Kato T, Namba K (2009) Specific arrangement of alpha-helical coiled coils in the core domain of the bacterial flagellar hook for the universal joint function. Structure 17:1485–1493
27. Lau WC, Rubinstein JL (2010) Structure of intact Thermus thermophilus V-ATPase by cryo-EM reveals organization of the membrane-bound V(O) motor. Proc Natl Acad Sci USA 107:1367–1372
28. Lau WC, Baker LA, Rubinstein JL (2008) Cryo-EM structure of the yeast ATP synthase. J Mol Biol 382:1256–1264
29. Rubinstein JL, Walker JE, Henderson R (2003) Structure of the mitochondrial ATP synthase by electron cryomicroscopy. EMBO J 22:6182–6192
30. Schmidt-Krey I, Rubinstein JL (2011) Electron cryomicroscopy of membrane proteins: specimen preparation for two-dimensional crystals and single particles. Micron 42: 107–116

31. Spence JCH (2008) High-resolution electron microscopy, 3rd edn. Oxford University Press, Oxford
32. Langmore JP, Smith MF (1992) Quantitative energy-filtered electron microscopy of biological molecules in ice. Ultramicroscopy 46:349–373
33. Bammes BE, Jakana J, Schmid MF, Chiu W (2010) Radiation damage effects at four specimen temperatures from 4 to 100 K. J Struct Biol 169:331–341
34. Comolli LR, Downing KH (2005) Dose tolerance at helium and nitrogen temperatures for whole cell electron tomography. J Struct Biol 152:149–156
35. Carragher B, Kisseberth N, Kriegman D, Milligan RA, Potter CS, Pulokas J, Reilein A (2000) Leginon: an automated system for acquisition of images from vitreous ice specimens. J Struct Biol 132:33–45
36. Zhang J, Nakamura N, Shimizu Y, Liang N, Liu X, Jakana J, Marsh MP, Booth CR, Shinkawa T, Nakata M, Chiu W (2009) JADAS: a customizable automated data acquisition system and its application to ice-embedded single particles. J Struct Biol 165:1–9
37. Booth CR, Jakana J, Chiu W (2006) Assessing the capabilities of a 4kx4k CCD camera for electron cryo-microscopy at 300 kV. J Struct Biol 156:556–563
38. Sander B, Golas MM, Stark H (2005) Advantages of CCD detectors for de novo three-dimensional structure determination in single-particle electron microscopy. J Struct Biol 151:92–105
39. Baker LA, Smith EA, Bueler SA, Rubinstein JL (2010) The resolution dependence of optimal exposures in liquid nitrogen temperature electron cryomicroscopy of catalase crystals. J Struct Biol 169:431–437
40. Henderson R, Cattermole D, McMullan G, Scotcher S, Fordham M, Amos WB, Faruqi AR (2007) Digitisation of electron microscope films: six useful tests applied to three film scanners. Ultramicroscopy 107:73–80
41. Faruqi AR, Henderson R (2007) Electronic detectors for electron microscopy. Curr Opin Struct Biol 17:549–555
42. Yonekura K, Braunfeld MB, Maki-Yonekura S, Agard DA (2006) Electron energy filtering significantly improves amplitude contrast of frozen-hydrated protein at 300 kV. J Struct Biol 156:524–536
43. Glaeser RM, Downing KH, DeRosier DJ, Chiu W, Frank J (2007) Electron crystallography of biological macromolecules. Oxford University Press, Oxford
44. DeRosier DJ (2000) Correction of high-resolution data for curvature of the Ewald sphere. Ultramicroscopy 81:83–98
45. Wolf M, DeRosier DJ, Grigorieff N (2006) Ewald sphere correction for single-particle electron microscopy. Ultramicroscopy 106:376–382
46. Bracewell RN (2000) The Fourier transform and its applications, 3rd edn. McGraw-Hill, New York
47. Orlova EV, Dube P, Harris JR, Beckman E, Zemlin F, Markl J, van Heel M (1997) Structure of keyhole limpet hemocyanin type 1 (KLH1) at 15 A resolution by electron cryomicroscopy and angular reconstitution. J Mol Biol 271:417–437
48. Crowther RA, Henderson R, Smith JM (1996) MRC image processing programs. J Struct Biol 116:9–16
49. Mindell JA, Grigorieff N (2003) Accurate determination of local defocus and specimen tilt in electron microscopy. J Struct Biol 142:334–347
50. Grigorieff N (2007) FREALIGN: high-resolution refinement of single particle structures. J Struct Biol 157:117–125
51. Ludtke SJ, Baldwin PR, Chiu W (1999) EMAN: semiautomated software for high-resolution single-particle reconstructions. J Struct Biol 128:82–97
52. Zhu Y, Carragher B, Glaeser RM, Fellmann D, Bajaj C, Bern M, Mouche F, de Haas F, Hall RJ, Kriegman DJ, Ludtke SJ, Mallick SP, Penczek PA, Roseman AM, Sigworth FJ, Volkmann N, Potter CS (2004) Automatic particle selection: results of a comparative study. J Struct Biol 145:3–14
53. Rosenthal PB, Henderson R (2003) Optimal determination of particle orientation, absolute hand, and contrast loss in single-particle electron cryomicroscopy. J Mol Biol 333:721–745
54. Frank J, Radermacher M, Penczek P, Zhu J, Li Y, Ladjadj M, Leith A (1996) SPIDER and WEB: processing and visualization of images in 3D electron microscopy and related fields. J Struct Biol 116:190–199
55. Tang G, Peng L, Baldwin PR, Mann DS, Jiang W, Rees I, Ludtke SJ (2007) EMAN2: an extensible image processing suite for electron microscopy. J Struct Biol 157:38–46
56. van Heel M, Harauz G, Orlova EV, Schmidt R, Schatz M (1996) A new generation of the IMAGIC image processing system. J Struct Biol 116:17–24
57. Sorzano CO, Marabini R, Velazquez-Muriel J, Bilbao-Castro JR, Scheres SH, Carazo JM,

Pascual-Montano A (2004) XMIPP: a new generation of an open-source image processing package for electron microscopy. J Struct Biol 148:194–204
58. Radermacher M (1988) Three-dimensional reconstruction of single particles from random and nonrandom tilt series. J Electron Microsc Tech 9:359–394
59. Leschziner AE, Nogales E (2006) The orthogonal tilt reconstruction method: an approach to generating single-class volumes with no missing cone for ab initio reconstruction of asymmetric particles. J Struct Biol 153: 284–299
60. Crowther RA, Amos LA, Finch JT, De Rosier DJ, Klug A (1970) Three dimensional reconstructions of spherical viruses by fourier synthesis from electron micrographs. Nature 226:421–425
61. Van Heel M (1987) Angular reconstitution: a posteriori assignment of projection directions for 3D reconstruction. Ultramicroscopy 21:111–123
62. van Heel M, Frank J (1981) Use of multivariate statistics in analysing the images of biological macromolecules. Ultramicroscopy 6:187–194
63. Shaikh TR, Trujillo R, LeBarron JS, Baxter WT, Frank J (2008) Particle-verification for single-particle, reference-based reconstruction using multivariate data analysis and classification. J Struct Biol 164:41–48
64. Scheres SH, Valle M, Carazo JM (2005) Fast maximum-likelihood refinement of electron microscopy images. Bioinformatics 21(suppl 2):ii243–ii244
65. Schmid MF, Booth CR (2008) Methods for aligning and for averaging 3D volumes with missing data. J Struct Biol 161:243–248
66. Walz J, Typke D, Nitsch M, Koster AJ, Hegerl R, Baumeister W (1997) Electron tomography of single ice-embedded macromolecules: three-dimensional alignment and classification. J Struct Biol 120:387–395
67. Shaikh TR, Gao H, Baxter WT, Asturias FJ, Boisset N, Leith A, Frank J (2008) SPIDER image processing for single-particle reconstruction of biological macromolecules from electron micrographs. Nat Protoc 3: 1941–1974
68. Coifman RR, Shkolnisky Y, Sigworth FJ, Singer A (2010) Reference free structure determination through eigenvectors of center of mass operators. Appl Comput Harmon Anal 28:296–312
69. Singer A, Coifman RR, Sigworth FJ, Chester DW, Shkolnisky Y (2010) Detecting consistent common lines in cryo-EM by voting. J Struct Biol 169:312–322
70. Baker LA, Rubinstein JL (2008) Angle determination for side views in single particle electron microscopy. J Struct Biol 162:260–270
71. Harauz G, Ottensmeyer FP (1984) Nucleosome reconstruction via phosphorus mapping. Science 226:936–940
72. Press WH, Teukolsky SA, Vetterling WT, Flannery BP (1992) Numerical recipes in Fortran 77, 2nd edn. Cambridge University Press, Cambridge
73. Stewart A, Grigorieff N (2004) Noise bias in the refinement of structures derived from single particles. Ultramicroscopy 102:67–84
74. Tagare HD, Barthel A, Sigworth FJ (2010) An adaptive expectation-maximization algorithm with GPU implementation for electron cryomicroscopy. J Struct Biol 171:256–265
75. Harauz G, van Heel M (1986) Exact filters for general geometry 3-dimensional reconstruction. Optik 73:146–156
76. Bottcher B, Wynne SA, Crowther RA (1997) Determination of the fold of the core protein of hepatitis B virus by electron cryomicroscopy. Nature 386:88–91
77. Rosenthal PB, Crowther RA, Henderson R (2003) An objective criterion for resolution assessment in single-particle electron microscopy (appendix). J Mol Biol 333:743–745
78. van Heel M, Schatz M (2005) Fourier shell correlation threshold criteria. J Struct Biol 151:250–262
79. Samso M, Wagenknecht T, Allen PD (2005) Internal structure and visualization of transmembrane domains of the RyRl calcium release channel by cryo-EM. Nat Struct Mol Biol 12:539–544
80. Sousa D, Grigorieff N (2007) Ab initio resolution measurement for single particle structures. J Struct Biol 157:201–210
81. Zhou ZH (2008) Towards atomic resolution structural determination by single-particle cryo-electron microscopy. Curr Opin Struct Biol 18:218–228
82. Saad A, Ludtke SJ, Jakana J, Rixon FJ, Tsuruta H, Chiu W (2001) Fourier amplitude decay of electron cryomicroscopic images of single particles and effects on structure determination. J Struct Biol 133:32–42
83. Fernandez JJ, Luque D, Caston JR, Carrascosa JL (2008) Sharpening high resolution information in single particle electron cryomicroscopy. J Struct Biol 164:170–175
84. Wriggers W, Birmanns S (2001) Using situs for flexible and rigid-body fitting of multiresolution

single-molecule data. J Struct Biol 133: 193–202

85. Schroder GF, Brunger AT, Levitt M (2007) Combining efficient conformational sampling with a deformable elastic network model facilitates structure refinement at low resolution. Structure 15:1630–1641
86. Trabuco LG, Villa E, Schreiner E, Harrison CB, Schulten K (2009) Molecular dynamics flexible fitting: a practical guide to combine cryo-electron microscopy and X-ray crystallography. Methods 49:174–180
87. Trabuco LG, Villa E, Mitra K, Frank J, Schulten K (2008) Flexible fitting of atomic structures into electron microscopy maps using molecular dynamics. Structure 16:673–683
88. Goddard TD, Huang CC, Ferrin TE (2007) Visualizing density maps with UCSF Chimera. J Struct Biol 157:281–287
89. Henderson R, Baldwin JM, Ceska TA, Zemlin F, Beckmann E, Downing KH (1990) Model for the structure of bacteriorhodopsin based on high-resolution electron cryo-microscopy. J Mol Biol 213:899–929
90. Henderson R, Baldwin JM, Downing KH, Lepault J, Zemlin F (1986) Structure of purple membrane from *Halobacterium halobrium*: recording, measurement and evaluation of electron micrographs at 3.5 Å resolution. Ultramicroscopy 19:147–178
91. Kunji ER, von Gronau S, Oesterhelt D, Henderson R (2000) The three-dimensional structure of halorhodopsin to 5 A by electron crystallography: a new unbending procedure for two-dimensional crystals by using a global reference structure. Proc Natl Acad Sci USA 97:4637–4642
92. Kubalek E, Ralston S, Lindstrom J, Unwin N (1987) Location of subunits within the acetylcholine receptor by electron image analysis of tubular crystals from Torpedo marmorata. J Cell Biol 105:9–18
93. Stahlberg H, Dubochet J, Vogel H, Ghosh R (1998) Are the light-harvesting I complexes from Rhodospirillum rubrum arranged around the reaction centre in a square geometry? J Mol Biol 282:819–831
94. Zeng X, Stahlberg H, Grigorieff N (2007) A maximum likelihood approach to two-dimensional crystals. J Struct Biol 160:362–374
95. Egelman EH (2000) A robust algorithm for the reconstruction of helical filaments using single-particle methods. Ultramicroscopy 85:225–234

Chapter 23

Electron Tomography of Paracrystalline 2D Arrays

Hanspeter Winkler, Shenping Wu, and Kenneth A. Taylor

Abstract

Paracrystalline arrays possess specific types of disorder that reduce the structural information as well as resolution when spatially averaged over repeating motifs. Electron tomography combined with motif classification and averaging can solve the heterogeneity problem and provide information on the structural elements that give rise to the disorder. This chapter describes procedures that would be used in a typical tomography application to identify and characterize a paracrystalline specimen. Particular emphasis is given to actively contracting insect flight muscle, a specimen with particularly difficult to characterize structural heterogeneity and 2D paracrystalline arrays of myosin-V, from which a particularly high resolution motif average was obtained. All aspects of the study are described including data collection, merging of micrographs to produce the tomogram, alignment to an invariant structural element, classification and averaging of heterogeneous structures, and reassembly of focused class averages into high signal-to-noise ratio representations of the original raw repeats. Particular emphasis is placed on limitations of the various processes to produce the final class averages.

Key words: Cryoelectron microscopy, Electron microscope tomography, Computer-assisted image processing, Three-dimensional imaging

1. Introduction

Electron tomography (ET) might seem a poor choice of technique for determining the structure of a crystalline specimen. Electron crystallography is a much more powerful method for obtaining near atomic resolution three-dimensional (3D) images of molecules. However, high resolution crystallography requires well-ordered arrays that are ordinarily obtained with difficulty. The more usual condition is arrays that are poorly ordered, which would happen if the molecules are heterogeneous in either conformation or composition or both, in which case the spatially averaged 3D image displays an average of heterogeneous structures. The heterogeneity

Ingeborg Schmidt-Krey and Yifan Cheng (eds.), *Electron Crystallography of Soluble and Membrane Proteins: Methods and Protocols*, Methods in Molecular Biology, vol. 955, DOI 10.1007/978-1-62703-176-9_23,

may reveal aspects of function but is lost unless the source of heterogeneity is first identified and similar structures are grouped before averaging.

ET on the other hand seldom achieves resolutions approaching 2 nm (1) and far more often achieves resolutions <4 nm. Despite its resolution limitation, ET can provide a powerful complementary method when intrinsic disorder is present in a two-dimensional (2D) array (2). The strength of ET is its ability to produce a 3D image of even unique objects and as such can provide 3D images of the individual molecules within a disordered array thus revealing the structural variations lost with spatial averaging.

Paracrystals (para—Greek prefix meaning faulty) are arrays with intrinsic disorder. Many kinds of disorder can make an array paracrystalline, such as mixtures of filaments with different structure, which is the case in insect flight muscle (IFM) (3), poor ordering of homogeneous filaments, which is the case in many actin bundles (4), poor order of an otherwise homogeneous protein, such as arrays of myosin-V (1), or compositional heterogeneity of a well ordered assembly to name just a few. We have applied ET to several of the types of paracrystalline specimens mentioned above. Most of the methods described here were developed initially for studies of IFM but have been applied and modified for broader application to a much wider variety of specimens such as the Env spikes of HIV/SIV viruses (5–7), integrins reconstituted into vesicles (8, 9), and bacterial flagellar motors (10).

We use the term "motif" as a synonym for subvolume because of the repetitive nature of paracrystalline arrays and equate it with the term "single-particle" as used in 3D reconstruction from projections. The structure determination of a paracrystalline specimen by ET would be similar to solving a single particle structure when preferred particle orientation is present. The high density of partially aligned motifs in paracrystals is a significant mitigating factor. Generally, the motifs in paracrystals are heterogeneous, otherwise the array would not be paracrystalline. Single-particle methods are most effective when the variation between particles arises primarily from orientation and are more difficult when heterogeneity in structure is also present. The presence of preferred orientation requires that the object be tilted to obtain 3D data, yet the close apposition of motifs within a paracrystal causes superposition when tilted so that a single particle approach to solving the structure using tilt series images is impractical.

ET largely overcomes these problems but with the qualification that the tomograms, especially of ice-embedded specimens have low signal-to-noise (S/N) ratio requiring that averages be obtained if molecular conformations are desired. When ET is applied to active muscle, which arguably lacks discrete structural states in the myosin cross-bridges, reduction in the number of structures as well as an accurate count of the different structures is imperative.

Improved resolution, improved S/N ratio, data reduction as well as quantification are goals achievable using the techniques of multivariate data analysis (MDA) (11, 12). However, processing of volume data presents additional challenges including angular alignment over three degrees of freedom and treatment of the missing wedge (or pyramid) that do not occur in 2D image classification.

In this chapter, we will describe approaches that we have found useful in structural work on disordered protein arrays. The chapter covers all aspects of the work from data collection to motif classification and averaging. The software used in these studies includes *protomo* for tilt series alignment (13, 14), *i3* for motif alignment and classification (15). Accessory programs have been written for specific functions and these have been combined into shell scripts to handle more complicated tasks that are commonly performed. Table 1 provides a list of *i3* modules and a brief description of their function so that comparable modules in other software packages can be substituted if desired. *protomo* can be downloaded free at http://www.electrontomography.org/.

2. Principles of Electron Tomography

Space limitations do not permit a detailed treatment of the principles of ET. For this, the interested reader is referred to two excellent books on ET (16, 17). Here we will limit the discussion to those aspects of ET that directly affect the visualization of molecules within a paracrystalline specimen. Molecular ET has many characteristics in common with single particle 3D image reconstruction, but differs in several key respects. Data are collected using tilt series, which must be aligned first to produce the tomogram. This is followed by alignment, classification, and averaging of motifs to improve the S/N ratio and determine heterogeneity. The treatment of defocus (18, 19) also differs but is not covered here.

Dose fractionation is one of the foundations of ET (20–22). Dose fractionation defines the relationship between the S/N ratio in the individual projections with the S/N ratio in the final tomogram. Simply put, if a tilt series can be recorded and aligned using the same total electron dose as a typical projection, the tilt series has the same statistical significance as that projection. However, the two are not equivalent. The tomogram also has higher contrast than its projection, a quality that can be deceptive as it suggests that motifs in tomograms are "better" than projections of single particles recorded with the same electron dose. In fact they are not better; they have the same statistical significance. Consequently, just as reference bias can lead to a faulty result in single particle reconstruction from noisy images, reference bias can also lead to a faulty result in motif averaging in ET.

Table 1
i3 Software modules and their functions

Multireference alignment scripts	
i3init.sh, i3make.sh	Create initial database, reads positions, transformations, and missing wedge geometry for data subsets. Select and define alignment set from subsets
i3align.in	Script template for multireference alignment. Each subset can be aligned concurrently. Calls the i3align executable
i3totsum.sh	Calculate global average. Calls the i3window executable
i3msa.sh	Compute eigenvalues/eigenimages for multivariate statistical analysis. Calls the i3window and i3svd executables
i3class.sh	Hierarchical ascendant classification, class averages and dendrograms. Calls i3hac and i3window executables
i3select.sh, i3selectalign.in	Select class averages, align the selected averages with respect to each other concurrently, based on the script template
i3selectapply.sh	Applies alignment transformation obtained by aligning class averages to the transformation of the raw subvolume, stored in the database
Executables that operate on whole data sets	
i3dataimport, i3datalist	Imports and exports positions and transformations into/out of the database
i3window	Extract motifs from raw tomograms based on positions and transformations stored in the database. Produces either image stacks or averages
i3align	Single- or multi-reference alignment
i3svd	Computes a singular value decomposition of image vectors for multivariate statistical analysis
i3hac	Hierarchical ascendant classification, based on the results of i3svd
Executables that operate on images	
i3cut	Extracts a window
i3paste	Inserts an image, overwriting pixels in the target image
i3mask	Apply masks to real space and Fourier space images
i3fourier	Fourier transform driver; calls selectable transform algorithms
i3compute	Perform arithmetic operations with images
i3project	Compute projection in x, y, or z-direction
i3resample	Apply a linear transformation to an image
i3concat	Creates an image stack from individual images
i3avg	Produces an average and/or variance image from an image stack
i3montage	Produces a montage of individual images or image stacks
i3stat	Print image statistics
i3display	Graphical display program and manual motif picking program

3. Tilt Series Data Collection

The collection of the tilt series is the most important aspect of ET. The results obtained by motif averaging and classification will be dependent on the quality of the tomogram that in turn will depend on the quality of the tilt series and the accuracy of the tilt series alignment. Several important factors impact the quality of the tomogram: the tilt increment, the choice of dual-axis or single-axis tilt series, defocus, and radiation damage.

3.1. Fixed Increment or Saxton (Cosine) Scheme

Data collection in both electron crystallography and ET involve tilting the specimen. The main difference is that in ET, all the images of the tilted specimen come from the same array; no assumption is made that any two arrays are identical. In electron crystallography, each array is assumed to be identical, but oriented variably with respect to the tilt azimuth. Typically in electron crystallography, it is assumed that the highest resolution data on each lattice line in the Fourier transform will be spaced at a distance no greater than the inverse of some multiple, which may be 1, of the crystal thickness. This distance in ET when expressed as an angle is often referred to as the tilt angle increment. The tilt angle increment is tightly correlated with the specimen thickness and the desired resolution. The number of views needed for a desired resolution is generally determined from the formula

$$N = 180^{\circ} D / d, \qquad (1)$$

where N is the number of images, D is the specimen diameter, and d is the desired resolution (23). An alternative formula makes use of the principle that the Fourier transform of an object changes at a rate that is proportional to the specimen thickness, $1/D$, and must be measured at least as frequently at the resolution desired. Thus,

$$\theta = \tan^{-1}(d / D), \qquad (2)$$

where θ is the increment between images of the tilt series. Both equations express the intuitive feeling that the thicker the specimen, the more rapidly it changes appearance with tilt angle and that fine features change their appearance more rapidly than do coarse features. Equation 1 is strictly speaking valid for spherical objects or cylindrical objects tilted about their axis; these objects do not change in thickness as the tilt angle increases. Hence, tilt series images can be collected in equally spaced increments. However, generally speaking, electron microscopy specimens such as paracrystals are slabs and thus increase in thickness as the specimen is tilted making Eq. 2 the more correct form. That the image changes more rapidly as the tilt angle increases can have important consequences if cross-correlation is used for the tilt series alignment, a procedure known as marker-free alignment. Thus, in our work we emphasize tilt increments determined by the Saxton scheme (24), which decreases the tilt increment by the cosine of the tilt angle exactly in tune with the increase in specimen thickness.

3.2. Single-Axis Versus Dual-Axis Tomography

It is intuitively obvious that more data will lead to better feature definition in the final reconstruction. Hence, dual-axis tilt series will generally lead to better final reconstructions than single-axis tilt series, especially when motif averaging is not anticipated. This fact has to be balanced with other considerations, such as radiation damage and defocus determination. Radiation damage is a serious limitation in ET of frozen-hydrated specimens, which will not tolerate excessive electron exposure. Negatively stained and plastic embedded sections *appear* more forgiving because they don't disappear with excessive electron exposure, and also are more limited with respect to ultimate resolution. Measurement of defocus in extremely low exposure images, such as those taken in cryoET, is very problematic. Thus, the extra images required for a dual-axis tomogram may be counterbalanced by the need to measure and correct for defocus to obtain the highest resolution.

In muscle fibrils, the filaments are parallel in a typical thin section. The grid, with its ribbon of sections can easily be positioned on the specimen rod with the filament axis roughly parallel to the tilt axis, which is the optimal orientation for filament visibility. Would this kind of specimen benefit from a dual-axis data collection scheme under these circumstances? A single-axis tilt series recorded about the filament axis provides the best rendition of the filament profiles. On the other hand, the cross-bridges connecting the thick and thin filaments are generally oriented perpendicular to the filament axis; an axial orientation is least optimal for their visualization. The most recent report on the structure of cross-bridges visualized in IFM using ET utilized dual-axis tilt series and showed the best cross-bridge resolution to date (3). This would suggest that there are some advantages to using dual-axis tilt series if they can be conveniently obtained.

Dual-axis tomograms of ice-embedded specimens have been obtained (25). However, if motif averages are required to obtain the desired information, and motifs can be obtained over a wide range of orientations within a set of tomograms, then single-axis tomography should be sufficient to obtain an *average* whose transform samples all of Fourier space or at worst leaves only a missing cone of data. This has been the case with many of the studies that have extracted molecular images from tomograms of frozen-hydrated specimens for classification and averaging (1, 10, 26, 27).

Special stages designed for side entry goniometers are available for the collection of dual-axis electron tomograms even of ice-embedded specimens (28) and some newer electron microscopes such as the FEI Polara and Titan-Krios have built-in dual-axis capability. However, when a rotation or "flip-flop" holder is unavailable, extracting the grid, rotating it 90° in the holder and then collecting a second tilt series is a simple solution. Remarkably, this has been done for frozen-hydrated specimens and an entire second tilt series collected *manually* (29). The separate tomograms can be merged into a single tomogram (30, 31) or the two tilt series images merged simultaneously (32). Conical tilting (33) is an alternative to dual-axis tilt series for plastic embedded and metal replicas, but not at this time for frozen-hydrated specimens.

3.3. Radiation Damage

From the earliest days of cryo-EM, it was realized that ice-embedded specimens suffer catastrophic damage (34, 35), later referred to as "bubbling", when the electron dose exceeded that which completely destroyed the crystalline diffraction (36). Thus, practitioners of cryoET typically try to approach the dose for terminal damage without exceeding it. Negatively stained specimens do not display the same terminal damage effect, but even they suffer from radiation damage (37). Plastic embedded specimens often "appear" more resistant to electron irradiation because of their inherently low resolution. However, even they display radiation damage effects, such as specimen thinning, that are more subtle (38). Progressive change in the thickness of the section during tilt series data collection will degrade the final tomogram. Specimen thinning is most rapid in the early stages of irradiation and after an initial rapid change, the rate becomes much slower and can be minimized using minimal electron exposure methods. However, in the dose range of ~50 $e^-/Å^2$, about 40% of the typical exposure used to collect a tilt series of ice-embedded specimens, there is virtually no shrinkage in the *z* direction (39).

More recently, Wu et al. (3, 40) repeated a study done in 1999 on isometrically contracting IFM (41) using motif classification and averaging. The new data were recorded automatically using a minimal dose procedure whereas the 1999 data were recorded manually while the specimen was continually irradiated resulting in a high total electron exposure. The recent results using motif averaging resolved the helical array of actin subunits on the thin filament. The same procedures carried out on the 1999 data could not, even though the sections were cut from the same specimen block. This observation suggests that even with plastic embedded specimens, continued electron irradiation will lead to progressive destruction of molecular details of the specimen.

Because of differences in the characteristics of CCD cameras, it is advisable to record an exposure series of frozen-hydrated specimens using the conditions that are judged suitable for the type of ET data desired. This involves simply recording images repeatedly of a suitable area of the specimen until "bubbling" is observed. If the specimen is suspended over a hole, bubbling is usually observed first on the surrounding matrix and only later within the hole itself.

4. Missing Wedge

The nature of electron microscope specimens, which are thin slabs of material, means that any tomogram will be missing some views of the specimen and thus the final 3D image comes from incomplete data. Single-axis tilt series have a missing wedge that is usually not larger than ±30° and usually not smaller than ±20°. This missing wedge can affect motif alignment, which is generally based on cross-correlation, thereby requiring special weighting schemes (42). It can also influence motif classification since it can affect the visibility of certain features, especially at low resolution, and therefore bias classification in favor of missing wedge orientation rather than structure. On the other hand, when motifs are obtained from multiple tomograms of paracrystals with different tilt azimuths, an average structure can be produced with only a missing cone of data rather than a missing wedge.

The goal of 3D motif classification is different from the goal of 2D motif classification in single particle reconstruction. Classification of projections in single particle reconstruction is largely used to identify different orientations of ostensibly identical objects. Because ET is usually applied to specimens that are heterogeneous, motif classification in ET is largely used to identify heterogeneity; if there were none, spatial averaging would be more appropriate. The power of 3D motif classification lies in the fact that far more of the Fourier transform of the motif is measured than is missing, so the structure and thus its variability is better defined. Because of the preferred orientation enforced by the array, when motif classification is applied to paracrystals, there will always be some regions of the transform that are missing from each motif and some regions that are held in common among all motifs. Thus, the transform of each motif will usually have in common the central section representing the in-plane projection and varying degrees of overlap of the rest of the transform according to the orientation and size of their missing wedges.

All atoms in the specimen contribute to all Fourier coefficients in the molecular transform. Thus, even with a ±30° missing wedge no specimen feature should be entirely missing from a tomogram. However, at low-resolution where ET is generally applied, the transform of certain low-resolution features may be concentrated in a small region of Fourier space. The two best-known examples are the density profile across a lipid bilayer and the mean radial density distribution of a filament. The membrane profile is a 1D projection of the density onto the normal to the membrane. Tomograms of lipid vesicles or native membranes usually reveal the lipid bilayer only along the lateral edges of the vesicle where the transform of the membrane profile falls within the region of measured data; the bilayer is invisible at the top and bottom (43) where the transform of the membrane profile comes to lie within the missing wedge. Thus, classification of integral membrane proteins can be strongly influenced by their orientation with respect to the missing wedge (15).

Information on the mean radial density distribution of a filament is concentrated in the equator of the Fourier transform, for which the power spectrum has the shape of a disk passing through the origin and oriented perpendicular to the filament axis. When the resolution is so low that the subunits of the filament are not visualized, the filament appears as either a stick (F-actin or intermediate filaments) or a straw (microtubules) and the equator contains all the information about the filament. When the filament equator falls within the missing wedge, which occurs when the filament axis is perpendicular to the tilt axis and the filament lies in the specimen plane, the filament will all but disappear in the tomogram (31). Much of the incentive for dual-axis (30, 31) and conical tilt (33) methods in cellular tomography arises from this effect.

In our IFM work, multiple tomograms are collected from identically oriented sections and thus, the missing wedge orientation is identical over the population of motifs effectively eliminating the missing wedge as a parameter in motif classification. In the other paracrystalline specimens described below, the visibility of the membrane has not been an issue as all have been planar. However, the mean radial density distribution across elongated structures has been found to have an effect (15).

The missing wedge also affects motif averages, particularly if the averages are computed in real space, where the point spread function corresponding to the missing wedge is convoluted over the motif. The missing wedge is defined in Fourier space and can thus be treated explicitly only if averages are computed in Fourier space (44). Averaging in real space effectively down weights amplitudes of measured Fourier coefficients; averaging in Fourier space can avoid this using proper weighting. The *i3* software counts the number of actual measurements contributing to each Fourier coefficient therefore accounting explicitly for the missing wedge.

5. Alignment of Tilt Series and Computation of the Tomogram

Alignment of tilt series using gold clusters added as fiducial markers is a popular method and can be done using software packages such as IMOD (45). An alternative method, known as "marker-free" alignment (reviewed in ref. 46), aligns the tilt series based on intrinsic specimen features and this is the method we use exclusively (2, 13, 14). Marker-free alignment works readily with thin specimens such as paracrystalline arrays which have the best potential for providing high resolution of macromolecular assemblies. We have also obtained excellent alignment for several different kinds of ice-embedded specimens up to ~200 nm thick (5–7). Thicker specimens are more difficult. A hybrid method proposed by (47) benefits from both techniques: an initial marker-based alignment is refined subsequently by marker-free alignment. This speeds up the more time-consuming iterative marker-free alignment.

The software implementation for our method of marker-free tilt series alignment (*protomo*) is a Python extension module that can be run from within the *SPARX* environment (48). It can be downloaded from http://www.electrontomography.org/software/. The steps used in the image processing of a typical specimen (Fig. 1) are described below. Steps 3–5 are iterated; step 4 is optional.

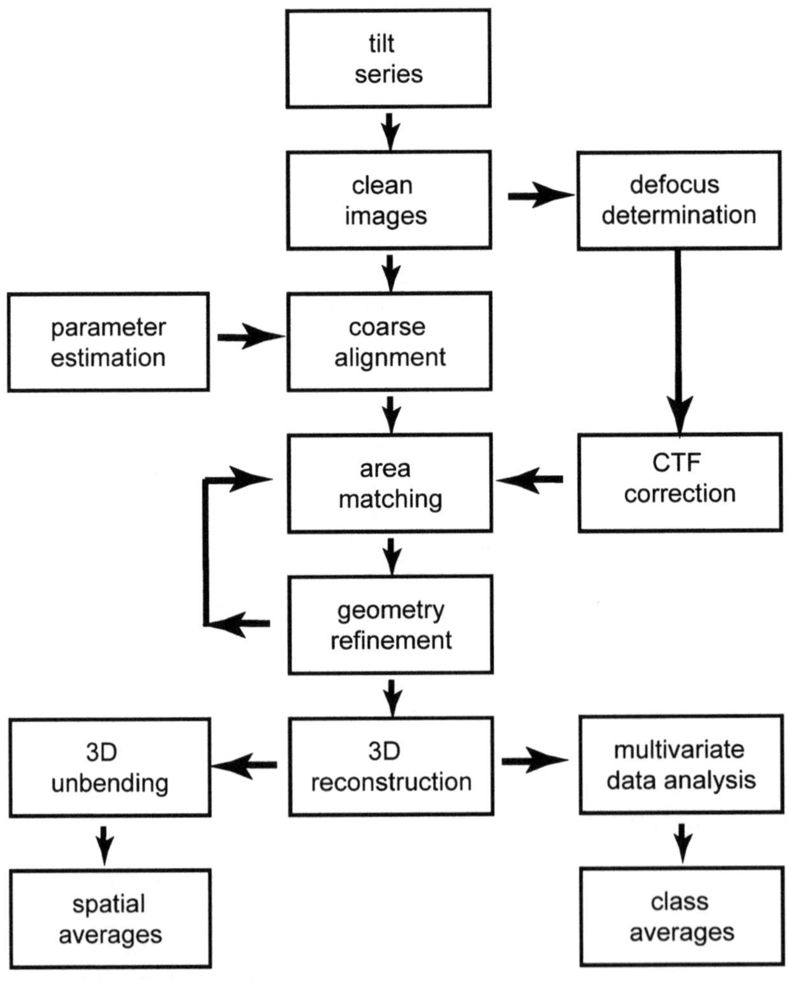

Fig. 1. Tomography flowchart. From ref. 13.

1. Raw images are initially cleaned to remove density gradients and density outliers (hot or cold pixels). They can optionally be binned to reduce the image size and speed up the alignment in the initial stages. Images are low pass filtered to the first zero in the contrast transfer function.
2. The initial alignment of the tilt series starts with the image of the untilted specimen as the reference and progresses successively to higher tilt using the neighboring image at the adjacent lower and higher tilt angle as a reference. For the first alignment the tilt angles from the goniometer and the tilt azimuth obtained from previous tomograms collected at this magnification are used as parameters. Coarse alignment takes only minutes after the last image in the tilt series is collected and is carried out either automatically for the whole tilt series, or manually with a graphical software tool (*tomoalign-gui*) for problematic images or tilt series (Fig. 2).
3. An "area matching" refinement scheme is undertaken in which each member of the tilt series is aligned to a reference obtained by reprojecting a preliminary back-projection. This process also starts with the 0° tilt and proceeds to lower and higher tilt angles while successively including each aligned image in the preliminary map computation.

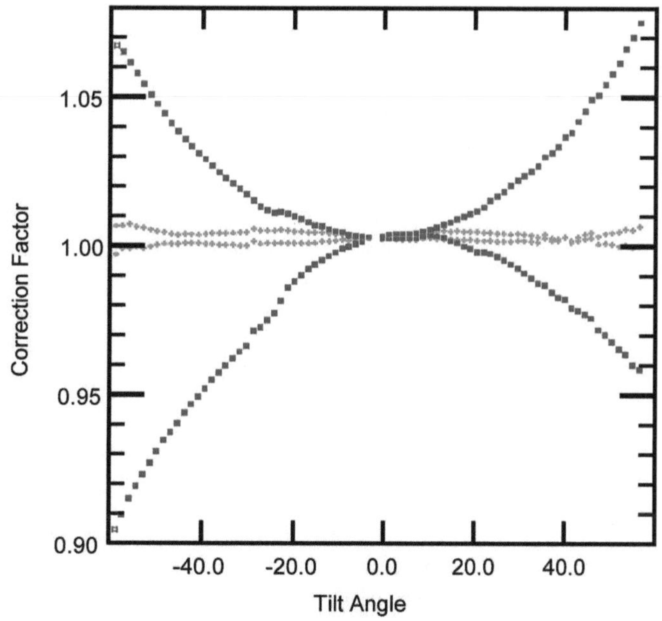

Fig. 2. Singular values of the correction matrix after the first iteration (*filled square*) and after the last iteration (*filled diamond*) for an ice-embedded specimen. Once an accurate estimate of the tilt azimuth has been obtained, the singular values usually deviate less than 1% from a value of 1.

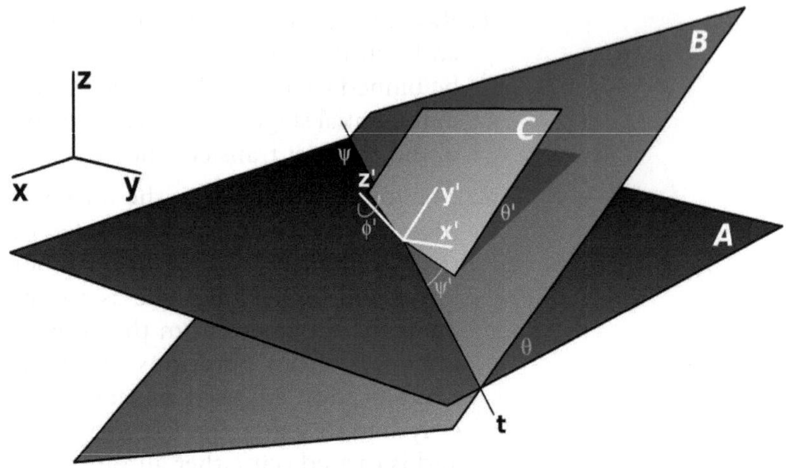

Fig. 3. Tilt geometry. (*x*, *y*, *z*): coordinate system fixed with respect to the microscope. *z* is the optical axis, *A* (the *x*-*y*-plane) is the image plane. (*x*′, *y*′, *z*′): coordinate system fixed with respect to the specimen. The transformation from (*x*, *y*, *z*) to (*x*′, *y*′, *z*′) consists of a tilt about the axis *t* and angle θ and an additional rotation (ψ′, θ′, ϕ′) which defines the orientation of the specimen (*C*) with respect to the specimen holder (*B*). From ref. 13.

4. A back-projection map can be calculated at this point to produce a 3D reconstruction. If this is an intermediate step, it is recommended to reduce the size of the map in order to speed up the calculation. The size need not be equal to the area that is used for area matching. This reconstruction can be examined to determine whether the specimen has the features desired. If so, the tomogram is further refined using area matching. A better estimate of the specimen thickness is also obtained from this back-projection and is needed as a parameter for the back-projection weighting algorithm.
5. Geometry fitting takes the alignment parameters obtained from the area matching of each image and performs a least squares fit to obtain a new estimate of the tilt series geometry in an attempt to determine the in-plane rotations of the images, the common tilt axis (tilt azimuth) and the three Euler angles that define the specimen orientation relative to the specimen holder more accurately (Fig. 3). Included in the rotation defined by the Euler angles is the deviation of the nominal untilted image from 0°.

Area matching determines four parameters of a linear transformation to match each image to the reference (49). The total transformation is split into two terms, the estimated transformation (which is an input to the area matching) and a correction term. Two parameters of the correction transformation represent the amount of stretch or compression in two orthogonal directions, the other two are the direction of the stretch/compression relative to the corrected and uncorrected image. The stretching/compression factors are plotted as a function of the image number and indicate the progress of the tilt series alignment after each iteration of area matching (Fig. 2). The plot of the stretching/compression factors should be flat across the tilt series when alignment and area matching are completed and ideally have a value of 1. In practice, good factors can usually be obtained for the lower tilt angles, while they occasionally deviate more at the high angles. The quality of the alignment is also assessed by the shape and height of the cross-correlation peaks. The peaks should be sharp and strong relative to the local background, but they usually broaden in a direction perpendicular to the tilt axis as the tilt angle increases due to the foreshortening of the projection in the specimen.

protomo uses as the reference image a reprojection of the most recently calculated back-projection. A reprojection is a much better reference, statistically speaking, than any single image in the tilt series and also prevents errors from propagating through the alignment process. Four different cross-correlation functions (CCFs) are available in *protomo*: (1) standard, (2) mutual (50), (3) phases only (51), and (4) phase doubled (52). Phase only CCF, in which amplitudes of all Fourier coefficients are set to 1.0, produces especially sharp correlation peaks even in spatially redundant specimens but requires good alignment before a peak can even be identified. The mutual CCF, which utilizes the square root of the amplitude in the product of the two transforms, is useful when the CCF is periodic or dominated by strong low frequency terms, e.g., muscle. The phase doubled CCF doubles the magnitude of the shift thereby increasing the accuracy.

6. 3D Unbending

One type of heterogeneity that plagues electron crystallography is specimen flatness (53–55). The tomogram contains information on the departure from flatness of a paracrystal since the z coordinate of the motif can be determined thereby providing the potential to correct specimen flatness computationally rather than experimentally. Specimen flatness is particularly likely to affect 2D arrays formed on lipid monolayers, which are most efficiently recovered over holes in reticulated carbon films (56, 57) and whose flatness will be affected by surface tension at the air–water interface. A 3D "unbending" scheme to correct for out-of-plane bending was developed for IFM (58) and used subsequently in a "column averaging" procedure to obtain spatial averages along only the filament axis (41). The approach has not yet been applied to well-ordered 2D crystalline arrays to correct for specimen flatness.

Table 2
Software modules for 3D unbending

3D unbending scripts	
tomoavg2.sh	Spatial averaging
tomomot.sh	Create reference motif for cross-correlation
tomocorr.sh	Calculate cross-correlation map
tomosearch.sh	Determine lattice from observed cross-correlation peaks
tomofit.sh	Generate an unbent grid from lattice points by spline fitting
tomounbend.sh	Reinterpolate raw map based on the grid
Executables that operate on images	
i3createmask	Create a periodic mask at reciprocal lattice points for filtering repetitive image structures
i3peaksearch	Peak search for lattices in cross-correlation maps
i3unbend3d	Spline fitting and reinterpolation

Correction for specimen flatness is also related to the problem of identifying motifs in the paracrystalline array using cross-correlation. The software used for 3D unbending is listed in Table 2 along with the functions. The following procedure can be used to calculate a cross-correlation map with the goal of correcting a 2D crystalline array for out-of-plane bending.

1. Lattice parameters are determined from a projection of the 3D map, possibly from a slice of the map to reduce noise contributions of the outer sections. Software for conventional 2D array processing such as the MRC package (59) or *2dx* (60) can be used for this purpose.
2. An averaged 3D image is calculated by filtering out aperiodic information in the Fourier transform. The filter function is generated using *i3createmask* by convoluting an array of delta-functions placed at the reciprocal lattice positions with a mask function of rectangular, ellipsoidal, or Gaussian shape, optionally apodized in the rectangular and ellipsoidal case. For arrays of helical filaments, where the transform is continuous in one direction (layer lines), the mask function is extended to the full width of the transform in two dimensions perpendicular to the filament axis.
3. From the averaged 3D image, a reference motif is selected comprising one or more unit cells. Optionally the reference can be generated by other means, in which case step 2 can be skipped. One such possibility would be the use of a raw motif extracted from the tomogram.

4. A cross-correlation map is computed from the tomogram and the generated reference using *tomocorr.sh*.
5. A peak search is carried out using *tomosearch.sh* to determine the location of the correlation peak maxima. This is done in two steps. First, a 2D search determines the x and y coordinates of the maxima from a projection of the correlation map, or better, from a central slice of prespecified thickness through the map. Then, a 3D search within a small window centered at the previously determined positions and a user-provided z-height locates the actual maxima in the 3D correlation map.
6. A smooth 2D surface is fitted to the measured correlation peaks using *tomofit.sh*. The fitted spline representation is evaluated on a regular grid of points corresponding to the original sampling points. Points above and below the fitted 2D surface are assumed to lie on lines parallel to the local surface normal in equidistant intervals, corresponding to the lateral sampling distance.
7. The final unbent map is computed by trilinear interpolation of the original map at the points computed in step 6 using *tomounbend.sh*.

7. Selecting Motifs

Selecting motifs from tomograms of paracrystals can be done by either using cross-correlation methods as shown above for 3D unbending or by manual picking. Because the paracrystal is likely to be thin with the motifs positioned within a flat plane, a cross-correlation map can usually be calculated from the tomogram projection and peaks fit to obtain a set of x, y coordinates using any of the lattice fitting programs used to analyze 2D crystals (60–63), or the first part of the unbending procedure described above. The z coordinate corresponding to the center of the tomogram will often suffice since the position will be refined later.

Tomograms of paracrystals often contain many separate domains. When this occurs, a CCF is calculated for each domain and a separate lattice fit. Even in IFM, where the motifs usually have at least some similarity, because they are the product of interactions between filaments arranged in parallel, motifs can be selected by fitting a 3D lattice to a cross-correlation map (40).

The use of cross-correlation to identify 3D motifs in paracrystals is straightforward, but is not easily applied to randomly oriented motifs, such as envelope spikes on a virus or large individual particles such as ribosomes. The correlation peak height is the measure of similarity between the reference and the “raw” motif,

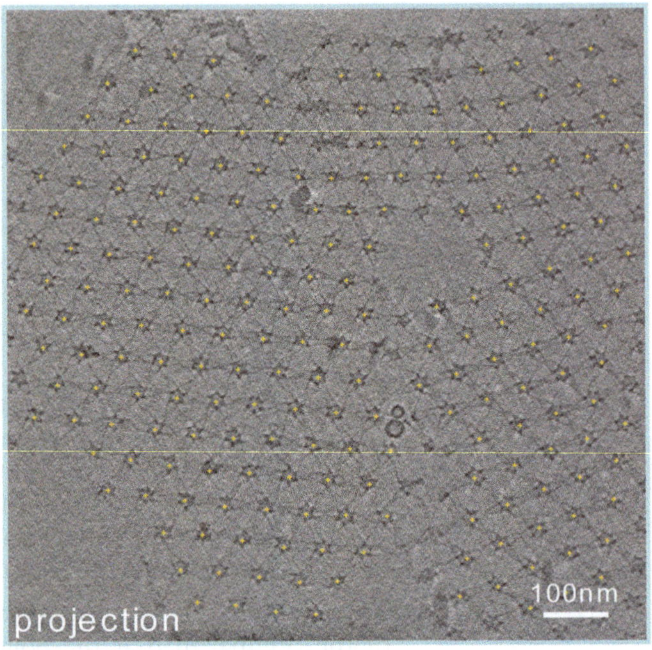

Fig. 4. Manual picking using *i3display*. The image shown is a projection from a tomogram of 2D paracrystalline arrays of myosin-V (1). The centers of so-called "flower motifs" are marked in *yellow*. Had the image been the actual tomogram, the *z*-coordinate of these points would be color coded: yellow for points at the same *z*-level as the currently displayed section, blue for points below and green for points above that level. Image courtesy of Dr. Jun Liu.

and as such depends on similarity in both orientation and structure as well as the local normalization, and this in turn is affected by the missing wedge (64). Thus, similarities in particle structure and alignment could be offset by missing wedge orientation. This should not present a problem for paracrystals because their preferred orientation within a narrow volume leads to less variability in orientation and thus less effect from the missing wedge and the local normalization.

The alternative approach to motif selection is a manual pick. The image display program, *i3display*, allows the user to select and mark points in a 3D map or a 2D image (Fig. 4). *i3display* is completely general and not specifically designed for paracrystalline arrays. Besides displaying grayscale images, it can thus be used for such diverse activities as picking randomly oriented particles, spikes on envelope viruses and filaments.

If the paracrystal is reasonably flat but too poorly ordered for fitting a CCF, the boxer utility of EMAN (65) is a good alternative for manual picking of motifs from the in-plane projection of the tomogram. A single value for the *z* coordinate can then be added to the list of particles.

8. Motif Alignment

Before classifying 3D images they must be aligned to an appropriate reference. Invariant functions such as the autocorrelation function that reduce the angular search are not generally available for volumes. Some specialized functions have been constructed, such as an adapted version of the double-autocorrelation method (66), or using spherical harmonics (67). The drawback of the latter is that phase information must be discarded to achieve rotational invariance, and as a consequence these methods, either in 2D or 3D are not as robust and effective as standard cross-correlation alignment.

8.1. Reference-Free Alignment

Reference bias must be avoided in order that the result would not depend on the choice of reference. To achieve this, we have adapted a method known as "alignment by classification" (68) to tomographic motifs (15). In this method, raw motifs are not aligned and instead averages obtained in a motif classification step are aligned by multireference alignment and the resulting transformation applied to the class members. For the first alignment step, a simple translational alignment is made to an initial global average. This average generally lacks fine detail. Subsequent alignment and classification will refine fine detail, but only if it appears in the class averages.

Procedures for the initial alignment depend on the orientational degrees of freedom of the motifs, of which the three angular degrees are of primary concern. Application of prior knowledge can greatly reduce the angular search. One example has been described by (15) where the orientation of Env spikes on SIV or HIV virions was computed by fitting an ellipsoidal surface to the manually picked spike positions, under the assumption that the spikes are oriented perpendicular to the surface. The result of the fit defines the geometrical center as well as diameters and orientations of the major axes from which the normal at the position of the spike can be computed. By fitting to an ellipsoid rather than a sphere, deformation of the virion envelope can be approximately accounted for. The program *ellipsofit* in *i3* produces this fit using as input the spike coordinates for individual virions.

Muscle sections are often cut in variable orientations through the filament even when care is used to obtain specific orientations. For example, 80 Å thick transverse sections through the sarcomere are often cut at an angle to the filament axis that can be as large as 20° without appearing to be oblique. In these cases, the following procedure has been found useful to define the initial orientation of the filaments and thus reduce the angular search range.

1. The start and end point of a typical filament, either thick or thin filament will do, is selected using *i3display*. It is not necessary to select two points for all filaments due to the paracrystalline nature of the muscle.

2. Using *i3resample* in concert with *i3display*, the three Euler angles that are needed are determined one by one. These angles can be computed from the two points but it is often essential to verify that the order is applied correctly and for this *i3display* is used. *i3resample* utilizes first a rotation about the *z*-axis, then a rotation around the resulting *x*-axis and a final rotation about the resulting *z*-axis.
3. The program *i3euler* is used to generate a 3×3 rotation matrix by giving it the three Euler angles in the same order as determined in step 2 above. This matrix is used as the starting orientation for the angular alignment.

The positions of all the motifs within the paracrystal, which can be the center of the filament along the *z-axis* or the center of a repeating motif, are computed and defined as the rotation center. The centers and rotation matrices are imported from a text file into a database which will be used throughout the motif alignment and classification. The alignment proceeds as follows.

1. Compute a global average of all picked motifs using *i3totsum.sh* after application of any prior orientation information.
2. Align by translation alone, all of the motifs to the global average using the alignment scripts generated from the template script *i3align.in*. The generated scripts can be run in parallel.
3. Classify all motifs using *i3msa.sh* and *i3class.sh*. For this, a mask must be constructed (see below).
4. Compute class averages.
5. Align all class averages with respect to each other using *i3selectalign.in*. From among all the multiple orientations for each class average, the alignment parameters are selected from the single class that provides the best overall alignment for all the other classes.
6. Apply the alignment transformations of each class to the respective class members using *i3selectapply.sh*.
7. Steps 4–6 are applied iteratively until one is satisfied that no further improvement can be obtained. Processing parameters are specified in a parameter file, which is applied to a single cycle (steps 4–6). Multiple parameter files can be used if parameters change.

No bias is introduced since no arbitrary reference selection takes place. Also, the alignment is carried out with class averages that have a higher S/N ratio, so that it is more robust than a multireference alignment of raw motifs. The number of classes computed in this procedure should be chosen as high as possible, because the goal is to capture as many different spatial orientations as possible in addition to the structural differences. After completion of the alignment procedure, the final classification is based on the expected number of structures.

A problem in the classification of tomographic data is the effect of the missing wedge, which tends to group motifs according to the orientation of the missing wedge rather than structural differences (44). One way to overcome this problem is the use of constrained cross-correlation as a similarity measure for the classification (42). A similar approach based on the signal overlap of a pair of motifs in reciprocal space has been described (67). Additionally, there is also evidence that classification without explicit missing wedge compensation does not necessarily lead to an orientation based grouping of the motifs. A judicious choice of the classification mask can essentially eliminate the effects of the missing wedge. For details see ref. 15.

8.2. Reference-Based Alignment

Although it is generally best to use a reference-free alignment scheme to avoid reference bias, in some cases, the S/N ratio of the raw tomogram may be sufficiently high and the heterogeneity in the motif structure comparatively low that a reference can be chosen from among the raw motifs with relatively little risk. Liu et al. (69) adopted a reference-based alignment scheme to classify motifs from rigor IFM so that actin subunits could be resolved. A similar approach has been used to visualize heterogeneous conformations in myosin and actin (70). In IFM, the least variable structure within each motif is likely to be the actin filament, which is reasonable at the resolutions typical of ET. Based on this assumption, Liu et al. (69) used the following approach.

1. A single actin filament is chosen as a reference from the partly aligned raw motifs selected using a cross-correlation map. This reference should have the features that you want enhanced after alignment, such as the staggered arrangement of actin subunits along the filament axis.
2. All the raw motifs are aligned to this reference using a small angular search within a cone of 4° half width about the filament axis.
3. This is followed by several cycles of MDA, using a mask that selects the actin filament, followed by multireference alignment.
4. The process is continued iteratively until the global average reveals the actin subunit structure along the filament.

The procedure is not guaranteed to resolve the actin subunits since that is also a function of preservation and staining.

9. Multivariate Data Analysis and Classification

Improvements in S/N ratio require averaging but this needs to be done within the context of identifying self-similar motifs so that S/N improvement is not achieved at the expense of the structural variation which may be at the heart of the investigation. Motif classification is the solution to this dilemma. Several classification schemes have been used in 2D image classification, among them K-means and hierarchical ascendant methods (HAC), e.g., (71) and Self Organizing Maps (SOM) (72). The classification of heterogeneous cross-bridge motifs in IFM have largely used hierarchical ascendant methods (73), but self-organizing maps have proven effective for this use because all motifs have the same missing wedge orientation. K-means classification has been applied to cluster 3D motifs of negatively stained integrins (74). More recently, maximum likelihood approaches have also been investigated (75).

In addition to improvement in S/N ratio, motif averaging can reduce the need for dual-axis tilt series as long as the motifs can be obtained over the largest possible range of tilt azimuth. Whatever method is used for classification one must be cognizant of the potential for missing wedge bias. Dual-axis tilt series reduce the missing wedge bias and perhaps the need for specialized classification algorithms, though it may not eliminate the problem.

9.1 Mask Construction

In motif classification using MDA, a Boolean mask is computed which defines the voxels that will be used in the classification. The mask is thus a critical component of the procedure. In alignment by classification, it is not necessary to use a single mask for the entire procedure. The mask can be varied throughout the procedure from a very simple one such as a sphere or a cylinder, to a more complex mask derived from the variance map of the global average or from the merged class averages.

Conceptually, volume masks are constructed the same way as projection masks but are somewhat complicated by the added dimension. One method (76) suggests computing a global average, thresholding the average, low-pass filtering, and thresholding again to produce the final Boolean mask. A variance map of the global average is a useful alternative and tends to concentrate the classification on the most variable regions. Masks based on the variance map have been used several times in motif classification (4, 40). Masks can also be computed from the variance maps of the individual class averages using the logical "OR" operation (66). The variance map approach will not be particularly useful until at least some partial alignment has been achieved and the variance map has some structure.

9.2. Number of Classes to Compute

The problem of motif classification particularly in contracting muscle is not one of binning a finite number of discrete states, but rather one of binning a continuum of varying structures. The problem begs the question of what is the most suitable number of bins. S/N improvement and recovery of structure variability are conflicting goals; S/N improvement is achieved by computing the smallest number of classes but recovery of structure variability

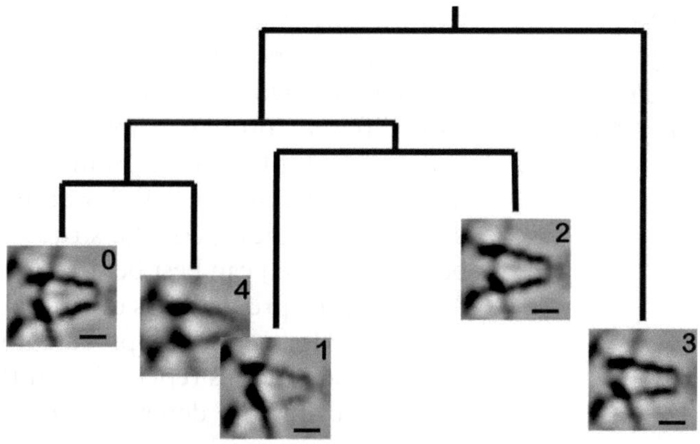

Fig. 5. Dendrogram showing the relationship between class averages of myosin-V petal motifs. See Fig. 8b legend for explanation of the structural elements. The class number is given in the *upper right hand corner*. Class 3 is the most isolated class because the myosin heads are more closely apposed than for the other four classes, all of which are related at one level. Bar equals 10 nm.

is achieved by computing the largest reasonable number of classes. HAC allows different numbers of classes to be computed from a single classification run but without an estimate of the optimal number of different classes needed to retain the image variance. SOMs are an alternative method to estimate the number of classes.

One useful tool in assessing the most correct number of classes to compute is the dendrogram, which is a graph produced during the classification by *i3class.sh*. The dendrogram shows the relationship between the different classes (Fig. 5). If the dendrogram shows a number of classes that are similar, then those classes can be merged into a single class. The example shown comes from a classification of myosin-V motifs (1,15) which have undergone MDA to produce five classes. Class 3 is noticeably different from the other four in that the myosin heads are closer together than for the other four classes. Classes 1 and 2, which the dendrogram shows are closely related, are obviously different. This difference will be explained below. Classes 0 and 4, which the dendrogram also shows are closely related, appear different because Class 0 has four times as many members as Class 4.

The number of factors (eigenimages) to use in the classification is also not readily defined because classifications optimizing a single feature can be performed using only a limited number of factors. Although the number of factors containing signal can be determined from a plot of the factors after sorting based on their eigenvalues, it is not necessary to use all of these factors for a classification. Factors that may bias the classification in favor of missing wedge orientation can be eliminated from the classification.

9.3. Example 1: Classifying Heterogeneity in Contracting Muscle

Contracting muscle is one challenging example of how image classification and alignment can resolve the highly heterogeneous set of actin–myosin interactions. Muscle has two sets of filaments with different protein compositions and different structures. To get the most information from tomograms of contracting muscle, multiple different sets of classifications would be necessary, each designed specifically for the structure of interest. How this might be accomplished can best be described using the following example. Although the procedure is described using utilities within *i3*, it was actually performed with a precursor version of *i3*. Where necessary, the original utility is replaced with the appropriate *i3* utility.

Wu et al. (40) described an approach for analysis of highly heterogeneous motifs in contracting muscle that may be applicable for other highly heterogeneous specimens. The approach aligned the set of motifs onto one of the common frames of reference, the actin filament, and utilized the knowledge that the alignment problem was largely one of determining the rotation about the filament axis, which can take on values of either 0° or 180° due to the helical structure of the thin filament. Motifs were defined as 38.7 nm segments of the actin filament centered between successive troponin complexes; successive motifs along each thin filament are rotated 180°. This rotation can be applied to motifs from individual filaments in the initial stages prior to alignment, but entire filaments are also rotated at random by 180° (77) and this has to be determined, but a validity check can be done knowing the rotation of their individual motifs. A multireference alignment scheme was implemented utilizing class averages that showed improved actin subunit detail. Finally, a single reference alignment was applied once the axial rotations were determined in order to place all of the motifs into a single frame of reference, this being necessary to facilitate quasi-atomic model building. Each of the 14 actin subunits in the repeat could then be examined for myosin head binding by a judicious choice of classification mask.

With a common frame of reference, the next problem was to obtain class averages of the myosin heads that were attached to the thin filament. Because the alignment procedure had minimized the variance within the actin filament, the classification mask could exclude those voxels that contained exclusively actin subunits, thereby facilitating a focused classification for specific regions of the motif repeat. A total of 12 different classification masks (Fig. 6) were constructed using the following general procedure.

1. Compute the variance map of the global average. The average and variance map are computed using *i3avg*.
2. Select an appropriate threshold. Usually, this is done by overlaying the global average and variance map in *chimera* (78) and adjusting the threshold. *Chimera* can also be used to edit the variance map to eliminate voxels in undesired regions, such as the thick filament backbone.

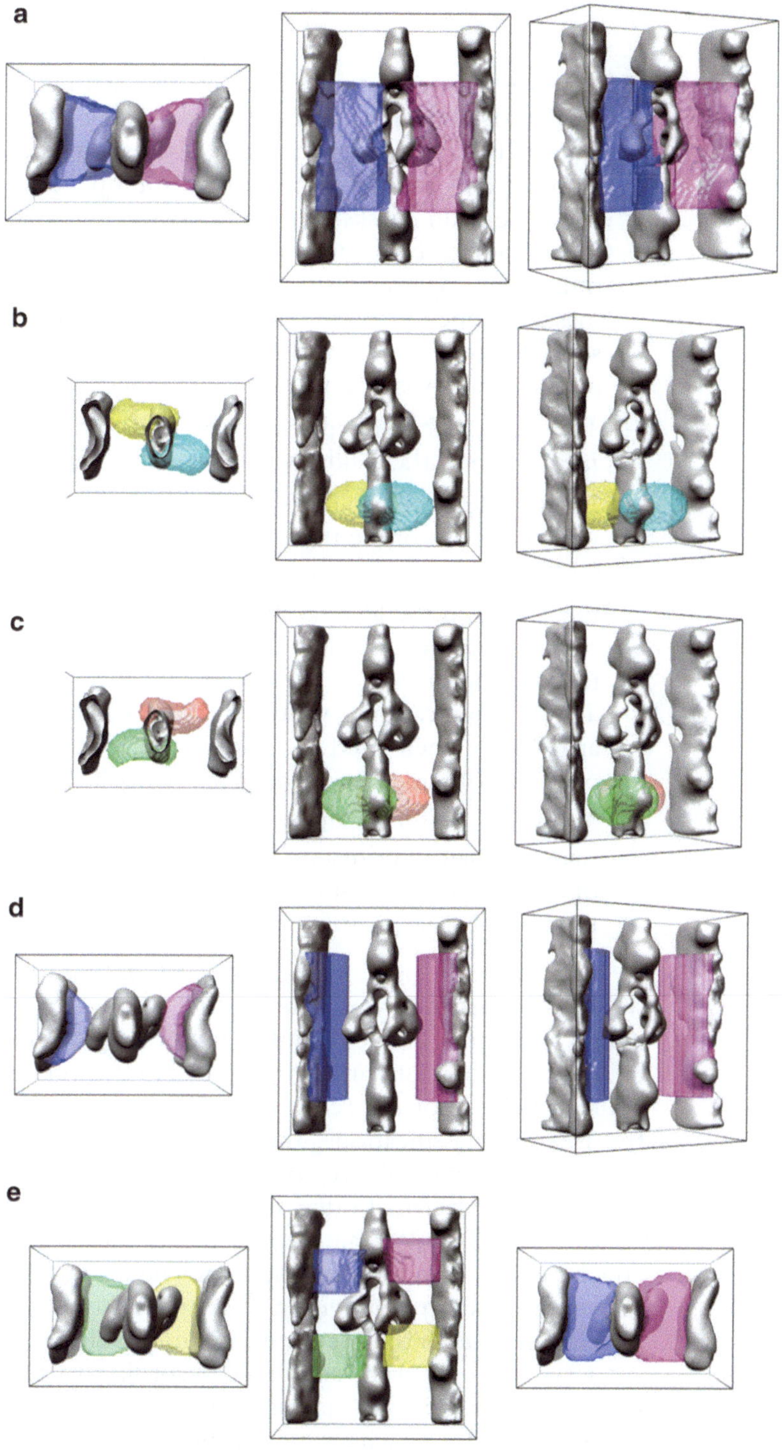

Fig. 6. MDA masks. Masks are shown as translucent, colored surfaces superimposed on the global average as a solid gray surface to provide a spatial reference point. (**a**) Surface display of *left* (*light purple*) and *right* (*magenta*) primary masks for primary cross-bridge classifications. The actin target zone, where most of the myosin heads bind, is positioned in the *middle* of the mask. View perspectives from *left* to *right*: top view (looking toward Z-line), front view (with the Z-line positioned toward the *bottom*), and tilted view. (**b**, **c**) Surface display of masks specific for troponin bridges. (**b**) Mask for back-left (*yellow*) and front-right (*cyan*) masks. (**c**) Mask for front-left (*green*) and back-right (*orange*) masks. (**d**) Surface display of masks specific for the myosin lever arms positioned near the thick filament surface. (**e**) Special mask to identify out-of-target-zone cross-bridges in four separate regions of the repeat. From ref. 40.

3. Remove the region occupied by the actin filament. Whole blocks of voxels can be removed using the *i3cut* utility. Alternatively, a complementary Boolean image can be constructed and the developing mask multiplied sequentially by these Boolean masks to remove entire regions, such as the thick filament backbone or the thin filament core. Images are multiplied together using the *i3compute* utility. This pair of masks, dubbed primary classification masks, is trimmed axially so that it covers all actin subunits except those to which a troponin complex is bound.
4. The mask is then divided into left and right hand sides using *i3cut* because these sides will be classified separately.
5. A set of four ellipsoidal masks are constructed covering the vicinal regions of the troponin complex.
6. To provide a cross-check on the primary mask classifications in the region of the lever arms, a key structural parameter of the contracting muscle, the primary classification mask was edited to eliminate the region occupied by the myosin motor domains. These classifications, which are based entirely on the lever arms, were then used to validate the lever arm's C-terminal positions obtained from the primary mask class averages.
7. A final set of four classification masks were constructed by trimming the pair of primary classification masks to cover only pairs of actin subunits outside of the target zone. These masks were used simply for quantifying numbers of attached cross-bridges.

The presence of numerous focused classifications required a reassembly procedure to make the results interpretable. We define reassembly as the recombination of regions from the different classifications into an improved S/N ratio version of the individual repeats. For the reassembly, Wu et al. (40) utilized six of the focused classifications, the two primary classification masks and the four troponin region masks. The primary classification masks captured cross-bridge structures in and around the target zone, while the troponin bridge masks identified myosin head attachments in and around troponin. The class averages only have meaning for the region within the mask. However, class averages were always computed for the entire motif volume.

Class averages were combined according to their class membership, which is to say that each motif was reassembled from the six class averages to which it contributed (Fig. 7). The procedure included the following steps.

1. The two primary mask class averages to which a particular repeat contributed are identified.
2. The left side of the left side class average and the right side of the right side class average were windowed separately using *i3cut*.

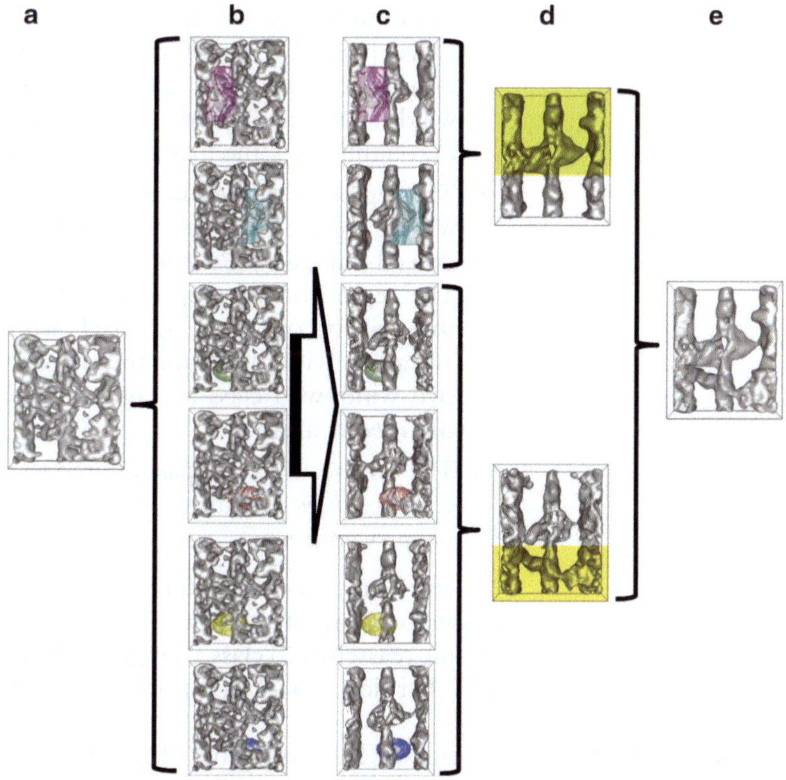

Fig. 7. Repeat reassembly. Different masked regions of each raw repeat are subjected to separate MDA and classification using left and right-side primary masks and four Tn region masks. Each classified region is represented by a corresponding class average. The original repeat is represented by a reassembled repeat from all these class averages. From *left* to *right*: column A: a raw repeat; column B: six different MDA masks, which are the same as depicted in Fig. 5, are applied to this repeat for separate MDA and classification with the mask superimposed; column C: class averages computed for each of the different masked regions of the raw repeat; column D: partially reassembled repeat, the *top* image combines the class averages from primary left- and right-side classifications and the *bottom* one combines the class averages from different Tn bridge mask classifications; column E: the whole repeat is reassembled by combining the *highlighted* portions in column D. Modified from ref. 40.

Floating and apodizing around the edges is not desirable here and voxels within the rejected region are simply set to zero.

3. The windowed left and right side class averages are then combined using the *i3paste* utility. If the raw repeat did not contribute to any troponin bridge class averages, the procedure is stopped here. If it contributed to troponin bridge classes, then the next few steps are followed.
4. Usually, there were no more than two troponin bridges in any single motif. However, because troponin bridges are distributed over a comparably large axial range, it was necessary to examine manually the axial extent of the volume that contained the

troponin bridge so this procedure could not be automated. Once the axial extent of the troponin bridge was determined, the information could be specified within *i3cut* to appropriately window out the left side and/or the right side troponin bridge in roughly the same manner as the primary mask class averages.

5. The region of the reassembled primary mask class average that will be replaced with the troponin bridge class average is windowed out using *i3cut* and the values of the voxels set to zero. The region of the troponin bridge reassembled class average that will be occupied by voxels from the primary mask class average is windowed out and the voxel values set to zero. The two windowed class averages are then added together using the *i3paste* utility.
6. The final result is low pass filtered to the same resolution as the class averages to remove any seams resulting from the reassembly procedure.

Implicitly, this procedure answers the question of how many classes to compute. If the data set consists of N motifs, then a unique reassembled class average can be obtained for each raw motif if $N^{1/2}$ left side and $N^{1/2}$ right side class averages are computed.

9.4. Example 2: 2D Arrays of Myosin-V

Several different isoforms of the motor protein myosin have been crystallized in 2D on lipid monolayers for structural analysis (1, 79–81). Most of these arrays were well enough ordered that spatial averaging could be used to solve the structure. However, 2D arrays of full-length myosin-V proved to be too poorly ordered for spatial averaging (Fig. 4). Images of negatively stained arrays show the presence of numerous vacancies within the lattice as well as points where insertions of new rows disrupted the alignment. This kind of disorder provided a perfect opportunity to see whether ET combined with motif averaging and classification could provide a higher resolution result.

The hexagonal unit cell was 653 Å in size and the transform of particularly extensive and well-ordered arrays extended to only the 5th or 6th order for a resolution of ~90 Å (Fig. 8a). The final result obtained by motif averaging from defocus corrected electron tomograms of frozen-hydrated specimens achieved 25 Å by an FSC criterion of 0.5. The cone-flower motif (Fig. 8b) showed sufficient detail that the calmodulin light chains could be uniquely positioned and the 15 Å wide coiled-coil dimerization domain was clearly seen.

For data collection, a Philips CM300-FEG electron microscope operated at 300 kV. Eight single-axis tilt series were collected at 43,200× magnification. The tilt series were collected using 3.5° cosine rule increments over an angular range of −70° to +70°. Each series consisted of about 60 images. According to Eq. 2, and an array thickness of ~100 Å this would be sufficient for

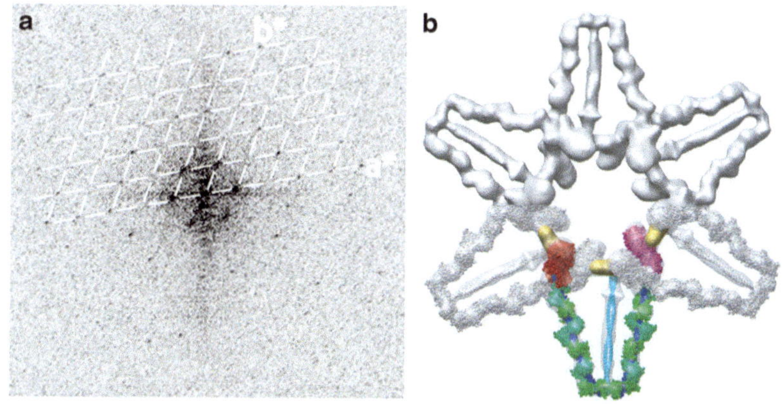

Fig. 8. Myosin-V paracrystalline arrays. (**a**) Power spectrum of a myosin-V array showing the resolution limitation. *White lines* define the reciprocal lattice. *Spots* extend to about the 5th order or a resolution of ~90 Å. (**b**) Molecular arrangement within the "cone-flower" motif. At the *top* is an opaque surface rendering viewed from the solvent phase onto the lipid monolayer; *bottom* is a translucent surface view with the myosin-V atomic model rendered in space filling. Color scheme for the *bottom* molecule: motor domains—*red* and *magenta*, light chains—*green*, heavy chain component of the lever arm—*blue*, coiled-coil domain—*cyan*, cargo-binding domain density envelope—*yellow*, adjacent molecules—*gray*. Note that the cargo binding domains have swapped binding partners and in the cone-flower motif bind the motor domains of adjacent myosin-V molecules. *Panel* **b** from ref. 1. *Panel* (**a**) courtesy of Dr. Jun Liu.

a resolution of ~6 Å so the tilt series itself was not resolution limiting. The pixel size at the specimen level, 5.56 Å, would have been resolution limiting for the tilt increment used, but the final resolution did not reach this limit.

The cumulative electron dose of each tilt series was about 30 $e^-/Å^2$ or about 0.5 $e^-/Å^2$ per image and ~15 e^-/pixel. At these doses and defoci and with an unstained specimen, the tilt series images themselves show literally nothing except at the highest tilt angles where the array just begins to be visible. However, the tomogram showed very distinctly the hexagonal array of cone-flower motifs in which six myosin-V molecules, the petals, are arranged symmetrically about a sixfold rotation axis.

Because a CTF correction was anticipated, the data were collected at three different defoci (−5, −8, and −12 μm). Defocus itself was determined with CTFIT (65). Focus gradient correction (18) was applied to each image taken at more than 30° tilt, and a Wiener filter defocus correction was used for images with less than 30° tilt. The defocus-corrected images replaced the low-pass-filtered micrographs for further tilt series refinement.

The first myosin-V structure (1) was obtained before the *i3* software was available and utilized both real space averaging without missing wedge compensation, and used a reference-based alignment. As part of the development of the *i3* software, the data were reanalyzed using reference-free alignment with missing wedge compensation (15). The following protocol is largely based on this reanalysis.

To obtain class averages of the myosin-V petal motif, the following procedure was used.

1. Cone-flower motifs are picked automatically from a cross-correlation map but using the projection of the tomogram. Each coneflower motif has *x*, *y*, *z* dimensions of 200 × 200 × 80 pixels. The *z* coordinate is initially assigned to be 0.
2. The first alignment cycle used projections rather than volumes to determine the single rotation angle and the *x*, *y* displacement.
3. All subsequent cycles of alignment by classification used the volume data and the angular search range over the three Euler angles was limited to a few degrees in order to take into account the wrinkling of the monolayers. Up to 60 class averages were used during the volume alignment step.
4. The final classification produced 20–80 classes but was ultimately reduced to five as the differences between many classes were slight.

The myosin-V study showed a clear effect of the missing wedge (Fig. 9). When carried out using volume data, Classes 1 and 2 had complementary distributions of orientations with respect to the tilt azimuth (Fig. 9a) that produced complementary effects on the density of the lever arm and cargo binding domains. Class 2 had higher density in the lever arm domain and lower density in the cargo binding domain, while Class 1 had higher density in the cargo binding domain and lower density in the lever arm. The cargo binding domains and the lever arms are approximately at right angles to each other. The orientations of these classes with respect to the tilt azimuth were such that the lever arms of one were parallel to the tilt azimuth and for the other perpendicular to the tilt azimuth with just the opposite arrangement for the cargo binding domain. This seems to be largely an effect on the average density of the lever arm and cargo binding domain, which have similar shapes overall and in the dendrogram (Fig. 5) showed a closer relationship to each other than to the other three classes. In other words, the missing wedge bias in the classification did not lead to a motif with a different overall shape.

This is exactly the missing wedge effect described above on elongated structures such as filaments. How can this be avoided? With paracrystals, it is quite likely that the motifs will have a limited range of orientations with respect to the optical axis of the microscope so that the projection of the motif can be used in place of the motif itself. When this is done, the classification does not show a missing wedge bias while at the same time, producing averages with the same kind of shape variation (Fig. 9b).

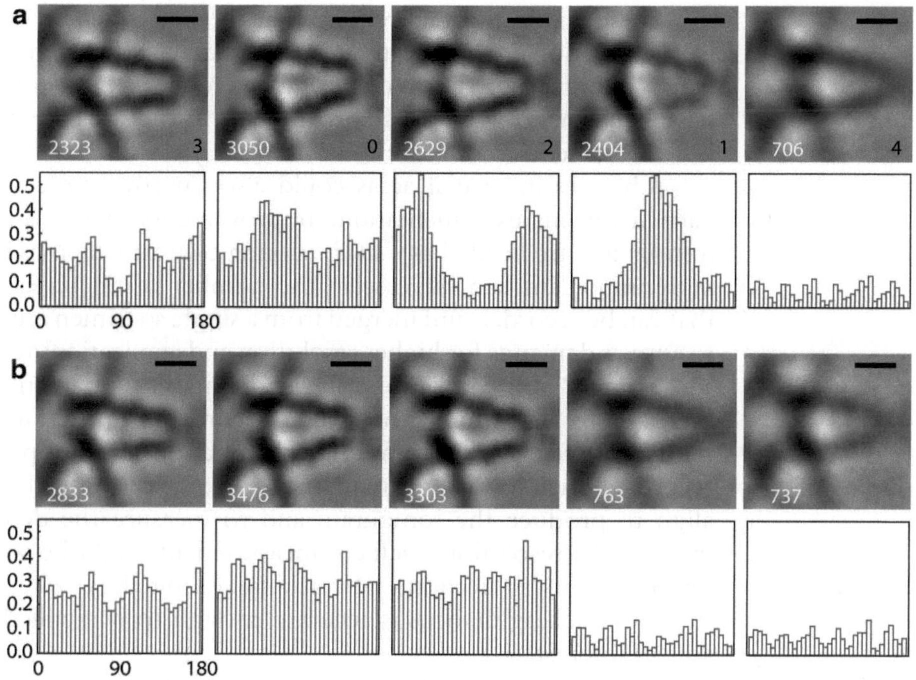

Fig. 9. Classification of myosin V petal motifs. Fig. 8b legend for explanation of the structural elements. Bar equals 10 nm. (**a**) Here the classification was carried out using aligned motifs to produce five classes (0–4, in the). The number of motifs () contributing to the class averages is given at the . The shows histograms of the tilt axis directions with respect to the petal motif with each bin representing the fraction of the total number of motifs that fall within a particular angle range. Note the complementarity of the angular distribution for classes 1 and 2. The complementarity is reflected in the density of the lever arms and the density attributable to the cargo binding domain. (**b**) Here the classification was carried out with projections of the motifs instead of the motifs themselves. Note that when using the projections the angular bias found for the volumes disappears. Modified from ref. 15.

10. Future Prospects

One of the main attractions of molecular ET is the opportunity it provides to look at a molecule's structure in situ or in vivo and compare it to the structure when removed from its original context. One of the most obvious applications here is the structure of enveloped viruses for which this method has found multiple applications (5, 27, 82–85), but it is also being used to visualize molecular complexes in situ that simply cannot be isolated. The bacterial flagellar motor is one such complex whose details are now being revealed in situ (10, 86, 87). ET also provides a means by which to study a molecular structure in a difficult-to-trap state, such as when it is generating tension in a contracting muscle. It also provides a means to visualize molecules in 3D in states or complexes that are very difficult to isolate with sufficient homogeneity for single particle reconstruction.

These applications all require that the heterogeneity inherent in visualization in situ be treated adequately. MDA as described here is but one approach to classifying heterogeneity. Other methods have been mentioned and improvements in this area are essential if heterogeneous motifs are to be accurately classified.

Advances in several areas could also improve molecular ET, particularly in instrumentation. Improvements in the quality of tomograms are needed as visualization in context becomes increasingly important. An increase in the number of tilt series images that can be recorded and merged from a single specimen are needed to answer demands for higher resolution and applications to thicker specimens. Because radiation damage provides a fundamental limit, increases in the number of images recorded from a single area require offsetting reductions in the number of electrons per image recorded. Such low exposure images become very difficult to both align to produce the tomogram and to measure the defocus to increase the resolution. Defocus measurement might be solved by improvements in accuracy of tracking during data collection so that the tracking defocus is accurate enough to be applied to the area of interest.

Phase plates (88–90) are one area where the contrast, and therefore S/N ratio in each image can be improved thereby providing a possible offset to a reduction in the number of electrons needed per image and resulting increase in the number of images. Phase plates would also allow reductions in defocus, so-called in-focus phase contrast, potentially removing the need to measure defocus in the tilt series.

As specimen thickness increases under the drive for more in situ imaging, inelastic scattering also increases and zero-loss electrons decrease. One solution to this problem is the use of an energy filter with the acceptance window set at the peak of the inelastic scattering (91). This method, known as most-probable loss imaging produces interpretable tomograms even in the near absence of zero-loss electrons. At some point, constructive use of inelastically scattered electrons is going to be a necessity and this will require a chromatic aberration corrector (92). Few of these improvements in instrumentation are widely available at the moment, but they are available in a few selected laboratories.

Acknowledgments

The research on insect flight muscle was supported by NIH Grant GM30598, the myosin-V reconstruction under NIH grant AR47421, and *protomo* is being developed under NIH grant GM82948.

References

1. Liu J, Taylor DW, Krementsova EB, Trybus KM, Taylor KA (2006) Three-dimensional structure of the myosin V inhibited state by cryoelectron tomography. Nature 442:208–211
2. Taylor KA, Tang J, Cheng Y, Winkler H (1997) The use of electron tomography for structural analysis of disordered protein arrays. J Struct Biol 120:372–386
3. Wu S, Liu J, Reedy MC, Tregear RT, Winkler H, Franzini-Armstrong C et al (2010) Electron tomography of cryofixed, isometrically contracting insect flight muscle reveals novel actin-myosin interactions. PLoS One 5:e12643
4. Hampton CM, Liu J, Taylor DW, DeRosier DJ, Taylor KA (2008) The 3D structure of villin as an unusual F-Actin crosslinker. Structure 16:1882–1891
5. Zhu P, Liu J, Bess J Jr, Chertova E, Lifson JD, Grise H et al (2006) Distribution and three-dimensional structure of AIDS virus envelope spikes. Nature 441:847–852
6. Zhu P, Winkler H, Chertova E, Taylor KA, Roux KH (2008) Cryoelectron tomography of HIV-1 envelope spikes: further evidence for tripod-like legs. PLoS Pathog 4:e1000203
7. Hu G, Liu J, Taylor KA, Roux KH (2011) Structural comparison of HIV-1 envelope spikes with and without the V1/V2 loop. J Virol 85:2741–2750
8. Ye F, Liu J, Winkler H, Taylor KA (2008) Integrin alpha IIb beta 3 in a membrane environment remains the same height after Mn^{2+} activation when observed by cryoelectron tomography. J Mol Biol 378:976–986
9. Ye F, Hu G, Taylor D, Ratnikov B, Bobkov AA, McLean MA et al (2010) Recreation of the terminal events in physiological integrin activation. J Cell Biol 188:157–173
10. Liu J, Lin T, Botkin DJ, McCrum E, Winkler H, Norris SJ (2009) Intact flagellar motor of *Borrelia burgdorferi* revealed by cryo-electron tomography: evidence for stator ring curvature and rotor/C-ring assembly flexion. J Bacteriol 191:5026–5036
11. Frank J, van Heel M (1982) Correspondence analysis of aligned images of biological particles. J Mol Biol 161:134–137
12. Van Heel M, Frank J (1981) Use of multivariate statistics in analysing the images of biological macromolecules. Ultramicroscopy 6:187–194
13. Winkler H, Taylor KA (2006) Accurate marker-free alignment with simultaneous geometry determination and reconstruction of tilt series in electron tomography. Ultramicroscopy 106:240–254
14. Winkler H (2006) 3D reconstruction and processing of volumetric data in cryo-electron tomography. J Struct Biol 157:126–137
15. Winkler H, Zhu P, Liu J, Ye F, Roux KH, Taylor KA (2009) Tomographic subvolume alignment and subvolume classification applied to myosin V and SIV envelope spikes. J Struct Biol 165:64–77
16. Frank J (ed) (2006) Electron tomography—methods for three-dimensional visualization of structures in the cell, 2nd edn. Springer, Berlin
17. McIntosh JR (ed) (2007) Cellular electron microscopy, vol 79. Academic, San Diego, CA
18. Winkler H, Taylor KA (2003) Focus gradient correction applied to tilt series image data used in electron tomography. J Struct Biol 143: 24–32
19. Fernandez JJ, Li S, Crowther RA (2006) CTF determination and correction in electron cryotomography. Ultramicroscopy 106:587–596
20. Hegerl R, Hoppe W (1976) Influence of electron noise on three-dimensional image reconstruction. Z Naturforsch 31a:1717–1721
21. Hoppe W, Hegerl R (1981) Some remarks concerning the influence of electron noise on 3D reconstruction. Ultramicroscopy 6: 205–206
22. McEwen BF, Downing KH, Glaeser RM (1995) The relevance of dose-fractionation in tomography of radiation-sensitive specimens. Ultramicroscopy 60:357–373
23. Crowther RA, Amos LA, Finch JT, De Rosier DJ, Klug A (1970) Three dimensional reconstructions of spherical viruses by Fourier synthesis from electron micrographs. Nature 226:421–425
24. Saxton WO, Baumeister W, Hahn M (1984) Three-dimensional reconstruction of imperfect two-dimensional crystals. Ultramicroscopy 13:57–70
25. Nickell S, Hegerl R, Baumeister W, Rachel R (2003) *Pyrodictium cannulae* enter the periplasmic space but do not enter the cytoplasm, as revealed by cryo-electron tomography. J Struct Biol 141:34–42
26. Beck M, Forster F, Ecke M, Plitzko JM, Melchior F, Gerisch G et al (2004) Nuclear pore complex structure and dynamics revealed by cryoelectron tomography. Science 306: 1387–1390
27. Förster F, Medalia O, Zauberman N, Baumeister W, Fass D (2005) Retrovirus envelope protein complex structure in situ studied by cryo-electron tomography. Proc Natl Acad Sci USA 102:4729–4734

28. Iancu CV, Wright ER, Benjamin J, Tivol WF, Dias DP, Murphy GE et al (2005) A "flip-flop" rotation stage for routine dual-axis electron cryotomography. J Struct Biol 151:288–297
29. Cheng Y, Boll W, Kirchhausen T, Harrison SC, Walz T (2007) Cryo-electron tomography of clathrin-coated vesicles: structural implications for coat assembly. J Mol Biol 365:892–899
30. Penczek P, Marko M, Buttle K, Frank J (1995) Double-tilt electron tomography. Ultramicroscopy 60:393–410
31. Mastronarde DN (1997) Dual-axis tomography: an approach with alignment methods that preserve resolution. J Struct Biol 120:343–352
32. Chen LF, Blanc E, Chapman MS, Taylor KA (2001) Real space refinement of acto-myosin structures from sectioned muscle. J Struct Biol 133:221–232
33. Lanzavecchia S, Cantele F, Bellon PL (2001) Alignment of 3D structures of macromolecular assemblies. Bioinformatics 17:58–62
34. Taylor KA, Glaeser RM (1976) Electron microscopy of frozen hydrated biological specimens. J Ultrastruct Res 55:448–456
35. Taylor KA, Glaeser RM (1973) Hydrophilic support films of controlled thickness and composition. Rev Sci Instrum 44:1546–1547
36. Glaeser RM, Taylor KA (1978) Radiation damage relative to transmission electron microscopy of biological specimens at low temperature: a review. J Microsc 112:127–138
37. Glaeser RM (1971) Limitations to significant information in biological electron microscopy as a result of radiation damage. J Ultrastruct Res 36:466–482
38. Luther PK (2006) Sample shrinkage and radiation damage of plastic sections. In: Frank J (ed) Electron tomography—methods for three-dimensional visualization of structures in the cell, 2nd edn. Springer, Berlin, pp 17–48
39. Braunfeld MB, Koster AJ, Sedat JW, Agard DA (1994) Cryo automated electron tomography: towards high-resolution reconstructions of plastic-embedded structures. J Microsc 174:75–84
40. Wu S, Liu J, Reedy MC, Winkler H, Reedy MK, Taylor KA (2009) Methods for identifying and averaging variable molecular conformations in tomograms of actively contracting insect flight muscle. J Struct Biol 168:485–502
41. Taylor KA, Schmitz H, Reedy MC, Goldman YE, Franzini-Armstrong C, Sasaki H et al (1999) Tomographic 3-D reconstruction of quick frozen, Ca^{2+}-activated contracting insect flight muscle. Cell 99:421–431
42. Förster F, Pruggnaller S, Seybert A, Frangakis AS (2008) Classification of cryo-electron subtomograms using constrained correlation. J Struct Biol 161:276–286
43. Dierksen K, Typke D, Hegerl R, Walz J, Sackmann E, Baumeister W (1995) Three-dimensional structure of lipid vesicles embedded in vitreous ice and investigated by automated electron tomography. Biophys J 68:1416–1422
44. Walz J, Typke D, Nitsch M, Koster AJ, Hegerl R, Baumeister W (1997) Electron tomography of single ice-embedded macromolecules: three-dimensional alignment and classification. J Struct Biol 120:387–395
45. Kremer JR, Mastronarde DN, McIntosh JR (1996) Computer visualization of three-dimensional image data using IMOD. J Struct Biol 116:71–76
46. Brandt SS (2006) Markerless alignment in electron tomography. In: Frank J (ed) Electron tomography—methods for three-dimensional visualization of structures in the cell, 2nd edn. Springer, Berlin, pp 187–216
47. Liu J, Wright ER, Winkler H (2010) 3D visualization of HIV virions by cryoelectron tomography. Methods Enzymol 483:267–290
48. Hohn M, Tang G, Goodyear G, Baldwin PR, Huang Z, Penczek PA et al (2007) SPARX, a new environment for Cryo-EM image processing. J Struct Biol 157:47–55
49. Amat F, Castano-Diez D, Lawrence A, Moussavi F, Winkler H, Horowitz M (2010) Alignment of cryo-electron tomography datasets. Methods Enzymol 482:343–367
50. Van Heel M, Schatz M, Orlova E (1992) Correlation functions revisited. Ultramicroscopy 46:307–316
51. Horner JL, Gianino PD (1984) Phase only matched filtering. Appl Opt 23:812–816
52. Saxton WO (1994) Accurate alignment of sets of images. J Microsc 174:61–68
53. Vonck J (2000) Parameters affecting specimen flatness of two-dimensional crystals for electron crystallography. Ultramicroscopy 85:123–129
54. Han BG, Wolf SG, Vonck J, Glaeser RM (1994) Specimen flatness of glucose-embedded biological materials for electron crystallography is affected significantly by the choice of carbon evaporation stock. Ultramicroscopy 55:1–5
55. Glaeser RM (1992) Specimen flatness of thin crystalline arrays: influence of the substrate. Ultramicroscopy 46:33–43
56. Kubalek EW, Kornberg RD, Darst SA (1991) Improved transfer of two-dimensional crystals

from the air/water interface to specimen support grids for high-resolution analysis by electron microscopy. Ultramicroscopy 35:295–304
57. Taylor DW, Kelly DF, Cheng A, Taylor KA (2007) On the freezing and identification of lipid monolayer 2-D arrays for cryoelectron microscopy. J Struct Biol 160:305–312
58. Winkler H, Taylor KA (1996) Three-dimensional distortion correction applied to tomographic reconstructions of sectioned crystals. Ultramicroscopy 63:125–132
59. Crowther RA, Henderson R, Smith JM (1996) MRC image processing programs. J Struct Biol 116:9–16
60. Gipson B, Zeng X, Zhang ZY, Stahlberg H (2007) 2dx—user-friendly image processing for 2D crystals. J Struct Biol 157:64–72
61. Henderson R, Baldwin JM, Downing KH, Lepault J, Zemlin F (1986) Structure of purple membrane from halobacterium halobium: recording, measurement and evaluation of electron micrographs at 3.5 Å resolution. Ultramicroscopy 19:147–178
62. Baldwin JM, Henderson R, Beckman E, Zemlin F (1988) Images of purple membrane at 2.8 Å resolution obtained by cryo-electron microscopy. J Mol Biol 202:585–591
63. Schmid M, Dargahi R, Tam M (1993) SPECTRA: a system for processing electron images of crystals. Ultramicroscopy 48:251–264
64. Frangakis AS, Bohm J, Forster F, Nickell S, Nicastro D, Typke D et al (2002) Identification of macromolecular complexes in cryoelectron tomograms of phantom cells. Proc Natl Acad Sci USA 99:14153–14158
65. Ludtke SJ, Baldwin PR, Chiu W (1999) EMAN: semiautomated software for high-resolution single-particle reconstructions. J Struct Biol 128:82–97
66. Winkler H, Taylor KA (1999) Multivariate statistical analysis of three-dimensional cross-bridge motifs in insect flight muscle. Ultramicroscopy 77:141–152
67. Bartesaghi A, Sprechmann P, Liu J, Randall G, Sapiro G, Subramaniam S (2008) Classification and 3D averaging with missing wedge correction in biological electron tomography. J Struct Biol 162:436–450
68. Dube P, Tavares P, Lurz R, van Heel M (1993) The portal protein of bacteriophage SPP1: a DNA pump with 13-fold symmetry. EMBO J 12:1303–1309
69. Liu J, Reedy MC, Goldman YE, Franzini-Armstrong C, Sasaki H, Tregear RT et al (2004) Electron tomography of fast frozen, stretched rigor fibers reveals elastic distortions in the myosin crossbridges. J Struct Biol 147:268–282
70. Burgess SA, Walker ML, Thirumurugan K, Trinick J, Knight PJ (2004) Use of negative stain and single-particle image processing to explore dynamic properties of flexible macromolecules. J Struct Biol 147:247–258
71. Frank J (1990) Classification of macromolecular assemblies studied as 'single particles'. Q Rev Biophys 23:281–329
72. Pascual-Montano A, Taylor KA, Winkler H, Pascual-Marqui RD, Carazo JM (2002) Quantitative self-organizing maps for clustering electron tomograms. J Struct Biol 138:114–122
73. van Heel M (1989) Classification of very large electron microscopical image data sets. Optik 82:114–126
74. Iwasaki K, Mitsuoka K, Fujiyoshi Y, Fujisawa Y, Kikuchi M, Sekiguchi K, Yamada T (2005) Electron tomography reveals diverse conformations of integrin alpha IIb beta3 in the active state. J Struct Biol 150:259–267
75. Stolken M, Beck F, Haller T, Hegerl R, Gutsche I, Carazo JM et al (2011) Maximum likelihood based classification of electron tomographic data. J Struct Biol 175:77–85
76. Frank J (1996) Three-dimensional elecron microscopy of macromolecular assemblies. Academic, San Diego, CA
77. Holmes KC, Tregear RT, Barrington Leigh J (1980) Interpretation of the low angle X-ray diffraction from insect muscle in rigor. Proc R Soc Lond B Biol Sci 207:13–33
78. Pettersen EF, Goddard TD, Huang CC, Couch GS, Greenblatt DM, Meng EC, Ferrin TE (2004) UCSF Chimera—a visualization system for exploratory research and analysis. J Comput Chem 25:1605–1612
79. Jontes JD, Milligan RA (1997) Three-dimensional structure of Brush Border Myosin-I at approximately 20 resolution by electron microscopy and image analysis. J Mol Biol 266:331–342
80. Wendt T, Taylor D, Trybus KM, Taylor K (2001) Three-dimensional image reconstruction of dephosphorylated smooth muscle heavy meromyosin reveals asymmetry in the interaction between myosin heads and placement of subfragment 2. Proc Natl Acad Sci USA 98:4361–4366
81. Liu J, Wendt T, Taylor DW, Taylor KA (2003) Refined model of the 10S conformation of smooth muscle myosin by cryoEM 3-D image reconstruction. J Mol Biol 329:963–972
82. Grunewald K, Desai P, Winkler DC, Heymann JB, Belnap DM, Baumeister W, Steven AC

(2003) Three-dimensional structure of herpes simplex virus from cryo-electron tomography. Science 302:1396–1398

83. Zanetti G, Briggs JA, Grunewald K, Sattentau QJ, Fuller SD (2006) Cryo-electron tomographic structure of an immunodeficiency virus envelope complex in situ. PLoS Pathog 2:e83
84. Liu J, Bartesaghi A, Borgnia MJ, Sapiro G, Subramaniam S (2008) Molecular architecture of native HIV-1 gp120 trimers. Nature 455:109–113
85. Dai W, Jia Q, Bortz E, Shah S, Liu J, Atanasov I et al (2008) Unique structures in a tumor herpes virus revealed by cryo-electron tomography and microscopy. J Struct Biol 161:428–438
86. Murphy GE, Leadbetter JR, Jensen GJ (2006) In situ structure of the complete *Treponema primitia* flagellar motor. Nature 442:1062–1064
87. Kudryashev M, Cyrklaff M, Wallich R, Baumeister W, Frischknecht F (2010) Distinct in situ structures of the Borrelia flagellar motor. J Struct Biol 169:54–61
88. Nagayama K, Danev R (2009) Phase-plate electron microscopy: a novel imaging tool to reveal close-to-life nano-structures. Biophys Rev 1:37–42
89. Murata K, Liu X, Danev R, Jakana J, Schmid MF, King J et al (2010) Zernike phase contrast cryo-electron microscopy and tomography for structure determination at nanometer and subnanometer resolutions. Structure 18:903–912
90. Danev R, Kanamaru S, Marko M, Nagayama K (2010) Zernike phase contrast cryo-electron tomography. J Struct Biol 171:174–181
91. Bouwer JC, Mackey MR, Lawrence A, Deerinck TJ, Jones YZ, Terada M et al (2004) Automated most-probable loss tomography of thick selectively stained biological specimens with quantitative measurement of resolution improvement. J Struct Biol 148:297–306
92. Freitag B, Kujawa S, Mul PM, Ringnalda J, Tiemeijer PC (2005) Breaking the spherical and chromatic aberration barrier in transmission electron microscopy. Ultramicroscopy 102:209–214

Chapter 24

High-Resolution Imaging of 2D Outer Membrane Protein F Crystals by Atomic Force Microscopy

Dimitrios Fotiadis and Daniel J. Müller

Abstract

In this chapter the methodological bases are provided to achieve subnanometer resolution on two-dimensional (2D) membrane protein crystals by atomic force microscopy (AFM). This is outlined in detail with the example of AFM studies of the outer membrane protein F (OmpF) from the bacterium *Escherichia coli* (*E. coli*). We describe in detail the high-resolution imaging of 2D OmpF crystals in aqueous solution and under near-physiological conditions. The topographs of OmpF, and stylus effects and artifacts encountered when imaging by AFM are discussed.

Key words: Atomic force microscopy, Biological membranes, High resolution, Outer membrane protein F, Stylus artifacts, Topography

1. Introduction

In more than 20 years of continuous technological and methodological development the AFM (1, 2) has become a powerful multifunctional nanoscopic tool to image and to sense a variety of cell and molecular biological systems (3–8). The unique features of imaging biological structures in their physiologically relevant buffer solution, temperature and pressure, the high lateral and vertical resolution, and the high signal-to-noise ratio of the AFM topographs make this microscope outstanding. Today, AFM is routinely applied to directly contour the topography of complex biological systems across dimensions ranging from tissues to cells and towards the substructure of single proteins. Whereas in the imaging mode the AFM stylus is used to contour the topography of the native biological sample, in the sensing mode the AFM stylus

Ingeborg Schmidt-Krey and Yifan Cheng (eds.), *Electron Crystallography of Soluble and Membrane Proteins: Methods and Protocols*, Methods in Molecular Biology, vol. 955, DOI 10.1007/978-1-62703-176-9_24,

is applied to probe chemical, biological, or physical interactions at molecular resolution. As a contouring and probing tool the AFM is particularly suited to image biological surfaces such as those of living cells, cell membranes, and reconstituted membrane proteins. When applied to biological membranes, high-resolution AFM has allowed the observation of the assembly, oligomeric state, diffusion, and function of native single membrane proteins. Complementary, the sensing mode AFM enables the user to spatially correlate and quantify biochemical and physical properties of membrane proteins, such as electrostatic potential, stability, flexibility, elasticity, conformational entropy, folding and unfolding pathways, and ligand- and inhibitor binding (for recent reviews see refs. 6, 9–12).

From this variety of possible applications of AFM to characterize membrane proteins (4, 13), this chapter focuses on the application of contact mode AFM imaging to acquire high-resolution structural information of native membrane proteins in buffer solution. In contact mode AFM, a nanometer-sized sharp stylus mounted on a 50–400 μm long flexible cantilever is raster scanned, thereby contouring the sample topography at constant force. The topography is reconstructed from the up-and-down displacements of the cantilever recorded as the stylus is scanned. To minimize possible damage of the biological specimen by the AFM stylus, soft cantilevers with spring constants around 0.1 N/m are preferred and scanning must be performed at minimal force applied to the stylus (~50–100 pN). Lateral resolutions better than 0.5 nm and vertical resolutions approaching 0.1 nm have been achieved on biological membranes in solution (7, 10, 11). Up to now this AFM imaging mode has provided the highest resolution with biological samples. Alternative AFM modes to record topographs are the oscillation mode (14–18) and the frequency modulation mode (19–21). The oscillation mode is frequently used to contour the surface topography of weakly attached biomolecules, i.e., single proteins, fibrils, and chromosomes, by vertically oscillating the AFM stylus while scanning the sample. Because the oscillating AFM stylus touches the sample surface only at the end of its downward movement, the frictional forces are reduced, avoiding the lateral deformation and displacement of the sample. In contrast to contact and oscillation mode, in the frequency modulation mode the AFM stylus hardly touches the sample surface as it senses interactions that are not based on the formation of a physical contact with the sample. Although having already provided promising results on membrane proteins (22), this AFM mode is still in its infancy for biological AFM imaging approaching a lateral resolution of <1 nm. In the last years, high-speed AFM has been introduced making it possible to observe proteins at high resolution, i.e., approaching 1 nm lateral resolution, and at scanning rates of up to 100 frames per second (23, 24). So far, for high-resolution imaging of biological membranes, AFM operated in the contact mode has shown to be the best choice provided the imaging parameters, e.g., applied force and imaging buffer, are adjusted correctly (11, 25). To prevent deformation of the relatively soft biological structures, the forces acting between AFM stylus and sample should not exceed 100 pN in contact mode (25). On the other hand, application of higher forces can sometimes be useful to mechanically dissect biological samples using the AFM stylus as a nanoscalpel (26).

The AFM imaging experiments presented here were carried out in buffer solution, at room temperature and under ambient pressure using 2D crystals of OmpF porin reconstituted into lipid membranes. This 39.3 kDa protein forms trimeric channels (Fig. 1) and is located in the outer membrane of the bacterium *E. coli*. *In vivo* OmpF together with other outer membrane proteins facilitates the uptake of nutrients into *E. coli*. The 16 antiparallel β-strands of the β-barrel forming the transmembrane pore are connected by short turns on the periplasmic surface and by long hydrophilic loops of variable length on the extracellular surface (27) (Fig. 1). The latter constitutes the rough side of the porin surface, the former the smooth side. When reconsituted in the presence of the lipid dimyristoyl phosphatidylcholine, OmpF porin assembles into various 2D crystal forms (28, 29). Such crystals exhibit a rectangular (a = 13.5 nm; b = 8.2 nm) or trigonal (a = b = 8.2 nm) packing arrangement and are frequently double-layered. Such

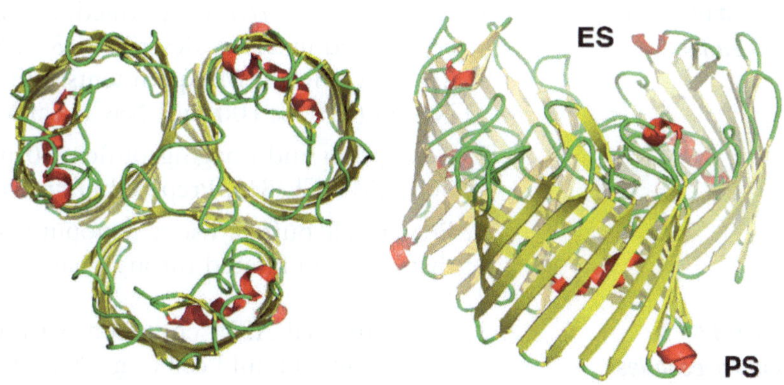

Fig. 1. Cartoon illustration of porin OmpF. *Left panel*: top view of the OmpF trimer as seen from the extracellular side. The channels are clearly visible as well as the pore constriction. *Right panel*: tilted side view of the OmpF trimer. The extracellular (ES) and periplasmic sides (PS) are indicated. The periplasmic surface is relatively smooth compared to the extracellular surface—small polypeptide loops protrude on the PS and large ones on the ES. These two illustrations of OmpF were calculated using the coordinates of Cowan et al. (1992; PDB accession code: 2OMF; see ref. 27) and the PyMol molecular graphics system (www.pymol.org).

double-layered crystals only expose the periplasmic side to the aqueous environment (30). OmpF porin is well-suited for our experiments since its structure has been well-characterized by X-ray crystallography (27), AFM (31–33), and electron microscopy (28–30, 34). In addition, 2D crystals of OmpF are highly ordered and flat, making them very convenient for high-resolution AFM.

2. Materials

2.1. Preparation of Mica Supports for Sample Immobilization

1. Inoxydable and magnetic steel disks of 12 mm diameter (Ted Pella, Inc., Redding, CA, USA).
2. Teflon sheets of ~0.2 mm thickness (Maag Technic AG, Birsfelden, Switzerland).
3. Mica sheets with a thickness between 0.3 and 0.6 mm (Mica House, 2A Pretoria, Street, Calcutta 700 071, India).
4. "Punch and die" set from Precision Brand Products, Inc. (Downers Grove, IL, USA).
5. Ethanol (concentration >96% (v/v)).
6. Loctite 406 superglue from KVT König, Dietikon, Switzerland.
7. Araldit Rapid: Two-component epoxy glue from Tesa AG, Bergdietikon, Switzerland.
8. Scotch tape.

2.2. OmpF Porin and Buffers

1. OmpF from *E. coli* was purified, reconstituted into lipids, and crystallized in 2D as described (30). Stock solution of OmpF 2D crystals for AFM experiments: 1 mg/ml in 20 mM HEPES–NaOH (pH 7), 100 mM NaCl, 2 mM $MgCl_2$. Store at 4°C.
2. Adsorption and imaging buffer: 20 mM Tris–HCl (pH 7.8), 300 mM KCl (31). Prepare buffers for AFM freshly.
3. Prepare all buffers using nanopure water (~18 MOhm/cm) that has been filtered through nanometer sized pores.

2.3. AFM and Accessories

1. A commercial AFM equipped with a >100 μm piezoelectric scanner and a liquid cell (e.g., NanoWizard, JPK Instruments, Berlin, Germany or Multimode, Veeco Metrology Group, Santa Barbara, CA, USA).
2. Oxide-sharpened Si_3N_4 microcantilevers of 100 μm length and a nominal spring constant of $k = 0.08$ N/m (Olympus Optical Co., LTD, Tokyo, Japan).

3. Method

3.1. Preparation of Mica Supports for Sample Immobilization

1. Punch mica disks of 6 mm and Teflon disks of 14 mm diameter using the "punch and die" set and a hammer.
2. Clean the Teflon and steel disks with ethanol and paper wipes.
3. Glue a Teflon disk on a steel disk with Loctite 406.
4. Glue a mica disk on the Teflon surface of the Teflon-steel disk using the two-component epoxy glue Araldit.
5. Let the glued supports strengthen for at least 1 day.

3.2. Adsorption of 2D OmpF Crystals to Mica

1. Dilute and mix 2 μl of 2D OmpF crystal stock solution with 40 μl adsorption buffer in an Eppendorf tube. Typical final protein concentrations are between 20 and 60 μg/ml.
2. Cleave mica with Scotch tape.
3. Pipet 20–30 μl of the diluted OmpF 2D crystals onto the freshly cleaved mica support.
4. Adsorb OmpF for 15–30 min.
5. Rinse the mica to remove weakly attached 2D crystals by aspirating approximately 2/3 of the fluid volume from the mica surface and re-adding the same amount of imaging buffer. Repeat this washing procedure at least three times.
6. Transfer and mount the mica support to the AFM.
7. Mount the cantilever to the fluid cell (preferably without o-ring seal to minimize drift) and mount the fluid cell to the AFM.
8. Fill the space between the mica surface and the fluid cell with imaging buffer to avoid dehydration of the sample.

3.3. Operation of the AFM

1. After thermal relaxation of the AFM, initial engagement of the AFM stylus is performed. Set the scan size to 0 to avoid that the approaching AFM stylus starts scanning and becomes contaminated by the biological specimen. Prior to scanning the sample surface, the operating point of the instrument is set to forces below 0.5 nN. For overview imaging at low magnification (frame size >1 μm) the forces are kept below 0.5 nN and for high-resolution imaging below 0.1 nN (frame size <500 nm). The forces must be corrected manually (or if available automatically) to compensate for thermal drift.
2. Record two frames of 512 by 512 pixel (or more pixel) simultaneously either showing topography or deflection (error) signal in trace or retrace scanning direction. Typically, the scan speed is set to 3.8–6.8 Hz (lines per second).
3. Optimize the feedback gains to reduce the deflection signal to a minimum. Usually deflection and height signals are recorded

at low magnification (frame size >1 μm), whereas height signals are acquired in both, trace and retrace scanning direction when approaching higher magnification (frame size <1 μm). Simultaneously recording two topographs in different scanning directions of the same specimen area allows to judge whether the sample is deformed by the scanning stylus in either one direction. Possible deformations are minimized by lowering the force applied to the AFM stylus and by optimizing the scanning speed.

4. At high magnification the vertical range of the piezoelectric scanner is reduced to enhance the vertical resolution by the digitalisation of the signal (AD-conversion).

3.4. Conclusions

In this chapter, we have provided a detailed description of how to contour at subnanometer resolution, and under near-physiological conditions the surface structures of OmpF by AFM. OmpF 2D crystals are suitable for beginners as well as for advanced microscopists, offering one easy (periplasmic side of OmpF) and one difficult (extracellular side of OmpF) surface structure for exercising AFM imaging. We have also shown that forces between AFM stylus and biological sample as well as the shape and geometry of the AFM stylus play an important role for successful imaging of the porin OmpF and biological samples in general.

4. Notes

4.1. Vibrations and Temperature

When scanning at minimal forces, smallest vibrations of the AFM can cause the AFM stylus to loose the contact to the sample surface. Thus, an efficient vibration isolated set-up of the AFM is crucial to record high-resolution topographs. Antivibration and damping tables, or heavy platforms supported by bungees offer excellent vibration damping. Acoustic isolation of the AFM can efficiently be achieved by a vacuum bell jar. Temperature-induced drift is another factor limiting the applicability of the AFM to reveal high-resolution topographs of native membrane proteins. To reduce thermal drift, the AFM room should be temperature stabilized, and all buffer solutions and parts (i.e., fluid cell, cantilever etc.) being in contact with buffer and AFM should be stored at the same temperature.

4.2. Choice of the Appropriate Support for Sample Immobilization

For high-resolution contact mode AFM imaging, membranes have to be immobilized on supporting surfaces to impede their lateral displacement while being contoured by the scanning AFM tip (35). Most frequently, immobilization is achieved by the adsorption of proteoliposomes or 2D crystals to mica. Mica is atomically flat and a relatively chemically inert and hydrophilic solid support (36). Such "solid-supported membranes" have made and will make possible the gain of important insights into the structure–function relationship of membrane proteins. In buffer solution, the gap between membrane and support is only 0.5–2 nm. This constraint is not of significance when imaging reconstituted membrane proteins. However, nonsupported membranes may be desirable for specific applications such as the study of membrane protein function and dynamics (37). Certain approaches make use of nanoscopic holes or spacers over which membranes are spanned to expose reconstituted membrane proteins to two different milieus simultaneously, e.g., solutions at different pHs or ionic strengths. Such setups allow the AFM imaging of proton or ion-dependent membrane proteins at work. Finally, immobilization of specific membranes may not work on mica and require coating of mica by ultrathin synthetic polymers or extracellular matrix proteins.

4.3. Buffer Conditions for High-Resolution AFM Imaging

High-resolution topographs of native membrane proteins approaching a lateral resolution of <1 nm (31, 38) can reproducibly be recorded with the AFM provided that sample preparation, imaging force, and buffer are adjusted correctly (11, 25, 36). In general, AFM contours the biological object at minimal forces applied to the AFM stylus to avoid friction and deformation. However, often even the smallest force adjustable by the instrument is too high for preventing deformation of the biomolecule.

The effective interaction force acting between AFM stylus and specimen is the sum of the force applied to the stylus, and the electrostatic repulsion and the van der Waals attraction between stylus and sample surface. By adjusting pH, electrolyte concentration and electrolyte valency of the imaging buffer, van der Waals attraction and electrostatic repulsion between the tip of the AFM stylus and the sample can be balanced. Under these conditions the scanning AFM stylus is assumed to surf on a cushion of electrostatic repulsion that minimizes the force interacting between the tip of the AFM stylus and the biomolecule. If this force, locally interacting between AFM tip and biomolecule, is sufficiently small, structural deformation can be avoided or mininimzed to a level below the resolution of the AFM. The best imaging conditions are determined by recording and analyzing force–distance curves between stylus and sample in different buffer solutions. Conditions that yield force–distance curves with a small repulsive peak (~100 pN) are ideal for high-resolution AFM imaging. When applying an imaging force of ~100 pN to the AFM stylus, only a small fraction of this force will interact locally between the tip of the AFM stylus and the biological sample, whereas most of the applied force is balanced by long range electrostatic interactions between AFM stylus and sample. By this compensation method the imaging buffer for OmpF mentioned in Subheading 2.2. was optimized. For further reading on this topic see refs. 11, 25.

4.4. Morphology of 2D OmpF Crystals

Figure 2 shows a typical overview (frame size ~15 μm) of 2D OmpF crystals adsorbed to freshly cleaved mica. The size of the OmpF crystals varies from reconstitution to reconstitution with sizes ranging between 0.5 and 5 μm. The number of adsorbed membrane patches is dependent on the adsorption buffer, time and the concentration of the OmpF 2D crystal solution deposited onto mica. To avoid contamination of the AFM stylus, 2D crystals may not be adsorbed too densely on the support.

As mentioned in the introduction, reconstituted *E. coli* porin OmpF 2D crystals form double layers exposing the periplasmic side. Therefore, the extracellular sides are rarely accessible to the AFM stylus. However, double layers often do not overlap perfectly and terraces with the extracellular side of single layers are exposed. This is nicely seen at higher magnification (Fig. 3; frame size ~7 μm): double layers exposing the periplasmic side (Fig. 3; 1) and single layers exposing the extracellular side (Fig. 3; 2). Multiple layers are also seen frequently (Fig. 3; 3). Single, double, and multiple layers can also easily be identified by measuring their height, e.g., with the "Section Analysis" tool of the AFM software. In the topograph displayed in Fig. 3 heights of ~5.5, ~11, and ~22 nm were measured for single (2), double (1), and multiple layers (3), respectively. As described by Müller et al. (1997; see ref. 39) electrostatic interactions can contribute to the height measured between biological samples with different surface properties and their support. This electrostatic repulsion contributing to the measured height depends on the surface charges of the membrane,

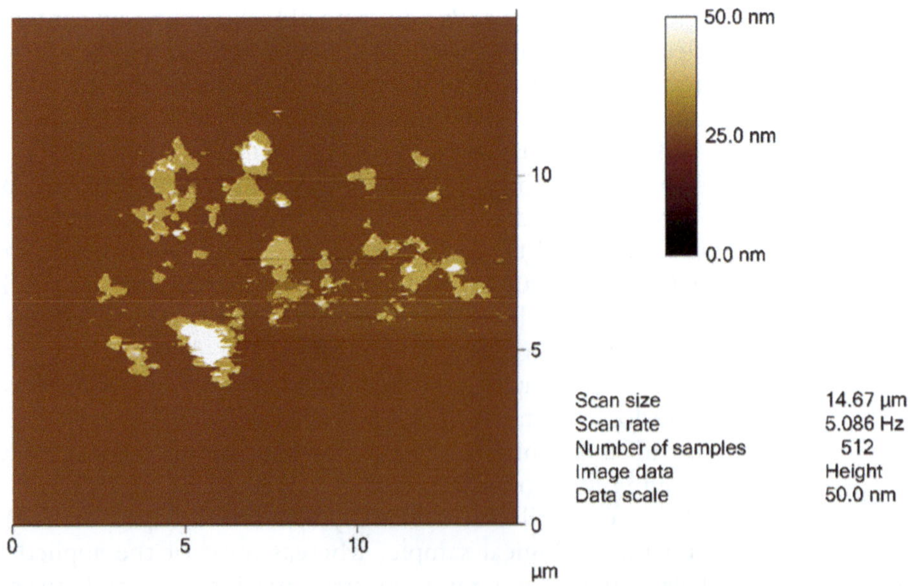

Fig. 2. Low magnification AFM topograph of 2D OmpF crystals adsorbed on mica. Crystals immobilized onto the atomically flat mica surface are clearly discerned.

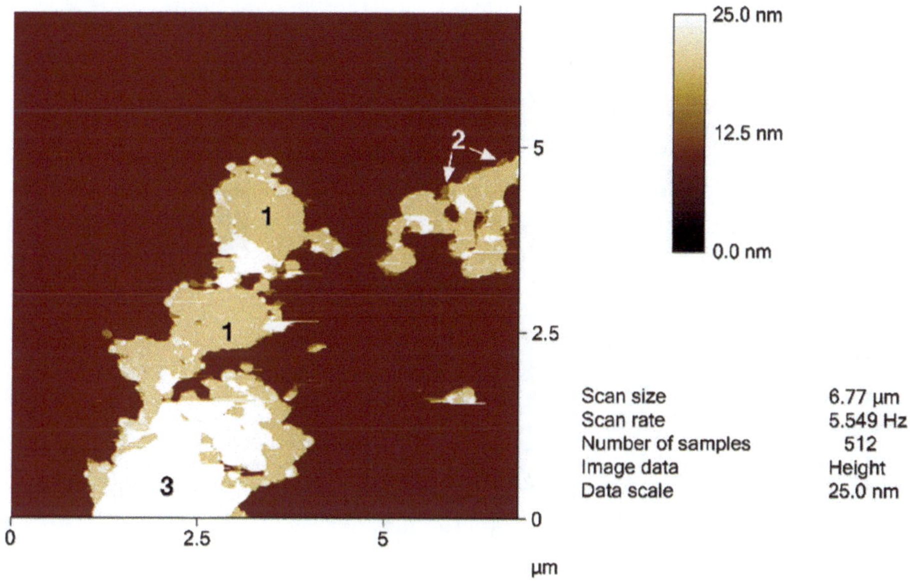

Fig. 3. Medium magnification AFM topograph of 2D OmpF crystals adsorbed on mica. Double layers exposing the periplasmic side (1), single layers exposing the extracellular side (2), and multiple layers are seen (3).

the AFM stylus and the supporting surface, and the ionic strength and pH of the buffer solution. At sufficiently high electrolyte concentration the electrostatic repulsion is screened and the height contributions become neglegible. The heights mentioned above correspond to heights measured in 20 mM Tris–HCl (pH 7.8), 300 mM KCl and may differ from heights measured under other buffer conditions.

4.5. The Periplasmic Surface of OmpF

At high magnification individual porin trimers are easily recognized on topographs (Fig. 4; frame size ~100 nm). On the periplasmic side porin trimers exhibit a prominent Y-shaped protrusion that separates the three transmembrane channels: the white circle in Fig. 4 marks one OmpF trimer. This elevation on the relatively smooth periplasmic surface protrudes less than 0.5 nm from the lipid bilayer (marked by an asterisk in the inset). Channels appear as black spots on AFM topographs. The rectangular unit cell of 2D crystals comprises two porin trimers which are related by a twofold symmetry axis (Fig. 4; white frame and inset). Because of the exquisite high signal-to-noise ratio in AFM topographs no processing of the raw data is necessary. Even missing trimers in the 2D crystal are detected (white arrowhead in Fig. 4). An estimate of the resolution of this topography is given in Subheading 4.7.

4.6. The Extracellular Surface of OmpF

The extracellular surface of porin OmpF shows domains which protrude by ~1 nm from the lipid bilayer (Fig. 5) covering to a great extent the channel entrance (see also Fig. 1). As a result of

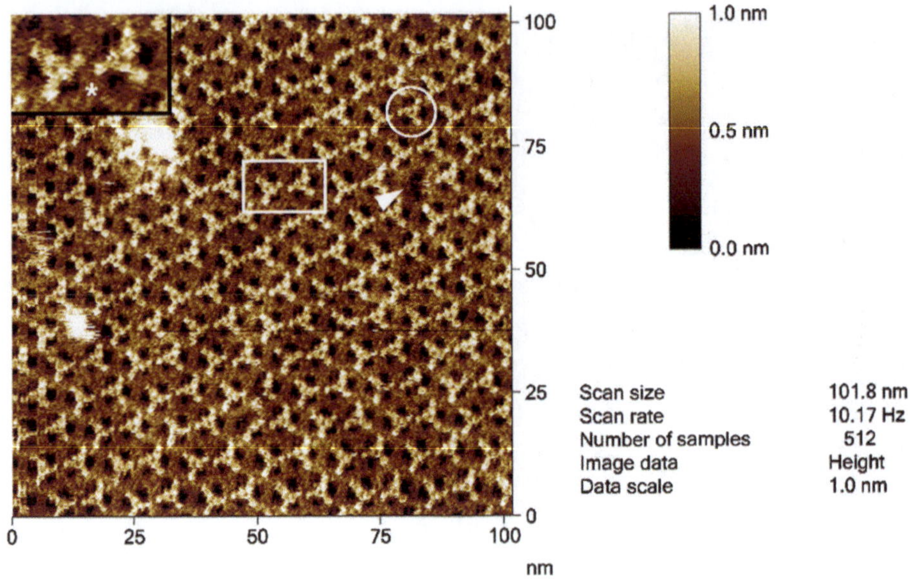

Fig. 4. High magnification AFM topograph of the periplasmic side of porin OmpF. To guide the eye, one single OmpF trimer is marked (*white* circle). The rectangular unit cell of such 2D crystals comprises two porin trimers which are related by a twofold symmetry axis (*white* frame and *inset*). The *arrowhead* indicates one missing porin trimer in the 2D crystal. The *asterisk* in the *inset* marks the lipid bilayer surface. Frame size of the *inset*: 16 × 9.8 nm.

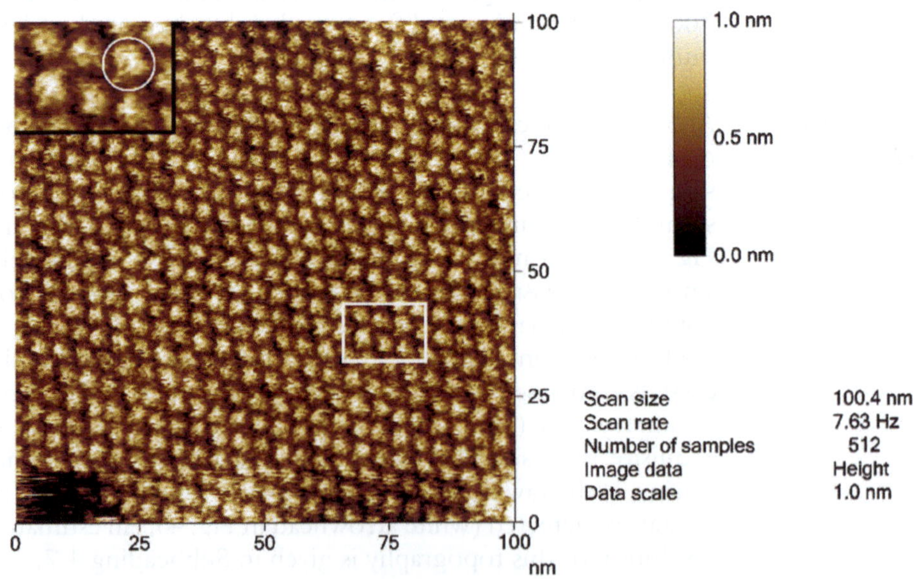

Fig. 5. High magnification AFM topograph of the extracellular side of porin OmpF. In order to have a detailed view on the extracellular surface of OmpF, the region marked by the *white* frame was magnified and is displayed in the *inset*. One protrusion (*white* circle) corresponds to one OmpF monomer. Frame size of the *inset*: 16 × 10.9 nm.

the large protruding domains on the extracellular side, trimers are difficult to distinguish and the pores can not be contoured by the scanning AFM stylus. Imaging of the extracellular surface of porin OmpF at subnanometer resolution is challenging, because of the long and flexible extracellular loops (31, 33).

4.7. Estimating the Resolution of AFM Topographs

The resolution of a topography containing a regular structure can be estimated from its powerspectrum. Software packages delivered with most AFMs provide the option to calculate powerspectra of topographs. Alternatively various public domain programs like NIH image from the National Institutes of Health (Bethesda, Maryland, USA, rsb.info.nih.gov/nih-image) or SXM from the University of Liverpool (Liverpool, UK, www.liv.ac.uk/~sdb/ImageSXM) have such algorithms implemented.

The powerspectrum calculated from the periplasmic surface of OmpF (Fig. 4) is shown in Fig. 6. The higher the order of the discernable spots, the higher is the resolution of the details preserved in the topograph. The encircled spot (–13,4) corresponds to a lateral resolution of 0.93 nm. To calculate the resolution at a selected diffraction spot (*h,k*) of such a rectangular lattice with lattice parameters: $a = 13.5$ nm; $b = 8.2$ nm, formula (1) can be used. This formula is applicable to all lattice types, e.g., for trigonal

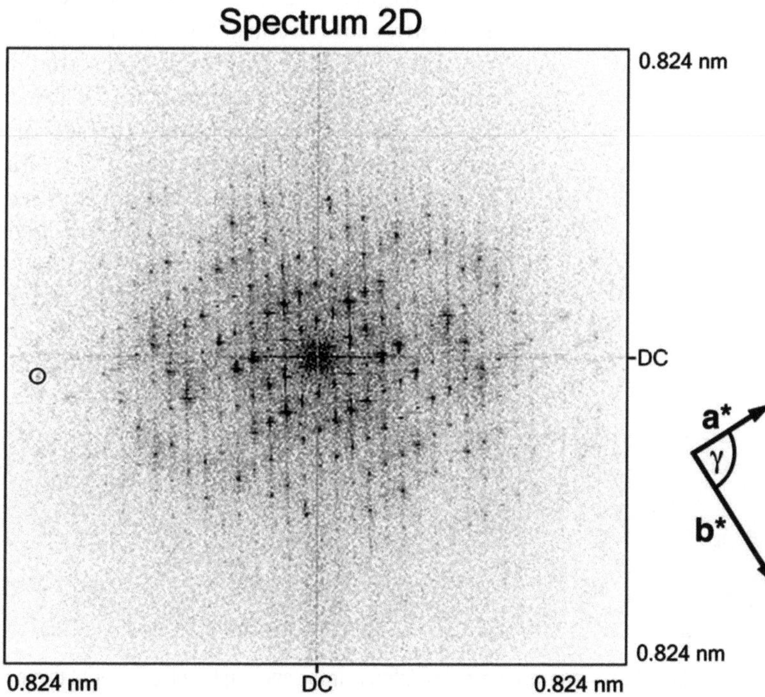

Fig. 6. Powerspectrum calculated from the AFM topograph displayed in Fig. 4 (periplasmic side of OmpF). Diffraction spot (–13,4) is marked by a small, *black* circle and represents a resolution of 0.93 nm.

and orthorhombic lattices, which also occur frequently in native and reconstituted 2D crystals of membrane proteins (for further reading on crystallography, see ref. 40).

$$\delta = \frac{1}{|\vec{r}|} = \frac{\sin\gamma}{\sqrt{\frac{h^2}{a^2} + \frac{2 \cdot h \cdot k \cdot \cos\gamma}{a \cdot b} + \frac{k^2}{b^2}}} \quad (1)$$

δ: resolution; a, b: basic lattice vectors; γ: angle between the basic lattice vectors; h, k: Miller indices; $\vec{r}$: vector from origin to diffraction spot (h,k).

4.8. Stylus Effects and Artifacts

Commercial AFM styluses are not ideal point probes. They show imperfect geometries in the subnanometer range from which imaging effects and artifacts arise that have to be considered when interpreting AFM topographs (for further reading on stylus effects and artifacts see refs. 41, 42). Such effects result when the stylus interacts with the sample at different sites simultaneously, instead of at only one site. Alternatively, the different sides of an assymetric stylus interact with the sample differently. This leads to an AFM topograph of the sample that is convoluted with the stylus shape. To be sure of having acquired the correct surface structure and not an artifact, the topography of the object being investigated has to be reproduced several times with different AFM styluses taken from different batches. To evaluate and confirm the observed AFM topography, structures of samples that have been determined by other methods, e.g., electron and X-ray crystallography for OmpF, can be used. The periplasmic surface of OmpF recorded with an artifact-free stylus is shown in Fig. 7a, while Fig. 7b–e display topographs acquired with artifacteous styluses. Some tentative explanations for the observed artifacts are given. However, the reader should consider that these are speculative. Compared to Fig. 7a, the trimers in Fig. 7b do not show the characteristic Y-shape. Substructure is missing and the pores are not evident. This AFM stylus was most probably blunt and thus unable to resolve small

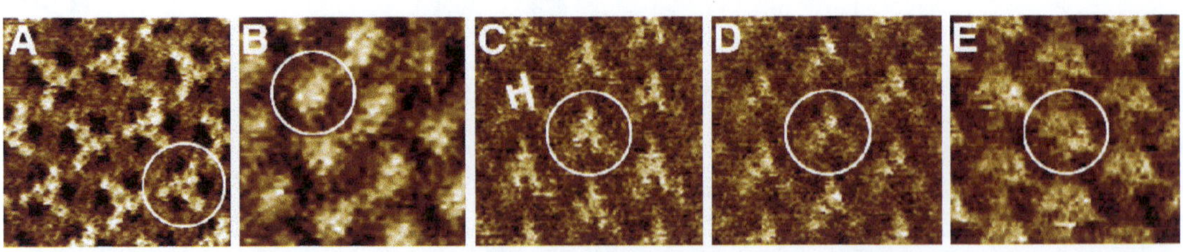

Fig. 7. Stylus effects and artifacts. (**a**) Artifact-free topography of the periplasmic side of OmpF. (**b–e**) Artifacteous topographs of the periplasmic surface of OmpF. In all topographs (**a–d**) one porin trimer is highligthed by a *white* circle. The topographs in (**a**) and (**b**) were recorded on 2D OmpF crystals with a rectangular unit cell, while those in panels (**c–e**) on crystals with trigonal symmetry. The frame sizes in (**a–e**) are 25.6 nm. The brightnesses of the topographs correspond to scales of 1 nm (**a**), 0.6 nm (**b**), 0.85 nm (**c** and **d**), and 0.7 nm (**e**).

elevations and indentations. In Fig. 7c the Y-shaped protrusion of the porin trimer appears more like the letter H with shorter legs and slighly tilted. Figure 7d is the retrace topograph of Fig. 7c. In this topograph OmpF has a dimeric appearance. The AFM stylus in Fig. 7c, d was most probably a double tip. The differences between trace (Fig. 7c) and retrace (Fig. 7d) indicate an asymmetry of the stylus (ideally a conus) in line with the prior assumption of a double tip. In Fig. 7e the porin trimer is imaged as an irregular object of almost constant height. This stylus geometry is difficult to explain but may be a combination of a blunt and an asymmetric tip. Proteins with symmetry are of great advantage to detect possible AFM stylus artifacts: in an artifact-free topography the monomers within the oligomer should all look the same.

Acknowledgments

We are grateful to Andreas Engel for his continous support. This work was supported by the Swiss National Foundation for Scientific Research (Grant 31003A_125150 to D.F.), the National Center of Competence in Research (NCCR) TransCure, and the European Science Foundation (Grant 09-EuroSYNBIO-FP-012 NANOCELL to D.F. and D.J.M.).

References

1. Binnig G, Quate CF, Gerber C (1986) Atomic force microscope. Phys Rev Lett 56:930–933
2. Gerber C, Lang HP (2007) How the doors to the nanoworld were opened. Nat Nanotec 1:3–5
3. Dufrêne YF (2008) Towards nanomicrobiology using atomic force microscopy. Nat Rev Microbiol 6:674–680
4. Engel A, Gaub HE (2008) Structure and mechanics of membrane proteins. Annu Rev Biochem 77:127–148
5. Frederix PLTM, Akiyama T, Staufer U, Gerber C, Fotiadis D, Müller DJ et al (2003) Atomic force bio-analytics. Curr Opin Chem Biol 7: 641–647
6. Müller DJ, Dufrêne YF (2008) Atomic force microscopy as a multifunctional molecular toolbox in nanobiotechnology. Nat Nanotechnol 3:261–269
7. Müller DJ, Helenius J, Alsteens D, Dufrêne YF (2009) Force probing surfaces of living cells to molecular resolution. Nat Chem Biol 5:383–390
8. Puchner EM, Gaub HE (2009) Force and function: probing proteins with AFM-based force spectroscopy. Curr Opin Struct Biol 19:605–614
9. Engel A, Lyubchenko Y, Müller DJ (1999) Atomic force microscopy: a powerful tool to observe biomolecules at work. Trends Cell Biol 9:77–80
10. Engel A, Müller DJ (2000) Observing single biomolecules at work with the atomic force microscope. Nat Struct Biol 7:715–718
11. Müller DJ, Engel A (2007) Atomic force microscopy and spectroscopy of native membrane proteins. Nat Protoc 2:2191–2197
12. Müller DJ, Wu N, Palczewski K (2008) Vertebrate membrane proteins: structure, function, and insights from biophysical approaches. Pharmacol Rev 60:43–78
13. Müller DJ (2008) AFM: a nanotool in membrane biology. Biochemistry 47:7986–7998
14. Han WH, Lindsay SM, Dlakic M, Harrington RE (1997) Kinked DNA. Nature 386:563
15. Han WH, Lindsay SM, Jing TW (1996) A magnetically driven oscillating probe microscope for operation in liquids. Appl Phys Lett 69:4111–4113

16. Hansma PK, Cleveland JP, Radmacher M, Walters DA, Hillner PE, Bezanilla M et al (1994) Tapping mode atomic force microscopy in liquids. Appl Phys Lett 64:1738–1740
17. Putman CAJ, Vanderwerf KO, de Grooth BG, Vanhulst NF, Greve J (1994) Tapping mode atomic-force microscopy in liquid. Appl Phys Lett 64:2454–2456
18. Zhong Q, Inniss D, Kjoller K, Elings VB (1993) Fractured polymer/silica fiber surface studied by tapping mode atomic force microscopy. Surf Sci Lett 290:L688–L692
19. Fukuma T (2008) Instrumentation and biological applications of high-resolution frequency modulation atomic force microscopy in liquid. J Nano Res 4:1–10
20. Fukuma T, Jarvis SP (2006) Development of liquid environment frequency modulation atomic force microscope with low noise deflection sensor for cantilevers of various dimensions. Rev Sci Inst 77:043701
21. Hoogenboom BW, Hug HJ, Pellmont Y, Martin S, Frederix PLTM, Fotiadis D et al (2006) Quantitative dynamic-mode scanning force microscopy in liquid. Appl Phys Letters 88:193109
22. Hoogenboom BW, Suda K, Engel A, Fotiadis D (2007) The supramolecular assemblies of voltage-dependent anion channels in the native membrane. J Mol Biol 370:246–255
23. Hansma PK, Schitter G, Fantner GE, Prater C (2006) Applied physics. High-speed atomic force microscopy. Science 314:601–602
24. Yokokawa M, Wada C, Ando T, Sakai N, Yagi A, Yoshimura SH et al (2006) Fast-scanning atomic force microscopy reveals the ATP/ADP-dependent conformational changes of GroEL. EMBO J 25:4567–4576
25. Müller DJ, Fotiadis D, Scheuring S, Müller SA, Engel A (1999) Electrostatically balanced subnanometer imaging of biological specimens by atomic force microscope. Biophys J 76: 1101–1111
26. Fotiadis D, Scheuring S, Müller SA, Engel A, Müller DJ (2002) Imaging and manipulation of biological structures with the AFM. Micron 33:385–397
27. Cowan SW, Schirmer T, Rummel G, Steiert M, Ghosh R, Pauptit RA et al (1992) Crystal structures explain functional properties of two *E. coli* porins. Nature 358:727–733
28. Dorset DL, Engel A, Häner M, Massalski A, Rosenbusch JP (1983) Two-dimensional crystal packing of matrix porin. A channel forming protein in *Escherichia coli* outer membranes. J Mol Biol 165:701–710
29. Hoenger A, Gross H, Aebi U, Engel A (1990) Localization of the lipopolysaccharides in metal-shadowed reconstituted lipid-porin membranes. J Struct Biol 103:185–195
30. Hoenger A, Pages JM, Fourel D, Engel A (1993) The orientation of porin OmpF in the outer membrane of *Escherichia coli*. J Mol Biol 233:400–413
31. Müller DJ, Engel A (1999) Voltage and pH-induced channel closure of porin OmpF visualized by atomic force microscopy. J Mol Biol 285:1347–1351
32. Schabert FA, Engel A (1994) Reproducible acquisition of *Escherichia coli* porin surface topographs by atomic force microscopy. Biophys J 67:2394–2403
33. Schabert FA, Henn C, Engel A (1995) Native *Escherichia coli* OmpF porin surfaces probed by atomic force microscopy. Science 268: 92–94
34. Engel A, Massalski A, Schindler H, Dorset DL, Rosenbusch JP (1985) Porin channel triplets merge into single outlets in *Escherichia coli* outer membranes. Nature 317:643–645
35. Tanaka M, Sackmann E (2005) Polymer-supported membranes as models of the cell surface. Nature 437:656–663
36. Müller DJ, Amrein M, Engel A (1997) Adsorption of biological molecules to a solid support for scanning probe microscopy. J Struct Biol 119:172–188
37. Müller DJ, Engel A (2008) Strategies to prepare and characterize native membrane proteins and protein membranes by AFM. Curr Opin Coll Int Sci 13:338–350
38. Müller DJ, Sass H-J, Müller S, Büldt G, Engel A (1999) Surface structures of native bacteriorhodopsin depend on the molecular packing arrangement in the membrane. J Mol Biol 285:1903–1909
39. Müller DJ, Engel A (1997) The height of biomolecules measured with the atomic force microscope depends on electrostatic interactions. Biophys J 73:1633–1644
40. Misell DL, Brown EB (1987) Electron diffraction: an introduction for biologists. In: Glauert AM (ed) Practical methods in electron microscopy, vol 2. Elsevier Science Publishers B.V., The Netherlands
41. Schwarz UD, Haefke H, Reimann P, Güntherodt H-J (1994) Tip artefacts in scanning force microscopy. J Microsc 173:183–197
42. Xu S, Arnsdorf MF (1994) Calibration of the scanning (atomic) force microscope with gold particles. J Microsc 173:199–210

Chapter 25

Determination of Soluble and Membrane Protein Structures by X-Ray Crystallography

Raquel L. Lieberman, Mary E. Peek, and J. Derrick Watkins

Abstract

X-ray crystallography is a technique used to determine the atomic-detail structure of a biological macromolecule. The method relies on the ability to generate a three-dimensional crystal of a highly purified protein or nucleic acid for diffraction by X-rays. The extent of scattering of X-rays by the crystal determines the accuracy of the resulting structural model. Unlike electrons, X-rays cannot be refocused after they have been scattered by their target. Thus, calculations are needed to reconstruct the image of the macromolecule that builds the crystal lattice. Tremendous advances over the past 60 years in recombinant expression and purification, crystal growth methods and equipment, X-ray sources, computer processing power, programs, and graphics have taken X-ray crystallography from a highly specialized field to one increasingly accessible to researchers in the biomedical sciences. In this chapter, we review the major concepts of macromolecular X-ray crystallography, focusing mainly on techniques for crystallizing soluble and membrane proteins, and provide a protocol for the crystallization of lysozyme as a model for the crystallization of other proteins.

Key words: Crystals, Vapor diffusion, Fourier transform, Model building, Refinement, Protein data bank, Membrane protein, Lysozyme

1. Introduction

X-ray crystallography is the predominant method used today to determine the molecular structure of proteins and nucleic acids. There is no known limit to the size of the biological macromolecule or complex as long as a crystal containing a high-quality three-dimensional lattice can be grown. Incredible advances over the past two generations have occurred in numerous fields that are indispensable to X-ray crystallography. In the early days of crystallography (1), proteins selected for structure determination were limited to molecules highly abundant in nature that could be

Ingeborg Schmidt-Krey and Yifan Cheng (eds.), *Electron Crystallography of Soluble and Membrane Proteins: Methods and Protocols*, Methods in Molecular Biology, vol. 955, DOI 10.1007/978-1-62703-176-9_25,

purified from their natural sources by simple procedures. Today, production of adequate quantities of low abundance proteins is readily accessible by recombinant expression techniques. Milligrams of many proteins can now be obtained from a recombinant bacterial, yeast, mammalian cell culture, or by in vitro transcription/ translation methods. Modern chromatography techniques using affinity purification, ion exchange and size exclusion chromatography take advantage of computer-controlled liquid chromatography systems that enable the isolation of highly purified and homogenous samples. Preparation of DNA and RNA samples for crystallization is accessible by solid phase phosphodiester chemical synthesis or by in vitro transcription (2). In addition, the commercialization of crystallization cocktails and peripheral consumables, as well as automated crystallization robots that generate cocktails, set crystallization drops, and document crystal growth, have reduced the volumes of sample needed and personal intervention during the course of a crystallization experiment.

In terms of the X-ray diffraction experiment, synchrotron radiation sources have been built around the world, with high flux undulating X-ray beamlines available for biological macromolecular structure determination, which are tunable to virtually any desired wavelength in the X-ray range of (~0.6–2.3 Å) (3, 4). Software programs controlling the beamlines have intuitive graphical user interfaces (5). Advances in robotics and off-site networking has made remote data collection possible at some synchrotron sources, relieving the inconvenience and expense of traveling to the site for data collection. Finally, personal computers equipped with software for automated or semi-automated processing, structure determination, as well as excellent graphics programs for model building and refinement, have significantly lowered the barrier for a newcomer to the field to solve a crystal structure of a macromolecule from virtually anywhere in the world. Many excellent resources are available for the aspiring X-ray crystallographer both online and in print, some of which are listed in the Appendix.

1.1. Crystal Growth and Harvesting

The starting entity of a crystal, the so-called asymmetric unit, is subjected to symmetry operations such as reflection, rotation, inversion, screw axis rotation, and glide plane translation. These operations on the asymmetric unit build up the unit cell. The unit cell is repeated in a periodic array that eventually yields the macroscopic crystalline object. Crystals are cataloged according to the space group and dimensions of their internal lattice structure, which is determined by the symmetry elements involved. Due to their chiral nature, biological macromolecules lack a center of inversion, and thus can crystallize using just 65 of the 230 possible

Table 1
Bravais lattice and space group possibilities for macromolecules

Crystal lattice system	Chiral space groups
Triclinic	P1
Monoclinic	P2, $P2_1$, C2
Orthorhombic	P222, $P222_1$, $P2_12_12$, $P2_12_12_1$, I222, $I2_12_12_1$, C222, $C222_1$, F222
Tetragonal	P4, $P4_1$, $P4_2$, $P4_3$, P422, $P42_12$, $P4_12_12$, $P4_222$, $P4_22_12$, $P4_322$, $P4_32_12$, I4, $I4_1$, I422, $I4_122$
Trigonal	P3, $P3_1$, $P3_2$, P312, P321, $P3_112$, $P3_121$, $P3_212$, $P3_221$, R3, R32
Hexagonal	P6, $P6_1$, $P6_2$, $P6_3$, $P6_4$, $P6_5$, P622, $P6_122$, $P6_222$, $P6_322$, $P6_422$, $P6_522$
Cubic	P23, $P2_13$, P432, $P4_132$, $P4_232$, $P4_332$, $P4_132$, I23, $I2_13$, I432, $I4_132$, F23, F432, $F4_132$

space groups (Table 1) (6). It is the precise and predictable arrangement of the molecules, combined with a very large number of repeating units, which gives crystals diffracting power.

In principle, growing crystals of a biological macromolecule is no different than growing crystals of table salt or other small organic molecule. A solution containing the molecule of interest is slowly concentrated until a metastable supersaturated solution is generated, which allows for nucleation and crystal growth. The crystallization process is governed by both thermodynamics and kinetics, which can be subtle and difficult to control or reproduce (7). Moreover, the cocktail of reagents that promote crystallization of a new macromolecule is not known in advance, nor can it be deduced, and multiple parameters must be tested to determine the appropriate conditions for crystallization. Macromolecular crystals typically contain 50% water, are highly fragile, and imperfections in the lattice result in crystal mosaicity (8). For these reasons, crystal growth is often a major limiting step to structure determination by X-ray crystallography (Fig. 1).

Several different methods exist for crystallization. Microbatch crystallization, used in the early days of crystallography, involved the addition of a precipitating agent into a large volume (>1 mL) of a protein solution in a vial in an attempt to grow crystals of the protein from a supersaturated solution. However, this technique, which requires a large amount of material, has largely been replaced with the more versatile technique of vapor diffusion (9). In vapor diffusion, a small volume (~0.5–10 μL) of a concentrated solution (~5–20 mg/mL) of a macromolecule is mixed with a precipitant solution, and then allowed to equilibrate against a reservoir that contains a higher concentration of the precipitant, called the mother liquor. Because the initial protein drop contains the precipitant at half the total concentration, the direction of equilibrium dictates that the drop will dehydrate. An appropriate combination

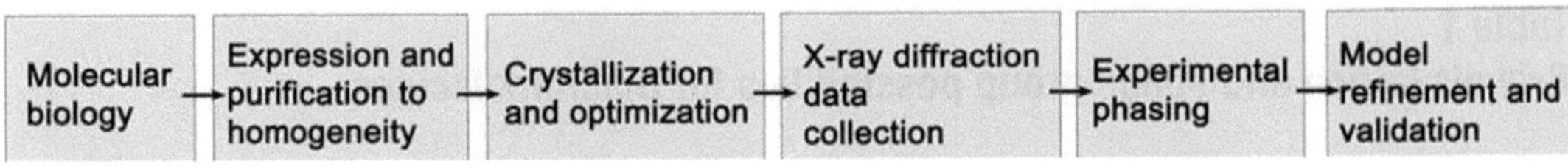

Fig. 1. Overall schematic for macromolecular structure determination by X-ray crystallography.

of analytes present at this increased concentration will promote crystal growth. Because of the small amount of protein used to test each condition, this technique allows for the screening of a much wider range of conditions. Numerous variations in geometries exist for vapor diffusion, but hanging and sitting drops are the most common (Fig. 2). The sitting drop geometry is particularly amenable to automated drop setting crystallization robots. An emerging alternative to vapor diffusion is free interface diffusion, particularly using microfluidic crystallization devices (10).

As a starting point for the crystallization of biological macromolecules, sparse matrix screens are available from commercial vendors (11). These screens contain cocktails that consist of a broad range of precipitant, pH, and salts in cocktails that have been successful in the crystallization of other macromolecules (12). The screens are typically tested at a constant temperature such as room temperature and/or a cooler temperature (4 or 10°C). The results of a sparse matrix screen are usually evaluated just after setup and at 24 h intervals thereafter. On occasion, sparse matrix screening results in crystals of suitable quality for structure determination in one or more conditions, but more commonly yields precipitation that needs to be evaluated for its promise to generate suitable crystals upon further parameterization. If most of the tray contains mainly clear drops or mostly heavy precipitate, strategies include increasing or decreasing the protein concentration of protein, respectively, changing the ionic strength of the buffer, or changing the sample buffer pH based on the isoelectric point of the macromolecule (13). Unless the protein is naturally colored, brown precipitate usually indicates a denatured protein.

Although the objective is to form three-dimensional crystals (Fig. 3, left), one-dimensional needles and fine white microcrystals are promising conditions that are obtained from sparse matrix screens. A series of optimization steps are typically undertaken to improve lattice structure and diffraction limit. Parameters to be tested include fine screening for precipitant concentration, pH, salt concentration and identity, low concentrations of additives such as organic solvents, as well as protein concentration, drop geometry, and incubation temperature (14, 15). Initial crystals can also be seeded into drops without crystals to better control nucleation (16, 17). Typical macroscopic size of crystals used for structure determination is 50–300 μm in each dimension, and the X-ray beam is usually of similar dimensions. In some cases, crystals as small as 5–10 μm in each dimension can be used in conjunction with specialized synchrotron mini-beamlines (18).

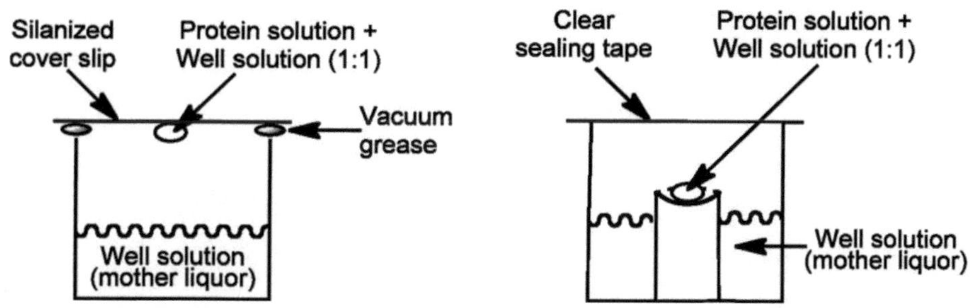

Fig. 2. Plates for crystallization by vapor diffusion. *Left*, hanging drop configuration in which the drop is placed on a siliconized cover slip and mounted upside down above the reservoir using vacuum grease. *Right*, sitting drop configuration in which the drops sit in a pedestal in the center. The plate is sealed with clear packing tape.

Crystals are harvested individually into nylon loops on the end of a pin and immediately plunged in liquid nitrogen, a method known as cryocooling. The pin is mounted on a base compatible with equipment used for data collection, so once cryocooled, crystals are kept at liquid nitrogen temperature through X-ray data collection. The advent of cryocooling has increased the resolution and data quality obtained from macromolecular crystals by reducing radiation damage during the X-ray diffraction experiments and limiting thermal vibrations and conformational changes (19). Crystals are transferred from their original crystallization solution to solutions known as cryoprotectants that form a glass upon cryocooling. These solutions often contain the mother liquor supplemented with ethylene glycol, glycerol, certain alcohols, or oil (20).

The vast majority of the macromolecular structures solved by X-ray crystallography are water soluble proteins, and their crystallization can be approached using the aforementioned screens and optimization steps. Well-folded, single domain, globular proteins typically contain adequate charged and polar residues on their surface for direct crystal contacts, or contacts mediated by metal ions (21). In these cases, the goal is to find the best combination of protein concentration, buffer, and precipitant for crystallization. The challenge of crystallization is greater for macromolecules with multiple domains or inherent flexibility, in which a compact, homogenous sample is not present in solution. For example, many multidomain proteins (22) and RNA (23) are difficult to crystallize. In these instances, a strategy might involve identifying a small molecule capable of locking the macromolecule into a single conformation, or generating individual well-defined domains for separate crystallization trials, either by recombinant expression (22) or by proteolysis (24). Natively unfolded macromolecules, which are significantly disordered, are better studied by solution methods such as nuclear magnetic resonance spectroscopy (25).

Integral membrane proteins, which comprise 1/3 of all sequenced genomes, are vastly underrepresented in the protein

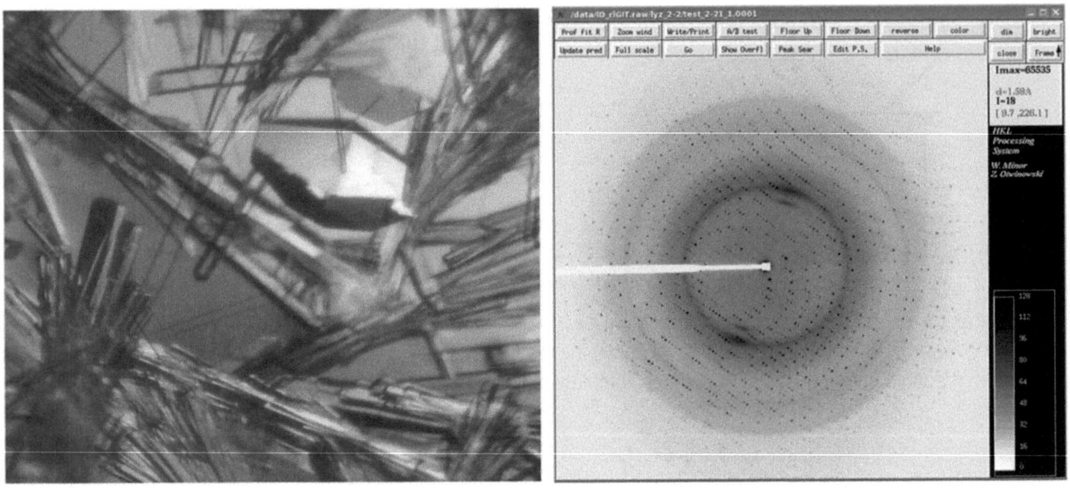

Fig. 3. Lysozyme crystals and sample diffraction pattern. *Left*, lysozyme crystals grown in our laboratory. Note different sizes and shapes. *Right*, diffraction pattern acquired from lysozyme crystals and visualized using HKL2000. Reflections are observed with high degree of symmetry predicted by the space group of the crystal.

databank (PDB) (26). Of the over 66,000 structures deposited into the PDB to date, just ~300 membrane protein structures have been solved by any method. Membrane proteins are notoriously difficult to study by X-ray crystallography for many reasons including bottlenecks at the stages of protein expression, purification, and crystallization (27). Like the early days of protein crystallography, membrane protein structures have typically been limited to highly abundant membrane proteins that can be isolated from natural sources. Lower yields of recombinant expression of membrane proteins in heterologous hosts can be overcome by large-scale fermentation. Membrane protein purification involves solubilization in specialized micelles of mild detergents, which is identified empirically for retention of structure and function (or enzyme activity). The detergent is present at a concentration typically two times its critical micellar concentration throughout the process of purification and crystallization.

There exist two major challenges to the crystallization of a purified membrane protein-detergent complex. First, an inherent issue is that membrane proteins have few hydrophilic or polar loops/domains that are critical to the formation of crystal contacts that stabilize a lattice. Second, detergents undergo phase changes as their concentration is increased in the vapor diffusion crystallization experiment. Thus, the successful crystallization cocktail favorably influences the detergent behavior in a manner that simultaneously promotes membrane protein crystallization and growth. The detergent ideal for solubilization may not be appropriate for crystallization, and each detergent to be tested involves rescreening using sparse matrix. In our experience, fine screening of cocktails is less productive in improving crystals than testing new detergents.

An emerging approach to overcome some of the limitations of crystal growth for so-called difficult proteins is cocrystallization of the target protein with a high affinity or tethered crystallizing agent (28). Ideally, the cocrystallization chaperone does not interfere with the function of the target protein, nor does it change its native conformation. Although the fusion of a crystallizable protein domain into a target protein of interest at the DNA level has enabled a few recalcitrant proteins to be crystallized, complexation with non-covalent binding partners for cocrystallization has been met with more widespread success. Traditional hybridoma screening has been used to identify a monoclonal antibody that recognizes an epitope in the native conformation of the protein. The antibody can be cleaved to yield the Fab fragment for complexation and crystallization (29), or the cDNA for the antibody can be used to clone just the Fv region (30). Phage display libraries of other easily crystallizable domains have also been screened to identify a high-affinity binder (28). Both of these approaches suffer from the lack of control over the epitope recognized, and a new binding partner must be identified for each new protein of interest. Thus, an emerging tactic to broaden the scope of these cocrystallization chaperones involves the use of a protein, such as an antibody fragment, with well-characterized crystallization properties that recognizes a short epitope sequence. The epitope sequence can be installed into any accessible position of protein of interest without disrupting the structure or function of the protein (31).

1.2. The X-ray Diffraction Experiment

The fundamental basis of X-ray crystallography includes concepts from the late nineteenth and early twentieth century such as the discovery of X-rays, the wave/particle duality of light, particularly the diffraction of light, and the complex interactions of electromagnetic radiation with electrons in matter. X-rays are high energy photons that have wavelengths from 100 to 0.1 Å (0.12–12 keV). Diffraction of light occurs when the size of an opening or slit that light passes through is on the same order as the wavelength of light. X-ray crystallography uses hard X-rays of in the range of 1 Å because this is the approximate distance of most covalent bonds in a macromolecule. The X-rays are scattered when their electric field vector interacts with the electrons of matter. Although the scatter of X-rays by a single molecule is very weak, molecules arranged in a periodic crystal lattice enables the amplification of the signal.

Laboratory X-ray generators typically use a rotating copper anode that is filtered to produce monochromatic X-rays of 1.54 Å. In house generators are limited, however, because this is the only single wavelength of X-ray available. Synchrotron beamlines, by contrast, are tunable to a wide range of wavelengths, have higher flux, and are therefore generally employed for high resolution data collection.

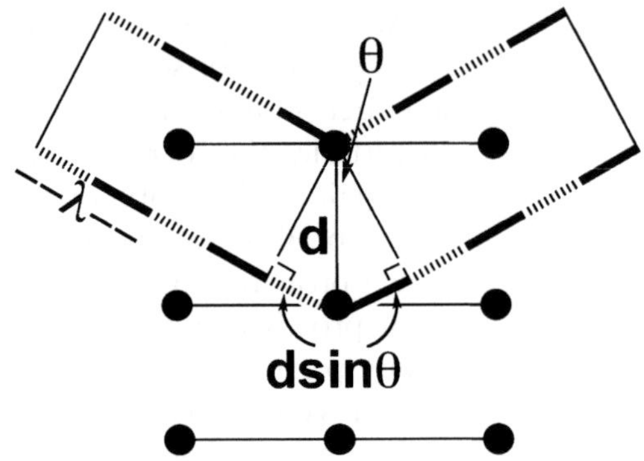

Fig. 4. Visualization of Bragg's law. The equation $(n)\lambda = 2d\sin\theta$ is the condition for constructive interference when an incident X-ray radiation (*solid*, *dashed* lines) is diffracted by atoms in the lattice.

The superposition of the diffracted X-rays by the molecules in a crystal is described by Bragg's law. Maximal constructive interference occurs when waves are in phase, namely when the equation $n\lambda = 2d\sin\theta$ is satisfied (Fig. 4). This requirement to satisfy Bragg's law results in discrete positions of constructively interfering waves. The positions and intensities (which are proportional to the amplitudes) of these waves are recorded on a detector and appear as a spot, or reflection, on a diffraction pattern (Fig. 3, right). The diffraction pattern comprises the data measured in the experiment leading to structure determination.

A complete X-ray diffraction data set contains accurate, and preferably redundant, measurements of all of the possible reflections (32). The number of possible reflection for a data set at given resolution is dictated by the space group, and redundancy of the data reduces the random errors associated with the measurements. The majority of data indexing, scaling, and phasing are simplified by the use of a mathematical construct called the reciprocal lattice. Fourier transforms are then used to convert from reciprocal space back to the real space coordinate system.

In practice, a complete data set is recorded by rotating a cryo-cooled crystal mounted in a loop attached to a goniometer head that is maintained at 100 K by a stream of nitrogen. The crystal is centered in the path of an X-ray beam, rotated up to 360° in small intervals, typically 0.5° or 1°, and a diffraction image is collected at each interval of rotation. The interval is set to minimize reflection overlaps. The total minimum rotation wedge that a crystal must be rotated through to yields a complete data set (i.e., all predicted reflections measured one time) depends on the space group.

Programs such as HKL2000 (33) among others are used to select and record the intensity of the reflections on each image, identify the space group and unit cell dimensions, and index the reflections (i.e., catalog each reflection according to Bragg's law). Reflections that have been cataloged with the same set of indices are then merged and scaled together. In all, data sets range from tens to hundreds of thousands of unique indexed reflections, and various statistics indicate the quality of the data set. Traditionally, 90–95% of the data collected are used in model building and refinement while 5–10%, respectively, are set aside as an unbiased "test set" (see below) (34).

1.3. Structure Solution

A common misconception is that once a crystal is obtained, a structure is imminent, but diffraction intensities do not by themselves provide the information to build a structural model. The intensities of the reflections from the diffracted X-ray beam do contain mathematically predictable contributions from each atom in its unique location in the macromolecule, but the detector is only capable of recording the intensity of the diffracted waves. Unlike other forms of electromagnetic radiation, X-rays cannot be refocused by any material after they have been scattered by their target and the phase information of the traveling wave is lost. Thus, to solve the "phase problem," i.e., retrieve the lost phase information, a variety of experimental and computational methods are used. The Fourier synthesis of all the amplitude/intensity and phase information, also called the structure factor, is what ultimately leads to an interpretable electron density for model building (35).

Here we will outline the several strategies used to obtain initial phases that can be used to solve a crystal structure but refrain from detailed mathematical treatment, which can be found in several excellent textbooks listed in the Appendix. We discuss three main approaches to solving the phase problem, namely, Molecular Replacement (MR), Multiple Isomorphous Replacement (MIR), and Anomalous Dispersion (MAD). The latter two techniques fall into the category of "substructure methods" because they involve identifying the real positions of just a few atoms in the lattice. These few atom positions, which yield initial phases, can often reveal interpretable electron density for the rest of the macromolecule when combined with density improvement protocols. In the rather rare case of atomic resolution data (defined as 1.2 Å resolution or better), ab inito or direct methods can also be used to solve the structure (36), a method beyond the scope of this chapter. A comparison of the methods is summarized in Table 2.

1.3.1. Molecular Replacement

In this method, a search model of known and similar predicted structure to the protein crystal of interest is used to obtain initial phase information (37). The major variable in the success of the MR procedure is the search model. Traditionally, the initial qualification for the search model is 30% identity or higher in the primary sequence between the search model and the protein of interest, and similar three-dimensional structures are predicted by using

Table 2
Summary of phasing methods used in macromolecular X-ray crystallography

Phasing strategy	Phasing entity	Comment
Direct methods	N/A	Small proteins, better than 1.2 Å resolution
Molecular replacement	Search model	Homologous protein
Multiple isomorphous replacement	Heavy atom ions, clusters, halides, naturally bound metal ion cofactors	High occupancy sites, isomorphous crystals, difference Patterson methods
Anomalous dispersion	Anomalous scatterers, often transition metal ions	Tunable X-ray radiation to absorption edge of metal ion, highly redundant data

proteins in the same family or that perform a similar function as the protein of interest. Search programs available on-line are useful in performing homology searches to identify suitable MR models. With the large number of proteins in the PDB, a three-dimensional structure of a suitable model can often be identified, and indeed ~80% of structures in the PDB are solved in this manner.

The coordinates of the model are typically reduced to just one copy of a single macromolecule plus any conserved prosthetic groups. Loop regions are often deleted and side chains removed to reduce model bias. Once diffraction data are collected and processed, the search model is positioned by rotation and translation functions until the calculated diffraction pattern of the search model matches that of the current observed data set, evaluated initially by a correlation coefficient. The main computational challenge is to position a real space probe given only amplitudes of reciprocal space data. Numerous excellent programs using somewhat different algorithms can be used for this calculation. Although the resulting solution is heavily biased towards the search model, a promising solution is one that results in new electron density in unmodeled regions, such as loops or side chains that are not in the original model (38). A detailed protocol of how to use this method in electron crystallography of 2D crystals is described in Chapter 14 of this book.

1.3.2. Multiple Isomorphous Replacement

In MIR, ions of heavy atoms such as lead, platinum, or mercury are introduced into the crystals typically either by soaking or cocrystallization (39, 40). Heavy metal ions contain many more electrons than the macromolecule's C, N, and O light atoms and contribute significantly to the intensity of the diffracted X-rays. The technique requires two or more crystals, one crystal without the heavy atoms present, called the native crystal, and one or more crystals that contains the heavy atoms, called the derivative(s). MIR further requires that the derivatized crystals be isomorphous with the native crystals: i.e. retain the same lattice parameters. As in MR, initially only amplitudes are known. Patterson maps, which provide relative vector relationships between pairs of atoms in the structure in reciprocal space, can enable the conversion to real and absolute positions of the heavy atoms with the correct handedness. In terms of computational methods, both Patterson and direct methods are employed in programs such as Solve/Resolve (41) and those found within the CCP4 suite (42).

1.3.3. Anomalous Dispersion (Single Wavelength: SAD, Multiple Wavelength: MAD)

Like MIR, the SAD/MAD method involves identifying the coordinates of metal ions in real space, but specifically employs those that can eject a core electron when bombarded with a suitable energy of the incident X-ray beam, so-called anomalous scatterers. Ions of elements such as bromine, gold, platinum, lead, and selenium can all absorb X-ray radiation near 1 Å, a common default wavelength of the X-ray beam. Tunable X-ray sources also enable ions of zinc, copper, iron, manganese, nickel, as well as sulfur (43) to serve as anomalous scatterers for MAD or SAD phasing. If the protein of interest is not already a metalloprotein, metal ions can be introduced using the same methods as in MIR. Alternatively, such as in the case of selenium, the metal ion can be incorporated using recombinant expression techniques (44) or by nucleic acid synthesis in the case of DNA/RNA structures (45). The binding sites for the metal ion should be fully occupied throughout the crystal. In general, there is no a priori knowledge of particular binding sites but amino acid composition and ligand preferences for a particular ion can guide selection.

When the doped macromolecular crystal is subjected to X-ray radiation at an energy tuned to the absorption edge of the anomalous scatterer, particular symmetry-related reflections called Bijvoet pairs no longer have the same intensity (46). The small differences in intensities among Bijvoet pairs at the peak of the anomalous absorption are utilized in solving the phase problem in the SAD protocol. When MAD phasing is employed, the data set collected just above the absorption edge is also compared to one collected at the inflection and/or a remote energy (47). The quality, resolution and redundancy of the data, as well as the number of absorbing ions, are critical to the success of these related methods. Computational protocols for SAD or MAD are found within CNS (48) and CCP4 (42).

MAD phasing has the advantage of over MIR in that only a single crystal is needed. The changes in the intensities of reflections used to solve for the phases are obtained by collecting data sets at different wavelengths from a single crystal. The use of a single crystal also removes the problem of isomorphism between the native crystal and the heavy atom derivative. Nevertheless, MIR diffraction data necessary to solve the phase problem by MIR can be collected using an in house generator, while the MAD technique requires the tunable wavelengths of a synchrotron source. MAD and MIR are much less biased phasing techniques than MR and, in the absence of a good search model for MR, MAD and MIR are the only options for solving the structure. Finally, it is often the case that the two techniques of MIR and MAD can be combined to give even greater phasing power.

1.4. Density Improvement, Model Building and Refinement

The substructure methods described in the previous section locate the coordinates of a selective group of atoms in real space. This information is then used to obtain an estimate of the phases. The estimated phases are combined with the experimentally measured amplitudes to produce an electron density map. If the initial estimates of the phases are a close approximation of the actual phases of the model, and the experimental data is of high enough quality, an electron density map for the entire macromolecule or major structural elements can be interpreted to build a model. Typically, initial phasing is conducted at lower resolution than the highest resolution data set, and the resulting map is of poor quality and a model is difficult to trace. However, identification of the position of atoms within the electron density maps through model building increases the accuracy of the phases, and lead to better and more easily interpretable maps.

Powerful computational protocols to improve the phases and thus the quality of the resulting electron density map include taking advantage of solvent boundaries (49) as well as averaging over non-crystallographic symmetry (NCS) elements, which are features of the macromolecule within the asymmetric unit, such as dimerization (50). Many of these protocols improve the reflection-to-parameter ratio. The reduction is advantageous because although diffraction data sets typically contain tens or hundred of thousands of observed reflections, the total is comparable to the degrees of freedom associated with the position of and the vibrational freedom for each C, N, and O atom in a typical macromolecule model.

Two electron density maps are used for modeling, which differ in the weighting of structure factors for the observed data (F_o) and that of the calculated data (F_c). The $2F_o - F_c$ map is used to guide model construction whereas the $F_o - F_c$ map is used to identify errors in the model (35). The maps and model are viewed and built using graphics programs such as Coot (51). Automated model building using ArpWarp (52) or Resolve (41) is also successful with good quality data. The model is refined using protocols within CCP4 or CNS, which differ in the computational algorithms used to minimize the differences in observed data and the model while retaining known geometries (53, 54). After a round of computational refinement, the model is further improved, typically by visual inspection of the maps and modification of the model one residue at a time. This iterative process improves the phases, making it easier to build correctly into the electron density. After the model has been optimized, additional density corresponding to waters or ions may be apparent and added to the model in CNS, and CCP4, as well as within Coot. Ordered water molecules can be validated by eye by measuring hydrogen-bonding interactions to the residues in the macromolecule. Larger ions, such as phosphate, sulfate, Mg^{2+}, Br^{-} may also appear bound to the macromolecule. It is not always straightforward to identify these ions, but the combination of knowledge of the crystallization cocktail and ligand preferences can provide clues. The reduced differences between data and model are evaluated by a decreasing residual value at each refinement step. Both the R_{work}, the difference between the model and the portion of the data being used for refinement, and the R_{free}, the residual calculated using the unused test set, are monitored. The addition of new atoms to the model usually decreases the R_{work}, but may not be correct. The test set and corresponding R_{free} were introduced to help monitor real improvements in the model; generally, a discrepancy of 5% between R_{work} and R_{free} is acceptable, and for a 2 Å resolution structure, the R_{free} for the final model is ~25% (34).

1.5. Crystallization of Lysozyme

Lysozyme is the most common protein structure deposited in the PDB. Lysozyme hydrolyzes acetal links between monomers in certain carbohydrates, and is known for its proclivity to crystallize. Indeed, lysozyme has been reported to crystallize in every lattice system due to the plethora of favorable residues found on the surface that enable the non-covalent interactions to form the crystal lattice.

The experiment we provide below falls under the "optimization" category of crystallization. A sparse matrix screen has identified a condition containing high concentrations of NaCl as being favorable for lysozyme crystallization, and the impact of ionic strength of a variety of chloride salts is tested on the rate of crystal growth and quality of the resultant crystals. Crystals grown by the method described below will exhibit varying shapes and sizes but their internal structure, identified by their space group, can only be determined by an X-ray diffraction experiment. In our hands, rates of crystal growth differed among the salts and were of different shapes (Fig. 3, left) yet chloride salts yield lysozyme in a tetrahedral space group configuration (55) reproducibly within 1 week.

2. Materials

1. Siliconized cover slips (for example, from Hampton Research, HR3-217).
2. Hanging drop plates (such as VDX plates from Hampton Research, HR3-142).

3. High Vacuum grease (Dow Corning).
4. Disposable syringe, 10 cm^3.
5. Mounting clay (available from art supply store).
6. Can of pressurized air.
7. Forceps.
8. Chicken Egg White Lysozyme (for example, Sigma L6876).
9. Sodium chloride.
10. Calcium chloride.
11. Potassium chloride.
12. Magnesium chloride.
13. Ammonium chloride.
14. Stereo microscope (such as Zeiss Stemi 2000) with digital camera for documentation.
15. 0.1 M sodium acetate buffer, pH 4.8.
16. 0.2 mL of 75 mg/mL lysozyme in acetate buffer, stored on ice.
17. Solutions of each salt at 1.5, 4, 6.5, 9% w/v in water.

3. Methods (see Subheading 6.2.1)

1. Apply a small amount of vacuum grease onto the mouth of each well you are going to use (see Note 1).
2. Pipette 1 mL of the appropriate precipitant into a well of the hanging drop plate according to Table 1 (see Note 4).
3. Remove dust particles from the cover slip using a pressurized duster.
4. Place cover slip onto a clean surface.
5. Pipette 2.5 μL of 75 mg/mL lysozyme into the center of the cover slip (see Note 4).
6. Pipette 2.5 μL of the appropriate precipitant into the lysozyme droplet (see Note 4).
7. Using foreceps, carefully invert the cover slip such that the droplet hangs upside down in the center of the cover slip.
8. Gently lower the cover slip onto the opening of the well. Do not allow grease to make contact with the protein droplet (see Note 5).
9. Press the cover slip down gently to seal the well (see Note 6).
10. Repeat this procedure until all wells for a given plate are set up according to Table 3 (see Note 6).
11. Using a stereomicroscope, record observations from each well (see Note 7).

Table 3
Plate setup for lysozyme crystallization with chloride salts

	1	2	3	4	5	6
A	*Drop*: 2.5 μL Lysozyme + 2.5 μL 1.5% NaCl *Reservoir*: 1.5% NaCl	*Drop*: 2.5 μL Lysozyme + 2.5 μL 1.5% $CaCl_2$ *Reservoir*: 1.5% $CaCl_2$	*Drop*: 2.5 μL Lysozyme + 2.5 μL 1.5% KCl *Reservoir*: 1.5% KCl	*Drop*: 2.5 μL Lysozyme + 2.5 μL 1.5% $MgCl_2$ *Reservoir*: 1.5% $MgCl_2$	*Drop*: 2.5 μL Lysozyme + 2.5 μL 1.5% NH_4Cl *Reservoir*: 1.5% NH_4Cl	X
B	*Drop*: 2.5 μL Lysozyme + 2.5 μL 4% NaCl *Reservoir*: 4% NaCl	*Drop*: 2.5 μL Lysozyme + 2.5 μL 4% $CaCl_2$ *Reservoir*: 4% $CaCl_2$	*Drop*: 2.5 μL Lysozyme + 2.5 μL 4% KCl *Reservoir*: 4% KCl	*Drop*: 2.5 μL Lysozyme + 2.5 μL 4% $MgCl_2$ *Reservoir*: 4% $MgCl_2$	*Drop*: 2.5 μL Lysozyme + 2.5 μL 4% NH_4Cl *Reservoir*: 4% NH_4Cl	X
C	*Drop*: 2.5 μL Lysozyme + 2.5 μL 6.5% NaCl *Reservoir*: 6.5% NaCl	*Drop*: 2.5 μL Lysozyme + 2.5 μL 6.5% $CaCl_2$ *Reservoir*: 6.5% $CaCl_2$	*Drop*: 2.5 μL Lysozyme + 2.5 μL 6.5% KCl *Reservoir*: 6.5% KCl	*Drop*: 2.5 μL Lysozyme + 2.5 μL 6.5% $MgCl_2$ *Reservoir*: 6.5% $MgCl_2$	*Drop*: 2.5 μL Lysozyme + 2.5 μL 6.5% NH_4Cl *Reservoir*: 6.5% NH_4Cl	X
D	*Drop*: 2.5 μL Lysozyme + 2.5 μL 9% NaCl *Reservoir*: 9% NaCl	*Drop*: 2.5 μL Lysozyme + 2.5 μL 9% $CaCl_2$ *Reservoir*: 9% $CaCl_2$	*Drop*: 2.5 μL Lysozyme + 2.5 μL 9% KCl *Reservoir*: 9% KCl	*Drop*: 2.5 μL Lysozyme + 2.5 μL 9% $MgCl_2$ *Reservoir*: 9% $MgCl_2$	*Drop*: 2.5 μL Lysozyme + 2.5 μL 9% NH_4Cl *Reservoir*: 9% NH_4Cl	X

12. Store the crystal plate at room temperature uninterrupted and view once on successive days, recording snapshots as desired (see Note 8).

4. Notes

1. A 10 mL syringe loaded with grease may be helpful for greasing the tray; alternatively, a toothpick may be employed.
2. All solutions should be prepared using sterile, ultrapure water.
3. Sterile filter all solutions except protein solution, which should be dissolved in previously filtered buffer.
4. Be careful not to introduce air bubbles into the droplet.
5. Leave a small gap in the circle of grease to allow for spreading once the cover slip is placed.
6. Be as gentle as possible during this process so as not to disturb the drops in any way.
7. Trays should be viewed immediately after setting up, as well as after 1 h.
8. Use the digital camera to record images of your observations. It is useful to turn off flash and auto-focusing, and turn on the timer feature.

5. Appendix

Macromolecular crystallographers worldwide continually provide assistance to fellow community members via listservs, wikis and web pages. A few of the many helpful resources online and in print are listed below.

Websites

http://ccp4wiki.org/~ccp4wiki/wiki/index.php?title=Main_Page#Content

http://www.ccp4.ac.uk/ccp4bb.php

https://solve.lanl.gov/

http://cns-online.org/v1.21/

http://www.ruppweb.org/

http://xray.bmc.uu.se/usf/

http://www.pdb.org/pdb/home/home.do

http://sage.ucsc.edu/~wgscott/xtal/wiki/index.php/Crystallography_on_OS_X

Books

Methods in Molecular Biology Vol 363 and 364, Macromolecular Crystallography Protocols Volumes 1 and 2. Editor: Sylvie Doublié, (Humana Press), 2007. ISBN 978-1617374753, 978-1617378362.

International Tables for Crystallography Volumes A-F. (Wiley), 2001–2004, ISBN 978-0792365914, 978-0792365921, 978-1402019005, 978-1402007149, 978-1402007156, 978-0792368571.

Methods in Enzymology Volumes 276, 277, 368, 374. Editors: Charles W. Cantor and Robert M. Sweet. (Elsevier), 1997 & 2003. ISBN: 978-0-12-182177-7, 978-0-12-182178-4, 978-0-12-182271-2, 978-0-12-182777-9.

Acknowledgments

This work was supported by grants to RLL from NSF (CAREER award 0845445) and the American Federation for Aging Research. Diffraction data were collected at the Southeast Regional Collaborative Access Team (Ser-CAT) beamline 22-BM at the Advanced Photon Source, Argonne National Laboratory. Supporting institutions may be found at www.ser-cat.org/members.html. Use of the Advanced Photon Source was supported by the U. S. Department of Energy, Office of Science, Office of Basic Energy Sciences, under Contract No. W-31-109-Eng-38. We acknowledge the feedback on lysozyme crystallization from undergraduate students enrolled in CHEM 4582, 4681, as well as visitors from Westlake High School. This manuscript is dedicated to the memory of biology teacher Ayesha Johnson.

References

1. Perutz M (1985) Early days of protein crystallography. Methods Enzymol 114:3–18
2. Ducruix A, Giege R (1999) Crystallization of nucleic acids and proteins. Oxford University Press, New York
3. Hendrickson WA (2000) Synchrotron crystallography. Trends Biochem Sci 25:637–643
4. Sweet RM (1998) The technology that enables synchrotron structural biology. Nat Struct Biol 5:654–656
5. McPhillips TM, McPhillips SE, Chiu HJ, Cohen AE, Deacon AM, Ellis PJ, Garman E, Gonzalez A, Sauter NK, Phizackerley RP, Soltis SM, Kuhn P (2002) Blu-Ice and the distributed control system: software for data aquisition and instrument control at macromolecular crystrallography beamlines. J Synchrotron Radiat 9:401–406
6. Burzlaff, H., and Zimmermann, H. (2006) *Bases, lattices, Bravais lattices and other classifications*, International Tables for Crystallography Vol. A. Chapter 9.1, pp 742–749
7. Newman J, Xu J, Willis MC (2007) Initial evaluations of the reproducibility of vapor-diffusion crystallization. Acta Cryst D63: 826–832
8. Matthews BW (1968) Solvent content of protein crystals. J Mol Biol 33:491–497

9. Chayen NE (1998) Comparative studies of protein crystallization by vapour-diffusion and microbatch techniques. Acta Cryst D54:8–15
10. Hansen CL, Skordalakes E, Berger JM, Quake SR (2002) A robust and scalable microfluidic metering method that allows protein crystal growth by free interface diffusion. Proc Natl Acad Sci U S A 99:16531–16536
11. Baldock P, Mils V, Stewart PS (1996) J Cryst Growth 168:170
12. Carter J, Charles W (1990) Efficient factorial designs and the analysis of macromolecular crystal growth conditions. Methods 1:12–24
13. Rupp B (2004) Protein isoelectric point as a predictor for increased crystallization screening efficiency. Bioinformatics 20:2162–2168
14. Cudney R, Patel S, Weisgraber K, Newhouse Y, McPherson A (1994) Screening and optimization strategies for macromolecular crystal growth. Acta Cryst D50:414–423
15. Prater BD, Tuller SC, Wilson LJ (1999) Simplex optimization of protein crystallization conditions. J Cryst Growth 196:674–684
16. D'Arcy A, Mac Sweeney A, Haber A (2003) Using natural seeding material to generate nucleation in protein crystallization experiments. Acta Cryst D59:1343–1346
17. Bergfors T (2003) Seeds to crystals. J Struct Biol 142:66–76
18. Baker M (2010) Structural biology: crystal-clear images. Nature 465:824–825
19. Garman EF, Nave C (2009) Radiation damage in protein crystals examined under various conditions by different methods. J Synchrotron Radiat 16:129–132
20. Petsko GA (1975) Protein crystallography at sub-zero temperatures: cryo-protective mother liquors for protein crystals. J Mol Biol 96:381–392
21. Carugo O, Argos P (1997) Protein-protein crystal-packing contacts. Protein Sci 6:2261–2263
22. Edwards AM, Arrowsmith CH, Christendat D, Dharamsi A, Friesen JD, Greenblat JF, Vegadi M (2000) Protein production: feeding the crystallographers and NMR spectroscopists. Nat Struct Mol Biol 7:970–972
23. Ke A, Doudna JA (2004) Crystallization of RNA and RNA-protein complexes. Methods 34:408–414
24. Wernimont A, Edwards A (2009) In situ proteolysis to generate crystals for structure determination: an update. PLoS One 4:e5094
25. Tompa P (2002) Intrinsically unstructured proteins. Trends Biochem Sci 27:527–533
26. Wiener MC (2004) A pedestrian guide to membrane protein crystallization. Methods 34:364–372
27. Loll PJ (2003) Membrane protein structural biology: the high throughput challenge. J Struct Biol 142:144–153
28. Warke A, Momany C (2007) Addressing the protein crystallization bottleneck by cocrystallization. Cryst Growth Des 7:2219–2225
29. Zhou YM-C, Morais-Cabral JH, Kaufman A, MacKinnon R (2001) Chemistry of ion coordination and hydration revealed by a K^+ channel-Fab complex at 2.0 A resolution. Nature 414:43–48
30. Ostermeier C, Essen LO, Michel H (1995) Crystals of an antibody Fv fragment against an integral membrane protein diffracting to 1.28 A resolution. Proteins 21:74–77
31. Pai JC, Culver JA, Drury JE, Lieberman RL, Maynard JA (2011) Peptide binding antibody fragment (scFv) chaperones for cocrystallization. Prot Engin Des Sel 24:419–428
32. Pflugrath JW (1999) The finer things in X-ray diffraction data collection. Acta Cryst D55:1718–1725
33. Otwinowski Z, Minor W (1997) Processing of x-ray diffraction data collected in oscillation mode. Methods Enzymol 276:307–326
34. Brunger AT (1997) Free R value: cross-validation in crystallography. Methods Enzymol 277:366–396
35. Rupp B (2010) Biomolecular crystallography. Garland Science, New York
36. Sheldrick GM (2008) A short history of SHELX. Acta Cryst A64:112–122
37. Evans P, McCoy A (2008) An introduction to molecular replacement. Acta Cryst D64:1–10
38. Dodson E (2008) The befores and afters of molecular replacement. Acta Cryst D64:17–24
39. Rould MA, Carter WC Jr (2003) Isomorphous difference methods. Methods Enzymol 374:145–163
40. Rould MA (1997) Screening for heavy-atom derivatives and obtaining accurate isomorphous differences. Methods Enzymol 276:461–472
41. Terwilliger T (2004) SOLVE and RESOLVE: automated structure solution, density modification, and model building. J Synchrotron Radiat 11:49–52
42. Collaborative Computational Project, Number 4 (1994) The CCP4 suite: programs for protein crystallography. Acta Cryst D50:760–763
43. Ramagopal UA, Dauter M, Dauter Z (2003) Phasing on anomalous signal of sulfurs: what is the limit? Acta Cryst D59:1020–1027
44. Hendrickson WA, Horton JR, Lemaster DM (1990) Selenomethionyl proteins produced for analysis by Multiwavelength Anomalous Diffraction (Mad) - a vehicle for direct deter-

mination of 3-dimensional structure. EMBO J 9:1665–1672

45. Sheng J, Huang Z (2008) Selenium derivatization of nucleic acids for phase and structure determination in nucleic acid X-ray crystallography. Int J Mol Sci 9:258–271
46. Blow DM (2003) How Bijvoet made the difference: the growing power of anomalous scattering. Methods Enzymol 374:3–22
47. Hendrickson WA, Ogata CM (1997) Phase determination from multiwavelength anomalous diffraction measurements. Methods Enzymol 276:494–523
48. Brunger AT, Adams PD, Clore GM, DeLano WL, Gros P, Grosse-Kunstleve RW, Jiang JS, Kuszewski J, Nilges M, Pannu NS, Read RJ, Rice LM, Simonson T, Warren GL (1998) Crystallography & NMR system: a new software suite for macromolecular structure determination. Acta Cryst D54:905–921
49. Kostrewa D (1997) Bulk solvent correction: practical application and effects in reciprocal and real space. CCP4 Newsletter Protein Crystallogr 34:9–22
50. Terwilliger TC (2002) Rapid automatic NCS identification using heavy-atom substructures. Acta Cryst D58:2213–2215
51. Emsley P, Cowtan K (2004) Coot: model-building tools for molecular graphics. Acta Cryst D60:2126–2132
52. Langer G, Cohen SX, Lamzin VS, Perrakis A (2008) Automated macromolecular model biolding for X-ray crystallography using ARP/wARP version 7. Nat Protoc 2:1171–1179
53. Engh RA, Huber R (1991) Accurate bond and angle parameters for x-ray protein structure refinement. Acta Cryst A47:392–400
54. Ramachandran GN, Ramakrishnan C, Saisekharan V (1963) Stereochemistry of polypeptide chain configurations. J Mol Biol 7:95–99
55. Ries-Krautt MM, Ducruix AF (1989) Relative effectiveness of various ions on the solubility and crystal growth of lysozyme. J Biol Chem 264:745–748

Chapter 26

Solution Nuclear Magnetic Resonance Spectroscopy

James J. Chou and Remy Sounier

Abstract

Solution nuclear magnetic resonance (NMR) spectroscopy has come a long way in characterizing the structure and function of biological molecules since the first one-dimensional spectrum of protein was recorded about 30 years ago. To date (September 1, 2012), there are 9,521 solution NMR structures in the Protein Data Bank, compared to 74,009 determined by crystallographic methods. Unlike X-ray and electron microscopy (EM) methods, which are based on the concepts of Fourier optics and image reconstruction, structure determination by NMR involves measuring structural restraints and finding structural solutions that satisfy the restraints. Although the NMR approach is much less direct in a physical sense, it has proven itself over the years to be capable of de novo structure determination at high precision. Moreover, the method is highly versatile and can be used in a variety of ways for addressing mechanistic questions. NMR measurements of protein internal dynamics and protein–protein or protein–ligand interaction are directly relevant to function in vivo because the molecules are often in physiological buffer conditions. The method can also be applied to investigate protein-folding intermediates, conformational changes, as well as intrinsically unfolded proteins. Recently, along with X-ray and EM, solution NMR has entered a state of rapid growth for structural studies of membrane proteins, already demonstrating its feasibility in de novo structure determination of membrane-embedded ion channels and receptors. As the hardware advances rapidly, especially in cryogenic probes that have much higher sensitivity, the sample concentration required for solution NMR investigation is decreasing, hopefully soon to a concentration level at which nonspecific protein aggregation is no longer an issue. After three decades of improvement in spectrometer technology, NMR pulse experiments, isotope labeling schemes, and structure determination software, we believe that solution NMR will truly enter the production phase in the next decade to answer biological questions of high impact, and to become more versatile than ever in complementing X-ray and EM in investigating protein structure and function.

Key words: NMR, Biomolecular application, Protein structure, Protein dynamics

1. Introduction

1.1. Solution NMR in Structure Determination of Biological Molecules

After Bloch and Purcell first observed the resonance spectrum of paraffin in solid phase (1, 2) and water in liquid phase (1, 2), liquid-state nuclear magnetic resonance (NMR) spectroscopy rapidly became an indispensable tool for characterizing small molecules in

Ingeborg Schmidt-Krey and Yifan Cheng (eds.), *Electron Crystallography of Soluble and Membrane Proteins: Methods and Protocols*, Methods in Molecular Biology, vol. 955, DOI 10.1007/978-1-62703-176-9_26,

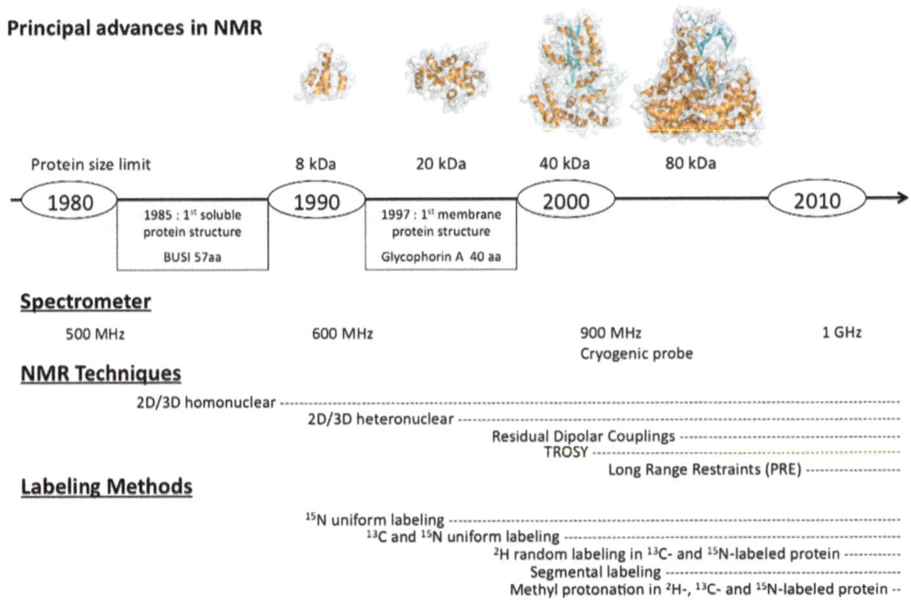

Fig. 1. A timeline summarizing the advances in NMR hardware, spectroscopic methods, and isotope labeling methods.

chemistry laboratories, and is now universally used in structural characterization of larger biomolecules such as proteins and nucleic acids as well as their complexes. A timeline that briefly describes the milestones in biomolecular NMR is shown in Fig. 1.

The obvious advantage of solution NMR is that biomolecules tumble freely in the more native aqueous environment, not having to form crystals. But the requirement for molecular tumbling in solution also happens to pose the major limitation of solution NMR. In NMR spectroscopy, the instruments generally excite spin ½ nuclides and detect their chemical shift evolution, with proton spin being the most popular one owing to its natural abundance and large gyromagnetic ratio (γ). For other biological atoms like carbon, nitrogen, and oxygen, the natural abundance of spin ½ nuclides is extremely low and isotope labeling is usually required for them to be NMR visible. The nuclear spins all have magnetic dipole moments and can thus interact with each other when they are close in space. The interaction between a pair of dipole moments is measured as the dipolar coupling constant, which is proportional to $3\cos^2\theta - 1$, where θ is the angle between the internuclear vector and the static magnetic field. When molecules are in solid phase, a spin experiences very strong dipolar couplings from others in the molecule (e.g., 0.1–50 kHz for ^{1}H), resulting in multiple levels of resonance splitting—the cause of low-resolution spectra of solids. These dipolar couplings, however, completely cancel out if the molecules in solution tumble fast enough. Proteins of average size (~40 kDa) reorient randomly on the nanosecond timescale, so fast that chemical-shift evolution, which occurs on the millisecond timescale, does not feel the dipolar interactions. Therefore NMR peaks of proteins in solution are sharp and generally have the perfect Lorentzian line-shape. Although dipolar coupling is averaged to zero on the timescale at which NMR peaks are recorded, the physical quantity of dipole–dipole interaction is still present and has a profound effect on the decay of spin coherence (generally described by the transverse relaxation rate R_2). The coherence relaxes faster for larger or slower tumbling molecules, thus setting a theoretical size limit of molecules that can be studied by solution NMR.

It is difficult to predict what exactly is the size limit because new methods are constantly being developed to push the size limit. For example, protein deuteration, or replacing the non-exchangeable aliphatic protons with deuterium, can dramatically slow down the transverse relaxation because deuterium has a sixfold smaller dipole moment than a proton. Furthermore many relaxation-optimized NMR experiments have been developed, such as the ^{1}H–^{15}N TROSY HSQC (3) and the methyl ^{1}H–^{13}C TROSY HMQC pulse schemes (4). The size of proteins amenable to solution NMR has increased from 8 kDa in 1990 to 82 kDa today (Fig. 1), and few in the field 20 years ago could have predicted such rapid progress. In addition to the size limit, resonance complexity is also a serious problem. Typical chemical shift dispersion of a protein is about 6, 30, and 20 ppm for ^1H_N, ^{15}N, and methyl ^{13}C, respectively. Increasing the number of peaks will inevitably result in greater resonance overlap that prohibits unambiguous assignment of protein resonances.

Structure determination by NMR uses an approach that is fundamentally different from that of diffraction methods such as X-ray and EM. Whereas the diffraction methods involve reconstructing images from diffraction data, NMR spectroscopy measures structural constraints (distances or angles) and finds a structural solution that is consistent with those experimental constraints. Therefore, the accuracy and precision of NMR structures depend strongly on the number of unambiguous restraints that can be collected for each residue. In general, the number of NMR restraints measurable for larger or more complex systems is smaller due to greater resonance overlap and lower spectral intensity, and thus precision (or resolution) is lower. The same is true for proteins that yield low-quality NMR spectra. Therefore, for proteins or nucleic acids that readily form well-diffracting crystals, it is usually more efficient to determine the structure by X-ray or electron crystallography.

1.2. Solution NMR in Functional Investigation of Biological Molecules

There are advantages of establishing an NMR system for a macromolecule even if its high-resolution structure has been determined by X-ray or electron crystallography, for further investigation of molecular interactions and dynamics. For example, once sequence-specific assignment of NMR resonances is achieved for a protein, it is very convenient to investigate its interaction with ligands or other proteins. By comparing the chemical shift of the assigned protein resonances with and without the bound ligand, one can quickly map the binding site onto the protein of known structure. This technique is commonly known as the *chemical shift perturbation* assay. Solution NMR can also be used to detect very weak interactions (with $K_D \sim 1$ mM), so weak that protein–protein or protein–ligand complexes cannot be isolated by gel filtration for crystallization trials. In those cases of weak interaction, the binding site can still be mapped using methods such as the *saturation transfer difference* method (5), in which nuclear spin states of the ligand are selectively saturated while observing dipolar cross relaxation that affects the resonances of the protein. Alternatively, one can introduce a paramagnetic label to the ligand and measure paramagnetic broadening of the protein resonances, or vice versa. These techniques are now routinely used for studying protein–protein and protein–ligand interactions (6–8).

Solution NMR is uniquely suited for characterizing the internal dynamics of a protein that is related to its function. Although high-resolution crystal structures also contain temperature factors that can be used to infer dynamics, an aspect of dynamics only visible to NMR is the timescale of protein internal motion. As described above, relaxation of NMR signals depends strongly on protein dynamics in solution. The dynamics are a combination of the overall molecular tumbling and the internal motion of the structural segments, and motions of various frequencies contribute differently to different types of relaxation processes. Longitudinal relaxation rate (R_1) is the rate at which the excited spin-state population returns to Boltzmann equilibrium and is dominated by the fast motions (nanosecond timescale; e.g., the rotational correlation time of a 30 kDa globular protein at 25°C is around 25 ns). The transverse relaxation rate (R_2) is the rate of dephasing of spin coherence and it is dominated by the slow motions. R_2 also has a strong contribution from chemical exchange (R_{ex}). If two conformational or chemical states have different chemical shift, exchange between them on a millisecond to microsecond timescale usually leads to substantial dephasing of spin coherence, thus effectively increasing R_2. Since most protein functional switches occur in this timescale, measurement of R_{ex} has become the most important aspect of dynamics studies by NMR.

One of the best examples of NMR dynamics measurements that led to answering important mechanistic question is the R_1 and R_2 measurement of calmodulin, a ubiquitous calcium (Ca^{2+}) sensor protein in cells. The crystal structure of Ca^{2+}-bound calmodulin shows a dumbbell-like structure in which the N- and C-terminal Ca^{2+}-binding domains are connected by a rigid helix (known as the central helix) (9), but the structure of the Ca^{2+}-calmodulin bound to the peptide from the smooth muscle myosin light-chain kinase (smMLCK) shows that the central helix is a loop, and that the N- and C-domains come together to wrap around the smMLCK helix (10). Dissolving the central helix to accommodate a ligand is an energy costly process and it was not clear why evolution selected such a mechanism. Measurement of ^{15}N R_1 and R_2 of Ca^{2+}-calmodulin showed that the central helix observed in the crystal is largely a flexible loop in solution (11). Therefore, the rigid central helix in the crystal structure likely represents only a small population of the conformer that was stabilized by crystal packing. More recently, NMR dynamics measurements of side-chain methyl groups have been demonstrated even for very large protein machineries such as the 650 kDa 20S core-particle proteasome, which identified dynamic residues inside the antechamber that facilitate the movement of substrates to the sites of proteolysis (12). The conventional R_1/R_2 measurement and analysis can only provide information on the very fast dynamics (nanosecond to picosecond), but many proteins undergo conformational switching on the millisecond to microsecond timescale. Chemical or conformational exchange in this frequency range strongly affects chemical shift evolution of nuclear spins (known as exchange broadening of NMR resonances), and this effect has been utilized by spectroscopists for extracting millisecond to microsecond internal dynamics of interesting protein systems. The exchange contribution to R_2 can be measured using the relaxation-compensated Carr–Purcell–Meiboom–Gill (CPMG) experiment (13, 14). A good example of this application is characterizing the rate of conformational switching of the enzyme cyclophilin A (15, 16). In another interesting study, the CPMG method was used to observe the opening or "unlocking" of the channel gate during proton conduction by the M2 proton channel of influenza virus at low pH (17).

2. NMR Structural Restraints and Structure Determination

2.1. Distance Restraints from ^1H–^1H NOE

Distance restraints derived from the ^1H–^1H nuclear overhauser effect (NOE) are the key restraints used in structural studies by solution NMR. The Overhauser effect was named after the physicist Albert Overhauser, who showed theoretically that spin polarization of electrons could be transferred to the nuclear spins via dipole–dipole cross relaxation through space (18). This effect is the basis of the dynamic nuclear polarization (DNP) technology that aims to increase the sensitivity of NMR signals by passing the strong electron spin magnetic moment to nuclear spins (19–21). In solution NMR, the NOE is generally detected between different ^1H nuclear spins in a molecule. After inverting two spin populations A and B so that they are away from equilibrium, the two spin populations do not relax toward equilibrium (governed by

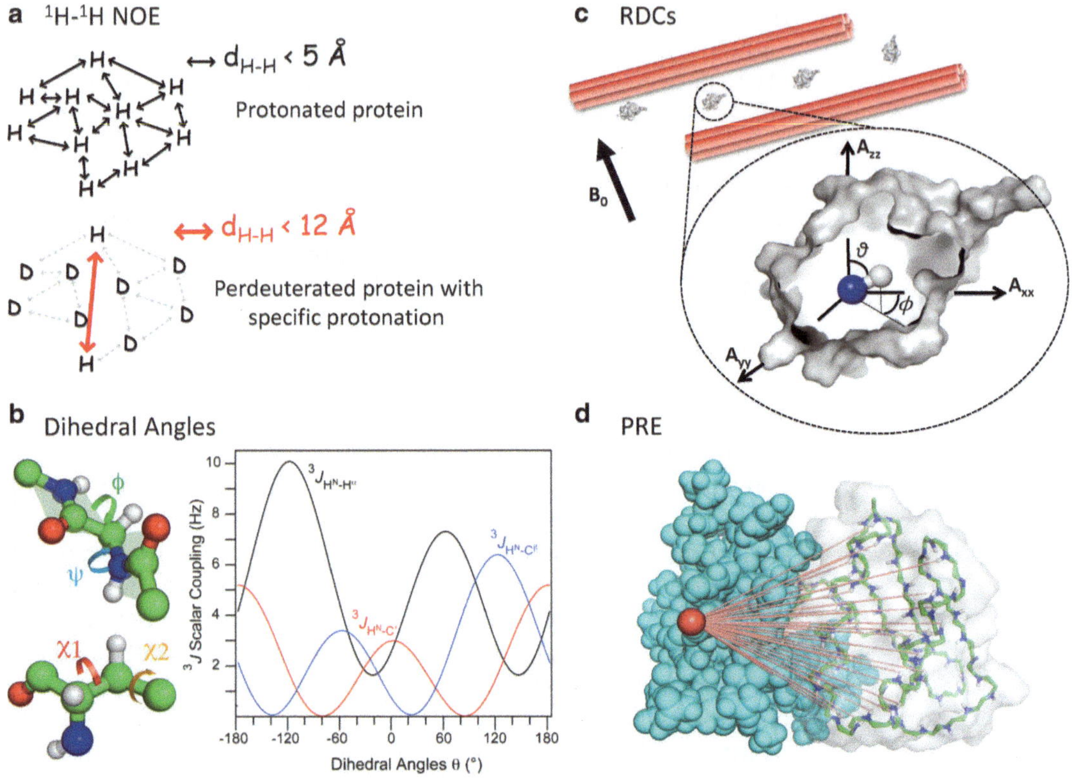

Fig. 2. Schematic illustration of the key structural restraints derived from NMR measurements, including (**a**) inter-proton distances from NOEs, (**b**) backbone and side chain dihedral angles from 3-bond *J* couplings, (**c**) internuclear vector orientation from dipolar couplings, and (**d**) long-range distances from PREs.

R_1) independently. If spins A and B are close in space, their magnetic dipoles interact with each other, giving rise to a cross relaxation term that transfers spin polarization between the two spin populations. This cross relaxation rate is dependent on the interatomic distance (r), and is proportional to $1/r^6$. The amount of NOE transfer depends on both the cross relaxation rate and the time of NOE mixing applied in the NMR experiment. Therefore, similar to the fluorescence resonance energy transfer (FRET), NOE is a very steep function of the distance. Using normal NOE mixing time (~100 ms), it is only detectable when two protons are within 5 Å (Fig. 2a). Using a long mixing time makes the NOE distance measurement less quantitative because other protons nearby can mediate spin diffusion. Longer distance NOE can however be achieved by selectively protonating two nuclear spin probes in an otherwise perdeuterated environment (because deuteration largely removes spin diffusion) (Fig. 2a). This approach has been used to measure NOEs between protons that are separated by as much as 12 Å (22).

NOEs can be measured using a variety of multidimensional experiments. One of the most powerful NOE experiments in protein NMR spectroscopy is the 3D or 4D ^{15}N-edited NOESY used for selectively observing NOEs between backbone amide protons and between amide protons and aliphatic protons (23, 24). Despite the upper bound of ~5 Å for conventional NOE detection, the distance restraints from this experiment are very effective in determining protein secondary structures. For example, in a β-sheet, the inter-strand H_N–H_N and H_N–H_α distances are <4 Å and yield measurable NOEs. Within an extended β strand, there is also very strong NOE between H_N of residue i ($H_N^{\,i}$) and H_α of residue $i-1$ ($H_\alpha^{\,i-1}$) (~2.2 Å). The distances that give rise to characteristic NOEs in an α helix are between $H_N^{\,i}$ and $H_\alpha^{\,i-3}$, and between $H_N^{\,i}$ and $H_\alpha^{\,i-1}$ (both at ~3.5 Å). Assigning tertiary distance restraints, e.g., those between two helices, is much more challenging. The inter-helical distances between backbone H_N and aliphatic protons are significantly longer and therefore it is usually necessary to assign NOEs between amino acid side chains such as the methyl and aromatic groups. For larger proteins, assignment of side chain resonances can be challenging due to higher spectral complexity. Table 1 lists the type of NOE restraints important in various types of protein structures.

2.2. Dihedral Angles from Chemical Shift and Scalar Coupling Constants

Chemical shift, or the frequency at which nuclear spins evolve under local magnetic field, depends on the local electronic and structural environment of the molecule. A complete understanding of how structure determines chemical shifts would in principle allow accurate prediction of macromolecular structures based on chemical shift values alone. In protein NMR spectroscopy, assigning residue-specific chemical shift of backbone ^{15}N, $^{13}C_\alpha$, ^{13}C′, $^{13}C_\beta$, $^{1}H_\alpha$, and $^{1}H_N$ is relatively straightforward, even for proteins as large as 82 kDa (25). Thus the notion of structure determination using chemical shift values is highly attractive and has been intensely pursued (26–29). Unfortunately, this approach is not yet feasible because the chemical shifts of backbone $^{1}H_N$, ^{15}N, and ^{13}C′ are very sensitive to buffer conditions such as pH and ionic strength, and to structural factors such as hydrogen bonding geometry, hydration, and intrinsic dynamics of molecules. However, the chemical shifts of some nuclides, such as $^{13}C_\alpha$, $^{13}C_\beta$, and $^{1}H_\alpha$, are not so sensitive to the above factors, and are largely determined by the local structure. Statistical analysis showed that their chemical shifts departing from the random coil values are strongly correlated to backbone dihedral angles of known secondary structural elements (30, 31). The empirical relation between chemical shift and backbone torsion angle has been implemented in programs such as TALOS/TALOS+ (26, 32) and DANGLE (33), which are widely used today in protein NMR for identifying regions of secondary structures. The procedure for assigning the secondary structure of proteins

Table 1
Typical NOEs observed in structure determination

NOE restraints	Distance (Å)	Types of structure
H_N^i–H_N^j	2.3	β-strand
H_N^i–H_α^j	3.2	β-strand
H_α^i–H_α^j	3.2	β-strand
H_N^i–H_N^{i-1}	2.8/4.3	α-helix/β-strand
H_N^i–H_N^{i-2}	4.2	α-helix
H_N^i–H_α^{i-1}	3.5/2.2	α-helix/β-strand
H_N^i–H_α^{i-2}	4.4	α-helix
H_N^i–H_α^{i-3}	3.4	α-helix
H_N^i–H_α^{i-4}	4.2	α-helix
H_N–$H_{aromatic}$	1.8–6	Long-range
H_N–H_{methyl}	1.8–6	Long-range
H_{methyl}–H_{methyl}	1.8–12	Long-range
H_{methyl}–$H_{aromatic}$	1.8–6	Long-range

based on chemical shift differences with respect to random coil values is generally known as Chemical Shift Index (CSI).

A more quantitative method of obtaining dihedral angles in NMR is measuring the 3-bond scalar coupling constants (J). The J coupling is indirect dipole–dipole interaction between two nuclear spins that is mediated by the bonding electrons. For two atoms separated by three chemical bonds, the J coupling between them (^3J) is a function of the dihedral angle and is given by the Karplus relation $^3J = A\cos^2\varphi + B\cos\varphi + C$, where φ is the dihedral angle and A, B, and C are constants (sometimes referred to as the Karplus parameters) (Fig. 2b). The A, B, and C constants have been characterized for many different types of dihedral angles in proteins and nucleic acids; they are usually determined semi-empirically by fitting experimental ^3J values to φ in high-resolution crystal structures (reviewed in ref. 34).

In the early days of protein NMR spectroscopy, the ^3J between backbone ^1H_N and $^1H_\alpha$ ($^3J_{HNH\alpha}$) and between the backbone $^1H_\alpha$ and the side chain $^1H_\beta$ ($^3J_{H\alpha H\beta}$) were commonly measured for the φ and χ_1 dihedral restraints, respectively (35, 36). Although these coupling constants could be conveniently measured for small proteins with slow R_2, they are difficult to obtain for larger proteins due to the requirement for a long lifetime of spin coherences in the

J-coupling experiments. For larger proteins, the backbone φ and ψ are commonly estimated using the CSI method, whereas side chain χ rotamers are extracted from ^3J between heavy atoms with a favorable relaxation property. For example, the side chain χ_1 rotamers of Val, Ile, and Thr can be extracted from $^3J_{C'C\gamma}$ and $^3J_{NC\gamma}$ coupling constants (37, 38). The χ_2 rotamers of Leu and Ile residues can be extracted from $^3J_{C\alpha C\delta}$ coupling constants (39). These coupling constants can be measured using the protein methyl groups as the NMR readout probes because methyl groups give strong signal and generally have slower R_2 compared to other groups in the protein. For amino acids that do not have methyl groups at γ or δ position, $^3J_{NC\gamma}$ couplings can be measured using deuterated proteins using the backbone amide as the readout nuclide (40, 41).

2.3. Orientation Restraints from Residual Dipolar Couplings

In a macromolecule marginally oriented in a magnetic field, measurable dipolar coupling between a pair of spin 1/2 nuclides encodes orientations of the internuclear vector connecting the two atoms (42). The orientations are defined relative to a common reference frame, known as the principal axes of the molecule's alignment tensor (43) (Fig. 2c). Hence, dipolar couplings provide the powerful global orientation restraints that had been absent in the traditional NOE-based structure determination. In solution NMR, dipolar interactions are essentially cancelled by random rotational diffusion of macromolecules on the nanosecond timescale, though there are a few isolated cases, e.g., paramyoglobin, in which the protein's own susceptibility can give rise to a few Hz of ^1H–^{15}N dipolar couplings (44). Most proteins however do not have large enough magnetic susceptibility anisotropy to be preferentially oriented by the magnetic field, and therefore must be forced to align by either physical interaction with large oriented particles or by fusion with a weakly aligned paramagnetic tag. The selection criteria in choosing an alignment medium are stringent. It must be compatible with most biological molecules in water. It must be highly ordered in the magnetic field and must be large enough so that collisions with proteins do not alter its orientation. Finally, for introducing sizable dipolar couplings for structure determination without introducing too much ^1H–^1H dipolar broadening of NMR resonances, macromolecules typically need to acquire an order parameter of ~10^{-3} (equivalently, a molecule is aligned for 0.1% of time).

About 10 years ago, it was first demonstrated for proteins that such order can be achieved in ~4–5 wt% DMPC/DHPC liquid crystalline medium, and that in the case of ubiquitin, optimal-sized ^1H–^{15}N residual dipolar couplings (RDCs) of about ±20 Hz can be measured (45). Over the past decade, spectroscopists have developed a number of different liquid crystalline media; each of them is suitable for aligning a particular type of proteins. These media include filamentous phage virus particles that form liquid crystals above a certain concentration (46, 47), the ternary mixtures of cetylpyridinium chloride (CPCl) or bromide (CPBr), hexanol, and NaCl or NaBr in water (48, 49), and the binary mixture of alkylpoly(ethylene glycol) and hexanol in water (50). Among them, the rodlike filamentous phage Pf1 is the most widely used because its uniform length and rigid assembly yield stable liquid crystal in concentration as low as 12 mg/ml.

Liquid crystals have delicate phase transitions and usually disintegrate under extreme sample conditions, such as high temperature, low pH, and presence of anionic detergent. A method of weakly orienting biological molecules without the use of liquid crystals is Strain-induced Alignment in polyacrylamide Gel (SAG) in polyacrylamide Gel (SAG) (51, 52). The advantage of polyacrylamide gel is obvious because the chemically cross-linked polymers can withstand very harsh sample conditions. The SAG method has been successfully applied in studies where the above liquid crystals are not applicable. These studies include measuring RDCs for Staphylococcal Nuclease partially denatured in 8 M Urea for studying folding intermediates (53) and measuring RDCs of membrane-associated proteins in the presence of high concentrations of detergent (54, 55). However, a fundamental problem of the SAG method is that it is difficult to soak high concentrations of protein into the gel because the gel pore size has a broad normal distribution. For larger protein–detergent or protein–protein complexes, only a small fraction of the pores can accommodate the macromolecules without obstructing their rotational diffusion. It is therefore more challenging to collect high-quality RDCs using the SAG method. The problem of measuring RDCs for membrane proteins is partially solved owing to the emergence of DNA-based liquid crystals that are resistant to detergents. One liquid crystal is formed with fabricated DNA nanotubes (56). The other medium is the liquid crystal of G-Tetrad DNA (57).

RDCs are extracted by subtracting J couplings acquired with the regular sample from the $J+D$ couplings of the weakly aligned sample. These couplings are typically measured using ^{15}N-, ^{13}C-, and 85% ^{2}H-labeled protein for decreasing ^{15}N and $^{13}C^{\alpha}$ R_2 and for reducing adverse ^{1}H–^{1}H dipolar interactions in the aligned sample. For medium-sized proteins smaller than 50 kDa, it is usually possible to acquire high-quality RDCs for backbone chemical bonds H_N–N, C′–C_α, and C′–N using the HNCO triple-resonance experiments (58–60). Probably the biggest advantage of RDC restraints is that they are completely quantitative and unambiguous. Being quantitative means the one-bond RDC value has a clean mathematical relation to the orientation of the chemical bond. RDC values are unambiguous because their assignment is based on the sequence-specific assignment of backbone resonances, which is usually very reliable in modern protein NMR. Therefore, RDC values can be used as numerical data for structure determination, similar to diffraction in X-ray crystallography. An active question is what is the most effective way to use RDCs in structure determination. Another shortcoming of RDCs is that they do not provide translational information of the corresponding chemical bonds and thus need to be used in combination with distance restraints for de novo structure determination.

2.4. Long-Range Distance from Paramagnetic Relaxation Enhancement Measurements

A good complement to the short-range ^{1}H–^{1}H NOEs is the long-range distance restraint derived from paramagnetic relaxation enhancement (PRE) measurements (61). Paramagnetic centers typically consist of one or more unpaired electrons, which have a very strong magnetic dipole moment. The interaction between the strong electron dipole and other nuclear spin dipoles in the molecule can strongly affect the relaxation rates of the nuclear spin populations. Similar to the rate of spin polarization transfer via NOE, the PRE of nuclear spin is proportional to $1/r^6$, where r is the distance between the nuclear spin and the paramagnetic center (Fig. 2d). The most commonly used paramagnetic labels in structural studies of proteins and nucleic acids are nitroxide compounds that have one unpaired electron. For proteins, the nitroxide compound often used is *S*-(2,2,5,5-tetramethyl-2,5-dihydro-1H-pyrrol-3-yl)methyl methanesulfonothioate (MTSL), which can be attached at cysteine positions in the protein by reacting with the thiol group of cysteine. The protein should only have one cysteine at a time to ensure that PRE-derived restraints are unambiguous. Similar nitroxide compounds have been used for spin labeling of DNA and RNA. For example, the MTSL-like compound 3-Iodomethyl-1-oxyl-2,2,5,5-tetramethylpyrroline can be attached to a phosphorothioate that is substituted at a specific backbone position during chemical synthesis of DNA or RNA (62).

The strong electron magnetic dipole moment (γ_e ~ 660 times that of ^{1}H) can broaden NMR line-width of nuclides much farther away (up to ~25 Å), but can also wipe out resonances of nuclides nearby (<12 Å). Therefore, the range of PRE-derived distances that can be quantified is 12–25 Å. Probably the biggest advantage of PRE over NOE, if the sequence-specific resonance assignments are known, is that the measurement and analysis are both simple and unambiguous. PRE measurement typically involves measuring (using either 2D or 3D experiments) ^{1}H $R_2 + R_2^{para}$ for ^{1}H resonances that are broadened by a particular spin label and ^{1}H R_2 after reducing the unpaired electron with ascorbic acid. R_2^{para} is then used to derive distance restraints based on the known calibrations (61). For resonances that are completely wiped out by the spin label, the distance is set to <12 Å. Therefore PRE restraints are unambiguous if the sequence-specific resonance assignments are unambiguous. The weakness of PRE is also glaring. The nitroxide spin label is ~8 Å long and is flexible like the arginine side chain, and thus PRE restraints have very large uncertainty, typically ±5 Å.

It is therefore not useful for determining local or secondary structures. Moreover, introducing an unnatural spin label could affect protein structure or function. Nonetheless, a large number of unambiguous PRE restraints can compensate for their low precision. With existing knowledge of the local structures from NOE and RDCs, PRE restraints are useful in providing a low-resolution global fold that can facilitate further NOE assignments. Another application of PRE mentioned above is mapping protein–protein or protein–ligand interactions that are too weak or complex to be measured by intermolecular NOEs. If the structures of two interacting proteins were known, it would be easy to identify positions in the proteins for attaching spin labels.

2.5. Structure Calculation

The objective of structure calculation in NMR is to find structural solutions that are consistent with all experimental restraints while not violating the standard covalent geometry of protein or nucleic acid polymers. All NMR restraints have experimental uncertainties and are included in structure calculation as ranges of allowed values (e.g., lower and upper bounds). Some NMR restraints are even ambiguous. It is thus important to calculate an ensemble of structures to account for the errors and ambiguity in the restraints. A number of methods have been developed for this purpose. In the early days of protein NMR, when computers were not so powerful, algebraic methods such as Distance Geometry (DG) were used. The first step of the DG method is extrapolating, using geometric inequality limits, a complete set of lower and upper limits on all the interatomic distances from the sparse set of experimental distance restraints. The next step is to choose a random distance matrix from within the complete set of experimental and extrapolated distances, and fit a set of atomic coordinates to it using the EMBED algorithm. By repeating this procedure with different random distance matrices, one obtains an ensemble of conformations that is consistent with experimentally derived restraints. The first de novo structure determination by NMR was achieved using the DG method (63). The advantage of the DG method is that it is rather deterministic. For example, in the EMBED algorithm, coordinates that are a best-fit to the distance matrix are calculated by eigenvalue methods, completely avoiding the local minima problem. Probably the biggest weakness of the DG method is that the approach relies almost completely on distance restraints and thus is difficult to be implemented to include other NMR data such as dipolar couplings, which encode bond orientation information, and chemical shifts, which encode dihedral angle information.

A more generally used structure calculation method is the restrained Molecular Dynamics (rMD) method (64, 65). Although this method is more computationally intensive, it can be used to minimize variable target functions of any type and is thus readily applicable for refining structures against different types of

NMR-derived restraints. The Molecular Dynamics (MD) part of the method involves numerically solving Newton's equations of motion for a many particle system in which the total potential energy is the sum of physical potentials (such as those for chemical bonds, angles, van der Waals, etc.) and pseudo potentials from experimental restraints (such as those for interatomic distances, dihedral angles, bond orientations, etc.). For macromolecules, the refinement energy landscape becomes very complicated, posing serious local minima problems. Therefore, the rMD calculation is done in combination with simulated annealing (heating the system followed by gradual cooling), an effective procedure for "jumping" out of false local minima (66, 67).

3. Solution NMR Studies of Membrane Proteins

Structure determination of membrane proteins by solution NMR is still in the exploratory phase. In principle, the NMR protocols that have been established for water-soluble proteins should be directly applicable to membrane proteins. These protocols however need to be tailored to account for the fundamental physical chemical differences between membrane proteins and water-soluble proteins and the imperative for use of a model membrane media. There are a number of issues that complicate solution NMR studies of membrane proteins. (1) Membrane proteins need to be solubilized in detergent micelles or detergent/lipid bicelles. We do not understand exactly how various types of detergents assemble a micelle around membrane proteins and thus cannot predict the effective size of a protein–micelle complex. Furthermore, the presence of very high concentrations of detergent requires methods to suppress NMR signals arising from the detergent. (2) Amino acid sequences of membrane proteins have been optimized in nature in the membrane environment, and thus it is unclear whether detergent micelles fully mimic the lateral lipid pressure that a protein experiences in a true membrane. For α-helical membrane proteins, insufficient lipid pressure in detergent micelles could result in weaker helix–helix packing or increased internal "breathing" in solution. The internal dynamics would pose a problem for measuring long-range NOEs. (3) In general, membrane proteins contain many more methyl-bearing amino acids than water-soluble proteins. Moreover, most of the hydrophobic residues are exposed to the dynamic detergent, unconfined to a unique chemical environment. These properties of membrane proteins result in a much smaller chemical shift dispersion of the methyl groups as compared to water-soluble proteins (in which hydrophobic residues are strongly packed in the protein core). The poor chemical shift dispersion poses a big

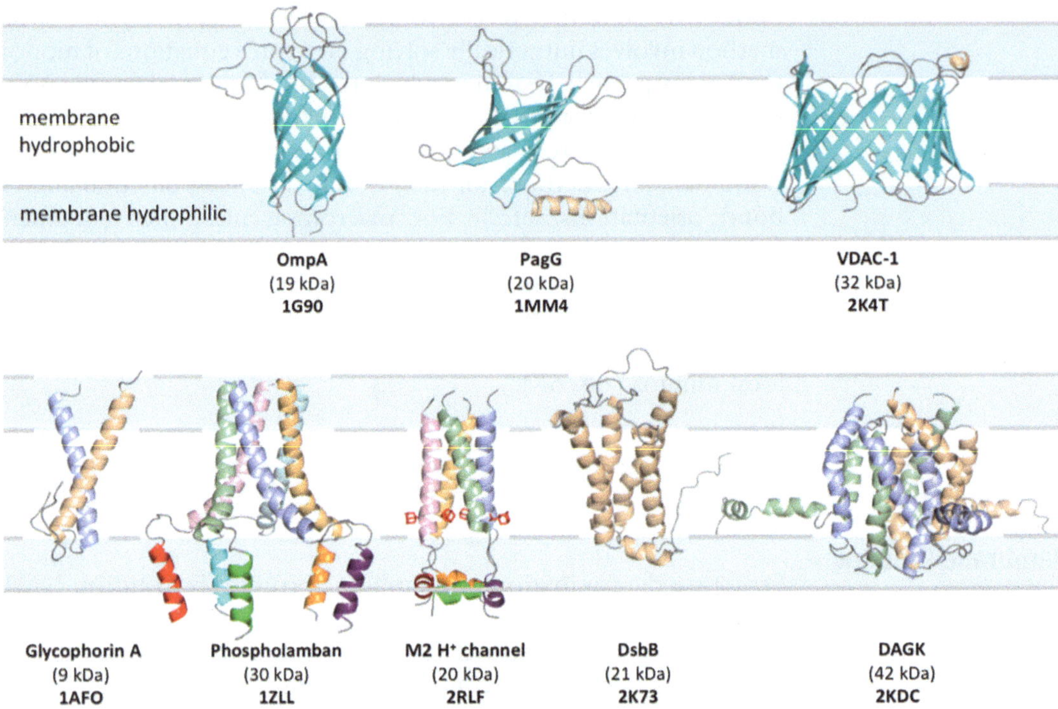

Fig. 3. Representative membrane protein structures determined using solution NMR spectroscopy. Below each ribbon structure is the protein name, molecular weight, and PDB code. On a scale relative to protein, the hydrophobic and hydrophilic regions of the membrane are ~30 Å and ~15 Å thick.

problem in assigning inter-helical NOEs, which typically requires unambiguous assignments of the methyl resonances.

Despite the above issues, individual laboratories have managed to solve new membrane protein structures by solution NMR in the past two decades, although these projects typically took longer than 4 years to complete. Figure 3 shows a number of membrane protein structures determined by solution NMR that had a high impact on the biological community, providing a rough assessment of the current capability of the technique. Although the number of structures solved is small, they cover a rather large range of membrane protein fold space, including oligomeric helical, polytopic helical, and β-barrel structures, with a substantial fraction of them revealing new structural features. The structures in Fig. 3 also span a wide range of sizes from 6 to 42 kDa. Therefore, the state-of-the-art technology in solution NMR is already capable of generating high-resolution structures of small- to medium-sized membrane proteins. The challenge in the field is to increase the speed and accuracy of structure determination.

There is a good opportunity for solution NMR to immediately contribute to the overall structural database for membrane proteins. Genomic analysis of the distribution of number of transmembrane helices (TMHs) in membrane protein families estimated that about 2/3 of the proteins span the membrane less than seven times (68). Moreover, the number of crystal structures of membrane proteins with less than 7 TMHs is roughly equal to those with greater than 7 TMHs, implying that crystallization of large and small proteins is equally challenging. The structures shown in Fig. 3 imply that it is possible to establish a robust solution NMR protocol for studying membrane proteins of sizes up to 50 kDa, which can cover the region of structural space up to 7 TMHs (assuming that the proteins are monomeric). Another motivation for establishing an NMR system for a membrane protein even if its crystal structure is available is that the system can be employed in studies to address function and mechanism (e.g., ligand binding of receptors or dynamics of ion transport).

3.1. Major Technical Challenges to Overcome

Probably the biggest challenge in solution NMR of membrane proteins is finding a solubilization condition that both supports the native fold of the protein and yields workable NMR spectra. Because integral membrane proteins are hydrophobic in nature, it is usually difficult to keep the proteins mono-dispersed in solution at concentrations higher than 0.5 mM—roughly the minimum concentration requirement for a full-scale NMR structure determination. Establishing an NMR system for membrane protein thus involves extensive screening of various detergents, lipids, and buffer conditions. Detergent selection is usually not based on rationale, although empirically detergents with phospho-head groups work better than those with sugar-head groups. A rather simple view of the problem of membrane protein solubilization is to find a detergent or detergent/lipid system that can form a tight micelle around the protein to prevent aggregation while still allowing the protein to tumble fast enough to yield good NMR spectrum. In general, detergents with a lower critical micelle concentration (c.m.c) are more stable. In some cases, doping the micelles with natural lipids could increase the overall stability of the protein–micelle complex. Introducing an anionic detergent to the micelles may reduce protein aggregation that is driven mainly by hydrophobic interactions. For example, LMPG (14:0 lyso phosphoglycerol) has become a popular detergent for solution NMR studies of membrane proteins, probably because it has a very low c.m.c and is negatively charged. It is obvious that there will not be a universal detergent/lipid system that will solve all the problems. As we experiment more with various combinations of detergents and lipids, we believe that there will soon be an optimized set of detergent/lipid systems for solution NMR studies of membrane proteins.

Another challenge widely recognized by the NMR community is finding sufficiently long-range NOEs for helical membrane proteins. The amino acid distribution of membrane proteins is very different from that of water-soluble proteins. Large hydrophobic residues show no preference for the protein core, and in β-barrels, the preference is opposite to that of water-soluble proteins (69). Therefore, the number of close methyl–methyl and methyl–aromatic contacts (<5 Å) is much smaller than in water-soluble proteins. Instead, residues like alanine, glycines, prolines, serines, and threonines are often found in the helix–helix interface and the protein core. It would thus be important to develop new isotope labeling strategies for assigning NOEs based on these small amino acids. The difficulty in finding long-range NOEs could also be due to dynamics. As mentioned above, membrane proteins in detergent micelles may adopt increased internal "breathing" because detergent micelles do not fully exert the lateral lipid pressure that a protein experiences in true membrane. Since the NOE is proportional to $1/r^6$, internal motion of TMHs relative to each other can substantially reduce inter-helical NOEs while not having a significant effect on intra-helical NOEs. One way to compensate for the lack of tertiary NOEs is measuring PRE restraints and this approach has been successfully used in the structure determination of DAGK (70) and OmpA (71).

Finally, structural investigation of membrane proteins in lipid bilayers has been a long-sought goal of structural biologists. Solid-state NMR (ssNMR) spectroscopy has been the much-anticipated technique for reaching this goal. While the technology has been progressing rapidly in the past decade, its capability is still far from de novo structure determination at high resolution—there is not yet a single example of de novo structure determination of a membrane protein at high resolution by ssNMR. The biggest problem of ssNMR is still poor spectral quality. Unlike in solution, the conformational inhomogeneity in the solid phase is not averaged out on the NMR timescale, giving rise to inhomogeneous peaks and low spectral resolution. Is solution NMR study of membrane proteins in a lipid bilayer possible? One approach is to reconstitute the protein in bicelles, which is a disc of lipid bilayer surrounded by a ring of detergents. The ratio of lipid to detergent (q) determines the size of the bicelles (72). Although at large q (>0.8) the lipid region of the bicelles is an excellent representation of the lipid bilayer, at smaller q (<0.5), the lipid bilayer is really a mixture of lipid and detergent. An intriguing alternative is the use of nanodiscs, which are self-assembled patches of lipid bilayer surrounded by a ring of amphipathic membrane scaffold protein such as apolipoprotein (73). The advantage of the nanodisc system is that there is no detergent involved. The large size of nanodiscs (~150 kDa) has, in the past, discouraged NMR spectroscopists. This system, however, has recently been revisited and showed promising NMR spectra for medium-sized membrane proteins (74). This result, though still preliminary, shows the potential of solution NMR to at least study membrane protein interactions in a true membrane environment.

4. The Future of Solution NMR

4.1. Isotope Labeling

The history of progress in biomolecular NMR tells us that isotope labeling and NMR techniques must go hand in hand in order to push the envelope of the technology farther. In the last two decades, two isotope-labeling schemes have fundamentally changed modern protein NMR spectroscopy. One is deuteration as described above in Subheading 1, which dramatically reduces R_2 by reducing dipolar interactions from protons. The other is selective methyl group protonation in deuterated protein, which enables high-resolution methyl spectroscopy in large proteins (75, 76). Growing *E. coli* in a deuterated medium can achieve protein deuteration (77). Selective protonation of leucine, valine, and isoleucine methyl sites can be done by adding selectively protonated a-keto acid precursors to a perdeuterated medium (78). This labeling scheme allowed full resonance assignment of a 723-residue single polypeptide protein, malate synthase G (79). Recently, stereospecific methyl labeling of valines and leucines has also been achieved (80). In addition to the branched methyl-baring amino acids, new protocols for labeling methyl groups have been developed for other amino acids, including methionine (81, 82) and alanine (83, 84). Moreover, the carbons of the protein backbone can be selectively labeled. For example, the metabolic precursors 1,3-$^{13}C_2$-glycerol and 2-$^{13}C_1$-glycerol can be used to produce protein that is selectively isotopic enriched at C′, C_α, H_α, and N positions, while the C_β position is selectively ^{13}C depleted and the adjacent protons are deuterated. This labeling scheme allows recording of a spectrum with very high resolution in the $^{13}C_\alpha$ dimension by removing the $^{13}C_\alpha$–$^{13}C_\beta$ coupling (85, 86). Owing to the complete characterization of the major metabolic pathways of amino acid synthesis in *E. coli*, it is possible to introduce various labeling strategies by adding isotope-labeled precursors and by manipulating components of the biosynthetic pathways. We anticipate many more new labeling schemes to be introduced in the near future that would fundamentally overcome some of the limitations in NMR spectroscopy of large proteins.

Another continuing development that would provide more labeling options is the cell-free expression of proteins. The open nature of the translation reactions allows the addition of many different compounds, such as protease and RNAse inhibitors, ligands, or chaperones, directly into the reaction. Owing to the lack of metabolic scrambling, amino acid type-specific labeling is possible in almost any combination (87). These methods are still very costly at present, but are expected to be increasingly affordable.

The above-described labeling schemes all selectively but uniformly label certain types of chemical groups in a polypeptide chain. There is great interest however in methods that selectively label only one or more segments of a protein. The basic approach involves splitting a polypeptide chain, expressing and labeling it separately, and splicing. Two intein-based approaches, Expressed Protein Ligation (EPL) and Protein Trans-Splicing (PTS), have been employed to produce segmental labeled protein (88). EPL is based on a reaction involving two protein fragments containing an α-thioester at the C-terminus of the first fragment and an α-cysteine at the N-terminus of the second fragment; a cysteine is required at the ligation site. EPL has been frequently used for segmental isotope labeling of proteins (89–92). In the PTS method, a particular intein is split into two fragments, which have no activity on their own. After mixing them together in solution, they become active and can perform a splicing reaction that results in the fused protein (93, 94).

4.2. NMR Instrument

As in the case of X-ray and electron crystallography, advance in instrumentation has been one of the main driving forces in enhancing the capability of solution NMR spectroscopy. High field instruments such as the 800 and 900 MHz NMR spectrometers are now common around the world, while construction of commercial 1.3 GHz spectrometers is under way. It is important to note however that higher field magnet does not necessarily yield better NMR data. In general, spectra recorded at higher magnetic field have better sensitivity and resolution if the protein is rigid. But this is often not the case because many interesting proteins currently being studied undergo conformational exchanges in solution, in particular those having multiple physiological states. For these molecular systems, high field magnets could further amplify resonance broadening due to chemical shift exchange. What has fundamentally boosted NMR capability in the past decade is the use of cryogenic probes. These "cold" probes, in which the receiver coils are kept at ~30 K, can provide up to fourfold gains in the signal-to-noise ratio (S/N) under low ionic strength condition. The "cold" probe technology fundamentally changed the protein concentration requirement for NMR structure determination from > 1 mM to 0.5 mM. Since a large portion of the probe electronics in the current generation of the cryoprobe is still at room temperature, we are hopeful that another two- to threefold gains in S/N may be achieved in the near future. Such improvement would dramatically increase the applicability of solution NMR in studying membrane proteins because many membrane proteins can be made soluble to about 0.2 mM.

4.3. Structure Calculation Program

The powerful combination of rMD and SA has been the dominant tool for generating new NMR structures in the past two decades. Despite its great success, we should not neglect the fact that the MD/SA protocol is still subject to the local minima problem for complex systems and thus potentially structure calculation artifacts. For example, the rmsd of an NMR ensemble can be affected by various parameters of the MD/SA procedure such as temperature, size of annealing steps, mass of atoms, and even mathematical properties of the target functions. As the computer processor speed is increasing rapidly, it is time for the NMR community to envision a new generation of structure calculation tools that exhaustively search for conformations that are consistent with experimental and/or knowledge-based restraints. The completeness of the search would ensure that structure calculation does not miss any native conformations, and assess the structural precision from data alone (complete data-driven structure determination). A comparison between a complete search method and MD/SA has been made for the homo-pentameric phospholamban protein, showing that rmsd of the structural ensemble obtained from a complete grid search is substantially larger (95). The complete search method is, however, not yet practical due to the unrealistically long computation time required.

An alternate, more feasible approach is to exhaustively searching the protein database for structural fragments that best fit experimental data, commonly known as the Molecular Fragment Replacement (MFR) method. The MFR method was first used in crystallography for building molecular fragments into low-resolution crystallographic density (96), and was subsequently applied in NMR to fragments that agree with RDCs (97) and other NMR data such as chemical shifts (98, 99). Although in principle the MFR approach limits conformational space to what has already been observed, in practice this approach demonstrated to be very effective because (1) the Protein Data Bank is so large and diverse that its content is an excellent representation of the conformational space of proteins and nucleic acids in nature and (2) the method greatly reduces the search space such that it is computationally affordable. Furthermore, the database of crystal structures is expanding at a rapid pace and thus the MFR approach would only become more powerful with time. Our expanding knowledge of structure will increasingly facilitate de novo NMR structure determination. The same is true for cryo-EM, where computer algorithms that effectively utilize the knowledge from the existing structure database have been largely responsible for dramatically increasing the resolution of cryo-EM structures (100). As solution NMR gradually collects structural restraints for larger proteins, there could be new opportunities for combining the RDC/chemical shift MFR method with low-resolution EM density, that is, to fit NMR-derived fragments into EM density. The EM density, although at low resolution, solves a big problem for NMR—determining the global fold or shape of a molecule by NOE restraints.

References

1. Bloch F, Hansen WW, Packard M (1946) The nuclear induction experiment. Phys Rev 70:474–485
2. Purcell EM, Torrey HC, Pound RV (1946) Resonance absorption by nuclear magnetic moments in a solid. Phys Rev 69:37–38
3. Pervushin K, Riek R, Wider G, Wuthrich K (1997) Attenuated T2 relaxation by mutual cancellation of dipole-dipole coupling and chemical shift anisotropy indicates an avenue to NMR structures of very large biological macromolecules in solution. Proc Natl Acad Sci USA 94(23):12366–12371
4. Tugarinov V, Hwang PM, Ollerenshaw JE, Kay LE (2003) Cross-correlated relaxation enhanced ^{1}H-^{13}C NMR spectroscopy of methyl groups in very high molecular weight proteins and protein complexes. J Am Chem Soc 125:10420–10428
5. Meyer B, Peters T (2003) NMR Spectroscopy techniques for screening and identifying ligand binding to protein receptors. Angew Chem Int Ed Engl 42(8):864–890
6. Battiste JL, Pestova TV, Hellen CU, Wagner G (2000) The eIF1A solution structure reveals a large RNA-binding surface important for scanning function. Mol Cell 5(1): 109–119
7. Gross JD et al (2003) Ribosome loading onto the mRNA cap is driven by conformational coupling between eIF4G and eIF4E. Cell 115(6):739–750
8. Marintchev A, Frueh D, Wagner G (2007) NMR methods for studying protein-protein interactions involved in translation initiation. Methods Enzymol 430:283–331
9. Babu YS, Bugg CE, Cook WJ (1988) Structure of calmodulin refined at 2.2 A resolution. J Mol Biol 204(1):191–204
10. Ikura M et al (1992) Solution structure of a calmodulin-target peptide complex by multidimensional NMR. Science 256(5057):632–638
11. Barbato G, Ikura M, Kay LE, Pastor RW, Bax A (1992) Backbone dynamics of calmodulin studied by 15N relaxation using inverse detected two-dimensional NMR spectroscopy: the central helix is flexible. Biochemistry 31(23):5269–5278
12. Sprangers R, Kay LE (2007) Quantitative dynamics and binding studies of the 20S proteasome by NMR. Nature 445:618–622
13. Loria JP, Rance M, Palmer AG 3rd (1999) A relaxation-compensated carr-purcell-meiboom-gill sequence for characterizing chemical exchange by NMR spectroscopy. J Am Chem Soc 121(10):2331–2332
14. Palmer AG 3rd, Kroenke CD, Loria JP (2001) Nuclear magnetic resonance methods for quantifying microsecond-to-millisecond motions in biological macromolecules. Methods Enzymol 339:204–238
15. Eisenmesser EZ, Bosco DA, Akke M, Kern D (2002) Enzyme dynamics during catalysis. Science 295(5559):1520–1523
16. Eisenmesser EZ et al (2005) Intrinsic dynamics of an enzyme underlies catalysis. Nature 438(7064):117–121
17. Schnell JR, Chou JJ (2008) Structure and mechanism of the M2 proton channel of influenza A virus. Nature 451(7178):591–595
18. Overhauser AW (1953) Polarization of nuclei in metals. Phys Rev 91(2):476
19. Carver TR, Slitcher CP (1953) Polarization of nuclear spins in metal. Phys Rev 92:212
20. Becerra LR, Gerfen GJ, Temkin RJ, Singel DJ, Griffin RG (1993) Dynamic nuclear polarization with a cyclotron resonance maser at 5 T. Phys Rev Lett 71(21):3561–3564
21. Hall DA et al (1997) Polarization-enhanced NMR spectroscopy of biomolecules in frozen solution. Science 276(5314):930–932
22. Sounier R, Blanchard L, Wu Z, Boisbouvier J (2007) High-accuracy distance measurement between remote methyls in specifically protonated proteins. J Am Chem Soc 129(3): 472–473
23. Zuiderweg ER, Fesik SW (1989) Heteronuclear three-dimensional NMR spectroscopy of the inflammatory protein C5a. Biochemistry 28(6):2387–2391
24. Kay LE, Clore GM, Bax A, Gronenborn AM (1990) 4-Dimensional heteronuclear triple-resonance NMR-spectroscopy of interleukin-1-beta in solution. Science 249(4967):411–414
25. Tugarinov V, Choy WY, Orekhov VY, Kay LE (2005) Solution NMR-derived global folded of a monomeric 82-kDA enzyme. Proc Natl Acad Sci USA 102(3):622–627
26. Cornilescu G, Delaglio F, Bax A (1999) Protein backbone angle restraints from searching a database for chemical shift and sequence homology. J Biomol NMR 13(3): 289–302
27. Bowers PM, Strauss CE, Baker D (2000) De novo protein structure determination using sparse NMR data. J Biomol NMR 18(4): 311–318
28. Cavalli A, Salvatella X, Dobson CM, Vendruscolo M (2007) Protein structure determination from NMR chemical shifts. Proc Natl Acad Sci USA 104(23):9615–9620

29. Gong H, Shen Y, Rose GD (2007) Building native protein conformation from NMR backbone chemical shifts using Monte Carlo fragment assembly. Protein Sci 16(8):1515–1521
30. Spera S, Bax A (1991) An empirical correlation between protein backbone conformation and Ca and Cb chemical shifts. J Am Chem Soc 113:5490–5492
31. Wishart DS, Sykes BD (1994) The 13C chemical-shift index: a simple method for the identification of protein secondary structure using 13C chemical-shift data. J Biomol NMR 4(2):171–180
32. Shen Y, Delaglio F, Cornilescu G, Bax A (2009) TALOS+: a hybrid method for predicting protein backbone torsion angles from NMR chemical shifts. J Biomol NMR 44(4):213–223
33. Cheung MS, Maguire ML, Stevens TJ, Broadhurst RW (2010) DANGLE: a Bayesian inferential method for predicting protein backbone dihedral angles and secondary structure. J Magn Reson 202(2):223–233
34. Bax A et al (1994) Measurement of homo- and heteronuclear J couplings from quantitative J correlation. Methods Enzymol 239: 79–105
35. Deber CM, Torchia DA, Blout ER (1971) Cyclic peptides. I. Cyclo(tri-l-prolyl) and derivatives. Synthesis and molecular conformation from nuclear magnetic resonance. J Am Chem Soc 93(19):4893–4897
36. Demarco A, Llinas M, Wuthrich K (1978) Analysis of H-1-NMR Spectra of ferrichrome peptides. 2. Amide resonances. Biopolymers 17(3):637–650
37. Vuister GW, Wang AC, Bax A (1993) Measurement of three-bond nitrogen-carbon J couplings in proteins uniformly enriched in nitrogen-15 and carbon-13. J Am Chem Soc 115(12):5334–5335
38. Grzesiek S, Vuister GW, Bax A (1993) A simple and sensitive experiment for measurement of Jcc couplings between backbone carbonyl and methyl carbons in isotopically enriched proteins. J Biomol NMR 3(4):487–493
39. MacKenzie KR, Prestegard JH, Engelman DM (1996) Leucine side-chain rotamers in a glycophorin A transmembrane peptide as revealed by three-bond carbon-carbon couplings and 13C chemical shifts. J Biomol NMR 7(3):256–260
40. Hu J-S, Grzesiek S, Bax A (1997) Chi1 angle information from a simple two-dimensional NMR experiment which identifies trans 3JNCg couplings in isotopically enriched proteins. J Biomol NMR 9:323–328
41. Hu JS, Bax A (1997) Determination of phi and chi(1) angles in proteins from C-13-C-13 three-bond J couplings measured by three-dimensional heteronuclear NMR. How planar is the peptide bond? J Am Chem Soc 119(27):6360–6368
42. Abragham A (1961) The principles of nuclear magnetism. Clarendon Press, Oxford, England
43. Saupe A, Englert G (1963) High-resolution nuclear magnetic resonance spectra of orientated molecules. Phys Rev Lett 11:462–464
44. Tolman JR, Flanagan JM, Kennedy MA, Prestegard JH (1995) Nuclear magnetic dipole interactions in field-oriented proteins – information for structure determination in solution. Proc Natl Acad Sci USA 92(20):9279–9283
45. Tjandra N, Bax A (1997) Direct measurement of distances and angles in biomolecules by NMR in a dilute liquid crystalline medium. Science 278(5340):1111–1114
46. Hansen MR, Mueller L, Pardi A (1998) Tunable alignment of macromolecules by filamentous phage yields dipolar coupling interactions. Nat Struct Biol 5(12):1065–1074
47. Clore GM, Starich MR, Gronenborn AM (1998) Measurement of residual dipolar couplings of macromolecules aligned in the nematic phase of a colloidal suspension of rod-shaped viruses. J Am Chem Soc 120(40):10571–10572
48. Prosser RS, Losonczi JA, Shiyanovskaya IV (1998) Use of a novel aqueous liquid crystalline medium for high-resolution NMR of macromolecules in solution. J Am Chem Soc 120(42):11010–11011
49. Barrientos LG, Dolan C, Gronenborn AM (2000) Characterization of surfactant liquid crystal phases suitable for molecular alignment and measurement of dipolar couplings. J Biomol NMR 16(4):329–337
50. Ruckert M, Otting G (2000) Alignment of biological macromolecules in novel nonionic liquid crystalline media for NMR experiments. J Am Chem Soc 122(32):7793–7797
51. Tycko R, Blanco FJ, Ishii Y (2000) Alignment of biopolymers in strained gels: a new way to create detectable dipole-dipole couplings in high-resolution biomolecular NMR. J Am Chem Soc 122(38):9340–9341
52. Sass HJ, Musco G, Stahl SJ, Wingfield PT, Grzesiek S (2000) Solution NMR of proteins within polyacrylamide gels: diffusional properties and residual alignment by mechanical stress or embedding of oriented purple membranes. J Biomol NMR 18(4):303–309
53. Shortle D, Ackerman MS (2001) Persistence of native-like topology in a denatured protein in 8 M urea. Science 293(5529):487–489

54. Chou JJ, Gaemers S, Howder B, Louis JM, Bax A (2001) A simple apparatus for generating stretched polyacrylamide gels, yielding uniform alignment of proteins and detergent micelles. J Biomol NMR 21(4):377–382
55. Chou JJ, Kaufman JD, Stahl SJ, Wingfield PT, Bax A (2002) Micelle-induced curvature in a water-insoluble HIV-1 Env peptide revealed by NMR dipolar coupling measurement in stretched polyacrylamide gel. J Am Chem Soc 124(11):2450–2451
56. Douglas SM, Chou JJ, Shih WM (2007) DNA-nanotube-induced alignment of membrane proteins for NMR structure determination. Proc Natl Acad Sci USA 104(16):6644–6648
57. Lorieau J, Yao L, Bax A (2008) Liquid crystalline phase of G-tetrad DNA for NMR study of detergent-solubilized proteins. J Am Chem Soc 130:7536–7537
58. Kontaxis G, Clore G, Bax A (2000) Evaluation of cross-correlation effects and measurement of one-bond couplings in proteins with short transverse relaxation times. J Magn Reson 143(1):184–196
59. Chou JJ, Delaglio F, Bax A (2000) Measurement of one-bond 15N-13C′ dipolar couplings in medium sized proteins. J Biomol NMR 18(2):101–105
60. Jaroniec CP, Ulmer TS, Bax A (2004) Quantitative J correlation methods for the accurate measurement of 13C′-13Calpha dipolar couplings in proteins. J Biomol NMR 30(2):181–194
61. Battiste JL, Wagner G (2000) Utilization of site-directed spin labeling and high-resolution heteronuclear nuclear magnetic resonance for global fold determination of large proteins with limited nuclear overhauser effect data. Biochemistry 39(18):5355–5365
62. Qin PZ et al (2007) Measuring nanometer distances in nucleic acids using a sequence-independent nitroxide probe. Nat Protoc 2(10):2354–2365
63. Williamson MP, Havel TF, Wüthrich K (1985) Solution conformation of proteinase inhibitor IIA from bull seminal plasma by ^{1}H nuclear magnetic resonance and distance geometry. J Mol Biol 182:295–315
64. Kaptein R, Zuiderweg ERP, Scheek RM, Boelens R, Vangunsteren WF (1985) A protein-structure from nuclear magnetic-resonance data – LAC repressor headpiece. J Mol Biol 182(1):179–182
65. Clore GM et al (1986) The 3-dimensional structure of alpha-1-purothionin in solution – combined use of nuclear-magnetic-resonance, distance geometry and restrained molecular-dynamics. EMBO J 5(10): 2729–2735
66. Clore GM, Brunger AT, Karplus M, Gronenborn AM (1986) Application of molecular-dynamics with interproton distance restraints to 3-dimensional protein-structure determination – a model study of crambin. J Mol Biol 191(3):523–551
67. Nilges M, Clore GM, Gronenborn AM (1988) Determination of 3-dimensional structures of proteins from interproton distance data by hybrid distance geometry-dynamical simulated annealing calculations. FEBS Lett 229(2):317–324
68. Liu Y, Engelman DM, Gerstein M (2002) Genomic analysis of membrane protein families: abundance and conserved motifs. Genome Biol 3(10):research0054
69. Ulmschneider MB, Sansom MS (2001) Amino acid distributions in integral membrane protein structures. Biochim Biophys Acta 1512(1):1–14
70. Van Horn WD et al (2009) Solution nuclear magnetic resonance structure of membrane-integral diacylglycerol kinase. Science 324(5935):1726–1729
71. Liang B, Bushweller JH, Tamm LK (2006) Site-directed parallel spin-labeling and paramagnetic relaxation enhancement in structure determination of membrane proteins by solution NMR spectroscopy. J Am Chem Soc 128(13):4389–4397
72. Vold RR, Prosser RS (1996) Magnetically oriented phospholipid bilayered micelles for structural studies of polypeptides. Does the ideal bicelle exist? J Magn Reson B 113(3):267–271
73. Bayburt TH, Sligar SG (2003) Self-assembly of single integral membrane proteins into soluble nanoscale phospholipid bilayers. Protein Sci 12(11):2476–2481
74. Raschle T et al (2009) Structural and functional characterization of the integral membrane protein VDAC-1 in lipid bilayer nanodiscs. J Am Chem Soc 131(49):17777–17779
75. Gardner KH, Kay LE (1998) The use of ^{2}H, ^{13}C, ^{15}N multidimensional NMR to study the structure and dynamics of proteins. Annu Rev Biophys Biomol Struct 27:357–406
76. Tugarinov V, Kay LE (2004) An isotope labeling strategy for methyl TROSY spectroscopy. J Biomol NMR 28:165–172
77. LeMaster DM, Richards FM (1988) NMR sequetial assignment of *E. coli* thioredoxin utilizing random fractional deuteration. Biochemistry 27:142–150

78. Tugarinov V, Kanelis V, Kay LE (2006) Isotope labeling strategies for the study of high-molecular-weight proteins by solution NMR spectroscopy. Nat Protoc 1(2): 749–754
79. Tugarinov V, Muhandiram DR, Ayed A, Kay LE (2002) Four-dimensional NMR spectroscopy of a 723-residue protein: chemical shift assignments and secondary structure of malate synthase G. J Am Chem Soc 124:10025–10035
80. Gans P et al (2010) Stereospecific isotopic labeling of methyl groups for NMR spectroscopic studies of high-molecular-weight proteins. Angew Chem Int Ed Engl 49(11):1958–1962
81. Fischer M et al (2007) Synthesis of a C-13-methyl-group-labeled methionine precursor as a useful tool for simplifying protein structural analysis by NMR spectroscopy. Chembiochem 8(6):610–612
82. Gelis I et al (2007) Structural basis for signal-sequence recognition by the translocase motor SecA as determined by NMR. Cell 131(4):756–769
83. Isaacson RL et al (2007) A new labelling method for methyl transverse relaxation-optimized spectroscopy NMR spectra of Alanine residues. J Am Chem Soc 129(50):15428–15429
84. Ayala I, Sounier R, Use N, Gans P, Boisbouvier J (2009) An efficient protocol for the complete incorporation of methyl-protonated alanine in perdeuterated protein. J Biomol NMR 43(2):111–119
85. LeMaster DM, Kushlan DM (1996) Dynamical mapping of E. coli thioredoxin via C-13 NMR relaxation analysis. J Am Chem Soc 118(39):9255–9264
86. Takeuchi K, Sun ZYJ, Wagner G (2008) Alternate C-13-C-12 labeling for complete mainchain resonance assignments using C alpha direct-detection with applicability toward fast relaxing protein systems. J Am Chem Soc 130(51):17210–17211
87. Kainosho M et al (2006) Optimal isotope labelling for NMR protein structure determinations. Nature 440:52–57
88. Skrisovska L, Schubert M, Allain FHT (2010) Recent advances in segmental isotope labeling of proteins: NMR applications to large proteins and glycoproteins. J Biomol NMR 46(1):51–65
89. Camarero JA et al (2002) Autoregulation of a bacterial sigma factor explored by using segmental isotopic labeling and NMR. Proc Natl Acad Sci USA 99(13):8536–8541
90. Skrisovska L, Allain FHT (2008) Improved segmental isotope labeling methods for the NMR study of multidomain or large proteins: application to the RRMs of Npl3p and hnRNP L. J Mol Biol 375(1):151–164
91. Vitali F et al (2006) Structure of the two most C-terminal RNA recognition motifs of PTB using segmental isotope labeling. EMBO J 25(1):150–162
92. Xu R, Ayers B, Cowburn D, Muir TW (1999) Chemical ligation of folded recombinant proteins: segmental isotopic labeling of domains for NMR studies. Proc Natl Acad Sci USA 96(2):388–393
93. Muralidharan V, Muir TW (2006) Protein ligation: an enabling technology for the biophysical analysis of proteins. Nat Methods 3(6):429–438
94. Xu MQ, Evans TC (2005) Recent advances in protein splicing: manipulating proteins in vitro and in vivo. Curr Opin Biotechnol 16(4):440–446
95. Oxenoid K, Chou JJ (2005) The structure of phospholamban pentamer reveals a channel-like architecture in membranes. Proc Natl Acad Sci USA 102(31):10870–10875
96. Jones TA, Thirup S (1986) Using known substructures in protein model-building and crystallography. EMBO J 5(4):819–822
97. Delaglio F, Kontaxis G, Bax A (2000) Protein structure determination using molecular fragment replacement and NMR dipolar couplings. J Am Chem Soc 122(9):2142–2143
98. Shen Y et al (2008) Consistent blind protein structure generation from NMR chemical shift data. Proc Natl Acad Sci USA 105(12):4685–4690
99. Raman S et al (2010) NMR structure determination for larger proteins using backbone-only data. Science 327(5968):1014–1018
100. Gonen T et al (2005) Lipid-protein interactions in double-layered two-dimensional AQP0 crystals. Nature 438(7068):633–638

Chapter 27

Structure–Function Insights of Membrane and Soluble Proteins Revealed by Electron Crystallography

Tina M. Dreaden, Bharanidharan Devarajan, Bridgette A. Barry, and Ingeborg Schmidt-Krey

Abstract

Electron crystallography is emerging as an important method in solving protein structures. While it has found extensive applications in the understanding of membrane protein structure and function at a wide range of resolutions, from revealing oligomeric arrangements to atomic models, electron crystallography has also provided invaluable information on the soluble α/β-tubulin which could not be obtained by any other method to date. Examples of critical insights from selected structures of membrane proteins as well as α/β-tubulin are described here, demonstrating the vast potential of electron crystallography that is first beginning to unfold.

Key words: Electron crystallography, Cryo-EM, Membrane protein, Soluble protein, Structure

1. Introduction

Advances over the last three decades have propelled electron crystallography to the forefront of biophysical methods for structure determination of membrane proteins. Since the first ground-breaking results in 1975 (1), dozens of membrane protein structures have been determined at a range of resolutions. Because of the near physiological environment provided in the two-dimensional (2D) proteoliposome crystals used for data collection, valuable information can be acquired at various resolution, ranging from the identification of the oligomeric nature over three-dimensional (3D) models of secondary structure to atomic models. Here we describe selected structures that demonstrate the capabilities and

Ingeborg Schmidt-Krey and Yifan Cheng (eds.), *Electron Crystallography of Soluble and Membrane Proteins: Methods and Protocols*, Methods in Molecular Biology, vol. 955, DOI 10.1007/978-1-62703-176-9_27,

potential of electron crystallography for answering fundamental structure–function questions, which are the chief aims for the structure determination of membrane proteins.

2. Bacteriorhodopsin

Membrane protein structure determination by electron crystallography was pioneered by Henderson and Unwin in 1975 with their work on the purple membrane of *Halobacterium salinarium* (1, 2). Bacteriorhodopsin (bR) is the light-driven proton pump in the membrane of purple bacteria. It consists of seven transmembrane α-helices and a covalently bound retinal chromophore that is responsible for the purple color of the membranes and initiation of the photocycle. In this work, defined transmembrane α-helices of the first membrane protein structure were seen at 7 Å resolution by electron microscopy of tilted, unstained specimens (1, 2). Critical advancements that enabled the influential milestone for the field of membrane protein structure determination included improvements in methods of specimen preservation by glucose embedding and data collection by low-dose imaging as well as the correction for specimen distortions by computational "unbending" of images (3, 4). With the development of electron cryo-microscopy (cryo-EM) and further improvements in data processing, the bR structure was solved to 3.5 Å resolution from 2D crystals in 1990, making it possible to build an atomic model (5). Interestingly, bR formed 2D crystalline arrays within membranes in vivo, rather than after the purification, delipidation, and reconstitution commonly used in 2D crystallization. These ordered arrays within native membranes were used for data collection because they formed the most stable and reproducible crystals (5). In the following years, structures by both electron and X-ray crystallography provided detailed, critical information on conformational changes in the photocycle of bR (6).

3. Plant Light Harvesting Complex II

Plant light harvesting complex II (LHCII) was the second membrane protein of which the structure was determined by electron crystallography. LHCII, the most abundant membrane protein in chloroplasts, binds chlorophyll and carotenoid pigments that function in energy capture and transfer to the reaction center during photosynthesis. LHCII also tends to form native crystalline arrays in vivo. It was possible to induce the formation of highly ordered arrays of the detergent-solubilized protein by incubation with divalent salts (7). These samples contained absolutely minimal amounts of essential co-purified lipid, eliminating the need for the addition of exogenous lipids. The 3D structure of LHCII at 3.4 Å resolution revealed three transmembrane α-helices and a fourth short 3-turn α-helix at the interface between the membrane interior and the thylakoid lumen (8); this fourth α-helix was not seen in the initial 3D structure at 6 Å resolution (9). The structure also revealed 12 non-covalently bound chlorophyll molecules and two carotenoid molecules per monomer (8).

4. α/β-tubulin

One of the first structures solved by electron crystallography, and the only one to date of soluble proteins providing an atomic model from high-resolution data by this method, is of α/β-tubulin (10), which plays roles in intracellular transport as well as cell division. A structure at 2.89 Å resolution of epothilone A bound to α/β-tubulin provided key data on the effect of epothilone A on the activity of α/β-tubulin and mutational resistance, with a comparison of the binding of the taxane cancer treatment Taxol allowing for critical conclusions on the binding pocket (11). To date, a comparable structure of α/β-tubulin has not been solved by any other method, including X-ray crystallography.

5. Aquaporins

The aquaporins are a family of membrane water channels present in numerous types of cells in a vast range of organs and organisms. The first atomic resolution structure of a member of this family was Aquaporin-1 (AQP1) (12, 13). In contrast to some other aquaporins that possess broad substrate specificity, AQP1 has a high selectivity for water and excludes other solutes and ions. The structure revealed a heterotetrameric organization of monomers, each containing six transmembrane α-helices as well as two half helices. These six α-helices result from two tandem sequence repeats, each with three membrane α-helices; this is a unique feature characteristic of the aquaporins. The overall shape of the pore and the identities of the amino acid side chains that lined the pore provided valuable insight into the structural basis for the functional water specificity. The pore architecture was unique in that it tapered from the surface into a narrow diameter of ~3–4 Å with a slight curvature, just slightly larger than a 2.8 Å diameter water molecule (12, 13). The majority of the side chains lining the pore were also hydrophobic in nature, which should exclude some solutes. The constricted pathway was reasoned to enhance interruption of a continuous, hydrogen-bonded network of water molecules. This in turn may facilitate the ability of water to hydrogen-bond with two asparagine residues located at the center of the pore, which would inhibit conduction of protons or other small, charged molecules (12, 13).

Aquaporin-0 (AQP0) differs from AQP1 in that it forms membrane junctions in vivo, as well as functions as a water pore. The protein serves as an essential component of the lens core system, which supplies underlying tissues with nutrients and water. At a resolution of 1.9 Å, the AQP0 structure remains the highest resolution structure of a membrane protein determined by electron crystallography to date (14, 15). Membrane junction formation by AQP0 is a consequence of natural proteolytic cleavage of a subset of proteins near the C-terminus in aging cells. While un-cleaved tetramers do not form junctions, partial proteolytic cleavage in a subpopulation from sheep lens tissue induced formation of double-layered 2D crystals in which tetramers interacted in opposing membranes through their extracellular surfaces (14, 15). The structures determined from these crystals thereby mimicked the natural membrane junctions. The conformational change induced by proteolytic cleavage to enable junction formation also induced a relative closed pore conformation. While seven water molecules were observed in non-junctional AQP0 (16), only three were present in the junctional AQP0 that had a much narrower pore at two constriction sites (15). Another interesting note about the AQP0 structure was that 2D crystals could form with a range of lipid types, which indicated a nonspecific requirement or low affinity binding to lipids, and nine lipid molecules were observed per monomer (15). These lipid molecules surrounded AQP0 tetramers and bridged lateral contacts between them in the membrane. Very impressively, the entire lipid bilayer could be resolved in addition to the protein (15).

Similar to AQP0, aquaporin-4 (AQP4) forms membrane junctions in vivo. AQP4 is the most prevalent water channel in the mammalian brain, and double-layered 2D crystals also formed from rat AQP4 expressed in insect cells (17). Of note, the tetramers in two interacting membranes in AQP0 are precisely in register (14, 15), while the tetramers in two layers of AQP4 are shifted half a unit cell with respect to one another (17).

6. The Nicotinic Acetylcholine Receptor

In contrast to the water permeable aquaporins, the nicotinic acetylcholine receptor is a neurotransmitter gated ion channel present at synapses between nerve and muscle cells. Upon acetylcholine release into the synapse, acetylcholine binds to the ligand-binding domain on the extracellular surface of the receptor. The binding induces a conformational change that opens the pore gate to allow cations to flow down their electrochemical gradient, resulting in a change in membrane potential. Although it was known that separate regions of the primary sequence composed the extracellular ligand binding domain and the membrane spanning pore, the 4 Å resolution structure depicted an abrupt transition between the two from a β-sheet fold to an α-helical fold (18). The interface at the transition was also ~10 Å away from the surface of the membrane, which was not predicted (18). Each subunit of the pentameric pore is composed of four α-helices, and the overall arrangement resembled a propeller blade (18). While one α-helix of each monomer (M2) tilted inward towards the central region of the pore, the other three α-helices tilted radially outward (18). The different structural components of the pore suggested that the α-helices in the inner ring would move during ion gating, but the outer helices would remain unchanged. For the first time, a structure-based mechanism was proposed for the pore opening mechanism induced by acetylcholine binding to the β-sheet binding domain. It was previously known that a 15 ° rotation of the β-sheets occurred upon acetylcholine-binding, which was also known to induce pore opening (19), but the basis for the allosteric control was unclear. In the structure, the only direct link between the ligand-binding domain and the helical pore was a valine residue on the β-sheet loop that had a "pin-into-socket" interaction with the M2 helix in the central movable region of the pore (18). It was proposed that this interaction enabled the communication and succeeding conformational change (18).

7. ATPases

While a broad range of important structural information had been gained of ATPases since the first successful 2D crystallization trials of Na^+, K^+-ATPase in the presence of the ligand vanadate (20), one of the major breakthroughs for the study of ATPases, and membrane proteins in general, was the first 3D structure showing the secondary arrangement of the transmembrane domain of Ca^{+2}-ATPase from rabbit sarcoplasmic reticulum (21). This information was obtained from helically arranged Ca^{+2}-ATPase, which also allowed the assignment of the transmembrane α-helices at 8 Å resolution and allowed for the proposal of Ca^{+2}–binding site as well as a mechanism for the release of Ca^{+2} while in the E_2 conformation. Rice et al. (22) were then able to relate the E_2 to E_1 conformational change of Ca^{+2}-ATPase to a structure of Na^+, K^+-ATPase at 11 Å resolution, demonstrating the power of docking and modeling of structures at, what could currently be described as, intermediated resolution. At this time, a substantial number of different types of membrane proteins in ordered arrays fall within this resolution range, and these ATPase illustrate the wealth of functional information that could be obtained without necessarily the highest resolution. The combination of electron crystallography and modeling then allowed for the modeling of the molecular interactions of phospholamban and Ca^{2+}-ATPase (23).

Critical functional insights could also be gained on the gastric H^+, K^+-ATPase by 3D structures from 2D crystals in the resolution range of 6.5–8.0 Å (24–26). It could be shown that the noncatalytic β-subunit blocks the catalytic α-subunit while in the E_2 conformation, precluding a reversal of the transport cycle (24). The action of an acid suppressant on the gastric H^+, K^+-ATPase showed distinct conformational rearrangements of specific α-helices on the luminal membrane side, locking the protein in an open conformation (26).

8. Conclusions

The field of electron crystallography has provided valuable information towards our understanding of many types of membrane protein structures. This is not only true of the few selected examples of high-resolution structures outlined here but also of critical information revealed from structures at more moderate resolution such as the structure–function studies of the ATPases (see Chapters 1, 27, and 30 for further examples). Since proteins are visualized in the context of a near native physiological environment within a lipid bilayer, electron crystallography can provide answers to structure–function questions that cannot be addressed by any other biophysical method to date. This is particularly valuable in the context of eukaryotic membrane proteins, which still provide many challenges in terms of obtaining sufficient quantities and their stability in a detergent environment. Although the time investment for structure determination by electron crystallography is currently longer than more established methods, such as X-ray crystallography, significant advances have been made in all aspects of the field. With the advantage in electron crystallography of working with smaller protein quantities (27) and often much faster time to crystallization which could recently be decreased substantially for a large range of membrane proteins (Schmidt-Krey, unpublished observations), the overall time frames can in some cases be comparable, with more time spent on the data collection than on the purification and crystallization in this case. Accordingly, the time required to determine a 3D structure will continue to decrease as knowledge on the possibly perceived or overcome bottleneck of 2D crystallization, increased use of the technique, and technology (see Chapter 20), currently in the form of automation (see Chapters 15–18), progress. With that said, electron crystallography should not be seen as competing with other structural methods such as X-ray crystallography, NMR, and modeling, but in fact data from these techniques is very frequently combined to elucidate the greatest possible detail in the understanding of structure–function of membrane and soluble proteins (28).

References

1. Henderson R, Unwin PNT (1975) Three-dimensional model of purple membrane obtained by electron microscopy. Nature 257:28–32
2. Unwin PNT, Henderson R (1975) Molecular structure determination by electron microscopy of unstained crystalline specimens. J Mol Biol 94:425–440
3. Henderson R, Baldwin JM, Downing KH, Lepault J, Zemlin F (1986) Structure of purple membrane from halobacterium halobium: recording, measurement and evaluation of electron micrographs at 3.5 Å resolution. Ultramicroscopy 19:147–178
4. Kunji ER, von Gronau S, Oesterhelt D, Henderson R (2000) The three-dimensional structure of halorhodopsin to 5 A by electron crystallography: a new unbending procedure for two-dimensional crystals by using a global reference structure. Proc Natl Acad Sci USA 97:4637–4642
5. Henderson R, Baldwin JM, Ceska TA, Zemlin F, Beckmann E, Downing KH (1990) Model for the structure of bacteriorhodopsin based on high-resolution electron cryo-microscopy. J Mol Biol 213:899–929
6. Hirai T, Subramaniam S (2009) Protein conformational changes in the bacteriorhodopsin photocycle: comparison of findings from electron and X-ray crystallographic analyses. PLoS One 4:5769
7. Wang DN, Kühlbrandt W (1991) High-resolution electron crystallography of light-harvesting chlorophyll *a/b*-protein complex in three different media. J Mol Biol 217:691–699
8. Kühlbrandt W, Wang DN, Fujiyoshi Y (1994) Atomic model of plant light-harvesting complex by electron crystallography. Nature 367:614–621
9. Kühlbrandt W, Wang DN (1991) Three-dimensional structure of plant light-harvesting complex determined by electron crystallography. Nature 350:130–134
10. Nogales E, Wolf SG, Downing KH (1998) Structure of the alpha beta tubulin dimer by electron crystallography. Nature 391:199–203
11. Nettles JH, Li H, Cornett B, Krahn JM, Snyder JP, Downing KH (2004) The binding mode of epothilone A on alpha, beta-tubulin by electron crystallography. Science 305:866–899
12. Murata K, Mitsuoka K, Hirai T, Walz T, Agre P, Heymann JB, Engel A, Fujiyoshi Y (2000) Structural determinants of water permeation through aquaporin-1. Nature 407:599–605
13. Ren G, Reddy VS, Cheng A, Melnyk P, Mitra AK (2001) Visualization of a water-selective pore by electron crystallography in vitreous ice. Proc Natl Acad Sci U S A 98:1398–1403
14. Gonen T, Sliz P, Kistler J, Cheng YF, Walz T (2004) Aquaporin-0 membrane junctions reveal the structure of a closed water pore. Nature 429:193–197
15. Gonen T, Cheng YF, Sliz P, Hiroaki Y, Fujiyoshi Y, Harrison SC, Walz T (2005) Lipid-protein interactions in double-layered two-dimensional aqpo crystals. Nature 438: 633–638
16. Harries WEC, Akhavan D, Miercke LJW, Khademi S, Stroud RM (2004) The channel architecture of aquaporin 0 at a 2.2-angstrom resolution. Proc Natl Acad Sci U S A 101: 14045–14050
17. Hiroaki Y, Tani K, Kamegawa A, Gyobu N, Nishikawa K, Suzuki H, Walz T, Sasaki S, Mitsuoka K, Kimura K, Mizoguchi A, Fujiyoshi Y (2006) Implications of the aquaporin-4 structure on array formation and cell adhesion. J Mol Biol 355:628–639
18. Miyazawa A, Fujiyoshi Y, Unwin N (2003) Structure and gating mechanism of the acetylcholine receptor pore. Nature 423:949–955
19. Unwin N, Miyazawa A, Li J, Fujiyoshi Y (2002) Activation of the nicotinic acetylcholine receptor involves a switch in conformation of the a subunits. J Mol Biol 319:1165–1176
20. Skriver E, Maunsbach AB, Jørgensen PL (1981) Formation of two-dimensional crystals in pure membrane-bound Na+, K+–ATPase. FEBS Lett 131:219–222
21. Zhang P, Toyoshima C, Yonekura K, Green NM, Stokes DL (1998) Structure of the calcium pump from sarcoplasmic reticulum at 8-Å resolution. Nature 392:835–839
22. Rice WJ, Young HS, Martin DW, Sachs JR, Stokes DL (2001) Structure of Na+, K+–ATPase at 11-A resolution: comparison with Ca2+–ATPase in E1 and E2 states. Biophys J 80:2187–2197
23. Stokes DL, Pomfret AJ, Rice WJ, Glaves JP, Young HS (2006) Interactions between Ca^{2+}-ATPase and the pentameric form of phospholamban in two-dimensional co-crystals. Biophys J 90:4213–4223
24. Abe K, Tani K, Nishizawa T, Fujiyoshi Y (2009) Inter-subunit interaction of gastric H+, K+–ATPase prevents reverse reaction of the transport cycle. EMBO J 28:1637–1643
25. Abe K, Tani K, Fujiyoshi Y (2010) Structural and functional characterization of H+, K+–ATPase with bound fluorinated phosphate analogs. J Struct Biol 170:60–68

26. Abe K, Tani K, Fujiyoshi Y (2011) Conformational rearrangement of gastric H(+), K(+)-ATPase induced by an acid suppressant. Nat Commun 2:155
27. Schmidt-Krey I (2007) Electron crystallography of membrane proteins: two-dimensional crystallization and screening by electron microscopy. Methods 41:417–426
28. Kim LY, Johnson MC, Schmidt-Krey I (2011) Cryo-EM in the study of membrane transport proteins. Compr Physiol 2:283–293

Chapter 28

Lipid Monolayer and Sparse Matrix Screening for Growing Two-Dimensional Crystals for Electron Crystallography: Methods and Examples

Mark Yeager, Kelly A. Dryden, and Barbie K. Ganser-Pornillos

Abstract

Electron microscopy provides an efficient method for rapidly assessing whether a solution of macromolecules is homogeneous and monodisperse. If the macromolecules can be induced to form two-dimensional crystals that are a single layer in thickness, then electron crystallography of frozen-hydrated crystals has the potential of achieving three-dimensional density maps at sub-nanometer or even atomic resolution. Here we describe the lipid monolayer and sparse matrix screening methods for growing two-dimensional crystals and present successful applications to soluble macromolecular complexes: carboxysome shell proteins and HIV CA, respectively. Since it is common to express recombinant proteins with poly-His tags for purification by metal affinity chromatography, the monolayer technique using bulk lipids doped with Ni^{2+} lipids has the potential for broad application. Likewise, the sparse matrix method uses screening conditions for three-dimensional crystallization and is therefore of broad applicability.

Key words: Lipid monolayers, Sparse matrix, Two-dimensional crystallization, Electron crystallography, Carboxysome proteins, Retroviral CA proteins

1. Introduction

As in any crystallization trial, one must refine the conditions for crystallization and specimen handling to maximize resolution (1). In general the variables to be tested to achieve two-dimensional (2D) crystallization are similar to those used for three-dimensional (3D) crystallization and include evaluation of divalent cations, pH, detergents, lipids, buffers, ionic strength, temperature, precipitants, ligands, and inhibitors. Purity and yield of the protein are maximized, and proteolysis is minimized. For membrane proteins

Ingeborg Schmidt-Krey and Yifan Cheng (eds.), *Electron Crystallography of Soluble and Membrane Proteins: Methods and Protocols*, Methods in Molecular Biology, vol. 955, DOI 10.1007/978-1-62703-176-9_28,

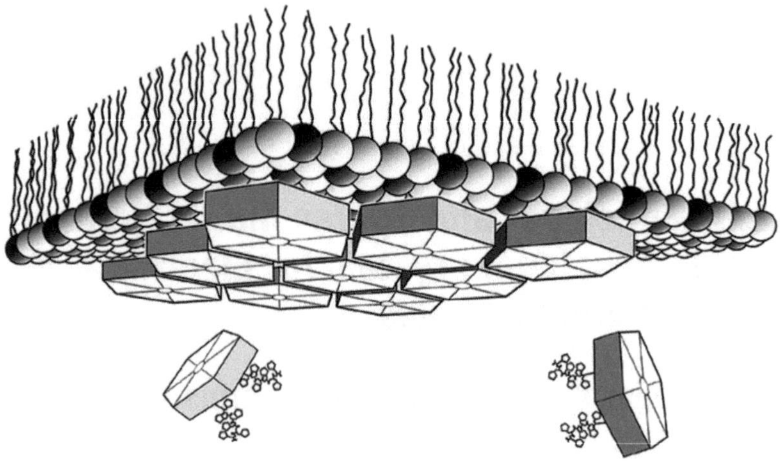

Fig. 1. Schematic depiction of 2D lipid monolayer crystallization. Carboxysome BMC shell protein hexamers were localized at the lipid-aqueous interface of a mixed lipid layer with 4:1 (wt:wt) l-α-phosphatidylcholine (**light** spheres) plus DOGS-NTA-Ni^{2+} (**dark** spheres) by means of polyhistidine tag chelation via Ni^{2+} bound within the polar lipid layer. Resulting 2D crystals were transferred to an EM grid and stained for image diffraction analysis (From ref. (13)).

in detergent micelles, 2D crystals are typically grown by reconstitution methods involving dialysis and/or the use of detergent binding agents (reviewed by Yeager et al., 1999 (2)). Here we will describe two methods for growing 2D crystals of soluble proteins and macromolecular complexes—lipid monolayer and sparse matrix screening. In all cases, the progress of 2D crystallization is assessed by negative-stain electron microscopy (reviewed by Adair and Yeager, 2007 (3)).

In the *lipid monolayer crystallization* technique (4–6), lipids are spread at an air–water interface so that the aliphatic chains extend into the air, and the lipid headgroups are exposed at the aqueous interface (Fig. 1). Soluble protein molecules present in the aqueous phase bind to the lipid headgroups via electrostatic interactions or by a specific affinity tag. Negatively charged lipids include stearic and oleic acid. DOPS and octadecylamine are positively charged. Affinity tags can exploit a receptor–ligand or enzyme–substrate interaction, such as cholera toxin binding to its receptor (7) and avidin binding to a biotinylated lipid (8). An interaction of general utility exploits the strategy of using recombinant His-tagged proteins (9, 10). The lipid monolayer can be doped with a synthetic lipid containing a chelating nickel headgroup that will bind to a protein having a polyhistidine tag. This method would also be applicable for the 2D crystallization of soluble ectodomains of membrane proteins generated by recombinant DNA technology or by release from the membrane by proteolysis. This method can also be used for membrane proteins solubilized in detergent micelles if the detergent is sufficiently mild so that the lipid monolayer is not perturbed. In those instances where the monolayer is disrupted by the detergent, the monolayer can be stabilized by the use of fluorinated lipids (11). In our laboratory, we have used the lipid monolayer technique for growing 2D crystals of a variety of His-tagged proteins: Fab fragments (2), the murine leukemia virus CA protein (12), and carboxysome shell proteins (13).

Carboxysomes are bacterial organelles that form microcompartments for performing enzymatic reactions. The polyhedral shells of carboxysomes (Fig. 2a) are formed by a single layer of CcmK proteins. The lipid monolayer crystallization method was used to grow 2D hexagonal crystals of CcmK proteins (Fig. 2b,c) that mimic the hexagonal lattice within a facet of the polyhedral shell. An aliquot of the protein (10 μl at a protein concentration of 0.5–1.0 mg/ml) was overlaid with 1 μl of a 1:1 chloroform/hexane solution containing 50 mg/ml L-α-phosphatidylcholine doped with 12.5 mg/ml nickel-charged lipid (DOGS-NTA Ni^{2+}), and the crystals grew in 4 h at room temperature. The diffraction pattern of the crystals (Fig. 2d) could be indexed with p1 unit cell parameters that were nearly hexagonal and in close agreement with the center-to-center spacings in the X-ray structures of CcmK proteins. 2D and 3D maps of CcmK proteins (Fig. 3), derived by electron crystallography and tilt reconstruction of the hexagonal crystals, revealed that the packing of the CcmK proteins recapitulated the packing within the planar facets of the polyhedral carboxysome shells.

In the *sparse matrix technique*, the protein is immersed in solutions typically used for screening for three-dimensional crystallization. After defined periods of time, an aliquot is tested for the formation of 2D crystals. This approach was particularly successful for growing subnanometer resolution crystals of the HIV CA protein (14). A particular mutant of CA (R18L) formed micron-sized, single-layered protein vesicles upon incubation of the protein at 32 mg/ml for 1 h at 37 °C in a buffer containing 17.5 % (w/v) PEG 20,000, 50 mM sodium cacodylate (pH 6.5), and 100 mM calcium acetate. When flattened onto the carbon substrate of an EM grid, the protein layer formed a planar hexagonal lattice (Fig. 4b) that was an in vitro mimic of the continuously variable hexagonal lattice within the HIV capsid (Fig. 4a). Computed diffraction patterns of frozen-hydrated crystals (Fig. 4c) displayed sharp spots to better than 10 Å resolution, so that the 3D map derived by tilt reconstruction revealed rods of density corresponding to α-helices (Fig. 5, gray-scale mesh). The α-helical rods served as fiducials for precise docking of atomic resolution structures of the N-terminal and C-terminal domains of CA. This hybrid approach that combined mutagenesis, EM and molecular modeling of high-resolution structures yielded the first atomic resolution model for the hexagonal lattice within HIV capsids.

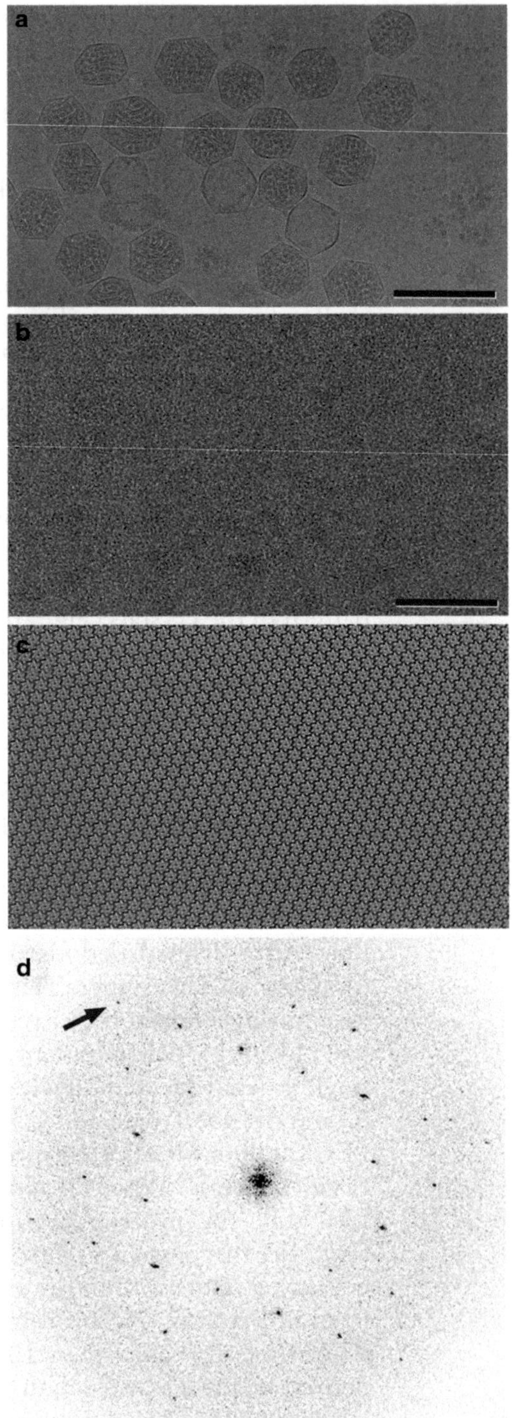

Fig. 2. (**a**) Electron cryomicrograph of carboxysomes, which are bacterial microcompartments formed by CcmK proteins that assemble as polyhedral shells containing key metabolic enzymes. Scale bar = 2,000 Å. (**b**) Image of negatively stained CcmK1 2D crystals. Scale bar = 500 Å. (**c**) The underlying order is observed when the image is Fourier filtered, and corrections are made for lattice distortions. (**d**) The Fourier transform of the negatively stained image displays hexagonal symmetry. The **arrow** identifies the (2,2) reflection at 17-Å resolution. Images were contrast stretched to emphasize salient features (From ref. (13)).

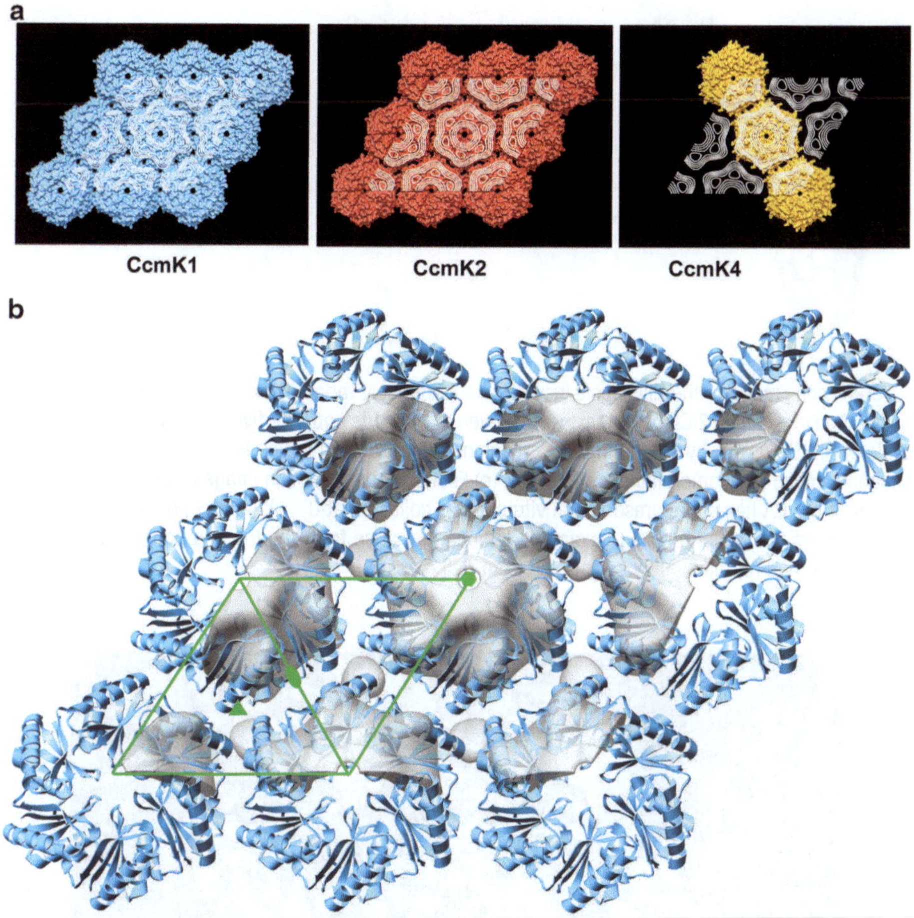

Fig. 3. (**a**) Projection maps from 2D crystals of three CcmK proteins (**white** contours) shown projected onto the molecular packings visualized in 3D crystals. In the case of CcmK4, hexamers in 3D crystals packed into uniformly ordered strips (illustrated) but not uniformly oriented layers. (**b**) 3D density map (tilted by 15 ° to enhance depth) showing the hexamer packing of CcmK1 in 2D lipid monolayer crystals (**gray** surface), which recapitulates the packing in the **ab** lattice plane in 3D crystals (**blue** ribbon) (pdb id. 3dn9). Center-to-center hexamer spacing is ~68 Å for the EM map and ~70 Å for the crystal structure. For comparison, the X-ray structure was manually docked into the EM density obtained from 3D reconstruction. The **green line** outlines the 2D unit cell with standard symbols at the axes of rotational symmetry (From ref. (13)).

If conditions for growing 3D crystals have been identified, a refinement of the sparse matrix approach is to directly apply an aliquot of the protein and precipitant solution to an EM grid and perform a time-course experiment to screen for nucleation of 2D crystals. Fine screening around the conditions for growing 3D crystals can also be performed. This approach was successful for generating 2D crystals of myosin II, in which the *a* and *b* lattice planes of the 2D crystals corresponded to similar planes in the 3D crystals (15).

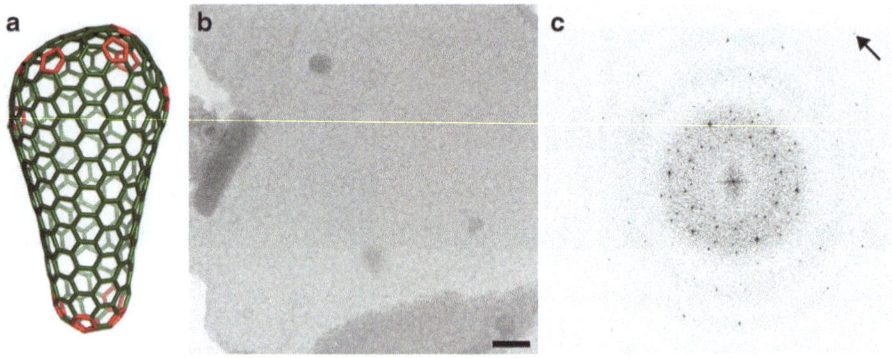

Fig. 4. (**a**) Fullerene cone model of the HIV capsid in which a continuously variable hexagonal lattice of CA hexamers (**green**) is closed by insertion of exactly 12 CA pentamers (**red**), seven at the **wide** end, and five at the **narrow** end of the cone (16). (**b**) 2D crystals of HIV CA grown by sparse matrix screening. The hexagonal lattice is an in vitro symmetric mimic of the variable hexagonal lattice in pleomorphic capsids of HIV. (**c**) Fourier transform of an image of a flattened sphere (**b**) formed by a single layer of HIV CA hexamers preserved in vitreous ice, collected at 0 ° tilt (**arrow** points to a reflection at 9.8 Å). The transform displays two superimposed hexameric lattices that arise from the top and bottom layers of the flattened sphere (From ref. (14) and reproduced by permission from Cell press).

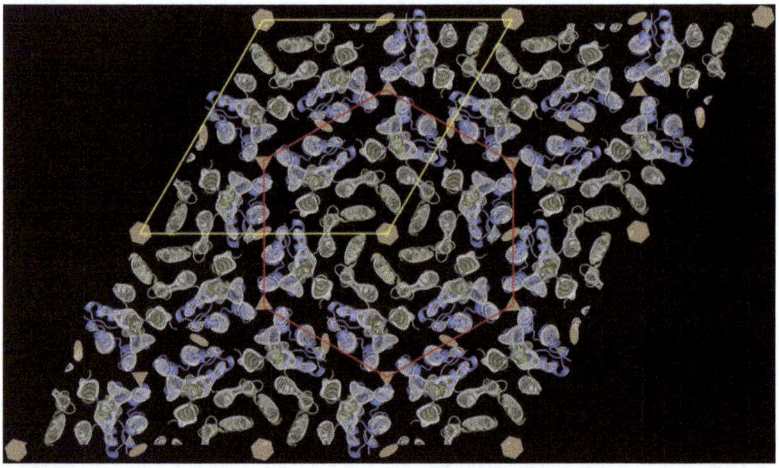

Fig. 5. The 3D cryoEM density map of HIV-1 CA (gray-scale, mesh) at 9-Å in-plane resolution serves as a template for docking the high-resolution structures of the N-terminal (**green**) and C-terminal (**blue**) domains. A hexamer of N-terminal domains is outlined in **red**, and the p6 unit cell is outlined in **yellow**. The hexamers are linked together by dimers of the C-terminal domains (**blue**). The six, three, and twofold symmetry axes are indicated by **hexagons**, **triangles**, and **ellipses**, respectively (From ref. (14) and reproduced by permission from Cell press).

2. Materials

2.1. Equipment

2.1.1. Lipid Monolayer 2D Crystallization

Teflon block (25 × 25 × 5 mm) with wells 3 mm in diameter (Fig. 6).

Hamilton syringe (10, 50, and 100 μl) (Fisher).

5 cm Whatman filter paper #2 (Fisher).

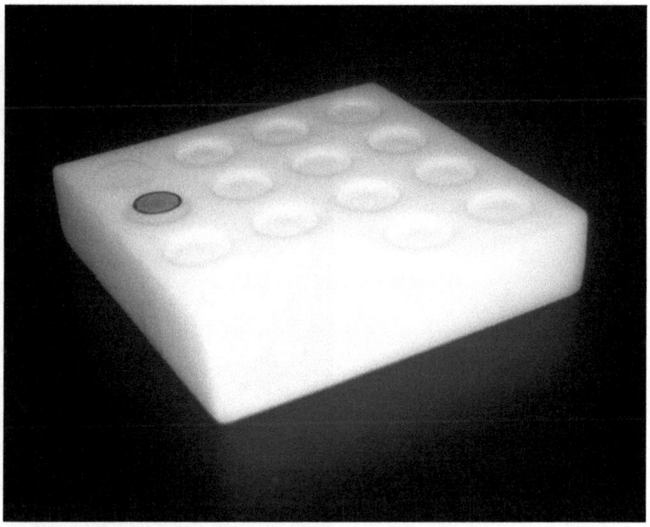

Fig. 6. Teflon block used for lipid monolayer crystallization. The wells are 3 mm in diameter to accommodate the EM grid and have a depth of 0.5 mm.

5 cm plastic Petri dish (Fisherbrand).

Parafilm™.

EM forceps (Dumont #5).

2.1.2. Sparse Matrix Screening

600 μl Eppendorf, 8-tube pcr strip (Fisher) (Fig. 7).

96-well plate (Fisherbrand).

EM forceps (Dumont #5).

2.2. Reagents and Solutions

2.2.1. Lipid Monolayer 2D Crystallization

Egg lipids or 10 % l-a-phosphatidylcholine (Avanti Polar Lipids) (see Note 1).

Nickel-charged 1,2-dioleoyl-*sn*-glycero-3-((*N*-(5-amino-1-carboxypentyl)iminodiacetic acid)succinyl) (DOGS-NTA Ni^{2+}) (Avanti Polar Lipids).

Hexane (Sigma).

Chloroform (Sigma).

Fisher Cleaning Solution.

7×, Linbro or Versa-Clean, Fisher.

Double-distilled, deionized water.

Uranyl acetate (Ted Pella).

3 mm Copper grids, 300 mesh (Ted Pella or EMS).

2.2.2. Sparse Matrix Screening

Solutions for 3D crystallization screens (e.g., Hampton Research, Wizard and Qiagen).

Uranyl acetate (Ted Pella).

3 mm Copper grids, 300 mesh (Ted Pella or EMS).

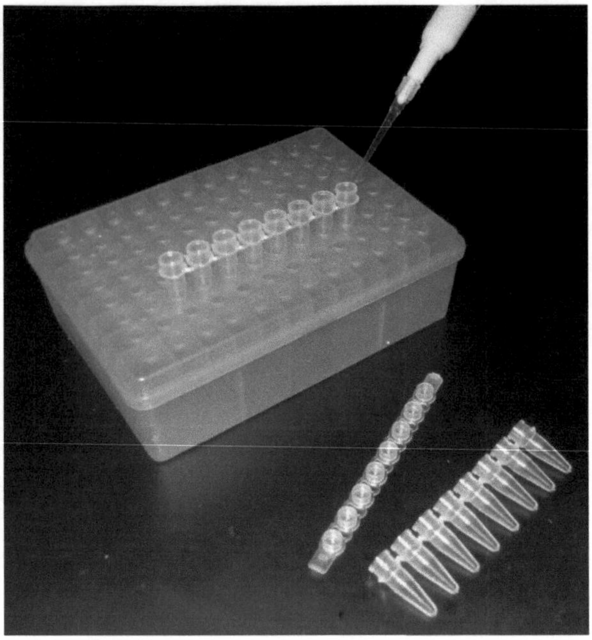

Fig. 7. 600 μl Eppendorf, 8-tube pcr strip, for sparse matrix screening. An aliquot of the precipitant solution is mixed with an aliquot of the solution containing the protein to be crystallized.

3. Methods

3.1. Lipid Monolayer 2D Crystallization

1. Fresh lipid solutions are prepared in chloroform and hexane from stock solutions of the special lipid (i.e., charged or liganded) and egg phosphatidylcholine and maintained on ice. Crystallization trials are performed in which the ratio of the special lipid to the bulk lipid is varied from 0 (as a control) to 1. A successful concentration for crystallization of MuLV CA was 50 mg/ml egg phosphatidylcholine and 12.5 mg/ml DOGS-NTA Ni_{2+} (12). Egg phosphatidylcholine is typically used because it remains fluid over a wide range of temperatures and compressions.
2. A teflon block (Fig. 1) that has wells 3 mm in diameter, corresponding to the diameter of an EM grid, is thoroughly cleaned in chromic and sulfuric acid (Fisher Cleaning Solution). The block is washed by sonication for 30 min in detergent solution (7×, Linbro or Versa-Clean, Fisher) and then rinsed for ~30 min in double-distilled, deionized H_2O.
3. To maintain a hydrated atmosphere, filter paper is placed on the bottom of a 5 cm Petri dish and thoroughly wetted with buffer. The teflon block is then inserted.

4. Use an Eppendorf pipette to place a 10 μl droplet of the protein solution at a concentration of 0.5–1.0 mg/ml in the center of the 0.5 mm deep well (see Note 2).
5. Use a glass micropipette (e.g., Hamilton syringe) to add ~0.5–1.0 μl of the lipid solution to the top of the drop, which will lower the surface tension and flatten the droplet, but should not overflow the well.
6. Cover the Petri dish and seal with Parafilm. Incubate at the desired temperature (e.g., room temperature or 4 °C).
7. 2D crystallization on the lipid monolayer can occur quickly (30–60 min) or take days.
8. EM grids with a hydrophobic carbon substrate are used since the attachment occurs via the lipid aliphatic chains.
9. EM forceps are used to carefully place the grid on top of the droplet with the carbon side facing the aliphatic chains on the droplet surface.
10. For some systems, a clue whether crystals have formed is that the grid will not rotate or spin when placed on the droplet.
11. The grid is immediately removed from the droplet or allowed to sit for 30 s and carefully lifted off (see Note 3). For staining, the grid is rinsed once on a droplet of distilled, deionized H_2O, blotted with filter paper, and stained. For freezing, the grid is immediately blotted and plunged into a liquid ethane slush.
12. Alternatively, the surface of the droplet can be picked up with a platinum/paladium wire loop via surface tension (17). The loop has a diameter slightly larger than the EM grid, so that transfer to the grid is accomplished by carefully passing the grid through the loop.

3.2. Sparse Matrix Screening for 2D Crystallization

1. Mix 5 μl protein solution with 5 μl of a given crystal screen in either a 600 μl Eppendorf, 8-tube pcr strip (Fig. 5), or a 96-well plate, and mix rapidly by pipetting up and down several times (see Note 4).
2. Incubate the reactions overnight at room temperature or 3–4 h at 37 °C. Obviously, time and temperature can be varied, but since there is no diffusion step, the reaction reaches completion (see Note 5).
3. Pipete 3 μl of sample onto an EM grid.
4. For staining, the grid is rinsed once on a droplet of either distilled, deionized H_2O or 0.1 M KCl (in the case of HIV CA), blotted with filter paper, and stained, or simply stained directly (see Note 6). For freezing, the grids need to be thoroughly washed to remove precipitants and buffers. In general, we use a 3 ml syringe equipped with a 0.2 μM filter to wash the grid with 5–10 drops of distilled, deionized H_2O or 0.1 M KCl.

4. Notes

4.1. Lipid Monolayer 2D Crystallization

1. In general, lipids are stored as 10× stock solutions (e.g., ~2 mg/ml of the egg lipids and 0.5 mg/ml of the nickel lipids). For our experiments on MuLV CA, lipids are mixed with equal amounts of hexane and chloroform (i.e., 10 μl of 10× egg lipid, 10 μl 10× nickel lipid, 40 μl hexane, and 40 μl chloroform) as the base.
2. 2D crystallization may be impeded if the sample buffer contains high concentrations of glycerol or salt. Hence, the sample may need to be diluted just before addition to the well in the teflon block.
3. It is essential that the grid be lifted from the droplet with as little disturbance as possible. Any rotation or translation of the grid will fracture the crystals.

4.2. Sparse Matrix Screening for 2D Crystallization

4. Due to volume restraints in robotic 3D crystallization, small amounts of the screening solutions are often discarded by users. These small aliquots are an excellent source for 2D screening buffers.
5. The protein should be concentrated to 2–30 mg/ml in a suitable buffer. Although low concentrations of buffer and salt are ideal, it is often not necessary for screening.
6. Some of the chemicals in the crystal screens will likely precipitate in the presence of uranyl acetate. Despite considerable precipitation, grids can still be screened and crystals still visualized.

Acknowledgments

Work in the Yeager laboratory is supported currently by NIH grants R01 HL48908, R01GM066087, R01 GM084545, and P50 GM082545. Parts of this review were abstracted from (2).

References

1. McPherson A (1982) Preparation and analysis of protein crystals. Wiley, New York
2. Yeager M, Unger VM, Mitra AK (1999) Three-dimensional structure of membrane proteins determined by two-dimensional crystallization, electron cryomicroscopy, and image analysis. Methods Enzymol 294:135–180
3. Adair BD, Yeager M (2007) Electron microscopy of integrins. Methods Enzymol 426:337–373
4. Hemming SA, Bochkarev A, Darst SA, Kornberg RD, Ala P, Yang DSC, Edwards AM (1995) The mechanism of protein crystal growth from lipid layers. J Mol Biol 246: 308–316
5. Brisson A, Olofsson A, Ringler P, Schmutz M, Stoylova S (1994) Two-dimensional crystallization of proteins on planar lipid films and structure determination by electron crystallography. Biol Cell 80:221–228

6. Chiu W, Avila-Sakar AJ, Schmid MF (1997) Electron crystallography of macromolecular periodic arrays on phospholipid monolayers. Adv Biophys 34:161–172
7. Ribi HO, Ludwig DS, Mercer KL, Schoolnik GK, Kornberg RD (1988) Three-dimensional structure of cholera toxin penetrating a lipid membrane. Science 239:1272–1276
8. Darst SA, Ahlers M, Meller PH, Kubalek EW, Blankenburg R, Ribi HO, Ringsdorf H, Kornberg RD (1991) Two-dimensional crystals of streptavidin on biotinylated lipid layers and their interactions with biotinylated macromolecules. Biophys J 59:387–396
9. Kubalek EW, Le Grice SFJ, Brown PO (1994) Two-dimensional crystallization of histidine-tagged, HIV-1 reverse transcriptase promoted by a novel nickel-chelating lipid. J Struct Biol 113:117–123
10. Barklis E, McDermott J, Wilkens S, Schabtach E, Schmid MF, Fuller S, Karanjia S, Love Z, Jones R, Rui Y, Zhao X, Thompson D (1997) Structural analysis of membrane-bound retrovirus capsid proteins. EMBO J 16: 1199–1213
11. Lebeau L, Lach F, Vénien-Bryan C, Renault A, Dietrich J, Jahn T, Palmgren MG, Kühlbrandt W, Mioskowski C (2001) Two-dimensional crystallization of a membrane protein on a detergent-resistant lipid monolayer. J Mol Biol 308:639–647
12. Ganser BK, Cheng A, Sundquist WI, Yeager M (2003) Three-dimensional structure of the M-MuLV CA protein on a lipid monolayer: a general model for retroviral capsid assembly. EMBO J 22:2886–2892
13. Dryden KA, Crowley CS, Tanaka S, Yeates TO, Yeager M (2009) Two-dimensional crystals of carboxysome shell proteins recapitulate the hexagonal packing of three-dimensional crystals. Protein Sci 18:2629–2635
14. Ganser-Pornillos BK, Cheng A, Yeager M (2007) Structure of full-length HIV-1 CA: a model for the mature capsid lattice. Cell 131:70–79
15. Turbedsky K, Pollard TD, Yeager M (2005) Assembly of *Acanthamoeba* myosin-II minifilaments. Model of anti-parallel dimers based on EM and X-ray diffraction of 2D and 3D crystals. J Mol Biol 345:363–373
16. Ganser-Pornillos BK, Yeager M, Sundquist WI (2008) The structural biology of HIV assembly. Curr Opin Struct Biol 18:203–217
17. Asturias FJ, Kornberg RD (1995) A novel method for transfer of two-dimensional crystals from the air/water interface to specimen grids. EM sample preparation/lipid-layer crystallization. J Struct Biol 114:60–66

Chapter 29

Processing of Electron Diffraction Patterns with the XDP Program

Kaoru Mitsuoka

Abstract

Electron crystallography using two-dimensional crystals of membrane protein can provide high-resolution structure of a membrane protein within a lipid bilayer. With this technique, it is advantageous to use electron diffraction patterns to collect accurate intensities of the structure factors at high resolution. Here we describe how to process diffraction patterns using the XDP program and show what parameters are used and how they are determined in the process. Finally, the procedures for merging the intensity data will be described briefly.

Key words: Electron crystallography, Diffraction pattern, Two-dimensional crystal, Image processing, Three-dimensional reconstruction

1. Introduction

Because the diffraction pattern is not modified by the contrast transfer function (CTF) and not sensitive to specimen movement during data collection, the derived intensities from the diffraction patterns are more accurate than those from images. Therefore, it is advantageous in electron crystallography to collect electron diffraction patterns from two-dimensional crystals of membrane protein for high-resolution structural analysis. Furthermore, the collected electron diffraction data allow for the use of molecular replacement methods in some cases, greatly reducing the time required for the structure determination by avoiding the time-consuming image collection and processing. However, relatively large crystals over 1 μm in diameter are needed to collect diffraction data and only a

Ingeborg Schmidt-Krey and Yifan Cheng (eds.), *Electron Crystallography of Soluble and Membrane Proteins: Methods and Protocols*, Methods in Molecular Biology, vol. 955, DOI 10.1007/978-1-62703-176-9_29,

handful of three-dimensional structures have been analyzed using electron diffraction data so far.

The electron crystallographic technique was developed to solve the high-resolution structure of bacteriorhodopsin (bR), a light-driven proton pump, and the atomic structure of bR was reported in 1990 (1). In the process, the programs of calculating intensities from electron diffraction patterns were created and improved (2, 3), and those programs were included in the so-called MRC package. From then, several atomic structures were solved by electron crystallography (4), and the MRC package was used to calculate almost all the atomic structures. However, because of the limited number of the solved structures, the algorithms are not fully automated and a trial-and-error approach to optimize the parameters is needed to use the package for the structural analysis of a target protein.

We started to use the MRC package from the structural analysis of light harvesting complex II (LHC-II) (5), and we added some GUIs (graphical user interfaces) to the programs to simplify the optimization of the parameters for our higher resolution structural analysis of bR (6). Then we continued to use the GUI program, called the XDP program, for the structural analysis of aquaporins (7) and MAPEG-family proteins (8). After we developed the XDP program, the algorithms of the programs in the MRC package were substantially modified for the auto-indexing and, therefore, the procedures of processing using the XDP program and the MRC package are now different.

Here, we show our typical procedures of processing using the XDP program. These protocols for the XDP program including the parameter optimization details will be helpful even for the researchers using the MRC package, which is widely used for structural analysis by electron microscopy. Because of the limited number of specimens used for the analysis, the parameter optimization is currently necessary for the appropriate results from a new specimen using both programs, and we believe the parameter optimization is easier using the GUI in the XDP program. It is highly likely that future efforts to improve the XDP program, or newly develop a more sophisticated program for electron diffraction processing, could be beneficial (9).

2. Materials

1. Films or CCD data (see Note 1) from EMMENU or Digital Micrograph, where the electron diffraction patters are recorded (see Note 2).
2. A scanner if films are used to record electron diffraction patterns (see Note 3). OD values are usually digitized using the scanner.

3. A look-up table for the digitized OD values for the estimated number of electrons. In the case of a CCD camera, the pixel values and the number of electrons to the pixel are mostly linear and so a linear look-up table can be used (see Note 4).
4. A computer (see Note 5). The XDP program and the MRC package are both developed on the VMS system initially and now for the UNIX system. Thus, a proper operating system must be installed on the computer.
5. An installed program package for processing. A program package for processing electron diffraction patterns and necessary utility programs like editors to modify scripts must be installed properly to start processing (see Note 6).

3. Methods

Here we describe how to process electron diffraction patterns by the XDP program. However, before starting the detailed description about the XDP program, we would like to explain about basic considerations in the processing, including the difference between the XDP program and the MRC package.

3.1. Basic Considerations

Processing of diffraction patterns requires adjustments of a wide range of parameters. One can use the same values for some parameters, like pixel sizes and unit cell dimensions, for the same type of samples and apparatus, while other parameters must be chosen for each pattern. Guidelines for the parameter selection will be given in the following sections, but the different specimen and apparatus can behave differently. Therefore, it is recommended to try a reasonable range of parameters to evaluate the performance of each algorithm at least once for a similar set of data. Then the chosen parameters can be used safely without deteriorating the quality of the calculated intensities.

For efficient processing of a large data set using appropriate parameters, a reasonable organization of the files is required. The choices of directory structure, timing of data backup, the job separation for error handling, and so forth, are important for proper processing. Filename convention is also important for data retrieval in the later stages. Improper handling of these organizational choices easily leads to data loss and wrong processing. It is recommended to determine the organization of the data in the early stage of the processing of a large data set, and some guidelines for the XDP program will be given in the following subsection.

Compared with the procedures described below for the XDP program, the procedures for the MRC package are simpler. The procedures for the MRC package consist of three steps: background estimation by BACKAUTO, indexing by AUTOINDEX, and the peak integration by PICKAUTO or PICKPROF. Expect some additional parameters in PICKPROF; the necessary parameters in the MRC package and the XDP program are similar.

3.2. Preparation of MRC Files for Processing

The data must be converted to an MRC-format file in the case of the MRC package and the XDP program. This procedure is called "C3" in the XDP program, while the procedures for processing by the XDP program are lined up in the menu from "C3" to "PICK" (see Fig. 1). A similar conversion is possible by the tif2mrc program in the MRC package, if you use TIFF files as your original data (see Note 7).

1. Make a directory and place the files for processing in the directory. Similar parameters can be used for data from the same crystalline batch and the same tilt angle, so those files are placed in the same directory (see Note 8).
2. Create parameter files, which record the parameters for processing of each file, by the "padd" command in the XDP program. In the following procedures, the parameter file is modified and the parameters in the file are used for processing.
3. In the "padd" command, you have to specify which CCD camera or films are used. If films are used, the lookup-table file

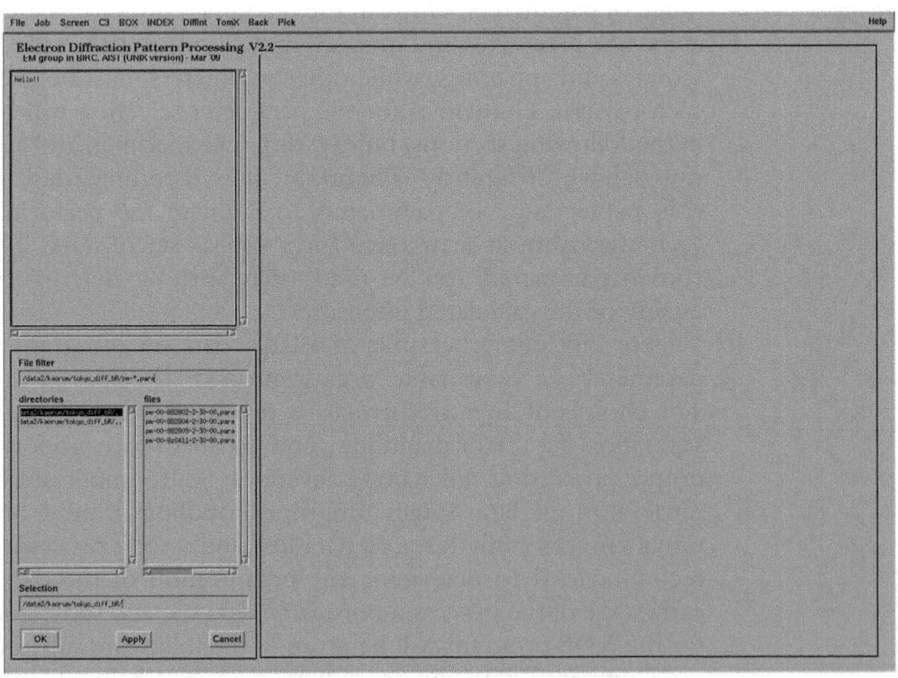

Fig. 1. Screen shot of the XDP program. The screen shot after the XDP program is started. The procedures for the processing are lined up in the menu from "C3" to "PICK".

is copied. If a CCD camera is used, then the linear lookup-table file is created and used.

4. In addition, you have to specify the sample, i.e., an abbreviation of the name of the protein, for processing by the two-letter sample code. Then the appropriate parameter file for the sample will be copied. The parameters conserved in the sample like lattice parameters are recorded in this file.
5. For future reference, tilt angle, film or file number, size and binning for CCD, scan number for films, and the exposure time are asked (see Note 9). These parameters are not used for processing.
6. Byte order and horizontal direction in the file are also asked (see Note 10). It is possible to compensate the fluctuation in the horizontal direction in the following procedures, which is implemented because of the fluctuation of the scanner light.
7. Then the actual processing to calculate the MRC files from each original file can be started from the "C3" menu. The log files are stored in the same directory with the data by the file extension, log (see Note 11).

3.3. Exclusion of the Beam-Stopper Area

To prevent blurring by the CCD camera, a beam stopper can be used. Then the procedure to exclude the beam-stopper area is needed for further processing. This procedure is called "BOX" in the XDP program.

1. Diffraction patterns can be displayed on the computer screen from the "Screen" menu (see Fig. 2). The file used for the display is a 1K×1K MRC-format file created by the file conversion by binning the original size of the MRC-format file (see Note 12).
2. Select four points to specify a box, which represent the beam-stopper area. The region inside the box will be excluded from the following processing by replacing the pixel values inside the box with values less than the minimum in the pattern (see Note 13).
3. Save the coordinates of the four points for the box in a text file and then start the actual replacement by executing a shell script from the "BOX" menu. The shell script is created by the XDP program and is stored in the same directory with the data as the file extension, com.
4. In addition, change the inner radius that is used to exclude the inner region from the following calculation. A scale, which can be used to determine the radius, can be displayed from the "Screen" menu. The inner radius and outer radius for processing are recorded in the parameter file and you can change the parameter file from the "File" menu (see Note 14).

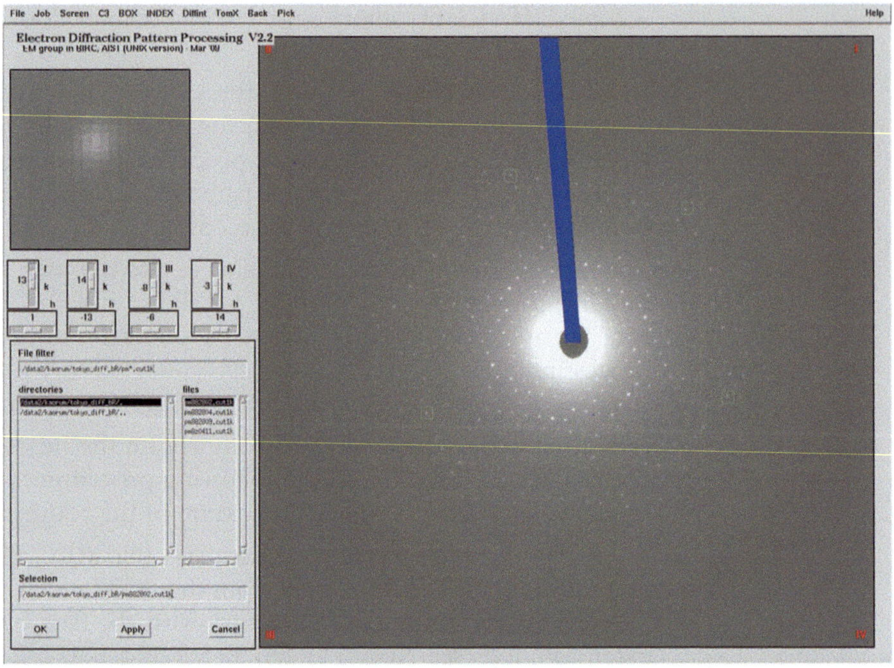

Fig. 2. Indexing of the four peaks in the XDP program. Four peaks in each sector were chosen and were indexed in the *left panels*. Each sector is indicated by Roman numerals from I to IV.

3.4. Indexing of the Peaks and Optimization of the Lattice Parameters

For further processing, you have to determine the lattice parameters to estimate all of the peak positions. To determine the lattice parameters, four peaks in each sector are selected and you have to index the four peaks in the "INDEX" menu. And then the lattice parameters are further optimized in the "Diffint" menu.

1. Display the diffraction pattern and adjust the brightness by selecting the proper minimum value from the "Screen" menu. Usually similar minimum values can be used for the diffraction patterns from the same grid and the same exposure time.
2. Select four sharp bright peaks at relative high resolution in each sector. Because the peak position is estimated by the center of gravity, a sharp bright spot can give a better estimate of the peak position. If you use the higher resolution peaks, the lattice vectors will be calculated more accurately from the peak positions.
3. Index the four peaks in each sector indicated by the Roman numerals I to IV (see Fig. 2). Then you can display the calculated peak positions from the "INDEX" menu. Check whether all the peaks are almost at the center of the calculated peak positions shown by boxes (see Note 15).
4. The mask boxes for lattice-parameter optimization can be displayed from the "Diffint" menu. Change the mask box size

and the minimum and maximum radius for the mask from the "File" menu, if needed. The mask size must be big enough to contain the entire spots inside the mask boxes and the maximum radius should be determined from the visible peaks.

5. Execute the script from the "Diffint" menu, and then the lattice parameters are optimized using the peak positions in the mask boxes (see Note 16). Spots with good signal-to-noise ratio, which are used for the lattice parameter optimization, can be shown using boxes from the "Diffint" menu.
6. The optimized parameters are recorded in the MRC-format file by the script created from the "TomX" menu. Because we modify the original file by this procedure, it is recommended to make a backup copy before the execution.

3.5. Background Subtraction and Peak Integration

After determining the lattice parameters, we use the background boxes between the peak positions to estimate the background values. The estimated background values are recorded in separate files and then they are used for background subtraction during the peak integration. The intensities are calculated by summing the values inside the peak boxes. These procedures are called "Back" and "Pick" in the XDP program.

1. At first, several parameters for background estimation should be determined in the popup panel from the "File" menu. Inner radius for the background estimation is determined to exclude the high-background region close to the center.
2. In the background estimation, radial average and a smoothed surface are calculated and recorded in a text file and an MRC-format file, respectively (see Note 17). For the smooth-surface calculation, the parameter, NCP, is used to specify the smoothness of the surface. After specifying these parameters, execute the background estimation.
3. There are three types of positions for background boxes and it can be specified by the parameter, NTYPE. Two types have six background boxes, while another type gives four background boxes at centers of lattice. This parameter can be determined from the crystal type and the tilt angle. Larger areas should be included for the estimation but overlapping of the boxes should be avoided.
4. The background box sizes are determined from the parameters, NXB and NYB. We usually use square boxes. When the specimen was tilted, the spots can become bigger in the perpendicular direction to the tilt axis, and then the parameters NXBT and NYBT are used to select the smaller size to exclude those bigger spots. These parameters to exclude the spots are determined by visual examination on the screen. Repeat the background estimation several times until you are satisfied with the results.

5. After the estimation of the background values, execute peak integration from the "Pick" menu. Additional parameters of peak box sizes, NXM and NYM, are first determined by visual inspection (see Notes 18 and 19). For the non-tilt specimen, the same NXMT and NYMT should be used, but these two parameters can be increased for tilted specimen.
6. In addition, use the two parameters, COUNT and STEP, to change the box sizes systematically. STEP is the increment size difference between each peak-integration execution and COUNT is the number of executions. For example, by STEP = 2 and COUNT = 5 with NXM = NYM = 12, you can check the peak box sizes of 8, 10, 12, 14, 16. The actual box size can be checked on the screen from the "Screen" menu.
7. Determine the appropriate values for the parameters, NXM and NYM, from the plot between the logarithm of average intensities and the resolution, which can be displayed from the "Screen" menu (see Fig. 3). From this plot, you can check the saturation of maximum intensity and slowing down of the increase of the sum of the intensities at the appropriate box sizes (see Table 1 and Note 20).
8. In addition, the outer radius for the peak integration should be determined from the linear part of this plot. The parameter,

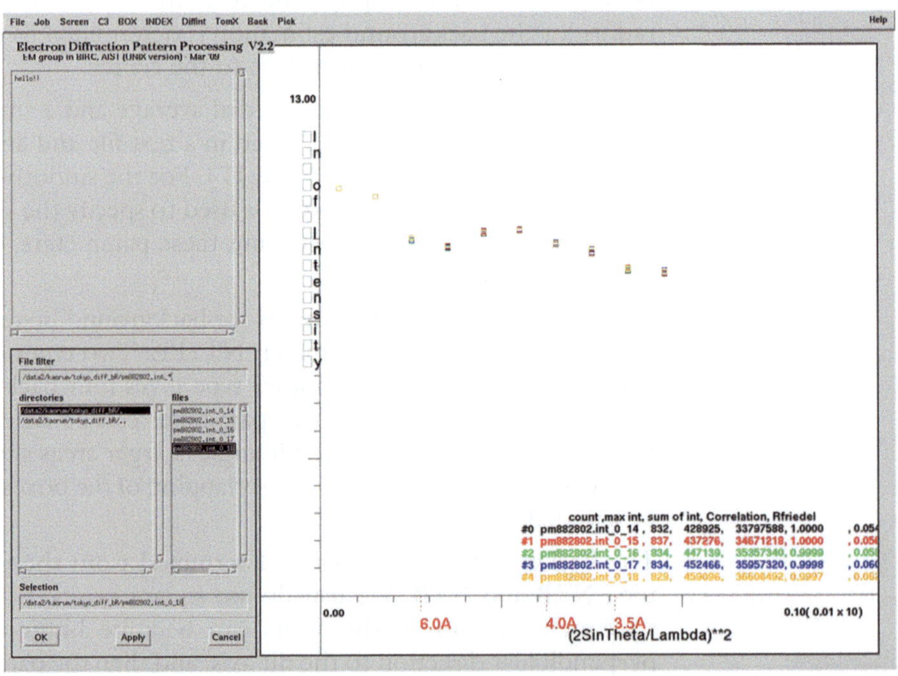

Fig. 3. Plot between the logarithm of average intensities and the resolution (Wilson plot). Plots between the logarithm of average intensities and the resolution for five different peak box sizes were shown. At those box sizes, the maximum intensities are already saturated and the increase of the sum of the intensities becomes slower from the box size 12. Thus, the appropriate box size was estimated as 12.

Table 1
Some statistical values calculated using the different sizes for the peak integration were listed

Peak box size	Number of reflections	Maximum intensity	Sum of intensities	Friedel R-factor
10	832	428,925	33,797,588	0.054
11	837	437,276	34,671,218	0.056
12	834	447,139	35,357,340	0.058
13	834	452,466	35,957,320	0.060
14	829	459,096	36,606,492	0.062

Please notice that the increase of maximum intensity become slower at the box size of 12

INTMAX, is used to specify the maximum value used for peak integration (see Note 21).

9. Final calculated intensities are checked by showing the intensities with corresponding box sizes on the background-subtracted diffraction patterns. In addition, the Friedel pair difference and signal-to-noise ratio of each spot can be shown and inspected visually from the "Screen" menu. Repeat the peak integration several times until you are satisfied with the results.

3.6. Merging the Intensity Data

For merging, the XDP program does not have a GUI but it provides semi-automated procedures for making scripts from a list of parameters. Here we describe how we change the list of parameters in the merging cycles.

1. Text files with indexes and the averaged Friedel-pair intensities for each diffraction pattern are used for the merging (see Note 22). The MERGEDIFF program in the MRC package is used to scale and merge the data.
2. The first scaling is done by using the best non-tilt diffraction pattern as the reference (see Note 23). Only low-resolution data, typically up to 5 Å resolution, are used for the scaling. Then all the non-tilt data are merged and used as a reference for the scaling of low-tilt data.
3. The scaled data are merged and then higher tilt data will be scaled using the merged data as a reference. After merging all the data, the systematic searches of the appropriate tilt angle and axis for low-tilt patterns up to 20° are applied (see Notes 24 and 25).
4. Gradually higher resolution data are included in the scaling with global temperature factors after several cycles of scaling

and merging. Anisotropic temperature factors to the tilt direction should be used for the high-tilt data.

5. In these cycles, the diffraction patterns showing large merging R-factors are excluded from the lists and thus comparable quality of each diffraction pattern can be assured. In addition, each intensity in a diffraction pattern should be carefully checked and some reflections with noise should be excluded.
6. Then the lattice line fitting by the SYNCFIT program in the MRC package is used to calculate the equally spaced intensities from the sampled data sets. The calculated lattice lines can be used by the MERGEDIFF program as a reference (see Note 26). The lattice lines can be plotted as a Postscript file from the SYNCFIT program. Several cycles to create the references by the SYNCFIT program are needed for refinement.
7. The intensity data can be converted to the amplitude data by the TRUNCATE program in the CCP4 package (10). Then the CCP4 package can be used to calculate the three-dimensional structure with phases from the images.

4. Notes

1. The wide dynamic range of slow-scan CCD cameras is beneficial. A CCD camera with a 1K× 1K pixel size can be used but a larger CCD camera will give more accurate intensities.
2. For recording diffraction pattern, a selective area aperture that fits the crystalline area is used to maximize the signal-to-noise ratio.
3. To scan diffraction patterns recorded on films, a point-illuminated scanner is preferable to digitize the small spots from diffraction patterns accurately.
4. Anti-blurring of some CCD cameras sets the maximum value below which a linear relationship can be assumed.
5. A recent-model personal computer with typical disk space is usually sufficient for the processing of electron diffraction patterns.
6. For the XDP program, Open Motif is needed in addition to the typical linux packages.
7. In the case of the XDP program, data-only files from the Perkin-Elmer scanner or Digital Micrograph and TIFF-format files from EMMENU can be converted.
8. Creating backup copies of the original files is recommended. Currently we use DVD-R for making backups, and just an additional hard disk might be convenient.

9. For example, the optimal exposure time is different depending on the crystal sizes but the longer exposure time could give bigger radiation damage. Thus, these parameters can be of interest in the later stage.
10. If the byte order or the horizontal direction is not correct, the converted file cannot give a proper view and therefore you will easily find the proper answer by trial-and-error.
11. The log files can be used to check the results, the error messages, etc.
12. A file with a smaller size is preferable to view on the screen promptly as well as for fast manipulation.
13. The pixel values outside the range specified will be excluded from the processing. The maximum is used to exclude the noise from cosmic rays and so on.
14. The outer radius should be determined according to the resolution of each diffraction pattern.
15. No box will appear when the sum of the *h* index or the *k* index is 0, because of our algorithm for the grid calculation. This is easily solved by selecting different peaks.
16. The parameters describing the pincushion effect due to the curvature of the Ewald sphere and random distortions are considered in the following procedures.
17. Both radial average and smooth surface calculations can be avoided by clicking the corresponding buttons in the popup panel.
18. A suggested value for peak box sizes is given by the AUTOINDEX program in the MRC package.
19. The same parameters as for the background estimation were also used for peak integration.
20. Rapid increase of the Friedel R-factor is another sign of inappropriate box size. The Friedel R-factors could be very different and may vary significantly depending on specific conditions, while the same type of crystals from a protein usually gives a similar value for the non-tilt specimen.
21. The boxes having larger pixel values than INTMAX are excluded from the calculation, because the very large values indicate noise from, e.g., X-rays.
22. The observed intensities may have small negative values. One source of this systematic error is thermal diffuse scattering (11).
23. Baldwin and Henderson (2) used the high-tilt data to obtain the initial reference, while Wang and Kühlbrandt (12) started from low-tilt data. We followed the latter strategy.
24. For non-tilt diffraction patterns, a 5°-step search of the tilt axis and 1°-step search of the tilt angle are needed. For 20° tilt data, a 1°-step search of the tilt axis was used.

25. You can also use the tilt angle and axis refinement option in the MERGEDIFF program with enough data after the systematic search.
26. It is also possible to use LATLINE (13) to combine the experimental phases from the images with the intensities from diffraction patterns simultaneously.

References

1. Henderson R, Baldwin JM, Ceska TA, Zemlin F, Beckmann E, Downing KH (1990) A model for the structure of bacteriorhodopsin based on high resolution electron cryo-microscopy. J Mol Biol 213:899–929
2. Baldwin JM, Henderson R (1984) Measurement and evaluation of electron diffraction patterns from two-dimensional crystals. Ultramicroscopy 14:319–336
3. Ceska TA, Henderson R (1990) Analysis of high-resolution electron diffraction patterns from purple membrane labelled with heavy-atoms. J Mol Biol 213:539–560
4. Hite RK, Raunser S, Walz T (2007) Revival of electron crystallography. Curr Opin Struct Biol 17:389–395
5. Kühlbrandt W, Wang DN, Fujiyoshi Y (1994) Atomic model of plant light-harvesting complex by electron crystallography. Nature 367:614–621
6. Kimura Y, Vassylyev DG, Miyazawa A, Kidera A, Matsushima M, Mitsuoka K, Murata K, Hirai T, Fujiyoshi Y (1997) Surface of bacteriorhodopsin revealed by high-resolution electron crystallography. Nature 389:206–211
7. Murata K, Mitsuoka K, Hirai T, Walz T, Agre P, Heymann JB, Engel A, Fujiyoshi Y (2000) Structural determinants of water permeation through aquaporin-1. Nature 407:599–605
8. Holm PJ, Bhakat P, Jegerschöld C, Gyobu N, Mitsuoka K, Fujiyoshi Y, Morgenstern R, Hebert H (2006) Structural basis for detoxification and oxidative stress protection in membranes. J Mol Biol 360:934–945
9. Philippsen A, Schenk AD, Signorell GA, Mariani V, Berneche S, Engel A (2007) Collaborative EM image processing with the IPLT image processing library and toolbox. J Struct Biol 157:28–37
10. French S, Wilson K (1978) Acta Crystalog Sect A 34:517–525
11. Grigorieff N, Ceska TA, Downing KH, Baldwin JM, Henderson R (1996) Electron-crystallographic refinement of the structure of bacteriorhodopsin. J Mol Biol 259: 393–421
12. Wang DN, Kühlbrandt W (1992) Three-dimensional electron diffraction of plant light-harvesting complex. Biophys J 61:287–297
13. Adard DA (1983) A least-squares method for determining structure factors in three-dimensional tilted-view reconstructions. J Mol Biol 167:849–852

Chapter 30

Future Directions of Electron Crystallography

Yoshinori Fujiyoshi

Abstract

In biological science, there are still many interesting and fundamental yet difficult questions, such as those in neuroscience, remaining to be answered. Structural and functional studies of membrane proteins, which are key molecules of signal transduction in neural and other cells, are essential for understanding the molecular mechanisms of many fundamental biological processes. Technological and instrumental advancements of electron microscopy have facilitated comprehension of structural studies of biological components, such as membrane proteins. While X-ray crystallography has been the main method of structure analysis of proteins including membrane proteins, electron crystallography is now an established technique to analyze structures of membrane proteins in the lipid bilayer, which is close to their natural biological environment. By utilizing cryo-electron microscopes with helium-cooled specimen stages, structures of membrane proteins were analyzed at a resolution better than 3 Å. Such high-resolution structural analysis of membrane proteins by electron crystallography opens up the new research field of structural physiology. Considering the fact that the structures of integral membrane proteins in their native membrane environment without artifacts from crystal contacts are critical in understanding their physiological functions, electron crystallography will continue to be an important technology for structural analysis. In this chapter, I will present several examples to highlight important advantages and to suggest future directions of this technique.

Key words: Structural physiology, Native membrane environment, Physiologically relevant structural information, Charge status analysis, 2D crystal contacts, Physiological lipid and ionic setting, Helium-cooled specimen stage, Experimental phase data

1. Introduction

In 1975, a real image of a membrane protein, bacteriorhodopsin (bR), was first generated by Henderson and Unwin using electron crystallography (1). Later, Henderson et al. determined the first atomic structure of bR by electron crystallography and demonstrated the power of this method (2). Whereas the method of

Ingeborg Schmidt-Krey and Yifan Cheng (eds.), *Electron Crystallography of Soluble and Membrane Proteins: Methods and Protocols*, Methods in Molecular Biology, vol. 955, DOI 10.1007/978-1-62703-176-9_30,

electron crystallography is extremely powerful for structural studies of membrane proteins, its spreading in the structural biology field has been rather slow because of several practical limitations. Firstly, unlike three-dimensional (3D) crystallization, in which one can easily check crystallization conditions by an optical microscope at a rate of a few second per condition, electron crystallography requires tedious treatments to find a good two-dimensional (2D) crystallization condition. Samples for 2D crystallization have to be stained on an EM grid and observed by an electron microscope (EM) for identifying the most optimal crystallization conditions. For an experienced scientist, it requires at least half an hour or more to check one crystallization condition for 2D crystals. Secondly, specimen preparation for cryo-EM requires many trials so that the condition can be optimized for high-resolution data collection. To find the best condition of specimen preparation, many specimen exchanges are inevitably required, which in turn requires a very efficient cryo-transfer system and also stable cryo-stage. Such a suitable instrumental setting is, however, rather difficult at this moment. Only a few instruments, some of them specifically developed for such a purpose, are available for high-resolution data collection (3). Thirdly, in order to reconstruct a 3D structure of a 2D crystal, it is necessary to tilt specimen to a high angle, such as over 60°, to minimize the influence of the missing core. It is, however, extremely difficult to collect images at highly tilted conditions. Lastly, it requires tremendous experience to properly analyze data collected from 2D crystals. The computer packages that are used for such analysis have not been as automated and easy to use as those used in X-ray crystallography (2). Furthermore, the computer analysis and the specimen preparation for data collection have to be tightly concerted for properly analyzing structures at the best possible resolution according to quality as well as type of 2D crystals.

Nevertheless, electron crystallography could become the core method for studying functional structures of membrane proteins because of a number of reasons. Most importantly, structures can be analyzed in membranes that provide a similar environment for membrane proteins as under native conditions. Also, structures can be analyzed even from small crystals, while the resolution is strongly related with the quality of crystals at this moment. Phases are determined directly from images, which give a better map than that of X-ray crystallography at similar resolution (Fig. 1) (4). Since the hydrophilic surfaces of membrane proteins within 2D crystals are exposed, there is no crystal packing artifact in the direction perpendicular to the membrane plane, which could be very important when studying structures such as gap junction channel. Membrane proteins in 2D crystals are also still accessible by ligand after crystal formation, as demonstrated by studies of gating

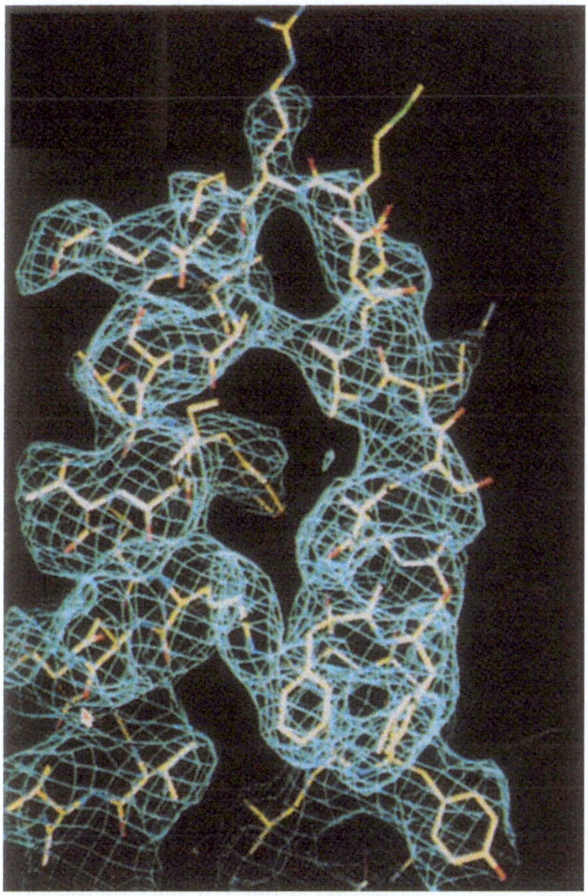

Fig. 1. Density map, indicated by the electron-potential map around the loop between the E and F helices of bacteriorhodopsin, was analyzed at 3 Å resolution by electron crystallography. The refined atomic model is represented by a ball-and-stick model.

mechanism of Nicotinic Acetylcholine receptor (AChR) by Unwin (5). Considering all these important features helps us to contemplate future directions of electron crystallography.

2. Obtaining Physiologically Relevant Structural Information

A main goal of electron crystallography, as of any other method, is to obtain physiologically relevant structural information of integral membrane proteins. An important advantage of electron crystallography is that integral membrane proteins under study are embedded in their native or near native membrane environment. The interaction between membrane lipids and integral membrane proteins is also a very important factor to consider when addressing mechanistic questions of integral membrane proteins. The importance of analyzing membrane protein in membrane could be clearly seen from examples of water channel studies.

Water is the most abundant molecule in any cell on the earth. The ordinary cell membranes, therefore, require an efficient water channel. The first water channel was identified and named aquaporin-1 (AQP1) (5). While the channel must be very efficient for water, it has to be impermeable to any ions including protons, because the cell membrane exquisitely regulates entry and exit of ions for controlling ion concentration and its dynamic changes which are strongly related with cell signaling functions. Similarly, pH regulation in a cell is well known to be crucially important for cell functions, such as, signal transduction, proliferation, and survival of cells. The AQP1 molecule was found to attain effective water selective transport by keeping strict water selectivity, providing many questions. An atomic model analyzed at 3.8 Å resolution by electron crystallography gave an answer to many of these questions and revealed a molecular contrivance of a water selective channel (6). The atomic model is interestingly the first structure of a membrane protein from a human source. For accomplishing effective water channel functions, the structure showed peculiar structural determinants including an unusual fold, which was named AQP fold (Fig. 2).

The critical function of AQP1 is a high water permeability of about three billion water molecules per channel per second. Almost all residues within a central 20 Å zone in the pore are highly hydrophobic. A narrow part of the pore with a diameter of about 3 Å is located in the middle of the membrane. Here loops B and E interact with each other, in particular proline 77 and 193 (AQP1) of the conserved NPA sequences in all water channels. Based on structure analysis by electron crystallography, a model named hydrogen bond isolation mechanism was proposed for explaining the high water selectivity while at the same time retaining the high speed water permeability (6). Contiguous hydrogen-bonded water chains are known to be efficient proton conductors. AQP must prevent proton conduction along its pore, to maintain the proton gradient across the cell membrane that serves as a major energy storage mechanism and also signal regulation mechanisms for the above-mentioned cell functions. Water regulation without ion and proton permeation is crucially important for every cell and therefore for all life forms on earth. Unexpected structural features, such as the AQP fold and the hydrophobic channel wall were revealed to facilitate the water transport through the channel. All of these were impossible to predict without structure analysis.

Another water channel, AQP4, is the predominant water channel in the mammalian brain. The AQP4 structure was analyzed by electron crystallography of double-layered, 2D crystals which were prepared from AQP4 molecules expressed in Sf9 cells (7). Analysis

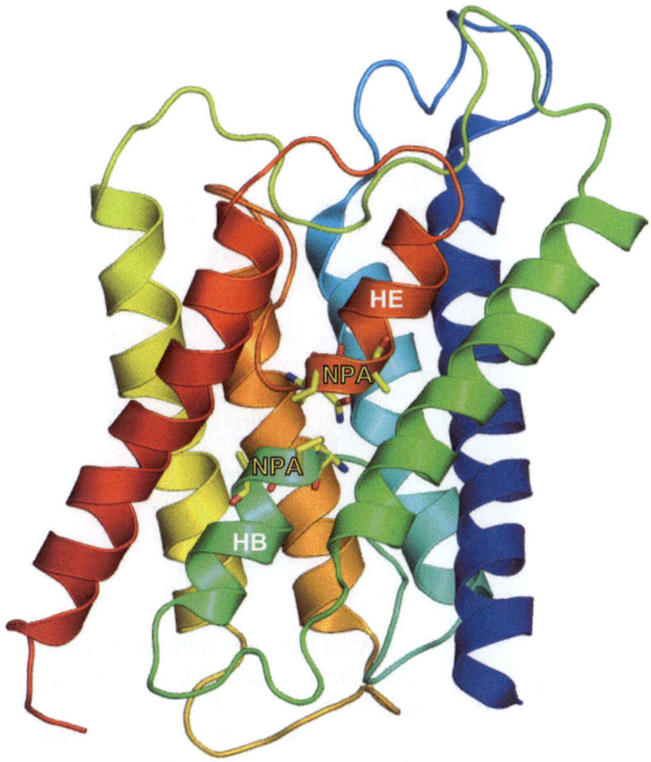

Fig. 2. Ribbon model of AQP1 structure. The side view of a monomer of the AQP1 molecule showing the two short helices formed in loop B and E.

of the AQP4 structure was complicated by variations in the double-layered 2D crystals in terms of lateral alignment and distance between the two layers. Electron diffraction data proved to not be sufficiently sensitivity to select one specific crystal type among different crystal variants, despite the high resolution of 3.2 Å. Phase data extracted from images even at medium resolution, on the other hand, were sensitive enough to discriminate between crystals with different arrangements of the two layers. The use of a helium-cooled electron microscope (8) and the carbon sandwich specimen preparation technique, which significantly increases the yield of good images (9), allowed us to collect a good image of each crystal that produced a high-resolution diffraction pattern. Classification based on the image data, which provided phase information to better than 6 Å resolution, identified one predominant crystal type, which accounted for about 70% of the analyzed crystals that yielded a high-quality electron diffraction pattern. This example illustrates another strength of electron crystallography which provides image data together with diffraction data, hence the best crystal type could be selected and resulted in a 3D structure even from a

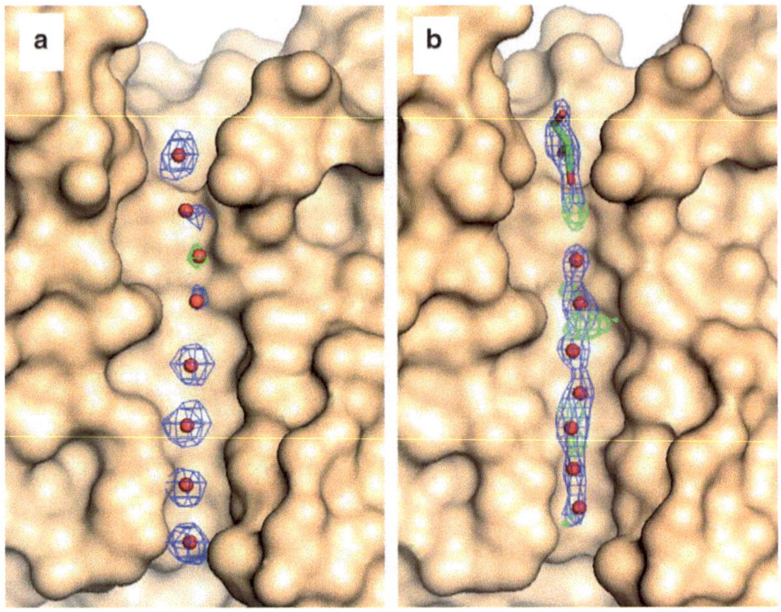

Fig. 3. Water molecules in the water channel AQP4 analyzed by electron crystallography at 2.8 Å resolution (**a**) and X-ray crystallography at 1.8 Å resolution (**b**). Electron crystallography results showed discriminate densities of water molecules in the channel, while X-ray could show only one continuous water density even at higher resolution, presumably because the structure of AQP4 molecules by electron crystallography was analyzed in a membrane and the 3D crystal structure was obtained from AQP4 in micelles. The lipid bilayer, which gives characteristic dielectric constant from $\varepsilon = 2$ to $\varepsilon = 80$ in 2D crystals, enhances the dipole moments of the two short helices, while the helical dipole moments are weaker in detergent micelles of 3D crystals. Therefore, oxygen atoms of water molecules in the channel are forced to face the PNA side by the strong electrostatic field in the channel, when the water channel is in a membrane. The counter-intuitive notion could be attributed to the difference of the helical dipole of these two short helices.

mixture of different types of crystals. This type of sample would provide a tremendous challenge for X-ray protein nanocrystallography (10).

The hydrogen bond isolation mechanism was proposed to explain how AQPs are impermeable to protons despite their very fast water conduction. However, inconsistent models were proposed and the mechanism is still controversial. Therefore, the resolution of the structure of AQP4 was improved to 2.8 Å, which was achieved by electron crystallography of double-layered two-dimensional crystals. Seven individual water molecules were identified in the channel as shown in Fig. 3a (11), while the map by X-ray crystallography showed only continuous densities of water molecules in the channel even at 1.8 Å resolution (Fig. 3b) (12). In addition, the Fo–Fc map of electron crystallography showed an additional spherical density at the ar/R constriction site, which forms the narrowest part of the channel (green in Fig. 3a). Since the side chains

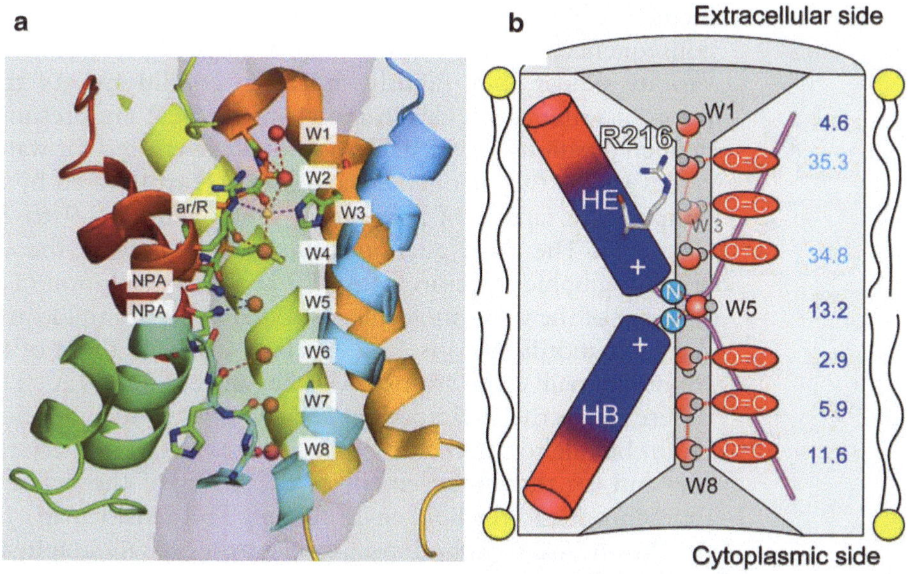

Fig. 4. Ribbon and schematic model including eight water molecules in the AQP4 channel based on the structural analysis by electron crystallography. The eight water molecules in the AQP4 channel are in a single-file arrangement as indicated by W1 to W8 (**a**). The strong dipole moments of the two short helices, HB and HE of AQP4, that are shown as *red* (negative potential) and *blue* (positive potential) in the lipid bilayer, which gives the characteristic dielectric constant, and carbonyl groups are schematically indicated together with eight water molecules (**b**). The temperature (B-) factors of the water molecules are indicated at the *right side* of these water molecules in the schematic figure (**b**). From the measured distances between successive water molecules in the channel, all water molecules appear to form hydrogen bonds with their neighbors (*red dotted lines*), except for the water molecule at the NPA site and the one below it. These two water molecules (W5 and W6) thus seem to be separated from the other water molecules in the channel, lending support to the hydrogen bond isolation mechanism.

of AQP4 around the ar/R region were represented by clear density and the atoms of the protein molecule had low temperature factors in this region, an eighth water molecule in the channel was assigned to the spherical density at the ar/R site (W3 in Fig. 4a). The narrow diameter at this constriction would make it an unfavorable position for a water molecule, potentially explaining the weak density for the water at this position (Fig. 3a). The two neighboring water molecules on either side of the ar/R constriction, which form hydrogen bonds with the unstable water molecule in the constriction, showed higher temperature factors (35 Å^2) compared to those of all the other water molecules in the channel (2.9–13.2 Å^2) which are shown in the right side of the schematic figure of the channel (Fig. 4b). The density map obtained by electron crystallography clearly, and counter-intuitively at such relatively low resolution, resolved eight spherical densities that could be assigned to water molecules. In the structure analysis by electron crystallography, more than 1,000 diffraction patterns were collected, but only the 199 very best patterns were selected, so that the final data set includes only information from the very best crystals (11). As a consequence, the map represents an average of the specimen under various similar, but slightly different, conditions. As the map resolved water molecules in the channel, these are presumably in defined positions where stable positions are created for water molecules within the channel. This notion is corroborated by the low temperature factors of the waters, ranging from 2.9 to 35.3 Å^2 (Fig. 4b). The same phenomenon was observed in the electron crystallographic structure of AQP0 at 1.9 Å resolution (13). The B-factor of the water molecule associated with the amide groups of the NPA motifs (W5) is 13.2 Å^2 (Fig. 4), while those of the surrounding main chain atoms are in the range of 30–40 Å^2. The large difference in the B factors between the water and protein atoms might be due to the characteristic features of electron crystallography and/or the enhancement of the effect of the helical dipole moments due to the lipid environment. The density map of AQP4, AQP0, and bacteriorhodopsin by electron crystallography actually resolved lipid molecules. The dipole moments of the two short helices, HB and HE of AQP4, that are schematically shown as red (negative potential) and blue (positive potential) in Fig. 4b, are stronger in the lipid bilayer which gives characteristic dielectric constant, such as $\varepsilon = 2 \sim 80$, while the helical dipole moments are weaker in detergent micelles. Therefore, oxygen atoms of water molecules in the channel are forced to face the PNA side by the strong electrostatic field in the channel as shown in Fig. 4b, when the water channel is in the membrane such as in electron crystallographic studies. The eight water molecules in the AQP4 channel are in a single-file arrangement as indicated by W1 to W8 in Fig. 4. The inside surface of water channels is largely hydrophobic except for a narrow hydrophilic belt or a stepping stone arrangement formed by the oxygen atoms of the main chain carbonyl groups of Gly209, Ala210, Ser211, His95, Gly94, and Gly93 and the nitrogen atoms of the side chain amide groups of Asn213 and Asn97 of the NPA motifs (11). The line of mechanically stable carbonyl groups provides "a guide rail or stepping stones" of hydrogen bonding partners for the hydrogen atoms of the permeating water molecules as illustrated in Fig. 4b. From the measured distances between successive water molecules in the channel, all water molecules appear to form hydrogen bonds with their neighbors (red dotted lines in Fig. 4), except for the water molecule at the NPA site and the one below it. These two water molecules (W5 and W6) thus seem to be separated from the other water molecules in the channel, lending support to the hydrogen bond isolation mechanism. Each water molecule in the single file can thus form two or three hydrogen bonds. Since water in bulk solution usually forms three or four hydrogen bonds with neighboring water molecules, water molecules entering the channel only have to sacrifice a single hydrogen bond, an energy cost of about 3 kcal. The arrangement of carbonyl and amide groups in the AQP4 channel thus dramatically lowers the energy barrier for water molecules which could possibly leap from stepping stone to stone at the positions of W1–W8 and allow for the very fast water permeation through the otherwise hydrophobic channel.

Glial cells contain characteristic orthogonal arrays in the plasma membrane, which are especially prominent in glial end feet surrounding vascular capillaries in the brain. Immunogold labeling experiments showed that these arrays consist of AQP4 (14). While AQP4 and AQP1 both function as very fast water-selective pores, AQP4 has distinctive biological characteristics as it forms orthogonal arrays in intact membranes. Furthermore, AQP4 exists in glial cells as a full-length protein starting with Met1 (AQP4M1) and an alternative shorter splicing isoform that starts with Met23 (AQP4M23) (15). Based on structural analysis by electron crystallography of AQP4M23 in a lipid bilayer, the characteristic intermolecular interactions were demonstrated to be mediated by Arg108 and Tyr250 and also by hydrophobic interactions of Gly157, Trp231, and Ile239. However, the structure of an AQP4S180D mutant analyzed at an improved resolution of 2.8 Å revealed the insertion of lipid molecules between the tetramers, and thus array formation could be mediated by hydrophobic interaction with lipid molecules. This lipid insertion is consistent with experimental data that crystallization of AQP4M23S180D required larger amounts of lipid molecules (11). The arrangement of AQP4 tetramers in 2D crystals has the same dimensions as the orthogonal array. These interactions analyzed by electron crystallography may suggest an array stabilization mechanism of AQP4. Revealing this type of mechanism of molecular assembly in biological cells is another strength of electron crystallography.

Certain characteristics of electron crystallography could be applied to the structural study of disordered domains. A typical example was yet again revealed in structural studies of the water channel AQP4. While the N-terminal tail structure of the AQP4M23S180D mutant was less clearly observed in the high-resolution map of 2.8 Å resolution, the density was more clearly observed in a lower resolution map at 10 Å resolution as shown in Fig. 5 (16). The density suggests the interaction among N-termini of AQP4 acts as a stabilizing factor for the formation of crystalline arrays which are characteristic orthogonal arrays formed by an isoform, AQP4M23. For understanding the array structure, the high-resolution structure of four tetramers of AQP4 including lipid molecules is shown together with the low resolution density map at the boundary of the tetramer (Fig. 5). The interacting bridge structure at the N-termini explains how lipid modification on Ser13 and Ser17 at the N-terminus of AQP4M1 effectively disrupts the arrays. The bridge formed by the N-termini is forced down to the lipid surface by the lipid modification on these serine residues and

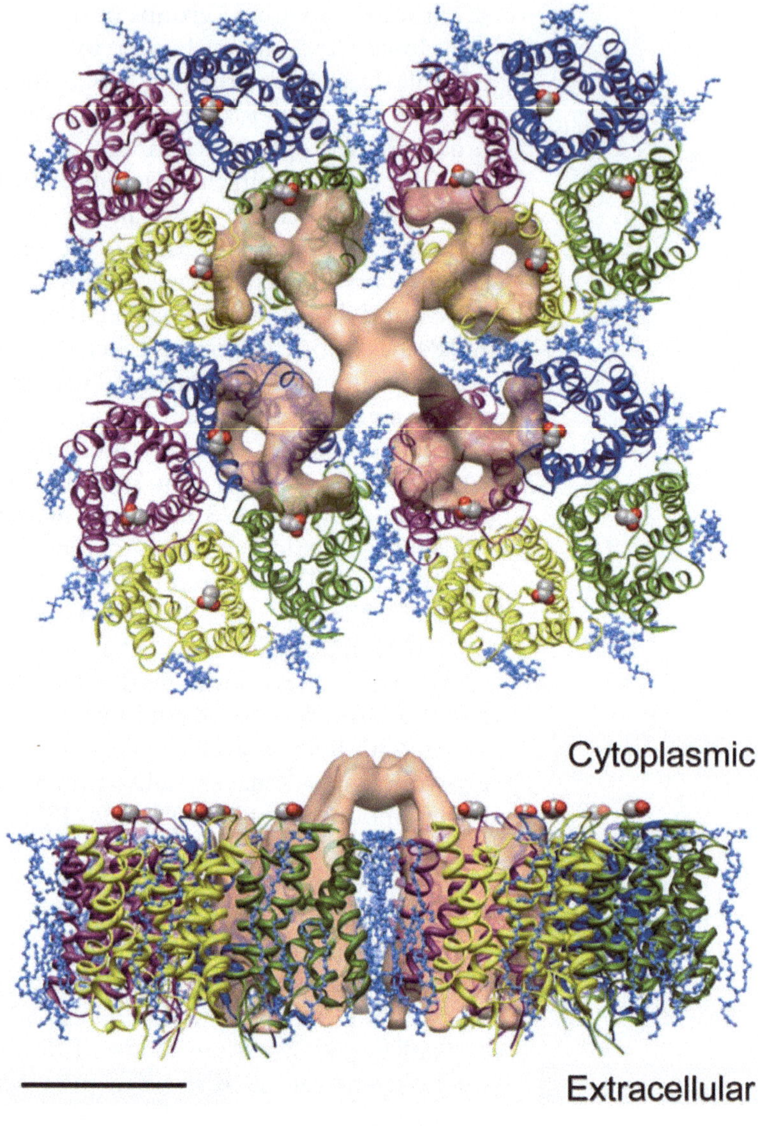

Fig. 5. Density map of the AQP4M23S180D mutant calculated based on image data at a lower resolution of 10 Å and structure of the mutant with lipid molecules analyzed by electron diffraction at 2.8 Å resolution. The density of the N-termini of AQP4 was observed in the lower resolution map, which is different from the higher resolution one. The density map of the interacting bridge formed by the N-termini was essentially free of noise even at a contour level of 0.65σ shown by the *pink* surface.

such a bridge movement pushes away the AQP4 tetramers and inhibits the tetramer–tetramer interaction, which is the key interaction for array formation. This feature is yet another strength of electron crystallography, in which phase data can be calculated directly from images as shown by the N-terminal bridge structure of AQP4 arrays (Fig. 5).

Structure analysis of AQP4 of double-layered 2D crystals revealed that the molecule contains a short 3_{10} helix in an extracellular loop, which mediates weak but specific interactions between AQP4 molecules in adjoining membranes. The analysis also revealed that lipid molecules of phosphatidylethanolamine, which ordinarily exist in the inner leaflet of the cell membrane, interact with the extracellular side of AQP4 molecules in the adjoining membrane and contribute to form double-layered packing of AQP4 (8). This finding by electron crystallography suggests a previously unexpected role for AQP4 in cell adhesion. This notion was corroborated by expression of AQP4 in L-cells, which resulted in clustering of the cells. Although further experiments are needed to elucidate the interplay of the two functions in AQPs and potentially other membrane channels with adhesive properties, we have proposed the name "adhennels" for adhesive water and ion channels (8). Analysis of membrane junctions could be achieved by electron crystallography. Therefore, electron crystallography could be a key technology for structural studies of membrane junctions, such as gap junction which will be discussed in a later section.

3. Structure Analysis from Image Data Without Diffraction Data

Another advantage of electron crystallography is that structures can be analyzed even from relatively small or poor quality crystals that give no electron diffraction patterns, although the resolutions obtained from such analyses are restricted by the quality of the crystals. A typical example is the structural analysis of H^+, K^+-ATPase as shown in Fig. 6, which shows the 3D structure of H^+, K^+-ATPase with bound acid suppressant at 7 Å resolution (17). It was difficult to grow large 2D crystals of these pump molecules, but rather easy to obtain tubular crystals which were too small for electron diffraction. Even with such small tubular crystals, several structures with inhibitors were effectively analyzed by electron crystallography (17–19). The gastric H^+, K^+-ATPase can generate a million-fold proton gradient across the gastric membrane. Based on structures determined only at 6.5 Å resolution, together with functional studies, the molecular mechanism for forming such a large proton gradient was explained by the finding of a ratchet function on the N-terminal part of the β-subunit (18). All of these analyses were accomplished only from image data without diffraction data.

This can be easily explained as direct calculation of phase data from images gives a better quality of the density map, when compared with map qualities based on indirect phase data calculated from diffraction data at the same resolution. This comparison is true for a large resolution range, from low to high resolution analysis, such as in the structure analyses of bR at 3 Å resolution (4). We anticipate more membrane protein structures analyzed only from images and without diffraction data.

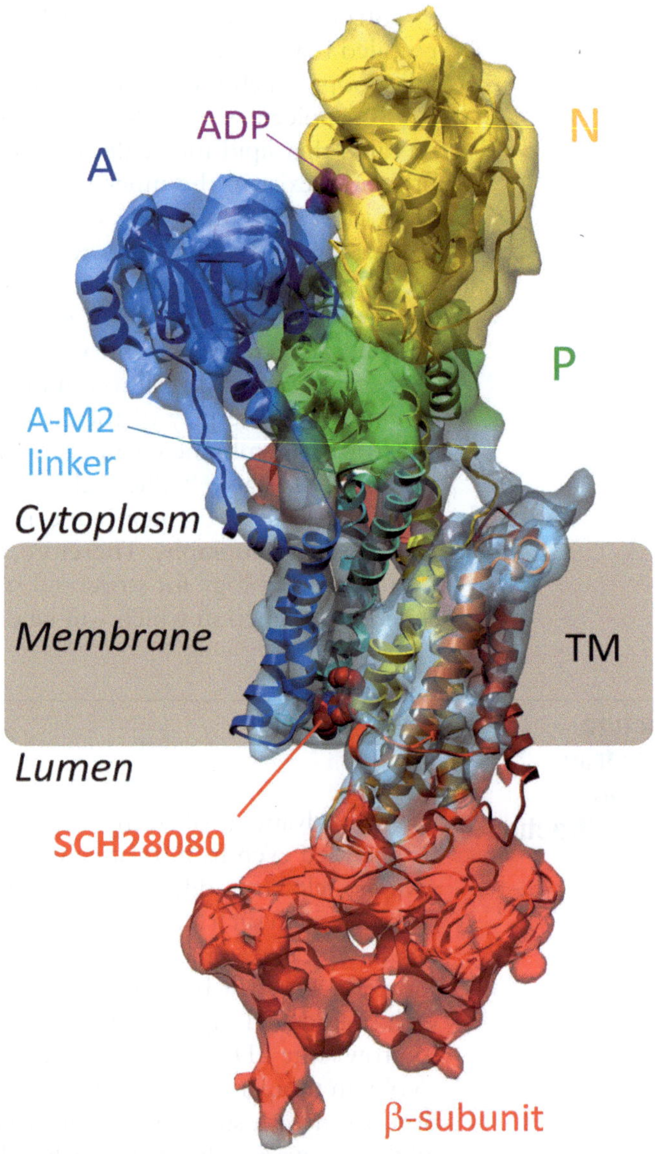

Fig. 6. An acid suppressant, SCH28080, bound to the E_2P structure of H^+, K^+-ATPase analyzed at 7 Å resolution. The molecular surface represents the EM density map (contoured at 1σ) with a superimposed homology model of H^+, K^+-ATPase (*ribbon*). Bound SCH28080 at the luminal cavity and ADP at the N domain are shown as *red* and *purple spheres*, respectively. The *yellow-brown box* indicates the approximate location of the lipid bilayer. Color code: A domain, *blue*; P domain, *green*; N domain, *yellow*; TM domain, *light blue* (surface); β-subunit, *red*. The color of the TM helices of the homology model gradually changes from M1 (*blue*) to M10 (*red*). The homology model does not contain the A-M2 linker (*dotted line in blue*). Bound SCH28080 (*red stick*, position predicted by the docking simulation) is found at the center of the luminal cavity, apparently blocking the K^+-entry pathway from the luminal side of the membrane.

4. Charge Status Analysis

Atomic scattering factors for electrons are strongly affected by the charge status of atoms, unlike the scattering factors by X-rays. Atomic scattering factors of neutral and charged atoms for electrons show largely different values especially below 4.5 Å resolution (Fig. 7a). For example, although neutral hydrogen atoms give small scattering factors, protons, which are positively charged hydrogen atoms, have larger scattering factors than that of neutral oxygen as shown in Fig. 7a.

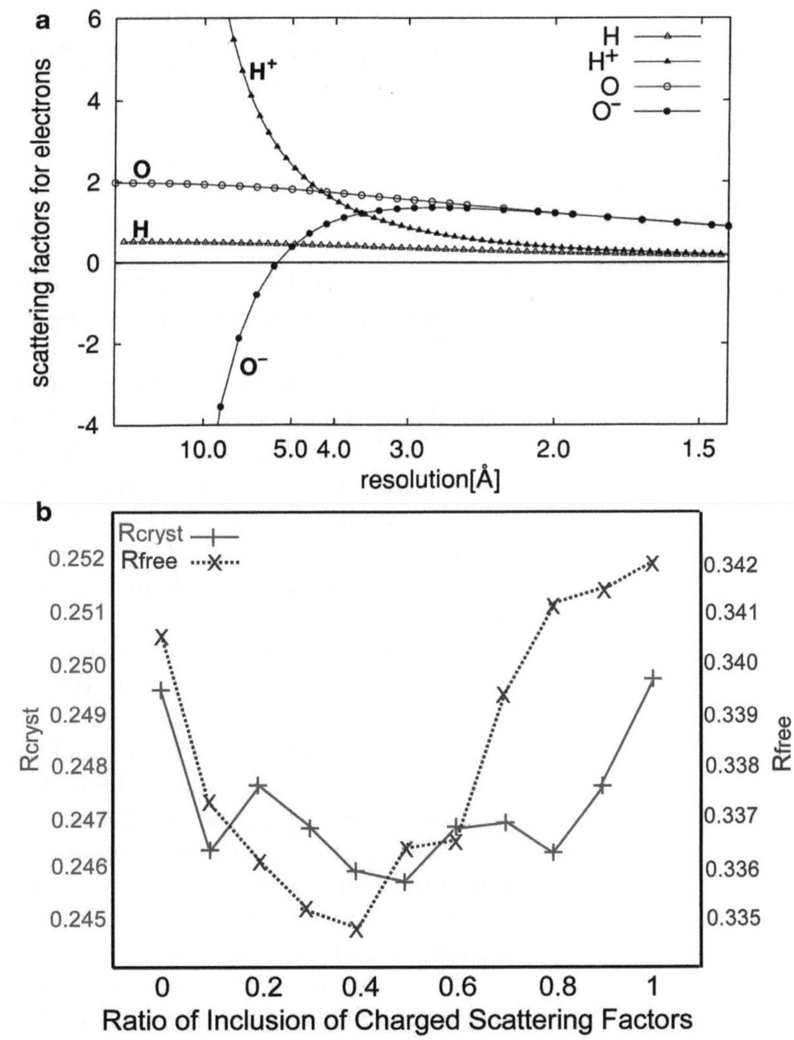

Fig. 7. Atomic scattering factors of charged and neutral hydrogen and oxygen atoms for electrons were shown in (**a**). R-factors and free R-factors calculated using scattering factors of partially charged atoms for carbon and oxygen atoms in the polypeptide carbonyl group. R-factors and free R-factors are shown in *red* and *blue*, respectively (**b**). In these calculations, the same partial charge values were assumed for both the carbon and oxygen atoms.

Comparison of refined results using various ratios of atomic scattering factors for neutral and charged atoms gave an interesting result. R-factors (red) and free R-factors (blue) were calculated in refinements using scattering factors of partially charged atoms for carbon and oxygen atoms in the peptide group. The minimum value for the R-factor and the free R-factor were at the charged ratios of 0.45 and 0.40, respectively (Fig. 7b). The results suggest polypeptide carbonyl groups are charge shifted by 40% or 45% in the peptide bonds (20). When high-resolution structure analysis is achieved, careful analyses of the |Fo|−|Fc| map with and without low-resolution data may give information about the charge status of atoms based on electron crystallography. While high-quality data as well as careful analysis are required, the charge status analysis could become one of the major applications of electron crystallography.

5. Structures with Less Artificial Crystal Contacts

Each membrane side of a 2D crystal is exposed to buffer and enables researchers to perform challenging dynamics studies by the freeze-trapping technique that was developed by N. Unwin to study the gating mechanism of the nicotinic acetylcholine receptor (AChR) (5). This is a powerful application of electron crystallography. Another important feature of electron crystallography of 2D crystal with buffer-exposed surfaces is that structures of membrane proteins could be analyzed with less artificial contacts, which are different from 3D crystals. One typical example is the structural analysis of a gap junction channel, which here is used to illustrate this point.

Gap junctions act as intercellular communication channels that allow a wide variety of solutes with different sizes to be transferred between the cytoplasm of adjacent cells. These solutes include ions, metabolites, nucleotides, peptides, and secondary messengers. Gap junction channels have critical roles in many biologically important processes including cardiac development, fertility, the immune system, and electrical signaling in the nervous system. The multiple gating mechanisms by voltage, calcium ions, phosphorylation, and pH are intricate and the types of regulation are also dependent on the type of connexin molecules.

The structure of human Connexin26 (hCx26) gap junction channel, which is diversely expressed and the second smallest member of the conserved mammalian gap junction protein family, was analyzed by using a site-specific Met34A mutant, hCx26M34A, because this mutant expresses in baculovirus-infected Sf9 cells at higher quantities than wild-type Cx26-infected cells. Remarkably, 2D crystals of hCx26M34A were

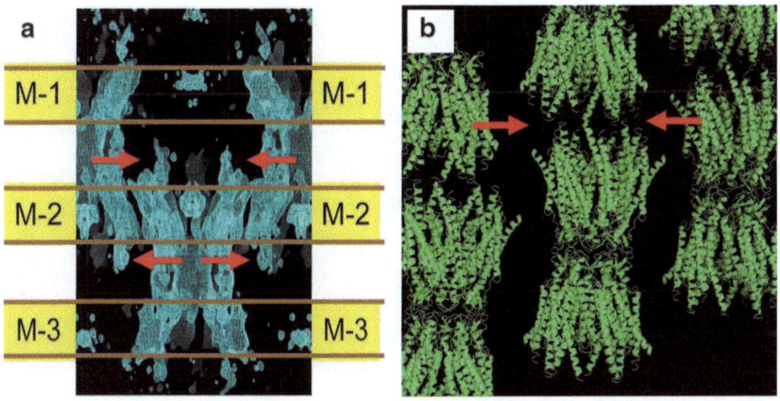

Fig. 8. The structure of three membrane-layered 2D crystals of Cx26 which are indicated as M-1, M-2, and M3 on both sides of the figure (**a**). Structure of Cx26 analyzed from a 3D crystal (**b**). *Red arrows* indicate the cytoplasmic domains of Cx26 molecules. While the cytoplasmic domains heavily interact in the 3D crystals, in the M-2 layer these domains are protected by the two "sandwiching" layers of M-1 and M-3 in the 2D crystals.

formed as three membrane layers. In bilayers of Mem-1 and Mem-3, the hemichannels show poorer structural density than in Mem-2 (Fig. 8a), presumably because the large cytoplasmic domains of the connexin subunits are variable and could easily be deformed by their contact with the carbon film to which the crystals are adsorbed in the sample preparation procedure for cryo-electron microscopy. By contrast, the hemichannels in Mem-2 are protected from any forces such as the surface tension upon specimen drying and mechanical interactions with the carbon film. Therefore, the structural features of the flexible cytoplasmic domains of the connexins of the hemichannels in Mem-2 should be the most accurate and, in particular, well preserved.

The 3D-map shows a novel density in the center of the pore, and this density was named a plug (21). Since the plug density was dramatically reduced in the structure of the N-terminal deletion mutant, it was confirmed that the density was formed by N-terminal helices of six subunits in the hemichannel. The plug density is located inside of the membrane layer and contacts the surrounding channel wall, which at the constricted part of the vestibule is formed by the innermost helices 1 of each subunit. The Cx26 gap junction crystal structure shows that the channel vestibule is blocked by a physical obstruction, which is the reason why the density was named a plug. A structure analyzed by X-ray crystallography at a higher resolution of the open state provides more structural details and supported the plug gating mechanisms in these widely expressed channels (22). The gap junction channel molecules are heavily interacting in 3D crystals, and the cytoplasmic domains were found to be seriously deformed (Fig. 8b). The map of hCx26M34A

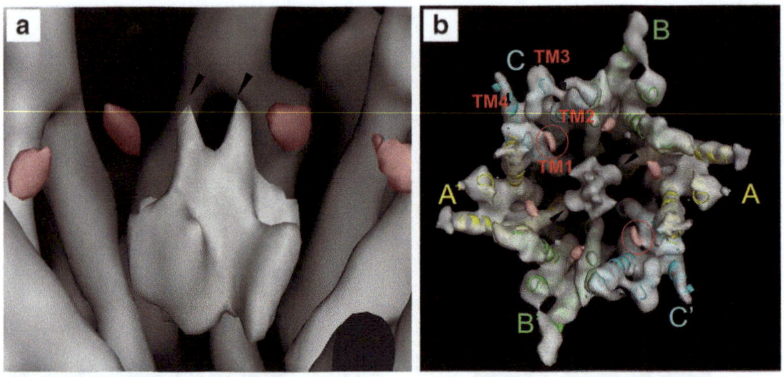

Fig. 9. The two-layer arrangement of the plug structure in the gap junction channel (**a**) and the gap junction structure with twofold symmetry observed from the cytoplasmic side (**b**). *Red* densities could be related with loops between helix 1 to the plug helix in each subunit. Electron crystallography revealed the complex structure of the plug as well as the cytoplasmic domains which could be analyzed presumably because of the lacking artificial interaction effects in the M-2 layer of these 2D crystals.

mutant was improved based on electron crystallography and the structure revealed the plug-forming helices are not arranged with sixfold symmetry at same height in the channel but form a two-layer arrangement (Fig. 9a) (23). The whole gap junction structure, especially the cytoplasmic domain, does not have sixfold symmetry but twofold symmetry. The cytoplasmic loop of the Cx26 molecule interacts with the plug loop and is therefore a key structural determinant for the regulation of the complex and multifunctional gating mechanism by gap junction channels (23). These results of a structure with less artificial influences by crystal packing clearly suggest the critical importance of cytoplasmic conformational changes (Fig. 9b). The cytoplasmic structure, key for understanding the function of the gap junction channel, could not be revealed without data by electron crystallography.

6. Conclusions

The merits of electron crystallography, which were discussed in terms of four examples in this chapter, suggest future directions suitable for this method. Electron crystallography could be recognized as a powerful method for acquiring detailed structural and functional information about membrane proteins in their physiological lipid and ionic settings with less artificial interactions. Development of user-friendly systems such as a cryo-electron microscope, software such as discussed in Chapters 10 and 11, and specimen preparation methods will be extending the possibilities of obtaining high-resolution information from increasingly complex protein–lipid arrays. Therefore, electron crystallography has the potential to provide a vast range of structural and functional information on membrane proteins, and the examples outlined here provide an indication of the power of the method that can relatively easily be extended to studies of further membrane proteins.

Acknowledgments

These studies were performed in nice collaboration with many researchers whose names were recorded as authors in each referenced paper and I appreciate the contribution by each researcher. This work was supported by Grants-in-Aid for Scientific Research (S) and the Japan New Energy and Industrial Technology Development Organization (NEDO).

References

1. Henderson R, Unwin PN (1975) Three-dimensional model of purple membrane obtained by electron microscopy. Nature 257:28–32
2. Henderson R, Baldwin JM, Ceska TA, Zemlin F, Beckmann E, Downing KH (1990) Model for the structure of bacteriorhodopsin based on high-resolution electron cryo-microscopy. J Mol Biol 213:899–929
3. Fujiyoshi Y (1998) The structural study of membrane proteins by electron crystallography. Adv Biophys 35:25–80
4. Kimura Y, Vassylyev DG, Miyazawa A, Kidera A, Matsushima M, Mitsuoka K, Murata K, Hirai T, Fujiyoshi Y (1997) Surface of bacteriorhodopsin revealed by high-resolution electron crystallography. Nature 389:206–211
5. Unwin N (1995) Acetylcholine receptor channel imaged in the open state. Nature 373:37–43
6. Preston GM, Carroll TP, Guggino WB, Agre P (1992) Appearance of water channels in Xenopus oocytes expressing red cell CHIP28 protein. Science 256:385–387
7. Murata K, Mitsuoka K, Hirai T, Walz T, Agre P, Heymann JB, Engel A, Fujiyoshi Y (2000) Structural determinants of water permeation through aquaporin-1. Nature 407:599–605
8. Hiroaki Y, Tani K, Kamegawa A, Gyobu N, Nishikawa K, Suzuki H, Walz T, Sasaki S, Mitsuoka K, Kimura K, Mizoguchi A, Fujiyoshi Y (2006) Implications of the aquaporin-4 structure on array formation and cell adhesion. J Mol Biol 355:628–639
9. Gyobu N, Tani K, Hiroaki Y, Kmegawa A, Mitsuoka K, Fujiyoshi Y (2004) Improved specimen preparation for cryo-electron microscopy using a symmetric carbon sandwich technique. J Struct Biol 146:325–333
10. Chapman HN et al (2011) Femtosecond X-ray protein nanocrystallography. Nature 470:73–77
11. Tani K, Mitsuma T, Hiroaki Y, Kamegawa A, Nishikawa K, Tanimura Y, Fujiyoshi Y (2009) Mechanism of aquaporin-4's fast and highly selective water conduction and proton exclusion. J Mol Biol 389(4):694–706
12. Ho JD, Yeh R, Sandstrom A, Chorny I, Harries WEC, Robbins RA, Miercke LJW, Stroud RM (2009) Crystal structure of human aquaporin 4 at 1.8 Å and its mechanism of conductance. Proc Natl Acad Sci USA 106:7437–7442
13. Gonen T, Cheng Y, Sliz P, Hiroaki Y, Fujiyoshi Y, Harrison SC, Walz T (2005) Aquaporin-0 membrane junctions reveal the structure of a closed water pore. Nature 438:633–638
14. Rash JE, Yasumura T, Hudson CS, Agre P, Nielsen S (1998) Direct immunogold labeling of aquaporin-4 in square arrays of astrocyte and ependymocyte plasma membranes in rat brain and spinal cord. Proc Natl Acad Sci U S A 95:11981–11986
15. Furman CS, Gorelick-Feldman DA, Davidson KG, Yasumura T, Neely JD, Agre

P, Rush JE (2003) Aquaporin-4 square array assembly: opposing actions of M1 and M23 isoforms. Proc Natl Acad Sci USA 100: 13609–13614

16. Mitsuma T, Tani K, Hiroaki Y, Kamegawa A, Suzuki H, Hibino H, Kurachi Y, Fujiyoshi Y (2010) Influence of the cytoplasmic domains of aquaporin-4 on water conduction and array formation. J Mol Biol 402:727–739
17. Abe K, Tani K, Fujiyoshi Y (2011) Conformational rearrangement of gastric H(+), K(+)-ATPase induced by an acid suppressant. Nat Commun 2:155
18. Abe K, Tani K, Fujiyoshi Y (2009) Intersubunit interaction of gastric H+, K+-ATPase prevents reverse reaction of the transport cycle. EMBO J 28:1637–1643
19. Abe K, Tani K, Fujiyoshi Y (2010) Structural and functional characterization of H+, K+-ATPase with bound fluorinated phosphate analogs. J Struct Biol 170:60–68
20. Mitsuoka K, Hirai T, Murata K, Miyazawa A, Kidera A, Kimura Y, Fujiyoshi Y (1999) The structure of bacteriorhodopsin at 3.0 Å resolution based on electron crystallography: implication of the charge distribution. J Mol Biol 286:861–882
21. Oshima A, Tani K, Hiroaki Y, Fujiyoshi Y, Sosinsky GE (2007) Three-dimensional structure of a human connexin26 gap junction channel reveals a plug in the vestibule. Proc Natl Acad Sci U S A 104:10034–10039
22. Maeda S, Nakagawa S, Suga M, Yamashita E, Oshima A, Fujiyoshi Y, Tsukihara T (2009) Structure of the connexin 26 gap junction channel at 3.5 Å resolution. Nature 458:597–602
23. Oshima A, Tani K, Toloue MM, Hiroaki Y, Smock A, Inukai S, Cone A, Nicholson BJ, Sosinsky GE, Fujiyoshi Y (2011) Asymmetric configurations and N-terminal rearrangements in connexin26 gap junction channels. J Mol Biol 405:724–735

INDEX

A

Aberration 174, 175, 356, 359, 363, 368–371, 456
Aberration corrector 356, 359, 363, 368–371, 456
Accelerating/acceleration voltage 174–176, 319, 326, 332, 336, 357, 372, 409–411
Acoustic
  noise 134, 135
  paneling 347, 348
AFM stylus 461–463, 465–469, 471–473
Aharonov-Bohm effect 363
Air delivery 346
Air–water interface 59–63, 66–68, 121, 280, 408, 439, 528
ALAT 205
Align/alignment
  image alignment
    alignment by classification 443, 446, 454
    alignment parameters 137, 200, 205, 234, 393, 416, 438, 444
    angular alignment 429, 444
  microscope column alignment 130–131
ALLSPACE 214, 223
Ammonium molybdate 78, 98, 406
Amore 245
Amplitude 7, 8, 10, 11, 154, 173–176, 179, 183–185, 188, 195–200, 202–204, 206–208, 212, 214, 216, 217, 219, 220, 222–224, 244, 246, 249, 253–255, 258, 266–268, 321, 323, 325, 328, 329, 358, 359, 370, 371, 376, 392, 393, 397, 409, 411, 419, 421, 435, 439, 482–486, 548
  contrast 359, 371, 376, 409, 411
Amplitudes and phases file (APH file) 183, 185, 190, 197–199, 203–205, 321
Amyl amine 83, 84, 294
Anamorphotic system 360, 363
Ancillary equipment 130, 133–134, 331, 333, 347, 349
Angle 25, 46, 133, 134, 139, 176, 177, 180, 219, 250, 276, 279, 280, 293, 307, 311, 318, 357, 369, 374, 405, 415, 418, 431, 438, 443, 454, 455, 472, 496, 501, 552
Angular range 374, 452
Anisotropic/anisotropically/anisotropy 121, 154, 213, 244, 246, 255, 259, 263, 327, 404, 413, 503, 548
Anisotropic distortions 327
Anti-contaminator/anticontaminator 131–132, 136, 146, 345
Aperture 6, 77, 86, 88, 109, 131, 138, 140, 141, 146, 154, 159–161, 339, 342, 343, 360, 387, 548
Apodizing 451
AQP fold 554
Aquaporin (AQP) 9, 12, 153, 158, 164, 165, 252, 521, 522, 554
Area matching 437–439
ar/R constriction site 556
Artifact 472, 473, 552
Astigmatic 176
Astigmatism
  condenser lens astigmatism 139
  objective lens astigmatism 413
Asymmetric particles 402, 415, 417
Asymmetric unit 98, 212, 213, 246, 249–251, 255–258, 262–264, 269, 390, 422, 476, 486
Atomic *B*-factor 249, 259
Atomic force microscopy (AFM) 99, 461–473
Atomic model 7–10, 117, 172, 245, 395, 397, 421, 448, 453, 519–521, 553, 554
Atomic resolution 2, 7, 8, 13, 105–107, 130, 133, 137, 139, 154, 226, 244, 245, 307, 376, 382, 401, 421, 427, 483, 521, 529
Atomic scattering factors 115, 563, 564
Autocorrelation 182, 443
AutoDock 232
Autoloader 289, 290, 299
Automated fitting 236
Automation 13, 14, 31, 34, 74, 187, 208, 209, 252, 258, 267, 274, 279, 298, 307–311, 313–329, 336, 374–375, 524
Average intensity 221

Ingeborg Schmidt-Krey and Yifan Cheng (eds.), *Electron Crystallography of Soluble and Membrane Proteins: Methods and Protocols*, Methods in Molecular Biology, vol. 955, DOI 10.1007/978-1-62703-176-9,

Averaging....104, 105, 117, 177, 198–199, 211, 214–216, 222, 223, 229, 307, 402, 412, 415, 416, 419, 428, 429, 431–435, 439, 440, 445, 446, 452, 453, 486
AVRGAMPHS....212, 214–216
AVRGAMPHSC....212
Axial distance....388
Axial rise (Δz)....390

B

Background
  amplitude....197, 202
  subtraction....164, 184, 545–547
Back-injection....120–124
Back projection....393, 437–439
Backscattering....350, 366
Bacteriorhodopsin (bR)....1, 4–9, 12, 13, 104, 116, 143, 145, 153, 208, 216, 373, 520, 540, 551, 553, 558
B_{add}....263
Balbes....252, 267, 268
β-barrel....9, 18, 463, 508, 509
Baverage....257
Beam brightness....130
Beam-induced movement....6, 137, 334
Beam/radiation damage....4, 6, 103–112, 131, 132, 137, 141, 159, 160, 354, 366, 373, 376, 402, 409, 412, 431–433, 456, 479, 549
Beam size....334
Beam stop....162–164, 166, 168, 339, 342
Beam stopper....154, 160, 161, 543–544
Beam tilt....141, 144, 175, 176, 309, 348, 413
Bessel function....388
B-factor....216, 219, 220, 224, 249, 255, 258–261, 558
Bilayer....48, 61, 88, 91, 105, 226, 251, 283, 285, 291, 292, 434
BioBeads....38, 55, 61, 64, 67–70, 279
Biotinylated lipid....70, 528
Blotting....85, 123, 126, 127, 144, 157, 158, 288, 295, 344, 386, 396, 407, 408
Blurring....113, 138, 402, 543, 548
Boersch, H. 358–360, 372
B_{off}....263
Boltzmann equilibrium....498
Bond angle....270
Bond length....270
Boolean mask....446, 450
Boundary conditions....216
Boxed reference structure....322
Boxer program....390
Box size....189, 215, 216, 390, 544–547, 549
Brightness/beam brightness....77, 130, 187, 188, 355, 369, 472, 544
Bsoft....173

C

Ca^{2+}-ATPase....230, 231, 233, 382, 524
Calidris software package....173
Camera length....142, 147, 154, 164–166
Cantilever
  arm....450, 453, 454
  spring....462
Carbon
  coating....76, 80–82, 333
  evaporation....75–76, 78–80, 120–122, 134, 294
  evaporator....75, 76, 80, 133, 158, 333, 342, 343, 350, 404, 405
  film....4–6, 57, 65, 67–69, 74, 78, 80–86, 88–91, 93, 98, 99, 112, 116, 119–127, 133, 134, 149, 157, 294, 333, 342, 343, 350, 351, 359, 360, 362, 404–406, 408, 439, 565
  floating device....76, 81
  rod....76, 78, 80, 120–122, 133, 278, 294
  sandwich
    specimen preparation technique....565
    technique....117, 121–125, 168
  support film....2, 113, 117, 119, 133, 144, 342, 404
Carboxysome proteins....529
Carr–Purcell–Meiboom–Gill (CPMG) experiment....499
Cartesian....199, 299
Cassette, film....339, 350
C_c correction/corrector....370, 371
CcmK proteins....529–531
CCP4
  CCP4i....245, 252–269
  CCP4i project....254
  CCP4 programs....185, 245, 252, 259
CCUNBEND....182
$C_{12}E_8$. *See* Dodecyl octaethylene glycol ether ($C_{12}E_8$)
Central section theorem....196, 197
Centroiding....367
Centrosymmetric constraint....212, 215, 216, 222–224
Centrosymmetric phase error....215, 222
Chain display....253
Chainsaw....249, 260, 267
Charge coupled device (CCD) camera....7, 77, 78, 88, 92–94, 96, 97, 99, 100, 132, 133, 138, 141, 148, 154, 166, 276, 315, 333, 335–339, 341, 342, 349, 350, 364, 365, 383, 433, 541–543, 548
Charge-mediated....61
Charge shifted....564
Charge status....563–564
Charging....6, 83, 97, 116, 138, 154, 244, 359, 361–364, 409
Chelating nickel headgroup/nickel headgroup....528
Chemical cross-linking....403
Chemical exchange....498

Chemical shift............ 496–49, 501–503, 506, 507, 512, 513
  dispersion.. 497, 507–508
  perturbation assay...498
Chemical shift index (CSI) 502, 503
Chimera............207, 232, 234, 236, 387, 394, 421, 422, 448
Chromatic aberration coefficient....................................368
Chromatic aberration corrector, 456. *See also* C_c correction/corrector
Class average 414–416, 419, 430, 443, 444, 446–448, 450–452, 454, 455
C2 lens..139
ClustalW.... 232
CMOS. *See* Complementary metal-oxide-semiconductor (CMOS)
CMOS direct electron detection camera........................341
CMOS pixel detector..7
Coherence/coherent 90, 130, 139, 155, 176, 184, 189, 327, 342, 355, 358, 368–370, 409, 497, 498, 502
Coil..6, 135, 354, 452, 453, 501, 502, 512
Colacor.....…..394
Cold stage.................................... 7, 334, 335, 343, 349, 354
Cold trap..6, 121, 131, 283
Column
  alignment.. 137, 146
  averaging...439
Coma...141, 144, 148, 348
Coma-free alignment 137, 141, 147, 149
Combined phase error 215, 216, 222, 223
Common-lines 415–418
Complementary metal-oxide-semiconductor (CMOS).......7, 133, 148, 315, 337, 338, 341–342, 365–368, 375, 376
Completeness 41, 207, 217, 221, 222, 225, 246, 255, 263, 264, 513
Computational biology.......... 8, 13, 245, 265, 362, 401, 402, 411–413, 422, 483–486, 520
Concentration/protein concentration 21–23, 25, 27, 28, 32, 34, 37–41, 44, 45, 51, 54, 60–63, 65, 66, 68, 69, 83–85, 98, 117, 125, 126, 274–277, 279–282, 284, 287, 288, 291–293, 374, 384, 385, 406, 409, 464, 467–469, 477–480, 504, 509, 512, 529, 534, 535, 554
Concrete plinth 346, 347
Condenser aperture 140, 146
Condenser lens 139, 342
Condenser lens astigmatism 140
Conductivity...6, 109, 110, 294, 362, 384
Conformation................13, 18, 53, 230, 233, 235, 266, 267, 427, 479, 481, 522–524
Conformational change.. 521, 523
Conformational flexibility 249, 402
Conical tilt/tilting... 433, 435
Contact mode 463
Contact mode AFM... 462, 466
Contamination/ice contamination.......................5, 6, 34, 44, 45, 47, 53, 75, 78, 91, 130, 131, 281, 283, 311, 332, 334, 345, 346, 361–363, 396, 404, 413, 468
Contour...... ....219, 220, 394, 395, 461, 462, 466, 467, 471, 531, 560, 562
Contrast............................13, 61, 74, 75, 80, 85, 87, 89–91, 96, 98, 130, 131, 138, 139, 141, 142, 147, 154, 175, 176, 185, 188, 192, 198, 199, 213, 216, 217, 222, 224, 244, 264–266, 299, 309, 322, 343, 354, 356–364, 368–371, 376, 387, 391, 397, 402–404, 408, 409, 411, 429, 437, 456, 462, 481, 521, 522, 530, 539, 565
Contrast reversal...176
Contrast transfer function (CTF)...................... 94, 96, 139, 154, 174–177, 183–185, 188, 191, 192, 198, 200, 203, 207, 213–216, 219, 223–224, 244, 325, 327, 357, 358, 360, 362, 368–370, 391–393, 397, 411, 413–414, 419, 437, 453, 539
Convert from mmCIF to mtz 254
Convert to/modify/extend MTZ...................................254
Cooling.. 6, 36, 108, 120, 134, 135, 336, 345, 346, 349, 354, 479, 507
Coot..232, 266, 269, 270, 486, 487
Co-purified lipid.................... 37, 41, 42, 49, 50, 53, 55, 521
Corrector................................356, 359, 363, 368–371, 456
Correlation coefficient............. 176, 236, 252, 263, 393, 484
Correlation peak.....................................182, 202, 439–441
Critical micelle concentration (CMC), 27, 38, 41, 54, 62, 63, 66, 83, 275, 276, 279, 280, 287, 288, 291, 298, 299, 409
Cropped reference ....…. 181, 182
Cross-correlation........................... 143, 145, 177, 184, 198, 201, 236, 388, 431, 434, 439–441, 443, 445
Cross-correlation coefficient..236
Cross-correlation functions (CCF)................. 439, 441, 442
Cross-correlation map (CC-map) 181–183, 189, 198, 201, 440, 441, 445, 454
Cross relaxation term...500
Cross-rotation function (RF) 250
Cross-validation 253
Cross-vector 246, 250
Cryo-blades 131
Cryo-box... .... 131, 146, 345
Cryo-cycling..136
Cryo-data collection...334
Cryo-electron tomography 331
Cryo-EM. *See* Electron cryo-microscopy (Cryo-EM)
CryoET.. 432, 433
Cryo-holder/cryo-specimen holder/cryo-transfer specimen holder......124, 125, 136, 137, 146, 343, 344, 354
Cryo protection 111, 131–132

Cryo stage....138, 244, 331, 332, 334, 552
Cryo-transfer.... 7, 105, 107, 108, 113–115, 354, 552
Crystal contact....11, 12, 479, 480, 564–566
Crystallization....1, 2, 4, 6–14, 21, 22, 24, 31–57, 59–70, 73–100, 104, 107, 108, 112, 113, 117, 119, 120, 127, 129–151, 153–158, 161, 164, 165, 171–192, 195–197, 199, 208, 209, 211–226, 244, 246–248, 251, 274, 278, 282–283, 297–304, 307–310, 314, 318–320, 324, 325, 327–329, 332–334, 336, 338, 342, 343, 351, 373, 382, 409, 410, 412, 422, 441, 463–466, 468–470, 472, 484, 520, 522–524, 527–536, 539, 552, 554–556, 559, 561, 564–566
Crystallization trial....26, 32, 37, 39–45, 50, 53, 54, 56, 64, 73, 88, 97, 99, 275, 284–288, 290, 291, 298, 336, 479, 498, 523, 527, 534
Crystallographic phase origin....212
Crystallography and NMR System (CNS).... 236, 237, 245, 485–487
Crystal quality....12, 77, 97, 173, 298, 303, 329, 336, 339, 375, 561
C_s correction....370
Cs value.... 174, 176, 319
CTF. *See* Contrast transfer function (CTF)
CTFAPPLY....185
CTF-corrected IQ-plot....213
CTF correction/defocus correction.... 184–185, 188, 200, 207, 213, 214, 219, 222, 327, 391–392, 414, 419, 453
CTF determination/CTF-assignment.... 392, 413–414
CTFFIND....150, 325
CTFFIND3.... 176, 177, 392
Ctfit/CTFIT....392, 453
CTFSEARCH....323
CTFTILT.... 177, 325, 392
Custom script....189, 200, 205, 323
Cut-on frequency.... 358, 359, 363
Cyclodextrin....274, 277, 279, 282, 286–288, 291, 292
Cytoplasmic domain....566

D

DANGLE....501
Dark room....339
Data collect
  automated.... 308, 404, 412
  cryo-EM.... 32, 77, 97, 98, 100, 209, 411–413
Data collection....8, 32, 51, 77, 97, 98, 100, 112, 113, 115, 116, 121, 126, 137, 154, 155, 157, 159–168, 196, 197, 208, 209, 216, 225, 244, 308, 321, 333, 334, 336–338, 345, 355, 375, 387, 404, 411–413, 429, 431–433, 452, 456, 476, 479, 481, 519, 520, 524, 539, 552
Data completeness....213, 255
Data processing.... 164, 186, 187, 196, 208, 309, 316, 328, 374, 520
Data quality.... 155, 161, 212, 213, 224, 255, 349, 479, 486, 564
Data reduction....254, 429
Data redundancy....213, 224
Dataset/data set....184, 190, 191, 198–202, 204–208, 212, 215–221, 223–226, 321–324, 329, 414–416
n-decyl-β-d-maltopyranoside (DM).... 19, 22, 23, 25, 26, 64, 65, 275, 280, 282, 284
Deflection signal....465
Defocus (ΔZ)....368
Defocus correction/CTF correction.... 184–185, 188, 200, 207, 213, 214, 219, 222, 327, 391–392, 414, 419, 453
Defocus range....213
Dehydration....4, 104–113, 117, 465
Dendrogram.... 430, 447, 454
Densitometer....275, 280
Density map.... 129, 213, 215, 217, 220, 226, 232–236, 240, 244, 245, 266, 267, 270, 371, 394, 395, 397, 421, 486, 532, 553, 557–562
Desiccator....76, 80, 85, 121, 122, 134, 277, 283, 339, 340
Detective quantum efficiency (DQE).... 7, 132, 133, 366, 367, 410, 412
Detector....7, 11, 13, 132–133, 138, 139, 141, 145, 147, 151, 176, 337, 338, 341–342, 349, 356, 364–368, 372, 374, 375, 377, 410, 412, 482, 483
Detergent
  free buffer....32, 37
  micelle....27, 291, 409, 420, 507, 510, 528, 556, 558
  solubilized....32, 37, 40, 45, 48, 274, 278, 294, 408, 520
Developer.... 339, 350, 383
Dialysis
  apparatus....35
  block....277, 283, 285, 286
  buffer....32, 34, 37, 39, 43–48, 98, 285–287, 291, 292
  buttons....34–36, 45–47, 52, 56, 277
  clips....46, 51
  tubing.... 34, 35, 37, 45–48, 51, 56, 83
Difference map.... 212, 217, 224
Difference mapping.... 212, 213, 216–217, 221, 223, 224
Differential phase residual....394
Diffraction
  data....129, 159, 164, 253, 259, 355, 482, 484–486, 497, 539, 555, 561–563 (*see also* Electron diffraction data)
  length....159–161, 163–166
  mode....140–142, 147, 159, 310
  pattern.... 97, 110, 139–142, 147, 162, 328, 336, 342, 359, 360, 363, 364, 389, 480, 482, 484, 529, 539, 544, 547–549, 555 (*see also* Electron diffraction pattern)
  peak....7, 178–180, 182–185, 190
  plane....339, 358–360, 362–364

ring......110
shift......142, 147, 160, 161, 163
Diffractogram......141
Digital image......74, 412
Digitize/digitising/digitizing......138, 333, 341, 355, 364, 387, 548
Dihedral angles......500–503, 507
1,2-dimyris toyl-*sn*-glycero-3-phosphocholine (DMPC), 31, 42, 43, 56, 276, 282–284, 288, 503
1,2-dioleoyl-*sn*-glycero-3-phosphocholine (DOPC)......63, 65, 70, 277, 282–284
1,2-Dioleoyl-*sn*-glycero-3-phospho-(10-racglycerol) (DOPG), 277, 282, 283
Dipolar coupling constant......496
Dipole-dipole cross-relaxation......499
Dipole-dipole interaction......497, 502
Direct electron detector......337, 410
Directory structure......197, 315, 541
Disordered domains......559
Distance geometry (DG)......506
Distance restraints......499–501, 505, 506
Distortion......142, 147, 174, 178, 180, 182–184, 189, 225, 310, 322, 325, 409
Distortion vector plot......184, 189
Divalent cations......69, 527
DMPC. *See* 1,2-Dimyristoyl-*sn*-glycero-3-phosphocholine (DMPC)
Docking......229–240, 382, 394–395, 421, 523, 529, 532, 562
n-dodecyl-β-d-maltopyranoside (DDM)......41, 63, 65, 275, 282, 284, 287, 291
Dodecylmaltoside (DDM)......41, 63, 65, 275, 282, 284, 287, 291
Dodecyl octaethylene glycol ether ($C_{12}E_8$)......275, 284
Domain interface......230
DOPC. *See* 1,2-Dioleoyl-*sn*-glycero-3-phosphocholine (DOPC)
DOPG. *See* 1,2-Dioleoyl-*sn*-glycero-3-phospho-(10-racglycerol) (DOPG)
Dose/electron dose......4, 6, 103–117, 132, 137–148, 159, 166–168, 195, 309–310, 334, 336–338, 387, 411, 429, 433, 453
Dose fractionation......429
Double-autocorrelation method......443
Double-layered 2D crystals......154, 172, 522, 554–556, 561
DQE. *See* Detective quantum efficiency (DQE)
Drift......7, 97, 100, 112, 135, 138, 143, 145, 149, 150, 154, 176, 244, 332, 334, 339, 349, 360, 362, 375, 411, 413, 465, 466
DropBox......293
Drop-box......275, 281
Dual axis tilt series......432, 433, 446
2dx
2dx package......173, 197
2DX robot......279, 287
2dx software system......171, 173–177, 184, 185, 188, 189, 197, 199, 202, 208
2dx_findlat......179, 180
2dx_getlat......179
2dx_image......174, 175, 177, 179, 181, 183–192, 197, 198, 200–203, 206, 314, 318–327
2dx_logbrowser......314
2dx_master.cfg......318, 319, 326, 327
2dx_merge......185, 197, 199–205, 208, 314–321, 323, 324, 326, 327, 329
2dx_merge.cfg......317–320, 326, 327
2DX_MERGE_COMBILEB......201
2dx_peaksearch......179
Dyads......251
Dynamic nuclear polarization (DNP)......499
Dynamic range......337, 338, 355, 364, 548
Dynamic transmission electron microscope (DTEM) 372, 373

E

E. coli. See Escherichia coli
Effective resolution......130, 212, 219–221, 225, 338
Effective resolution cutoff......219, 221
Eigenimage......430, 447
Eigenvalues......430, 447
Einzel-lens......359, 360, 362
Elastically scattered electron......356, 410, 456
Elastic/elastic scattering......371, 374
Electrical field......347
Electron beam......88, 103, 104, 108, 109, 111, 126, 130, 131, 140, 148, 166, 175, 176, 196, 309, 358, 366
Electron cryo-microscopy (Cryo-EM)......1, 4, 5, 14, 32, 73, 77, 78, 83, 93, 96–98, 100, 104, 108, 112–115, 117, 119–127, 155–157, 161, 178, 208, 209, 244, 313, 314, 317, 321, 328, 331, 334, 336, 339, 345, 346, 356, 357, 371, 382, 386, 387, 390, 391, 395, 396, 401–406, 408–413, 421, 422, 433, 513, 520, 552
Electron density......245, 249, 421, 483, 484, 486
Electron diffraction (ED)......4, 5, 7–11, 14, 97, 99, 108, 111, 114, 153–168, 172, 195, 196, 216, 224, 243–270, 337, 338, 355, 358, 364, 373, 539–550, 555, 560, 561
Electron diffraction data......9, 153–168, 243–271
Electron diffraction pattern......4, 5, 7, 8, 11, 108, 111, 114, 161, 165, 172, 337, 358, 539–550, 555, 561
Electron dose/dose......4, 6, 103–117, 132, 137–148, 166–168, 195, 309–310, 334, 336–338, 387, 429, 433, 453
Electron-hole pair......366, 368
Electron imaging film......350

Electron magnetic dipole moment 505
Electron micrograph........................... 2, 3, 5, 11, 68, 69, 93, 97, 114, 129, 132
Electron microscope room......................................346–347
Electron microscopy (EM) 1–5, 7, 8, 10, 13, 14, 33, 34, 42, 49, 50, 57, 60, 63, 68, 73–100, 103–105, 107, 108, 112, 120, 122, 133–135, 142, 149, 154–158, 171, 177, 198, 208, 229, 230, 232, 234–237, 239, 240, 244, 274, 278, 288–291, 295, 308, 331–336, 339, 341, 345–347, 349–351, 353–356, 364, 365, 368, 373–375, 385–386, 394, 396, 401–422, 431, 464, 497, 513, 520, 528, 529, 531, 533–535, 540, 552, 562
Electron optical/optic........................11, 341, 362, 368, 372
Electron scattering....................................75, 196, 365, 366
Electron source......................................130, 154, 168, 244, 364, 369, 372, 409
Electron tomography (ET).....................1, 6, 104, 114, 177, 307, 331, 333, 415, 416, 418, 421, 427–456
Electron wavelength...............................175, 368, 411, 414
Electrostatic charging...6
Electrostatic lens..358
Ellipsoidal mask 450
EMAN................383, 388, 390, 392, 414, 417, 419, 421, 442
EMBED...506
Embedding................................ 6, 104, 107, 112, 120, 123, 125, 144, 155–157, 161–163, 244, 355, 386, 520
Embedding medium...............................123, 125, 144, 355
EM column 346, 347, 359
EM map..... .... 8, 230–237, 239, 402, 420–422, 531
EM room..135, 333, 347, 351
Energy filter...335, 356, 410, 456
Envelope/envelope function 139, 175, 176, 234, 357, 368–370, 391, 404, 421, 441–443, 453, 455, 511
ERROR field.. 322, 323
Escherichia coli................ 18, 19, 24, 27, 28, 63, 65, 282–284, 292, 384, 463, 464, 468, 511
Escherichia coli lipid extract 283
Eucentric/eucentricity 114, 141, 144, 147, 159, 160, 300, 301, 309
Eukaryotic.. .... 2, 8, 9, 13, 18, 28, 524
Euler angle 392, 415–417, 438, 444, 454
Evaluate lattice 319
Ewald sphere 196, 411, 549
Excited spin-state population..498
Experimental constraints..497
Experimental phase 214, 220, 225, 550
Experimental phase value 214, 225
Exposure
  mode.........................137, 138, 143, 148, 150, 159, 160, 334, 387
  time....................138, 143, 149, 160, 161, 168, 304, 309, 355, 387, 411, 543, 544, 549
Expression.. ....14–29, 274, 291, 315, 318, 384, 476, 479, 480, 485, 511, 561
Expression system...17, 18
Extramembranous domains...................................121, 125

F

$|F|^2$, 246
Fab fragments 529
Fasta... 253, 268
Fast Fourier transform (FFT).................74, 77, 92–96, 100, 149, 150, 175, 181–183, 187, 188, 191, 256, 325
Fast freezing device 333
Feedback gain..465
FFTIR..324
Fiducial marker...435
Field emission gun (FEG)......................113, 130, 142, 154, 159, 332, 334–336, 345, 346, 348, 349, 355, 376, 383, 409, 452
Figure Of Merit (FOM) 185, 211, 214, 215, 219, 220, 223, 226, 417, 419
File format...175, 200, 207, 254–255, 268, 315
File structure 314–315
Film density...341
Film desiccator 339
Film development...333, 335
Film emulsion..339
Film processing chemicals 333
Film rack.... 339
Film scanner...175, 333, 341, 410
Filter mask radius 322
Fitting........................... 179, 207, 217, 218, 229–240, 276, 292, 392, 395, 414, 420, 421, 438, 440–443, 502, 548
Fixer.. 339, 350, 383
Flatbed scanner 341
Flatness
  carbon-film flatness 113, 120, 133, 439
  crystal flatness.. 113, 161–163
  specimen flatness 7, 119, 439, 440
Floating...... .... 64, 78, 81, 158, 175, 303, 346, 404, 405, 451
Flow rate..349
Fluorinated lipid.. 63, 65, 529
Focal plane...140, 141, 176, 339
Focus gradient correction 453
Focus mode.................... 137, 138, 143–145, 147, 149, 150, 159, 334, 387
Fogging..350
2-Fold axis.. .... 8, 248, 251
FOM. *See* Figure Of Merit (FOM)
FOMSTATS 212, 215–217, 223
Forbidden reflection 22, 215
Fortran-77.. ....172, 323

Fourier........ ....2, 4, 7, 14, 74, 77, 132, 133, 138, 141, 143, 149, 150, 156, 158, 173–176, 178, 179, 181–185, 188–190, 192, 195–199, 204, 207, 225, 244, 246, 256, 290, 322, 325, 336, 338, 358, 388–390, 392, 394, 413, 416, 417, 419–421, 430–432, 434, 435, 439, 440, 482, 483, 530, 532
Fourier-Bessel...390
Fourier coefficient.. 434, 435, 439
Fourier filter mask radius...322
Fourier mask..322
Fourier ring correlation (FRC)...................................394
Fourier shell correlation (FSC).......................335, 336, 394, 419, 420, 452
Fourier space........................ 4, 138, 173, 175, 176, 179, 185, 195–199, 204, 207, 417, 420, 430, 432, 434, 435
Fourier transform (FT) 2, 7, 14, 74, 77, 132, 141, 143, 149, 156, 158, 173–175, 179, 181, 183, 184, 188, 189, 196, 199, 246, 256, 290, 322, 325, 336, 338, 358, 388–390, 392, 413, 416, 417, 431, 434, 435, 440, 482, 530, 532
FRC. *See* Fourier ring correlation (FRC)
Frealign.. 414, 419, 421
Free electron laser... 12, 373
FreeR.. 254, 255
Free R-factor.. 212, 563, 564
Frequency modulation mode..462
Fresnel fringes 370
Frozen-hydrated/vitreous 8, 68, 107, 126, 148, 195, 307, 308, 343, 350, 386, 402–404, 406, 410, 432, 433, 452, 529, 532
FSC. *See* Fourier shell correlation (FSC)
f term......... 392

G

Gap junction channel.................................... 552, 564–566
Generator... 391, 481, 485
Geometry fitting...438
Geometry restraints..506
Get Spacegroup & PhaseOrigin......................................200
GIXD. *See* Grazing incidence X-ray diffraction (GIXD)
GLIDE...232
Global average.................................430, 443–446, 448, 449
Glow discharge..................................76–77, 83, 84, 99, 278, 288, 294, 333, 343, 350, 383, 385, 386, 406–408
Glow discharge unit 83, 333, 343, 350
Glucose.......................4–6, 8, 11, 19, 85, 104, 107, 120, 125, 143–145, 155, 157, 168, 520
Grazing incidence X-ray diffraction (GIXD)....................61
Grid
atlas... 302–304, 311
bar..86, 110, 123, 124
copper... .... 76, 80, 120, 133, 155, 157, 168, 383, 405, 533
freezing robot.. 406, 408
gold...140
molybdenum........120, 121, 123–125, 133, 155, 157, 168
potentials 237
square........................74, 78, 86, 87, 92, 98, 99, 144, 149, 290, 294, 302, 308, 404, 408, 412
Grid-loading robot 299–302
Grid preparation
Cryo.................................... 83, 117, 119–127, 155, 313
negative stain 75, 83–85
GRIP system 173
Gyromagnetic ratio...496

H

Handedness...............203, 389, 394, 397, 418, 420, 421, 485
Heat conductivity 109
Height signal 465, 466
Helical........ ... 2, 10, 11, 33, 88, 226, 229, 230, 381–397, 422, 433, 440, 448, 501, 508–510, 523, 556, 558
crystals2..............10, 11, 33, 88, 120, 230, 381–397, 422
dipole moment.. 556, 558
filament...440
indexing 388–390
reconstruction 229, 230, 390
repeat distance 226, 388
selection rule..388
symmetry 387, 389–391, 393
Helium-cooled stage (helium-cooled electron microscope), 244, 555
Helix.................................... 2, 226, 388–390, 394, 397, 498, 499, 501, 507, 510, 523, 561, 566
α-Helix... 502, 521, 523
Helixboxer.. 388
Heterogeneity......................... 402, 403, 414, 415, 427–429, 434, 439, 445, 448–452, 456
Heterogeneous samples 402
Hierarchical ascendant methods (HAC).........................446
High frequency.. 139, 421
High-resolution.............................. 4, 5, 7–11, 14, 104, 105, 107, 112, 113, 115–117, 119, 120, 129–151, 153–168, 172, 173, 175, 178, 179, 183, 208, 244, 259, 307, 309, 313, 327, 329, 332–334, 336, 338, 342, 347–349, 351, 354, 356, 365, 373, 386, 387, 402, 410, 411, 419–421, 461–473, 497, 498, 502, 508, 511, 521, 524, 529, 532, 539, 540, 552, 555, 559, 564, 567
High-resolution imaging......... 105, 130, 139, 145, 461–473
High resolution structure....... 7–11, 104, 115, 116, 175, 244, 259, 356, 497, 508, 524, 529, 532, 540, 559, 564
High tension (HT)............................77, 146, **277**, 335, 385
High-throughput.......................34, 208, 209, 273–295, 374
High voltage tank..346
Hilbert/Hilbert contrast/Single sideband.............. 359, 360, 362, 363
His-tag (His-tagged).......................24, 60–63, 69, 528, 529

H^+,K^+-ATPase 524, 561, 562
^1H-^{15}N TROSY HSQC 497
Holey carbon film/perforated carbon film 5, 6, 112, 120, 343, 350, 351, 404, 408
Homologous structure 233, 237, 238, 248, 249, 258, 267
Homology model 233–238, 245, 562
Human embryonic kidney (HEK)cells 18, 19, 24–25
Humidity 64, 126, 339, 343, 346, 348, 386, 407
Hybrid pixel device 365, 368
Hydrated/hydration 4, 8, 37, 46, 53, 68, 107, 108, 112, 113, 138, 148, 156, 195, 308, 354, 386, 403, 404, 406, 411, 432, 433, 452, 501, 529, 534
Hydrogen bonding geometry 501
Hygroscopic 43, 55
Hysteresis effects 144, 148

I

i3align.in 430, 444
i3avg 430, 448
Ice contamination/contamination 5, 6, 34, 44, 45, 47, 53, 75, 78, 91, 130, 131, 281, 283, 311, 332, 334, 345, 346, 361–363, 396, 404, 413, 468
ICE package 172
Ice thickness 112, 308, 371, 404, 408, 412
i3class.sh 430, 444, 447
ICM 232, 234, 235, 237, 238, 240
i3compute 430, 450
i3createmask 440
i3cut 430, 450, 452
IDEAL 213, 219
i3display 430, 442–444
i3euler 444
IHRSR. *See* Iterative real-space helical reconstruction (IHRSR)
i3/i3 software 429, 430, 435, 443, 448, 453
Illumination 6, 105, 109, 130, 139–142, 145, 147, 148, 150, 196, 281, 334, 348, 357, 367, 369
Illumination spot 140
Image/Imaging
  acquisition 5, 6, 92, 94, 100, 104, 105, 130, 132, 133, 135, 137–141, 143, 145, 147, 148, 159, 172, 175, 224, 244, 274, 275, 278, 281, 290, 294, 297–304, 307–310, 334, 338, 339, 350, 354, 355, 359, 362, 363, 365, 372–374, 387, 402, 409–411, 414, 456, 461–473, 520
  blurring 138
  classification 429, 446, 448, 555
  data 97, 116, 117, 129, 137–150, 195–209, 212, 216, 314, 336, 338, 555, 560–563
  detection system 331, 333, 336–342
  directory 187, 189, 197, 202, 318, 319, 323, 327
  format 132, 315, 338, 421
  intensifier 88, 99, 338
  low dose imaging 5, 135, 137, 139, 140, 143, 151, 334, 350, 354, 387, 520
  mode 141, 159, 310, 334, 461, 462
  number 203, 315, 318, 320, 439
  shift 116, 117, 143, 145, 198, 304, 309, 334, 349, 354
  shift coil 354
Image processing/image analysis 2, 4, 7–11, 32, 63, 75, 93, 96, 97, 99, 100, 133, 138, 141, 145, 154, 171–192, 197, 198, 216, 313–329, 338, 383, 387–390, 392, 397, 403, 412–422, 436
Image Processing Library and Toolbox (IPLT) 173, 244, 328
Image processing parameters 173, 316, 324, 327
Image quality (IQ)/Image quality (IQ) value 97, 149, 150, 183–185, 189, 190, 198, 199, 202, 203, 207, 213–215, 325, 339, 341, 348, 349, 357, 358, 364, 402
IMAGIC software suite 231
imgccf 388
Import images 174, 197, 291, 315, 318–320, 326
IMQ 217, 218
IMQ value 217
i3msa.sh 430, 444
Inclusion bodies 18, 20–22, 27
Incubator 36, 46, 47, 277
Indexing 164, 166, 179, 190, 246, 248, 254, 388–390, 482, 540, 542, 544–545
Inelastic 109, 356, 370, 371, 374, 376, 410, 456
Inelastically scattered electron 356, 410, 456
In-focus phase contrast 357–364, 456
INHERITABLE_UPON_INIT 317
INHERIT_UPON_INIT 319
Init program 318
In-plane resolution 10, 219, 221, 225, 226, 532
Instability 131, 244, 335, 349, 359, 369, 370
Instrumentation 2, 297, 353–377, 401, 456, 512
Intensity 77, 94, 96, 135, 139–142, 144, 147, 149, 155, 162, 163, 166–168, 184, 188, 190, 191, 212, 221, 246, 254, 279, 341, 342, 364, 367–369, 372, 388, 392, 482–485, 497, 546–548
Intensity statistics 246
Inter-atomic vector 246, 248, 249
Interference 374, 409, 482
Internuclear vector 496, 500, 503
Intrinsic disorder 428
Inverse transform 219, 220
Ion pump 346
i3paste 430, 451, 452
IQ-plot 213
i3resample 430, 444
Irradiation 111, 362, 373, 404, 433
i3selectalign.in 430, 444

i3selectapply.sh 430, 444
Isotropic/isotropically 154, 244
ISWRONG 317
Iterative 2D/3D merging 180, 197, 199, 204, 207, 314, 324, 328, 329, 419
Iterative marker free alignment 436
Iterative real-space helical reconstruction (IHRSR) 381–397
Iterative reference refinement 322
Iterative refinement 391–394
i3totsum.sh 430, 444

J

J coupling 500, 502–504

K

k_{50} 217, 218
Karplus parameters 502
Karplus relation 502
keV, kV 77, 130, 146, 154, 159, 332–337, 345–347, 349, 350, 356, 357, 359, 365, 368, 376, 409–412, 414, 452, 481
K-means [methods]/K-means classification 446
Köhler illumination 139
k_{50}-value 217
120 kV, 120keV 77, 333, 336, 337, 347, 368, 414
200 kV, 200 keV 130, 146, 154, 159, 356, 357, 359, 411, 412, 414
300 kV, 300 keV 154, 332, 334, 335, 345–347, 350, 365, 376, 409–412, 414, 452

L

LaB_6 .... 130, 333, 336, 337, 383
Laplacian filter 395
Laser 12, 13, 80, 97, 187, 279, 339, 342, 360, 364, 372, 373
Lateral resolution 462, 467, 471
LATLINE 207, 550
LATLINEK 213, 218, 219, 225
LATLINPRESCAL 207, 218
Lattice
- distortion 7, 11, 174, 179, 180, 182, 183, 189, 191, 322, 323, 325, 327, 530
- distortion vectors 174, 189
- fringes 370
- heat capacities 109
- indexing 179, 190
- line
  - fitting 207, 548
  - interpolation 208, 328, 329
  - plot 218
- parameters 180, 326, 440, 471, 485, 543–545
- refinement 188
- refinement mode 188, 190
- vector 180, 388, 397, 472, 544

Laue group 246
Layer line 388, 389, 397, 440
Layer line index/indices 388
Least-square approximation 207
Leginon 145, 290, 291, 298–304, 307–311, 321, 355, 410
Lens 6, 77, 100, 131, 135, 139–147, 150, 154, 165, 174–176, 252, 339, 342, 354, 358–360, 362, 363, 365, 369, 370, 376, 411, 413, 522
Lens aberration .... 359, 376. *See also* Aberration
Lens coupled 365
Lens pole piece 354
Ligand 10, 13, 26, 29, 37, 53, 60–63, 65–69, 237, 240, 249, 270, 271, 283, 285, 372, 373, 415, 462, 485, 487, 498, 499, 506, 509, 511, 522, 523, 527, 528, 552
Ligand-mediated 61
Light-harvesting complex II (LHC-II) 8, 9, 53, 540
Lighting 347
Light microscope/light microscopy 14, 343, 357, 367
Light scattering 279, 284, 286–288
Linearity 355, 364
Lipid .... 42, 61, 62, 75, 88. *See also* Phospholipid
Lipid bilayer 12, 21, 32, 41, 67, 244, 274, 278, 287, 434, 469, 470, 510, 522, 524, 556–559, 562
Lipid fluid phase 33
Lipid monolayer/monolayer 32, 59–70, 75, 308, 408, 439, 452, 453, 527–536
Lipid-protein interaction 9, 244
Lipid stocks 36, 38, 39, 43–45, 55, 56, 282, 283
Lipid-to-protein ratios (LPRs) 31, 32, 37, 38, 40–45, 49–51, 53–55, 283, 285
Liposome(s) 42
Liquid condense phase 62
Liquid ethane 5, 6, 13, 68, 120, 144, 157, 333, 343, 386, 396, 407, 408, 535
Liquid expanded (LE) phase 62, 63
Liquid-handling robots 279, 285, 286, 297
Liquid helium 6, 8, 10, 105, 108, 109, 112, 326, 332
Liquid nitrogen 5–8, 13, 25, 48, 112, 120–125, 127, 131, 133, 134, 137, 146, 157, 334, 336, 345–347, 351, 386, 402, 407–409, 479
Liquid nitrogen temperature 5–8, 112, 120, 131, 133, 336, 402, 409, 479
List reflections into logfile (ILIST) 202
Location of fast FFTIR disk 324
Location of fast SCRATCH disk 324
LOCKED 317
Locked cross rotation function (LRF) 252
Locked translation function (LTF) 252

Longitudinal relaxation rate 498
Loop....... 68–69, 122, 123, 125, 231, 234, 235, 240, 249, 260, 261, 264, 266, 309, 463, 471, 479, 480, 482, 484, 499, 523, 535, 553–555, 561, 566
Lorentzian line-shape....... 497
Low-dose....... 5, 77, 103–117, 130, 135, 137–151, 154, 159, 309–311, 334, 339, 341, 350, 354, 355, 387, 411, 520
Low dose image....... 5, 137, 151, 334, 350
Low dose image acquisition system....... 334
Lowest energy solution....... 237
Low frequency....... 357, 359, 361, 363, 439
Low pass filtering/filter 138, 391, 446
Low-resolution....... 9, 13, 100, 107, 116, 179, 198, 208, 223, 246, 257, 270, 325, 402, 410, 411, 413, 419, 421, 433, 434, 496, 506, 513, 547, 557, 559, 564
Low-resolution spectrum/spectra....... 496
LPRs. *See* Lipid-to-protein ratios (LPRs)

M

Macromolecular assemblies 401, 436
Magnetic dipole moment 496, 505
Magnetic field....... 349, 363, 496, 501, 503, 512
Magnification 22, 24, 68, 74, 77, 78, 85–97, 99, 132, 138, 139, 141–148, 159, 160, 174, 179, 180, 186, 189, 190, 290, 294, 298, 300, 302–304, 308–311, 319, 326, 334, 337, 338, 341, 387, 412, 437, 452, 465, 466, 468–470
  mode....... 138, 141–143, 147, 159, 160, 174, 310, 334, 387
Maintenance....... 53, 54, 129, 131, 331–351
MAKETRAN....... 184, 321
Manual fitting....... 234–237
Map....... 4, 55, 63, 97, 129, 172, 196, 212, 229, 244, 321, 371, 391, 401, 437, 485, 498, 529, 552
Map quality 218, 420, 561
Mapslicer....... 257
Marker based alignment....... 436
Marker-free alignment/protomo 429, 431, 435, 436, 439
Mask construction 446
Mask function 199, 440
Mask/MASK
  projection mask....... 446
  radius....... 182, 183, 189, 322, 545
  volume mask....... 446
Matthews' coefficient....... 255–256
Maximum likelihood (ML) 173, 192, 266, 328, 415, 419, 422, 446
Maximum tilt angle....... 219, 225
MDA mask....... 449, 451
MDS. *See* Minimum dose system (MDS)
MD/SA....... 512, 513
Mean radial density distribution....... 434, 435
Mechanical instability 335, 349
Mechanical vibration....... 112, 346
Membrane
  morphology 42, 50, 51, 74, 75, 78, 86, 88, 89, 92, 93, 96
  protein....... 1, 17, 31, 59, 88, 104, 121, 129, 153, 172, 195, 216, 243, 274, 328, 356, 382, 408, 435, 461, 475, 504, 519, 527, 539, 551
  protein complex 408–409, 420
  size....... 49–51, 93
Merge....... 91, 92, 198, 200, 202, 204, 206, 207, 318, 319, 326, 418, 547
Merged dataset....... 184, 190, 191, 198, 200, 202, 204–206, 208, 321, 323, 324
Merge directory....... 318, 319, 326
Merge Once 200, 202, 206
Merging
  in 2D....... 197, 199–204, 314, 324, 329
  in 3D....... 180, 197, 199, 204–208, 314, 324, 328, 329
Meridian....... 388
Methylamine tungstate....... 406
Methyl ^{1}H-^{13}C TROSY HMQC....... 497
MFR. *See* Molecular fragment replacement (MFR)
Mica....... 74–76, 78–82, 91, 120–125, 127, 133, 134, 155, 157, 158, 294, 342, 404, 405, 464–469
Micelle/detergent micelle 280, 408, 409, 420, 507, 509
Micrograph....... 2, 3, 5, 6, 11, 68, 69, 74, 75, 92, 93, 96, 97, 100, 114, 129, 132, 176, 192, 202, 207, 290, 341, 388, 390, 392, 413, 414, 453, 540, 548
Miller index....... 179, 185, 188, 197
Minimum dose system (MDS)....... 105–107
Missing cone 4, 199, 208, 209, 219, 225, 329, 382, 432, 434
Missing wedge....... 416, 429, 430, 434–435, 442, 445–447, 453, 454
ML. *See* Maximum likelihood (ML)
MMBOX....... 183, 185, 325
mmCIF....... 253, 254
Model building....... 234, 239, 240, 269, 270, 448, 476, 483, 486–487
Modeling....... 172, 229–240, 245, 486, 523, 524, 529
Modeller....... 232, 234, 236
Modeller Web server (Modweb) 232, 234
Modulation transfer function (MTF)....... 132, 133, 176, 367, 368, 410, 412
Modus of Merging 199, 204, 205
Molar LPR....... 40–42, 44, 50, 55
Molecular ET 429, 455, 456
Molecular fragment replacement (MFR) 513
Molecular replacement (MR)....... 8, 9, 154, 172, 243–271, 483–486, 539
Molecular transform....... 434

Molrep........245, 251, 252, 258, 260–269
Molsoft........232
Molybdenum........120, 121, 123–125, 133, 155, 157, 168
Molybdenum grid........123, 124
Monolayer/lipid monolayer........32, 59–70, 75, 308, 408, 439, 452–454, 527–536
Morphology........22, 32, 42, 43, 50, 51, 74, 75, 78, 85–93, 96, 468–469
Mosaicity........7, 50, 92–97, 477
Motif alignment........429, 434, 443–455
Motif averaging........429, 431–433, 446, 452
Motif classification........429, 433–435, 443, 446
Motif/subvolume........231, 389, 428–435, 439–446, 448, 450–455, 548
MR. *See* Molecular replacement (MR)
MrBUMP........252, 267, 268
MRC format........231, 388, 542, 543, 545
MRC programmes/MRC package/MRC Programme Suite........172, 173, 177, 182–185, 197, 198, 200–202, 214, 244, 314, 321–323, 325, 328, 388, 440, 540–542, 547–549
MSI applications........298, 304
MTZ/MTZ file........200, 205, 207, 254–258, 261, 262, 268, 269, 321
MULTALIN........232
Multi-domain model........249
Multi-reference alignment........392, 393, 430
Multi-scaled imaging (MSI)........298, 300, 303, 304, 310
Multivariate data analysis (MDA)........415, 429, 445–456
m values........390

N

Na^+,K^+-ATPase........523
Nano-probe........139, 140
Navigator........187, 188
NCS. *See* Non-crystallographic symmetry (NCS)
NCS restraint........270
Negative........74, 75, 77, 78, 83–85, 95–99, 104, 155, 156, 175, 176, 184, 185, 257, 270, 278, 288, 289, 295, 298, 317, 339, 385, 389, 402–406, 408, 411, 413, 419, 528, 549, 557, 558
Negative stain/staining........20, 74, 75, 77, 78, 83–86, 95–99, 103, 104, 107, 155, 156, 278, 279, 288, 289, 295, 297, 302, 317, 402–406, 408, 411, 418, 445, 528, 535
Nickel headgroup/chelating nickel headgroup........528
Nicotinic acetylcholine receptor (AChR)........6, 10, 522–523, 553, 564
Ni^{2+}-NTA........60, 62, 63
NMR instrument/NMR spectrometer........512
NMR peak........497
n number........390
Node........182, 189, 301–304, 310, 311
NOE. *See* Nuclear Overhauser effect (NOE)
NOE restraint........499, 501, 502, 513
NOESY........501
Noise........7, 94, 134, 135, 138, 166, 176, 181, 183, 184, 189, 222, 342, 347, 348, 364–366, 368, 371, 386, 387, 391, 392, 397, 402, 417, 419, 420, 440, 548, 549, 560
Nominal resolution cutoff........219, 220, 225
Nominal tilt category........320
Non-crystallographic symmetry (NCS)...252, 258, 270, 486
Non-crystallographic translation........246, 248, 256, 257
Non-parallel illumination........196
Non-parallelism........139
Non-tilted reference........205
Nuclear magnetic resonance (NMR) spectroscopy/Solution NMR spectroscopy........1, 259, 401, 403, 479, 495–513
Nuclear Overhauser effect (NOE)........499–503, 505, 506, 510, 513
Nuclear spins........496, 498–502, 505
Nyquist/Nyquist frequency/Shannon-Nyquist........189, 410, 412, 413

O

Objective aperture/objective lens........6, 77, 86, 88, 100, 109, 131, 138–142, 147, 150, 174–176, 359, 387, 411, 413
Objective astigmatism........144, 149, 334
Objective lens astigmatism........100, 175, 413
Observations........50, 55, 57, 65, 67–68, 100, 112, 113, 124, 126, 214, 222, 309, 433, 462, 490, 524
n-octyl-β-d-glucopyranoside (OG)........19, 21, 26, 275, 276, 280, 282, 293
n-octyl-β-d-thioglucopyranoside (OTG)........275, 284
Olap........390
Optical density........175, 540, 541
Optical diffraction........65, 68, 69, 74, 92, 93, 96–97, 413
Optical diffractometer........77, 78, 333, 339, 342
Optics........168, 362, 364–366, 372
Optimization parameter........322, 540, 544, 545
Order........8, 12, 32, 45, 50, 51, 54, 56, 57, 62, 66, 68–70, 74, 75, 78, 82, 83, 87, 88, 93, 95–97, 105, 119, 122, 125, 130, 131, 134, 137, 141, 143, 144, 148, 156, 158, 175, 178, 198, 204, 224, 230, 232, 234, 235, 244, 246, 258, 261, 282, 283, 290, 291, 294, 302, 320, 322, 326, 333, 336, 338, 347, 354, 358, 359, 364, 367, 372, 373, 375, 394, 403, 405, 409, 410, 413, 415, 428, 438, 443, 444, 448, 452–454, 470, 471, 481, 503, 511, 530, 543, 549, 552
ORIGTILT........200–202, 206
origtiltk........202
Orthogonal tilt reconstruction (OTR)........416, 421
Oscillation mode........462
OTR. *See* Orthogonal tilt reconstruction (OTR)
Out-of-plane bending........439, 440

Out-of-plane tilt angle 393
Overall phase error 213, 214
Over-expression 17, 18, 32
Over-focus 142
Overhead gantry crane 347
Oxygen sensor 347

P

Packing function (PF) 263
1-Palmitoyl-2-oleoyl-sn-glycero-3-phosphatidic acid (POPA) 277
1-Palmitoyl-2-oleoyl-sn-glycero-3-phosphocholine (POPC) 42, 63, 65, 276, 283, 284
1-Palmitoyl-2-oleoyl-sn-glycero-3-phosphoethanolamine (POPE) 277
Paracrystal/paracrystalline/paracrystal array 427–456
Parallel illumination 6, 139–141
Paramagnetic broadening 498
Paramagnetic label 498, 505
Paramagnetic relaxation enhancement (PRE) measurements 505–506, 510
Parameter refinement 184
Parameter synchronization 327
Parent image 304, 311
Particle selection 413–414
Patterson function 245–246, 248–251, 257
Patterson map 256–257, 485
Patterson vector 250
PCO. *See* Projective constraint optimization (PCO)
PDB. *See* Protein data bank (PDB)
Peak 4, 7, 11, 23, 26, 138, 167, 175, 178–180, 182–185, 188–190, 198, 201, 202, 219, 220, 250, 256–258, 263–266, 278, 384, 439–441, 456, 457, 467, 485, 497, 510, 542, 544–547, 549
Peak height 184, 257, 258, 266, 441
Peakmax 257
Pentaethylene glycol monooctyl ether (C_8E_5) 275–276, 284
Perforated carbon film/holey carbon film 112, 120, 343, 351, 404, 408
PhaOri 202, 203
PhaOri Change 202, 203
PhaRes 202, 203
Phase
- angle 219
- contrast 343, 356–364, 370, 409, 411, 456
- data 196, 208, 216, 223, 555, 560, 561
- deviation 214
- error 213–217, 222
- origin 200–203, 205, 206, 212
- origin search 200–202, 205, 206
- plate 11, 13, 357–359, 361–363, 376, 456
- residual 200–204, 214, 219, 221, 394
- reversal 358, 392
- shift 139, 198, 203, 206, 357–364, 376
- statistics 212, 213
- transition temperature 33, 43, 91, 92, 287
- variables 198

Phase contrast transfer function. *See* Contrast transfer function (CTF)
Phase-flipping/flipped 176, 206
Phase object/weak phase object 357, 359, 370
Phaser 245, 252, 266, 268
Phosphatidylethanolamine 561
Phospholipid bilayer 61, 75, 88
Phospholipid/lipid 2, 9, 12–14, 21, 22, 24, 29, 31–46, 49–51, 53, 59–70, 75, 78, 88, 91, 92, 94, 99, 117, 172, 244, 270, 274–279, 281–284, 286–288, 291–294, 374, 434, 439, 452, 453, 463, 464, 469, 470, 507, 509, 510, 521, 522, 524, 527–536, 553, 556–562, 566
Phospholipid monolayer 61
Phosphotungstic acid (PA) 78, 406
Photographic film 336–342, 355, 410
Pipetting robot 279
Pitch 388
Pixel 2, 7, 132, 138, 174, 175, 177, 179, 181–183, 186, 188–189, 198, 199, 319, 322, 337–339, 341, 364, 365, 367, 368, 374, 375, 387, 397, 412, 413, 430, 437, 453, 454, 465, 541, 543, 548, 549
Pixel size 138, 174, 179, 186, 189, 337, 341, 364, 365, 367, 368, 453, 541, 548
Planar-tubular 9, 10, 33, 50, 74, 88–91, 93
Plane group 212, 214, 215, 218, 221–223, 225, 246–248
Plinth 346, 347
PLOTALL 215–217
PLOTALLK 212
PLTILT 207, 213, 217, 218, 220, 221, 225, 226
PLTILTK 213, 217, 218, 220, 221, 225, 226
Plunge-freezer 343
Plunge-freezing 120, 372, 383, 386, 396
Point group symmetry 251, 252, 390
Point resolution 368
Point spread function (PSF) 176, 212, 219–221, 225, 226, 338, 365, 367, 435
Pointspread peak 220
Polar space group 247, 251
Pole piece 354
Ponderomotive force 363
Poor quality crystals 561
POPA. *See* 1-Palmitoyl-2-oleoyl-sn-glycero-3-phosphatidic acid (POPA)
POPC. *See* 1-Palmitoyl-2-oleoyl-sn-glycero-3-phosphocholine (POPC)
POPE. *See* 1-Palmitoyl-2-oleoyl-sn-glycero-3-phosphoethanolamine (POPE)

Post-dialysis........47–49
PowerBins........191
Power cables........347
Power spectrum........145, 150, 155, 176, 179, 188, 190, 392, 414, 421, 435, 453, 471
Power supply........275, 368
Pre-dialysis........44–47
Prefabricated grids........404, 407
Pre-irradiation........334
PREPNTSPREAD........213
(Re)Process all images........204, 320, 321
Processing parameters........173, 174, 186, 187, 314, 316, 318, 321, 324, 325, 327, 444
Project default........174, 314, 319, 321, 324, 326
Projection
  data........97, 212–216, 221
  map........9, 55, 63, 97, 174, 185, 192, 196, 203, 204, 213, 321, 438, 531
  matching........393, 417, 419
Projective constraint optimization (PCO)........207–209, 329
Projector lens pole piece........354
Prokaryotic........18
Protein complex........63, 177, 400, 402, 408–409, 411, 413, 415, 420, 504
Protein concentration........21–23, 28, 37, 39, 40, 44, 54, 61, 62, 68, 69, 83, 279, 385, 403, 465, 478, 479, 512, 529
Protein data bank (PDB)........11, 231, 233, 234, 249, 252, 253, 256, 258–260, 266–269, 394, 463, 480, 484, 487, 508, 513
Protein deuteration........497, 511
Protein dynamics in solution........498
Protein-protein docking........232, 233, 237, 240
Protein resonances........497, 498
Protein stability........37, 38, 41, 43, 274, 281
Proteoliposomes........32, 33, 42, 43, 50, 52, 54, 57, 74, 86, 88–93, 96, 98, 99, 466, 519
Protomo/marker free tilt series alignment........429, 436, 439
Pseudoatomic model........382, 394, 395
Pseudo-translation........255
PSF. *See* Point spread function (PSF)
PTT........33
P-type ATPases........230, 234, 239
Pumping station........343, 344
Purification........17–29, 32, 37, 38, 40–43, 49, 53–55, 115, 126, 275, 281, 291, 292, 384, 403, 476, 480, 520, 524
PyMol........232, 266, 463

Q

QUADSERCH........182, 189, 322
Quality factor IMQ........217
Quality value (QVal)........184, 190, 200, 321
Quarter wave phase shift........364
QVal1........190
QVal2........190, 200, 321
QValS........190

R

Radiation/beam damage........4, 6, 103–112, 131, 132, 137, 141, 160, 354, 363, 373, 376, 402, 409, 412, 431–433, 456, 479, 549
Radiation tolerance........354
Radius........181–183, 189, 246, 250, 257, 258, 263, 266, 322, 359, 388, 389, 394, 397, 414, 543, 545, 546, 549
RADLIM........182
Raleigh definition of resolution........225
Random conical tilting (RCT)........414, 416, 418, 421
Raw image........181–183, 322, 391, 437
RCSB. *See* Research consortium for structural bioinformatics (RCSB)
RCT. *See* Random conical tilting (RCT)
RDCs. *See* Residual dipolar couplings (RDCs)
Real space........94–96, 117, 179, 181, 185, 189, 190, 199, 207, 225, 319, 322, 327, 381–397, 416, 417, 422, 430, 435, 453, 482, 484–486
Real-space lattice........190, 319
Reciprocal lattice........2, 179, 180, 183, 190, 207, 440, 453, 482
Reciprocal space (Fourier space)........4, 138, 173, 175, 176, 179, 185, 195–199, 204, 207, 222, 417, 420, 430, 432, 434, 435, 445, 482, 484, 485
Reconstitution........25–26, 31–57, 61, 63, 65, 67, 69, 75, 88, 91, 99, 279, 285, 288, 291, 416, 418, 468, 520, 528
Reconstitution lipid........42
Reconstruction........32, 99, 132, 171, 196, 199, 203, 205–209, 211–213, 219, 225, 226, 229–231, 274, 321, 323, 329, 372, 376, 381–397, 415, 428, 429, 432, 434, 438, 455, 529, 531
Reconstruction album........203, 206
Redundancy........104, 213, 217, 222, 224, 482, 485
Reference
  based alignment........445, 453
  bias........429, 443, 445
  data set........198, 202, 205, 220, 323
  image........149, 177, 198, 439
  motif........440
  refinement........322
  structure/volume........322, 390, 393
Reference-free alignment........443–445, 453
Refine defocus from merged data........323
Refinement parameters........184, 269, 270
Refinement/structure refinement........417–419
Refine Once........200–202, 205, 206
Reflection
  data utilities........254
  peak amplitude........198

Refmac......266–270
Refolding......21, 28
Relaxation-compensated Carr–Purcell–Meiboom–Gill (CPMG) experiment......499
Relaxation of NMR signals......498
Relaxation-optimized NMR......497
Re-processing......204, 321
Reprojection......439
Research consortium for structural bioinformatics (RCSB)......231
Residual dipolar couplings (RDCs)......503–506, 513
RESMAX......317
RESMIN......317
Resolution bins......214, 218, 222
Resolution cutoff......164, 166, 219–221, 223, 225, 226, 317
Resonance overlap......497
Resonance splitting......496
Restrained molecular dynamics (rMD)......506, 507, 512
Restrained refinement......262, 265–270
Retroviral CA proteins......536
R-factor......212, 266, 267, 547–549, 563, 564
R_{free}......267, 270, 487
Rf/sigma......263
Rhodopsin......18, 19, 23–26, 28, 29
Rigid body......234, 237, 262, 269
  docking......236, 421
  refinement......265, 266, 268, 269
rMD. *See* Restrained molecular dynamics (rMD)
Robot......277, 279, 285–290, 297–303, 374, 406, 408, 476, 478
Robotics......274, 288, 476, 536
Room noise......347, 348
Room requirements......346
Rotational symmetry......250, 251, 257, 531
Rotation between subunits (Δϕ)......390
R_1/R_2 measurement......499
R values......270, 271
R_{work}......270, 487

S

SA aperture/select area aperture/selected area aperture......146, 154, 159–161
Sample flatness......223
Sample holder......134, 137, 289
Sample preparation......83, 120, 154–158, 161, 230, 345, 396, 467, 476, 565
Sampling......48, 56, 126, 196, 207, 212, 213, 218, 222, 223, 226, 232, 283, 298, 302, 412, 413, 441
Sarcoplasmic reticulum (SERCA)......231, 233–235, 239, 240
Saturation transfer difference method......498
Save as Project Default......174, 319, 324
Saxton/Saxton scheme......431–432
Scalar coupling constants......501–503
Scaling......220, 263, 482, 547
Scanner......138, 175, 275, 333, 339, 341, 349, 350, 383, 387, 410, 464, 466, 540, 543, 548
Scanning transmission electron microscopy (STEM)......374
Scattered/scattering......7, 75, 109, 115, 196, 279, 284, 286–288, 356–360, 362, 365–367, 370, 371, 373, 374, 376, 409, 410, 421, 456, 481, 483, 549, 563, 564
Scherzer/Scherzer focus......105, 130, 370
Scintillator......132, 365–367, 410
SCRATCH......324
Screening......2, 14, 32, 34, 42, 47–51, 73–100, 144, 274, 278, 288, 297–304, 307, 333, 334, 336, 338, 347, 349, 374, 385, 478, 481, 509, 527–536
Screw symmetry......390
Script......141, 145, 172, 175–177, 180, 183–192, 199–206, 236, 245, 253, 290, 316, 318–323, 325, 390, 391, 393, 416, 429, 430, 440, 444, 541, 543, 545, 547
Sculptor......236
Search
  mode......137, 141–145, 147–150, 159–161, 334, 387
  model......245, 246, 248–253, 257–261, 264–268, 483, 484, 486
Sections......130, 148, 155, 156, 161, 185–188, 196, 197, 213, 236, 255, 257–259, 262, 263, 279, 290, 301, 320, 323, 324, 356, 360, 409, 432–435, 440, 442, 443, 468, 486, 541, 561
Segmentation......390–391, 414, 421
Selection rule......388, 390
Self-assembly......63
Self-difference map......217
Self organizing map (SOM)......446
Self-pressurizing container......345
Self-rotation function......250, 252, 255, 257–258, 264, 266
Self-vectors......246, 249, 250
Sequence......18, 27, 28, 53, 61, 137, 138, 144–145, 150, 232–235, 239, 240, 245, 248, 249, 252, 253, 256, 258, 260–262, 265, 267–269, 288, 291, 393, 396, 481, 483, 497, 504, 505, 507, 521, 522, 554
SERCA. *See* Sarcoplasmic reticulum (SERCA)
SerialEM......310, 355
Service......2, 299, 349, 350
Service contract......349
Set phase origin manually......200
set RESMAX......317
SFall......267
Sfcheck......255, 256
Shadowing......103, 397
Shannon-Nyquist. *See* Nyquist/Nyquist frequency/Shannon-Nyquist

Sheet..........1, 7–11, 33, 50, 51, 55, 74, 75, 78, 79, 88–92, 99, 116, 120–125, 155, 156, 277, 286, 290, 333, 339, 382, 405, 464
β-Sheet..........238, 259, 501, 523
Shift
beam shift..........142–144, 146–148
gun shift..........144
image shift..........116, 117, 131, 143, 145, 198, 304, 309, 334, 349, 354, 419
Short-range ^{1}H-^{1}H NOEs..........505
Shot noise..........176
Sidechain χ rotamer..........503
Side-entry
EM..........134, 299
holder..........134, 344, 432
Signal-to-noise ratio/SNR, S/N..........74, 94–96, 100, 132, 138, 139, 171, 177, 178, 183, 184, 198, 199, 211, 213, 218, 253, 266, 309, 334, 356, 361, 366, 387, 402, 410–412, 414, 415, 419, 428, 429, 444–446, 450, 456, 461, 469, 512, 545, 547, 548
SINC function..........199, 207
Single axis tilt series..........431, 432, 434, 452
Single-electron detection..........366, 368
Single molecule/single particle..........1, 6, 10, 13, 83, 104, 114, 132, 138, 139, 148, 173, 177, 192, 198, 229, 274, 307, 328, 331, 333, 357, 361, 369, 373, 376, 390, 394, 401–422, 428, 429, 434, 455, 481
Single particle averaging..........229, 307
Single-particle cryo-EM..........114, 132, 139, 148, 331, 403, 404, 409
Single particle electron microscopy..........177, 198, 331, 333, 401–422
Single particle reconstructions..........229, 274, 394, 429, 434, 455
Single sideband, Hilbert..........359, 360, 362, 363
Situs..........236, 383, 394
Size exclusion chromatography..........278, 285, 476
Slow-scan CCD camera..........338, 355, 364, 375, 548
Small or poor quality crystals..........561
SNAP-shot method..........104
Software..........77, 86, 93, 99, 141, 143, 145, 148, 172–174, 179, 192, 197, 231, 232, 234, 236, 240, 245, 252, 254, 260, 275, 278–281, 290, 293, 297, 299, 307–310, 314, 315, 317, 321, 323, 328, 329, 338, 342, 355, 356, 374, 375, 382, 383, 387, 414, 416, 417, 419, 421, 429, 430, 435–437, 440, 453, 468, 471, 476, 566
Solid phase..........476, 495, 496, 510
Solubilization/solubilized..........22–23, 25, 28, 32, 37, 38, 40, 41, 44, 45, 48, 53, 54, 56, 61–63, 65, 66, 68, 83, 274, 275, 278, 283, 284, 294, 408, 480, 507, 509, 520, 528
Solubilizing detergent..........37, 40, 41, 44, 54
Soluble protein..........27, 60, 61, 70, 382, 408, 409, 479, 507, 509, 519–524, 528
Solution nuclear magnetic resonance (NMR) spectroscopy..........1, 259, 401, 403, 479, 495–513
Solvent..........65, 66, 237, 256, 266, 281, 282, 453, 478, 486
Solvent-exposed charges..........237
SOM. *See* Self organizing map (SOM)
Sonicator..........38, 45, 56, 66, 76, 80, 293
Space group..........200, 221, 246–248, 250–252, 254, 256, 262, 266, 267, 270, 317, 327, 476, 477, 480, 482, 487
Sparking..........120, 122, 125, 133, 134, 294
Sparse matrix screening..........478, 527–536
Spatial average/averaging..........428, 434, 439, 440, 452
Spatial coherence..........130, 176, 358, 368, 369, 409
Spatial frequency..........358, 420
Special translation function (STF)..........251
Specifications..........135, 346, 347, 351
Specimen airlock..........335, 336
Specimen flatness..........7, 113, 119, 120, 133, 161–163, 223, 293, 294, 439, 440
Specimen holder..........14, 124, 125, 131, 134, 145, 343, 354, 374, 383, 411, 438
Specimen preparation..........7, 14, 98, 113–115, 119, 333, 342–346, 348, 373, 375, 385–386, 404–406, 408, 552, 555, 566
Specimen preservation..........354, 520
Specimen thickness/thick specimen..........94, 221, 376, 409, 431, 432, 438, 456
Specimen thinning..........433
Specimen tilt..........174, 176–177, 184, 187, 552
SPECTRA software..........172
Spherical harmonics..........443
Spherical, spherical aberration coefficient (Cs)..........174–176, 249, 250, 319, 368–370, 413, 431, 443, 556, 557
Spider..........173, 185, 231, 383, 392–394, 414, 416, 417, 421
Spin diffusion..........500
Spin polarization..........499, 500, 505
Spin population..........499, 500, 505
Spin probe..........500
Spin-1/2 nuclide..........496, 503
Spot..........2, 6, 50, 77, 93–96, 107, 113, 114, 116, 119, 132, 140–144, 147, 148, 156–166, 178, 179, 183–185, 188–190, 197, 214, 281, 321, 325, 334, 355, 364, 411, 453, 469, 471, 472, 482, 529, 544, 545, 547, 548
Spotlist/spot list..........174, 183, 189, 322
SPOTS..........191
Spot-scan imaging..........6, 132, 325, 334, 355
Spot-scan/spotscan..........6, 132, 325, 334, 355
Spot size..........6, 77, 140, 141, 144, 147, 159, 160
Spot-splitting..........191
Spring constant..........462, 464
SpSplit..........191

Stability............ 32, 34, 37, 38, 40, 41, 43, 55, 56, 60, 61, 69, 134, 135, 137, 138, 274, 278, 281, 285, 334, 336, 362, 375, 395, 396, 462, 509, 524
Stacking............ 57, 96
Stage.......... 6, 7, 13, 24, 32, 40, 42, 48, 50, 56, 78, 79, 83, 88, 105, 108, 109, 112–115, 117, 131, 132, 134, 135, 137, 138, 142, 145, 148, 149, 157, 158, 160, 161, 206, 244, 270, 274, 275, 287, 293, 298–301, 311, 326, 331, 332, 334–336, 343, 346, 349, 354, 372, 373, 375, 412, 416, 418, 419, 432, 433, 437, 448, 480, 541, 549, 552
Staining, negative stain............ 20, 75, 77, 78, 83–86, 98, 103, 104, 107, 155, 278, 279, 288, 289, 297, 302, 403–406, 445, 535
Standard script............ 187, 189, 200, 202, 316, 320, 322, 325
Standard scripts panel............ 187, 189, 325
1-start helix (n=1)............ 390
Status panel............ 189–191
STEM. *See* Scanning transmission electron microscopy (STEM)
Stepsize............ 201, 202, 206
Stepsize of the phase origin search............ 201, 202, 206
STF. *See* Special translation function (STF)
Strength of electron crystallography............ 244, 555, 559, 560
Stroboscopic imaging............ 372
Structural constraint............ 497
Structure determination............ 11, 12, 17, 18, 53, 89, 99, 153, 154, 172, 208, 209, 243–271, 314, 324, 356, 361, 396, 428, 475–478, 482, 495–497, 499–510, 512, 513, 519, 520, 524, 539
Structure factor............ 214–216, 218, 219, 222–225, 246, 253, 255, 258, 261, 268, 483, 486
Structure modeling............ 229–240
Structure refinement............ 51, 182, 184, 188–190, 201, 202, 211, 236, 237, 243–271, 298, 322, 354, 383, 391–394, 401, 410, 414, 416–422, 437, 453, 476, 483, 486–487, 507, 531, 548, 550, 564
Stylus artifacts............ 473
Stylus effect............ 472–473
Sublimating............ 110
Subphase............ 61–63, 70
Subvolume, motif............ 428, 430
Sugar embedding............ 6, 120, 156
Surface residues............ 249
Surflex............ 232
Symmetrization............ 200, 207, 224
Symmetry
  axis............ 248, 251, 469, 470
  constraints............ 215, 219
  restraints............ 270
Synchronization............ 321, 327
SYNC_WITH_UPPER_LEVEL............ 317, 327
SYN_maska............ 322
SYN_maskb............ 322
Synthetical reference............ 322
Synthetic unbend............ 184, 190, 322
Synthetic unbend scripts............ 184, 322
SYN_UNBENDING............ 322, 323
Systematic absences............ 179, 222, 246, 247

T

TALOS/TALOS+............ 501
TANGL............ 180, 191, 204, 323
TAXA............ 180, 191, 204
Teflon............ 64–66, 280, 282, 292, 293, 464, 465, 532–534, 536
TEM. *See* Transmission electron microscopy (TEM)
Temperature cycling............ 285
Temperature factors............ 176, 498, 547, 548, 557, 558
Temperature fluctuation............ 135, 346
Template............ 10, 61, 187, 231–235, 238, 413, 430, 444, 532
Template model............ 238
Template structure............ 232, 234
Temporal coherence............ 368–370
Tertiary distance restraint............ 501
Test specimen............ 348, 351
TFcnt............ 263–265
Tf/sig............ 263
Theoretical magnification............ 189
Thermal cycling............ 285
Thermal drift............ 7, 244, 465, 466
Thermal relaxation............ 465
Thermal stabilization............ 145
Thin film device............ 358, 362
Thin layer chromatography (TLC)............ 42, 49, 275, 279, 281, 282, 287, 293
Thin section............ 432
Thon rings............ 97, 100, 175, 176, 188, 203, 325, 358, 413
Three-dimensional (3D) correlation map............ 441
Three-dimensional (3D) crystallization............ 12–14, 55, 278, 291, 527, 529, 533, 536, 552
Three-dimensional (3D) data............ 212–213, 217–221, 428
Three-dimensional (3D) density map............ 129, 217, 220, 244, 392–395, 531
Three-dimensional (3D) Fourier space............ 197, 199, 204, 207
Three-dimensional (3D) Fourier space representation............ 199
Three-dimensional (3D) image............ 427–429, 434, 440, 443
Three-dimensional (3D) reconstruction............ 32, 132, 171, 196, 199, 205–209, 211, 219, 329, 382, 387–394, 428, 438, 531
Three-dimensional (3D) structure............ 4, 9, 10, 70, 99, 104, 213, 229, 231, 313, 324, 390, 415, 416, 422, 521, 523, 524, 552, 555, 561
Three-dimensional (3D) transform............ 217, 223
Thresholding............ 367, 446
TIFF............ 175, 315, 318, 320, 387, 542, 548

Tilt angle......... 121, 133, 139, 162, 177, 180, 195, 197, 205, 207–209, 219, 225, 226, 310–311, 314, 315, 318, 319, 323–327, 329, 334, 393, 416, 418, 431, 432, 437, 439, 453, 542, 543, 545, 547, 549, 550
Tilt angle increment...431
Tilt axis/azimuth 7, 116, 119, 148, 150, 162, 163, 177, 180, 204, 217, 218, 244, 311, 349, 431–435, 437–439, 446, 452, 454, 455, 545, 549
Tilt category...320
Tilted
  crystal..................161, 191, 197, 205–206, 212, 213, 321
  data...157, 168, 199, 205–207
  image.........70, 121, 197, 204–206, 208, 311, 314, 321, 323–325, 327, 328, 416
  sample...148, 176, 199, 204
  TTF...327
Tilt geometry/nominal tilt category 176, 177, 179, 180, 184, 186, 190, 191, 196, 204, 212, 320, 323, 325, 438
Tilt increment 431, 432, 453
Tilt series........................ 150, 274, 361, 415, 428–439, 446, 452, 453, 456
  alignment..429, 431, 436, 439
  data collection...431–433
  geometry...438
Time resolution range 372
Titled transfer function (TTF)............................... 191, 327
TLC. *See* Thin layer chromatography (TLC)
TLTANG... 177, 180, 191
TLTAXA... 180, 191
TLTAxis/TLTAXIS 177, 180, 191
TLTPLOT...207
Tomoalign-gui... 437
Tomocorr.sh... 440, 441
Tomofit.sh.. 440, 441
Tomogram........................ 361, 415, 418, 428–442, 445, 448, 452–454, 456
Tomogram projection...441
Tomographic motif...443
Tomographic reconstruction..274
Tomography 1, 6, 104, 114, 177, 229, 307, 310, 331, 333, 371, 375, 376, 415, 416, 418, 421, 427–456
Tomosearch.sh... 440, 441
Tomounbend.sh... 440, 441
TOM Toolbox... 310, 355
Top-entry
  EM.. 113–115
  holder...134
Topograph/topography............................461, 462, 465–473
Translation function (TF)............................. 250–252, 256, 261–266, 484
Transmission electron microscopy (TEM) 73–100, 124, 125, 159, 171, 175, 176, 297–300, 304, 310, 331–333, 371–374
Transverse relaxation rate 497, 498
Trehalose......... .... 6, 107, 112, 120, 121, 125, 144, 155, 157, 168
Trilinear interpolation 441
Triton X-100 .... 19, 21, 41, 61–63, 65, 275
TTF. *See* Titled transfer function (TTF)
Tube..3, 10, 21, 22, 38, 45, 47, 48, 50, 52, 55, 64, 66, 67, 74, 83–85, 88, 89, 91, 127, 277, 282, 285–290, 292, 294–296, 350, 355, 360, 362, 382, 405, 465, 533–535
Tubular
  assemblies 382, 387, 388, 395, 396
  crystal.................. 9–11, 89, 91, 230, 350, 381–397, 561
Tubulin......... .... 8, 11, 172
Tungsten... 130, 349
Turbidity...287
Turbomolecular pump 131, 136, 137
Turn..4, 33, 79, 80, 83, 123–125, 133, 134, 142, 159, 160, 211, 224, 386, 388, 431, 442, 463, 490, 521, 552
TV-rate...148–150, 154, 355
Twinning......... .... 246, 255
Two-dimensional crystal/crystallization (2D crystal/crystallization) 1, 2, 4, 6–14, 21, 22, 24, 31–57, 59–70, 73–100, 104, 107, 108, 112, 113, 117, 119, 120, 127, 129–151, 153–158, 161, 164, 165, 171–192, 195–197, 199, 208, 209, 211–226, 244, 246–248, 251, 274, 278, 282–283, 297–304, 307–310, 314, 318–320, 324, 325, 327–329, 332–334, 336, 338, 342, 343, 351, 373, 382, 409, 410, 412, 422, 441, 463–466, 468–470, 472, 484, 520, 522–524, 527–536, 539, 552, 554–556, 559, 561, 564–566
Two-fold axis... 8, 248, 251
Two-pass screening...300–303
Two-sided plane group...221
TYPE...317

U

UCSF DOCK...232
UCSFImage 145, 146, 148–150
Unambiguous restraints...497
UNBEND 1 .... 174, 182, 183, 189, 190, 322
UNBEND II 174, 182, 189, 190, 321, 322
Unbending...................................... 7, 11, 13, 158, 173, 174, 179, 181–185, 189, 190, 319–323, 325, 356, 422, 439–441, 520
Unbending algorithm 173, 181–183, 322
Unbend synthetic 184, 190, 322
Unbent image 174, 182, 322, 422
Unbent map...441
Under-focus...142, 177, 221, 411
Unique reflections.................... 213, 214, 216, 219, 222, 225

Unit....7, 11–13, 83, 91, 94–96, 98, 100, 132, 178, 181, 182, 187, 192, 198, 200, 201, 203, 205, 212–214, 221, 223, 246, 248–252, 254–258, 262–264, 269, 270, 319, 322, 331, 333, 343, 344, 347, 349, 350, 370, 388, 390, 422, 440, 452, 469, 470, 472, 476, 482, 486, 522, 529, 531, 532, 541
Unit cell
  dimensions....94–96, 100, 200, 213, 221, 254, 482, 541
  parameter....214, 221, 529
Unscattered....357–360, 362, 363, 374, 409
Uranyl acetate....68, 75, 77, 78, 278, 288, 385, 406, 533, 536
Uranyl formate....78, 98, 155, 406
Use synthetical reference if possible?....322

V

Vacuum
  pump....133, 283, 340, 342, 346
  system....130–131, 136
Validation....237, 253, 255, 419–421
Variance map....446, 448
Vector averaging....222
Vector field....182, 481
Vector map....246
Ventilation....135, 346
Vertical resolution....217, 219, 220, 225, 226, 461, 462, 466
Vesicle....9, 10, 33, 50, 55, 74, 88–91, 93, 126, 155, 156, 188, 284, 293, 300, 428, 434, 529
Vibration....76, 79, 112, 134, 135, 176, 342, 346, 349, 466, 479, 486
Vitreous/frozen hydrated....8, 68, 107, 113, 126, 138, 148, 195, 307, 308, 343, 350, 386, 402–404, 406, 410, 411, 432, 433, 452, 529, 532
Vitrification....6, 120, 156, 343, 350
Vitrified....6, 144, 396, 404
V_{mol}....262, 264
Voxel....420, 446, 448, 450–452

W

Water cooler....346, 347
Water molecule....84, 115, 120, 130, 131, 136, 137, 172, 244, 249, 259, 487, 521, 522, 554, 556–559
Weak phase object/phase object....357, 359, 370
Weighted phase residual....219, 221
Wet-cell stage....354
Wiener filter/filtration....185, 392, 395, 453
Wilson/Wilson plot....255, 257, 259, 546
Wind sock....346, 348

X

XDP software....328
Xdroptrace....281, 293
Ximdisp....172, 179
XMIPP....173, 414, 419
X-ray/X-ray crystallography....1, 2, 4, 8–14, 61, 104, 115, 117, 154, 165, 212, 230, 232, 233, 237, 238, 243–245, 248, 252, 259, 260, 274, 278, 291, 356, 373, 374, 382, 401, 403, 420, 421, 464, 472, 475–491, 497, 505, 512, 520, 521, 524, 529, 531, 549, 552, 556, 563, 565
X,Y peak....198
X/Y plane/x,y-plane....196, 225, 311, 438
X,Y shift....198

Z

z*....197, 199, 204–208, 218, 219, 222, 223, 225, 226
Z-axis....4, 60, 61, 222, 225, 444
z*-bin....218
z* coordinate....197, 206–208
z* dimension....204
Zemlin tableau....141
Zernike-type phase contrast/Zernike phase contrast....356, 361
Zero-loss electrons....456
Zstarwin....205, 206
z* tolerance....205, 206

MIX
Papier aus verantwortungsvollen Quellen
Paper from responsible sources
FSC® C105338

If you have any concerns about our products,
you can contact us on
ProductSafety@springernature.com

In case Publisher is established outside the EU,
the EU authorized representative is:
Springer Nature Customer Service Center GmbH
Europaplatz 3, 69115 Heidelberg, Germany

Printed by Libri Plureos GmbH
in Hamburg, Germany